Country	Currency	Symbol	Country	Currency	Symbol
Macao	pataca	P	Spain	peseta	Ptas
Malawi	kwacha	MK	Sri Lanka	rupee	SL Rs
Malaysia	ringgit	M$	Sweden	krona	SKr
Malta	lira	Lm	Switzerland	franc	SFr
Mauritius	rupee	Mau Rs	Taiwan	dollar	NT$
Mexico	peso	Mex$	Thailand	baht	B
Morocco	dirham	DH	Trinidad and Tobago	dollar	TT$
Namibia	rand (S.Afr.)	R			
Netherlands	guilder	$f.	Tunisia	dinar	D
Netherlands Antilles	guidler	NA. f	Turkey	lira	LT
			Ukraine	ruble	rub
New Zealand	dollar	$NZ	United Arab Emirates	dirham	Dh
Nigeria	naira	₦			
Norway	krone	NKr	United Kingdom	pound	£ or £ stg.
Oman	rial Omani	RO			
Pakistan	rupee	PRs	Uruguay	new peso	NUr$
Panama	balboa	B	Vanuatu	vatu	VT
Papua New Guinea	kina	K	Venezuela	bolívar	Bs
			Vietnam	dong	D
Paraguay	guaraní	₲	Western Samoa	tala	WS$
Peru	new sol	S/.	Zaire	zaire	Z
Philippines	peso	₱	Zambia	kwacha	K
Portugal	escudo	Esc	Zimbabwe	dollar	Z$
Qatar	riyal	QR			
Russia	ruble	Rb			
Saudi Arabia	riyal	SRIs			
Senegal	franc	CFAF			
Singapore	dollar	S$			
Somalia	shiling	So. Sh.			
So. Africa	rand	R			

ALAN C. SHAPIRO
University of Southern California

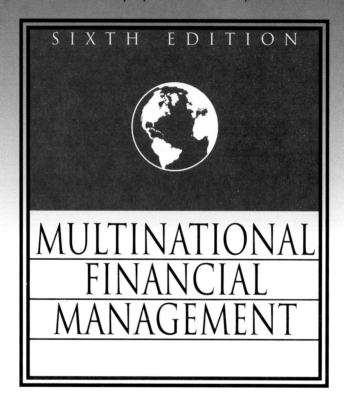

SIXTH EDITION

MULTINATIONAL
FINANCIAL
MANAGEMENT

Prentice Hall, Upper Saddle River, NJ 07458

Senior Acquisitions Editor: Paul Donnelly
Assistant Editor: Gladys Soto
Editorial Assistant: Jodi Hirsh
Marketing Manager: Lori Braumberger
Editorial Director: James C. Boyd
Production Editor: Lynda P. Hansler
Permissions Coordinator: Monica Stipanov
Senior Manufacturing Supervisor: Paul Smolenski
Manufacturing Manager: Vincent Scelta
Design Manager: Patricia Smythe
Interior Design: Jill Yutjowitz
Cover Design: Cheryl Asherman
Cover Photo: John Still/Photonica, Inc.
Composition: Progressive Publishing Alternatives

© 1999, 1996 by Prentice-Hall, Inc.
Upper Saddle River, New Jersey 07458

Library of Congress Cataloging-in-Publication Data

Shapiro, Alan C.
    Multinational financial management / Alan C. Shapiro. —6th ed.
      p.   cm.
    Includes bibliographical references and index
    ISBN 0-13-010142-7
    1. International business enterprises—Finance.   I. Title.
  HG4027.5.S47   1999
  658.15′99—dc21   98-50571
                                                                              CIP

Prentice-Hall International (UK) Limited, London
Prentice-Hall of Australia Pty. Limited, Sydney
Prentice-Hall Canada, Inc., Toronto
Prentice-Hall Hispanoamericana, S.A., Mexico
Prentice-Hall of India Private Limited, New Delhi
Prentice-Hall of Japan, Inc., Tokyo
Pearson Education Asia Pte. Ltd., Singapore
Editora Prentice-Hall do Brasil, Ltda., Rio de Janeiro

Printed in the United States of America

10 9 8 7 6 5 4 3 2 1

To my parents,
Hyman and Lily Shapiro,
*for their encouragement,
support, and love*

BRIEF CONTENTS

CONTENTS

PREFACE

APPROACH

The basic thrust of this sixth edition of *Multinational Financial Management* is to provide a conceptual framework within which the key financial decisions of the multinational firm can be analyzed. The approach is to treat international financial management as a natural and logical extension of the principles learned in the foundations course in financial management. Thus, it builds on and extends the valuation framework provided by domestic corporate finance to account for dimensions unique to international finance.

Multinational Financial Management focuses on decision making in an international context. Analytical techniques help translate the often vague rules of thumb used by international financial executives into specific decision criteria. The book offers a variety of real-life examples, both numerical and institutional, that demonstrate the use of financial analysis and reasoning in solving international financial problems. These examples have been culled from the thousands of illustrations of corporate practice that I have collected over the years from business periodicals and my consulting practice. Examples scattered throughout the text show students the value of examining decision problems with the aid of a solid theoretical foundation. Seemingly disparate facts and events can then be interpreted as specific manifestations of more general financial principles.

All the traditional areas of corporate finance are explored, including working capital management, capital budgeting, cost of capital, and financial structure. However, these areas are explored from the perspective of a multinational corporation, concentrating on those decision elements that are rarely, if ever, encountered by purely domestic firms. These elements include multiple currencies with frequent exchange rate changes and varying rates of inflation, differing tax systems, multiple money markets, exchange controls, segmented capital markets, and political risks such as nationalization or expropriation. Throughout the book, I try to demystify and simplify multinational financial management by showing that its basic principles rest on the same foundation as the principles of corporate finance.

The emphasis throughout this book is on taking advantage of being multinational. Too often, companies focus on the threats and risks inherent in venturing abroad rather than on the opportunities that are available to multinational firms. These opportunities include the ability to obtain a greater degree of international diversification than security purchases alone can provide, as well as the ability to arbitrage between imperfect capital markets, thereby obtaining funds at a lower cost than could a purely domestic firm.

AUDIENCE

Multinational Financial Management is designed for use in masters-level courses. It is also suitable for use in bank management and other executive development programs.

FEATURES

Multinational Financial Management presumes a knowledge of basic corporate finance, economics, and algebra. However, it assumes no prior knowledge of international economics or international finance and is, therefore, self-contained in that respect. For those who are not familiar with *Multinational Financial Management*, here are some of the distinctive features that have led to its widespread adoption.

Distinctive topic coverage includes the following:

- Role of expectations in determining exchange rates (Chapter 2)
- Discussions of the equilibrium approach to exchange rate determination (Chapter 2) and the monetary approach to exchange rate determination (Chapter 7)
- Analysis of the Mexican and Asian currency crises (Chapter 2)
- Discussion of currency boards and the role of central bank credibility in the context of the dramatic changes made recently in Argentina and New Zealand (Chapter 2)
- Analysis of the September 1992 and August 1993 currency crises in the European Monetary System (Chapter 3)
- Discussion of European Monetary Union, the Maastricht criteria, and optimum currency areas (Chapter 3)
- Use of the balance-of-payments framework to assess the economic links among nations (Chapter 4)
- Comprehensive discussion of the foreign exchange market and its institutions and mechanisms, including electronic trading (Chapter 5)
- Discussion of covered interest arbitrage with and without transaction costs (Chapter 5)
- Understanding of currency futures and options contracts (Chapter 6) and their use in exchange risk management (Chapter 9)
- Demonstration of put-call option parity (Chapter 6)
- Discussion of the key parity conditions in international finance, with numerous real-world applications (Chapter 7)
- Discussion of the distinction between real and nominal exchange rates and its significance for exchange risk management (Chapters 7 and 10)
- Use of currency risk sharing in international contracts (Chapter 9)
- Identification of the economic, as opposed to accounting, aspects of foreign exchange risk (Chapter 10)
- Development of marketing and production strategies to cope with exchange risk, including a discussion of how Japanese firms coped with a strong yen (Chapter 11)
- Role of countertrade in financing international trade (Chapter 12)
- Evaluation of foreign currency-denominated debt (Chapter 13)
- Costs and benefits of managing interaffiliate fund flows on a global basis (Chapter 15)
- Integration of tax management with financial management based on the Tax Reform Act of 1986 and the 1993 Tax Act (Chapters 14, 16, and 18)
- Extensive discussion of the nature of national and international capital markets and their role in gathering and allocating capital (Chapter 15)
- Discussion of international differences in corporate governance and their consequences (Chapter 15)

- Use of interest rate and currency swaps, interest rate derivatives, leasing, and less-developed country debt-equity swaps (Chapter 16)
- Discussion of new perspectives on the international debt crisis and country risk analysis (Chapter 17)
- Assessment of the cost of capital for foreign investments and extended discussion on alternative approaches to estimating this figure (Chapter 18)
- Understanding of the nature and consequences of international portfolio investment, including investments in emerging markets (Chapter 19)
- Development of global strategies of multinational corporations (Chapter 20), including joint ventures (Appendix 20A) and the strategy for Europe 1992 (Appendix 20B)
- Discussion of the Japanese strategy for global expansion (Chapter 20)
- Analysis of foreign investments, including assessing the true profitability of a foreign operation, calculating adjusted net present values, and factoring in various economic and political risks (Chapter 21)
- Valuation of the growth options often associated with foreign investments (Chapter 21)
- Understanding of political risk measurement and management, with a discussion of the dramatic changes taking place throughout Latin America, Eastern Europe, and China (Chapter 22)

CHANGES TO THE SIXTH EDITION

Changes that have been made to the sixth edition include the following:

- Appendix 1A (The Origins and Consequences of International Trade) has been added to provide a more extensive discussion of the basic concepts of absolute and comparative advantage as well as some insight into the gains and income redistributions associated with free trade and the consequences of barriers to trade.
- The section on interest rate parity theory has been transferred from Chapter 7 (Parity Conditions in International Finance) to Chapter 5 (The Foreign Exchange Market) where it provides the link between the spot and forward exchange markets and money markets.
- The section on put-call option interest rate parity has been transferred from Chapter 7 (Parity Conditions in International Finance) to Chapter 6 (Currency Futures and Options Markets).
- Part III (Working Capital Management) has been rearranged and shortened. It now begins with Chapter 12 on Financing Foreign Trade. The former Chapter 12 (Short-Term Financing) has been consolidated with the previous Chapter 14 (Current Asset Management) into a new Chapter 13 (Current Asset Management).
- The old Part V (Financing Foreign Operations) has been shifted to come before the old Part IV (Foreign Investment Analysis, which is now Part V). The aim was to deal with the financing subject matter, including the cost of capital, prior to discussing the investment decision.
- Part VI (International Banking) has been eliminated. Instead, the two chapters comprising it have been shortened and consolidated into one new Chapter 17 (International Banking Trends and Strategies).
- Chapter 23 (Designing a Global Financing Strategy) has been eliminated, with most of its material consolidated into one new Chapter 18 (The Cost of Capital for Foreign Investments) and transferred to Part IV. Parts of the former Chapter 23 have also been shifted to Chapter 16 (Special Financing Vehicles).
- Chapter 21 (International Tax Management) has been deleted. Faculty wishing to still use it can receive permission to do so free of charge by contacting me directly.

At the same time that the book has been updated, reorganized, and shortened, new material has been added, including the following:

- Many new solved numerical problems in the body of the chapters to illustrate the application of the various concepts and techniques presented in the text along with a number of new end-of-chapter problems
- Extensive discussion of the Asian currency crisis in 1997–1998 (Chapters 1 and 2)
- A section on valuing the ECU and calculating central cross rates in the exchange-rate mechanism (Chapter 3)
- Demonstration of double-entry bookkeeping in balance-of-payments accounting (Chapter 4)
- Presentation on how to compute gains, losses, and maintenance margins on a futures contract (Chapter 6)
- A discussion of currency spreads and knockout (or barrier) options (Chapter 6)
- A discussion of recent research on biases in the forward exchange rate and possible causes, including the peso problem (Chapter 7)
- Discussion on linking currency forecasts to specific decision rules (Chapter 7)
- Discussion and comparison of the three basic types of exposure—accounting exposure, transaction exposure, and operating exposure (Chapter 8)
- Extended discussion on the design of a hedging strategy (Chapter 9)
- Discussion of managing the risk management function, including lessons learned from some highly publicized cases of derivatives-related losses (Chapter 9)
- Comparing hedging alternatives when there are transaction costs (Chapter 9)
- Discussion of cross-hedging using a simple regression analysis (Chapter 9)
- Discussion of how to structure and use currency collars (or range forwards) and currency cylinders to hedge exchange risk (Chapter 9)
- Discussion of international differences in corporate governance and the economic consequences of those differences (Chapter 15)
- Discussion of how to calculate effective costs of Eurocurrency loans and Eurocommercial paper and all-in costs of Eurobonds (Chapter 15)
- Discussion of fixed-for-fixed and fixed-for-floating currency interest rate swaps (Chapter 16)
- Discussion of structured notes, such as inverse floaters and step-up and step-down notes (Chapter 16)
- Discussion of interest rate risk management techniques, including forward rate agreements, forward forwards, and Eurodollar futures (Chapter 16)
- Discussion of the home bias in international portfolio investing and the effects of hedging on the efficient frontier for internationally diversified portfolios (Chapter 19)
- Addition of three new cases, including a comprehensive case on currency forecasting and exchange risk management (Case of the Depreciating Indian Rupee), a second on Indonesia's decision of whether to establish a currency board (Rescuing the Indonesian Rupiah with a Currency Board), and a third on Brazil's decision to defend its currency (Brazil Fights a *Real* Battle)

The book also contains many new charts and illustrations of corporate practice that are designed to highlight specific techniques or teaching points. Again, the emphasis is on reinforcing and making more relevant the concepts developed in the body of each chapter. To make the text more suitable as a teaching vehicle, I have added numerous questions and problems, most of which are based on up-to-date information and real-life situations, to the ends of the chapters.

PEDAGOGY

The pedagogical thrust of the book is greatly enhanced by the following learning and teaching aids:

1. *Focus on corporate practice* Throughout the text, there are numerous real-world examples and vignettes that provide actual applications of financial concepts and theories. They show students that the issues, tools, and techniques discussed in the book are being applied to day-to-day financial decision making.

2. *Extensive use of examples and illustrations* Numerous short illustrations and examples of specific concepts and techniques are scattered throughout the body of most chapters.

3. *Lengthier illustrations of corporate practice* There are nine longer illustrations of actual company practices, at the end of key chapters, that are designed to demonstrate different aspects of international financial management.

4. *Problems and discussion questions* Hundreds of realistic end-of-chapter questions and problems offer practice in applying the concepts and theories being taught. Many of these questions and problems are related to actual situations and companies.

6. *Glossary* A glossary at the back of the book defines key terms in the text.

7. *Supplements* Ancillary materials are available for adopters of *Multinational Financial Management* to supplement the text. These include

   ■ *Instructor's Manual with Solutions and Test Bank* This all-inclusive ancillary contains, for every chapter: lecture outline and objectives, key points, suggested answers to questions and cases, and solutions to all end-of-chapter problems. This manual also includes a test bank completely revised and updated for the sixth edition by Joseph Greco of California State, Fullerton.

   ■ *PowerPoint Lecture Presentation* Also prepared by Joseph Greco, the PowerPoint Presentation can be downloaded from The Prentice Hall Finance Center.

THANKS

I have been greatly aided in developing *Multinational Financial Management* by the helpful suggestions of the following reviewers: Robert Aubey, University of Wisconsin; James Baker, Kent State University; Donald T. Buck, Southern Connecticut State University; C. Edward Chang, Southwest Missouri State University; Jay Choi, Temple University; Robert C. Duvic, University of Texas, Austin; Janice Wickstead Jadlow, Oklahoma State University; Steve Johnson, University of Texas at El Paso; Boyden C. Lee, New Mexico State University; Marc Lars Lipson, Boston University; Richard K. Lyons, University of California, Berkeley; Dileep Mehta, Georgia State University; Margaret Moore, Franklin University; William Pugh, Auburn University; Bruce Seifert, Old Dominion University; Jay Sultan, Bentley College; Paul J. Swanson, Jr., University of Cincinnati; and Steve Wyatt, University of Cincinnati. I am very appreciative to the editorial staff at Prentice Hall for the effort they put into this edition. To Gladys Soto, the Assistant Editor for Prentice Hall Finance, who was so helpful in pulling together the supplements which accompany this new edition, I offer a resounding thank you. I am also grateful to Jodi Hirsh, whose help in the review process was professional and timely. I offer sincere thanks, as well, to Lynda Hansler for her dedication in managing the production of the book.

My family, especially my wife, Diane, as well as my mother and three brothers, have provided me (once again) with continual support and encouragement during the writing of this book. I appreciate the (usual) cheerfulness with which Diane and my children, Thomas and Kathryn, endured the many hours I spent writing the sixth edition of this text.

A.C.S.
Pacific Palisades

INTRODUCTION: MULTINATIONAL ENTERPRISE AND MULTINATIONAL FINANCIAL MANAGEMENT

What is prudence in the conduct of every private family can scarce be folly in that of a great kingdom. If a foreign country can supply us with a commodity cheaper than we ourselves can make it, better buy it of them with some part of the produce of our own industry employed in a way in which we have some advantage.

Adam Smith (1776)

International business activity is not new. The transfer of goods and services across national borders has been taking place for thousands of years, antedating even Joseph's advice to the rulers of Egypt to establish that nation as the granary of the Middle East. Since the end of World War II, however, international business has undergone a revolution out of which has emerged one of the most important economic phenomena of the latter half of the twentieth century: the multinational corporation.

1.1 THE RISE OF THE MULTINATIONAL CORPORATION

A *multinational corporation* (MNC) is a company engaged in producing and selling goods or services in more than one country. It ordinarily consists of a parent company located in the home country and at least five or six foreign subsidiaries, typically with a high degree of strategic interaction among the units. Some MNCs have upwards of 100 foreign subsidiaries scattered around the world. The United Nations estimates that at least 35,000 companies around the world can be classified as multinational.

Based in part on the development of modern communications and transportation technologies, the rise of the multinational corporation was unanticipated by the classical theory of international trade as first developed by Adam Smith and David Ricardo. According to this theory, which rests on the doctrine of comparative advantage, each nation should specialize in the production and export of those goods that it can produce with highest relative efficiency and import those goods that other nations can produce relatively more efficiently.

Underlying this theory is the assumption that goods and services can move internationally but factors of production, such as capital, labor, and land, are relatively immobile. Furthermore, the theory deals only with trade in commodities—that is, undifferentiated products; it ignores the roles of uncertainty, economies of scale, transportation costs, and technology in international trade; and it is static rather than dynamic. For all these defects, however, it is a valuable theory, and it still provides a well-reasoned theoretical foundation for free-trade arguments (see Appendix 1A). But the growth of the MNC can be understood only by relaxing the traditional assumptions of classical trade theory.

Classical trade theory implicitly assumes that countries differ enough in terms of resource endowments and economic skills for those differences to be at the center of any analysis of corporate competitiveness. Differences among individual corporate strategies are considered to be of only secondary importance; a company's citizenship is the key determinant of international success in the world of Adam Smith and David Ricardo.

This theory, however, is increasingly irrelevant to the analysis of businesses in the countries currently at the core of the world economy—the United States, Japan, the nations of Western Europe, and, to an increasing extent, the most successful East Asian countries. Within this advanced and highly integrated core economy, differences among corporations are becoming more important than aggregate differences among countries. Furthermore, the increasing capacity of even small companies to operate in a global perspective makes the old analytical framework even more obsolete.

Not only are the "core nations" more homogeneous than before in terms of living standards, lifestyles, and economic organization, but their factors of production tend to move more rapidly in search of higher returns. Natural resources have lost much of their previous role in national specialization as advanced, knowledge-intensive societies move rapidly into the age of artificial materials and genetic engineering. Capital moves around the world in massive amounts at the speed of light; increasingly, corporations raise capital simultaneously in several major markets. Labor skills in these countries no longer can be considered fundamentally different; many of the students enrolled in American graduate schools are foreign, while training has become a key dimension of many joint ventures between international corporations. Technology and "know-how" are also close to becoming a global pool. Trends in protection of intellectual property and export controls clearly have less impact than the massive development of the means to communicate, duplicate, store, and reproduce information.

Against this background, the ability of corporations of all sizes to use these globally available factors of production is a far bigger factor in international competitiveness than broad macroeconomic differences among countries. Contrary to the postulates of Smith and Ricardo, the very existence of the multinational enterprise is based on the international mobility of certain factors of production. Capital raised in London on the Eurodollar market may be used by a Swiss-based pharmaceutical firm to finance the acquisition of German equipment by a subsidiary in Brazil. A single Barbie doll is made in 10 countries—designed in California, with parts and clothing from Japan, China, Hong Kong, Malaysia, Indonesia, Korea, Italy, and Taiwan, and assembled in Mexico—and sold in 144 countries. Information technology also makes it possible for worker skills to flow with little regard to borders. In the semiconductor industry, the leading companies typically locate their design facilities in high-tech corridors in the United States, Japan, and Europe. Finished designs are transported quickly by computer networks to manufacturing plants in countries with more-advantageous cost structures. In effect, the traditional world economy in which products are exported has been replaced by one in which value is added in several different countries.

The value added in a particular country—product development, design, production, assembly, or marketing—depends on differences in labor costs and unique national attributes or skills. Although trade in goods, capital, and services, and the ability to shift production act to limit these differences in costs and skills among nations, differences nonetheless remain based on cultural predilections, historical accidents, and government policies. Each of these factors can affect the nature of the competitive advantages enjoyed by different nations and their companies. For example, at the moment, the United States has some significant competitive advantages. For one thing, individualism and entrepreneurship—characteristics that are deeply ingrained in the American spirit—are increasingly a source of competitive advantage as the creation of value becomes more knowledge-intensive. When inventiveness and entrepreneurship are combined with abundant risk capital, superior graduate education, and an inflow of foreign brainpower, it is not surprising that U.S. companies dominate world markets in software, biotechnology, microprocessors, aerospace, and entertainment. Also, U.S. firms are moving rapidly forward to construct an information superhighway and related multimedia technology, whereas their European and Japanese rivals face continued regulatory and bureaucratic roadblocks.

Recent experiences also have given the United States a significant competitive advantage. During the 1980s and 1990s, fundamental political, technological, regulatory, and economic forces radically changed the global competitive environment. A brief listing of some of these forces would include the following:

- Massive deregulation
- The collapse of Communism
- The sale of hundreds of billions of dollars of state-owned firms around the world in massive privatizations designed to shrink the public sector
- The revolution in information technologies
- The rise in the market for corporate control with its waves of takeovers, mergers, and leveraged buyouts

- The jettisoning of statist policies and their replacement by free-market policies in Third World nations
- The unprecedented number of nations submitting themselves to the exacting rigors and standards of the global marketplace.

These forces have combined to usher in an era of brutal price and service competition. The United States is further along than other nations in adapting to this new world economic order, largely because its more open economy has forced its firms to confront rather than hide from competitors. Facing vicious competition at home and abroad, U.S. companies—including such corporate landmarks as IBM, General Motors, Walt Disney, Xerox, American Express, Coca-Cola, and Kodak—have been restructuring and investing heavily in new technologies and marketing strategies to boost productivity and expand their markets. In addition, the United States has gone further than any other industrialized country in deregulating its financial services, telecommunications, airlines, and trucking industries. The result: Even traditionally sheltered U.S. industries have become far more competitive in recent years, and so has the U.S. workforce. The heightened competitiveness of U.S. firms has, in turn, compelled European and Japanese rivals to undergo a similar process of restructuring and renewal.

The prime transmitter of competitive forces in this global economy is the multinational corporation. What differentiates the multinational enterprise from other firms engaged in international business is the globally coordinated allocation of resources by a single centralized management. Multinational corporations make decisions about market-entry strategy; ownership of foreign operations; and production, marketing, and financial activities with an eye to what is best for the corporation as a whole. The true multinational corporation emphasizes group performance rather than the performance of its individual parts.

Evolution of the Multinational Corporation

Every year, *Fortune* publishes a list of the 10 most-admired U.S. corporations. Year in and year out, most of these firms are largely multinational in philosophy and operations. In contrast, the least-admired tend to be national firms with much smaller proportions of assets, sales, or profits derived from foreign operations. Although multinationality and economic efficiency do not necessarily go hand in hand, international business is clearly of great importance to a growing number of U.S. and non-U.S. firms. The list of large American firms that receive 50% or more of their revenues and profits from abroad and that have a sizable fraction of their assets abroad reads like a corporate *Who's Who*: IBM, Gillette, Compaq, Dow Chemical, Colgate-Palmolive, 3M, Xerox, and Hewlett-Packard. Coca-Cola earns more than 80% of its beverage profit overseas and makes more money in Japan alone than it does in the United States.

Industries differ greatly in the extent to which foreign operations are of importance to them. For example, oil companies and banks are far more heavily involved overseas than are packaged food companies and automakers. Even within industries, companies differ markedly in their commitment to international business. For example, in 1995 Exxon had 59% of its assets, 78% of its sales, and 77% of its profits abroad. The corresponding figures for Atlantic Richfield are 26%, 21%, and 8%. Similarly, General Motors generated 54% of its income overseas, in contrast to 14% for Ford and less than 8% for Chrysler.

GENERAL ELECTRIC GLOBALIZES ITS MEDICAL SYSTEMS BUSINESS

A critical element of General Electric's (GE) global strategy is to be first or second in the world in a business or to exit that business. For example, in 1987, GE swapped its RCA consumer electronics division for Thomson CGR, the medical equipment business of Thomson SA of France, to strengthen its own medical unit. Together with GE Medical Systems Asia (GEMSA) in Japan, CGR makes GE No. 1 in the world market for X-ray, CAT scan, magnetic resonance, ultrasound, and other diagnostic imaging devices, ahead of Siemens (Germany), Philips (Netherlands), and Toshiba (Japan).

General Electric's production also is globalized, with each unit exclusively responsible for equipment in which it is the volume leader. Hence, GE Medical Systems now makes the high end of its CAT scanners and magnetic resonance equipment near Milwaukee (its headquarters) and the low end in Japan. The middle market is supplied by General Electric CGR SA in France. Engineering skills pass horizontally from the United States to Japan to France and back again. Each subsidiary supplies the marketing skills to its own home market.

Recently, in line with GE's decision to shift its corporate center of gravity from the industrialized world to the emerging markets of Asia and Latin America, Medical Systems has set up joint ventures in India and China to make low-end CAT scanners and various ultrasound devices for sale in their local markets. These machines were developed in Japan with GE's 75% joint-venture partner, Yokogawa Medical Systems, but the design work was turned over to India's vast pool of inexpensive engineers. At the same time, engineers in India and China are developing low-cost products that could serve markets in Latin America and the United States, where there is a demand from a cost-conscious medical community for cheaper machines.

The degree of internationalization of the American economy is often surprising. For example, analysts estimate that about 60% of the U.S. film industry's revenues came from foreign markets in 1997. The film industry illustrates other dimensions of internationalization as well, many of which are reflected in *Total Recall*, a film that was made by a Hungarian-born producer and a Dutch director, starred an Austrian-born leading man and a Canadian villain, was shot in Mexico, and was distributed by a Hollywood studio owned by a Japanese firm.

Appendix 1B provides further evidence of the growing internationalization of American business. It presents data on the size and scope of overseas investment by U.S. firms and U.S. investment by foreign firms. The numbers involved are in the hundreds of billions of dollars. Moreover, these investments have grown steadily over time, facilitated by a combination of factors: falling regulatory barriers to overseas investment; rapidly declining telecommunications and transport costs; and freer domestic and international capital markets in which vast sums of money can be raised, companies can be bought, and currency and other risks can be hedged. These factors have made it easier for companies to invest abroad, to do so more cheaply, and to experience less risk than ever before. A brief taxonomy of the MNC and its evolution is given in the sections that follow.

RAW-MATERIALS SEEKERS ▪ Raw-materials seekers were the earliest multinationals, the villains of international business. They are the firms—the British, Dutch, and French East India Companies, the Hudson's Bay Trading Company, and the Union Miniere Haut-Katanga—that first grew under the protective mantle of the British, Dutch, French, and Belgian colonial empires. Their aim was to exploit the raw materials that could be found overseas. The modern-day counterparts of these firms, the multinational oil and mining companies, were the first to make large foreign investments, beginning during the early years of the 20th century. Hence, large oil companies such as British Petroleum and the Standard Oil companies, which went where the dinosaurs died, were among the first true multinationals. Hard-mineral companies such as International Nickel, Anaconda Copper, and Kennecott Copper were also early investors abroad.

MARKET SEEKERS ▪ The market seeker is the archetype of the modern multinational firm that goes overseas to produce and sell in foreign markets. Examples include IBM, Volkswagen, and Unilever. Similarly, branded consumer-products companies such as Nestlé, Levi Strauss, MacDonald's, Procter & Gamble, and Coca-Cola have been operating abroad for decades and maintain vast manufacturing, marketing, and distribution networks from which they derive substantial sales and income.

Although there are some early examples of market-seeking MNCs (for example, Colt Firearms, Singer, Coca-Cola, N.V. Philips, and Imperial Chemicals), the bulk of **foreign direct investment**, which is the acquisition abroad of physical assets such as plant and equipment, took place after World War II. This investment was primarily a one-way flow—from the United States to Western Europe—until the early 1960s. At that point, the phenomenon of *reverse foreign investment* began, primarily with Western European firms acquiring U.S. firms. More recently, Japanese firms have begun investing in the United States and Western Europe, largely in response to perceived or actual restrictions on their exports to these markets.

COST MINIMIZERS ▪ Cost minimizer is a fairly recent category of firms doing business internationally. These firms seek out and invest in lower-cost production sites overseas (for example, Hong Kong, Taiwan, and Ireland) to remain cost-competitive both at home and abroad. Many of these firms are in the electronics industry. Examples include Texas Instruments, Atari, and Zenith.

1.2 THE PROCESS OF OVERSEAS EXPANSION

Studies of corporate expansion overseas indicate that firms become multinational by degree, with foreign direct investment being a late step in a process that begins with exports. For most companies, the globalization process does not occur through conscious design, at least in the early stages. It is the unplanned result of a series of corporate responses to a variety of threats and opportunities appearing at random abroad. From a broader perspective, however, the globalization of firms is the inevitable outcome of the competitive strivings of members of oligopolistic industries. Each member tries both to create and to exploit monopolistic product and factor advantages internationally while simultaneously attempting to reduce the competitive threats posed by other industry members.

EXHIBIT 1.1 Typical Foreign Expansion Sequence

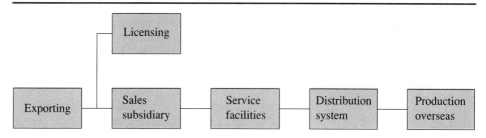

To meet these challenges, companies gradually increase their commitment to international business, developing strategies that are progressively more elaborate and sophisticated. The sequence normally involves exporting, setting up a foreign sales subsidiary, securing licensing agreements, and eventually establishing foreign production. This evolutionary approach to overseas expansion is a risk-minimizing response to operating in a highly uncertain foreign environment. By internationalizing in phases, a firm can gradually move from a relatively low-risk, low-return, export-oriented strategy to a higher-risk, higher-return strategy emphasizing international production. In effect, the firm is investing in information, learning enough at each stage to improve significantly its chances for success at the next stage. Exhibit 1.1 depicts the usual sequence of overseas expansion.

Exporting

Firms facing highly uncertain demand abroad typically will begin by exporting to a foreign market. The advantages of exporting are significant: Capital requirements and start-up costs are minimal, risk is low, and profits are immediate. Furthermore, this initial step provides the opportunity to learn about present and future supply and demand conditions, competition, channels of distribution, payment conventions, financial institutions, and financial techniques. Building on prior successes, companies then expand their marketing organizations abroad, switching from using export agents and other intermediaries to dealing directly with foreign agents and distributors. As increased communication with customers reduces uncertainty, the firm might set up its own sales subsidiary and new service facilities, such as a warehouse, with these marketing activities culminating in the control of its own distribution system.

Overseas Production

A major drawback to exporting is the inability to realize the full sales potential of a product. By manufacturing abroad, a company can more easily keep abreast of market developments, adapt its products and production schedules to changing local tastes and conditions, fill orders faster, and provide more comprehensive after-sales service. Many companies also set up research and development (R&D) facilities along with their foreign operations; they aim to pick the best brains, wherever they are. The results help companies keep track of the competition and design new products. For example, the Japanese subsidiary of Loctite, a U.S. maker of engineering adhesives, devised several new applications for sealants in the electronics industry.

Setting up local production facilities also shows a greater commitment to the local market, a move that typically brings added sales and provides increased assurance of supply stability. Certainty of supply is particularly important for firms that produce intermediate goods for sale to other companies. A case in point is SKF, the Swedish ball-bearing manufacturer. It was forced to manufacture in the United States to guarantee that its product, a crucial component in military equipment, would be available when needed. The Pentagon would not permit its suppliers of military hardware to be dependent on imported ball bearings, because imports could be halted in wartime and are always subject to the vagaries of ocean shipping.

Thus, most firms selling in foreign markets eventually find themselves forced to manufacture abroad. Foreign production covers a wide spectrum of activities from repairing, packaging, and finishing to processing, assembly, and full manufacture. Firms typically begin with the simpler stages—for example, packaging and assembly—and progressively integrate their manufacturing activities backward—to production of components and subassemblies.

Because the optimal entry strategy can change over time, a firm must continually monitor and evaluate the factors that bear on the effectiveness of its current entry strategy. New information and market perceptions change the risk-return trade-off for a given entry strategy, leading to a sequence of preferred entry modes, each adapted on the basis of prior experience to sustain and strengthen the firm's market position over time.

Associated with a firm's decision to produce abroad is the question of whether to *create* its own affiliates or to *acquire* going concerns. A major advantage of an acquisition is the capacity to effect a speedy transfer overseas of highly developed but underutilized parent skills, such as a novel production technology. Often the local firm also provides a ready-made marketing network. This network is especially important if the parent is a late entrant to the market. Many firms have used the acquisition approach to gain knowledge about the local market or a particular technology. The disadvantage, of course, is the cost of buying an ongoing company. In general, the larger and more experienced a firm becomes, the less frequently it uses acquisitions to expand overseas. Smaller and relatively less-experienced firms often turn to acquisitions.

Regardless of its preferences, a firm interested in expanding overseas may not have the option of acquiring a local operation. Michelin, the French manufacturer of radial tires, set up its own facilities in the United States because its tires are built on specially designed equipment; taking over an existing operation would have been out of the question.[1] Similarly, companies moving into developing countries often find they are forced to begin from the ground up because their line of business has no local counterpart.

Licensing

An alternative, and at times a precursor, to setting up production facilities abroad is to *license* a local firm to manufacture the company's products in return for royalties and other forms of payment. The principal advantages of licensing are the minimal investment required, faster market-entry time, and fewer financial and legal risks involved.

---

[1] Once that equipment became widespread in the industry, Michelin was able to expand through acquisition (which it did, in 1989, when it acquired Uniroyal-Goodrich).

But the corresponding cash flow is also relatively low, and there may be problems in maintaining product quality standards. The licensor may also face difficulty controlling exports by the foreign licensee, particularly when, as in Japan, the host government refuses to sanction restrictive clauses on sales to foreign markets. Thus, a licensing agreement may lead to the establishment of a competitor in third-country markets, with a consequent loss of future revenues to the licensing firm. The foreign licensee may also become such a strong competitor that the licensing firm will face difficulty entering the market when the agreement expires, leading to a further loss of potential profits.

For some firms, licensing alone is the preferred method of penetrating foreign markets. Other firms with diversified innovative product lines follow a strategy of trading technology for both equity in foreign joint ventures and royalty payments.

A Behavioral Definition of the Multinational Corporation

Regardless of the foreign entry or global expansion strategy pursued, the true multinational corporation is characterized more by its state of mind than by the size and worldwide dispersion of its assets. Rather than confine its search to domestic plant sites, the multinational firm asks, Where in the world should we build that plant? Similarly, multinational marketing management seeks global, not domestic, market segments to penetrate, and multinational financial management does not limit its search for capital or investment opportunities to any single national financial market. Hence, the essential element that distinguishes the true multinational is its commitment to seeking out, undertaking, and integrating manufacturing, marketing, R&D, and financing opportunities on a global, not domestic, basis. For example, IBM's superconductivity project was pioneered in Switzerland by a German scientist and a Swiss scientist who shared a Nobel Prize in physics for their work on the project.

Necessary complements to the integration of worldwide operations are flexibility, adaptability, and speed. Indeed, speed has become one of the critical competitive weapons in the fight for world market share. The ability to develop, make, and distribute products or services quickly enables companies to capture customers who demand constant innovation and rapid, flexible response. Exhibit 1.2 illustrates the combination of globally integrated activities and rapid response times of The Limited, a 3,200-store clothing chain headquartered in Columbus, Ohio.

Another critical aspect of competitiveness in this new world is focus. *Focus* means figuring out and building on what a company does best. This process typically involves divesting unrelated business activities and seeking attractive investment opportunities in the core business.

In this world-oriented corporation, a person's passport is not the criterion for promotion. Nor is a firm's citizenship a critical determinant of its success. Success depends on a new breed of business person: the global manager.

The Global Manager

In a world in which change is the rule and not the exception, the key to international competitiveness is the ability of management to adjust to change and volatility at an ever faster rate. In the words of General Electric Chairman Jack Welch, "I'm not here to

EXHIBIT 1.2 How the Limited Cuts the Fashion Cycle to 60 Days

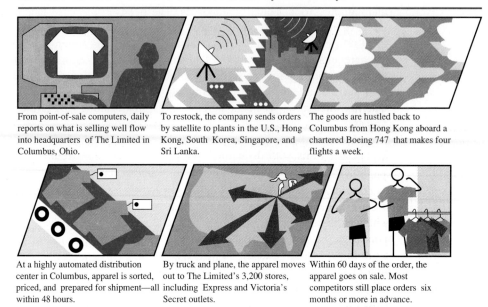

From point-of-sale computers, daily reports on what is selling well flow into headquarters of The Limited in Columbus, Ohio.

To restock, the company sends orders by satellite to plants in the U.S., Hong Kong, South Korea, Singapore, and Sri Lanka.

The goods are hustled back to Columbus from Hong Kong aboard a chartered Boeing 747 that makes four flights a week.

At a highly automated distribution center in Columbus, apparel is sorted, priced, and prepared for shipment—all within 48 hours.

By truck and plane, the apparel moves out to The Limited's 3,200 stores, including Express and Victoria's Secret outlets.

Within 60 days of the order, the apparel goes on sale. Most competitors still place orders six months or more in advance.

Source: Fortune, September 26, 1988, p. 56. Chart by Renee Klein, *Fortune*, © 1988 Time, Inc. Reprinted with permission by *Fortune* magazine.

predict the world. I'm here to be sure I've got a company that is strong enough to respond to whatever happens."[2]

The rapid pace of change means that new global managers need detailed knowledge of their own operations. Global managers must know how to make the products, where the raw materials and parts come from, how they get there, the alternatives, where the funds come from, and what their changing relative value does to their bottom lines. They must also understand the political and economic choices facing key nations and how those choices will affect the outcomes of their decisions.

In making decisions for the global company, managers search their array of plants in various nations for the most cost-effective mix of supplies, components, transport, and funds. All this is done with the constant awareness that the choices change and must be made again and again.

The problem of constant change disturbs some managers. It always has. But today's global managers have to anticipate it, understand it, deal with it, and turn it to their company's advantage. The payoff to thinking globally is a quality of decision making that enhances the firm's prospects for survival, growth, and profitability in the evolving world economy.

---

[2] Quoted in Ronald Henkoff, "How to Plan for 1995," *Fortune*, December 31, 1990, p. 70.

ARCO CHEMICAL DEVELOPS A WORLDWIDE STRATEGY

In the 1980s, ARCO Chemical shed its less successful product lines. At one point, revenue shrank from $3.5 billion annually to $1.5 billion. But by stripping down to its most competitive lines of business, ARCO could better respond to the global political and economic events constantly buffeting it. Around the world, it now can take advantage of its technological edge within its narrow niche—mostly intermediate chemicals and fuel additives. This strategy has paid off: In 1992, more than 40% of ARCO's $3 billion in sales was abroad, and it now makes about half of its new investment outside the United States. It also claims half the global market for the chemicals it sells.

ARCO Chemical went global because it had to. The company's engineering resins are sold to the auto industry. In the past, that meant selling exclusively to Detroit's Big Three in the U.S. market. Today, ARCO Chemical sells to Nissan, Toyota, Honda, Peugeot, Renault, and Volkswagen in Japan, the United States, and Europe. It also deals with Ford and General Motors in the United States and Europe. ARCO must be able to deliver a product anywhere in the world or lose the business.

Global operations also have meant, however, that ARCO Chemical faces increasingly stiff competition from abroad in addition to its traditional U.S. competitors such as Dow Chemical. European companies have expanded operations in America, and Japanese competitors also began to attack ARCO Chemical's business lines. For example, in 1990 Japan's Asahi Glass began a fierce price-cutting campaign in both Asia and Europe on products in which ARCO Chemical is strong.

In response, ARCO set up production facilities around the world and entered into joint ventures and strategic alliances. It counterattacked Asahi Glass by trying to steal one of Asahi's biggest customers in Japan. ARCO's joint-venture partner, Sumitomo Chemical, supplied competitive intelligence, and its knowledge of the Japanese market was instrumental in launching the counterattack.

Political and Labor Union Concerns about Global Competition

Politicians and labor leaders, unlike corporate leaders, usually take a more parochial view of global investment flows. Many instinctively denounce local corporations that invest abroad as job "exporters," even though most welcome foreign investors in their own countries as job creators. However, a growing number of U.S. citizens today view the current tide of American asset sales to foreign companies as a dangerous assault on U.S. sovereignty. They are unaware, for example, that foreign-owned companies account for more than 20% of industrial production in Germany and more than 50% in Canada, and neither of those countries appears to have experienced the slightest loss of sovereignty. Regardless of their views, however, the global rationalization of production will continue, as it is driven by global competition. The end result will be improvements in private sector efficiency and higher living standards.

Despite the common view that U.S. direct investment abroad comes at the expense of U.S. exports and jobs, the evidence clearly shows the opposite. By enabling MNCs to

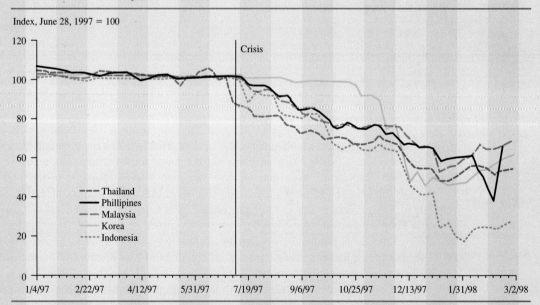

ILLUSTRATION

THE ASIAN TIGERS
FALL PREY TO WORLD
FINANCIAL MARKETS

For years, the nations of East Asia were held up as economic icons. Their typical blend of high savings and investment rates, autocratic political systems, export-oriented businesses, restricted domestic markets, government-directed capital allocation, and controlled financial systems were hailed as the ideal recipe for strong economic growth, particularly for developing nations. However, by the summer of 1997, the financial markets became disenchanted with this region, beginning with Thailand. Waves of currency selling left the Thai baht down 40% and the stock market down 50%. Thailand essentially went bankrupt. Its government fell and the In-

ternational Monetary Fund put together a $17 billion bailout package, conditioned on austerity measures. What the financial markets had seen that others had not was the rot at the core of Thailand's economy. Thais had run up huge debts, mostly in dollars, and were depending on the stability of the baht to repay these loans. Worse, Thai banks, urged on by the country's corrupt political leadership, were shoveling loans into money-losing ventures that were controlled by political cronies. As long as the money kept coming, Thailand's statistics on investment and growth looked good, but the result was a financially troubled economy that could not generate the income necessary to repay its loans.

Investors then turned to other East Asian economies and saw similar flaws there. One by one, the dominoes fell, from Bangkok to

EXHIBIT 1.3A Currency Devaluations

Index, June 28, 1997 = 100

Source: Southwest Economy, March/April 1998, Federal Reserve Bank of Dallas.

EXHIBIT 1.3B Stock Market Drop

Index, June 28, 1997 = 100

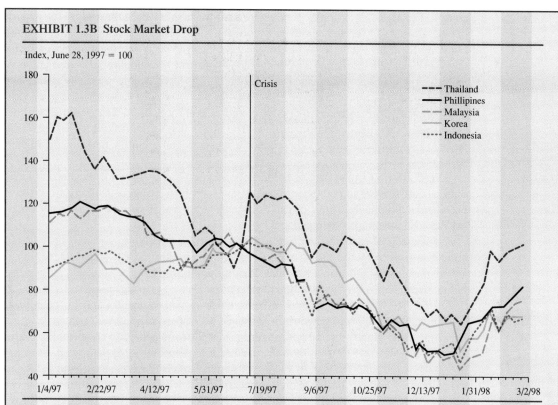

Source: *Southwest Economy*, March/April 1998, Federal Reserve Bank of Dallas.

Kuala Lumpur, Jakarta to Manila, Singapore to Taipei, Seoul to Hong Kong. The Asian tigers were humbled as previously stable currencies were crushed (see Exhibit 1.3a), local stock markets crashed (see Exhibit 1.3b), interest rates soared, banking systems tottered, economies contracted, bankruptcies spread, and governments were destabilized. The international bailout for the region grew to over $150 billion, crowned by $60 billion for South Korea, as the United States and other developed nations poured in funds for fear that the events in East Asia would spin out of control, threatening the world financial system with ruin and leading to a global recession. How to stave off such crises? The answer is financial markets that are open and transparent, leading to investment decisions that are based on sound economic principles rather than cronyism or political considerations.

What is the bright side of the awesome power wielded by the global financial markets? Simply this: These markets bring economic sanity even to nations run by corrupt elites. They have no tolerance for regimes that suppress enterprise, reward cronies, or squander resources on ego-building but economically dubious, grandiose projects.

expand their toeholds in foreign markets, such investments tend to increase U.S. exports of components and services and create more and higher-paying jobs in the United States. Ford and IBM, for example, would be generating less U.S.-based employment today had they not been able earlier to invest abroad—both by "outsourcing" the production of parts to low-wage countries such as Mexico and by establishing assembly plants and R&D centers in Europe and Japan.

Similarly, the argument that poor countries drain jobs from rich countries and depress wages for all is demonstrably false. The fact is that as poor countries prosper, they buy more of the advanced goods produced by the richer countries that support higher-paying jobs.

The growing irrelevance of borders for corporations will force policymakers to rethink old approaches to regulation. For example, corporate mergers that once would have been barred as anticompetitive might make sense if the true measure of a company's market share is global rather than national.

International economic integration also reduces the freedom of governments to determine their own economic policy. If a government tries to raise tax rates on business, for example, it is increasingly easy for business to shift production abroad. Similarly, nations that fail to invest in their physical and intellectual infrastructure—roads, bridges, R&D, education—will likely lose entrepreneurs and jobs to nations that do invest. Capital—both financial and intellectual—will go where it is wanted and stay where it is well treated. In short, economic integration is forcing governments, as well as companies, to compete. For example, after America's 1986 tax reform that slashed income tax rates, virtually every other nation in the world followed suit. In a world of porous borders, governments found it difficult to ignore what worked. Similarly, big U.S. mutual funds are wielding increasing clout in developing nations, particularly in Latin America and Asia. In essence, the funds are trying to do overseas what they are already doing domestically: pressure management (in this case governments) to adopt policies that will maximize returns. The carrot is more money; the stick is capital flight. Simply put, the globalization of trade and finance has created an unforgiving environment that penalizes economic mismanagement and allots capital and jobs to the nations delivering the highest risk-adjusted returns. As markets become more efficient, they are quicker to reward sound economic policy and swifter to punish the profligate. Their judgments are harsh and they cannot be appealed.

Paradoxically, however, even as people are disturbed at the thought of their government losing control of events, they have lost faith in government's ability to solve many of their problems. One result has been the collapse of Communism in Eastern Europe and the spread of free-market economics in developed and developing countries alike. Rejecting the statist policies of the past, they are shrinking, closing, pruning, or privatizing state-owned industries and subjecting their economies to the rigors of foreign competition. In response to these changes, developing countries in 1996 received over $240 trillion in new foreign investment. Five years earlier, by contrast, they were exporting savings, as they paid service costs on their large foreign debts and as local capital fled hyperinflation and confiscatory tax and regulatory regimes. These dramatic shifts in policy—and the rewards they have brought to their initiators—have further strengthened the power of markets to set prices and priorities around the world.

The stresses caused by global competition have stirred up protectionists and given rise to new concerns about the consequences of free trade. For example, the sudden entry

JAPANESE COMPETITION AFFECTS THE U.S. AUTO INDUSTRY

Until recently, Japanese competition steadily eroded the influence of the Big Three U.S. automakers in the auto industry. During the 1980s, Japanese auto companies raised their U.S. market share 8 points, to 28%, versus 65% for Detroit and 5% for Europe.

The tough Japanese competition was a big factor in the sales and profit crunch that hit the Big Three. General Motors, Ford, and Chrysler responded by shutting down U.S. plants and acting to curb labor costs. Thus, Japanese competition has limited the wages and benefits that United Auto Workers (UAW) union members can earn, as well as the prices that U.S. companies can charge for their cars. Both unions and companies understand that in this competitive environment, raising wages and car prices leads to fewer sales and fewer jobs. One solution, which allows both the Big Three and the UAW to avoid making hard choices — sales volume versus profit margin, and jobs versus wages and benefits — is political: Limit Japanese competition through quotas, tariffs, and other protectionist devices, and thereby control its effects on the U.S. auto industry. Unfortunately, American consumers get stuck with the tab for this apparent free lunch in the form of higher car prices and less choice.

The best argument against protectionism, however, is long-term competitiveness. It was, after all, cutthroat competition from the Japanese that forced Detroit to get its act together. The Big Three swept away layers of unneeded management, raised productivity, and dramatically increased the quality of their cars and trucks. They also shifted their focus toward the part of the business in which the Japanese did not have strong products but which just happened to be America's hottest and fastest growing automotive segment — light trucks, which includes pickups, minivans, and sport-utility vehicles. Combined with a strong yen and higher Japanese prices, these changes helped Detroit pick up three percentage points of market share in 1992 and 1993 alone, mostly at the expense of Japanese nameplates. By 1994, the Japanese share of the U.S. auto market, which peaked at 29% in 1991, had fallen to 25%.

The inescapable fact is that Japanese automakers forced Detroit to make better cars at better prices. Handicapping the Japanese couldn't possibly have had the same effect.

of 3 billion people from low-wage countries such as China, Mexico, Russia, and India into the global marketplace is provoking anxiety among workers in the old industrial countries about their living standards. As the illustration of the U.S. auto industry above indicates, companies and unions are quite rational in fearing the effects of foreign competition. It disrupts established industry patterns and limits the wages and benefits of workers by giving more choice to consumers. The U.S.-Canada trade agreement, which ends tariff barriers by the year 2000, has caused major disruption to Canada's manufacturing industry. Plants are closing, mergers are proliferating, and both domestic and multinational companies are adjusting their operations to the new continental market. Similarly, the North American Free Trade Agreement (NAFTA), which created a giant free-trade area from the Yukon to the Yucatán, has forced formerly sheltered companies, especially in Mexico, to cut costs and change their way of doing business. It led U.S. companies to shift production both into and out of Mexico, while confronting American and Canadian workers with a new pool of lower-priced (but also less productive) labor.

ꜱꜱʟᴜꜱᴛʀᴀᴛɪᴏɴ

Rᴏꜱꜱ Pᴇʀᴏᴛ Fɪɢʜᴛꜱ NAFTA, ᴀɴᴅ Cʟɪɴᴛᴏɴ Rᴇꜱᴘᴏɴᴅꜱ

In November 1993, the North American Free Trade Agreement (NAFTA) was signed into law, but not before it stirred spirited opposition among unions and politicians. The best-known critic of NAFTA was Ross Perot, the billionaire Texan who launched a multimillion-dollar campaign against the free-trade treaty. He claimed that if NAFTA were ratified, the United States would hear a giant sucking sound as businesses rushed to Mexico to take advantage of its lower wages, putting nearly 6 million U.S. jobs at risk.

This argument ignores the economic theory of trade as well as its reality. If it is true that American factory workers are paid about eight times as much as their Mexican counterparts, it is also true that they are about eight times as productive. As Mexican workers become more productive, their pay will rise proportionately. This prediction is borne out by recent economic history. Like critics of NAFTA in 1993, many in 1986 feared a giant sucking sound from south of another border—the Pyrenees. Spain, with wages less than half those of its northern neighbors, and Portugal, with wages about a fifth Europe's norm, were about to join the European Community. Opponents said their low wages would drag down wages or take away jobs from French and German workers.

What happened? Job creation in France and Germany exceeded job creation in Spain and Portugal. More important, workers got to trade up to better jobs because opening trade allows all countries to specialize where their advantage is greatest. Specialization raises incomes. It is the reason all parties benefit from trade. Again the evidence bears this assertion out: By 1993, French and German wages had doubled; Spanish and Portuguese wages increased slightly faster. Put simply, countries do not grow richer at each other's expense. If allowed to trade freely, they grow richer together—each supplying the other with products, markets, and the spur of competition.

After numerous appeals by NAFTA supporters that he speak out in favor of the treaty, President Bill Clinton finally responded. On September 14, 1993, he gave the following eloquent argument for open borders and open markets:

I want to say to my fellow Americans, when you live in a time of change, the only way to recover your security and to broaden your horizons is to adapt to the change, to embrace it, to move forward. Nothing we do in this great capital can change the fact that factories or information can flash across the world, that people can move money around in the blink of an eye. Nothing can change the fact that technology can be adopted, once created, by people all across the world and then rapidly adapted in new and different ways by people who have a little different take on the way that technology works.

For two decades, the winds of global competition have made these things clear to any American with eyes to see. The only way we can recover the fortunes of the middle class in this country so that people who work harder and smarter can at least prosper more, the only way we can pass on the American dream of the past 40 years to our children and their children for the next 40 is to adapt to the changes that are occurring.

In a fundamental sense, this debate about NAFTA is a debate about whether we will embrace these changes and create the jobs of tomorrow or try to resist these changes, hoping we can preserve the economic structures of yesterday. I tell you, my fellow Americans, that if we learn anything from the collapse of the Berlin Wall and the fall of the governments in Eastern Europe, even a totally controlled society cannot resist the winds of change that economics and technology and information flow have imposed in this world of ours. That is not an option. Our only realistic option is to embrace these changes and create the jobs of tomorrow.

So it is all the more encouraging that political leaders keep trying to stretch borders. The world's long march toward a global economy should accelerate considerably in the next few years if the U.S.-Canada-Mexico free-trade pact and the European Community's drive to create a truly common market proceed as planned. The greater integration of national economies is likely to continue despite the stresses it causes as politicians worldwide increasingly come to realize that they either accept this integration or watch their respective nations fall behind.

1.3 MULTINATIONAL FINANCIAL MANAGEMENT: THEORY AND PRACTICE

Although all functional areas can benefit from a global perspective, this book concentrates on developing financial policies that are appropriate for the multinational firm. The main objective of multinational financial management is to maximize shareholder wealth as measured by share price. This means making financing and investment decisions that add as much value as possible to the firm. It also means that companies must manage effectively the assets under their control.

The focus on shareholder value stems from the fact that shareholders are the legal owners of the firm and management has a fiduciary obligation to act in their best interests. Although other stakeholders in the company do have rights, these are not coequal with the shareholders' rights. Shareholders provide the risk capital that cushions the claims of alternative stakeholders. Allowing alternative stakeholders coequal control over capital supplied by others is equivalent to allowing one group to risk someone else's capital. This undoubtedly would impair future equity formation and produce numerous other inefficiencies.

A more compelling reason for focusing on creating shareholder wealth is that those companies who don't are likely to be prime takeover targets and candidates for a forced corporate restructuring. Conversely, maximizing shareholder value provides the best defense against a hostile takeover: a high stock price. Companies that build shareholder value also find it easier to attract equity capital. Equity capital is especially critical for companies that operate in a riskier environment and for companies that are seeking to grow.

Last, but not least, shareholders are not the only beneficiaries of corporate success. By forcing managers to evaluate business strategies based on prospective cash flows, the shareholder-value approach favors strategies that enhance a company's cash-flow generating ability—which is good for everyone, not just shareholders. Companies that create value have more money to distribute to all stakeholders, not just shareholders. Put another way, you have to create wealth before you can distribute it. Thus, there is no inherent economic conflict between shareholders and stakeholders. Indeed, most financial economists believe that maximizing shareholder value is not merely the best way, it is the *only* way to maximize the economic interests of *all* stakeholders over time.

Although an institution as complex as the multinational corporation cannot be said to have a single, unambiguous will, the principle of shareholder wealth maximization provides a rational guide to financial decision making. However, other financial goals that reflect the relative autonomy of management and external pressures also are examined here.

The Multinational Financial System

From a financial management standpoint, one of the distinguishing characteristics of the multinational corporation, in contrast to a collection of independent national firms dealing at arm's length with one another, is its ability to move money and profits among its affiliated companies through internal transfer mechanisms, the aggregate of which comprise the **multinational financial system**. These mechanisms include transfer prices on goods and services traded internally, intercompany loans, dividend payments, leading (speeding up) and lagging (slowing down) intercompany payments, and fee and royalty charges. They lead to patterns of profits and movements of funds that would be impossible in the world of Adam Smith.

Financial transactions within the MNC result from the internal transfer of goods, services, technology, and capital. These product and factor flows range from intermediate and finished goods to less tangible items such as management skills, trademarks, and patents. Those transactions not liquidated immediately give rise to some type of financial claim, such as royalties for the use of a patent or accounts receivable for goods sold on credit. In addition, capital investments lead to future flows of dividends and interest and principal repayments. Exhibit 1.4 depicts some of the myriad financial linkages possible in the MNC.

Although all the links portrayed in Exhibit 1.4 can and do exist among independent firms, the MNC has greater control over the mode and timing of these financial transfers.[3]

MODE OF TRANSFER ■ The MNC has considerable freedom in selecting the **financial channels** through which funds, allocated profits, or both are moved. For example, patents and trademarks can be sold outright or transferred in return for a contractual stream of royalty payments. Similarly, the MNC can move profits and cash from one unit to another by adjusting **transfer prices** on intercompany sales and purchases of goods and services. With regard to investment flows, capital can be sent overseas as debt with at least some choice of interest rate, currency of denomination, and repayment schedule, or as equity with returns in the form of dividends. Multinational firms can use these various channels, singly or in combination, to transfer funds internationally, depending on the specific circumstances encountered. Furthermore, within the limits of various national laws and with regard to the relations between a foreign affiliate and its host government, these flows may be more advantageous than those that would result from dealings with independent firms.

TIMING FLEXIBILITY ■ Some of the internally generated financial claims require a fixed payment schedule; others can be accelerated or delayed. This **leading and lagging** is most often applied to interaffiliate trade credit, where a change in open account terms, say from 90 to 180 days, can involve massive shifts in liquidity. (Some nations have regula-

[3] See Donald R. Lessard, "Transfer Prices, Taxes, and Financial Markets: Implications of Internal Financial Transfers within the Multinational Firm," in *The Economic Effects of Multinational Corporations*, Robert G. Hawkins, ed. (Greenwich, Conn.: JAI Press, 1979); and David P. Rutenberg, "Maneuvering Liquid Assets in a Multinational Company," *Management Science*, June 1970, pp. B-671–684. This section draws extensively from Lessard's article.

EXHIBIT 1.4 The Multinational Corporate Financial System

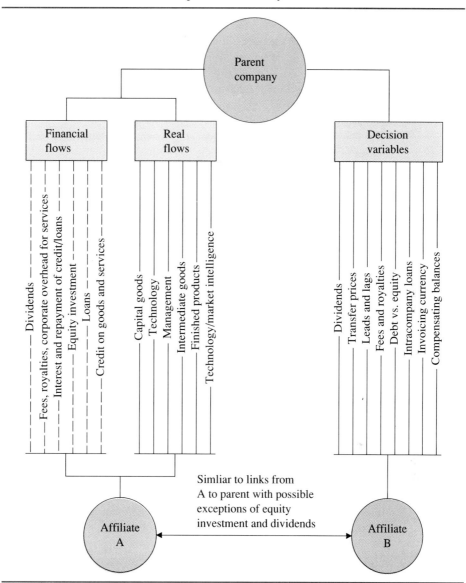

Similar to links from A to parent with possible exceptions of equity investment and dividends

Source: Adapted from Figure 1 in Donald R. Lessard, "Transfer Prices, Taxes, and Financial Markets: Implications of Internal Financial Transfers within the Multinational Firm," in *The Economic Effects of Multinational Corporations*, Robert G. Hawkins, ed. (Greenwich, Conn.: JAI Press, 1979), by permission of the author and the publisher.

tions about the repatriation of the proceeds of export sales. Thus, there is typically not complete freedom to move funds by leading and lagging.) In addition, the timing of fee and royalty payments may be modified when all parties to the agreement are related. Even if the contract cannot be altered once the parties have agreed, the MNC generally has latitude when the terms are initially established.

In the absence of **exchange controls**—government regulations that restrict the transfer of funds to nonresidents—firms have the greatest amount of flexibility in the timing of equity claims. The earnings of a foreign affiliate can be retained or used to pay dividends that in turn can be deferred or paid in advance.

Despite the frequent presence of government regulations or limiting contractual arrangements, most MNCs have some flexibility in the timing of fund flows. This latitude is enhanced by the MNC's ability to control the timing of many of the underlying real transactions. For instance, shipping schedules can be altered so that one unit carries additional inventory for a sister affiliate.

VALUE ■ By shifting profits from high-tax to lower-tax nations, the MNC can reduce its global tax payments. Similarly, the MNC's ability to transfer funds among its several units may allow it to circumvent currency controls and other regulations and to tap previously inaccessible investment and financing opportunities. However, because most of the gains derive from the MNC's skill at taking advantage of openings in tax laws or regulatory barriers, governments do not always appreciate the MNC's capabilities and global profit-maximizing behavior. Thus, controversy has accompanied the international orientation of the multinational corporation.

Criticisms of the Multinational Corporation

Critics of the MNC liken its behavior to that of an octopus with tentacles extended, squeezing the nations of the world to satisfy the apparently insatiable appetite of its center. Its defenders claim that only by linking activities globally can world output be maximized. According to this view, greater profits from overseas activities are the just reward for providing the world with new products, technologies, and know-how.

This book's focus is on multinational financial management, so it does not directly address this controversy. It concentrates instead on the development of analytical approaches to deal with the major environmental problems and decisions involving overseas investment and financing. In carrying out these financial policies, though, conflicts between corporations and nation-states will inevitably arise.

A classic case is that of General Motors-Holden's Ltd. The General Motors (GM) wholly owned Australian affiliate was founded in 1926 with an initial equity investment of A\$3.5 million. The earnings were reinvested until 1954, at which time the first dividend, for A\$9.2 million, was paid to the parent company in Detroit. This amount seemed reasonable to GM management, considering the 28 years of forgoing dividends, but the Australian press and politicians denounced a dividend equal to more than 260% of GM's original equity investment as economic exploitation and imperialism.[4]

---

[4] Reported in, among other places, Sidney M. Robbins and Robert B. Stobaugh, *Money in the Multinational Enterprise* (New York: Basic Books, 1973), p. 59.

More recently, Brazil, facing one of its periodic balance-of-payments crises, chose to impose stringent controls on the removal of profits by MNCs, thereby affecting the financial operations of such firms as Volkswagen and Scott Paper. In addition, companies operating in countries as diverse as Canada and Chile, Italy and India, and the United States and Uruguay have faced various political risks, including price controls and confiscation of local operations. This book examines the modification of financial policies to align better with national objectives in an effort to reduce such risks and minimize the costs of the adjustments.

This text also considers the links between financial management and other functional areas. After all, the analysis of investment projects is dependent on sales forecasts and cost estimates, and the dispersal of production and marketing activities affects a firm's ability to flow funds internationally, as well as its vulnerability to expropriation.

Functions of Financial Management

Financial management traditionally is separated into two basic functions: the acquisition of funds and the investment of those funds. The first function, also known as the *financing decision*, involves generating funds from internal sources or from sources external to the firm at the lowest long-run cost possible. The *investment decision* is concerned with the allocation of funds over time in such a way that shareholder wealth is maximized. Many of the concerns and activities of multinational financial management, however, cannot be categorized so neatly.

Internal corporate fund flows such as loan repayments often are undertaken to access funds that are already owned, at least in theory, by the MNC itself. Other flows, such as dividend payments, may take place to reduce taxes or currency risk. Capital structure and other financing decisions frequently are motivated by a desire to reduce investment risks as well as financing costs. Furthermore, exchange risk management involves both financing decisions and investment decisions. Throughout this book, therefore, the interaction between these decisions is stressed because the right combination of these decisions is the key to maximizing the value of the firm to its shareholders.

Theme

Financial executives in multinational corporations face many factors that have no domestic counterparts. These factors include exchange and inflation risks; international differences in tax rates; multiple money markets, often with limited access; currency controls; and political risks, such as sudden and creeping expropriation.

When companies consider the unique characteristics of multinational financial management, they understandably emphasize the additional political and economic risks faced when going abroad. But a broader perspective is necessary if firms are to take advantage of being multinational.

The ability to move people, money, and material on a global basis enables the multinational corporation to be more than the sum of its parts. By having operations in different countries, the MNC can access segmented capital markets to lower its overall cost of capital, shift profits to lower its taxes, and take advantage of **international diversification** of markets and production sites to reduce the riskiness of its earnings. Multinationals have taken the old adage of "don't put all your eggs in one basket" to its logical conclusion.

Operating globally confers other advantages as well: It increases the bargaining power of multinational firms when they negotiate investment agreements and operating conditions with foreign governments and labor unions; it gives MNCs continuous access to information on the newest process technologies available overseas and the latest research and development activities of their foreign competitors; and it helps them diversify their funding sources by giving them expanded access to the world's capital markets.

In summary, this book emphasizes the many opportunities associated with being multinational without neglecting the corresponding risks. To properly analyze and balance these international risks and rewards, we must use the lessons to be learned from domestic corporate finance.

Relationship to Domestic Financial Management

In recent years, there has been an abundance of new research in the area of international corporate finance. The major thrust of this work has been to apply the methodology and logic of financial economics to the study of key international financial decisions. Critical problem areas, such as foreign exchange risk management and foreign investment analysis, have benefited from the insights provided by **financial economics**—a discipline that emphasizes the use of economic analysis to understand the basic workings of financial markets, particularly the measurement and pricing of risk and the intertemporal allocation of funds.

By focusing on the behavior of financial markets and their participants rather than on how to solve specific problems, we can derive fundamental principles of valuation and develop from them superior approaches to financial management, much as a better understanding of the basic laws of physics leads to better-designed and better-functioning products. We also can better gauge the validity of existing approaches to financial decision making by seeing whether their underlying assumptions are consistent with our knowledge of financial markets and valuation principles.

Three concepts arising in financial economics have proved to be of particular importance in developing a theoretical foundation for international corporate finance: arbitrage, market efficiency, and capital asset pricing. Throughout the remainder of the book, we rely on these concepts, which are briefly described in the next sections.

ARBITRAGE ■**Arbitrage** traditionally has been defined as the purchase of securities or commodities on one market for immediate resale on another in order to profit from a price discrepancy. In recent years, however, arbitrage has been used to describe a broader range of activities. **Tax arbitrage**, for example, involves the shifting of gains or losses from one tax jurisdiction to another to profit from differences in tax rates. In a broader context, **risk arbitrage**, or speculation, describes the process that leads to equality of risk-adjusted returns on different securities, unless market imperfections that hinder this adjustment process exist. In fact, it is the process of arbitrage that ensures **market efficiency**.

MARKET EFFICIENCY ■An **efficient market** is one in which the prices of traded securities readily incorporate new information. Numerous studies of U.S. and foreign capital markets have shown that traded securities are correctly priced in that trading rules based on past prices or publicly available information cannot consistently lead to profits (after adjusting for transaction costs) in excess of those due solely to risk taking.

The predictive power of markets lies in their ability to collect in one place a mass of individual judgments from around the world. These judgments are based on current information. If the trend of future policies changes, people will revise their expectations, and prices will change to incorporate the new information.

To say that markets are efficient, however, is not to say that they never blunder. Swept up in enthusiasms or urged on by governments, investors appear to succumb periodically to herd behavior and go to excess, culminating in a financial crisis. In the 1980s, for example, there was the international banking crisis stemming from overly optimistic lending to developing nations, and in the 1990s the Asian crisis was associated with overly optimistic lending to the rapidly growing Asian tigers. To date, these crises have been resolved, albeit with much pain. Between crisis and resolution, however, is always uncharted territory, with the ever-present potential of panic feeding on itself and spreading from one nation to another, leading to global instability and recession. What we can say about markets, however, is that they are self-correcting; unlike governments, when investors spot problems, their instinct is to withdraw funds, not add more. At the same time, if a nation's economic fundamentals are basically sound, investors will eventually recognize that and their capital will return.

CAPITAL ASSET PRICING ■ **Capital asset pricing** refers to the way in which securities are valued in line with their anticipated risks and returns. Because risk is such an integral element of international financial decisions, this book briefly summarizes the results of more than two decades of study on the pricing of risk in capital markets. The outcome of this research has been to posit a specific relationship between risk (measured by return variability) and required asset returns, now formalized in the **capital asset pricing model (CAPM)** and the more general arbitrage price theory (APT).

Both the CAPM and the APT assume that the total variability of an asset's returns can be attributed to two sources: (1) marketwide influences that affect all assets to some extent, such as the state of the economy, and (2) other risks that are specific to a given firm, such as a strike. The former type of risk is usually termed **systematic**, or **nondiversifiable**, **risk**, and the latter, **unsystematic**, or **diversifiable**, **risk**. Unsystematic risk is largely irrelevant to the highly diversified holder of securities because the effects of such disturbances cancel out, on average, in the portfolio. On the other hand, no matter how well diversified a stock portfolio is, systematic risk, by definition, cannot be eliminated, and thus the investor must be compensated for bearing this risk. This distinction between systematic risk and unsystematic risk provides the theoretical foundation for the study of risk in the multinational corporation and is referred to throughout the book.

The Importance of Total Risk

Although the message of the CAPM and the APT is that only the systematic component of risk will be rewarded with a risk premium, this does not mean that total risk—the combination of systematic and unsystematic risk—is unimportant to the value of the firm. In addition to the effect of systematic risk on the appropriate discount rate, total risk may have a negative impact on the firm's *expected* cash flows.[5]

---

[5] The effect of total risk is discussed in Alan C. Shapiro and Sheridan Titman, "An Integrated Approach to Corporate Risk Management," *Midland Corporate Finance Journal*, Summer 1985, pp. 41–56.

The inverse relation between risk and expected cash flows arises because financial distress, which is most likely to occur for firms with high total risk, can impose costs on customers, suppliers, and employees and thereby affect their willingness to commit themselves to relationships with the firm. For example, potential customers will be nervous about purchasing a product they might have difficulty getting serviced if the firm goes out of business. Similarly, a firm struggling to survive is unlikely to find suppliers willing to provide it with specially developed products or services, except at a higher-than-usual price. The uncertainty created by volatile earnings and cash flows also may hinder management's ability to take a long view of the firm's prospects and make the most of its opportunities.

In summary, total risk is likely to affect a firm's value adversely by leading to lower sales and higher costs. Consequently, any action taken by a firm that decreases its total risk will improve its sales and cost outlooks, thereby increasing its expected cash flows.

These considerations justify the range of corporate hedging activities that multinational firms engage in to reduce total risk. This text focuses on those risks that appear to be more international than national in nature, including inflation risk, exchange risk, and political risk. As we will see, however, appearances can be deceiving, because these risks also affect firms that do business in only one country. Moreover, international diversification may actually allow firms to reduce the total risk they face. Much of the general market risk facing a company is related to the cyclical nature of the domestic economy of the home country. Operating in several nations whose economic cycles are not perfectly in phase should reduce the variability of the firm's earnings. Thus, even though the riskiness of operating in any one foreign country may be greater than the risk of operating in the United States (or other home country), diversification can eliminate much of that risk.

What is true for companies is also true for investors. International diversification can reduce the riskiness of an investment portfolio because national financial markets tend to move somewhat independently of each other.

The Global Financial Marketplace

Market efficiency has been greatly facilitated by the marriage of computers and telecommunications. The resulting electronic infrastructure melds the world into one global market for ideas, data, and capital, all moving at almost the speed of light to any part of the planet. Today there are more than 200,000 computer terminals in hundreds of trading rooms, in dozens of nations, that light up to display an unending flow of news. Only about two minutes elapse between the time a president, a prime minister, or a central banker makes a statement and the time traders buy or sell currency, stocks, and bonds according to their evaluation of that policy's effect on the market.

The result is a continuing global referendum on a nation's economic policies, which is the final determinant of the value of its currency. Just as we learn from television the winner of a presidential election weeks before the electoral college even assembles, so also we learn instantly from the foreign exchange market what the world thinks of our announced economic policies even before they are implemented. In a way, the financial market is a form of economic free speech. Although many politicians do not like what it is saying, the market presents judgments that are clear-eyed and hard-nosed. It knows that there are no miracle drugs that can replace sound fiscal and monetary policies. Thus, cosmetic political fixes will exacerbate, not alleviate, a falling currency.

The Role of the Financial Executive in an Efficient Market

The basic insight into financial management that we can gain from recent empirical research in financial economics is the following: *Attempts to increase the value of a firm by purely financial measures or accounting manipulations are unlikely to succeed unless there are capital market imperfections or asymmetries in tax regulations.*

Rather than downgrading the role of the financial executive, the net result of these research findings has been to focus attention on those areas and circumstances in which financial decisions can have a measurable impact. The key areas are capital budgeting, working capital management, and tax management. The circumstances to be aware of include **capital market imperfections**, primarily caused by government regulations, and asymmetries in the tax treatment of different types and sources of revenues and costs.

The value of good financial management is enhanced in the international arena because of the much greater likelihood of market imperfections and multiple tax rates. In addition, the greater complexity of international operations is likely to increase the payoffs from a knowledgeable and sophisticated approach to internationalizing the traditional areas of financial management.

1.4 OUTLINE OF THE BOOK

This book is divided into five parts.

- Part I: Environment of International Financial Management
- Part II: Foreign Exchange Risk Management
- Part III: Multinational Working Capital Management
- Part IV: Financing Foreign Operations
- Part V: Foreign Investment Analysis

The following sections briefly discuss these parts and their chapters.

Environment of International Financial Management

Part I examines the environment in which international financial decisions are made. Chapter 2 discusses the basic factors that affect currency values. It also explains the basics of central bank intervention in foreign exchange markets, including the economic and political motivations for such intervention. Chapter 3 describes the international monetary system and shows how the choice of system affects the determination of exchange rates. Chapter 4 analyzes the balance of payments and the links between national economies, and chapter 5 describes the foreign exchange market and how it functions. Foreign currency futures and options contracts are discussed in chapter 6. Chapter 7 is a crucial chapter because it introduces four key equilibrium relationships—among inflation rates, interest rates, and exchange rates—in international finance that form the basis for much of the analysis in the remainder of the text.

Foreign Exchange Risk Management

Part II discusses foreign exchange risk management, a traditional area of concern that is receiving even more attention today. Chapter 8 discusses the likely impact that an exchange rate change will have on a firm (its exposure) from an accounting perspective; chapter 9 analyzes the costs and benefits of alternative financial techniques to hedge against those exchange risks. Chapter 10 examines exposure from an economic perspective. As part of the analysis of economic exposure, the relationship between inflation and currency changes and its implications for corporate cash flows is recognized. Chapter 11 develops marketing, logistic, and financial policies to cope with the competitive consequences of currency changes.

Multinational Working Capital Management

Part III examines working capital management in the multinational corporation. The subject of trade financing is covered in chapter 12. Chapter 13 discusses current asset management in the MNC, including the management of cash, inventory, and receivables. It also deals with current liability management, presenting the alternative short-term financing techniques available and showing how to evaluate their relative costs. Chapter 14 describes the mechanisms available to the MNC to shift funds and profits among its various units, while considering the tax and other consequences of these maneuvers. The aim of these maneuvers is to create an integrated global financial planning system.

Financing Foreign Operations

Part IV focuses on laying out and evaluating the medium- and long-term financing options facing the multinational firm, then developing a financial package that is tailored to the firm's specific operating environment. Chapter 15 describes the alternative external, medium- and long-term debt financing options available to the multinational corporation. These options include an outline of the international capital markets—namely, the Eurocurrency and Eurobond markets. Chapter 16 discusses special financing vehicles available to the MNC, including interest-rate and currency swaps, structured notes, interest rates forwards and futures, international leasing, and debt-equity swaps. Chapter 17 discusses the development and expansion of international banking activities and the international debt crisis. It also shows how to analyze country risk—the credit risk on loans to a foreign nation—a topic of great concern these days. Chapter 18 seeks to determine the cost-of-capital figure(s) that MNCs should use in evaluating foreign investments, given the funding sources actually employed.

Foreign Investment Analysis

Part V analyzes the foreign investment decision process. Chapter 19 begins by discussing the nature and consequences of international portfolio investing—the purchase of foreign stocks and bonds. In chapter 20, the strategy of foreign direct investment is discussed, including an analysis of the motivations for going abroad and those factors that have contributed to business success overseas. Chapter 21 presents techniques for evaluating foreign investment proposals, emphasizing how to adjust cash flows for the various political and economic risks encountered abroad, such as inflation, currency fluctuations, and expropriations. Chapter 22 discusses the measurement and management of political risk. It identifies political risks and then shows how companies can control these risks by appropriately structuring the initial investment and by making suitable modifications to subsequent operating decisions.

➤ Questions

1. a. What are the various categories of multinational firms?
   b. What is the motivation for international expansion of firms within each category?
2. a. How does foreign competition limit the prices that domestic companies can charge and the wages and benefits that workers can demand?
   b. What political solutions can help companies and unions avoid the limitations imposed by foreign competition?
   c. Who pays for these political solutions? Explain.
3. a. What is the internal financial transfer system of the multinational firm?
   b. What are its distinguishing characteristics?
   c. What are the different modes of internal fund transfers available to the MNC?
4. How does the internal financial transfer system add value to the multinational firm?
5. a. Why do companies generally follow a sequential strategy in moving overseas?
   b. What are the pluses and minuses of exporting? of licensing? of foreign production?
6. What is an efficient market?
7. In seeking to predict tomorrow's exchange rate, are you better off knowing today's exchange rate or the exchange rates for the past 100 days?

8. a. What is the capital asset pricing model?
   b. What is the basic message of the CAPM?
   c. How might a multinational firm use the CAPM?
9. Why might setting up production facilities abroad lead to expanded sales in the local market?
10. A memorandum by Labor Secretary Robert Reich to President Bill Clinton suggests that the government penalize U.S. companies that invest overseas rather than at home. According to Reich, this kind of investment hurts exports and destroys well-paying jobs. Comment on this argument.
11. a. Are multinational firms riskier than purely domestic firms?
    b. What data would you need to address this question?
    c. Is there any reason to believe that MNCs may be less risky than purely domestic firms? Explain.
12. In what ways do financial markets grade government economic policies?
13. a. How might total risk affect a firm's production costs and its ability to sell? Give some examples of firms in financial distress that saw their sales drop.
    b. What is the relation between the effects of total risk on a firm's sales and costs and its desire to hedge foreign exchange risk?

♦APPENDIX 1A The Origins and Consequences of International Trade

Underlying the theory of international trade is the doctrine of comparative advantage. This doctrine rests on certain assumptions:

1. Exporters sell undifferentiated (commodity) goods and services to unrelated importers.
2. Factors of production cannot move freely across countries. Instead, trade takes place in the goods and services produced by these factors of production.

As noted at the beginning of chapter 1, the doctrine of comparative advantage also ignores the roles of uncertainty, economies of scale, and technology in international trade; and it is static rather than dynamic. Nonetheless, this theory helps to explain why nations trade with each other and forms the basis for assessing the consequences of international trade policies.

To illustrate the main features of the doctrine of comparative advantage and to distinguish this concept from that of absolute advantage, suppose the United States and the United Kingdom produce the same two products, wheat and coal, according to the following production schedules, where the units referred to are units of production (labor, capital, land, and technology):[6]

	Wheat	Coal
U.S.	2 units/ton	1 unit/ton
U.K.	3 units/ton	4 units/ton

These figures show clearly that the United States has an **absolute advantage** in both mining coal and growing wheat. That is, the United States is more efficient than the United Kingdom in producing both coal and wheat. However, although the United Kingdom is at an absolute disadvantage in both products, it has a **comparative advantage** in producing wheat. Put another way, the United Kingdom's absolute disadvantage is less in growing wheat than in mining coal. This lesser disadvantage can be seen by redoing the production figures above to reflect the output per unit of production for both countries:

	Wheat	Coal
U.S.	0.5 tons/unit	1 ton/unit
U.K.	0.33 tons/unit	0.25 tons/unit

Productivity for the United States relative to the United Kingdom in coal is $4:1$ $(1/.25)$, whereas it is "only" $1.5:1$ $(.5/.33)$ in wheat.

In order to induce the production of both wheat and coal prior to the introduction of trade, the profitability of producing both commodities must be identical. This condition is satisfied only when the return per unit of production is the same for both wheat and coal

---

[6] The traditional theory of international trade ignores the role of technology in differentiating products but it leaves open the possibility of different production technologies to produce commodities.

in each country. Hence, prior to the introduction of trade between the two countries, the exchange rate between wheat and coal in the United States and the United Kingdom must be as follows:

U.S.   1 ton wheat = 2 tons coal
U.K.   1 ton wheat = 0.75 tons coal

THE GAINS FROM TRADE

Based on the relative prices of wheat and coal in both countries, there will be obvious gains to trade. By switching production units from wheat to coal, the United States can produce coal and trade with the United Kingdom for more wheat than those same production units can produce at home. Similarly, by specializing in growing wheat and trading for coal, the United Kingdom can consume more coal than if it mined its own. This example demonstrates that trade will be beneficial even if one nation (the United States here) has an absolute advantage in everything. As long as the degree of absolute advantage varies across products, even the nation with an across-the-board absolute disadvantage will have a comparative advantage in making and exporting some goods and services.

The gains from trade for each country depend on exactly where the exchange rate between wheat and coal ends up following the introduction of trade. This exchange rate, which is known as the *terms of trade,* depends on the relative supplies and demands for wheat and coal in each country. However, any exchange rate between 0.75 and 2.0 tons of coal per ton of wheat will still lead to trade because trading at that exchange rate will allow both countries to improve their ability to consume. By illustration, suppose the terms of trade end up at 1 : 1 — that is, 1 ton of wheat equals 1 ton of coal.

Each unit of production in the United States can now provide its owner with either 1 ton of coal to consume or 1 ton of wheat or some combination of the two. By producing coal and trading for wheat, each production unit in the United States now enables its owner to consume twice as much wheat as before. Similarly, by switching from mining coal to growing wheat and trading for coal, each production unit in the United Kingdom will enable its owner to consume 0.33 tons of coal, 33% (0.33/0.25 = 133%) more than before.

SPECIALIZED FACTORS OF PRODUCTION

So far, we have assumed that the factors of production are unspecialized. That is, they can easily be switched between the production of wheat and coal. However, suppose that some factors such as labor and capital are specialized (that is, relatively more efficient) in terms of producing one commodity rather than the other. In that case, the prices of the factors of production that specialize in the commodity that is exported (coal in the United States, wheat in the United Kingdom) will gain because of greater demand once trade begins, whereas those factors that specialize in the commodity that is now imported (wheat in the United States, coal in the United Kingdom) will lose because of lower demand. This conclusion is based on the economic fact that the demand for factors of production is derived from the demand for the goods those factors produce.

The gains and losses to the specialized factors of production will depend on the magnitude of the price shifts after the introduction of trade. To take an extreme case, suppose the terms of trade become 1 ton of wheat equals 1.95 tons of coal. At this exchange rate, trade is still beneficial for both countries but far more so for the United Kingdom than the United States. The disparity in the gains from trade can be seen as follows: By producing coal and trading it for wheat, the United States can now consume approximately 2.5% more wheat than before per ton of coal.[7] On the other hand, the United Kingdom gains enormously. Each unit of wheat traded for coal will now provide 1.95 tons of coal, a 160% (1.95/0.75) increase relative to the earlier ratio.

These gains are all to the good. However, with specialization comes costs. In the United States, the labor and capital that specialized in growing wheat will be hurt. If they continue to grow wheat (which may make sense because they cannot easily be switched to mining coal), they will suffer an approximate 2.5% loss of income because the wheat they produce now will buy about 2.5% less coal than before (1.95 tons instead of 2 tons). At the same time, U.S. labor and capital that specializes in mining coal will be able to buy about 2.5% more wheat. Although gains and losses for specialized U.S. factors of production exist, they are relatively small. The same cannot be said for U.K. gains and losses.

As we saw earlier, U.K. labor and capital that specialize in growing wheat will be able to buy 160% more coal than before, a dramatic boost in purchasing power. Conversely, those factors that specialize in mining coal will see their wheat purchasing power plummet, from 1.33 (4/3) tons of coal before trade to 0.5128 (1/1.95) tons of wheat now. These figures translate into a drop of about 62% (0.5128/1.33 = 38%) in wheat purchasing power.

This example illustrates a general principle of international trade: *The greater the gains from trade for a country overall, the greater the cost of trade to those factors of production that specialize in producing the commodity that is now imported.* The reason is that in order for trade to make sense, imports must be less expensive than the competing domestic products. The less expensive these imports are, the greater will be the gains from trade. By the same token, however, less expensive imports drive down the prices of competing domestic products, thereby reducing the value of those factors of production that specialize in their manufacture.

It is this redistribution of income   from factors specializing in producing the competing domestic products to consumers of those products—that leads to demands for protection from imports. However, protection is a double-edged sword. These points are illustrated by the experience of the U.S. auto industry.

As discussed in the chapter, the onslaught of Japanese cars in the U.S. market drove down the price and quantity of cars sold by American manufacturers, reducing their return on capital and forcing them to be much tougher in negotiating with the United Auto Workers. The end result was better and less expensive cars for Americans but lower profits for Detroit automakers, lower wages and benefits for U.S. autoworkers, and fewer jobs in the U.S. auto industry.

The U.S. auto industry responded to the Japanese competition by demanding, and receiving, protection in the form of a quota on Japanese auto imports. The Japan-

---

[7] With trade, the United States can now consume $1/1.95 = 0.5128$ tons of wheat per ton of coal, which is 2.56% more wheat than the 0.5 tons it could previously consume (0.5128/0.50 = 1.0256).

ese response to the quota—which allowed them to raise their prices (Why cut prices when you can't sell more cars anyway?) and increase their profit margins—was to focus on making and selling higher quality cars in the U.S. market (as these carried higher profit margins) and shifting substantial production to the United States. In the end, protection did not help U.S. automakers nearly as much as did improving the quality of their cars and reducing their manufacturing costs. Indeed, to the extent that protection helped delay the needed changes while boosting Japanese automaker profits (thereby giving them more capital to invest), it may well have hurt the U.S. auto industry.

MONETARY PRICES

So far, we have talked about prices of goods in terms of each other. To introduce monetary prices into the example we have been analyzing, suppose that before the opening of trade between the two nations, each production unit costs $30 in the United States and £10 in the United Kingdom. In this case, the prices of wheat and coal in the two countries will be as follows:

	Wheat	Coal
U.S.	$60/ton	$30/ton
U.K.	£30/ton	£40/ton

These prices are determined by taking the number of required production units and multiplying them by the price per unit.

Following the introduction of trade, assume the same 1:1 terms of trade as before and that the cost of a unit of production remains the same. The prices of wheat and coal in each country will settle at the following, assuming that the exported goods maintain their prices and the prices of the goods imported adjust to these prices so as to preserve the 1:1 terms of trade:

	Wheat	Coal
U.S.	$30/ton	$30/ton
U.K.	£30/ton	£30/ton

These prices present a potential problem in that there will be equilibrium at these prices only if the exchange rate is $1 = £1. Suppose, however, that before the introduction of trade, the exchange rate is £1 = $3. In this case, dollar-equivalent prices in both countries will begin as follows:

	Wheat	Coal
U.S.	$60/ton	$30/ton
U.K.	$90/ton	$120/ton

This is clearly a disequilibrium situation. Once trade begins at these initial prices, the British will demand both U.S. wheat and coal whereas Americans will demand no British coal or wheat. Money will be flowing in one direction only (from the United

Kingdom to the United States to pay for these goods), and goods will flow only in the opposite direction. The United Kingdom will run a massive trade deficit, matched exactly by the U.S. trade surplus. Factors of production in the United Kingdom will be idle (because there is no demand for the goods they can produce) whereas U.S. factors of production will experience an enormous increase in demand for their services.

This is the nightmare scenario for those concerned with the effects of free trade: One country will sell everything to the other country and demand nothing in return (save money), leading to prosperity for the exporting nation and massive unemployment and depression in the importing country. However, such worries ignore the way markets work.

Absent government interference, a set of forces will swing into play simultaneously. The British demand for dollars (to buy U.S. coal and wheat) will boost the value of the dollar, making U.S. products more expensive to the British and British goods less expensive to Americans. At the same time, the jump in demand for U.S. factors of production will raise their prices and hence the cost of producing U.S. coal and wheat. The rise in cost will force a rise in the dollar price of U.S. wheat and coal. Conversely, the lack of demand for U.K. factors of production will drive down their price and hence the pound cost of producing British coal and wheat. The net result of these adjustments in the pound : dollar exchange rate and the cost of factors of production in both countries is to make British products more attractive to consumers and U.S. products less competitive. This process will continue until both countries can find their comparative advantage and the terms of trade between coal and wheat are equal in both countries (say, at 1 : 1).

TARIFFS

Introducing tariffs (taxes) on imported goods will distort the prices at which trade takes place and will reduce the quantity of goods traded. In effect, tariffs introduce a wedge between the prices paid by domestic customers and the prices received by the exporter, reducing the incentive of both to trade. To see this, suppose Mexican tomatoes are sold in the United States at a price of $0.30 per pound. If the United States imposes a tariff of, say, $0.15 per pound on Mexican tomatoes, then Mexican tomatoes will have to sell for $0.45 per pound to provide Mexican producers with the same pretariff profits on their tomato exports to the United States. However, it is likely that at this price, some Americans will forgo Mexican tomatoes and either substitute U.S. tomatoes or do without tomatoes in their salads. More likely, competition will preclude Mexican tomato growers from raising their price to $0.45 per pound. Suppose, instead, that the price of Mexican tomatoes, including the tariff, settles at $0.35 per pound. At this price, the Mexican tomato growers will receive only $0.20 per pound, reducing their incentive to ship tomatoes to the American market. At the same time, the higher price paid by American customers will reduce their demand for Mexican tomatoes. The result will be fewer Mexican tomatoes sold in the U.S. market. Such a result will benefit American tomato growers (who now face less competition and can thereby raise their prices) and farmworkers (who can now raise their wages without driving their employers out of business) while harming U.S. consumers of tomatoes (including purchasers of Campbell's tomato soup, Ragu spaghetti sauce, Progresso ravioli and Heinz ketchup).

➤ Questions

1. In a satirical petition on behalf of French candlemakers, Frederic Bastiat, a French economist, called attention to cheap competition from afar: sunlight. A law requiring the shuttering of windows during the day, he suggested, would benefit not only candlemakers but "everything connected with lighting" and the country as a whole. He explained: "As long as you exclude, as you do, iron, corn, foreign fabrics, in proportion as their prices approximate to zero, what inconsistency it would be to admit the light of the sun, the price of which is already at zero during the entire day!"

   a. Is there a logical flaw in Bastiat's satirical argument?

   b. Do Japanese automakers prefer a tariff or a quota on their U.S. auto exports? Why? Is there likely to be consensus among the Japanese carmakers on this point? Might there be any Japanese automakers that are likely to prefer U.S. trade restrictions? Why? Who are they?

   c. What characteristics of the U.S. auto industry have helped it gain protection? Why does protectionism persist despite the obvious gains to society from free trade?

   d. Review the arguments both pro and con on NAFTA. What is the empirical evidence so far?

2. Given the resources available to them, countries A and B can produce the following combinations of steel and corn.

	COUNTRY A		COUNTRY B	
Steel (tons)	Corn (bushels)	Steel (tons)	Corn (bushels)
36	0	54	0
30	3	45	9
24	6	36	18
18	9	27	27
15	12	18	36
6	15	9	45
0	18	0	54

   a. Do you expect trade to take place between countries A and B? Why?

   b. Which country will export steel? Which will export corn? Explain.

◆APPENDIX 1B Size and Scope of Multinational Corporations Abroad

This appendix presents data on direct foreign investment by U.S. firms and on the U.S. investment position of foreign firms. It also discusses some recent changes in the overall U.S. foreign investment position.

U.S. DIRECT INVESTMENT ABROAD

Exhibit 1B.1 shows the foreign direct investment positions of U.S. firms, broken down by major areas of the world, for the years 1977–1995. The *foreign direct investment position* is defined as the book value of the equity in and net loans outstanding to foreign businesses in which Americans own or control, directly or indirectly, at least 10% of the voting securities. The data reveal the strong preference exhibited by U.S. firms for investment in developed countries; historically, the fraction of U.S. direct investment going to developed countries has remained at a stable 73%–75% of the total, with Canada alone accounting for just under 20% of total U.S. direct investments abroad. But this pattern is changing. The share of U.S. foreign direct investment in the developing countries of Latin America, Africa, the Middle East, Asia, and the Pacific has increased from slightly less than 25% to almost 30% currently, while the share in the developed countries has fallen to around 70%. This change is largely accounted for by a jump in U.S. foreign direct

EXHIBIT 1B.1 U.S. Direct Investment Abroad by Major Regions, 1977–1996

IN BILLIONS OF DOLLARS	1977	1978	1979	1980	1981	1982	1983
Developed Countries							
Canada	35	36	41	45	47	46	48
Twelve European Union Countries*	49	56	66	78	81	78	79
Other Europe	13	14	17	19	21	22	24
Japan	5	5	6	6	7	7	8
Australia, New Zealand, and South Africa	8	9	10	11	12	12	12
Total developed countries	110	120	140	159	168	165	171
Developing Countries							
Latin America	28	32	35	39	39	33	30
Africa	2	3	3	4	4	5	5
Middle East	−3	−3	−1	2	2	2	3
Asia and Pacific	6	6	7	9	11	12	13
Total developing countries	33	38	44	54	56	52	51
International	4	4	4	4	5	5	6
Total for all countries	147	162	188	217	229	222	228

AS PERCENT OF TOTAL	1977	1978	1979	1980	1981	1982	1983
Developed Countries							
Canada	24	22	22	21	21	21	21
Twelve European Union Countries*	33	35	35	36	35	35	35
Other Europe	9	9	9	9	9	10	11
Japan	3	3	3	3	3	3	4
Australia, New Zealand, and South Africa	5	6	5	5	5	5	5
Total developed countries	75	74	74	73	73	74	75
Developing Countries							
Latin America	19	20	19	18	17	15	13
Africa	1	2	2	2	2	2	2
Middle East	−2	−2	−1	1	1	1	1
Asia and Pacific	4	4	4	4	5	5	6
Total developing countries	22	23	23	25	24	23	22
International	3	2	2	2	2	2	3
Total for all countries	100	100	100	100	100	100	100

NOTE: The numbers may not sum exactly because of rounding error.

*Twelve European Union nations include Belgium, Denmark, France, Germany, Greece, Ireland, Italy, Luxembourg, Netherlands, Portugal, Spain, and the United Kingdom. In 1995, Austria, Finland, and Sweden joined the European Union.

Source: Survey of Current Business, U.S. Department of Commerce, various issues.

1984	1985	1986	1987	1988	1989	1990	1991	1992	1993	1994	1995	1996
47	47	51	58	63	67	67	69	69	70	75	85	92
70	81	99	120	131	150	178	225	207	240	261	304	337
22	24	22	26	26	27	34	36	39	45	49	57	63
8	9	12	15	18	19	21	23	27	31	37	38	40
11	10	12	13	15	16	19	19	21	23	25	31	36
157	172	196	232	253	279	318	335	363	410	446	516	567
25	28	37	45	51	61	72	77	91	101	112	128	144
4	4	4	5	4	4	4	4	4	4	5	5	6
5	5	5	5	4	4	4	5	6	7	7	8	9
15	15	15	17	19	21	23	25	33	40	48	58	67
49	53	61	72	78	91	102	112	134	151	171	199	226
5	5	5	5	3	4	4	3	3	3	3	3	4
211	230	262	309	334	373	424	450	500	564	621	718	797

1984	1985	1986	1987	1988	1989	1990	1991	1992	1993	1994	1995	1996
22	20	19	19	19	18	16	15	14	12	12	12	12
33	35	38	39	39	40	42	50	41	43	42	42	42
10	10	8	8	8	7	8	8	8	8	8	8	8
4	4	5	5	5	5	5	5	5	6	6	5	5
5	4	5	4	4	4	4	4	4	4	4	4	4
74	75	75	75	76	75	75	74	73	73	72	72	71
12	12	14	15	15	16	17	17	18	18	18	18	18
2	2	2	2	1	1	1	1	1	1	1	1	1
2	2	2	2	1	1	1	1	1	1	1	1	1
7	7	6	6	6	6	5	6	7	7	8	8	8
23	23	23	23	23	24	24	25	27	27	28	28	28
2	2	2	2	1	1	1	1	1	1	1	0	1
100	100	100	100	100	100	100	100	100	100	100	100	100

EXHIBIT 1B.2 U.S. Direct Investment Abroad by Industrial Sector and Region, 1996
(U.S. $ billions)

Region	PETROLEUM		MANUFACTURING		OTHER		TOTAL	
	Amount	%	Amount	%	Amount	%	Amount	%
Developed Countries								
Canada	11	12	44	48	37	40	92	100
Europe	29	7	135	34	236	59	400	100
Japan	5	12	17	42	18	46	40	100
Australia, New Zealand, and South Africa	3	9	11	31	22	61	36	100
Total	48	8	206	36	313	55	567	100
Developing Countries								
Latin America	7	5	41	28	97	67	144	100
Other	21	25	26	30	39	45	86	100
Total	28	12	67	29	136	59	230	100
International	2	45	0	0	2	55	4	100
Total All Countries	76	9%	273	34%	448	56%	797	100%

Source: Survey of Current Business, U.S. Department of Commerce, July 1997, p. 36.

investment in the Asia/Pacific nations. According to the data in Exhibit 1B.1, direct investment in Latin America declined substantially in absolute and percentage terms beginning in 1982. This decline coincided with the Latin American debt crisis and reflected both the poorer economic prospects of Latin American countries and the additional constraints imposed on MNCs' ability to repatriate profits from their Latin American units. As Latin American prospects have improved in recent years so has the flow of direct investment, in both absolute and percentage terms. By 1993, the share of U.S. foreign direct investment in Latin America—18%—had returned to its 1980 level.

The basic generalization we can draw from the data is that American investment flows to those nations with the largest economies and best economic prospects. The most striking departure from this generalization is Japan. Although the Japanese economy is the second largest in the world, the share of U.S. direct investment in Japan is now about 6 percent. This amount is less even than U.S. direct investment in Switzerland and about one-third of such investment in Great Britain. These differences are strong evidence of the large barriers, both formal and informal, to American direct investment in Japan.

The pattern of U.S. foreign direct investment is also noteworthy for what it does not indicate. Despite claims by U.S. labor unions, overseas investment by American multinationals does not appear to be driven solely by the search for low-cost manufacture. U.S. investment in Mexico fell 17% between 1980 and 1987 despite the growth of low-cost, labor-intensive *maquiladora* factories on the American border.

Exhibit 1B.2 shows the U.S. direct foreign investment position cross-classified by industrial sector and region of the world at the end of 1996. The industrial sectors are petroleum, manufacturing, and other (mining, trade, banking and finance, and other industries).

EXHIBIT 1B.3 Rates of Return on U.S. Direct Investment Abroad: 1988–1992 (percent)

	1988	1989	1990	1991	1992
Developed Countries					
Canada	12.1	10.6	11.2	13.1	12.4
Petroleum	5.0	4.2	5.3	4.8	4.5
Manufacturing	14.0	13.9	15.0	14.6	14.2
Other	13.4	9.5	9.9	11.5	12.5
Europe	17.5	16.2	16.8	17.2	17.6
Petroleum	16.8	18.6	17.2	18.1	17.9
Manufacturing	20.0	18.6	19.2	19.5	19.8
Other	14.8	12.9	12.5	13.6	13.4
Other	16.5	14.4	15.0	15.6	14.9
Petroleum	18.9	8.7	9.5	12.0	14.3
Manufacturing	17.5	17.8	18.0	18.5	17.8
Other	13.7	13.3	13.5	13.9	14.0
Developed Countries—All	16.0	14.6	15.2	15.8	14.8
Petroleum	13.8	12.9	12.8	13.5	13.7
Manufacturing	18.2	17.3	17.0	18.1	17.6
Other	14.4	12.2	12.5	12.9	13.4
Developing Countries					
Latin America	8.9	14.2	13.0	12.5	9.8
Petroleum	7.5	12.5	8.2	10.2	12.4
Manufacturing	18.0	18.5	18.0	18.6	18.5
Other	3.1	11.8	8.5	11.0	9.6
Other	23.4	23.3	24.0	23.5	23.4
Petroleum	17.9	25.4	18.5	20.4	18.0
Manufacturing	32.0	24.3	26.0	28.6	28.4
Other	23.8	20.5	21.3	22.5	20.1
Developing Countries—All	13.9	17.2	14.3	14.0	16.5
Petroleum	14.4	21.2	18.5	15.6	17.2
Manufacturing	21.8	20.0	20.0	20.4	21.4
Other	8.4	14.0	9.2	10.5	13.5
All Countries	15.4	15.2	14.8	15.0	15.6
Petroleum	13.4	14.6	13.5	14.1	14.3
Manufacturing	18.8	17.8	17.5	17.6	18.2
Other	12.7	12.7	12.3	12.0	12.2

Source: U.S. Department of Commerce, *Survey of Current Business*, various issues.

    Manufacturing is the most important sector, accounting for 34% of total foreign direct investment by U.S. companies. Far behind is petroleum, at 9%. Although the data are not presented here, the three most important manufacturing industries in terms of foreign direct investment are chemicals, nonelectrical equipment, and transportation equipment.

Rates of Return on U.S. Direct Investment Abroad

The U.S. Department of Commerce calculates the rate of return on U.S. foreign direct investments annually. These data are contained in Exhibit 1B.3 for the years 1988 to 1992. The rate of return is estimated as net income (including interest income) before withholding tax divided by the average of the beginning- and end-of-year direct investment position. There are several biases in these data, however. Because assets are carried at historical costs instead of their current values, the return on investment is overstated. On the other hand, the estimate of income excludes fees and royalties and charges for other services. Inclusion of fees, royalties, and other charges would have increased estimated income by 29% in 1996. These biases downgrade the estimated return on foreign investment. Moreover, these estimates are not adjusted for varying debt ratios or risk.

Historically, direct investment in less-developed countries (LDCs), particularly in Latin America, has been relatively unprofitable compared with similar investments in developed countries. That fact helps account for the diminishing share of investment directed toward LDCs during most of the 1980s. However, the data in Exhibit 1B.3 show that this situation is now changing, with returns from LDCs on a par with those from developed countries. Petroleum investments historically have earned higher returns overseas than manufacturing investments. They no longer do, probably because of the sharp drop in oil prices in recent years.

Capital Expenditures by Majority-Owned Foreign Affiliates of U.S. Companies, 1977–1995

Exhibit 1B.4 shows the annual amount of capital expenditures made by majority-owned foreign affiliates of U.S. companies between 1977 and 1995 by area and industry (data for 1995 are estimated). Capital expenditures include all expenditures made to acquire, add to, or improve property, plant, and equipment and that are charged to a capital account. The resulting figure understates the amounts invested overseas because it does not include investments in research and development or additions to working capital.

The drop in overseas investment in the early 1980s reflects the severity of the worldwide recession. In general, the growth of foreign direct investment by U.S. companies has slowed substantially since the 1970s. However, it picked up again in the late 1980s as world economic growth strengthened, and then it rose further in the 1990s to take advantage of new opportunities in Europe, Latin America, and the Asia/Pacific region. Although petroleum and manufacturing investments historically have run neck and neck in importance from one year to the next, recent investments in manufacturing have consistently outpaced those in petroleum. This trend reflects the lower returns from petroleum investments, a casualty of the fall in oil prices.

New investment is primarily concentrated in Europe and Canada, with Latin America in third place. The slowdown in new investment in Canada in the early 1980s reflects both the world recession and the increasingly nationalistic posture of the Trudeau government. The replacement of the Trudeau government in 1984 by a more conservative one that welcomed foreign corporate investors reversed that trend. The effects of the debt crisis are evident in the steep decline in Latin American investments after 1982. Conversely, the promise of Europe 1992—an integrated market of 320 million wealthy consumers—has led to a dramatic rise in European investments by U.S. multinationals. By 1990, Europe accounted for more than two-thirds of U.S. overseas manufacturing investment. That percentage has fallen in recent years with the rapid growth in U.S. manufacturing investment in Asia and Latin America.

EXHIBIT 1B.4 Capital Expenditures by Majority-Owned Foreign Affiliates of U.S. Companies, 1977–1995

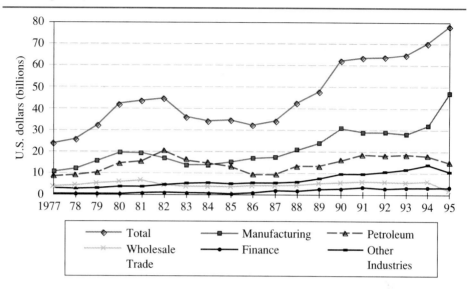

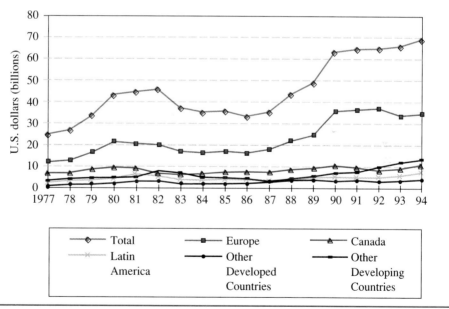

Source: U.S. Department of Commerce, *Survey of Current Business*, various issues.

FOREIGN DIRECT
INVESTMENT IN THE UNITED STATES

The United States itself is an increasingly attractive source for foreign direct investment prospects. Foreigners now own Firestone Tire & Rubber, Rockefeller Center, Columbia Pictures, MCA/Universal (first Japan's Matsushita and now Canada's Seagrams), TV Guide, A&P, 20th Century Fox, Brooks Brothers, CBS Records, RCA, Pillsbury, and about 30% of all office space in downtown Los Angeles.

Exhibit 1B.5 shows that the net amount of foreign direct investment in the United States grew by 12% in 1996—to $630 billion. This growth represents a continuation of the rapid increase in foreign direct investment inflows to the United States that began in 1994, coincident with the rising U.S. economy. The pickup in foreign investment coincided with a lower dollar, leading to the charge that foreigners have been buying American assets on the cheap. Of course, the foreigners also get a dollar earnings stream that is worth less in their currencies. Note that although current growth in foreign investment inflows to the United States is high, it still falls below the rates of growth during 1982–1990, when annual increases averaged 16%, and in the four years before 1982, when the average annual growth rate was 30%.

More likely reasons for foreign firms' increasing their investment in the United States in recent years are the expanding U.S. economy (following the recession in the

EXHIBIT 1B.5 Foreign Direct Investment (FDI) in the United States, 1996 (U.S. $ billions)

By Country	Total End 1995	1996 Inflow	Percent Increase	Total End 1996	Percent of Total FDI
United Kingdom	$126	$16	13	$143	23
Japan	108	10	9	118	19
Netherlands	66	0	0	74	12
Canada	48	6	11	54	9
Germany	49	13	26	62	10
France	39	11	28	49	8
Switzerland	36	−1	−1	35	6
Netherland Antilles	8	0	−6	8	1
Other	81	6	8	87	14
Total	$561	$69	12%	$630	100%

By Industry	Total End 1996	Percent of Total
Manufacturing	$234	37
Trade	93	15
Petroleum	42	7
Real estate	30	5
Insurance	60	10
Banking	32	5
Finance	70	11
Other	69	11
Total	$630	100%

EXHIBIT 1B.6 Japanese-Owned or Joint Venture Auto Plants in the United States

Year Completed	Company	Location	Annual Production Capacity
1982	Honda Motor Co.	Marysville, Ohio	360,000
1983	Nissan Motor Co.	Smyrna, Tennessee	250,000*
1984	Toyota-General Motors	Fremont, California	240,000
1987	Mazda Motor Co.	Flat Rock, Michigan	240,000
1988	Mitsubishi-Chrysler	Normal, Illinois	240,000
1988	Toyota Motor Co.	Georgetown, Kentucky	218,000
1989	Subaru-Isuzu	Lafayette, Indiana	120,000*
1989	Honda Motor Co.	East Liberty, Ohio	150,000
1992	Nissan Motor Co.	Smyrna, Tennessee	190,000
1993	Toyota Motor Co.	Georgetown, Kentucky	200,000
		Total capacity, 1993	2,208,000

*Includes light trucks.

early 1990s), the passage of NAFTA, the loosening of regulations, and the weak dollar. The weakness of the dollar against several foreign currencies, especially the Japanese yen and the Deutsche mark, made it less expensive for some companies to manufacture in the United States than to export to the U.S. market. The stronger dollar in 1996 has not yet slowed down foreign investment.

The importance of foreign investment in the United States is indicated by a few facts: Four of America's six major record concerns are now foreign-owned; Goodyear is the last major American-owned tire manufacturer; as of 1998, Japanese auto companies were able to build more than 2 million cars a year in their U.S. plants (see Exhibit 1B.6 for a list of these plants), over 25% of the total in a typical sales year; four of Hollywood's largest film companies are foreign-owned; and in July 1995, Zenith, the last of the 21 companies manufacturing televisions in the United States that was American-owned, was purchased by the South Korean firm LG Group.

European corporate parents accounted for 65% of the year-end 1996 foreign investment position. Canadian and Japanese parents accounted for another 9% and 19%, respectively. By industry of the U.S. affiliates, 37% of the position was in manufacturing, 15% in trade, 7% in petroleum, and 41% in other industries, mainly real estate, banking, finance (nonbanking), and insurance.

Two countries—the United Kingdom ($143 billion) and Japan ($118 billion)—account for 42% of the foreign direct investment in the United States. The Dutch, who were surpassed in 1987 by the Japanese as the second-largest foreign direct investors in the United States, had invested $74 billion by the end of 1996. Although not displayed here, a striking aspect of the returns on U.S. investments earned by foreign companies is how low they are. For example, during the 10-year period 1985 through 1994, the average return on assets for nonfinancial U.S. affiliates was only 4.3% (it was even lower for the Japanese), less than half the 8.3% return on assets enjoyed by their domestic U.S. peers. Apparently, foreign firms are willing to invest large sums in the United States, earning only marginal returns compared with what could be earned in Treasury bills or other riskless instruments. Of course, the low rates of return could just reflect the possibility that foreign firms overpaid for their acquisitions of U.S. companies.

Motives for Foreign Direct Investment in the United States

Investment by foreign multinationals in the United States can be attributed to many of the same factors that propelled U.S. firms abroad: the size and growth potential of the market, the fear of future protectionism (especially true for Japanese firms), the desire to compete with and learn from rivals on their home ground, and lower relative production costs in some industries. Foreign firms also may seek access to new technology and operations that complement existing product lines in order to improve their global market positions. The positive aspects of foreign direct investment in the United States are that it often brings technology, spurs competition, and provides access to foreign markets.

THE NET INTERNATIONAL WEALTH OF THE UNITED STATES

Direct investment is only one part of capital flows. Even more important is the flow of portfolio investment and bank lending overseas. The net of U.S. investment abroad (U.S. claims against foreigners) and foreign investment domestically (foreign claims against the United States) makes up the net international wealth of the United States. In 1981, the net U.S. international wealth peaked at $141 billion. Beginning in 1982, net U.S. international wealth began a long-term decline. Sometime during 1985, the continued decline in net international wealth turned negative. Exhibit 1B.7 depicts the changing net international wealth position of the United States from 1970 through 1997. The net international wealth position of the United States reached −$1.57 *trillion* by the end of 1997.

Note that these figures are expressed on a historical cost basis. The average age of U.S. investments abroad exceeds the average age of foreign investments in the United States, so historical cost figures will tend to understate significantly the true U.S. net international wealth position. Calculated on a current cost basis, the U.S. net international wealth position is less negative, −$1.01 trillion at the end of 1996, about three-quarters the historical cost figure of −$1.33 trillion in that year.

Historical Perspective

The United States was a net debtor to the rest of the world until World War I. In the early stages of its industrial development, the United States depended heavily on foreign capital; in building up its industries, it "mortgaged" part of its wealth to foreigners. In 1900, for example, when the total wealth of the United States, including land and reproducible assets, was an estimated $88 billion, net liabilities to foreigners were $2.5 billion; these liabilities far exceeded U.S. claims on foreigners. On the eve of World War I, total foreign investment in the United States amounted to $7.2 billion—nearly twice the $3.7 billion Americans had invested abroad. The effects of the war reversed this situation: By the end of 1919, U.S. claims on foreigners exceeded foreign claims on the United States by $3.7 billion.

Net U.S. international wealth continued to increase throughout the 1920s, and by 1930, it reached a peak of about $9 billion. During the depression-ridden 1930s, the value of U.S. assets abroad declined while the flight of capital from war-threatened Europe to the United States increased total foreign claims on the United States. These factors contributed to a slightly negative net U.S. foreign investment position.

EXHIBIT 1B.7 International Investment Position of the United States, 1970–1997

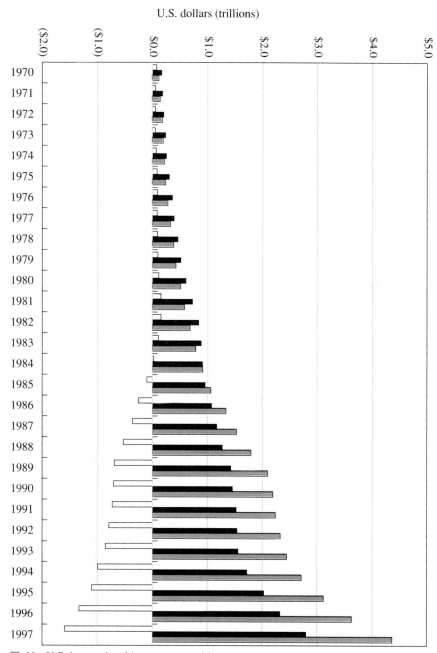

U.S. dollars (trillions)

☐ Net U.S. international investment position
■ U.S. assets abroad
▨ Foreign assets in the United States

*Data for 1997 are preliminary.

Source: Data from U.S. Department of Commerce, *Survey of Current Business*, various issues.

The Post-World War II Period

After the end of World War II, U.S. wealth abroad increased dramatically, first as a result of the vast flow of aid-related credits provided by the U.S. government to war-devastated countries, then later as U.S. companies sharply expanded their investment in foreign countries. By 1970 the U.S. international investment position showed a surplus of about $60 billion.

During the 1970s, the basic trend in the net U.S. international investment position was sharply upward. Between 1971 and 1979, U.S. foreign assets rose from about $165 billion to $510 billion, and foreign assets in the United States rose from $107 billion to $416 billion. These capital flows boosted the net U.S. international wealth position to $94 billion by 1980, as shown in Exhibit 1B.7.

The 1980s and 1990s

From 1980 through 1986, U.S. foreign assets rose at an average annual rate of 11%, bringing total U.S. assets abroad to $1.068 trillion. At the same time, however, U.S. liabilities to foreigners were rising at an annual rate of 18%. In 1985, the inevitable consequence of these trends finally occurred: The United States became a net international debtor, reverting to the position it was in at the turn of the century. This decline in net international wealth reflects the large trade deficits the United States has been running in recent years (in effect, the trade deficits have been financed by the sale of U.S. assets to foreigners). Indeed, as of 1998, the United States had run a trade deficit for 17 straight years, ever since 1982, and the deficit shows no signs of ending. The trade deficits in turn result from basic shifts in savings and investment behavior in the United States and abroad. Chapter 4 discusses these factors at length.

The Consequences

The consequences of a reduction in a nation's net international wealth depend on the nature of the foreign capital inflows that cause the erosion. If the capital inflows finance new investments that enhance the nation's productive capacity, they are self-financing in that they will eventually generate the necessary resources for their repayment. But if the investment flows finance current consumption, their repayment will eventually reduce the nation's standard of living below where it would have been in the absence of such inflows. The need to reduce future living standards to pay back foreign debts is a problem facing many of the debtor Latin American nations.

In the case of the United States, a good portion of the inflows seems to have been of the "productive" variety, but a growing share of capital inflows, particularly since 1982, might be classified as "consumption" in nature. If they are, the erosion of net U.S. international wealth, traceable in part to the large U.S. federal budget deficits during the 1980s and early 1990s (see Chapter 4), will inflict new burdens on the U.S. economy in the future.

➤ Bibliography

EAKER, MARK R. "Teaching International Finance: An Economist's Perspective." *Journal of Financial and Quantitative Analysis*, November 1977, pp. 607–608.

FOLKS, WILLIAM R., JR. "Integrating International Finance into a Unified Business Program." *Journal of Financial and Quantitative Analysis*, November 1977, pp. 599–600.

LESSARD, DONALD R. "Transfer Prices, Taxes, and Financial Markets: Implications of International Financial Transfers within the Multinational Firm." In *The Economic Effects of Multinational Corporations*, edited by Robert G. Hawkins, Greenwich, CT: JAI Press, 1979.

ENVIRONMENT OF INTERNATIONAL FINANCIAL MANAGEMENT

THE DETERMINATION OF EXCHANGE RATES

Experience shows that neither a state nor a bank ever have had the unrestricted power of issuing paper money without abusing that power.

David Ricardo (1817)

$\mathcal{E}$conomic activity is globally unified today to an unprecedented degree. Changes in one nation's economy are rapidly transmitted to that nation's trading partners. These fluctuations in economic activity are reflected, almost immediately, in fluctuations in currency values. Consequently, multinational corporations, with their integrated cross-border production and marketing operations, continually face devaluation or revaluation worries somewhere in the world. The purpose of this chapter and the next one is to provide an understanding of what an exchange rate is and why it might change. Such an understanding is basic to dealing with currency risk.

This chapter first describes what an exchange rate is and how it is determined in a *freely floating exchange rate* regime—that is, in the absence of government intervention. The chapter next discusses the role of expectations in exchange rate determination. It also examines the different forms and consequences of central bank intervention in the foreign exchange market. Chapter 3 describes the political aspects of currency determination under alternative exchange rate systems and presents a brief history of the international monetary system.

Before proceeding further, here are definitions of several terms commonly used to describe currency changes. Technically, a **devaluation** refers to a decrease in the stated par value of a **pegged currency**, one whose value is set by the government; an increase in par value is known as a **revaluation**. By contrast, a **floating currency**—one whose value is set primarily by market forces—is said to depreciate if it loses value and to appreciate if it gains value. However, discussions in this book will use the terms *devaluation* and **depreciation** and *revaluation* and **appreciation** interchangeably.

2.1 SETTING THE EQUILIBRIUM SPOT EXCHANGE RATE

An exchange rate is, simply, the price of one nation's currency in terms of another. For example, the yen/dollar exchange rate is just the number of yen that one dollar will buy. Equivalently, the dollar/yen exchange rate is the number of dollars one yen will buy. To understand how exchange rates are set it helps to recognize that they are market-clearing prices that equilibrate supplies and demands in foreign exchange markets. The determinants of currency supplies and demands are first discussed with the aid of a two-country model featuring the United States and Germany. Later, the various currency influences will be studied more closely.

The demand for the Deutsche mark (DM), Germany's currency, in the foreign exchange market (which in this two-country world is equivalent to the supply of dollars) derives from the American demand for German goods and services and DM-denominated financial assets. German prices are set in DM, so in order for Americans to pay for their German purchases they must first exchange their dollars for DM. That is, they will demand DM.

An increase in the DM's dollar value is equivalent to an increase in the dollar price of German products. This higher dollar price normally will reduce the U.S. demand for German goods, services, and assets. Conversely, as the dollar value of the DM falls, Americans will demand more DM to buy the less-expensive German products, resulting in a downward-sloping demand curve for Deutsche marks. As the dollar cost of the DM (the exchange rate) falls, Americans will tend to buy more German goods and so will demand more DM.

Similarly, the supply of Deutsche marks (which for the model is equivalent to the demand for dollars) is based on German demand for U.S. goods and services and dollar-denominated financial assets. In order for Germans to pay for their U.S. purchases, they must first acquire dollars. As the dollar value of the Deutsche mark increases, thereby lowering the DM cost of U.S. goods, the increased German demand for U.S. goods will cause an increase in the German demand for dollars and, hence, an increase in the amount of Deutsche marks supplied.[1]

In Exhibit 2.1, *e* is the spot exchange rate (dollar value of one DM), and Q is the quantity of Deutsche marks supplied and demanded. The DM supply (S) and demand (D) curves intersect at e_0, the equilibrium exchange rate. The foreign exchange market is said to be in equilibrium at e_0 because both the demand for DM and the supply of DM at this price are Q_0.

---

[1] This statement holds provided the price elasticity of German demand, E, is greater than 1. In general, $E = -(\Delta Q/Q)/(\Delta P/P)$, where Q is the quantity of goods demanded, P is the price, and ΔQ is the change in quantity demanded for a change in price, ΔP. If $E > 1$, then total spending goes up when price declines.

EXHIBIT 2.1 Equilibrium Exchange Rates

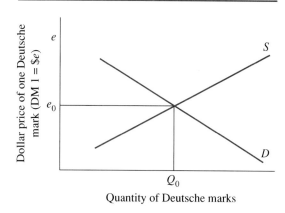

Suppose that the supply of dollars increases relative to its demand. This excess growth in the money supply will cause inflation in the United States, which means that U.S. prices will begin to rise relative to prices of German goods and services. German consumers are likely to buy fewer U.S. products and begin switching to German substitutes, leading to a decrease in the amount of Deutsche marks supplied at every exchange rate. The result is a leftward shift in the DM supply curve to S' as shown in Exhibit 2.2. Similarly, higher prices in the United States will lead American consumers to substitute German imports for U.S. products, resulting in an increase in the demand for Deutsche marks as depicted by D'. In effect, both Germans and Americans are searching for the best deals worldwide and will switch their purchases accordingly. Hence, a higher rate of inflation in the United States than in Germany will simultaneously increase German exports to the United States and reduce U.S. exports to Germany.

EXHIBIT 2.2 Impact of U.S. Inflation on the Equilibrium Exchange Rate

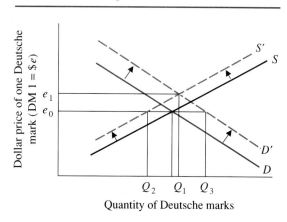

A new equilibrium rate $e_1 > e_0$ results. In other words, a higher rate of inflation in the United States than in Germany will lead to a depreciation of the dollar relative to the Deutsche mark or, equivalently, to an appreciation of the mark relative to the dollar. In general, a nation running a relatively high rate of inflation will find its currency declining in value relative to the currencies of countries with lower inflation rates. This relationship will be formalized in Chapter 7 as purchasing power parity (PPP).

Depending on the current value of the Deutsche mark relative to the dollar, the amount of DM appreciation or depreciation is computed as the fractional increase or decrease in the dollar value of the DM. For example, if the DM/$ exchange rate goes from DM 1 = $0.64 to DM 1 = $0.68, the DM is said to have appreciated by $(0.68 - 0.64)/0.64 = 6.25\%$.

The general formula by which we can calculate the DM's appreciation or depreciation is as follows:

$$\text{Amount of DM appreciation (depreciation)} = \frac{\text{New dollar value of DM} - \text{Old dollar value of DM}}{\text{Old dollar value of DM}} \quad (2.1)$$

$$= \frac{e_1 - e_0}{e_0}$$

Substituting in the numbers from the previous example (with $e_0 = \$0.64$ and $e_1 = \$0.68$) yields the 6.25% DM appreciation. Alternatively, the dollar is said to have depreciated (appreciated) by the fractional decrease (increase) in the DM value of the dollar:

$$\text{Amount of dollar depreciation (appreciation)} = \frac{\text{New DM value of dollar} - \text{Old DM value of dollar}}{\text{Old DM value of dollar}} \quad (2.2)$$

$$= \frac{1/e_1 - 1/e_0}{1/e_0} = \frac{e_0 - e_1}{e_1}$$

Equation 2.2 relies on the fact that if e equals the dollar value of a DM (dollars per DM), then the DM value of a dollar (DM per dollar) must be the reciprocal, or $1/e$. Em-

ILLUSTRATION

CALCULATING THE AMOUNT OF YEN DEPRECIATION AGAINST THE DOLLAR

During 1995, the yen went from $0.0125 to $0.0095238. By how much did the yen depreciate against the dollar?

Solution. Using Equation 2.1, the yen has depreciated against the dollar by an amount equal to $(0.0095238 - 0.0125)/0.0125 = -23.81\%$.

By how much has the dollar appreciated against the yen?

Solution. An exchange rate of ¥1 = $0.0125 translates into an exchange rate of $1 = ¥80 (1/0.0125 = 80). Similarly, the exchange rate of ¥1 = $0.0095238 is equivalent to an exchange rate of $1 = ¥105. Using Equation 2.2, the dollar has appreciated against the yen by an amount equal to $(105 - 80)/80 = 31.25\%$.

ILLUSTRATION

CALCULATING DOLLAR APPRECIATION AGAINST THE THAI BAHT
On July 2, 1997, the Thai baht fell 17 percent against the U.S. dollar. By how much has the dollar appreciated against the baht?

Solution. If e_0 is the initial dollar value of the baht and e_1 is the post-devaluation exchange rate, then we know from Equation 2.1 that $(e_1 - e_0)/e_0 = -17\%$. Solving for e_1 in terms of e_0 yields $e_1 = 0.83e_0$. From Equation 2.2, we know that the dollar's appreciation against the baht equals $(e_0 - e_1)/e_1$ or $(e_0 - 0.83e_0)/0.83e_0 = 0.17/.83 = 20.48\%$.

ploying Equation 2.2, we can find the increase in the DM exchange rate from \$0.64 to \$0.68 to be equivalent to a dollar depreciation of 5.88% [(0.64 − 0.68)/0.68 = −0.0588]. (Why don't the two exchange rate changes equal each other?)[2]

Interest rate differentials will also affect the equilibrium exchange rate. A rise in U.S. interest rates relative to German rates, all else being equal, will cause investors in both nations to switch from DM- to dollar-denominated securities to take advantage of the higher dollar rates. The net result will be depreciation of the DM in the absence of government intervention. Similarly, because a stronger economy attracts capital, economic growth should lead to a stronger currency. Empirical evidence supports this hypothesis.

Other factors that can influence exchange rates include political and economic risks. Investors prefer to hold lesser amounts of riskier assets; thus, low-risk currencies—those associated with more politically and economically stable nations—are more highly valued than high-risk currencies.

ILLUSTRATION

CALCULATING YUGOSLAV DINAR DEVALUATION
April 1, 1998, was a significant date in Yugoslavia. On that day, the government devalued the Yugoslav dinar, setting its new rate at 10.92 dinar to the dollar, from 6.0 dinar previously. By how much has the dinar devalued against the dollar?

Solution. The devaluation lowered the dinar's dollar value from \$0.1667 (1/6) to \$0.0916 (1/10.92). According to Equation 2.1, the dinar has devalued by (0.0916 − 0.1667)/0.1667 = 45%.

By how much has the dollar appreciated against the dinar?
Solution. Applying Equation 2.2, the dollar has appreciated against the dinar by an amount equal to (10.92 − 6)/6 = 82%.

---

[2] The reason the DM appreciation is unequal to the amount of dollar depreciation depends on the fact that the value of one currency is the inverse of the value of the other one. Hence, the percentage change in currency value differs because the base off which the change is measured differs.

ILLUSTRATION

EAST-WEST POLITICS AFFECT THE DEUTSCHE MARK

After the reform movement that swept Eastern Europe in late 1989, West Germany's stock market soared. The rise in the West German stock market reflected the general belief among investors that West German industry stood to benefit significantly from the opening of Eastern Europe. For example, economists estimated that the modernization of East Germany's economy alone would cost up to $300 billion over the coming decade, with much of the reconstruction to be performed by West German industry.

As foreign capital flowed into West Germany, attracted by its strong growth prospects in Eastern Europe, the demand for Deutsche marks in the foreign exchange market led to a steady increase in the DM's value. On January 30, 1990, however, the DM fell about 3 percent as panicked traders and investors dumped Deutsche marks and other foreign currencies and bought dollars after an unconfirmed report on Cable News Network that Mikhail Gorbachev, the architect of the Soviet Union's reform policies, might resign as head of the Soviet Communist Party. The dollar traditionally has been considered a safe haven for investors during periods of international turmoil.

The foreign exchange consequences of the CNN report, which was televised just after 2 P.M. EST and raised the specter of an unraveling of the reform movement in the Soviet Union and Eastern Europe, are vividly illustrated in Exhibit 2.3. By later in the day, the dollar retreated after the White House said that it had no information to support the CNN report. Over the next two days, the DM rose another 1.4 percent when Mr. Gorbachev denied the report. These currency fluctuations illustrate the importance of political factors in the foreign exchange market.

EXHIBIT 2.3 How Political Factors Affect Exchange Rates

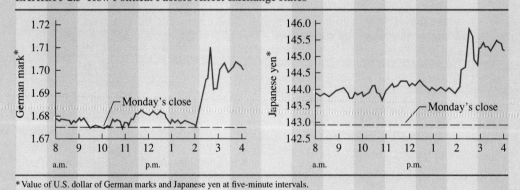

* Value of U.S. dollar of German marks and Japanese yen at five-minute intervals.

2.2 EXPECTATIONS AND THE ASSET MARKET MODEL OF EXCHANGE RATES

Although currency values are affected by current events and current supply and demand flows in the foreign exchange market, they also depend on expectations about future exchange rate movements. And exchange rate expectations are influenced by every conceivable economic, political, and social factor.

The role of expectations in determining exchange rates depends on the fact that currencies are financial assets and that an exchange rate is simply the relative price of two financial assets—one country's currency in terms of another's. Thus, currency prices are determined in the same manner that the prices of assets such as stocks, bonds, gold, or real estate are determined. Unlike the prices of services or products with short storage lives, asset prices are influenced comparatively little by current events. Rather, they are determined by people's willingness to hold the existing quantities of assets, which in turn depends on their expectations of the future worth of these assets. Thus, for example, frost in Florida can bump up the price of oranges, but it should have little impact on the price of the citrus groves producing the oranges; instead, longer-term expectations of the demand and supply of oranges governs the values of these groves.

Similarly, the value today of a given currency, say, the dollar, depends on whether or not—and how strongly—people still want the amount of dollars and dollar-denominated assets they held yesterday. According to this view—known as the asset market model of exchange rate determination—the exchange rate between two currencies represents the price that just balances the relative supplies of, and demands for, assets denominated in those currencies. Consequently, shifts in preferences can lead to massive shifts in currency values.

A brief examination of the dollar's ups and downs, using some stylized facts, provides further insight into the relation of exchange rate movements to the asset-like character of money. In the early 1960s, the United States was clearly the dominant nation in the world. The U.S. political system was stable and its economic health was good: U.S. inflation averaged about 1% annually; economic growth averaged almost 5% annually. Not surprisingly, the world was willing to hold a large fraction of its wealth in the form of dollars and dollar-denominated assets.

In the late 1960s and 1970s, problems appeared. The Vietnam War was sharply divisive, and the U.S. political system was shaken. One president was driven from office, another was forced to resign, and two others lost the confidence of the American people. Monetary growth accelerated, leading to double-digit inflation and sharply higher effective tax rates as inflation forced everyone into higher tax brackets. Economic growth slowed to about 3% annually, compared with about 4% in other developed nations. These factors, combined with increased government regulation, lowered the attractiveness of holding dollars and U.S. assets relative to foreign assets. The world decided to hold a much smaller fraction of its wealth in the form of dollars and dollar-denominated assets. This massive portfolio shift caused a sharp decline in the value of the dollar.

Then, in the early 1980s, Americans elected a new president who was committed to cutting inflation, taxes, and government regulation. The authority of the presidency was restored, taxes and inflation were sharply reduced, and following a brief recession, economic growth accelerated and reached 6.5% in 1984, compared with average growth of 0.9% in other developed countries. Capital was attracted to the United States by the strength of its economy, the high after-tax real (inflation-adjusted) rate of return, and the favorable political climate—conditions superior to those attainable elsewhere. Foreigners once again found the United States to be a safer and more rewarding place in which to invest, so they added many more U.S. assets to their portfolios. The dollar rose sharply.

As U.S. growth slowed down in 1985 and foreign growth accelerated, foreign assets became attractive again. At the same time, foreigners fled U.S. assets when they perceived that the U.S. government itself wanted the dollar to decline. The world's decreased willingness to hold U.S. assets led to another massive portfolio rebalancing; this time, investors substituted foreign assets for U.S. assets. The dollar fell once again.

The desire to hold a currency today depends critically on expectations of the factors that can affect the currency's future value, therefore, what matters is not only what is happening today but what markets expect will happen in the future. Thus, currency values are forward looking; they are set by investor expectations of their issuing countries' future economic prospects rather than by contemporaneous events alone. Moreover, in a world of high capital mobility, the difference between having the right policies and the wrong ones has never been greater. This point is illustrated by the Asian currency crisis of 1997.

The Nature of Money and Currency Values

To understand the factors that affect currency values, it helps to examine the special character of money. To begin, money has value because people are willing to accept it in exchange for goods and services. The value of money, therefore, depends on its purchasing power. Money also provides **liquidity**—that is, you can readily exchange it for goods or other assets, thereby facilitating economic transactions. Thus, money represents both a *store of value* and a *store of liquidity*. The demand for money, therefore, depends on money's ability to maintain its value and on the level of economic activity. Hence, the lower the expected inflation rate, the more money people will demand. Similarly, higher economic growth means more transactions and a greater demand for money to pay bills.

The demand for money is also affected by the demand for assets denominated in that currency. The higher the expected real return and the lower the riskiness of a country's assets, the greater is the demand for its currency to buy those assets. In addition, as people who prefer assets denominated in that currency (usually residents of the country) accumulate wealth, the value of the currency rises.

Because the exchange rate reflects the relative demands for two moneys, factors that increase the demand for the home currency should also increase the price of home currency on the foreign exchange market. In summary, the economic factors that affect a currency's foreign exchange value include its usefulness as a store of value, determined by its expected rate of inflation; the demand for liquidity, determined by the volume of transactions in that currency; and the demand for assets denominated in that currency, determined by the risk-return pattern on investment in that nation's economy and by the wealth of its residents. The first factor depends primarily on the country's future monetary policy, whereas the latter two factors depend largely on expected economic growth and political and economic stability. All three factors ultimately depend on the soundness of the nation's economic policies. The sounder these policies, the more valuable the nation's currency will be; conversely, the more uncertain a nation's future economic and political course, the riskier its assets will be, and the more depressed and volatile its currency's value.

The asset market view that sound economic policies strengthen a currency is challenged by critics who point to the links between the huge U.S. budget deficits, high real interest rates, and the strong dollar in the early 1980s. Others find this argument unconvincing, especially since the dollar fell in the late 1980s while the deficit rose further. If the big deficits were responsible for high real U.S. interest rates, there would be more evidence that private borrowing was being crowded out in interest-sensitive areas, such as

ILLUSTRATION

ASIAN CURRENCIES SINK DURING 1997

During the second half of 1997, beginning in Thailand, currencies and stock markets plunged across East Asia, while hundreds of banks, builders, and manufacturers went bankrupt. The Thai baht, Indonesian rupiah, Malaysian ringgit, Philippine peso, and South Korean won depreciated by 40% to 80% apiece. All this happened despite the fact that Asia's fundamentals looked good: low inflation, balanced budgets, well-run central banks, high domestic savings, strong export industries, a large and growing middle class, a vibrant entrepreneurial class, and industrious, well-trained, and often well-educated work forces paid relatively low wages. But investors were looking past these positives to signs of impending trouble. What they saw was that many East Asian economies were locked on a course that was unsustainable, characterized by large trade deficits,[3] huge short-term foreign debts, overvalued currencies, and financial systems that were rotten at their core. Each of these ingredients played a role in the crisis and its spread from one country to another.

To begin, most East Asian countries depend on exports as their engines of growth and development. Along with Japan, the United States is the most important market for these exports. Partly because of this, many of them had tied their currencies to the dollar. This tie served them well until 1995, promoting low inflation and currency stability. It also boosted exports at the expense of Japan as the dollar fell against the yen, forcing Japanese companies to shift production to East Asia to cope with the strong yen. Currency stability also led East Asian banks and companies to finance themselves with dollars, yen, and marks—some $275 billion worth, much of it short term—because dollar and other foreign currency loans carried lower interest rates than did their domestic currencies. The party ended in 1995, when the dollar began recovering against the yen and other currencies. By mid-1997, the dollar had risen by over 50% against the yen and by 20% against the mark. Dollar appreciation alone would have made East Asia's exports less price competitive. But their competitiveness problem was greatly exacerbated by the fact that during this period, the Chinese yuan depreciated by about 25% against the dollar.[4] China exported similar products, so yuan devaluation raised China's export competitiveness at East Asia's expense. The loss of export competitiveness slowed down Asian growth and caused utilization rates—and profits—on huge investments in production capacity to plunge. It also gave the Asian central banks a mutual incentive to devalue their currencies. According to one theory, recognizing these altered incentives, speculators attacked the East Asian currencies almost simultaneously and forced a round of devaluations.[5]

Another theory suggests that **moral hazard**—the tendency to incur risks that one is protected against—lies at the heart of Asia's financial problems. Specifically, most Asian banks and finance companies operated with implicit or explicit government guarantees. For

---

[3] The trade balance as used here refers to trade in both goods and services and is also known as the current-account balance (see Chapter 4).

[4] For a discussion of the role that Chinese yuan devaluation played in the Asian crisis, see Kenneth Kasa, "Export Competition and Contagious Currency Crises," *Economic Letter*, Federal Reserve Bank of San Francisco, January 16, 1998.

[5] See C. Hu and Kenneth Kasa, "A Dynamic Model of Export Competition, Policy Coordination, and Simultaneous Currency Collapse," working paper, Federal Reserve Bank of San Francisco, 1997.

example, the South Korean government directed the banking system to lend massively to companies and industries that it viewed as economically strategic, with little regard for their profitability. When combined with poor regulation, these guarantees distorted investment decisions, encouraging financial institutions to fund risky projects in the expectation that they would enjoy any profits, while sticking the government with any losses. (These same perverse incentives underlie the savings and loan fiasco in the United States during the 1980s.) In Asia's case, the problem was compounded by the crony capitalism that is pervasive throughout the region, with lending decisions often dictated more by political and family ties than by economic reality. Billions of dollars in easy-money loans were made to family and friends of the well-connected. Without market discipline or risk-based bank lending, the result was overinvestment—financed by vast quantities of debt—and inflated prices of assets in short supply, such as land.[6]

This financial bubble persists as long as the government guarantee is maintained. The inevitable glut of real estate and excess production capacity leads to large amounts of nonperforming loans and widespread loan defaults. When reality strikes, and investors realize that the government doesn't have the resources to bail out everyone, asset values plummet and the bubble bursts. The decline in asset values triggers further loan defaults, causing a loss of the confidence on which economic activity depends. Investors also worry that the government will try to inflate its way out of its difficulty. The result is a self-reinforcing downward spiral and capital flight. As foreign investors refuse to renew loans and begin to sell off shares of overvalued local companies, capital flight accelerates and the local currency falls, increasing the cost of servicing foreign debts. Local firms and

banks scramble to buy foreign exchange before the currency falls further, putting even more downward pressure on the exchange rate. This story explains why stock prices and currency values declined together and why Asian financial institutions were especially hard hit. Moreover, this process is likely to be contagious, as investors search for other countries with similar characteristics. When such a country is found, everyone rushes for the exit simultaneously and another bubble is burst, another currency is sunk.

The standard approach of staving off currency devaluation is to raise interest rates, thereby making it more attractive to hold the local currency and increasing capital inflows. However, this approach was problematic for Asian central banks. Raising interest rates boosted the cost of funds to banks and made it more difficult for borrowers to service their debts, thereby further crippling an already sick financial sector. Higher interest rates also lowered real estate values, which served as collateral for many of these loans, and pushed even more loans into default. Thus, Asian central banks found their hands were tied and investors recognized that.

These two stories—loss of export competitiveness and moral hazard in lending—combine to explain the severity of the Asian crisis. Appreciation of the dollar and depreciation of the yen and yuan slowed down Asian economic growth and hurt corporate profits. These factors turned ill-conceived and overleveraged investments in property developments and industrial complexes into financial disasters. The Asian financial crisis then was touched off when local investors began dumping their own currencies for dollars and foreign lenders refused to renew their loans. It was aggravated by politicians, such as in Malaysia and South Korea, who preferred to blame foreigners for their problems rather than seek structural reforms of their economies. Both

---

[6] This explanation for the Asian crisis is set forth in Paul Krugman, "What Happened to Asia?" MIT working paper, 1998.

foreign and domestic investors, already spooked by the currency crisis, lost yet more confidence in these nations and dumped more of their currencies and stocks, driving them to record lows.

This synthesized story is consistent with the experience of Taiwan, which is a net exporter of capital and whose savings are largely invested by private capitalists without government direction or guarantees. Taiwanese businesses also are financed far less by debt than by equity. In contrast to its Asian competitors, Taiwan suffered minimally during 1997, with the Taiwan dollar (NT$) down by a modest 15% (to counteract its loss of export competitiveness to China and Japan) and its stock market actually up by 17% in NT$ terms.

"The way out," said Confucius, "is through the door." The clear exit strategy for East Asian countries is to restructure their ailing financial systems by shutting down or selling off failing banks (for example, to healthy foreign banks) and disposing of the collateral (real estate and industrial properties) underlying their bad loans. The result should be fewer, but stronger and better-capitalized, banks and restructured and consolidated industries and a continuation of East Asia's strong historical growth record. At the same time, governments must step aside and allow those who borrow too much or lend too foolishly to fail. Ending government guarantees and politically-motivated lending will transform Asia's financial sector and force cleaner and more transparent financial transactions. The result will be better investment decisions—ones driven by market forces rather than personal connections or government whim—and healthier economies, which will quickly bring back foreign investors attracted by Asia's solid fundamentals.

fixed business investment and residential construction. However, the rise in real interest rates coincided with rapid growth in capital spending.

An alternative explanation for high real interest rates is that a vigorous U.S. economy, combined with a major cut in business taxes in 1981, raised the after-tax profitability of business investments. The result was a capital spending boom and a strong dollar.

Indeed, if large government deficits and excessive government borrowing cause currencies to strengthen, then the Russian ruble should be one of the strongest currencies in the world. Instead, it is among the weakest currencies because Russia's flawed economic policies scare off potential investors.

Central Bank Reputations and Currency Values

As the example of Mexico indicates, another critical determinant of currency values is central bank behavior. A **central bank** is the nation's official monetary authority; its job is to use the instruments of monetary policy, including the sole power to create money, to achieve one or more of the following objectives: price stability, low interest rates, or a target currency value. As such, the central bank affects the risk associated with holding money. This risk is inextricably linked to the nature of a **fiat money**, which is nonconvertible paper money. Until 1971, every major currency was linked to a commodity. Today no major currency is linked to a commodity. With a commodity base, usually gold, there was a stable, long-term anchor to the price level. Prices varied a great deal in the short term, but they eventually returned to where they had been.

ILLUSTRATION

THE PESO PROBLEM

On December 20, 1994, Mexico devalued its peso by 12.7%. Two days later, the government was forced to let the peso float freely, whereupon it quickly fell an additional 15%. By March 1995, the peso had fallen over 25% more, a total of 50% altogether (see Exhibit 2.4). Even President Clinton's dramatic rescue package involving $52 billion in loans and loan guarantees from the United States and various international financial institutions could only halt the freefall temporarily. The story of the peso's travails illustrates the importance of credibility in establishing currency values. This credibility depends, in part, on the degree of consistency between the government's exchange rate policy and its other macroeconomic objectives.

Until the devaluation, Mexico had a system under which the peso was allowed to fluctuate within a narrow band against the dollar. Pegging the peso to the dollar helped stabilize Mexico's economy against hyperinflation. The credibility of this exchange rate regime depended on people believing that Banco de Mexico (Mexico's central bank) would defend the currency to keep it within this band. As long as investors had confidence in the country's economic future, this policy worked well. However, that confidence was shaken during 1994 by an armed uprising in the state of Chiapas, assassinations of leading Mexican politicians (including the front-running presidential candidate), and high-level political resignations. Another source of concern was the enormous trade deficit, which was about 8% of gross domestic product (GDP) for 1994.

The trade deficit jeopardized future growth because, to attract the dollars needed to finance this deficit, the government had to keep interest rates high, especially because interest rates were rising in the United States and around the world. Foreign investors began to bet that this situation

EXHIBIT 2.4 The Peso's Plunge

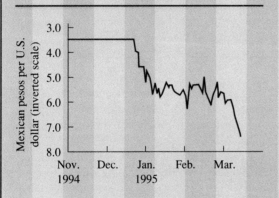

With a fiat money, there is no anchor to the price level—that is, there is no standard of value that investors can use to find out what the currency's future value might be. Instead, a currency's value is largely determined by the central bank through its control of the money supply. If the central bank creates too much money, inflation will occur and the value of money will fall. *Expectations* of central bank behavior also will affect exchange rates today; a currency will decline if people *think* the central bank will expand the money supply in the future.

Viewed this way, money becomes a brand name product whose value is backed by the reputation of the central bank that issues it. And just as reputations among automobiles vary—from Mercedes Benz to Yugo—so currencies come backed by a range of quality was unsustainable, that in order to continue to finance the deficit Mexico would have to raise interest rates so much that it would damage its economy. Such a rise was unlikely given the political difficulties the government was already facing. At the same time, under pressure from an administration facing a tough election, the central bank permitted a monetary expansion of more than 20% during 1994, leading to fears of rising inflation. Sensing that something had to give, many investors ran for the exits, draining Banco de Mexico's dollar reserves.

Here is where Mexico made a fundamental error. The central bank didn't allow the supply of pesos to fall even though the various political shocks—and the economic uncertainties they created—reduced the demand for pesos. As investors sold pesos to Banco de Mexico for dollars, reducing the supply of pesos to the level actually demanded, the central bank—fearing that a reduced supply of pesos would cause interest rates to rise (a politically costly step)—put these pesos back into circulation by buying an offsetting amount of government notes and bonds from the public (a process known as *sterilization*; see Section 2.3). The result was a continuing excess supply of pesos that the central bank kept buying up with its shrinking dollar reserves. Despite this inherent conflict between Mexico's monetary policy and its exchange rate policy, many investors trusted the government's

adamant promise to maintain the peso's link with the dollar.

Mexico's devaluation, therefore, represented an enormous gamble that foreign investors would not lose confidence in the country's financial markets. The payoff was swift and bloody: The Mexican stock market plunged 11% and interest rates soared as investors demanded higher returns for the new risk in peso securities. At the same time, investors rushed to cash in their pesos, causing Banco de Mexico to lose half its dollar reserves in one day. The next day, the government caved in and floated the peso. It also announced a tightened monetary policy to bolster the peso's value, along with a package of market-oriented structural reforms to enhance Mexican competitiveness and restore investor confidence. Despite the soundness of these new policies, they were too late; the government's loss of credibility was so great that the peso's fall continued until the U.S. rescue plan. Simply put, with the devaluation, Mexico sacrificed its most valuable financial asset— market confidence.

Mexico's peso problem quickly translated into faltering investor confidence in other countries that, like Mexico, suffer from political turmoil and are dependent on foreign investors to finance their deficits—including Canada, Italy, and other Latin American nations viewed as having overvalued currencies.

reputations—from the U.S. dollar, Deutsche mark, Swiss franc, and Japanese yen on the high side to the Mexican peso, Thai baht, and Russian ruble on the low side. Underlying these reputations is trust in the willingness of the central bank to maintain price stability.

The high-quality currencies are those expected to maintain their purchasing power because they are issued by reputable central banks. A reputable central bank is one that the markets trust to do the right thing, and not merely the politically expedient thing, when it comes to monetary policy. This trust, in turn, comes from past history: Reputable banks, like the Bundesbank, have developed their credibility by having done hard, cruel, and painful things for years in order to fight inflation. In contrast, the low-quality currencies are those that bear little assurance that their purchasing power will be maintained. As in the car market, high-quality currencies sell at a premium, and low-quality currencies sell at a discount (relative to their values based on economic fundamentals alone). That is, investors demand a risk premium to hold a riskier currency, whereas safer currencies will be worth more than their economic fundamentals would indicate.

Because good reputations are slow to build and quick to disappear, many economists recommend that central banks should adopt rules for price stability that are verifiable, unambiguous, and enforceable—along with the independence and accountability necessary to realize this goal.[7] Focus is also important. A central bank whose responsibilities are limited to price stability is more likely to achieve this goal. For example, the Bundesbank—a model for many economists—has managed to maintain such a low rate of German inflation because of its statutory commitment to price stability, a legacy of Germany's bitter memories of hyperinflation in the 1920s, which peaked at 200 billion percent in 1923. Absent such rules, the natural accountability of central banks to government becomes an avenue for political influence. For example, even though the U.S. Federal Reserve is an independent central bank, its legal responsibility to pursue both full employment and price stability (aims that conflict in the short term) hinders its effectiveness in fighting inflation. The greater scope for political influence in central banks such as the Federal Reserve that don't have a clear mandate to pursue price stability will, in turn, add to the perception of inflation risk.

This perception stems from the fact that government officials and other critics routinely exhort the central bank to follow "easier" monetary policies, by which they mean boosting the money supply to lower interest rates. These exhortations arise because many people believe (1) that the central bank can "trade off" a higher rate of inflation for more economic growth, and (2) that the central bank determines the rate of interest independently of the rate of inflation and other economic conditions. Despite the questionable merits of these beliefs, central banks—particularly those that are not independent—often respond to these demands by expanding the money supply.

Central banks that lack independence are also often forced to *monetize the deficit*, which means financing the public-sector deficit by buying government debt with newly created money. Whether monetary expansion stems from economic stimulus or deficit financing, it inevitably leads to higher inflation and a devalued currency.

The link between central bank independence and sound monetary policies is borne out by the empirical evidence.[8] Exhibit 2.5a shows that countries whose central banks are less subject to government intervention tend to have lower and less volatile inflation rates and vice versa. The central banks of Germany, Switzerland, and the United States, identified as the most independent in the post-World War II era, also showed the lowest inflation rates from 1951 to 1988. Least independent were the central banks of Italy, New Zealand, and Spain, countries wracked by the highest inflation rates in the industrial world. Moreover, Exhibit 2.5b indicates that this lower inflation rate is not achieved at the expense of economic growth; rather central bank independence and economic growth seem to go together.

---

[7] See, for example, W. Lee Hoskins, "A European System of Central Banks: Observations from Abroad," *Economic Commentary*, Federal Reserve Bank of Cleveland, November 15, 1990.

[8] See, for example, Alberto Alesina, "Macroeconomics and Politics," in *NBER Macroeconomic Annual, 1988* (Cambridge, Mass.: MIT Press), 1988.

EXHIBIT 2.5 Central Bank Independence, Inflation, and Economic Growth

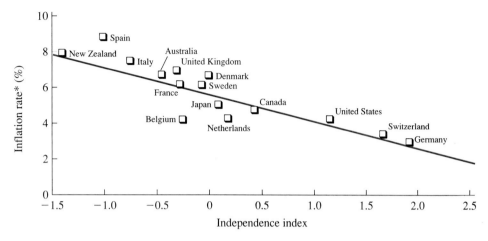

(a) Central Bank Independence versus Inflation

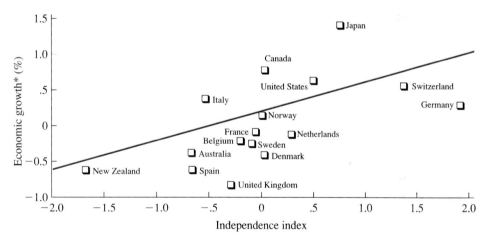

(b) Central Bank Independence versus Economic Growth

* Inflation and economic growth rates calculated for the period 1951–1988.

Source: Adapted from J. Bradford DeLong and Lawrence H. Summers, "Macroeconomic Policy and Long-Run Growth," *Economic Review*, Federal Reserve Bank of Kansas City, Fourth Quarter 1992, pp. 14, 16.

The idea that central bank independence can help establish a credible monetary policy is being put into practice today, as countries that have been plagued with high inflation rates are enacting legislation to reshape their conduct of monetary policy. For example, New Zealand, England, Mexico, Canada, Chile, and Bolivia all have passed laws that mandate an explicit inflation goal or that give their central banks more independence.

INFLATION DIES DOWN UNDER

In Germany and Switzerland, long seen as bastions of sound money, inflation rose during the early 1990s.

However Australia and New Zealand, so often afflicted by high inflation, boasted the lowest rates among the nations comprising the Organization for Economic Cooperation and Development (OECD), which consists of all the industrialized nations in the world (see Exhibit 2.6).

The cure was simple: Restrict the supply of Australian and New Zealand dollars. To increase the likelihood that it would stick to its guns, the Reserve Bank of New Zealand was made fully independent in 1990 and its governor, Donald Brash, was held accountable for cutting inflation to 0%–2% by December 1993. Failure carried a high personal cost: He would lose his job. Exhibit 2.6 shows why Mr. Brash kept his job; by 1993, inflation had fallen to 1.3% and has since then held at about 2%. At the same time, growth has averaged a rapid 4% a year.

The job of New Zealand's central banker was made easier by the government's decision to dismantle one of the OECD's most taxed, regulated, protectionist, and comprehensive welfare states and transform it into one of the most free-market oriented. By slashing welfare programs and stimulating economic growth through its market reforms and tax and tariff cuts, the government converted its traditionally large budget deficit into a budget surplus and ended the need to print money to finance it. To ensure continued fiscal sobriety, in 1994, parliament passed the "Fiscal Responsibility Act," which mandates budgetary balance over the business cycle.

EXHIBIT 2.6 Inflation Dies Down Under

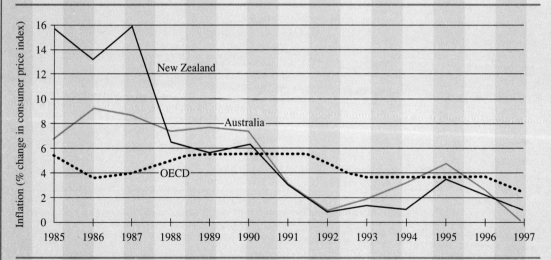

Source: International Financial Statistics and OECD, various years.

ILLUSTRATION

THE BANK OF ENGLAND GAINS INDEPENDENCE
On May 6, 1997, within days of the Labour Party's landslide victory, Britain's new Chancellor of the Exchequer announced a policy change that he described as "the most radical internal reform to the Bank of England since it was established in 1694." The reform granted the Bank of England independence from the government in the conduct of monetary policy, meaning that it is now free to pursue its policy goals without political interference, and charged it with the task of keeping inflation to 2.5%. The decision was a surprise, coming as it did from the Labour Party, a party with a strong socialist history that traditionally was unsympathetic to low-inflation policies, which it viewed as destructive of jobs. Investors responded to the news by revising downwards their expectations of future British inflation. This favorable reaction can be seen by examining the performance of index-linked gilts. Index-linked gilts are British government bonds that pay an interest rate that varies with the British inflation

EXHIBIT 2.7 British Inflation Expectations Fall as the Bank of England Gains Its Independence

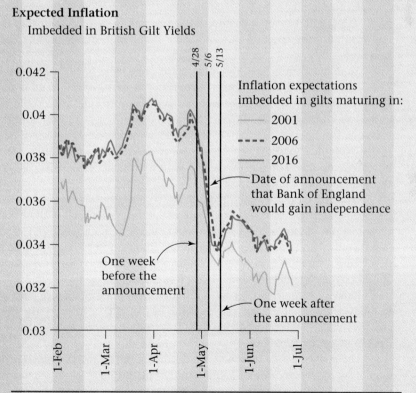

Expected Inflation
   Imbedded in British Gilt Yields

Inflation expectations imbedded in gilts maturing in:
—— 2001
- - - 2006
—— 2016

Date of announcement that Bank of England would gain independence

One week before the announcement

One week after the announcement

Source: Mark M. Spiegel, "British Central Bank Independence and Inflation Expectations," *FRBSF Economic Letter*, Federal Reserve Bank of San Francisco, November 28, 1997.

rate. One can use the prices of these gilts to estimate the inflation expectations of investors.[9] Exhibit 2.7 shows how the expected inflation rate embodied in three different index-linked gilts—maturing in 2001, 2006, and 2016—responded to the Chancellor's announcement of independence. Over the two-week period surrounding the announcement, the expected inflation rate dropped by 0.60% for the 2016 gilt

and by somewhat less for the shorter-maturity gilts. These results indicate that the market perceived that enhanced central bank independence would lead to lower future inflation rates. Consistent with our earlier discussion on the inverse relation between inflation and currency values, the British pound jumped in value against the U.S. dollar and the Deutsche mark on the day of the announcement.

[9]The methodology used to compute these inflation expectations is described in detail in Mark M. Spiegel, "Central Bank Independence and Inflation Expectations: Evidence from British Index-Linked Gilts," *Economic Review*, Federal Reserve Bank of San Francisco, 1998, forthcoming.

Evidence that even the announcement of greater central bank independence can boost the credibility of monetary policy comes from England. This example shows that institutional change alone can have a significant impact on future expected inflation rates.

Some countries, like Argentina, have gone even further and established what is in effect a currency board. Under a **currency board** system, there is no central bank. Instead, the currency board issues notes and coins that are convertible on demand and at a fixed rate into a foreign reserve currency. As reserves, the currency board holds high-quality, interest-bearing securities denominated in the reserve currency. Its reserves are equal to 100%, or slightly more, of its notes and coins in circulation. The board has no discretionary monetary policy. Instead, market forces alone determine the money supply.

Over the past 150 years, more than 70 countries (mainly former British colonies) have had currency boards. As long as they kept their boards, all of those countries had the same rate of inflation as the country issuing the reserve currency and successfully maintained convertibility at a fixed exchange rate into the reserve currency; no board has ever devalued its currency against its anchor currency. Currency boards are successfully operating today in Argentina, Estonia, Hong Kong, and Lithuania.

In addition to promoting price stability, a currency board also compels government to follow a responsible fiscal policy. If the budget is not balanced, the government must convince the private sector to lend to it; it no longer has the option of forcing the central bank to monetize the deficit.

The downside of a currency board is that a run on the currency forces a sharp contraction in the money supply and a jump in interest rates. High interest rates slow economic activity, increase bankruptcies, and batter financial markets. For example, Hong Kong's currency board weathered the Asian storm and delivered a stable currency but at the expense of high interest rates and plummeting stock and real estate markets.

ARGENTINA REFORMS ITS CURRENCY

Argentina, once the world's seventh-largest economy, has long been considered one of Latin America's worst basket cases. Starting with Juan Peron, decades of profligate government spending financed by a compliant central bank that printed money to cover the chronic budget deficits had triggered a vicious cycle of inflation and devaluation. High taxes and excessive controls compounded Argentina's woes and led to an overregulated, arthritic economy. However, in 1991, after 50 years of economic mismanagement, President Carlos Menem and his fourth Minister of Economy, Domingo Cavallo, launched the Convertibility Act. (The first Minister of Economy, Miguel Roig, took one look at the economy and died of a heart attack six days into the job.) This act made the austral (the Argentine currency) fully convertible at a fixed rate of 10,000 australs to the dollar, and by law the monetary supply must now be 100% backed by gold and foreign currency reserves, mostly dollars. This link to gold and the dollar imposes a straitjacket on monetary policy. If, for example, the central bank has to sell dollars to support the currency, the money supply automatically shrinks. Better still, the government can no longer print money to finance a budget deficit. In January 1992, the government knocked four zeros off the austral and renamed it the peso, worth exactly $1.

By effectively locking Argentina into the U.S. monetary system, the Convertibility Act has had remarkable success in restoring confidence in the peso and providing an anchor for inflation expectations. Inflation fell from more than 2,300% in 1990 to 170% in 1991 and 4% in 1994 (see Exhibit 2.8). By 1997, the inflation rate was 0.4%, among the lowest in the world. Argentine capital transferred overseas to escape Argentina's hyperinflation has started to come home. It has spurred rapid economic growth and led to a rock-solid currency. In response to the good economic news, stock prices quintupled, in dollar terms,

EXHIBIT 2.8 Argentina Ends Hyperinflation

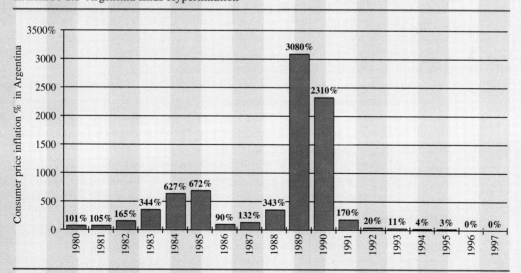

Source: International Financial Statistics, various issues.

during the first year of the plan. And the price of Argentina's foreign debt rose from 13% of its face value in 1990 to 45% in 1992.

The likelihood that the Convertibility Act marks a permanent change in Argentina and will not be revoked at a later date—an important consideration for investors—is increased by the other economic actions the Argentine government has taken to reinforce its commitment to price stability and economic growth: It has deregulated its economy, sold off money-losing state-owned businesses to the private sector, cut taxes and red tape, opened its capital markets, and lowered barriers to trade. In September 1994, the Argentine government announced a sweeping privatization plan designed to sell off all remaining state-owned enterprises—including the national mint, the postal service, and the country's main airports.

A byproduct of the Mexican peso crisis was a questioning by investors of the use of exchange-rate policy to achieve price stability. Despite investor concerns that the Argentine peso might be next to devalue, however, the nature of Argentina's currency-board arrangement prevented a speculative attack on the currency. In early January 1995, Domingo Cavallo announced that Argentina had $17.8 billion of reserves to cover a monetary base of $16.2 billion, more than enough reserves for every Argentine to convert all their pesos into dollars. The experience of Argentina suggests that countries such as Mexico and Indonesia can restore their credibility in the financial markets by instituting their own currency boards. Three years later, the Argentine peso also withstood an assault brought about by the Asian currency crisis.

The importance of expectations and central bank reputations in determining currency values was dramatically illustrated on June 2, 1987, when the financial markets learned that Paul Volcker was resigning as chairman of the Federal Reserve Board. Within seconds after this news appeared on the ubiquitous video screens used by traders to watch the world, both the price of the dollar on foreign exchange markets and the prices of bonds began a steep decline. By day's end, the dollar had fallen 2.6% against the Japanese yen, and the price of Treasury bonds declined 2.3%—one of the largest one-day declines ever. The price of corporate bonds fell by a similar amount. All told, the value of U.S. bonds fell by more than $100 billion.

The response by the financial markets reveals the real forces that are setting the value of the dollar and interest rates under our current monetary system. On that day, there was no other economic news of note. There was no news about American competitiveness. There was no change in Federal Reserve (Fed) policy or inflation statistics; nor was there any change in the size of the budget deficit, the trade deficit, or the growth rate of the U.S. economy.

What actually happened on that announcement day? Foreign exchange traders and investors simply became less certain of the path U.S. monetary policy would take in the days and years ahead. Volcker was a known inflation fighter. Alan Greenspan, the incoming Fed chairman, was an unknown quantity. The possibility that he would emphasize growth over price stability raised the specter of a more expansive monetary policy. Because the natural response to risk is to hold less of the asset whose risk has risen, investors tried to reduce their holdings of dollars and dollar-denominated bonds, driving down their prices in the process.

PRESIDENT CLINTON SPOOKS THE CURRENCY MARKETS In early 1994, the U.S. dollar began a steep slide, particularly against the yen (see Exhibit 2.9), that "baffled" President Clinton. According to him, the U.S. economy was stronger than it had been in decades and therefore the dollar's weakness was a market mistake. "In the end, the markets will have to respond to the economic realities," the president said. His critics, however, described the dollar's travails as a global vote of "no confidence" in his policies. They pointed to President Clinton's erratic handling of foreign affairs (e.g., Bosnia, Haiti, Somalia, North Korea, Rwanda) and threatened trade sanctions against Japan and China, along with his administration's tendency to use a weak dollar to bludgeon Japan into opening its markets without any concern that dollar weakness could boost inflation. Investors also noted White House resistance to the Federal Reserve Board's raising interest rates to stem incipient inflation as well as

President Clinton's appointment of two suspected inflation doves to the Federal Reserve Board. Even worse, the Clinton Administration did not appear to be particularly bothered by the dollar's drop. In June 1994, the administration did and said nothing to support the dollar as it fell to a 50-year low against the yen. At a meeting with reporters on June 21, for example, Treasury Secretary Lloyd Bentsen rebuffed three attempts to get him to talk about the dollar; he wouldn't even repeat the usual platitudes about supporting the dollar.

One investment banker summed up the problem. In order to reverse the dollar's decline, he said, "The U.S. administration must convince the market that it doesn't favor a continuing dollar devaluation and that it won't use the dollar as a bargaining chip in trade negotiations with Japan or other countries in the future."[10] Simply put, the administration needed to make credible its belated claim that it saw no advantage in a lower dollar. Finally, investors were not pleased with President Clinton's domestic economic policy, a policy that sought to sharply boost taxes, spending

EXHIBIT 2.9 The Clinton Dollar

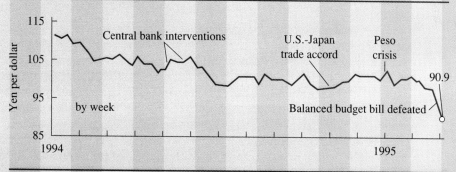

Source: Data from Bloomberg Business News, Nihon Ketzaí Shimbun, Mitsubishi Bank in *Investor's Business Daily,* March 13, 1995, p. B1. Reprinted by permission of *Investor's Business Daily.*

---

[10] Quoted in the *Wall Street Journal,* June 24, 1994, p. C1.

(on a huge new health-care entitlement program), and regulation. Such a policy was unlikely to encourage the high savings and investment and reduced government spending
necessary for low inflation and vigorous long-
term U.S. economic growth.

By mid-1995, the Clinton administration,
pushed by the Republican takeover of Congress
in November 1994, shifted its economic policies
to favor a balanced budget and a stable dollar
and away from talk of a trade war with Japan.
At the same time, rapid growth combined with
low inflation made the United States a magnet
for capital. In contrast, Japan and Europe exhibited feeble growth. The result was a dramatic
turnaround in the dollar's fortunes.

The import of what happened on June 2, 1987, is that prices of the dollar and those
billions of dollars in bonds were changed by nothing more or less than investors changing
their collective assessment of what actions the Fed would or would not take. A critical
lesson for business people and policymakers alike surfaces: A shift in the trust that people
have for a currency can change its value now by changing its expected value in the future.
The level of interest rates also is affected by trust in the future value of money. All else
being equal, the greater the trust in the promise that money will maintain its purchasing
power, the lower interest rates will be. This theory will be formalized in chapter 7 as the
Fisher effect.

2.3 THE FUNDAMENTALS OF CENTRAL BANK INTERVENTION

The exchange rate is one of the most important prices in a country because it links the domestic economy and the rest-of-world economy. As such, it affects relative national competitiveness.

We already have seen the link between exchange rate changes and relative inflation
rates. The important point for now is that an appreciation of the exchange rate beyond
that necessary to offset the inflation differential between two countries raises the price of
domestic goods relative to the price of foreign goods. This rise in the **real** or **inflation-
adjusted exchange rate**—measured as the nominal exchange rate adjusted for changes
in relative price levels—proves to be a mixed blessing. For example, the rise in the value
of the U.S. dollar from 1980 to 1985 translated directly into a reduction in the dollar
prices of imported goods and raw materials. As a result, prices of imports and of products
that compete with imports began to ease. This development contributed significantly to
the slowing of U.S. inflation in the early 1980s.

However, the rising dollar had some distinctly negative consequences for the U.S.
economy as well. Declining dollar prices of imports had their counterpart in the increasing foreign currency prices of U.S. products sold abroad. As a result, American exports
became less competitive in world markets, and American-made import substitutes became less competitive in the United States. Domestic sales of traded goods declined, gen-

EXHIBIT 2.10 Real Value of the U.S. Dollar: 1970–1998*

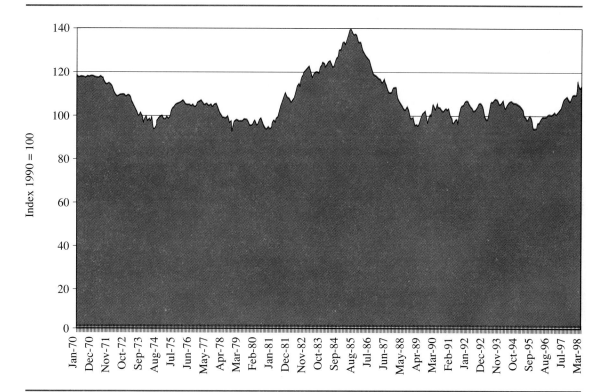

*Through March 1998.

Source: Morgan Guaranty Trust Company of New York, Data and Systems Group, Economic Research.

erating unemployment in the traded-goods sector and inducing a shift in resources from the traded- to the nontraded-goods sector of the economy.

Alternatively, home currency depreciation results in a more competitive traded-goods sector, stimulating domestic employment and inducing a shift in resources from the nontraded- to the traded-goods sector. The bad part is that currency weakness also results in higher prices for imported goods and services, eroding living standards and worsening domestic inflation.

From its peak in mid-1985, the U.S. dollar fell by more than 50% during the next few years, enabling Americans to experience the joys and sorrows of both a strong and a weak currency in less than a decade. The weak dollar made U.S. companies more competitive worldwide; at the same time, it lowered the living standards of Americans who enjoyed consuming foreign goods and services. The dollar hit a low point in 1995 and then began to strengthen, largely based on the substantial success that the United States has had in taming inflation and the budget deficit and generating strong economic growth. Exhibit 2.10 charts the real value of the U.S. dollar from 1970 to 1998. Despite its recent rise, though, the dollar is still below its level back in 1970.

Depending on their economic goals, some governments will prefer an overvalued domestic currency, whereas others will prefer an undervalued currency. Still others just want a correctly valued currency, but economic policymakers may feel that the rate set by the market is irrational; that is, they feel they can better judge the correct exchange rate than the marketplace can.

No matter what category they fall in, most governments will be tempted to intervene in the foreign exchange market to move the exchange rate to the level consistent with their goals or beliefs. **Foreign exchange market intervention** refers to official purchases and sales of foreign exchange that nations undertake through their central banks to influence their currencies.

For example, review section 2.1, Setting the Equilibrium Spot Exchange Rate. Now suppose the U.S. and German governments decide to maintain the old exchange rate e_0 in the face of the new equilibrium rate e_1. According to Exhibit 2.2, the result will be an excess demand for Deutsche marks equal to $Q_3 - Q_2$; this DM shortage is the same as an excess supply of $(Q_3 - Q_2)e_0$ dollars. Either the Federal Reserve (the American central bank), or the Bundesbank (the German central bank), or both will then have to intervene in the market to supply this additional quantity of Deutsche marks (to buy up the excess supply of dollars). Absent some change, the United States will face a perpetual balance-of-payments deficit equal to $(Q_3 - Q_2)e_0$ dollars, which is the dollar value of the German balance-of-payments surplus of $(Q_3 - Q_2)$ Deutsche marks.

Foreign Exchange Market Intervention

Although the mechanics of central bank intervention vary, the general purpose of each variant is basically the same: to increase the market demand for one currency by increasing the market supply of another. To see how this purpose can be accomplished, suppose in the previous example that the Bundesbank wants to reduce the value of the DM from e_1 to its previous equilibrium value of e_0. To do so, the Bundesbank must sell an additional $(Q_3 - Q_2)$ DM in the foreign exchange market, thereby increasing the supply of DM in the marketplace and eliminating the shortage of DM that would otherwise exist at e_0. This sale of DM (which involves the purchase of an equivalent amount of dollars) will also eliminate the excess supply, $(Q_3 - Q_2)e_0$, of dollars that now exists at e_0. The simultaneous sale of DM and purchase of dollars will balance the supply and demand for DM (and dollars) at e_0.

If the Fed also wants to raise the value of the dollar, it will buy dollars with Deutsche marks. Regardless of whether the Fed or the Bundesbank initiates this foreign exchange operation, the net result is the same: The U.S. money supply will fall, and Germany's money supply will rise.

Sterilized versus Unsterilized Intervention

The two examples just discussed are instances of **unsterilized intervention**; that is, the monetary authorities have not insulated their domestic money supplies from the foreign exchange transactions. In both cases, the U.S. money supply will fall, and the German

EXHIBIT 2.11 Mexico and Argentina Follow Different Monetary Policies

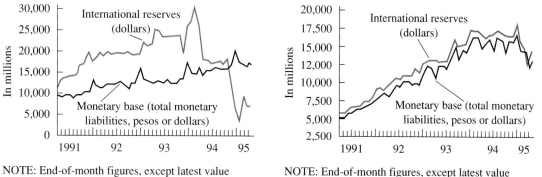

NOTE: End-of-month figures, except latest value
        plotted April 12, 1995

NOTE: End-of-month figures, except latest value
        plotted April 11, 1995

(a) International Reserves and Monetary Base in Mexico

(b) International Reserves and Monetary Base in Argentina

money supply will rise. As noted earlier, an increase (decrease) in the supply of money, all other things held constant, will result in more (less) inflation. Thus, the foreign exchange market intervention will not only change the exchange rate, it will also increase German inflation, while reducing U.S. inflation. Recall it was the jump in the U.S. money supply that caused this inflation. These money supply changes will also affect interest rates in both countries.

To neutralize these effects, the Fed and/or the Bundesbank can **sterilize** the impact of their foreign exchange market intervention on the domestic money supply through an **open-market operation**, which is just the sale or purchase of U.S. Treasury securities. For example, the purchase of U.S. Treasury bills by the Fed supplies reserves to the banking system and increases the U.S. money supply. After the open-market operation, therefore, the public will hold more cash and bank deposits and fewer Treasury securities. If the Fed buys enough T-bills, the U.S. money supply will return to its preintervention level. Similarly, the Bundesbank could neutralize the impact of intervention on the German money supply by subtracting reserves from its banking system through sales of German Treasury securities.

The net result of sterilization should be a rise or fall in the country's foreign exchange reserves but no change in the domestic money supply. These effects are shown in Exhibit 2.11a, which displays a steep decline in Mexico's reserves during 1994 while its money supply, measured by its **monetary base** (currency in circulation plus bank reserves), followed its usual growth path with its usual seasonal variations. As mentioned earlier, Banco de Mexico sterilized its purchases of pesos by buying back government securities. Conversely, Argentina's currency board has precluded its ability to sterilize changes in reserves, forcing changes in Argentina's monetary base to closely match changes in its dollar reserves, as can be seen in Exhibit 2.11b.

The Effects of Foreign Exchange
Market Intervention

The basic problem with central bank intervention is that it is likely to be either ineffectual or irresponsible. Because sterilized intervention entails a substitution of foreign currency-denominated securities for domestic currency securities,[11] the exchange rate will be permanently affected only if investors view domestic and foreign securities as being imperfect substitutes. If this is the case, then the exchange rate and relative interest rates must change to induce investors to hold the new portfolio of securities.

If investors consider these securities to be perfect substitutes, however, then no change in the exchange rate or interest rates will be necessary to convince investors to hold this portfolio. In this case, sterilized intervention is ineffectual. This conclusion is consistent with the experiences of Mexico as well as those of the United States and other industrial nations in their intervention policies. Between March 1973 and April 1983, industrial nations bought and sold a staggering $772 billion of foreign currencies to influence exchange rates. Foreign central banks bought more than $100 billion in 1987 alone to support the dollar. More recently, the U.S. government bought more than $36 billion of foreign currencies between July 1988 and July 1990 to stem a rise in the dollar's value. Despite these large interventions, exchange rates appear to have been moved largely by basic market forces. Similarly, Mexico ran through about $25 billion in reserves in 1994 and Asian nations ran through more than $100 billion in reserves in 1997 to no avail.

Sterilized intervention could affect exchange rates by conveying information or by altering market expectations. It does this by signaling a change in monetary policy to the market, not by changing market fundamentals, so its influence is transitory.

On the other hand, unsterilized intervention can have a lasting effect on exchange rates, but insidiously—by creating inflation in some nations and deflation in others. In the example presented above, Germany would wind up with a permanent (and inflationary) increase in its money supply, and the United States would end up with a deflationary decrease in its money supply. If the resulting increase in German inflation and decrease in U.S. inflation were sufficiently large, the exchange rate would remain at e_0 without the need for further government intervention. But it is the money supply changes, and not the intervention by itself, that affect the exchange rate. Moreover, moving the *nominal* (or actual) *exchange rate* from e_1 to e_0 should not affect the real exchange rate because the change in inflation rates offsets the nominal exchange rate change.

If forcing a currency below its equilibrium level causes inflation, it follows that devaluation cannot be much use as a means of restoring competitiveness. A devaluation improves competitiveness only to the extent that it does not cause higher inflation. If the devaluation causes domestic wages and prices to rise, any gain in competitiveness is quickly eroded. For example, Mexico's peso devaluation led to a burst of inflation, driving the peso still lower and evoking fears of a continuing inflation-devaluation cycle.

---

[11] Central banks typically hold their foreign exchange reserves in the form of foreign currency bonds. Sterilized intervention, therefore, involves selling off some of the central bank's foreign currency bonds and replacing them with domestic currency ones. Following the intervention, the public will hold more foreign currency bonds and fewer domestic currency bonds.

Britain Pegs the Pound to the Mark

In early 1987, Nigel Lawson, Britain's Chancellor of the Exchequer, began pegging the pound sterling against the Deutsche mark. Unfortunately, his exchange rate target greatly undervalued the pound. In order to prevent sterling from rising against the DM he had to massively intervene in the foreign exchange market, by selling pounds to buy marks. The resulting explosion in the British money supply reignited the inflation that Prime Minister Margaret Thatcher had spent so long subduing. With high inflation, the pound fell against the mark and British interest rates surged. The combination of high inflation and high interest rates led first to Mr. Lawson's resignation in October 1989 and then to Mrs. Thatcher's resignation a year later, in November 1990.

Of course, when the world's central banks execute a coordinated surprise attack, the impact on the market can be dramatic—for a short period. Early in the morning on February 27, 1985, for example, Western European central bankers began telephoning banks in London, Frankfurt, Milan, and other financial centers to order the sale of hundreds of millions of dollars; the action—joined a few hours later by the Federal Reserve in New York—panicked the markets and drove the dollar down by 5% that day.

But keeping the market off balance requires credible repetitions. Shortly after the February 27 blitzkrieg, the dollar was back on the rise. The Fed intervened again, but it wasn't until clear signs of a U.S. economic slowdown emerged that the dollar turned down in March.

Open-Market Operations

To summarize, exchange market intervention will have a lasting influence on exchange rates only if the intervention leads to permanent changes in relative money supplies. Thus, if the DM appreciation is viewed as resulting from a shortage of DM, an alternative to foreign exchange market intervention is for the Bundesbank to expand the supply of DM by buying Treasury bills from the public. Some of these added DM will be spent on foreign goods, services, and assets. More DM will be sold in the foreign exchange market to buy dollars to carry out these planned purchases. If the Bundesbank expands the money supply sufficiently through its open-market operations, the increased supply of DM in the foreign exchange market will cause the equilibrium exchange rate to return to e_0.

Conversely, if the depreciating dollar is traced to an excess of dollars, the Fed could reduce the U.S. money supply by selling Treasury bills to the public. As the U.S. money supply declines, fewer dollars will be sold in the foreign exchange market. If the drop in the U.S. money supply is large enough, the demand for DM will drop sufficiently to return the equilibrium exchange rate to e_0.

Open-market operations affect the equilibrium exchange rate in a manner analogous to unsterilized intervention—primarily through their impact on inflation rates. In the case of Bundesbank open-market operations, the increased supply of DM will raise the German inflation rate. If enough DM are created, German inflation will rise to the

U.S. INTERVENTION
FAILS TO HALT
DOLLAR'S SLIDE
On June 24, 1994, the dollar tumbled as an alliance of 17 nations failed in its bid to buoy the beleaguered U.S. currency. This failure stemmed, in part, from the view among traders that political pressure on the Federal Reserve by the Clinton Administration and the Congress would make it unwilling to raise interest rates to curb U.S. money supply growth.

In addition, the appearance of weak governments in Japan and the United Kingdom took credibility away from central bank efforts to stabilize the dollar and emboldened speculators to gamble that the dollar's plunge would continue. According to one money manager, "People are saying how can these weakened governments support the dollar? They know there isn't the political will to do this on a sustained basis."[12] Moreover, the Clinton Administration's unrelenting trade war with Japan had no doubt diminished that nation's enthusiasm for cooperation with the United States.

[12]Quoted in the *Los Angeles Times*, June 25, 1994, p. A23.

level of U.S. inflation, and the old equilibrium exchange rate e_0 will be restored. Alternatively, if the U.S. money supply is reduced sufficiently, the initial inflation that shifted the equilibrium exchange rate will end, and e_0 will again be the equilibrium rate. In either event, the real exchange rate and relative competitiveness should remain unchanged because the shift in inflation rates offsets the nominal currency change.

To summarize the empirical evidence, *real* exchange rates are primarily determined by real economic variables, such as relative national incomes and interest rates between countries; real exchange rates, in turn, determine the magnitudes and the direction of flows of goods, services, and capital among countries. A change in the supply-demand relationship in the foreign exchange market that is brought about by intervention may temporarily influence the movement of the real exchange rate. However, unless the underlying economic variables that typically give rise to broadly based, market-generated supply-and-demand forces change, these forces will eventually swamp the impact of the intervention. Thus, nations intent on fixing their exchange rate in defiance of market forces must ultimately bow to those forces or else resort to currency controls.

2.4 THE EQUILIBRIUM APPROACH TO EXCHANGE RATES

We have seen that changes in the nominal exchange rate are largely affected by variations or expected variations in relative money supplies. These nominal exchange rate changes are also highly correlated with changes in the real exchange rate. Indeed, many commentators believe that nominal exchange rate changes *cause* real exchange rate changes. As defined earlier, the *real* exchange rate is the price of domestic goods in terms of foreign goods. Thus, changes in the nominal exchange rate, through their impact on the real exchange rate, are said to help or hurt companies and economies.

One explanation for the correlation between nominal and real exchange rate changes is supplied by the disequilibrium theory of exchange rates.[13] According to this view, various frictions in the economy cause goods prices to adjust slowly over time, whereas nominal exchange rates adjust quickly in response to new information or changes in expectations. As a direct result of the differential speeds of adjustment in the goods and currency markets, changes in nominal exchange rates caused by purely monetary disturbances are naturally translated into changes in real exchange rates and can lead to exchange rate "overshooting," whereby the short-term change in the exchange rate exceeds, or overshoots, the long-term change in the equilibrium exchange rate (see Exhibit 2.12). The sequence of events associated with overshooting is as follows:

- *The central bank expands the domestic money supply.* In response, the price level will eventually rise in proportion to the money supply increase. However, because of frictions in the goods market, prices do not adjust immediately to their new equilibrium level.

- *This monetary expansion depresses domestic interest rates.* Until prices adjust fully, households and firms will find themselves holding more domestic currency than they want. Their attempts to rid themselves of excess cash balances by buying bonds will temporarily drive down domestic interest rates (bond prices and interest rates move inversely).

EXHIBIT 2.12 Exchange Rates Overshooting According to the Disequilibrium Theory of Exchange Rates

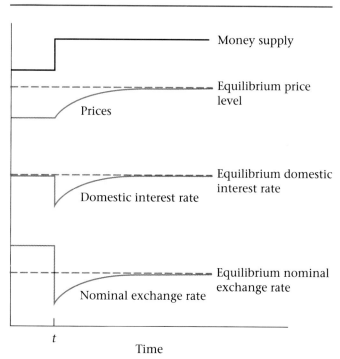

---

[13] The most elegant presentation of a disequilibrium theory is in Rudiger Dornbusch, "Expectations and Exchange Rate Dynamics," *Journal of Political Economy*, December 1976, pp. 1161–1176.

■ *Capital begins flowing out of the country because of the lower domestic interest rates, causing an instantaneous and excessive depreciation of the domestic currency.* In order for the new, lower domestic interest rates to be in equilibrium with foreign interest rates, investors must expect the domestic currency to appreciate to compensate for lower interest payments with capital gains. Future expected domestic currency appreciation, in turn, requires that the exchange rate temporarily overshoot its eventual equilibrium level. After initially exceeding its required depreciation, the exchange rate will gradually appreciate back to its new long-run equilibrium.

This view underlies most popular accounts of exchange rate changes and policy discussions that appear in the media. It implies that currencies may become "overvalued" or "undervalued" relative to equilibrium, and that these disequilibria affect international competitiveness in ways that are not justified by changes in comparative advantage.

However, the disequilibrium theory has been criticized by some economists in recent years, in part because one of its key predictions has not been upheld. Specifically, the theory predicts that as domestic prices rise, with a lag, so should the exchange rate. However, the empirical evidence is inconsistent with this predicted positive correlation between consumer prices and exchange rates. In place of the disequilibrium theory, these economists have suggested an equilibrium approach to exchange rate changes.[14] The basis for the equilibrium approach is that markets clear—supply and demand are equated—through price adjustments. Real disturbances to supply or demand in the goods market cause changes in relative prices, including the real exchange rate. These changes in the real exchange rate often are accomplished, in part, through changes in the nominal exchange rate. Repeated shocks in supply or demand thereby create a correlation between changes in nominal and real exchange rates.

The equilibrium approach has three important implications for exchange rates. First, exchange rates do not "cause" changes in relative prices but are part of the process through which the changes occur in equilibrium; that is, changes in relative prices and in real exchange rates occur simultaneously, and both are related to more fundamental economic factors.

Second, attempts by government to affect the real exchange rate via foreign exchange market intervention will fail. The direction of causation runs from the real exchange rate change to the nominal exchange rate change, and not vice versa; changing the nominal exchange rate by altering money supplies will affect relative inflation rates in such a way as to leave the real exchange rate unchanged.

Finally, there is no simple relation between changes in the exchange rate and changes in international competitiveness, employment, or the trade balance. With regard to the latter, trade deficits do not "cause" currency depreciation, nor does currency depreciation by itself help reduce a trade deficit.

Some of the implications of the equilibrium approach may appear surprising. They conflict with many of the claims that are commonly made in the financial press and by politicians; they also seem to conflict with experience. But according to the equilibrium view of exchange rates, many of the assumptions and statements commonly made in the media are simply wrong, and experiences may be very selective.

---

[14] See, for example, Alan C. Stockman, "The Equilibrium Approach to Exchange Rates," *Economic Review*, Federal Reserve Bank of Richmond, March/April 1987, pp. 12–30. This section is based on his article.

Econometric testing of these models is in its infancy, but there is some evidence that supports the equilibrium models, although it is far from conclusive. According to the disequilibrium approach, sticky prices cause changes in the nominal exchange rate to be converted into changes in the real exchange rate. But as prices eventually adjust toward their new equilibrium levels, the real exchange rate should return to its equilibrium value. Monetary disturbances, then, should create temporary movements in real exchange rates. Initial decreases in the real exchange rate stemming from a rise in the money supply should be followed by later increases as nominal prices rise to their new equilibrium level.

Statistical evidence, however, indicates that changes in real exchange rates tend, on average, to be nearly permanent or to persist for very long periods of time. The evidence also indicates that changes in nominal exchange rates—even very short-term day-to-day changes—are largely permanent. This persistence is inconsistent with the view that monetary shocks, or even temporary real shocks, cause most of the major changes in real exchange rates. On the other hand, it is consistent with the view that most changes in real exchange rates are due to real shocks with a large permanent component. Changes in real and nominal exchange rates are also very highly correlated and have similar variances, supporting the view that most changes in nominal exchange rates are due to largely permanent, real disturbances.

An alternative explanation is that we are seeing the effects of a sequence of monetary shocks, so that even if any given exchange rate change is temporary, the continuing shocks keep driving the exchange rate from its long-run equilibrium value. Thus, the sequence of these temporary changes is a permanent change. Moreover, if the equilibrium exchange rate is itself constantly subject to real shocks, we would not expect to see reversion in real exchange rates. The data do not allow us to distinguish between these hypotheses.

Another feature of the data is that the exchange rate varies much more than the ratio of price levels. The equilibrium view attributes this "excess variability" to shifts in demand and/or supply between domestic and foreign goods; the shifts affect the exchange rate but not relative inflation rates. Supply-and-demand changes also operate indirectly to alter relative prices of foreign and domestic goods by affecting the international distribution of wealth.

Although the equilibrium theory of exchange rates is consistent with selected empirical evidence, it may stretch its point too far. Implicit in the equilibrium theory is the view that money is just a unit of account—a measuring rod for value—with no intrinsic value. However, because money is an asset it is possible that monetary and other policy changes, by altering the perceived usefulness and importance of money as a store of value or liquidity, could alter real exchange rates. The evidence presented earlier that changes in anticipated monetary policy can alter real exchange rates supports this view. Moreover, the equilibrium theory fails to explain a critical fact: The variability of real exchange rates has been much greater when currencies are floating than when they are fixed. This fact is easily explained, if we view money as an asset, by the greater instability in relative monetary policies in a floating rate system. The real issue then is not whether monetary policy—including its degree of stability—has any impact at all on real exchange rates but whether that impact is of first- or second-order importance.

Despite important qualifications, the equilibrium theory of exchange rates provides a useful addition to our understanding of exchange rate behavior. Its main contribution is to suggest an explanation for exchange rate behavior that is consistent with the notion that markets work reasonably well if they are permitted to work.

ᴥ SUMMARY AND CONCLUSIONS

This chapter studied the process of determining exchange rates under a floating exchange rate system. We saw that in the absence of government intervention, exchange rates respond to the forces of supply and demand, which in turn are dependent on relative inflation rates, interest rates, and GDP growth rates. Monetary policy is crucial here. If the central bank expands the money supply at a faster rate than the growth in money demand, the purchasing power of money declines both at home (inflation) and abroad (currency depreciation). In addition, the healthier the economy is, the stronger the currency is likely to be. Exchange rates also are crucially affected by expectations of future currency changes, which depend on forecasts of future economic and political conditions.

In order to achieve certain economic or political objectives, governments often intervene in the currency markets to affect the exchange rate. Although the mechanics of such interventions vary, the general purpose of each variant is basically the same: to increase the market demand for one currency by increasing the market supply of another. Alternatively, the government can control the exchange rate directly by setting a price for its currency and then restricting access to the foreign exchange market.

A critical factor that helps explain the volatility of exchange rates is that with a fiat money there is no anchor to a currency's value, nothing around which beliefs can coalesce. In this situation, where people are unsure of what to expect, any new piece of information can dramatically alter their beliefs. Thus, if the underlying domestic economic policies are unstable, exchange rates will be volatile as traders react to new information.

➤ Questions

1. Suppose prices start rising in the United States relative to prices in Japan. What would we expect to see happen to the dollar:yen exchange rate? Explain.

2. If a foreigner purchases a U.S. government security, what happens to the supply of, and demand for, dollars?

3. Describe how these three typical transactions should affect present and future exchange rates:
   a. Seagram imports a year's supply of French champagne. Payment in French francs is due immediately.
   b. American Motors sells a new stock issue to Renault, the French car manufacturer. Payment in dollars is due immediately.
   c. Korean Airlines buys five Boeing 747s. As part of the deal, Boeing arranges a loan to KAL for the purchase amount from the U.S. Export-Import Bank. The loan is to be paid back over the next seven years with a two-year grace period.

4. In 1987, the British government cut taxes significantly, raising the after-tax return on investments in Great Britain. What would be the likely consequence of this tax cut on the equilibrium value of the British pound?

5. Some economists have argued that a lower government deficit could cause the dollar to drop by reducing high real interest rates in the United States. What does the asset view of exchange rates predict will happen if the United States lowers its budget deficit? What is the evidence from countries such as Mexico and Brazil?

6. The maintenance of money's value is said to depend on the monetary authorities. What might the monetary authorities do to a currency that would cause its value to drop?

7. For each of the following six scenarios, say whether the value of the dollar will appreciate, depreciate, or remain the same relative to the Japanese yen. Explain each answer. Assume that exchange rates are free to vary and that other factors are held constant.
   a. The growth rate of national income is higher in the United States than in Japan.
   b. Inflation is higher in the United States than in Japan.
   c. Prices in Japan and the United States are rising at the same rate.
   d. Real interest rates are higher in the United States than in Japan.
   e. The United States imposes new restrictions on the ability of foreigners to buy American companies and real estate.
   f. U.S. wages rise relative to Japanese wages, while American productivity falls behind Japanese productivity.

8. The Fed adopts an easier monetary policy. How is this likely to affect the value of the dollar and U.S. interest rates?

9. What is there about a fiat money that makes its exchange rate especially volatile?

10. Comment on the following headlines in the *Wall Street Journal*.
    a. "Sterling Drops Sharply Despite Good Health of British Economy: Oil Price Slump Is Blamed" (January 17, 1985).
    b. "Dollar Surges as Coup in Soviet Union Revives Unit's Appeal as a Safe Haven" (August 20, 1991).
    c. "Dollar Plummets on Soviet Coup Failure" (August 22, 1991).
    d. "Dollar Falls Across the Board as Fed Cuts Discount Rate to 6.5% From 7%" (December 19, 1990).
    e. "Canadian Dollar Likely to Fall Further On Recession and Constitutional Crisis" (September 28, 1992).
    f. "Dollar Soars on U.S. and Iraqi Tension, Hints of Possible Lower German Rates" (December 12, 1992).
    g. "Dollar Slips Against Mark and Yen As Iraq Orders a Troop Withdrawal" (October 11, 1994).
    h. "Inflation, Slow Growth Seen Spurring Latin America to Devalue Currencies" (January 22, 1990).

11. Comment on the following headline from *The New York Times*. "Germany Raises Interest Rate, and Value of Dollar Declines" (October 10, 1997).

12. Suppose a new Russian government makes threatening moves against Western Europe. How is this threat likely to affect the dollar's value? Why?

13. On May 11, 1995, the House Budget Committee approved a plan to slash federal spending through the year 2002 and thereby end the persistent U.S. budget deficits. How do you think the dollar responded to this news?

14. In the 1995 election for the French presidency, the Socialist candidate, Lionel Jospin, vowed to halt all privatizations, raise taxes on business, spend heavily on job creation, and cut the workweek without a matching pay cut. At the time Mr. Jospin made this vow, he was running neck-and-neck with the conservative Prime Minister Jacques Chirac, who espoused free-market policies.
    a. How do you think the French franc responded to Mr. Jospin's remarks?
    b. In the event, Mr. Chirac won the election. What was the franc's likely reaction?
    c. In a surprise ending to the 1997 French Parliamentary elections, the Socialist party won and Mr. Jospin became Prime Minister. Given that Mr. Jospin still espoused the same policies as before, what was the likely reaction of the French franc to his election?

15. Comment on the following statement: "One of the puzzling aspects of central bank intervention is how those who manage our economic affairs think they know what is the 'right' price for a dollar in terms of francs, pounds, yen, or Deutsche marks. And if they do know, why do they keep changing their minds?"

16. In a widely anticipated move, on August 30, 1990, the Bank of Japan raised the discount rate (the rate it charges on loans to financial institutions) to 6% from 5.25% in a move to reduce inflationary pressures in Japan. Many currency traders had expected the Japanese central bank to raise its rate by more than 0.75%. What was the likely consequence of this interest rate rise on the yen:dollar exchange rate?

17. On November 28, 1990, Federal Reserve Chairman Alan Greenspan told the House Banking Committee that despite possible benefits to the U.S. trade balance, "a weaker dollar also is a cause for concern." This statement departed from what appeared to be an attitude of benign neglect by U.S. monetary officials toward the dollar's depreciation. He also rejected the notion that the Fed should aggressively ease monetary policy, as some Treasury officials had been urging. At the same time, Mr. Greenspan did not mention foreign exchange market intervention to support the dollar's value.
    a. What was the likely reaction of the foreign exchange market to Mr. Greenspan's statements? Explain.
    b. Can Mr. Greenspan support the value of the U.S. dollar without intervening in the foreign exchange market? If so, how?

18. In the late 1980s, the Bank of Japan bought billions of dollars in the foreign exchange market to prop up the dollar's value against the yen. What were the likely consequences of this foreign exchange market intervention for the Japanese economy?

19. Countries with high inflation need to keep devaluing their currencies to maintain competitiveness. But countries that try to maintain their competitiveness by devaluing their currencies only end up with even higher inflation. Discuss.

20. The Russian government is trying to figure out how to stabilize the value of its currency. What advice would you offer to it?

21. "Unsterilized interventions are just open market operations conducted through the foreign exchange market rather than through the U.S. government securities market." Comment on this statement.

22. As 1992 began, the Russian government and the central bank tightened credit in an attempt to slow the growth in the supply of rubles. However, the moves were not popular with the country's giant state-run industrial enterprises, which are still dependent on official subsidies and cheap credit. In July 1992, the Russian Parliament appointed Viktor Gerashchenko head of the central bank. One of his first acts was to say that he did not think the time was right to make the ruble convertible. Then he said that he would continue to extend credits to bankrupt and inefficient state enterprises.
    a. How independent is the Russian central bank likely to be? What political pressures is it facing?
    b. What is the likely effect of Mr. Gerashchenko's statements on inflationary expectations in Russia?
    c. How do you think the ruble : dollar exchange rate was affected by these statements?
    d. In 1995, the Russian Central Bank signed an agreement with the International Monetary Fund not to issue cheap credits to state enterprises. How should the ruble react if the Central Bank sticks to its agreement with the IMF?

23. In January 1991, President Mikhail Gorbachev banned all 50-ruble and 100-ruble bills, while permitting Soviet citizens to change only 1,000 rubles in these large bills into smaller denominations. In addition, savings-bank accounts were frozen for six months. The object of these measures was to strip the country's powerful black marketeers of their operating capital, driving as many as possible out of business, and to reduce inflation, which had been running at about 80% a year. The official Russian news agency Tass reported that the government had "clearly decided that the confiscation version of monetary reform was the most efficient and least expensive version at its disposal."
    a. Were these measures likely to achieve President Gorbachev's objectives?
    b. How do you think the ruble's exchange rate responded to President Gorbachev's initiative? Explain.

24. On October 29, 1995, the Mexican government announced a new economic plan, which called for the government to boost the economy by cutting taxes and spending. The plan also included an agreement among business, labor, and government representatives to limit wage and price increases. How do you think the peso responded to this announcement? What about the Mexican stock market? Explain.

25. Under the Convertibility Act, Argentina's central bank is allowed to count dollar-denominated bonds issued by the Argentine government as part of its "foreign" reserve assets. What potential problem do you see with this rule?

26. After the Mexican devaluation, investors questioning Argentina's ability to maintain currency convertibility began pulling their money out of Argentina. In response, the Argentine government took extraordinary steps to maintain its exchange rate at $1/peso.
    a. What were the likely consequences of this capital flight for Argentina's peso money supply? for Argentine peso interest rates? for economic growth?
    b. Why was the Argentine government so reluctant to devalue the peso?
    c. As U.S. interest rates rise, what is likely to happen to Argentine rates? Why?

27. One recommended approach to strengthen the dollar against the yen is for the U.S. Treasury to issue about $70 billion a year (the Japanese share of the U.S. trade deficit) in yen-denominated bonds. How might this move help the dollar?

28. In 1993, President Carlos Salinas de Gortari proposed a bill that would formally grant the Bank of Mexico, Mexico's central bank, autonomy vis-à-vis the central government. As an investor, how would you view such a proposal? What other changes might help to amplify the signals sent by this proposal?

29. The People's Bank of China, China's central bank, is run by bureaucrats whose prime objective seems to be funding loss-making state-owned firms. What is your prediction about the inflation outlook for China and the value of its currency, the yuan? Explain.

30. In August 1994, Alan Blinder, the recently appointed vice chairman of the Federal Reserve Board, gave a talk in which he argued that the Fed should be willing to tolerate somewhat higher inflation in order to spur economic growth and job creation. The dollar fell, almost immediately. Explain the link between Dr. Blinder's views and the value of the dollar.

31. Describe the chief differences between a currency board and a central bank with a nominal exchange rate target.

32. Many Asian governments have attempted to promote their export competitiveness by holding down the values of their currencies through foreign exchange market intervention.
    a. What is the likely impact of this policy on Asian foreign exchange reserves? on Asian inflation? on Asian export competitiveness? on Asian living standards?
    b. Some Asian countries have attempted to sterilize their foreign exchange market intervention by selling bonds. What are the likely consequences of sterilization on interest rates? on exchange rates in the longer term? on export competitiveness?

33. As mentioned in the chapter, Hong Kong has a currency board that fixes the exchange rate between the U.S. and H.K. dollars.
   a. What is the likely consequence of a large capital inflow for the rate of inflation in Hong Kong? For the competitiveness of Hong Kong business? Explain.
   b. Given a large capital inflow, what would happen to the value of the Hong Kong dollar if it were allowed to float freely? What would be the effect on the competitiveness of Hong Kong business? Explain.
   c. Given a large capital inflow, will Hong Kong business be more or less competitive under a currency board than a freely floating currency? Explain.

34. In 1994, an influx of drug money to Colombia coincided with a sharp increase in its export earnings from coffee and oil.
   a. What was the likely impact of these factors on the value of the Colombian peso and the competitiveness of Colombia's legal exports? Explain.
   b. In 1996, Colombia's president, facing charges of involvement in his country's drug cartel, sought to boost his domestic popularity by pursuing more expansionary monetary policies. Standing in the way was Colombia's independent central bank—Banco de la Republica. In response, the president and his supporters discussed the possibility of returning central bank control to the executive branch. Describe the likely economic consequences of ending Banco de la Republica's independence.

35. In 1994, China sought to boost its foreign exchange reserves and stabilize the yuan (which was under pressure to appreciate) by mandating that Chinese enterprises sell all their foreign exchange to the country's commercial banks. The People's Bank of China, in turn, was forced to buy surplus foreign exchange with yuan.
   a. What are the likely consequences of this policy for China's rate of inflation? Explain.
   b. What alternatives are open to China to achieve its aim to simultaneously hold down the value of the yuan and curb inflation?

36. In the midst of the Asian financial crisis, Malaysia's Prime Minister Mahathir Mohamad accused an international cabal of Jewish financiers of deliberately provoking the crisis to wreck Malaysia's economy. "Jews are not happy to see Muslims prosper," he said. Following his remarks, Malaysia's financial markets and its currency, the ringgit, plunged to record lows. Explain.

➤ Problems

1. In the second half of 1997, the Indonesian rupiah devalued by 84% against the U.S. dollar. By how much has the dollar appreciated against the rupiah?

2. On February 1, the French franc is worth $0.1984. By May 1, it has moved to $0.2057.
   a. By how much has the franc appreciated or depreciated against the dollar during this three-month period?
   b. By how much has the dollar appreciated or depreciated against the franc during this period?

3. During 1997, the U.S. dollar appreciated by 104% against the South Korean won. By how much did the won depreciate against the U.S. dollar during the year?

4. On January 1, 1975, the Mexican peso/U.S.$ exchange rate was Ps 12.5 = $1. By 1985, the exchange rate stood at Ps 208.9 = $1.
   a. By how much has the Mexican peso appreciated or depreciated against the dollar over this 10-year period?
   b. By how much has the dollar appreciated or depreciated against the peso over this period?

5. Suppose the Russian ruble devalues by 75% against the dollar. What is the percentage appreciation of the dollar against the ruble?

6. Suppose the dollar appreciates by 500% against the Russian ruble. How much has the ruble devalued against the dollar?

7. In 1993, the Brazilian cruzeiro lost 95% of its dollar value. What happened to the cruzeiro value of the dollar during 1993?

8. Between 1988 and 1991, the price of a room at the Milan Hilton rose from Lit 346,400 to Lit 475,000. At the same time, the exchange rate went from Lit 1,302:$1 in 1988 to Lit 1,075:$1 in 1991.
   a. By how much has the dollar cost of a room at the Milan Hilton changed over this three-year period?
   b. What has happened to the lira's dollar value during this period?

9. On the day that the Chancellor of the Exchequer announced independence for the Bank of England, the British stock market rose 1.4%, and British gilts rose about 2.1%. The pound rose to $1.6333 from $1.6223 and to DM2.8165 from DM2.8006.
   a. Why did the British stock and bond markets jump on the news?

b. Why did the British pound appreciate on the news?

c. What was the pound's percentage appreciation against the U.S. dollar?

d. What was the pound's percentage appreciation against the DM?

10. During the currency crisis of September 1992, the Bank of England borrowed DM 33 billion from the Bundesbank when a pound was worth DM 2.78 or $1.912. It sold these DM in the foreign exchange market for pounds in a futile attempt to prevent a devaluation of the pound. It repaid these DM at the post-crisis rate of DM 2.50:£1. By then, the dollar:pound exchange rate was $1.782:£1.

a. How much had the pound sterling devalued in the interim against the Deutsche mark? Against the dollar?

b. What was the cost of intervention to the Bank of England in pounds? In dollars?

c. Who won?

➤ Bibliography

BATTEN, DALLAS S., and MACK OTT. "What Can Central Banks Do About the Value of the Dollar?" *Federal Reserve Bank of St. Louis Review*, May 1984, pp. 16–26.

DORNBUSCH, RUDIGER. "Expectations and Exchange Rate Dynamics." *Journal of Political Economy*, December 1976, pp. 1161–1176.

FRENKEL, JACOB A., and HARRY G. JOHNSON, eds. *The Economics of Exchange Rates*. Reading, Mass.: Addison-Wesley, 1978.

LEVICH, RICHARD M. "Empirical Studies of Exchange Rates: Price Behavior, Rate Determination and Market Efficiency." In *Handbook of International Economics*, vol. II. RONALD W. JONES and PETER B. KENEN, eds. Netherlands: Elsevier B.V., 1985, pp. 980–1040.

MARRINAN, JANE. "Exchange Rate Determination: Sorting Out Theory and Evidence." *New England Economic Review*, November/December 1989, pp. 39–51.

STOCKMAN, ALAN C. "The Equilibrium Approach to Exchange Rates." *Economic Review*, Federal Reserve Bank of Richmond, March/April 1987, pp. 12–30.

THE INTERNATIONAL MONETARY SYSTEM

The monetary and economic disorders of the past fifteen years . . . are a reaction to a world monetary system that has no historical precedent. We have been sailing on uncharted waters and it has been taking time to learn the safest routes.

Milton Friedman
Winner of Nobel Prize in Economics

The currency problems faced by firms today have been exacerbated by the breakdown of the postwar international monetary system established at the Bretton Woods Conference in 1944. The main features of the **Bretton Woods system** were the relatively fixed exchange rates of individual currencies in terms of the U.S. dollar and the convertibility of the dollar into gold for foreign official institutions. These fixed exchange rates were supposed to reduce the riskiness of international transactions, thus promoting growth in world trade.

However, in 1971, the Bretton Woods system fell victim to the international monetary turmoil it was designed to avoid. It was replaced by the present regime of rapidly fluctuating exchange rates, resulting in major problems and opportunities for multinational corporations. The purpose of this chapter is to help managers, both financial and nonfinancial, to understand what the international monetary system is and how the choice of system affects currency values. It also provides a historical background of the international monetary system to enable managers to gain perspective when trying to interpret the likely consequences of new policy moves in the area of international finance. After all, although the types of government foreign-exchange policies may at times appear to be limitless, they are all variations on a common theme.

85

3.1 ALTERNATIVE EXCHANGE RATE SYSTEMS

The **international monetary system** refers primarily to the set of policies, institutions, practices, regulations, and mechanisms that determine the rate at which one currency is exchanged for another. This section considers five market mechanisms for establishing exchange rates: free float, managed float, target-zone arrangement, fixed-rate system, and the current hybrid system.

As we shall see, each of these mechanisms has costs and benefits associated with it. Nations prefer economic stability and often equate this objective with a stable exchange rate. However, fixing an exchange rate often leads to currency crises if the nation attempts to follow a monetary policy that is inconsistent with that fixed rate. At the same time, a nation may decide to fix its exchange rate in order to limit the scope of monetary policy, as in the case of currency boards seen in the previous chapter. On the other hand, economic shocks can be absorbed more easily when exchange rates are allowed to float freely, but freely floating exchange rates may exhibit excessive volatility and hurt trade and stifle economic growth. The choice of a particular exchange rate mechanism depends on the relative importance that a given nation at a given point in time places on the trade-offs associated with these different systems.

Free Float

We already have seen that free-market exchange rates are determined by the interaction of currency supplies and demands. The supply-and-demand schedules, in turn, are influenced by price level changes, interest differentials, and economic growth. In a **free float**, as these economic parameters change—for example, because of new government policies or acts of nature—market participants will adjust their current and expected future currency needs. In the two-country example of Germany and the United States, the shifts in the Deutsche mark supply-and-demand schedules will, in turn, lead to new equilibrium positions. Over time, the exchange rate will fluctuate randomly as market participants assess and react to new information, much as security and commodity prices in other

EXHIBIT 3.1 Supply and Demand Curve Shifts

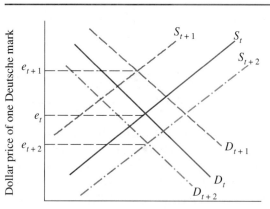

Quantity of Deutsche marks

EXHIBIT 3.2 Fluctuating Exchange Rate

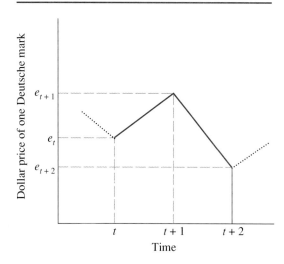

financial markets respond to news. These shifts and oscillations are illustrated in Exhibits 3.1 and 3.2; D_t and S_t are the hypothetical DM demand and supply curves, respectively, for period t. Such a system of freely floating exchange rates is usually referred to as a **clean float**.

Managed Float

Not surprisingly, few countries have been able to resist for long the temptation to intervene actively in the foreign exchange market in order to reduce the economic uncertainty associated with a clean float. The fear is that too abrupt a change in the value of a nation's currency could imperil its export industries (if the currency appreciates) or lead to a higher rate of inflation (if the currency depreciates). Moreover, the experience with floating rates in the 1980s was not encouraging. Instead of reducing economic volatility, as they were supposed to do, floating exchange rates appear to have increased it. Exchange rate uncertainty also reduces economic efficiency by acting as a tax on trade and foreign investment. Therefore, most countries with floating currencies have attempted, through central bank intervention, to smooth out exchange rate fluctuations. Such a system of managed exchange rates, called a **managed float**, is also known as a **"dirty float."**

Managed floats fall into three distinct categories of central bank intervention. The approaches, which vary in their reliance on market forces, are as follows:

1. *Smoothing out daily fluctuations.* Governments following this route attempt only to preserve an orderly pattern of exchange rate changes. They occasionally enter the market on the buy or sell side to ease the transition from one rate to another, rather than resist fundamental market forces, tending to bring about longer-term currency appreciation or depreciation. One variant of this approach is the "crawling peg" system used in some countries, such as Poland, Russia, and Brazil. Under a crawling peg, the local currency depreciates against a reference currency or currency basket on a regular, controlled basis. For example, the Brazilian *real* and Russian ruble are allowed to depreciate monthly by about 0.6% and 0.5%, respectively, against the dollar. Similarly, the Polish zloty depreciates by 1% a month against a basket of currencies.

2. *"Leaning against the wind".* This approach is an intermediate policy designed to moderate or prevent abrupt short- and medium-term fluctuations brought about by random events whose effects are expected to be only temporary. The rationale for this policy—which is primarily aimed at delaying, rather than resisting, fundamental exchange rate adjustments—is that government intervention can reduce for exporters and importers the uncertainty caused by disruptive exchange rate changes. It is questionable, though, whether governments are more capable than private forecasters of distinguishing between fundamental and temporary (irrational) values.

3. *Unofficial pegging.* This strategy evokes memories of a fixed-rate system. It involves resisting fundamental upward or downward exchange rate movements for reasons clearly unrelated to exchange market forces. Thus, Japan historically has resisted revaluation of the yen for fear of its consequences for Japanese exports. With unofficial pegging, however, there is no publicly announced government commitment to a given exchange rate level.

Target-Zone Arrangement

Many economists and policymakers have argued that the industrialized countries could minimize exchange rate volatility and enhance economic stability if the United States, Germany, and Japan linked their currencies in a target-zone system. Under a **target-zone arrangement**, countries adjust their national economic policies to maintain their exchange rates within a specific margin around agreed-upon, fixed central exchange rates. Such a system already exists for the major European currencies participating in the European Monetary System, which is discussed later in this chapter.

Fixed-Rate System

Under a fixed-rate system, such as the Bretton Woods system, governments are committed to maintaining target exchange rates. Each central bank actively buys or sells its currency in the foreign exchange market whenever its exchange rate threatens to deviate from its stated par value by more than an agreed-on percentage. The resulting coordination of monetary policy ensures that all member nations have the same inflation rate. Put another way, for a fixed-rate system to work, each member must accept the group's joint inflation rate as its own. A corollary is that monetary policy must become subordinate to exchange rate policy. In the extreme case, those who fix their exchange rate via a currency board system surrender all control of monetary policy. The money supply is determined solely by people's willingness to hold the domestic currency.

With or without a currency board system, there is always a rate of monetary growth (it could be negative) that will maintain an exchange rate at its target level. If it involves monetary tightening, however, maintaining the fixed exchange rate could mean a high interest rate and a resultant slowdown in economic growth and job creation.

Under the Bretton Woods system, whenever the commitment to the official rate became untenable, it was abruptly changed and a new rate was announced publicly.

EXHIBIT 3.3 Typical Currency Control Measures

- Restriction or prohibition of certain remittance categories such as dividends or royalties
- Ceilings on direct foreign investment outflows (e.g., the elaborate U.S. Office of Foreign Direct Investment controls in effect 1968–1975)
- Controls on overseas portfolio investments
- Import restrictions
- Required surrender of hard-currency export receipts to central bank
- Limitations on prepayments for imports
- Requirements to deposit in interest-free accounts with central bank, for a specified time, some percentage of the value of imports and/or remittances
- Foreign borrowings restricted to a minimum or maximum maturity
- Ceilings on granting of credit to foreign firms
- Imposition of taxes and limitations on foreign-owned bank deposits
- Multiple exchange rates for buying and selling foreign currencies, depending on category of goods or services each transaction falls into

---

Currency devaluation or revaluation, however, was usually the last in a string of temporizing alternatives for solving a persistent balance-of-payments deficit or surplus. These alternatives, which are related only in their lack of success, ranged from foreign borrowing to finance the balance-of-payments deficit, wage and price controls, and import restrictions to exchange controls. The latter have become a way of life in most developing countries. Nations with overvalued currencies ration foreign exchange, whereas countries facing revaluation, such as Germany and Switzerland, may restrict capital inflows.

In effect, government controls supersede the allocative function of the foreign exchange market. The most drastic situation occurs when all foreign exchange earnings must be surrendered to the central bank, which, in turn, apportions these funds to users on the basis of government priorities. The buying and selling rates need not be equal, nor need they be uniform across all transaction categories. Exhibit 3.3 lists the most frequently used currency control measures. These controls are a major source of market imperfection, providing opportunities as well as risks for multinational corporations.

Austerity brought about by a combination of reduced government expenditures and increased taxes can be a permanent substitute for devaluation. By reducing the nation's budget deficit, austerity will lessen the need to monetize the deficit. Lowering the rate of money supply growth, in turn, will bring about a lower rate of domestic inflation (disinflation). Disinflation will strengthen the currency's value, ending the threat of devaluation. However, disinflation often leads to a short-run increase in unemployment, a cost of austerity that politicians today generally consider to be unacceptable.

The Current System of Exchange Rate Determination

The current international monetary system is a hybrid, with major currencies floating on a managed basis, some currencies freely floating, and other currencies moving in and out of various types of pegged exchange rate relationships. Exhibit 3.4 presents a currency map that describes the various zones and blocs linking the world's currencies as of December 31, 1997.

EXHIBIT 3.4 Exchange Rate Arrangements (as of December 31, 1997)[1]

			FLEXIBILITY LIMITED IN TERMS OF A SINGLE CURRENCY OR GROUP OF CURRENCIES	
CURRENCY PEGGED TO				
U.S. Dollar	French Franc	Other Currency	Single Currency[3]	Cooperative Arrangements[4]
Angola	Benin	Bhutan	Bahrain	Austria
Antigua & Barbuda	Burkina Faso	(Indian rupee)	Qatar	Belgium
Argentina	Cameroon	Bosnia and	Saudi Arabia	Denmark
Bahamas, The	C. African Rep.	Herzegovina	United Arab	Finland
Barbados	Chad	(deutsche mark)	Emirates	France
Belize	Comoros	Brunei Darussalam		Germany
Djibouti	Congo. Rep. of	(Singapore		Greece
Dominica	Côte d'Ivoire	dollar)		Ireland
Grenada	Equatorial	Bulgaria		Italy
Iraq	Guinea	(deutsche mark)		Luxembourg
Lithuania	Gabon	Cape Verde		Netherlands
Marshall Islands	Guinea-Bissau	(Portuguese		Portugal
Micronesia,	Mali	escudo)		Spain
Fed. States of	Niger	Estonia		
Oman	Senegal	(deutsche mark)		
Palau	Togo	Kiribati		
Panama		(Australian		
St. Kitts & Nevis		dollar)		
St. Lucia		Lesotho		
St. Vincent and the	**SDR**	(South African		
Grenadines	Jordan	rand)		
Syrian Arab Rep.	Latvia	Namibia		
	Libya	(South African		
	Myanmar	rand)		
		Nepal		
		(Indian rupee)		
Other Composite[2]		San Marino		
		(Italian lira)		
Bangladesh	Malta	Swaziland		
Botswana	Morocco	(South African		
Burundi	Samoa	rand)		
Cyprus	Seychelles			
Fiji	Slovak Republic			
Iceland	Tonga			
Kuwait	Vanuatu			

[1] For members with dual or multiple exchange markets, the arrangements shown is that in the major market.
[2] Comprises currencies that are pegged to various "baskets" of currencies of the members own choice, as distinct from the SDR basket.
[3] Exchange rates of all currencies have shown limited flexibility in terms of the U.S. dollar.
[4] Refers to the cooperative arrangement maintained under the European Monetary System.
[5] Starting May 24, 1994, the Azerbaijan authorities ceased to peg the manat to the Russian ruble and the exchange arrangement was reclassified to "Independently floating."

Source: International Monetary Fund, International Financial Statistics, March 1998.

3.2 A BRIEF HISTORY OF THE INTERNATIONAL MONETARY SYSTEM

Almost from the dawn of history, gold has been used as a medium of exchange because of its desirable properties. It is durable, storable, portable, easily recognized, divisible, and easily standardized. Another valuable attribute of gold is that short-run changes in its

EXHIBIT 3.4 (continued)

MORE FLEXIBLE

Other Managed Floating

Algeria	Iran. I. R. of	Romania
Belarus	Israel	Russia
Bolivia	Kazakhstan	Singapore
Brazil	Kenya	Slovenia
Cambodia	Kyrgyz Rep.	Solomon Islands
Chile	Lao P.D. Rep.	Sri Lanka
China, P.R	Macedonia,	Sudan
Colombia	FYR of	Suriname
Costa Rica	Malawi	Tajikistan, Rep. of
Croatia	Malaysia	Thailand
Czech Republic	Maldives	Tunisia
Dominican Rep.	Mauritania	Turkey
Ecuador	Mauritius	Turkmenistan
Egypt	Nicaragua	Ukraine
El Salvador	Nigeria	Uruguay
Ethiopia	Norway	Uzbekistan
Georgia	Pakistan	Venezuela
Honduras	Poland	Vietnam
Hungary		

Independently Floating[5]

Afghanistan,	Japan	Sierra Leone
Islamic State of	Korea	Somalia
Albania	Lebanon	South Africa
Armenia	Liberia	Sweden
Australia	Madagascar	Switzerland
Azerbaijan	Mexico	Tanzania
Canada	Moldova	Trinidad and
Congo, Dem. Rep.	Mongolia	Tobago
Eritrea	Mozambique	Uganda
Gambia, The	New Zealand	United Kingdom
Ghana	Papua New Guinea	United States
Guatemala	Paraguay	Yemen, Republic of
Guinea	Peru	Zambia
Guyana	Philippines	Zimbabwe
Haiti	Rwanda	
India	São Tomé and	
Indonesia	Príncipe	
Jamaica		

stock are limited by high production costs, making it costly for governments to manipulate. Most important, because gold is a commodity money, it ensures a long-run tendency toward price stability. The reason is that the purchasing power of an ounce of gold, or what it will buy in terms of all other goods and services, will tend toward equality with its long-run cost of production.

For these reasons, most major currencies, until fairly recently, were on a gold standard, which defined their relative values or exchange rates. The **gold standard** essentially involved a commitment by the participating countries to fix the prices of their domestic currencies in terms of a specified amount of gold. The countries maintained these prices by being willing to buy or sell gold to anyone at that price. For example, from 1821 to 1914, Great Britain maintained a fixed price of gold at £4.2474 per ounce. The United States, during the 1834–1933 period, maintained the price of gold at $20.67 per ounce (with the exception of the Greenback period from 1861–1878). Thus, over the period 1834–1914 (with the exception of 1861–1878), the dollar:pound exchange rate, referred to as the par exchange rate, was perfectly determined at:

$$\frac{\$20.67/\text{ounce of gold}}{£4.2474/\text{ounce of gold}} = \$4.8665/£1$$

The value of gold relative to other goods and services does not change much over long periods of time, so the monetary discipline imposed by a gold standard should ensure long-run price stability for both individual countries and groups of countries. Indeed, there was remarkable long-run price stability in the period before World War I, during which most countries were on a gold standard. As Exhibit 3.5 shows, price levels at the start of World War I were roughly the same as they had been in the late 1700s before the Napoleonic Wars began.

This record is all the more remarkable when contrasted with the post-World War II inflationary experience of the industrialized nations of Europe and North America. As shown in Exhibit 3.6, 1995 price levels in all these nations were several times as high as they were in 1950. Even in Germany, the value of the currency in 1995 was only one-quarter of its 1950 level, whereas the comparable magnitude was less than one-tenth for France, Italy, and the United Kingdom. Although there were no episodes of extremely rapid inflation, price levels rose steadily and substantially.

The Classical Gold Standard

A gold standard is often considered an anachronism in our modern, high-tech world because of its needless expense; on the most basic level, it means digging up gold in one

EXHIBIT 3.5 Wholesale Price Indices: Pre-World War I (1913 = 100)

YEAR	BELGUIM	BRITAIN	FRANCE	GERMANY	UNITED STATES
1776	na	101	na	na	84
1793	na	120	na	98	100
1800	na	186	155	135	127
1825	na	139	126	76	101
1850	83	91	96	71	82
1875	100	121	111	100	80
1900	87	86	85	90	80
1913	100	100	100	100	100

EXHIBIT 3.6 Consumer Price Indices (CPI): Post-World War II

NATION	CPI, 1950	CPI, 1995	LOSS OF PURCHASING POWER DURING PERIOD (%)
Belgium	100	578	82.7
France	100	1294	92.3
Germany	100	388	74.2
Italy	100	2163	95.4
Netherlands	100	603	83.4
United Kingdom	100	1617	93.8
United States	100	622	83.9

corner of the globe for burial in another corner. Nonetheless, discontent with the current monetary system, which has produced more than two decades of worldwide inflation and widely fluctuating exchange rates, has prompted interest in a return to some form of a gold standard.

To put it bluntly, calls for a new gold standard reflect a fundamental distrust of government's willingness to maintain the integrity of fiat money. **Fiat money** is nonconvertible paper money backed only by faith that the monetary authorities will not cheat (by issuing more money). This faith has been tempered by hard experience; the 100% profit margin on issuing new fiat money has proved to be an irresistible temptation for most governments.

By contrast, the net profit margin on issuing more money under a gold standard is zero. The government must acquire more gold before it can issue more money, and the government's cost of acquiring the extra gold equals the value of the money it issues. Thus, expansion of the money supply is constrained by the available supply of gold. This fact is crucial in understanding how a gold standard works.

Under the classical gold standard, disturbances in the price level in one country would be wholly or partly offset by an automatic balance-of-payments adjustment mechanism called the **price-specie-flow mechanism**. (*Specie* refers to gold coins.) To see how this adjustment mechanism worked to equalize prices among countries and automatically bring international payments back in balance, consider the example that is illustrated in Exhibit 3.7.

Suppose a technological advance increases productivity in the non-gold-producing sector of the U.S. economy. This productivity will lower the price of other goods and services relative to the price of gold, and the U.S. price level will decline. The fall in U.S. prices will result in lower prices of U.S. exports; export prices will decline relative to import prices (determined largely by supply and demand in the rest of the world). Consequently, foreigners will demand more U.S. exports, and Americans will buy fewer imports.

Starting from a position of equilibrium in its international payments, the United States will now run a balance-of-payments surplus. The difference will be made up by a flow of gold into the United States. The gold inflow will increase the U.S. money supply (under a gold standard, more gold means more money in circulation), reversing the initial decline in prices. At the same time, the other countries will experience gold outflows,

EXHIBIT 3.7 The Price-Specie-Flow Mechanism

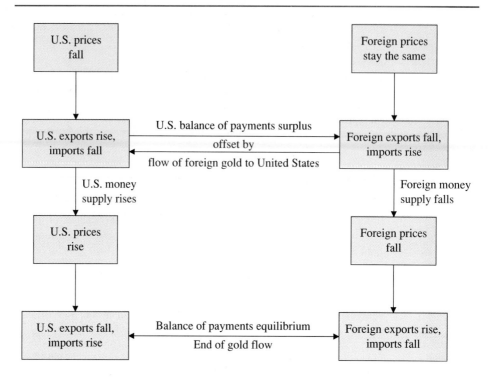

reducing their money supplies (less gold, less money in circulation) and, thus, their price levels. In final equilibrium, price levels in all countries will be slightly lower than they were before, because of the increase in the worldwide supply of other goods and services relative to the supply of gold. Exchange rates will remain fixed.

Thus, the operation of the price-specie-flow mechanism tended to keep prices in line for those countries that were on the gold standard. As long as the world was on a gold standard, all adjustments were automatic, and although many undesirable things might have happened under a gold standard, enduring inflation was not one of them.

Gold does have a cost, though—the opportunity cost associated with mining and storing it. By the late 1990s, with inflation on the wane worldwide, the value of gold as an inflation hedge has declined. Central banks also have begun selling their gold reserves and replacing them with U.S. Treasury bonds that, unlike gold, pay interest. The reduced demand for gold has lowered its price and its usefulness as a monetary asset.

How the Classical Gold Standard Worked in Practice: 1821–1914

In 1821, after the Napoleonic Wars and their associated inflation, England returned to the gold standard. From 1821 to 1880, more and more countries joined the gold standard. By 1880, most nations of the world were on some form of gold standard. The period from 1880 to 1914, during which the classical gold standard prevailed in its most pristine form, was a remarkable period in world economic history. The period was characterized by a rapid expansion of virtually free international trade, stable exchange rates and prices, a free flow of labor and capital across political borders, rapid economic growth, and, in general, world peace. Advocates of the gold standard harken back to this period as illustrating the standard's value.

Opponents of a rigid gold standard, in contrast, point to some less-than-idyllic economic conditions during this period: a major depression during the 1890s, a severe economic contraction in 1907, and repeated recessions. Whether these sharp ups and downs could have been prevented under a fiat money standard cannot be known.

The Gold Exchange Standard: 1925–1931

The gold standard broke down during World War I and was briefly reinstated from 1925 to 1931 as the Gold Exchange Standard. Under this standard, the United States and England could hold only gold reserves, but other nations could hold both gold and dollars or pounds as reserves. In 1931, England departed from gold in the face of massive gold and capital flows, owing to an unrealistic exchange rate, and the Gold Exchange Standard was finished.

After the devaluation of sterling, 25 other nations devalued their currencies to maintain trade competitiveness. These **"beggar-thy-neighbor" devaluations**, in which nations cheapened their currencies to increase their exports at others' expense and to reduce imports, led to a trade war. Many economists and policymakers believed that the protectionist exchange rate and trade policy fueled the global depression of the 1930s. To avoid such destructive economic policies in the future, the Allied nations agreed to a new postwar monetary system at a conference held in Bretton Woods, New Hampshire, in 1944. The conference also created two new institutions – the **International Monetary Fund (IMF)** and the **International Bank for Reconstruction and Development (World Bank)** – to implement the new system and to promote international financial stability. The IMF was created to promote monetary stability; the World Bank was set up to lend money to countries so they could rebuild their infrastructure that had been destroyed during the war.

Both agencies have seen their roles evolve over time. The IMF now oversees exchange-rate policies in member countries (currently totaling 181 nations) and advises developing countries about how to turn their economies around. In the process, it has become the **lender of last resort** to countries that get into serious financial trouble. It currently is exploring new ways to monitor member nations' financial health so as to prevent another Mexico-like surprise. Despite these efforts, the IMF was blindsided by the Asian crisis and wound up leading a $118 billion attempt to shore up Asian financial systems. Critics argue that by bailing out careless lenders and imprudent nations, IMF rescues make it too easy for governments to persist with bad policies and for investors to ignore the risks these policies create. In the long run, by removing from governments and investors the prospect of failure — which underlies the market discipline that encourages sound policies — these rescues magnify the problem of moral hazard and so make imprudent policies more likely to recur.[1]

---

[1] As economist Allan Meltzer puts it, "Capitalism without failure is like religion without sin."

The World Bank is looking to expand its lending to developing countries and to provide more loan guarantees for businesses entering new developing markets. But here too there is controversy. Specifically, critics claim that World Bank financing allows projects and policies to avoid being subject to the scrutiny of financial markets and permits governments to delay enacting the changes necessary to make their countries more attractive to private investors.

Another key institution is the **Bank for International Settlements (BIS)**, which acts as the central bank for the industrial countries' central banks. The BIS helps central banks manage and invest their foreign exchange reserves, and, in cooperation with the IMF and the World Bank, helps the central banks of developing countries, mostly in Latin America and Eastern Europe. The BIS also holds deposits of central banks so that reserves are readily available.

The Bretton Woods System: 1946–1971

Under the Bretton Woods Agreement, implemented in 1946, each government pledged to maintain a fixed, or pegged, exchange rate for its currency vis-à-vis the dollar or gold. As one ounce of gold was set equal to $35, fixing a currency's gold price was equivalent to setting its exchange rate relative to the dollar. For example, the Deutsche mark was set equal to 1/140 of an ounce of gold, meaning it was worth $0.25 ($35/140). The exchange rate was allowed to fluctuate only within 1% of its stated par value (usually less in practice).

The fixed exchange rates were maintained by official intervention in the foreign exchange markets. The intervention took the form of purchases and sales of dollars by foreign central banks against their own currencies whenever the supply and demand conditions in the market caused rates to deviate from the agreed-on par values. The IMF stood ready to provide the necessary foreign exchange to member nations defending their currencies against pressure resulting from temporary factors. Any dollars acquired by the monetary authorities in the process of such intervention could then be exchanged for gold at the U.S. Treasury, at a fixed price of $35 per ounce.

These technical aspects of the system had important practical implications for all trading nations participating in it. In principle, the stability of exchange rates removed a great deal of uncertainty from international trade and investment transactions, thus promoting their growth for the benefit of all the participants. Also, in theory, the functioning of the system imposed a degree of discipline on the participating nations' economic policies.

For example, a country that followed policies leading to a higher rate of inflation than that experienced by its trading partners would experience a balance-of-payments deficit as its goods became more expensive, reducing its exports and increasing its imports. The necessary consequences of the deficit would be an increase in the supply of the deficit country's currency on the foreign exchange markets. The excess supply would depress the exchange value of that country's currency, forcing its authorities to intervene. The country would be obligated to "buy" with its reserves the excess supply of its own currency, effectively reducing the domestic money supply. Moreover, as the country's reserves were gradually depleted through intervention, the authorities would be forced, sooner or later, to change economic policies to eliminate the source of the reserve-draining deficit. The reduction in the money supply and the adoption of restrictive policies would reduce the country's inflation, thus bringing it in line with the rest of the world.

In practice, however, governments perceived large political costs accompanying any exchange rate changes. Most governments also were unwilling to coordinate their monetary policies, even though this coordination was necessary to maintain existing currency values.

The reluctance of governments to change currency values or to make the necessary economic adjustments to ratify the current values of their currencies led to periodic foreign exchange crises. Dramatic battles between the central banks and the foreign exchange markets ensued. Those battles invariably were won by the markets. However, because devaluation or revaluation was used only as a last resort, exchange rate changes were infrequent and large.

In fact, Bretton Woods was a fixed exchange-rate system in name only. Of 21 major industrial countries, only the United States and Japan had no change in par value during the period 1946–1971. Of the 21 countries, 12 devalued their currencies more than 30% against the dollar, 4 had revaluations, and 4 were floating their currencies by mid-1971 when the system collapsed. The deathblow for the system came on August 15, 1971, when President Richard Nixon, convinced that the "run" on the dollar was reaching alarming proportions, abruptly ordered U.S. authorities to terminate convertibility even for central banks. At the same time, he devalued the dollar to deal with America's emerging trade deficit.

The fixed exchange-rate system collapsed along with the dissolution of the gold standard. There are two related reasons for the collapse of the Bretton Woods system. First, inflation reared its ugly head in the United States. In the mid-1960s, the Johnson administration financed the escalating war in Vietnam and its equally expensive Great Society programs by, in effect, printing money instead of raising taxes. This lack of monetary discipline made it difficult for the United States to maintain the price of gold at $35 an ounce.

Second, the fixed exchange-rate system collapsed because some countries—primarily West Germany, Japan, and Switzerland—refused to accept the inflation that a fixed exchange rate with the dollar would have imposed on them. Thus, the dollar depreciated sharply relative to the currencies of those three countries.

The Post-Bretton Woods System: 1971 to the Present

In December 1971, under the **Smithsonian Agreement**, the dollar was devalued to 1/38 of an ounce of gold, and other currencies were revalued by agreed-on amounts vis-à-vis the dollar. After months of such last-ditch efforts to set new fixed rates, the world officially turned to floating exchange rates in 1973.

OPEC AND THE OIL CRISIS OF 1973–1974 ■ October 1973 marked the beginning of successful efforts by the Organization of Petroleum Exporting Countries (OPEC) to raise the price of oil. By 1974, oil prices had quadrupled. Nations responded in various ways to the vast shift of resources to the oil-exporting countries. Some nations, such as the United States, tried to offset the effect of higher energy bills by boosting spending, pursuing expansionary monetary policies, and controlling the price of oil. The result was high inflation, economic dislocation, and a misallocation of resources without bringing about the real economic growth that was desired. Other nations, such as Japan, allowed the price of oil to rise to its market level and followed more prudent monetary policies.

The first group of nations experienced balance-of-payments deficits because their governments kept intervening in the foreign exchange market to maintain overvalued currencies; the second group of nations, along with the OPEC nations, wound up with balance-of-payments surpluses. These surpluses were recycled to debtor nations, setting the stage for the international debt crisis of the 1980s.

U.S. DOLLAR CRISIS OF 1977–1978 ■ During 1977–1978, the value of the dollar plummeted, and U.S. balance-of-payments difficulties were exacerbated as the Carter administration pursued an expansionary monetary policy that was significantly out of line with other strong currencies. The turnaround in the dollar's fortunes can be dated to October 6, 1979, when the Fed (under its new chairman, Paul Volcker) announced a major change in its conduct of monetary policy. From here on, in order to curb inflation, it would focus its efforts on stabilizing the money supply, even if that meant more volatile interest rates. Before this date, the Fed had attempted to stabilize interest rates, indirectly causing the money supply to be highly variable.

THE RISING DOLLAR: 1980–1985 ■ This shift had its desired effect on both the inflation rate and the value of the U.S. dollar. During President Ronald Reagan's first term in office (1981–1984), inflation plummeted and the dollar rebounded extraordinarily. This rebound has been attributed to vigorous economic expansion in the United States and to high real interest rates (owing largely to strong U.S. economic growth) that combined to attract capital from around the world.

THE SINKING DOLLAR: 1985–1987 ■ The dollar peaked in March 1985 and then began a long downhill slide. The slide is largely attributable to changes in government policy and the slowdown in U.S. economic growth relative to growth in the rest of the world.

By September 1985, the dollar had fallen about 15% from its March high, but this decline was considered inadequate to dent the growing U.S. trade deficit. In late September of 1985, the Group of Five, or **G-5 nations** (the United States, France, Japan, Great Britain, and West Germany), met at the Plaza Hotel in New York City. The outcome was the **Plaza Agreement**, a coordinated program designed to force down the dollar against other major currencies and thereby improve American competitiveness.

The policy to bring down the value of the dollar worked too well. The dollar slid so fast during 1986 that the central banks of Japan, West Germany, and Britain reversed their policies and began buying dollars to stem the dollar's decline. Believing that the dollar had declined enough, and in fact showed signs of "overshooting" its equilibrium level, the United States, Japan, West Germany, France, Britain, Canada, and Italy–also known as the Group of Seven, or **G-7 nations**–met again in February 1987 and agreed to

EXHIBIT 3.8 Effects of Government Actions and Statements on the Value of the 1987 Dollar

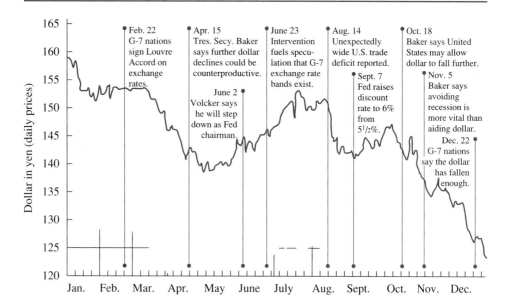

an ambitious plan to slow the dollar's fall. The **Louvre Accord**, named for the Paris land-mark where it was negotiated, called for the G-7 nations to support the falling dollar by pegging exchange rates within a narrow, undisclosed range, while they also moved to bring their economic policies into line.

As always, however, it proved much easier to talk about coordinating policy than to change it. The hoped-for economic cooperation faded, and the dollar continued to fall (see Exhibit 3.8).

RECENT HISTORY: 1988–1998 ■Beginning in early 1988, the U.S. dollar rallied some-what and then maintained its strength against most currencies through 1989. It fell sharply again in 1990 but then stayed basically flat in 1991 and 1992, while posting sharp intrayear swings. It began falling again in 1993, particularly against the yen and DM, and fell throughout most of 1994 and 1995 before rallying again in 1996 and continuing its upward direction as of mid-1998. Its future course is unpredictable given the absence of an anchor for its value.

Assessment of the Floating-Rate System

At the time floating rates were adopted in 1973, proponents said that the new system would reduce economic volatility and facilitate free trade. In particular, floating exchange rates would offset international differences in inflation rates so that trade, wages, employ-ment, and output would not have to adjust. High-inflation countries would see their currencies depreciate, allowing their firms to stay competitive without having to cut

IRAQ INVADES KUWAIT

After Iraq invaded Kuwait in August 1990, investors initially sought refuge in the dollar, and it surged in value. Then they had second thoughts, switched out of dollars, and the dollar slumped. The reversal reflected investor nervousness about the possibility that the United States would repeat its past mistakes. In particular, the reaction of many Americans to the rise in the price of oil after the invasion evoked memories of the 1970s. Large oil companies—attacked unmercifully by politicians and the media throughout the 1970s—were once again in the spotlight, accused of profiteering on the Persian Gulf crisis and threatened with new price controls. At the same time, U.S. Treasury policymakers began pushing again for a falling dollar to improve American trade competitiveness. In other words, the dollar's fall reflected lack of confidence in the ability of the U.S. political system to deal with economic crises and its willingness to protect the integrity of the dollar. The subsequent war in the Persian Gulf and its successful conclusion brought a reminder of American strength and renewed prospects for the U.S. economy—and a stronger dollar. The dollar's strength was reinforced by doubts about how well German economic integration was working.

wages or employment. At the same time, currency appreciation would not place firms in low-inflation countries at a competitive disadvantage. Real exchange rates would stabilize, even if permitted to float in principle, because the underlying conditions affecting trade and the relative productivity of capital would change only gradually; and if countries would coordinate their monetary policies to achieve a convergence of inflation rates, then nominal exchange rates would also stabilize.

The experience to date, however, is disappointing. The dollar's ups and downs have had little to do with actual inflation and a lot to do with expectations of future government policies and economic conditions. Put another way, real exchange rate volatility has increased, not decreased, since floating began. This instability reflects, in part, nonmonetary (or real) shocks to the world economy, such as changing oil prices and shifting competitiveness among countries, but these real shocks were not obviously greater during the 1980s than they were in earlier periods. Instead, uncertainty over future government policies has increased.

Given this evidence, a number of economists and others have called for a return to fixed exchange rates. To the extent that fixed exchange rates more tightly constrain the types of monetary and other policies governments can pursue, this approach should make expectations less volatile and, hence, should reduce fluctuations in the real exchange rate.

Although history offers no convincing model for a system that will lead to long-term exchange rate stability, it does point to two basic requirements. First, the system must be credible. If the market expects an exchange rate to be changed, the battle to keep it fixed is already lost. Second, the system must have price stability built into its very core. Without price stability the system will not be credible. Recall that under a fixed-rate system, each member must accept the group's inflation rate as its own. Only a zero rate of inflation will be mutually acceptable. If the inflation rate is much above zero, prudent governments will defect from the system.

Even with tightly coordinated monetary policies, freely floating exchange rates would still exhibit some volatility because of real economic shocks. However this volatility is not necessarily a bad thing because it could make adjustment to these shocks easier. For example, it has been argued that flexible exchange rates permitted the United States to cope with the buildup in defense spending in the early 1980s and the later slowdown in defense spending. Increased U.S. defense spending expanded aggregate U.S. demand and shifted output toward defense. The stronger dollar attracted imports and thereby helped satisfy the civilian demand. As the United States cut back on defense spending, the weakening dollar helped boost U.S. exports, making up for some of the decline in defense. To the extent that this argument is correct, the dollar's movements helped buffer the effects of defense spending shifts on American living standards.

3.3 THE EUROPEAN MONETARY SYSTEM

The **European Monetary System** (EMS) began operating in March 1979. Its purpose is to foster monetary stability in the *European Community* (EC, or the *Common Market*). As part of this system, the members have established the European Currency Unit, which plays a central role in the functioning of the EMS. The **European Currency Unit**, or ECU, is a composite currency that consists of fixed amounts of the 12 European Community member currencies. The quantity of each country's currency in the ECU reflects that country's relative economic strength in the European Community. The ECU functions as a unit of account, as a means of settlement, and as a reserve asset for the members of the EMS.

Exhibit 3.9 shows the amount of each member currency that goes into the ECU and the ECU's dollar value as of March 25, 1997. Although the ECU was worth $1.1546 on that date, its value changes continually in line with the market values of its 12 component

EXHIBIT 3.9 Composition and Value of the European Currency Unit

Currency	Amount of Currency in the ECU (1)	U.S. $ Value of Currency on March 25, 1997 (2)	Value of Component Currency (U.S. $) (3) = (1) × (2)	Currency Weight in ECU as of March 25, 1997 (4) = (3)/1.154634
Deutsche mark	0.6242	0.5918	0.369402	31.99%
French franc	1.332	0.1755	0.233766	20.25%
British pound	0.08784	1.6193	0.142239	12.32%
Dutch guilder	0.2198	0.5263	0.115681	10.02%
Italian lira	151.8	0.000591	0.089714	7.77%
Belgian franc	3.301	0.02865	0.094574	8.19%
Spanish peseta	6.885	0.006953	0.047871	4.15%
Danish krone	0.1976	0.155	0.030628	2.65%
Irish punt	0.008552	1.5728	0.013451	1.16%
Greek drachma	1.44	0.003754	0.005406	0.47%
Portuguese escudo	1.393	0.005871	0.008178	0.71%
Luxembourg franc	0.13	0.02865	0.003725	0.32%
ECU			1.154634	100.00%

Source: Currency values appeared in the *Wall Street Journal*, March 26, 1997.

currencies. The weight of each currency in the ECU equals the dollar value of its currency units in the ECU divided by the dollar value of the ECU. So, for example, with a dollar value of $0.2338, the French franc makes up 0.2338/1.1546 = 20.25% of the ECU's value. These currency weightings vary with the relative strength of the ECU's currencies.

To find the ECU's value in terms of another currency—say, the DM—is straightforward. According to Exhibit 3.9, on March 25, the DM was worth $0.5918. Hence, each ECU is worth 1.1546/0.5918 = DM 1.9510.

At the heart of the system is an **exchange-rate mechanism (ERM)**, which allows each member of the EMS to determine a mutually agreed-on central exchange rate for its currency; each rate is denominated in currency units per ECU. These central rates attempt to establish equilibrium exchange values, but members can seek adjustments to the central rates. Exhibit 3.10 shows the central rates against the ECU as of March 20, 1997. This exhibit, based on data appearing in the *Financial Times* of London, shows the EMS currencies ranked in descending order of strength. A positive deviation from the central rate (currency units/ECU), shown in the third column, indicates currency weakness. The fourth column contains the percentage spread of each currency's ECU value from the weakest currency in the EMS (the one with the greatest positive deviation from its central value, which at this time was the Italian lira). The greater the spread between the strongest (the Irish punt) and weakest (the lira) currencies, the greater the pressure on the weakest currency to devalue. Finally, the last column contains the divergence indicator, which measures the deviation of a currency's market rate from its ECU central rate relative to the maximum permitted deviation for that currency from its central rate.

Central rates establish a grid of bilateral cross-exchange rates between the currencies. For example, 1.92573 Deutsche marks per ECU, divided by 7.34555 Danish kroner (DKr) per ECU, equals 0.26216 DM per krone, which also implies DKr 3.81442 per DM. Nations participating in the ERM pledge to keep their currencies within a 15% margin on either side of these central cross-exchange rates (±2.25% for the DM/guilder cross rate). The upper and lower intervention levels for each currency pair can be found by applying

EXHIBIT 3.10 EMS Currency Central Rates Versus the ECU and Indicators of Strength

	ECU CENTRAL RATE (CURRENCY UNIT PER ECU)	MARKET RATE VS. ECU	DEVIATION (+/−) FROM CENTRAL RATE (%)	SPREAD VS. WEAKEST CURRENCY (%)	DIVERGENCE INDICATOR
Ireland	0.798709	0.737606	−7.65	10.95	52
Portugal	197.398	195.89	−0.76	3.25	5
Finland	5.85424	5.85182	−0.04	2.51	0
Spain	163.826	165.167	0.82	1.63	−6
Netherlands	2.16979	2.19163	1.01	1.44	−7
Germany	1.92573	1.94617	1.06	1.39	−10
Austria	13.5485	13.6967	1.09	1.36	−8
Belgium	39.7191	40.1582	1.11	1.34	−8
Denmark	7.34555	7.429	1.14	1.31	−8
France	6.45863	6.56701	1.68	0.77	−14
Italy	1906.46	1953.48	2.47	0	−18

Source: Data are from the *Financial Times*, March 21, 1997.

the appropriate margin to their central cross-exchange rate. For example, the upper and lower limits of the DM in units of krone can be approximated as[2]

$$\text{DKr/DM upper limit} = \text{Central rate} \times (1 + 0.15)$$

$$= 3.81442 \times 1.15 = 4.38659$$

$$\text{DKr/DM lower limit} = \text{Central rate} \times (1 - 0.15)$$

$$= 3.81442 \times 0.85 = 3.24226$$

Thus, if the Danish krone should strengthen against the Deutsche mark and reach the (approximate) lower intervention limit, the Bundesbank will sell kroner to commercial banks in exchange for DM at DKr 3.24226/DM, and Denmark's National Bank will buy DM at DM 0.30843/DKr [0.26216/0.85 = 1/3.24226].

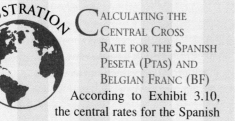

CALCULATING THE CENTRAL CROSS RATE FOR THE SPANISH PESETA (PTAS) AND BELGIAN FRANC (BF) According to Exhibit 3.10, the central rates for the Spanish and Belgian currencies on March 20, 1997, were Ptas 163.826/ECU and BF 39.7191/ECU. These central rates imply a central cross rate between the two currencies of Ptas 4.1246/BF (163.826/39.7191), or equivalently, BF 0.242447/Ptas (39.7191/163.826)

The original intervention limits were set at 2.25% above and below the central cross rates. (Spain and Britain had 6% margins.) Despite good intentions, the ERM came unglued in a series of speculative attacks that began in 1992. By 1993, the EMS had slipped into a two-tiered system. One tier consists of a core group of currencies tightly anchored by the DM. That tier includes the Dutch guilder, the French, Belgian, and Luxembourg francs, and possibly the Danish krone. The other tier consists of weaker currencies such as those of Spain, Portugal, Britain, Italy, and Ireland. However, the future of the ERM is very much in question.

A review of the European Monetary System and its history provides valuable insights into the operation of a target-zone system and illustrates the problems that such mechanisms are likely to encounter. Perhaps the most important lesson the EMS illustrates is that the exchange rate stability afforded by any target-zone arrangement requires a coordination of economic policy objectives and practices. Nations should achieve convergence of those economic variables that directly affect exchange rates— variables such as fiscal deficits, monetary growth rates, and real economic growth differentials.

---

[2] This formula is an approximation only because it ignores the fact that as a currency's value changes, its weight in the ECU changes, thereby affecting the ECU values of the other currencies as well.

Although the system helped keep its member currencies in a remarkably narrow zone of stability between 1987 and 1992, it has had a history of ups and downs. In its early years, the exchange-rate mechanism offered little anti-inflationary discipline, with Italy and France undergoing regular devaluations to offset higher inflation than in West Germany. By January 12, 1987, when the last realignment before September 1992 occurred, the values of the EMS currencies had been realigned 12 times despite heavy central bank intervention. Relative to their positions in March 1979, the Deutsche mark and the Dutch guilder soared, while the French franc and the Italian lira nose-dived. Between 1979 and 1988, the franc devalued relative to the DM by more than 50%.

The reason for the past failure of the European Monetary System to provide the currency stability it promised is straightforward: Germany's economic policymakers, responding to an electorate hypersensitive to inflation, have put a premium on price stability; in contrast, the French have historically pursued a more expansive monetary policy in response to high domestic unemployment. Neither country has been willing to permit exchange rate considerations to override political priorities.

The experience of the EMS also demonstrates once again that foreign exchange market intervention not supported by a change in a nation's monetary policy has only a limited influence on exchange rates. For example, the heavy intervention before the January 12, 1987, EMS realignment was generally not accompanied by changes in national monetary policies. West Germany, in particular, made only small adjustments to monetary policy in response to its increasingly undervalued currency. Consequently, the intervention failed to contain speculation that the DM would revalue, and a realignment became unavoidable.

The Currency Crisis of September 1992

The same attempt to maintain increasingly misaligned exchange rates in the EMS occurred again in 1992. And once again, the system broke down—in September 1992. The catalyst for the September currency crisis was the Bundesbank's decision to tighten monetary policy and force up German interest rates both to battle inflationary pressures associated with the spiraling costs of bailing out the former East Germany and to attract the inflows of foreign capital needed to finance the resulting German budget deficit. To defend their currency parities with the DM, the other member countries had to match the high interest rates in Germany (see Exhibit 3.11). The deflationary effects of high interest rates were accompanied by a prolonged economic slump and higher unemployment in Britain, France, Italy, Spain, and most other EMS members.

As the costs of maintaining exchange rate stability rose, the markets began betting that some countries with weaker economies would devalue their currencies or withdraw them from the ERM altogether rather than maintain painfully high interest rates at a time of rising unemployment.

The mere possibility of a currency realignment meant that nations had to raise their interest rates dramatically to halt speculative attacks on their currencies: 15% in Britain and Italy, 13.75% in Spain, 13% in France, and an extraordinary *500%* in Sweden. They also intervened aggressively in the foreign exchange markets. British, French, Italian, Spanish, and Swedish central banks together spent the equivalent of roughly $100 billion trying to prop up their currencies, with the Bank of England reported to have spent $15

EXHIBIT 3.11 Defending the ERM Required High
Interest Rates in Europe

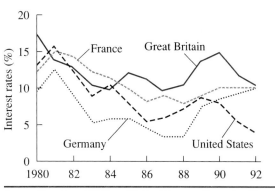

Source: Data from OECD and Federal Reserve, September 30, 1992.

billion to $20 billion in just one day to support the pound. The Bundesbank spent another $50 billion in DM to support the ERM. All to no avail.

On September 14, the central banks capitulated, but not before losing an estimated $4 billion to $6 billion in their mostly futile attempt to maintain the ERM. For example, the Bank of England borrowed DM 33 billion from the Bundesbank when a pound was worth DM 2.78. It had to repay this money at the less favorable rate. At a rate of DM 2.50:£1, a devaluation of 10%, the extra cost would be £1.3 billion (33 billion/2.50 − 33 billion/2.78). As noted earlier, despite these costly efforts, Britain and Italy were forced to drop out of the ERM, and Spain, Portugal, and Ireland devalued their currencies within the ERM. In addition, Sweden, Norway, and Finland were forced to abandon their currencies' unofficial links to the ERM.

The Exchange-Rate Mechanism Is Abandoned in August 1993

The final straw was the currency crisis of August 1993, which actually was touched off on July 29, 1993, when the Bundesbank left its key lending rate, the discount rate, unchanged. Traders and investors had been expecting the Bundesbank to cut the discount rate to relieve pressure on the French franc and other weak currencies within the ERM. As had happened the year before, however, the Bundesbank largely disregarded the pleas of its ERM partners and concentrated on reining in 4.3% German inflation and its fast-growing money supply. Given the way the ERM works, and the central role played by the Deutsche mark, other countries could not both lower interest rates and keep their currencies within their ERM bands unless Germany did so.

The French franc (FF) was the main focus of the ERM struggle. With high real interest rates, recession, and unemployment running at a post-World War II high of 11.6%, speculators doubted that France had the willpower to stay with the Bundesbank's tight monetary policy and keep its interest rates high, much less raise them to defend the franc. Speculators reacted logically: They dumped the French franc and other European currencies and bought DM, gambling that economic pressures, such as rising unemployment and deepening recession, would prevent these countries from keeping their interest rates well above those in Germany. In other words, speculators bet—rightly, as it turned out—that domestic priorities would ultimately win out despite governments' pledges to the contrary.

Despite heavy foreign exchange market intervention (the Bundesbank alone spent $35 billion trying to prop up the franc), the devastating assault by speculators on the ERM forced the franc to its ERM floor of DM 0.29150. Other European central banks also intervened heavily to support the Danish krone, Spanish peseta, Portuguese escudo, and Belgian franc, which came under heavy attack as well.

It was all to no avail, however. Without capital controls or a credible commitment to move to a single currency in the near future, speculators could easily take advantage of a one-sided bet. The result was massive capital flows that overwhelmed the central banks' ability to stabilize exchange rates. Over the weekend of July 31–August 1, the EC finance ministers agreed essentially to abandon the defense of each other's currencies. Under the new plan, announced on August 2, seven of the nine currencies could trade within a band of 15% above or below their central rates. The band for the peseta and the escudo was set at ±6%, while Germany and the Netherlands agreed to maintain the old 2.25% fluctuation range between their currencies. The immediate effect of the radically wider bands on the DM:FF exchange rate is shown in Exhibit 3.12. Because the new trading bands are so wide, the European Monetary System has become a floating-rate system in name only. Its preservation in form, but not substance, provides a fig leaf for politicians to say the system is still in place.

The currency turmoil of 1992–1993 showed once again that a genuinely stable European Monetary System, and eventually a single currency, requires the political will to direct fiscal and monetary policies at that European goal and not at purely national ones.

EXHIBIT 3.12 France Widens the ERM Band

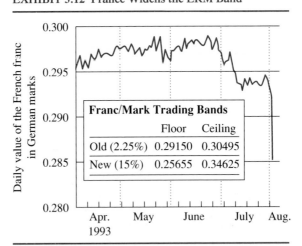

Source: Data from *Tradeline International* in the *Wall Street Journal*, August 3, 1993, p. C1. Used with permission of the *Wall Street Journal*, © Dow Jones & Company, Inc., 1993. All rights reserved worldwide.

In showing that they lacked that will, European governments proved once again that allowing words to run ahead of actions is a recipe for failure.

On the other hand, despite its recent problems, the EMS did achieve some significant success. By improving monetary policy coordination among its member states, the EMS has succeeded in narrowing inflation differentials in Europe. In 1980, the gap between the highest inflation rate (Italy's 21.2%) and the lowest (West Germany's 5.2%) was 16 percentage points. By 1990 the gap had narrowed to less than four percentage points. The narrowing of inflation rates, in turn, reduced exchange rate volatility until 1992. Indeed, from January 1987 to September 1992, currencies remained fixed. Moreover, Germany's importance to the European economy and the Bundesbank's unwillingness to compromise its monetary policy forced other members of the EMS to adjust their monetary policies to more closely mimic Germany's low-inflation policy. As a result, inflation rates have tended to converge toward Germany's lower rate. For example, in 1990 the Netherlands, France, Belgium, Denmark, and Ireland all had inflation rates within one percentage point of Germany's (see Exhibit 3.13). In effect, the EMS has moved to a Deutsche mark standard. Conversely, the gap between British and German inflation widened. In recognition of the benefits to be derived from a strong and stable monetary system, Prime Minister Margaret Thatcher finally relented on "monetary sovereignty" in late 1990 and entered Britain in the EMS (only to pull out two years later).

To summarize, the EMS was based on Germany's continuing to deliver low inflation rates and low real interest rates. As long as Germany lived up to its end of the bargain, the benefits to other EMS members of following the Bundesbank's policies would exceed the costs. But once the German government broke that compact by running huge and inflationary deficits, the costs to most members of following a Bundesbank monetary policy designed to counter the effects of the government's fiscal policies exceeded the benefits. Put another way, the existing exchange rates became unrealistic given what would have been required of the various members to maintain those exchange rates. In the end, there was no real escape from market forces.

EXHIBIT 3.13 The European Monetary System Forced Convergence toward Germany's Inflation Rate

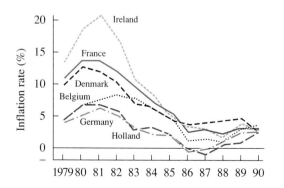
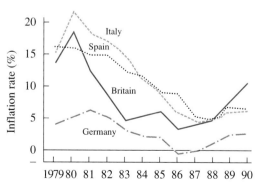

Monetary Union

Many politicians and commentators point to the recent turmoil in the EMS as increasing the need for the European Community to move toward **monetary union**. Under this scenario, formalized in the **Maastricht Treaty**, the EC nations would establish a single central bank with the sole power to issue a single European currency called the **euro** as of January 1, 1999. On that date, conversion rates would lock in for member currencies, and the euro would become a currency, although euro coins and bills would be unavailable until 2002. Francs, marks, guilders, schillings, and other currencies would be phased out, to be replaced, on January 1, 2002, by the euro. Member countries no longer would be able to create money as of January 1, 1999; only the new **European Central Bank** would be able to do so. Governments would be able to issue bonds denominated in euros, just as individual American states can issue dollar bonds. However, like California or New York, member nations would be unable to print the currency needed to service their debts. Instead, they would have to attract investors by convincing them they have the financial ability (through taxes and other revenues) to generate the euros to repay their debts.

In order to join the **European Monetary Union** (EMU), European nations were supposed to meet tough standards on inflation, currency stability, and deficit spending. Specifically, government debt could be no more than 60% of gross domestic product (GDP); the government budget deficit could not exceed 3% of GDP; the inflation rate could not be more than 1.5 percentage points above the average rate of Europe's three lowest-inflation nations; and long-term interest rates could not be more than two percentage points higher than the average interest rate in the three lowest-inflation nations. These standards are summarized in Exhibit 3.14, along with relevant financial statistics as of year-end 1997 for some of the potential participants. It should be noted that most countries, including Germany, fudged some of their figures through one-time maneuvers (redefining government debt or selling off government assets) or fudged the criteria (check out Italy's debt/GDP ratio) in order to qualify. Nonetheless, on May 2, 1998, the European Parliament formally approved the historic decision to launch the euro with 11 founder nations—Germany, France, Italy, Spain, the Netherlands, Belgium, Finland, Portugal, Austria, Ireland, and Luxembourg. Britain, Sweden, and Denmark opted out of the launch, whereas Greece intends to join in 2001, by which time it expects to be able to meet the economic convergence criteria.

The truth is that monetary union is as much about reining in the expensive European welfare state and its costly regulations as it is about currency stability. Owing to high taxes, generous social welfare and jobless benefits, and costly labor-market regulations—all of which reduce incentives to work, save, invest, and create jobs—and diminished competitiveness—fostered by onerous regulations on business as well as state subsidies and government protection to ailing industries—job growth has been stagnant throughout Western Europe for three decades. (Western Europe failed to create a single net new job from 1973 to 1994, a period in which the United States generated 38 million net new jobs.) As a result, the European unemployment rate has been averaging about 12% (in contrast to less than 5% for the United States). However, although crucial for strong and sustained economic growth, limiting the modern welfare state is politically risky; too many people live off the state. Enter the Maastricht Treaty. European govern-

EXHIBIT 3.14　Maastricht Criteria for European Monetary Union and Year-End 1997 Figures
for Potential Participants

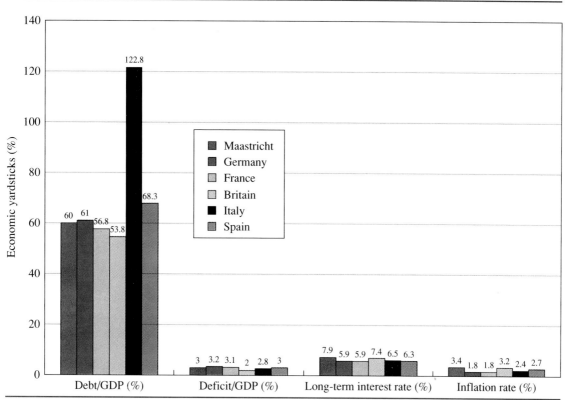

Source: Organization for Economic Cooperation and Development. The Maastricht inflation and interest rate criteria are based
on year-end 1997 data.

ments can blame the strict Maastricht criteria they must meet to enter EMU for the need
to take the hard steps that most economists believe are necessary for ending economic
stagnation: curbing social-welfare expenditures, reducing costly regulations on business,
increasing labor market flexibility (primarily by lowering the cost to companies of hiring
and firing workers), cutting taxes, and selling off state-owned enterprises (a process
known as **privatization**). Thus, the greatest benefit from monetary union likely will be
the long-term economic gains that will come from the fiscal discipline required for entry.
If, however, Europe does not take this opportunity to reform its economic policies, the
costs of monetary union will be high because member nations no longer will be able to
use currency or interest-rate adjustments to compensate for the pervasive labor market
rigidities that characterize the modern European economy.

　　Business clearly would benefit from EMU through lower cross-border currency
conversion costs. For example, Philips, the giant Dutch electronics company, estimates
that a single European currency would save it $300 million a year in currency transaction
costs. Overall, the EC Commission estimates that businesses in Europe spend $13 billion
a year converting money from one EC currency to another. Ordinary citizens also bear

BUDGET DISPUTE TOPPLES ITALY'S GOVERNMENT

In October 1997, Italian Prime Minister Romani Prodi, a Socialist, resigned after his allies in the Communist Party refused to tolerate welfare cuts the government said were vital for Italy's entry into the European Monetary Union. The fall of Prodi's 17-month-old government ended Italy's first leftist-dominated government since the end of World War II. Although the Communist Party backed down after a national outcry, and Prodi's government was reconstituted, this episode reveals the difficulties European governments face in getting their citizens to accept painful budget cuts to conform to the Maastricht criteria.

some substantial currency conversion costs. A tourist who left Paris with 1,000 francs and visited the other 11 EC countries, exchanging her money for the local currency in each country but not spending any of it, would find herself with fewer than 500 francs when she returned to Paris. Multinational firms would also find corporate planning, pricing, and invoicing easier with a common currency.

Adopting a common currency also would eliminate the risk of currency fluctuations and facilitate cross-border price comparisons. Lower risk and improved price transparency would encourage the flow of trade and investments among member countries and bring about greater integration of Europe's capital, labor, and commodity markets and a more efficient allocation of resources within the region as a whole. Increased trade and price transparency, in turn, would intensify Europewide competition in goods and services and spur a wave of corporate restructurings and mergers and acquisitions.

Moreover, monetary union—such as exists among the 50 states of the United States, where the exchange rate between states is immutably set at 1—would provide the ultimate in coordination of monetary policy. Inflation rates under monetary union would converge, but not in the same way as in the current system. The common inflation rate would be decided by the monetary policy of the European central bank. It would tend to reflect the average preferences of the people running the bank, rather than giving automatic weight to the most anti-inflationary nation as in the current system. Thus, for the European Monetary Union to be an improvement over the current state of affairs, the new European Central Bank (ECB) must be as averse to inflation as Europe's current de facto central bank—the Bundesbank.

To ensure the European Monetary Union's inflation-fighting success, the new central bankers would have to be granted independence along with a statutory duty to devote monetary policy to keeping the price level stable. One way to protect the central bank's independence is to appoint the governing body, as in the United States and Germany, with long-term contracts that cut across national electoral cycles. Only then would a European central bank be able to do what William McChesney Martin, a former Fed chairman, said a sound central bank must always do: "Take away the punch bowl just when the party gets going."

Even at this late date, independence of the European Central Bank is an unsettled issue. Germans, who favor a strong, fiercely independent ECB modeled on the Bundesbank, fear the French will politicize it by using it to push job-creation and other schemes requiring an expansionist (and, hence, inflationary) monetary policy. Many French see the Germans as favoring price stability over compassion for the unemployed. This dispute points out a hard reality: The ECB will find it difficult to be tough on inflation without the benefit of a uniformly prudent fiscal policy across all its member states and create jobs.

Another important issue in forming a monetary union is who gets the benefits of **seigniorage**—the profit to the central bank from money creation. In other words, who gets to spend the proceeds from printing money? In the United States, the answer is the federal government. In the case of Europe, however, this issue has not been resolved.

ILLUSTRATION

GERMAN MONETARY UNION

As a first step toward economic reunification, East and West Germany formed a monetary union. On July 2, 1990, the East German mark, or Ostmark, was replaced by the Deutsche mark as the single German currency. It was agreed that East Germans would trade their Ostmarks to the West German Bundesbank for Deutsche marks at an exchange rate fixed in advance. In addition, the East German central bank had to cede its power to issue currency or create bank balances; otherwise, the Bundesbank would lose control over the money supply. At the same time, restrictions on movements of capital across the border were eliminated, making it possible for East Germans to hold bank accounts in West Germany and for West Germans to invest freely in East Germany. All transactions would be denominated in DM.

Most economists believed that a key to smooth transition toward German monetary union—and ultimately economic integration—was to get the exchange rate between Deutsche marks and Ostmarks right. Valuing the DM too high would destroy the worth of East German savings. Overvaluing the Ostmark would keep East German wages and goods from being competitive. Businesses in East Germany would suffer, leading to a further exodus of workers to West Germany.

Most economists felt that the best approach would be to err on the side of undervaluation, making industry in eastern Germany more competitive and attracting investors from West Germany and other countries. As it turned out, Chancellor Helmut Kohl decided on political grounds to convert East German savings, debts, wages, and prices from Ostmarks into DM at vastly overvalued rates ranging between one and two Ostmarks to one Deutsche mark. The resulting expansion in the DM money supply helped contribute to the inflationary pressures facing Germany. At the same time, the combination of western German wages and eastern German productivity (about 25% of western Germany's) laid waste much of the East's industry and put most of the citizenry on welfare. The former East Germany now consumes about twice what it produces, creating a regional shortfall that is met by transferring DM 175 billion a year from the former West Germany. As we have already seen, the Bundesbank responded to unification-induced inflation and deficits by raising real interest rates, leading to the breakdown of the EMS.

An unspoken reason for strong business support for European Monetary Union is to boost growth by breaking the lock hold of government and unions on European economies. As described earlier, meeting the Maastricht criteria—particularly the one dealing with the reduced budget deficit—would help to diminish the role of the state in

Europe and its tax-financed cradle-to-grave benefits. Many economists believe that only by cutting back on government and its generous—and increasingly unaffordable—social welfare programs and costly business and labor market regulations can the stagnant economies of Western Europe start to grow again and create jobs.

Optimum Currency Area

Most discussion of European monetary union has highlighted its benefits, such as eliminating currency uncertainty and lowering the costs of doing business. The potential costs of currency integration have been overlooked. In particular, as the discussion of U.S. military spending shifts indicated, it may sometimes pay to be able to change the value of one currency relative to another. Suppose, for example, that the worldwide demand for French goods falls sharply. To cope with such a drop in demand, France must make its goods less expensive and attract new industries to replace its shrinking old ones. The quickest way to do this is to reduce French wages, thereby making its workers more competitive. But this reduction is unlikely to be accomplished quickly. Eventually, high unemployment might persuade French workers to accept lower pay. But in the interim, the social and economic costs of reducing wages by, say, 10% will be high. In contrast, a 10% depreciation of the franc would achieve the same thing quickly and relatively painlessly.

Conversely, a worldwide surge in demand for French goods could give rise to French inflation, unless France allowed the franc to appreciate. In other words, currency changes can substitute for periodic bouts of inflation and deflation caused by various economic shocks. Once France has entered monetary union, it no longer has the option of changing its exchange rate to cope with these shocks. This option would have been valuable to the ERM, which instead came unglued because of the huge economic shock to its fixed parities brought about by the absorption of East Germany into the German economy.

Taking this logic to its extreme would imply that not only should each nation have its own currency, but so should each region within a nation. Why not a southern California dollar, or indeed a Los Angeles dollar? The answer is that having separate currencies brings costs as well as benefits.

The more currencies there are, the higher the costs of doing business and the more currency risk exists. Both factors impair the functions of money as a medium of exchange and a store of value, so maintaining more currencies acts as a barrier to international trade and investment, even as it reduces vulnerability to economic shocks.

According to the theory of the **optimum currency area**, this trade-off becomes less and less favorable as the size of the economic unit shrinks. So how large is the optimum currency area? No one knows. But some economists argue that Europe might be better off with four or five regional currencies than with only one.[3] Similarly, some have argued that the United States too might do better with several regional currencies to cushion shocks such as those that afflicted the Midwest and the Southwest during the 1980s and the Northeast and California in the 1990s. Nonetheless, the experience with floating exchange rates since the early 1970s will likely give pause to anyone seriously

---

[3] See, for example, Geoffrey M. B. Tootell, "Central Bank Flexibility and the Drawbacks to Currency Unification," *New England Economic Review*, May–June 1990, pp. 3–18; and Paul Krugman, "A Europe-Wide Currency Makes No Economic Sense," *Los Angeles Times*, August 5, 1990, p. D2.

thinking of pushing that idea further. That experience suggests that exchange rate changes can add to economic volatility as well as absorb it. At the same time, economic flexibility—especially of labor markets—is critical to reducing the costs associated with currency union. This flexibility can only be attained through further deregulation, privatization, freer trade, labor market and social welfare reform, and a reduction in economic controls, state subsidies, and business regulations. Absent these changes, especially to reduce the rigidities of Europe's labor market, European Monetary Union will intensify economic shocks because their effects can no longer be mitigated by exchange rate adjustments.

❧ 3.4 SUMMARY AND CONCLUSIONS

This chapter studied the process of exchange rate determination under five market mechanisms: free float, managed float, target-zone system, fixed-rate system, and the current hybrid system. In the last four systems, governments intervene in the currency markets in various ways to affect the exchange rate.

Regardless of the form of intervention, however, fixed rates do not remain fixed for long. Neither do floating rates. The basic reason that exchange rates do not remain fixed in either a fixed- or floating-rate system is that governments subordinate exchange rate considerations to domestic political considerations.

We saw that the gold standard is a specific type of fixed exchange-rate system, one that required participating countries to maintain the value of their currencies in terms of gold. Calls for a new gold standard remind us of the fundamental lack of trust in fiat money due to the historical unwillingness of the monetary authorities to desist from tampering with the money supply.

Finally, we concluded that intervention to maintain a disequilibrium rate is generally ineffective or injurious when pursued over lengthy periods of time. Seldom have policymakers been able to outsmart for any extended period the collective judgment of buyers and sellers. The current volatile market environment, a consequence of unstable U.S. and world financial conditions, cannot be arbitrarily directed by government officials for long.

Examining U.S. experience since the abandonment of fixed rates, we found that free-market forces did correctly reflect economic realities thereafter. The dollar's value dropped sharply between 1973 and 1980 when the United States experienced high inflation and weakened economic conditions. Beginning in 1981, the dollar's value rose when American policies dramatically changed under the leadership of the Federal Reserve and a new president, but fell when foreign economies strengthened relative to the U.S. economy. Nonetheless, the resulting shifts in U.S. cost competitiveness have led many to question the current international monetary system.

The principal alternative to the current system of floating currencies with its economic volatility is a fixed exchange-rate system. History offers no entirely convincing model for how such a system should be constructed, but it does point to two requirements. To succeed in reducing economic volatility, a system of fixed exchange rates must be credible, and it must have price stability built into its very fabric. Otherwise, the market's expectations of exchange rate changes combined with an unsatisfactory rate of inflation will lead to periodic battles among central banks and between central banks and the financial markets. The recent experiences of the European Monetary System point to the costs associated with the maintenance of exchange rates at unrealistic levels. These experiences also point out that, in the end, there is no real escape from market forces.

A final lesson learned is that one must be realistic in what one can expect from a currency system. In particular, no currency system can achieve what many politicians seem to expect of it—a way to keep all the benefits of economic policy for their own nation while passing along the costs to foreigners (who don't vote) or to future generations (who don't vote yet).

Questions

1. Have exchange rate movements under the current system of managed floating been excessive? Explain.

2. Why has speculation failed to smooth exchange rate movements?

3. Is a floating-rate system more inflationary than a fixed-rate system? Explain.

4. Find a recent example of a nation's foreign exchange market intervention and note what the government's justification was. Does this justification make economic sense?

5. Gold has been called "the ultimate burglar alarm." Explain what this expression means.

6. Since 1979, the price of gold has fallen by more than 60%. What could explain such a steep price decline? Consider the roles of inflation and new financial instruments such as swaps and options that can provide lower-cost inflation hedges.

7. Comment on the following statement: "A system of floating exchange rates fails when governments ignore the verdict of the exchange markets on their policies and resort to direct controls over trade and capital flows."

8. Suppose nations attempt to pursue independent monetary and fiscal policies. How will exchange rates behave?

9. Will coordination of economic policies make exchange rates more or less stable? Explain.

10. The experiences of fixed exchange-rate systems and target-zone arrangements have not been entirely satisfactory.
    a. What lessons can economists draw from the breakdown of the Bretton Woods system?
    b. What lessons can economists draw from the exchange rate experiences of the European Monetary System?

11. Despite official parity between the Deutsche mark and the Ostmark, the black market rate in early 1990 was about 10 Ostmarks for one Deutsche mark. What problems might setting the exchange rate at one DM for each Ostmark create for Germany?

12. How did the European Monetary System limit the economic ability of each member nation to set its interest rate to be different from Germany's?

13. Historically, Spain has had high inflation and has seen its peseta continuously depreciate. In 1989, though, Spain joined the EMS and pegged the peseta to the DM. According to a Spanish banker, EMS membership means that "the government has less capability to manage the currency but, on the other hand, the people are more trusting of the currency for that reason."
    a. What underlies the peseta's historical weakness?
    b. Comment on the banker's statement.
    c. What are the likely consequences of EMS membership on the Spanish public's willingness to save and invest?

14. When Britain announced its entry into the exchange-rate mechanism of the EMS on October 5, 1990, the price of British gilts (long-term government bonds) soared and sterling rose in value.
    a. What might account for these price jumps?
    b. Sterling entered the ERM at a central rate against the DM of DM 2.95, and it was allowed to move within a band of plus and minus 6% of this rate. What were sterling's upper and lower rates against the DM?

15. What potential costs might be associated with the decision to widen the margins within which some currencies in the ERM can float?

16. In discussing European Monetary Union, a recent government report stressed a need to make the central bank accountable to the "democratic process." What are the likely consequences for price stability and exchange rate stability in the EMS if the "Eurofed" becomes accountable to the "democratic process"?

17. Comment on the following headline in the *Wall Street Journal* (January 11, 1993): "Germany's Rate Cut Takes Pressure Off French Franc, and the Rest of the EMS."

18. The French franc was the main target of speculators during the August 1993 assault on the EMS even though France was running a 2% inflation rate while Germany had a 4.3% inflation rate. Why might this be?

19. In early 1996, in response to growing doubts about the ability of EC nations to meet the Maastricht criteria and move toward monetary union by the 1999 deadline, yields on European bonds jumped. What is the likely link between the doubts on Maastricht and the EC bond yield increases?

20. For a fixed exchange rate system to work, the government must be able to make tight budget and monetary policies stick from the outset. Comment on this statement.

21. In 1996, Chancellor of the Exchequer Kenneth Clarke called for a national debate on whether Britain should join the European Monetary Union. Discuss the pros and cons for Britain of joining EMU.

22. Upon taking office in October 1993, the Bundes-

bank's new president, Hans Tietmeyer, said, "Forced reductions in central bank interest rates which are contrary to stability policies can solve neither economic nor structural problems. But they would undermine trust in currency values, drive long-term interest rates higher, and delay necessary corrections in the real economy." Explain the context in which Mr. Tietmeyer made these comments. Do you agree or disagree with his comments? Explain.

Problems

1. On Friday, September 13, 1992, the lira was worth DM 0.0013065. Over the weekend, the lira devalued against the DM to DM 0.0012613.
   a. By how much had the lira devalued against the DM?
   b. By how much had the DM appreciated against the lira?
   c. Suppose Italy borrowed DM 4 billion, which it sold to prop up the lira. What were the Bank of Italy's lira losses on this currency intervention?
   d. Suppose Germany spent DM 24 billion in an attempt to defend the lira. What were the Bundesbank's DM losses on this currency intervention?

2. Suppose the central rates within the ERM for the French franc and DM are FF 6.90403 : ECU 1 and DM 2.05853 : ECU 1, respectively.
   a. What is the cross-exchange rate between the franc and the mark?
   b. Under the former 2.25% margin on either side of the central rate, what were the approximate upper and lower intervention limits for France and Germany?
   c. Under the new 15% margin on either side of the central rate, what are the current approximate upper and lower intervention limits for France and Germany?

3. Exhibit 3.9 valued the ECU as of March 25, 1997.
   a. Using current spot exchange rates appearing in the *Wall Street Journal*, find the ECU's dollar value today.

   b. What is the ECU's DM value today? Its yen value today?
   c. What are the weights associated with each currency?

4. A Dutch company exporting to France has FF 3 million due in 90 days. Suppose that the current exchange rate is FF 1 = DFl 0.3291.
   a. Based on the data in Exhibit 3.10, what is the central cross-exchange rate between the two currencies?
   b. Based on the answer to Part a, what is the most the Dutch company could lose on its French franc receivable, assuming that France and the Netherlands stick to the ERM with a 15% band on either side of their central cross rate?
   c. Redo Part b, assuming the band were narrowed to 2.25%.
   d. Redo Part b, assuming you know nothing about the current cross-exchange rate.

5. Panama adopted the U.S. dollar as its official paper money in 1904. Currently, $400 million to $500 million in U.S. dollars is circulating in Panama. If interest rates on U.S. Treasury securities are 7%, what is the value of the seigniorage that Panama is forgoing by using the U.S. dollar instead of its own-issue money?

6. By some estimates, $185 billion to $260 billion in currency is held outside the United States.
   a. What is the value to the United States of the seigniorage associated with these overseas dollars? Assume that dollar interest rates are about 6%.
   b. Who in the United States realizes this seigniorage?

Bibliography

BORDO, MICHAEL DAVID. "The Classical Gold Standard: Some Lessons for Today." *Federal Reserve Bank of St. Louis Review*, May 1981, pp. 2–17.

COOMBS, CHARLES A. *The Arena of International Finance.* New York: John Wiley & Sons, 1976.

FRIEDMAN, MILTON, and ROBERT V. ROOSA. "Free versus Fixed Exchange Rates: A Debate." *Journal of Portfolio Management*, Spring 1977, pp. 68–73.

MUNDELL, ROBERT A. "A Theory of Optimum Currency Areas." *American Economic Review*, September 1961, pp. 657–663.

THE BALANCE OF PAYMENTS AND INTERNATIONAL ECONOMIC LINKAGES

I had a trade deficit in 1986 because I took a vacation in France. I didn't worry about it; I enjoyed it.

> Herbert Stein
> Chairman of the Council
> of Economic Advisors under
> Presidents Nixon and Ford

We have almost a crisis in trade and this is the year Congress will try to turn it around with trade legislation.

> Lloyd Bentsen
> Former U.S. Senator from Texas

Despite all the cries for protectionism to cure the trade deficit, protectionism will not lower the trade deficit.

> Phil Gramm
> U.S. Senator from Texas

A key theme of this book is that companies today operate within a global market-place, and they can ignore this fact only at their peril. In line with that theme, the purpose of this chapter is to present the financial and real linkages between the domestic and world economies and examine how these linkages affect business viability. The chapter identifies the basic forces underlying the flows of goods, services, and capital between countries and relates these flows to key political, economic, and cultural factors.

Politicians and the business press realize the importance of these trade and capital flows. They pay attention to the balance of payments, on which these flows are recorded, and to the massive and continuing U.S. trade deficits. As we saw in chapter 3, government foreign exchange policies are often geared toward dealing with balance-of-payments problems. However, as indicated by the three quotations that opened the chapter, many people disagree on the nature of the trade deficit problem and its solution. In the process of studying the balance of payments in this chapter, we will sort out some of these issues.

4.1 BALANCE-OF-PAYMENTS CATEGORIES

The **balance of payments** is an accounting statement that summarizes all the economic transactions between residents of the home country and residents of all other countries. Balance-of-payments statistics are published quarterly in the United States by the Commerce Department and include such transactions as trade in goods and services, transfer payments, loans, and short- and long-term investments. The statistics are followed closely by bankers and business people, economists, and foreign exchange traders; the publication affects the value of the home currency if these figures are more, or less, favorable than anticipated.

Currency inflows are recorded as *credits*, and outflows are recorded as *debits*. Credits show up with a plus sign, and debits have a minus sign. There are three major balance-of-payments categories:

■ **Current account**, which records flows of goods, services, and transfers
■ **Capital account**, which shows public and private investment and lending activities
■ **Official reserves account**, which measures changes in holdings of gold and foreign currencies—*reserve assets*—by official monetary institutions.

Exports of goods and services are credits; imports of goods and services are debits. Interest and dividends are treated as services because they represent payment for the use of capital. Capital inflows appear as credits because the nation is selling (exporting) to foreigners valuable assets—buildings, land, stock, bonds, and other financial claims—and receiving cash in return. Capital outflows show up as debits because they represent purchases (imports) of foreign assets. The increase in a nation's official reserves also shows up as a debit item because the purchase of gold and other reserve assets is equivalent to importing these assets.

The balance-of-payments statement is based on double-entry bookkeeping; every economic transaction recorded as a credit brings about an equal and offsetting debit entry, and vice versa. According to accounting convention, a source of funds (either a decrease in assets or an increase in liabilities) is a credit, and a use of funds (either an increase in assets or a decrease in liabilities or net worth) is a debit. Suppose a U.S. company exports machine tools to Switzerland at a price of 2,000,000 Swiss francs (SFr). At the current exchange rate of SFr 1 = $0.75, this order is worth $1,500,000. The Swiss importer pays for the order with a check drawn on its Swiss bank account. A credit is recorded for the increase in U.S. exports (a reduction in U.S. goods—a source of funds) and, because the exporter has acquired a Swiss franc deposit (an increase in a foreign asset—a use of funds), a debit is recorded to reflect a private capital outflow:

	Credit	Debit
U.S. exports	$1,500,000	
Private foreign assets		$1,500,000

Suppose the U.S. company decides to sell the Swiss francs it received to the Federal Reserve for dollars. In this case, a private asset would have been converted into an official (government) liability. This transaction would show up as a credit to the private asset account (as it is a source of funds) and a debit to the official assets account (as it is a use of funds):

	Credit	Debit
Private assets	$1,500,000	
Official assets		$1,500,000

Similarly, if a German sells a painting to a U.S. resident for $1,000,000, with payment made by issuing a check drawn on a U.S. bank, a debit is recorded to indicate an increase in assets (the painting) by U.S. residents, which is a use of funds, and a credit is recorded to reflect an increase in liabilities (payment for the painting) to a foreigner, which is a source of funds:

	Credit	Debit
Private liabilities to foreigners	$1,000,000	
U.S. imports		$1,000,000

In the case of **unilateral transfers**, which are gifts and grants overseas, the transfer is debited because the donor's net worth is reduced, whereas another account must be credited: exports, if goods are donated; services, if services are donated; or capital, if the recipient receives cash or a check. Suppose the American Red Cross donates $100,000 in goods for earthquake relief to Nicaragua. The balance of payments entries for this transaction would appear as follows:

	Credit	Debit
U.S. exports	$100,000	
Unilateral transfer		$100,000

Because double-entry bookkeeping ensures that debits equal credits, the sum of all transactions is zero. That is, the sum of the balance on the current account, the capital account, and the official reserves account must equal zero:

Current account balance + Capital account balance + Official reserves account

$$= Balance\ of\ payments = 0$$

These features of balance-of-payments accounting are illustrated in Exhibit 4.1, which shows the U.S. balance of payments for 1997, and in Exhibit 4.2, which gives examples of entries in the U.S. balance-of-payments accounts.

ot available.

EXHIBIT 4.1 The U.S. Balance of Payments for 1997[1] (U.S. $ Billions)

CREDITS		DEBITS	
a: Exports of civilian goods	$678.3	b: Imports of civilian goods	$877.3
c: Military sales abroad	15.2	d: Military purchases abroad	11.3
	Trade balance	$= a + c - (b + d)$	
		= Deficit of $195.0	
e: Exports of services (investment income and fees earned, foreign tourism in United States, etc.)	474.1	f: Imports of services (investment income and fees paid out, U.S. tourism abroad, etc.)	406.9
	Services balance	= Surplus of $67.2	
		g: Net unilateral transfers (gifts)	38.5
	Current-account balance	$= a + c + e - (b + d + f + g)$	
		Deficit of $166.4	
h: Foreign private investment in the United States	672.3	i : U.S. private investment overseas	426.1
j: Foreign official lending in the United States	18.2	k: U.S. government lending overseas*	−0.2
	Capital-account balance	$= h + j - (i + k)$	
		= Surplus of $264.6	
l: Net decrease in U.S. official reserves			1
	Official reserves balance	= Deficit of $1.0	
m: Statistical discrepancy	−97.1		

[1] Balance of payments data are preliminary. Numbers may not sum exactly owing to rounding.

*Negative sign indicates net repayment of U.S. government loans.

Source: Data from the Bureau of Economic Analysis, U.S. Department of Commerce, as published on their web page, April 17, 1998.

EXHIBIT 4.2 Examples of Entries in the U.S. Balance-of-Payments Accounts

CREDITS	DEBITS
Current Account	
a: Sales of wheat to Great Britain; sales of computers to Germany	b: Purchases of oil from Saudi Arabia; purchases of Japanese automobiles
c: Sales of Phantom jets to Canada	d: Payments to Filipino workers at U.S. bases in the Philippines
e: Interest earnings on loans to Argentina; profits on U.S.-owned auto plants abroad; licensing fees earned by Lotus 1-2-3; spending by Japanese tourists at Disneyland	f: Profits on sales by Nestlé's U.S. affiliate; hotel bills of U.S. tourists in Paris
	g: Remittances by Mexican Americans to relatives in Mexico; Social Security payments to Americans living in Italy; economic aid to Pakistan
Capital Account	
h: Purchases by the Japanese of U.S. real estate; increases in Arab bank deposits in New York banks; purchases by the French of IBM stock; investment in plant expansion in Ohio by Honda	i: New investment in a German chemical plant by Du Pont; increases in U.S. bank loans to Mexico; deposits in Swiss banks by Americans; purchases of Japanese stocks and bonds by Americans
j: Purchases of U.S. Treasury bonds by Bank of Japan; increases in holdings of New York bank deposits by Saudi Arabian government	k: Deposits of funds by the U.S. Treasury in British banks; purchases of Swiss-franc bonds by the Federal Reserve
Official Reserve Account	
	l: Purchases of gold by the U.S. Treasury; increases in holdings of Japanese yen by the Federal Reserve

Current Account

The balance on current account reflects the net flow of goods, services, and unilateral transfers. It includes exports and imports of merchandise (trade balance), military transactions, and service transactions (invisibles). The service account includes investment income (interest and dividends), tourism, financial charges (banking and insurance), and transportation expenses (shipping and air travel). Unilateral transfers include pensions, remittances, and other transfers overseas for which no specific services are rendered. In 1997, for example, the U.S. balance of trade registered a deficit of $195 billion, whereas the overall current-account deficit was $166.4 billion. The difference was accounted for by a $67.2 billion *surplus* on the services account and a $38.5 billion deficit in unilateral transactions. Even more striking, in 1991 the U.S. trade deficit was $73.4 billion, whereas the current-account deficit was $3.7 billion. The difference of $69.7 billion was largely

EXHIBIT 4.3 Current-Account Balances as a Percent of GDP (1998 estimated)

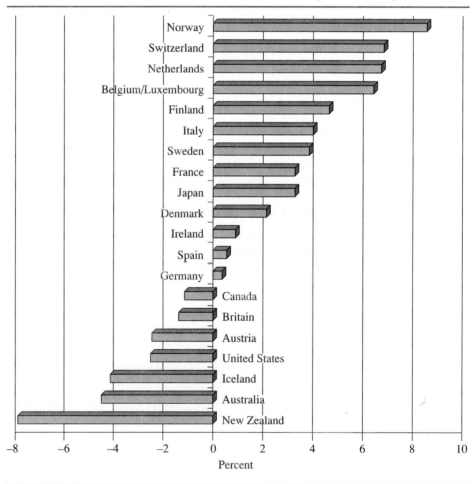

Source: OECD.

accounted for by the $42 billion in contributions that the United States received from other countries to help pay for the Gulf War.

The U.S. current-account deficit at $166.4 billion in 1997 was the world's largest, but as a percent of GDP, the deficit was just 2.1%—two-thirds of its peak value in 1987. However, as exhibit 4.3 shows, only three industrial countries were expected to have bigger deficits than the United States as a percent of GDP in 1998.

Capital Account

Capital-account transactions affect a nation's wealth and net creditor position. These transactions are classified as either portfolio, direct, or short-term investments. *Portfolio investments* are purchases of financial assets with a maturity greater than one year; *short-term investments* involve securities with a maturity of less than one year. *Direct investments* are those where management control is exerted, defined under U.S. rules as ownership of at least 10% of the equity. Government borrowing and lending are included in the balance-on-capital account. As shown in Exhibit 4.1, the U.S. balance-on-capital account in 1997 was a surplus of $264.6 billion.

Official Reserves Account

The change in official reserves measures a nation's surplus or deficit on its current- and capital-account transactions by netting reserve liabilities from reserve assets. For example, a surplus will lead to an increase in official holdings of foreign currencies or gold or both; a deficit will normally cause a reduction in these assets. However, U.S. balance-of-payments deficits have not been matched exactly by net changes in reserve assets because foreigners have been willing to hold many billions of dollars (estimated at more than $100 billion) for liquidity and other purposes. Instead of being converted into foreign currencies, many dollars have been placed on deposit in the Eurodollar market. For most countries, though, there is a close correlation between balance-of-payments deficits and reserve declines. A drop in reserves will occur, for instance, when a nation sells gold to acquire foreign currencies that it can then use to meet a deficit in its balance of payments. In 1997, the United States saw a decline in its official reserves even though it ran a surplus on its combined current- and capital-account balance. This oddity may stem from erroneous data, as reflected in the statistical discrepancy for that year.

Balance-of-Payments Measures

There are several balance-of-payments definitions. The *basic balance* focuses on transactions considered to be fundamental to the economic health of a currency. Thus, it includes the balance on current account and long-term capital, but it excludes such ephemeral items as short-term capital flows—mainly bank deposits—that are heavily influenced by temporary factors—short-run monetary policy, changes in interest differentials, and anticipations of currency fluctuations.

The **net liquidity balance** measures the change in private domestic borrowing or lending that is required to keep payments in balance without adjusting official reserves. Nonliquid, private, short-term capital flows and errors and omissions are included in the balance; liquid assets and liabilities are excluded.

The **official reserve transactions balance** measures the adjustment required in official reserves to achieve balance-of-payments equilibrium. The assumption here is that official transactions are different from private transactions.

Each of these measures has shortcomings, primarily because of the increasing complexity of international financial transactions. For example, changes in the official reserve balance may now reflect investment flows as well as central bank intervention. Similarly, critics of the basic balance argue that the distinction between short- and long-term capital flows has become blurred. Direct investment is still determined by longer-term factors, but investment in stocks and bonds can be just as speculative as bank deposits and sold just as quickly. The astute international financial manager, therefore, must analyze the payments figures rather than rely on a single summarizing number.

The Missing Numbers

In going over the numbers in Exhibit 4.1, you will note an item referred to as a **statistical discrepancy**. This number reflects errors and omissions in collecting data on international transactions. In 1997, that item was a mighty −$97.1 billion. (A negative figure reflects a mysterious outflow of funds; a positive amount reflects an inflow.)

Typically, the statistical discrepancy is positive. For example, in 1990 it was +$66.8 billion. This discrepancy coincided with such worrisome foreign events as the Iraqi invasion of Kuwait, turmoil in Iran, unrest in Central and Latin America, and the upheaval in the Soviet Union. Many experts believe that the statistical discrepancy in that year was primarily the result of foreigners' surreptitiously moving money into what they deemed to be a safe political haven—the United States.

4.2 THE INTERNATIONAL FLOW OF GOODS, SERVICES, AND CAPITAL

This section provides an analytical framework that links the international flows of goods and capital to domestic economic behavior. The framework consists of a set of basic macroeconomic accounting identities that link domestic spending and production to savings, consumption, and investment behavior, and thence to the capital-account and current-account balances. By manipulating these equations, we can identify the nature of the links between the U.S. and world economies and assess the effects on the domestic economy of international economic policies, and vice versa. As we see in the next section, ignoring these links leads to political solutions to international economic problems—such as the trade deficit—that create greater problems. At the same time, authors of domestic policy changes are often unaware of the effect these changes can have on the country's international economic affairs.

Domestic Savings and Investment and the Capital Account

The national income and product accounts provide an accounting framework for recording the national product and showing how its components are affected by international transactions. This framework begins with the observation that *national income*, which is the same as *national product*, is either spent on consumption or saved:

$$National\ income = Consumption + Savings \tag{4.1}$$

Similarly, *national expenditure*, the total amount that the nation spends on goods and services, can be divided into spending on consumption and spending on domestic real investment. *Real investment* refers to plant and equipment, research and development, and other expenditures designed to increase the nation's productive capacity. This equation provides the second national accounting identity:

$$National\ spending = Consumption + Investment \tag{4.2}$$

Subtracting Equation 4.2 from Equation 4.1 yields a new identity:

$$\frac{National}{income} - \frac{National}{spending} = Savings - Investment \tag{4.3}$$

This identity says that if a nation's income exceeds its spending, savings will exceed domestic investment, yielding surplus capital. The surplus capital must be invested overseas (if it were invested domestically there would not be a capital surplus). In other words, savings equals domestic investment plus net foreign investment. Net foreign investment equals the nation's net public and private capital outflows plus the increase in official reserves. The net private and public capital outflows equal the capital-account deficit if the outflow is positive (a capital-account surplus if negative); the net increase in official reserves equals the balance on the official reserves account. In a freely floating exchange rate system—that is, no government intervention and no official reserve transactions—excess savings will equal the capital-account deficit. Alternatively, a national savings deficit will equal the capital-account surplus (net borrowing from abroad); this borrowing finances the excess of national spending over national income.

Here is the bottom line: A nation that produces more than it spends will save more than it invests domestically and will have a net capital outflow. This capital outflow will appear as some combination of a capital-account deficit and an increase in official reserves. Conversely, a nation that spends more than it produces will invest domestically more than it saves and have a net capital inflow. This capital inflow will appear as some combination of a capital-account surplus and a reduction in official reserves.

The Link Between the Current and Capital Accounts

Beginning again with national product, we can subtract from it spending on domestic goods and services. The remaining goods and services must equal exports. Similarly, if we subtract spending on domestic goods and services from total expenditures, the remaining spending must be on imports. Combining these two identities leads to another national income identity:

$$\frac{National}{income} - \frac{National}{spending} = Exports - Imports \tag{4.4}$$

EXHIBIT 4.4 The Trade Balance Falls as Spending Rises Relative to GNP

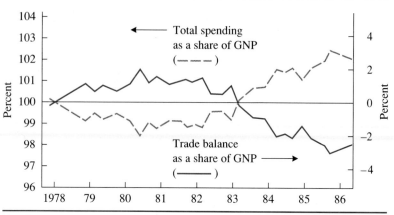

Note: Total spending is "Gross Domestic Purchases"; trade balance is "Net Exports of Goods and Services."

Equation 4.4 says that a current-account surplus arises when national output exceeds domestic expenditures; similarly, a current-account deficit is due to domestic expenditures' exceeding domestic output. Exhibit 4.4 illustrates this latter point for the United States. Moreover, when Equation 4.4 is combined with Equation 4.3, we have a new identity:

$$Savings - Investment = Exports - Imports \qquad (4.5)$$

According to Equation 4.5, if a nation's savings exceed its investment, that nation will run a current-account surplus. This equation explains the Japanese current-account surplus: The Japanese have an extremely high savings rate, both in absolute terms and relative to their investment rate. Conversely, a nation such as the United States, that saves less than it invests, must run a current-account deficit. Noting that savings minus investment equals net foreign investment, we have the following identity:

$$\frac{Net\ foreign}{investment} = Exports - Imports \qquad (4.6)$$

Equation 4.6 says that the balance on the current account must equal the net capital outflow; that is, any foreign exchange earned by selling abroad must be either spent on imports or exchanged for claims against foreigners. The net amount of these IOUs equals the nation's capital outflow. If the current account is in surplus, the country must be a net exporter of capital; a current-account deficit indicates that the nation is a net capital importer. This equation explains why Japan, with its large current-account surpluses, is a major capital exporter, whereas the United States, with its large current-account deficits, is a major capital importer. Bearing in mind that trade is goods plus services, to say that the United States has a trade deficit with Japan is simply to say that the United States is buying more goods and services from Japan than Japan is buying from the United States,

---

EXHIBIT 4.5 Linking National Economic Activity with Balance-of-Payments Accounts: Basic Identities

	Our national product (Y)	$-$	Our total spending (E)	
minus	Our spending for consumption		minus	Our spending for consumption

$=$ Our national savings (S) $-$ Our investment in new real assets (I_d)

$=$ Net foreign investment, or the net increase in claims on foreigners and official reserve assets, e.g., gold (I_f)

	Our national product (Y)	$-$	Our total spending (E)	
minus	Our spending on our own goods and services		minus	Our spending on our own goods and services

$=$ Our exports of goods and services (X) $-$ Our imports of goods and services (M)

$=$ Balance on current account

$=$ $-$(Balance on capital and official reserves accounts)

$$Y - E = S - I_d = X - M = I_f$$

Conclusions: A nation that produces more than it spends will save more than it invests, export more than it imports, and wind up with a capital outflow. A nation that spends more than it produces will invest more than it saves, import more than it exports, and wind up with a capital inflow.

---

and that Japan is investing more in the United States than the United States is investing in Japan. Between the United States and Japan, any deficit in the current account is exactly equal to the surplus in the capital account. Otherwise, there would be an imbalance in the foreign exchange market, and the exchange rate would change.

Another interpretation of Equation 4.6 is that the excess of goods and services bought over goods and services produced domestically must be acquired through foreign trade and must be financed by an equal amount of borrowing from abroad (the capital-account surplus and/or official reserves deficit). Thus, in a freely floating exchange rate system, the current-account balance and the capital-account balance must exactly offset each other. With government intervention in the foreign exchange market, the sum of the current-account balance plus the capital-account balance plus the balance on the official reserves account must be zero. These relations are shown in Exhibit 4.5.

These identities are useful because they allow us to assess the efficacy of proposed "solutions" for improving the current-account balance. It is clear that a nation can neither reduce its current-account deficit nor increase its current-account surplus unless it meets two conditions: (1) Raise national product relative to national spending, and (2) increase savings relative to domestic investment. A proposal to improve the current-account balance by reducing imports (say, via higher tariffs) that does not affect national output/spending and national savings/investment leaves the trade deficit the same; and the proposal cannot achieve its objective without violating fundamental accounting identities. With regard to Japan, a clear implication is that chronic Japanese trade surpluses are not reflective of unfair trade practices and restrictive import policies but rather are the natural effect of differing cultures and philosophies regarding saving and consumption. As long as the Japanese prefer to save and invest rather than consume, the imbalance will persist.

These accounting identities also suggest that a current-account surplus is not necessarily a sign of economic vigor nor is a current-account deficit necessarily a sign of weakness or of a lack of competitiveness. Indeed, economically healthy nations that provide good investment opportunities tend to run trade deficits, because this is the only way to run a capital-account surplus. The United States ran trade deficits from early Colonial times to just before World War I, as Europeans sent investment capital to develop the continent. During its 300 years as a debtor nation—a net importer of capital—the United States progressed from the status of a minor colony to the world's strongest power. Conversely, it ran surpluses while the infamous Smoot-Hawley tariff helped sink the world into depression. Similarly, during the 1980s, Latin America ran current-account surpluses because its dismal economic prospects made it unable to attract foreign capital. As Latin America's prospects improved in the early 1990s, money flowed in and it began running capital-account surpluses again—matched by offsetting current-account deficits.

Note, too, that nations that grow rapidly will import more goods and services at the same time that weak economies will slow down or reduce their imports, because imports are positively related to income. As a result, the faster a nation grows relative to other economies, the larger its current-account deficit (or smaller its surplus). Conversely, slower-growing nations will have smaller current-account deficits (or larger surpluses). Hence, current-account deficits may reflect strong economic growth or a low level of savings, and current-account surpluses can signify a high level of savings or a slow rate of growth.

Government Budget Deficits and Current-Account Deficits

Up to now, government spending and taxation have been included in aggregate domestic spending and income figures. By differentiating between the government and private sectors, we can see the effect of a government deficit on the current-account deficit.

National spending can be divided into household spending plus private investment plus government spending. Household spending, in turn, equals national income less the sum of private savings and taxes. Combining these terms yields the following identity:

$$\frac{National}{spending} = \frac{Household}{spending} + \frac{Private}{investment} + \frac{Government}{spending} \tag{4.7}$$

$$= \frac{National}{income} - \frac{Private}{savings} - Taxes + \frac{Private}{investment} + \frac{Government}{spending}$$

Rearranging Equation 4.7 yields a new expression for excess spending:

$$\frac{National}{spending} - \frac{National}{income} = \frac{Private}{investment} - \frac{Private}{savings} + \frac{Government}{budget\ deficit} \tag{4.8}$$

where the **government budget deficit** equals government spending minus taxes. Equation 4.8 says that excess national spending is composed of two parts: the excess of private domestic investment over private savings and the total government (federal, state, and local) deficit. Because national spending minus national product equals the net capital inflow, Equation 4.8 also says that the nation's excess spending equals its net borrowing from abroad.

Rearranging and combining Equations 4.4 and 4.8 provides a new accounting identity:

$$\underset{\text{balance}}{\text{Current-account}} = \underset{\text{surplus}}{\text{Savings}} - \underset{\text{budget deficit}}{\text{Government}} \qquad (4.9)$$

Equation 4.9 reveals that a nation's current-account balance is identically equal to its private savings-investment balance less the government budget deficit. According to this expression, a nation running a current-account deficit is not saving enough to finance its private investment and government budget deficit. Conversely, a nation running a current-account surplus is saving more than is needed to finance its private investment and government deficit.

In 1993, for example, private savings in the United States totaled $1,002 billion; private investment equaled $882 billion; and the government budget deficit amounted to $215 billion. Excess domestic spending thus equaled $95 billion, and the United States experienced a $104 billion current-account deficit. The $9 billion discrepancy reflects errors and omissions in the measurements of international transactions plus other small adjustments.

Exhibit 4.6 presents similar data for the United States between 1980 and 1994, along with the averages for the period 1973 to 1979. The data are expressed as percent-

EXHIBIT 4.6 U.S. National Income Accounts and the Current-Account Deficit, 1973–1994 (Percent of GNP)

YEAR	GROSS PRIVATE SAVINGS	GROSS PRIVATE INVESTMENT	SAVINGS LESS INVESTMENT	TOTAL GOVERNMENT DEFICIT	CURRENT-ACCOUNT BALANCE[1]
1973–79 (average)	18.0%	16.8%	1.2%	−0.9%	0.1%
1980	17.7%	16.6%	1.1%	−1.3%	0.1%
1981	18.6%	17.7%	0.9%	−1.0%	0.2%
1982	18.8%	15.4%	3.5%	−3.3%	−0.3%
1983	18.1%	15.4%	2.7%	−3.9%	−1.3%
1984	18.9%	18.3%	0.6%	−2.8%	−2.5%
1985	17.5%	17.0%	0.5%	−3.0%	−3.0%
1986	16.3%	16.2%	0.1%	−3.3%	−3.4%
1987	15.5%	15.9%	−0.4%	−2.4%	−3.6%
1988	15.8%	15.7%	0.2%	−1.9%	−2.5%
1989	15.0%	15.3%	−0.2%	−1.5%	−1.9%
1990	14.9%	14.0%	0.9%	−2.4%	−1.6%
1991	15.8%	12.6%	3.2%	−3.1%	−0.1%
1992	15.7%	12.6%	3.1%	−4.1%	−1.1%
1993	15.3%	13.4%	1.8%	−3.3%	−1.6%
1994	14.5%	15.0%	−0.5%	−1.3%	−2.2%

[1] The sum of the savings/investment balance and the government budget deficit should equal the current-account balance. Any discrepancy between these figures is due to rounding or minor data adjustments.

Source: Economic report of the President, February 1995.

ages of gross domestic product (GDP) to facilitate comparisons over time. The table shows that the increase in the U.S. current-account deficit during the 1980s was associated with an increase in the total government budget deficit and with a narrowing in private savings relative to private investment. As savings relative to investment rose beginning in 1989, the current-account deficit narrowed, even as the government deficit continued to grow. The growth in the total government budget deficit reflects the huge federal budget deficit, as state and local governments typically run surpluses. Conversely, even as strong economic growth during the 1990s eventually turned the federal deficit into a surplus, the current-account deficit continued to grow. In general, a current-account deficit represents a decision to consume, both publicly and privately, and to invest more than the nation currently is producing.

The purpose of Exhibit 4.6 is not to specify a channel of causation but simply to show a tautological relationship among private savings, private investment, the government budget deficit, and the current-account balance. Nevertheless, the important implication is that steps taken to correct the current-account deficit can be effective only if they also change private savings, private investment, and/or the government deficit. Policies or events that fail to affect both sides of the relationship shown in Equation 4.9 will not alter the current-account deficit.

4.3 COPING WITH THE CURRENT-ACCOUNT DEFICIT

Conventional wisdom suggests some oft-repeated solutions to a current-account deficit. The principal suggestions are currency devaluation and protectionism. There are important, though subtle, reasons, however, why neither is likely to work.

Currency Depreciation

An overvalued currency acts as a tax on exports and a subsidy to imports, reducing the former and increasing the latter. The result, as we saw in chapter 2, is that a nation maintaining an overvalued currency will run a trade deficit. Permitting the currency to return to its equilibrium level will help reduce the trade deficit.

Many academics, politicians, and business people also believe that devaluation can reduce a trade deficit in a floating-rate system. Key to the effectiveness of devaluation is sluggish adjustment of nominal prices, which translates changes in nominal exchange rates into changes in real (inflation-adjusted) exchange rates. This view of exchange rate changes implies a systematic relation between the exchange rate and the current-account balance. For example, it implies that the current U.S. trade deficit will be reduced eventually by a fall in the value of the dollar.

By contrast, we saw in chapter 2 that all exchange rates do is to equate currency supplies and demands; they do not determine the distribution of these currency flows between trade flows (the current-account balance) and capital flows (the capital-account balance). This view of exchange rates predicts that there is no simple relation between the exchange rate and the current-account balance. Trade deficits do not "cause" currency depreciation, nor does currency depreciation by itself help reduce a trade deficit: Both exchange rate changes and trade balances are determined by more fundamental economic factors.

These diametrically opposed theories can be evaluated by studying evidence on trade deficits and exchange rate changes. A good place to start is with recent U.S. experience.

From 1976 to 1980, the value of the dollar declined as the current-account deficit for the United States first worsened and then improved; but from 1980 to 1985, the dollar strengthened even as the current account steadily deteriorated. Many analysts attributed the rise in the U.S. trade deficit in the early 1980s to the sharp rise in the value of the U.S. dollar over that period. As the dollar rose in value against the currencies of America's trading partners, fewer dollars were required to buy a given amount of foreign goods and more foreign currencies were needed to buy a fixed amount of U.S. goods. Responding to these price changes, Americans bought more foreign goods; and foreign consumers reduced their purchases of U.S.-made goods. U.S. imports increased and exports declined.

After reaching its peak in early March 1985, the value of the dollar began to decline. This decline was actively encouraged by the United States and by several foreign governments in hope of reducing the U.S. trade deficit. Conventional wisdom suggested that the very same basic economic forces affecting the trade account during the dollar's run-up would now be working in the opposite direction, reducing the U.S. trade (current-account) deficit. As Exhibit 4.7 documents, however, the theory did not work. The U.S. trade deficit kept rising, reaching new record levels month after month. By 1987, it had risen to $167 billion. The 1988 trade deficit fell to $128 billion, but this figure still exceeded the $125 billion trade deficit in 1985. Even more discouraging from the standpoint of those who believe that currency devaluation should cure a trade deficit is that between 1985 and 1987 the yen more than doubled in dollar value without reducing the U.S. trade deficit with Japan. What went wrong?

LAGGED EFFECTS ■ The simplest explanation is that time is needed for an exchange rate change to affect trade. Exhibit 4.8 shows that when the exchange rate is lagged two years

EXHIBIT 4.7 The Dollar and the Deficit

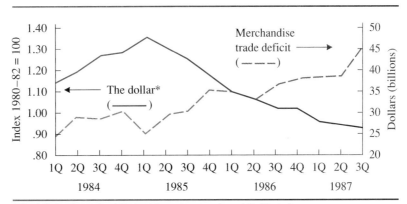

*The dollar's value against 15 industrial-country currencies weighted by trade.

EXHIBIT 4.8 The U.S. Current Account Balance Versus the Dollar: 1970–1997

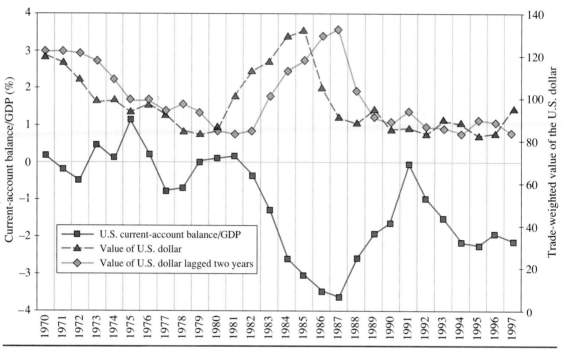

Source: Economic Report of the President, February 1998.

(that is, the current-account balance for 1987 is matched against the value of the dollar in 1985), there is a closer correspondence between the current-account balance as a percent of GDP and the exchange rate. Despite this closer correspondence, however, in some years the dollar is falling and the current-account balance is worsening, and in other years the dollar is strengthening and the current account balance is improving. Overall, changes in the dollar's value explain less than 4% of the variation in the U.S. current-account balance as a percent of GDP over the period 1970–1997.

J-CURVE THEORY ■ Another explanation, which is consistent with the presence of lagged effects, is based on the J-curve theory, illustrated in Exhibit 4.9. The letter J describes a curve that, viewed from left to right, goes down sharply for a short time, flattens out, then rises steeply for an extended period. That's how J-curve proponents have been expecting the U.S. trade deficit to behave. According to the **J-curve theory**, a country's trade deficit worsens just after its currency depreciates because price effects will dominate the effect on volume of imports in the short run: That is, the higher cost of imports will more than offset the reduced volume of imports. Thus, the J-curve says that a decline in the value of the dollar should be followed by a temporary worsening in the trade deficit before its longer-term improvement.

The initial worsening of the trade deficit occurred as predicted in 1985 but not until four years later, in 1989, was the trade deficit below where it had been in 1985. Moreover,

EXHIBIT 4.9 The Theoretical J-Curve

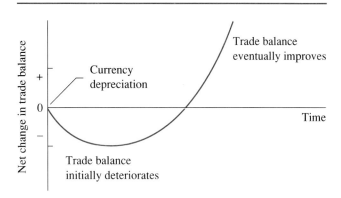

the improvement that occurred between 1985 and 1989 may owe more to the $59 billion drop in the federal budget deficit over this period than to depreciation of the dollar. Similarly, between 1970 and 1995, the yen rose from ¥360 to the dollar to ¥95 to the dollar. At the same time, America's trade deficit with Japan kept rising. In other words, the upturn of the J-curve has proved elusive. Part of the answer in the case of Japan is that Japanese manufacturers responded to currency appreciation by cutting costs and profit margins enough to keep their goods competitive abroad. Another confounding factor is that a strong yen makes Japanese raw material imports cheaper, offsetting some of their cost disadvantage.

The Japanese experience has been similarly disappointing to advocates of using currency changes to cure current-account imbalances. Between 1991 and 1995, when the yen was rising rapidly, Japan's current-account surplus widened. Since 1995, however, Japan's surplus has narrowed, even as the yen has fallen in value. One explanation for the recent turn of events is that the Japanese government's budget has moved into sizable deficit, reducing Japan's total domestic savings. Other things being equal, lower domestic savings will reduce a current-account surplus.

DEVALUATION AND INFLATION ■ Devaluing to gain trade competitiveness can also be self-defeating as a weaker currency tends to result in higher domestic inflation, offsetting the benefits of devaluation. For example, higher inflation brought about by a lower dollar will make U.S. exports more expensive abroad and imports more competitive in the U.S. market.

U.S. DEFICITS AND THE DEMAND FOR U.S. ASSETS ■ Another possible reason for the failure of dollar devaluation to cure the persistent U.S. trade deficit is that the earlier analysis mixed up cause and effect. The argument that the strong dollar was the main culprit of the massive U.S. trade deficit rested on the obvious fact that the dollar's high price made importing cheaper than exporting. But that was not the complete picture. The dollar's price was not a cause or even a symptom of the problem. It is axiomatic that price is a reflection of a fundamental value in the market. To argue that the high dollar hurt the U.S. economy does not explain how or why the price got there.

One plausible explanation is that owing to an increasingly attractive investment climate in the United States and added political and economic turmoil elsewhere in the world, foreign investors in the early 1980s wanted to expand their holdings of U.S. assets. They bid up the value of the dollar to a level at which Americans were willing to exchange their assets for foreign goods and services. The result was a capital-account surplus balanced by a current-account deficit. In effect, the capital-account surplus drove the current-account deficit. The net result was excess American spending financed by borrowing from abroad.

Adding fuel to the current-account deficit, particularly after 1982, was the growth of the federal government's budget deficit. The U.S. budget deficit could be funded in only one of three ways: restricting investment, increasing savings, or exporting debt. The United States rejected the first alternative, was unable to accomplish the second, and thus relied heavily on the third. Accordingly, the trade deficit was the equilibrating factor that enabled the United States to satisfy its extra-large debt appetite. The price of the dollar determined the terms on which the rest of the world was willing to finance that deficit. When foreigners wanted to hold U.S. assets, the terms were quite attractive (they were willing to pay a high price for dollars); when foreigners no longer found U.S. assets so desirable, the financing terms became more onerous (they reduced the price they were willing to pay for dollars).

This analysis suggests that the current-account deficit will disappear only if the U.S. savings rate rises significantly or the rate of investment falls. With the federal deficit disappearing but the current-account deficit growing, currency devaluation will work only if some mechanism is in place that leads to a rise in private savings, a cut in private investment, or a further increase in the government budget surplus.

Protectionism

Another response to a current-account deficit is **protectionism**—that is, the imposition of tariffs, quotas, or other forms of restraint against foreign imports. A **tariff** is essentially a tax that is imposed on a foreign product sold in a country. Its purpose is to increase the price of the product, thereby discouraging purchase of that product and encouraging the purchase of a substitute, domestically produced product.[1] A **quota** specifies the quantity of particular products that can be imported to a country, typically an amount that is much less than the amount currently being imported. By restricting the supply relative to the demand, the quota causes the price of foreign products to rise. In both cases, the results are ultimately a rise in the price of products consumers buy, an erosion of purchasing power, and a collective decline in the standard of living.

These results present a powerful argument against selective trade restrictions as a way to correct a nation's trade imbalance. An even more powerful argument is that such restrictions do not work. Either other imports rise or exports fall. This conclusion follows from the basic national-income accounting identity: Savings − Investment = Exports − Imports. Unless saving or investment behavior changes, this identity says that a $1 reduction in imports will lead to a $1 decrease in exports.

---

[1] The incidence of a tariff—that is, who pays it—depends on the relative elasticities of supply and demand. For example, the more elastic the demand, the more of the tariff that will be absorbed by the exporter. On the other hand, an elastic supply means that more of the tariff will be paid by the consumer.

The mechanism that brings about this result depends on the basic market forces that shape the supply and demand for currencies in the foreign exchange market. For example, when the U.S. government imposes restrictions on steel imports, the reduction in purchases of foreign steel effectively reduces the U.S. demand for foreign exchange. Fewer dollars tendered for foreign exchange means a higher value for the dollar. The higher-valued dollar raises the price of U.S. goods sold overseas and causes proportionately lower sales of U.S. exports. A higher-valued dollar also lowers the cost in the United States of foreign goods, thereby encouraging the purchase of those imported goods on which there is no tariff. Thus, any reduction in imports from tariffs or quotas will be offset by the reduction in exports and increase in other imports.

Restrictions on importing steel will also raise the price of steel, reducing the competitiveness of U.S. users of steel, such as automakers and capital goods manufacturers. Their ability to compete will be constrained both at home and abroad. Ironically, protectionism punishes the most efficient and most internationally competitive producers—those who are exporting or are able to compete against imports—while it shelters the inefficient producers.

Ending Foreign Ownership of Domestic Assets

One approach that would eliminate a current-account deficit is to forbid foreigners from owning domestic assets. If foreigners cannot hold claims on the nation, they will export an amount equal in value only to what they are willing to import, ending net capital inflows. The microeconomic adjustment mechanism that will balance imports and exports under this policy is as follows.

The cessation of foreign capital inflows, by reducing the available supply of capital, will raise real domestic interest rates. Higher interest rates will stimulate more savings because the opportunity cost of consumption rises with the real interest rate; higher rates also will cause domestic investment to fall because fewer projects will have positive net present values. The outcome will be a balance between savings/investment and elimination of the excess domestic spending that caused the current-account deficit in the first place. Although such an approach would work, most observers would consider the resulting slower economic growth too high a price to pay to eliminate a current-account deficit.

Many observers are troubled by the role of foreign investors in U.S. financial markets, but as long as Americans continue to spend more than they produce, there will be a continuing need for foreign capital. This foreign capital is helping to improve America's industrial base, while at the same time providing capital gains to those Americans who are selling their assets to foreigners. Foreign investors often introduce improved management, better production skills, or new technology that increases the quality and variety of goods available to U.S. consumers, who benefit from lower prices as well. Investment—even foreign investment—also makes labor more productive. And higher labor productivity leads to higher wages, whether the factory's owners live across town or across the Pacific. A recent study indicates that foreign direct investment is an especially powerful engine for stimulating the productivity of domestic industries.[2] The study demonstrates that domestic workers could achieve productivity levels on a par with leading foreign

---

[2] "Manufacturing Productivity," McKinsey Global Institute, October 1993.

workers, and foreign investment spreads good practices as employees move from the foreign firm to local ones. For example, Japanese auto transplants have provided a closeup learning lab for the Big Three U.S. automakers to grasp concepts such as "lean production" and "just-in-time" component delivery. Restrictions on foreign investment will eliminate such productivity gains and may also provoke reciprocal restrictions by foreign governments.

ILLUSTRATION

JAPAN'S TRANSPLANTED AUTO PARTS SUPPLIERS RAISE U.S. PRODUCT QUALITY

Japanese auto plants located in the United States buy about 60% of their parts from American sources. This figure includes purchases from Japanese component makers that have followed their customers to the United States. Japan's transplanted parts makers allow the Japanese carmakers to be very choosy customers. Many American parts suppliers have been rejected as not producing to the requisite quality. The good news is that the tough, new competition has forced many U.S. companies to raise their standards and cut their costs.

Japanese buyers put would-be suppliers through exhausting qualification trials, which often require suppliers to make fundamental improvements in their manufacturing. They are also sharing valuable know-how with American suppliers. Auto suppliers form a pyramid, so those feeding auto manufacturers at the top can meet stringent requirements only with better performance from their own suppliers, and so on, down the chain. In this way, demands from quality-conscious Japanese customers ripple down to the base of the industrial supplier infrastructure. For example, when Honda demanded a smoother steel coating, Inland Steel had to insist on better zinc from its suppliers. Moreover, suppliers to the Japanese-owned auto plants are also supplying Detroit's Big Three. As the supplier base improves, so do the components that make up a Ford, a Chevy, or a Chrysler.

Boosting the Savings Rate

We have seen that a low savings rate tends to lead to a current-account deficit. Thus, another way to reduce the current-account deficit would be to stimulate savings behavior. The data, however, indicate that the rate of U.S. private savings has declined over time. In particular, personal savings, which averaged about 8% of disposable income in the 1970s, slid to 4.3% in 1987 and stayed below 5% through 1998. Future U.S. wealth will be impaired if the low U.S. personal savings rate persists.

One possible explanation for the decline in personal savings is that Social Security benefits expanded greatly during the 1970s. By attenuating the link between savings behavior and retirement income, Social Security may have reduced the incentive for Americans to save for retirement. By contrast, Japan—which has only a rudimentary social security system and no welfare to speak of—has an extraordinarily high personal savings rate. Presumably, the inability of the Japanese to throw themselves on the mercy of the state has affected their willingness to save for a rainy day.

Similarly, changes in tax regulations and tax rates may greatly affect savings and investment behavior and, therefore, the nation's trade and capital flows. Thus, purely domestic policies may have dramatic—and unanticipated—consequences for a nation's international economic transactions. The lesson is clear: In an integrated world economy, everything connects to everything else; politicians can't tinker with one parameter without affecting the entire system.

Current-Account Deficits and Unemployment

One rationale for attempting to eliminate a current-account deficit is that such a deficit leads to unemployment. Underlying this rationale is the notion that imported goods and services are substituting for domestic goods and services and costing domestic jobs. For example, some have argued that every million-dollar increase in the U.S. trade deficit costs about 33 American jobs, assuming that the average worker earns about $30,000 a year ($1,000,000/$30,000 = 33). Hence, it is claimed, reducing imports would raise domestic production and employment. However, the view that reducing a current-account deficit promotes jobs is based on single-entry bookkeeping.

If a country buys fewer foreign goods and services, it will demand less foreign exchange. As discussed above, this result will raise the value of the domestic currency, thereby reducing exports and encouraging the purchase of other imports. Jobs are saved in some industries, but other jobs are lost by the decline in exports and rise in other imports. According to this line of reasoning, the net impact of a trade deficit or surplus on jobs should be nil.

The claims of some politicians, however, would lead one to believe that the economic performance of the United States has been dismal because of its huge current-account deficits. However, if the alternative story is correct—that a current-account deficit reflects excess spending and has little to do with the health of an economy—then there should be no necessary relation between economic performance and the current-account balance.

The appropriate way to settle this dispute is to examine the evidence. Recent research that examined the economic performance of the 23 OECD (Organization for Economic Cooperation and Development) countries during a 38-year period found no systematic relationship between trade deficits and unemployment rates.[3] This result is not surprising for those who have looked at U.S. economic performance over the past 20 years or so. During this period, the trade deficit soared, but the United States created jobs three times as fast as Japan and 20 times as fast as Germany. Also, in the same time period, America's GDP grew 43% faster than that of Japan or Germany, even though both nations had huge trade surpluses with the United States.

In general, no systematic relationship between net exports and economic growth should be expected—and none is to be found.[4] The evidence shows that current-account surpluses—in and of themselves—are neither good nor bad. They are not correlated with jobs, growth, decline, competitiveness, or weakness. What matters is why they occur.

[3] See David M. Gould, Roy J. Ruffin, and Graeme L. Woodbridge, "The Theory and Practice of Free Trade," Federal Reserve Bank of Dallas *Economic Review*, Fourth Quarter, 1993, pp. 1–16.

[4] See, for example, David M. Gould and Roy J. Ruffin, "Trade Deficits: Causes and Consequences," Federal Reserve Bank of Dallas *Economic Review*, Fourth Quarter, 1996, pp. 10–20. After analyzing a group of 101 countries over a 30-year period, they conclude (p. 17) that "trade imbalances have little effect on rates of economic growth once we account for the fundamental determinants of economic growth."

ILLUSTRATION

GERMANY'S CURRENT-ACCOUNT SURPLUS TURNS INTO A DEFICIT

The process of international payments equilibrium and its tie to both macroeconomic factors and exchange rates is shown in the case of Germany. With its mature industrial economy and aging population, western Germany has long run a current-account surplus. Modernization of the eastern German economy changed that. To finance re-unification, Germany had to shift from being a large capital exporter to being a capital importer. Consequently, Germany had to increase its imports of goods and services relative to exports. This it did, as shown in Exhibit 4.10. For example, Germany's current account moved from a surplus of DM 70 billion (about 3% of GDP) in 1990 to a deficit of DM 30 billion (about 1.5% of GDP) in 1991. Such a drastic change could be brought about in one of two ways.

One possibility would have been to reduce domestic demand in Germany by private and government consumption or investment. However, cuts in capital spending would be unwise because they would mean less productive capacity in the future. Private consumption would have to be curbed through tax increases, which—even if temporary—would be unpopular, to say the least. Cutting non-unification-related government expenditures would have been politically difficult as well, given the strong constituencies behind that spending.

The other alternative was to reduce demand for German exports and, at the same time, rely to a greater extent on imports to meet domestic demand. This path—which Germany chose—required its goods to become relatively more expensive to its trading partners (mostly fellow members of the EMS) and foreign goods relatively less expensive to Germans. This shift in relative prices could be accomplished through some combination of higher inflation in Germany, appreciation of the DM, depreciation of foreign currencies relative to the DM, or deflation in fellow EMS members.

As we saw in chapter 3, Germany was unwilling to accept inflation, and the ERM prevented exchange rate changes. That left only deflation in other EMS countries, until it too proved unacceptable and the system broke down in September 1992.

EXHIBIT 4.10 Reunification Turns Germany's Current-Account Surplus into a Deficit that Exceeds the U.S. Deficit (% of GDP)

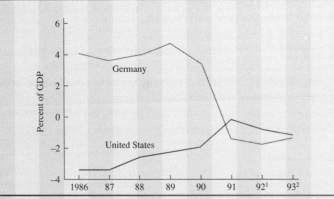

The Bottom Line on Current-Account Deficits and Surpluses

To summarize the previous discussion of current-account deficits and surpluses, consider the following stylized facts. Suppose the United States is a country whose citizens, for one reason or another, have a low propensity to save. And suppose also that for a variety of reasons, the United States is an attractive place to invest. Finally, suppose that in the rest of the world, people have high propensities to save but their opportunities for investment are less attractive. In these circumstances, there will be a flow of capital from the rest of the world and a corresponding net inflow of goods and services to the United States. The United States would have a current-account deficit.

In the above situation, the current-account deficit would not be viewed as a problem. Rather, it would be viewed as an efficient adaptation to different savings propensities and investment opportunities in the United States and the rest of the world. From this perspective, a current-account deficit becomes a solution, not a problem. This was the situation confronting the United States early in its history, when it ran almost continual trade deficits for its first 100 years. Nobody viewed that as a problem back then, and it still isn't one.

The real problem, if there is one, is either too much consumption, and thus too little savings, or too much investment. Regardless of what one's opinions are, the situation confronting the United States and the rest of the world is an expression of national preferences, to which trade flows have adjusted in a timely manner. An economist has no further wisdom to shed on this matter.

What are the long-term consequences for the United States or for any other nation that runs a current-account deficit? Here, an economist can speak with some authority. If the current-account deficit (and resulting capital-account surplus) finances productive domestic investment, then the nation is better off; the returns from these added investments will service the foreign debts with income left over to increase living standards. Conversely, a capital-account surplus that finances consumption will increase the nation's well-being today at the expense of its future well-being. That trade-off, however, has little to do with the balance of payments per se.

❧ 4.4 SUMMARY AND CONCLUSIONS

The balance of payments is an accounting statement of the international transactions of one nation over a specific period. The statement shows the sum of economic transactions of individuals, businesses, and government agencies located in one nation with those located in the rest of the world during the period. Thus, the U.S. balance of payments for a given year is an accounting of all transactions between U.S. residents and residents of all other countries during that year.

The statement is based on double-entry bookkeeping; every economic transaction recorded as a credit brings about an equal and offsetting debit entry, and vice versa. A debit entry shows a purchase of foreign goods, services, or assets, or a decline in liabilities to foreigners. A credit entry shows a sale of domestic goods, services, or assets, or an increase in liabilities to foreigners. For example, if a foreign company sells a car to a U.S. resident, a debit is recorded to indicate an increase in purchases made by the United States (the car); a credit is recorded to reflect an increase in liabilities to the foreigner (payment for the car).

The balance of payments is often divided into several components. Each shows a particular kind of transaction, such as merchandise exports or foreign purchases of U.S. government securities. Transactions that represent purchases and sales of goods and services in the current period are called the current account; those that represent capital transactions are called the capital account. Changes in official reserves appear on the official reserves account.

Double-entry bookkeeping ensures that debits equal credits, therefore the sum of all transactions is zero. In the absence of official reserve transactions, a capital-account surplus must offset the current-account deficit; and a capital-account deficit must offset a current-account surplus.

The United States is currently running a large current-account deficit. Much public discussion about why the United States imports more than it exports has focused on claims of unfair trading practices or on the high value of the dollar. However, economic theory indicates that the total size of the current-account deficit is a macroeconomic phenomenon; there is a basic accounting identity that a nation's current-account deficit reflects excess domestic spending. Equivalently, a current-account deficit equals the excess of domestic investment over domestic savings. Explicitly taking government into account yields a new relation: The domestic spending balance equals the private savings–investment balance minus the government budget deficit. As private savings and investment have historically been in balance for the United States, the trade deficit until recently could be traced to the federal budget deficit. With the federal budget now in balance, the culprit today appears to be a low U.S. savings rate.

We saw that failure to consider the elementary economic accounting identities can mislead policymakers into relying on dollar depreciation, trade restrictions, or trade subsidies in order to reduce U.S. trade deficits without doing anything about excess domestic spending. The current-account deficit can be reduced only if domestic savings rise, private investment declines, or the government deficit is reduced. Absent any of those changes, the current-account deficit will not diminish, regardless of the imposition of trade barriers or the amount of dollar depreciation.

➤Questions

1. In a freely floating exchange rate system, if the current account is running a deficit, what are the consequences for the nation's balance on capital account and its overall balance of payments?

2. As the value of the U.S. dollar rises, what is likely to happen to the U.S. balance on current account?

3. A current-account surplus is not always a sign of health; a current-account deficit is not always a sign of weakness. Comment.

4. Suppose Lufthansa buys $400 million worth of Boeing jets in 1996 and is financed by the U.S. Eximbank with a five-year loan that has no principal or interest payments due until 1997. What is the net impact of this sale on the U.S. current account, capital account, and overall balance of payments for 1996?

5. How does a trade deficit affect the current-account balance?

6. On which balance-of-payments account does tourism show up?

7. Suppose the United States expropriates all foreign holdings of American assets. What will happen to the U.S. current-account deficit? What will likely happen to U.S. savings and investment? Why?

8. What happens to Mexico's ability to repay its foreign loans if the United States restricts imports of Mexican agricultural produce?

9. For Brazil to service its foreign debts without borrowing more money, what must be true of its trade balance?

10. Suppose the United States imposes import restrictions on Japanese steel. What is likely to happen to the U.S. current-account deficit? What else is likely to happen?

11. Suppose Brazil starts welcoming foreign investment with open arms. How is this policy likely to affect the value of the Brazilian *real*? The Brazilian current-account balance?

12. According to popular opinion, U.S. trade deficits indicate any or all of the following: a lack of U.S. competitiveness owing to low productivity or low-quality products and/or lower wages, superior technology, and unfair trade practices by foreign countries. Which of those factors is likely to underlie the persistent U.S. trade deficits? Explain.

13. During the 1990s, Mexico and Argentina went from economic pariahs with huge foreign debts to countries posting strong economic growth and welcoming foreign investment. What would you expect these changes to do to their current-account balances?

14. Suppose the trade imbalances of the 1980s largely disappear during the 1990s. What is likely to happen to the huge global capital flows of the 1980s? What is the link between the trade imbalances and the global movement of capital?

15. In 1965, about 34% of all adult workers were under the age of 34, compared with almost 47% by 1980. Meanwhile, the share of the workforce between 35 years and 59 years of age shrank from about 60% to 49%. What impact might this dramatic shift in the age distribution of the U.S. workforce have had on the U.S. current-account balance over this 15-year period? (*Hint*: Consider the difference in savings behavior between younger and older workers.)

16. In 1990, Japan's Ministry of International Trade and Investment (MITI) proposed that firms be given a tax credit equal to 5% of the value of the country's increased imports. The purpose of this tax subsidy was to encourage Japanese imports of foreign products and thereby reduce Japan's persistent trade surplus. At the same time, the Japanese government announced that it would reduce its budget deficit during the coming year.
    a. What are the likely consequences of the tax-subsidy plan to Japan's trade balance, the value of the yen, and the competitiveness of Japanese firms?
    b. What are the likely consequences of a lower Japanese budget deficit on Japan's trade balance?

17. Currently, social security is minimal in Japan. Suppose Japan institutes a comprehensive social security system. How would this policy switch likely affect Japan's trade surplus?

18. During 1992, Japan entered a recession. At the same time, however, its current-account surplus hit a record. Is there a contradiction between Japan's large trade surplus and a weak national economy? Explain.

19. What will strong economic growth do for the U.S. balance on current account? What will a U.S. recession do?

20. In the early 1990s, Japan underwent a recession that brought about a prolonged slump in consumer spending and capital investment (some estimates were that in 1994 only 65% of Japan's manufactur-ing capacity was being used). At the same time, the U.S. economy emerged from its recession and began expanding rapidly. Under these circumstances, what would you predict would happen to the U.S. trade deficit with Japan?

21. In the early 1990s, interest rates fell worldwide. As the United States is a net debtor nation, how should this decline affect the U.S. current-account balance?

22. "The U.S. trade deficit is a consequence of the unwillingness of the current generation of American taxpayers to pay fully for the goods and services they want from government." Comment.

23. The devastating earthquake that hit Kobe, Japan, on January 17, 1995, was estimated to have caused about $100 billion in damage to the Japanese economy. What is the likely effect of this earthquake on Japan's 1995 current account? On its capital account? Explain.

24. In 1990, Germany's current-account surplus was more than $50 billion. However, by some estimates, the process of reunification will require Germany to invest several hundred billion dollars in its eastern states over the coming decade.
    a. What implications does this huge investment have for Germany's current-account balance in the future? Explain.
    b. How should the Deutsche mark's value change to facilitate the necessary shift in Germany's economy?

25. According to the *World Competitiveness Report 1994*, with freer markets, Third World nations now are able to attract capital and technology from the advanced nations. As a result, they can achieve productivity close to Western levels while paying low wages. Hence, the low-wage Third World nations will run huge trade surpluses, creating either large-scale unemployment or sharply falling wages in the advanced nations. Comment on this apocalyptic scenario.

26. On June 23, 1997, Japanese Prime Minister Ryutaro Hashimoto spooked Wall Street. At a Columbia University luncheon, he appeared to warn that the Japanese might sell U.S. Treasury bills unless the United States helped stabilize exchange rates. The Dow Jones Industrial Average fell 192 points.
    a. Why might the stock market have fallen on such a remark? Trace the causal links.
    b. How much substance is there to the possibility that the Japanese might sell off their U.S. investments? Explain.

➤ Problems

1. How would each of the following transactions show up on the U.S. balance of payments accounts?
   a. Payment of $50 million in Social Security to U.S. citizens living in Costa Rica
   b. Sale overseas of 125,000 Elvis Presley CDs
   c. Tuition receipts of $3 billion received by American universities from foreign students
   d. Payment of $1 million to U.S. consultants A.D. Little by a Mexican company
   e. Sale of a $100 million Eurobond issue in London by IBM
   f. Investment of $25 million by Ford to build a parts plant in Argentina
   g. Payment of $45 million in dividends to U.S. citizens from foreign companies

2. Set up the double-entry accounts showing the appropriate debits and credits associated with the following transactions:
   a. ConAgra, a U.S. agribusiness, exports $80 million of soybeans to China and receives payment in the form of a check drawn on a U.S. bank.
   b. The U.S. government provides refugee assistance to Somalia in the form of corn valued at $1 million.
   c. Dow Chemical invests $500 million in a chemical plant in Germany financed by issuing bonds in London.
   d. General Motors pays $5 million in dividends to foreign residents, who choose to hold the dividends in the form of bank deposits in New York.
   e. The Bank of Japan buys up $1 billion in the foreign exchange market to hold down the value of the yen and uses these dollars to buy U.S. Treasury bonds.
   f. Cemex, a Mexican company, sells $2 million worth of cement to a Texas company and deposits the check in a bank in Dallas.
   g. Colombian drug dealers receive $10 million in cash for the cocaine they ship to the U.S. market. The money is smuggled out of the United States and then invested in U.S. corporate bonds on behalf of a Cayman Islands bank.

3. Suppose Patagonia has a government surplus of $10 billion. At the same time, private investment in Patagonia exceeds private savings by $15 billion. What can you conclude about Patagonia's balance on current account?

4. During the year, Japan had a current-account surplus of $98 billion and a capital-account deficit of $67 billion.

   a. Assuming the preceding data are measured with precision, what can you conclude about the change in Japan's foreign exchange reserves during the year?
   b. What is the gap between Japan's national expenditure and its national income?
   c. What is the gap between Japan's savings and its domestic investment?
   d. What was Japan's net foreign investment for the year?
   e. Suppose the Japanese government's budget ran a $22 billion surplus during the year. What can you conclude about Japan's private savings-investment balance for the year?

5. The following transactions (expressed in U.S. $ billions) take place during a year. Calculate the U.S. merchandise-trade, current-account, capital-account, and official reserves balances.
   a. The United States exports $300 of goods and receives payment in the form of foreign demand deposits abroad.
   b. The United States imports $225 of goods and pays for them by drawing down its foreign demand deposits.
   c. The United States pays $15 to foreigners in dividends drawn on U.S. demand deposits here.
   d. American tourists spend $30 overseas using traveler's checks drawn on U.S. banks here.
   e. Americans buy foreign stocks with $60, using foreign demand deposits held abroad.
   f. The U.S. government sells $45 in gold for foreign demand deposits abroad.
   g. In a currency support operation, the U.S. government uses its foreign demand deposits to purchase $8 from private foreigners in the United States.

6. Ruritania is calculating its balance of payments for the year. As usual, its data are perfectly accurate. All the transactions for the year are listed below (in Rur$ millions).
   a. Ruritania received weapons worth $200 from the United States under its military aid program; no payment is necessary.
   b. A Ruritanian firm exported $400 of cloth and received an IOU from the foreign importer.
   c. A Ruritanian resident paid $10 in interest on a loan from a foreigner; the check was drawn on a domestic Ruritanian bank.
   d. Foreign tourists visited Ruritania and spent $100 in traveler's checks drawn on foreign banks.
   e. The Ruritanian central bank sold $60 in gold to a

foreign government and received U.S. Treasury bills in return.

f. A foreign central bank deposited $120 in a private domestic Ruritanian bank and paid with a check drawn on a private bank in the United States.

Fill in the correct number for each balance-of-payments account for items a through j.

1. Exports
   a. goods
   b. services
   Imports
   c. goods
   d. services
   e. unilateral transfers
2. Ruritanian assets abroad
   f. privately owned
   g. officially owned
3. Foreign assets in Ruritania
   h. privately owned
   i. officially owned
   j. current account

7. During the Reagan era, 1981–1988, the U.S. current account moved from a tiny surplus to a large deficit. The following table provides U.S. macroeconomic data for that period.
   a. Based on these data, to what extent would you attribute the changes in the U.S. current-account balance to a decline in the U.S. private savings-investment balance?
   b. To what extent would you attribute the changes in the U.S. current-account balance to an increase in the U.S. government budget deficit?
   c. Based on these data, what was the excess of national spending over national income during this period?

Year	1980	1981	1982	1983	1984	1985	1986	1987	1988
Private Savings	500	586	617	641	743	736	721	731	802
Private Investment	468	558	503	547	719	715	718	749	794
Government Budget Deficit	−35	−30	−109	−140	−109	−125	−147	−112	−98
Current-Account Balance	2	5	−11	−45	−100	−125	−151	−167	−129

8.[5] Select a country and analyze that country's balance of payments for 8 to 12 years, subject to availability of data. The analysis must include examinations (presentation of statistical data with discussion) of the trade balance, current-account balance, capital-account balance, basic balance, and overall balance. Your report should also address the following issues:
   a. What accounts for swings in these various balances over time?
   b. What is the relationship between shifts in the current-account balance and changes in savings and investment? Include an examination of government budget deficits and surpluses, explaining how they are related to the savings and investment and current-account balances.

9. For the country selected in problem 8, analyze the exchange rate against the dollar during the same period.
   a. Is there any observable relationship between the balance-of-payments accounts and the exchange rate?
   b. Provide a possible explanation for your observations in (a) above.

[5]Project suggested by Donald T. Buck.

THE FOREIGN EXCHANGE MARKET

The Spaniards coming into the West Indies had many commodities of the country which they needed, brought unto them by the inhabitants, to who when they offered them money, goodly pieces of gold coin, the Indians, taking the money, would put it into their mouths, and spit it out to the Spaniards again, signifying that they could not eat it, or make use of it, and therefore would not part with their commodities for money, unless they had such other commodities as would serve their use.

Edward Leigh (1671)

$\mathcal{T}$he volume of international transactions has grown enormously over the past 50 years. Exports of goods and services by the United States now total more than 10% of gross domestic product. For both Canada and Great Britain, this figure exceeds 25%. Imports are about the same size. Similarly, annual capital flows involving hundreds of billions of dollars occur between the United States and other nations. International trade and investment of this magnitude would not be possible without the ability to buy and sell foreign currencies. Currencies must be bought and sold because the U.S. dollar is not the acceptable means of payment in most other countries. Investors, tourists, exporters, and importers must exchange dollars for foreign currencies, and vice versa.

The trading of currencies takes place in foreign exchange markets whose primary function is to facilitate international trade and investment. Knowledge of the operation and mechanics of these markets, therefore, is important for any fundamental understanding of international financial management. This chapter provides this information. It discusses the organization of the most important foreign exchange market—the interbank market—including the spot market, the market in which currencies are traded for immediate delivery, and the forward market, in which currencies are traded for future delivery.

Chapter 6 examines the currency futures and options markets.

5.1 ORGANIZATION OF THE FOREIGN EXCHANGE MARKET

If there were a single international currency, there would be no need for a foreign exchange market. As it is, in any international transaction, at least one party is dealing in a foreign currency. The purpose of the *foreign exchange market* is to permit transfers of purchasing power denominated in one currency to another—that is, to trade one currency for another currency. For example, a Japanese exporter sells automobiles to a U.S. dealer for dollars, and a U.S. manufacturer sells machine tools to a Japanese company for yen. Ultimately, however, the U.S. company will likely be interested in receiving dollars, whereas the Japanese exporter will want yen. Similarly, an American investor in Swiss-franc-denominated bonds must convert dollars into francs, and Swiss purchasers of U.S. Treasury bills require dollars to complete these transactions. It would be inconvenient, to say the least, for individual buyers and sellers of foreign exchange to seek out one another, so a foreign exchange market has developed to act as an intermediary.

Most currency transactions are channeled through the worldwide **interbank market**, the wholesale market in which major banks trade with one another. This market, which accounts for about 95% of foreign exchange transactions, is normally referred to as *the* foreign exchange market. It is dominated by about 20 major banks. In the *spot market*, currencies are traded for immediate delivery, which is actually within two business days after the transaction has been concluded. In the *forward market*, contracts are made to buy or sell currencies for future delivery. Spot transactions account for about 60% of the market, with forward transactions accounting for another 10%. The remaining 30% of the market consists of *swap* transactions, which involve a package of a spot and a forward contract.[1]

The foreign exchange market is not a physical place; rather, it is an electronically linked network of banks, foreign exchange brokers, and dealers whose function is to bring together buyers and sellers of foreign exchange. The foreign exchange market is not confined to any one country but is dispersed throughout the leading financial centers of the world: London, New York City, Paris, Zurich, Amsterdam, Tokyo, Hong Kong, Toronto, Frankfurt, Milan, and other cities.

Trading is generally done by telephone, telex, or the SWIFT system. **SWIFT (Society for Worldwide Interbank Financial Telecommunications)**, an international bank-communications network, electronically links all brokers and traders. Foreign exchange traders in each bank usually operate out of a separate foreign exchange trading room. Each trader has several telephones and is surrounded by terminals displaying up-to-the-minute information. It is a hectic existence, and many traders burn out by age 35. Most transactions are based on verbal communications; written confirmation occurs later. Hence, an informal code of moral conduct has evolved over time in which the foreign exchange dealers' word is their bond.

Although one might think that most foreign exchange trading is derived from export and import activities, this turns out not to be the case. In fact, trade in goods and

[1] These volume estimates appear in Hendrik Bessembinder, "Bid-Ask Spreads in the Interbank Foreign Exchange Markets," *Journal of Financial Economics*, June 1994, pp. 317–348.

services accounts for less than 5% of foreign exchange trading. More than 95% of foreign exchange trading relates to cross-border purchases and sales of assets, that is, to international capital flows.

The Participants

The major participants in the foreign exchange market are the large commercial banks; foreign exchange brokers in the interbank market; commercial customers, primarily multinational corporations; and central banks, which intervene in the market from time to time to smooth exchange rate fluctuations or to maintain target exchange rates. Central bank intervention involving buying or selling in the market is often indistinguishable from the foreign exchange dealings of commercial banks or of other private participants.

Only the head offices or regional offices of the major commercial banks are actually marketmakers—that is, they actively deal in foreign exchange for their own accounts. These banks stand ready to buy or sell any of the major currencies on a more or less continuous basis. A large fraction of the interbank transactions in the United States is conducted through **foreign exchange brokers**, specialists in matching net supplier and demander banks. These brokers receive a small commission on all trades (traditionally, 1/32 of 1% in the U.S. market, which translates into $312.50 on a $1 million trade). Some brokers tend to specialize in certain currencies, but they all handle major currencies such as the pound sterling, Canadian dollar, Deutsche mark, and Swiss franc. Brokers supply information (at which rates various banks will buy or sell a currency); they provide anonymity to the participants until a rate is agreed to (because knowing who the other party is may give dealers an insight into whether that party needs or has a surplus of a particular currency); and they help banks minimize their contacts with other traders (one call to a broker may substitute for half a dozen calls to traders at other banks).

Commercial and central bank customers buy and sell foreign exchange through their banks. However, most small banks and local offices of major banks do not deal directly in the interbank market. Rather, they typically will have a credit line with a large bank or with their home office. Thus, transactions with local banks will involve an extra step. The customer deals with a local bank that in turn deals with its head office or a major bank. The various linkages between banks and their customers are depicted in Exhibit 5.1. Note that the diagram includes linkages with currency futures and options markets, which we will examine in the next chapter.

The major participants in the forward market can be categorized as arbitrageurs, traders, hedgers, and speculators. *Arbitrageurs* seek to earn risk-free profits by taking advantage of differences in interest rates among countries. They use forward contracts to eliminate the exchange risk involved in transferring their funds from one nation to another.

Traders use forward contracts to eliminate or cover the risk of loss on export or import orders that are denominated in foreign currencies. More generally, a forward-covering transaction is related to a specific payment or receipt expected at a specified point in time.

Hedgers, mostly multinational firms, engage in forward contracts to protect the home currency value of various foreign currency-denominated assets and liabilities on their balance sheets that are not to be realized over the life of the contracts.

EXHIBIT 5.1 Structure of Foreign Exchange Markets

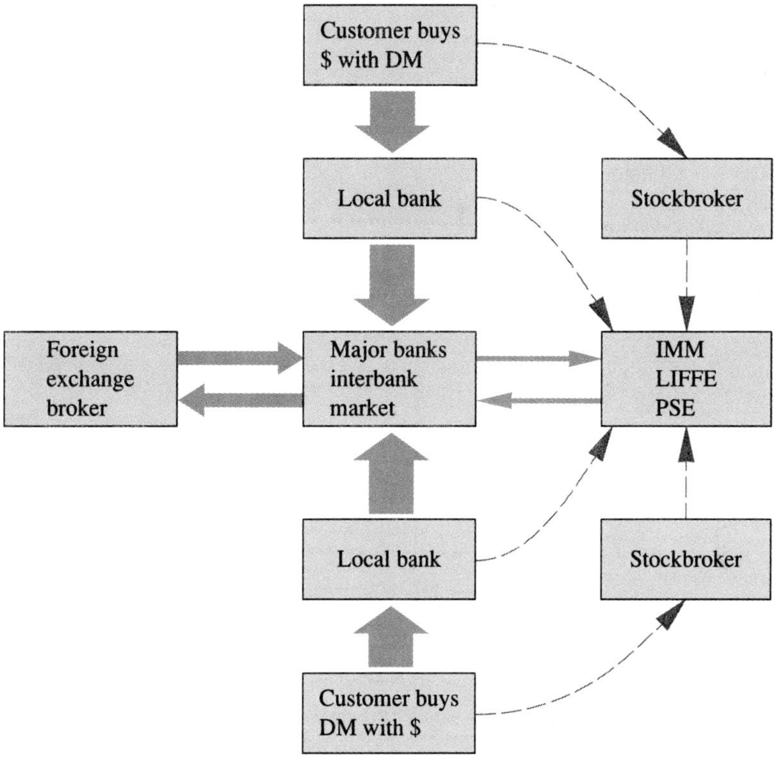

Note: The International Money Market (IMM) Chicago trades foreign exchange futures and DM futures options. The London International Financial Futures Exchange (LIFFE) trades foreign exchange futures. The Philadelphia Stock Exchange (PSE) trades foreign currency options.

Source: Federal Reserve Bank of St. Louis, *Review*, March 1984, p. 9.

Arbitrageurs, traders, and hedgers seek to reduce (or eliminate, if possible) their exchange risks by "locking in" the exchange rate on future trade or financial operations.

In contrast to these three types of forward market participants, *speculators* actively expose themselves to currency risk by buying or selling currencies forward in order to profit from exchange rate fluctuations. Their degree of participation does not depend on their business transactions in other currencies; instead, it is based on prevailing forward rates and their expectations for spot exchange rates in the future.

THE CLEARING SYSTEM ■ Technology has standardized and sped up the international transfer of funds, which is at the heart of clearing, or settling, foreign exchange transactions. In the United States, where all foreign exchange transactions involving dollars are cleared, electronic funds transfers take place through the **Clearing House Interbank Payments System**, or **CHIPS**. CHIPS is a computerized network developed by the New York Clearing House Association for transfer of international dollar payments, linking about 150 depository institutions that have offices or affiliates in New York City. Currently, CHIPS handles about 105,000 interbank transfers daily valued at $350 billion. The transfers represent about 90% of all interbank transfers relating to international dollar payments.

The New York Fed has established a settlement account for member banks into which debit settlement payments are sent and from which credit settlement payments are disbursed. Transfers between member banks are netted out and settled at the close of each business day by sending or receiving **FedWire** transfers of **fed funds** through the settlement account. Fed funds are deposits held by member banks at Federal Reserve branches.

The FedWire system is operated by the Federal Reserve and is used for domestic money transfers. FedWire allows almost instant movement of balances between institutions that have accounts at the Federal Reserve Banks. A transfer takes place when an order to pay is transmitted from an originating office to a Federal Reserve Bank. The account of the paying bank is charged, and the receiving bank's account is credited with fed funds.

To illustrate the workings of CHIPS, suppose Fuji Bank has sold U.S. $15 million to Citibank in return for ¥1.5 billion to be paid in Tokyo. In order for Fuji Bank to complete its end of the transaction, it must transfer $15 million to Citibank. To do this, Fuji Bank enters the transaction into its CHIPS terminal, providing the identifying codes for the sending and receiving banks. The message—the equivalent of an electronic check—is then stored in the CHIPS central computer.

As soon as Fuji Bank approves and releases the "stored" transaction, the message is transmitted from the CHIPS computer to Citibank. The CHIPS computer also makes a permanent record of the transaction and makes appropriate debits and credits in the CHIPS accounts of Fuji Bank and Citibank, as both banks are members of CHIPS. Immediately after the closing of the CHIPS network at 4:30 P.M. (Eastern Standard Time), the CHIPS computer produces a settlement report showing the net debit or credit position of each member bank.

Member banks with debit positions have until 5:45 P.M. (Eastern Standard Time) to transfer their debit amounts through FedWire to the CHIPS settlement account on the books of the New York Fed. The Clearing House then transfers those fed funds via FedWire out of the settlement account to those member banks with net creditor positions. The process usually is completed by 6:00 P.M. (Eastern Standard Time).

ELECTRONIC TRADING ■ In April 1992, Reuters, the news company that supplies the foreign exchange market with the screen quotations used in telephone trading, introduced a new service, Dealing 2000-Phase 2. This service adds automatic execution to the package, thereby creating a genuine screen-based market. Other quote vendors, such as EBS, Telerate, and Quotron have their own automatic systems. These **electronic trading systems** offer automated matching. Traders can enter buy and sell orders directly into their terminals on an anonymous basis, and these prices will be visible to all market participants. Another trader, anywhere in the world, can execute a trade by simply hitting two buttons.

The introduction of automated trading should reduce the cost of trading, partly by eliminating foreign exchange brokers. The new systems will replicate the order matching and the anonymity that brokers offer, and they will do it more cheaply. For example, small banks can now deal directly with each other instead of having to channel trades through larger ones. At the same time, automated systems threaten the oligopoly of information that has underpinned the profits of those who now do most foreign exchange business. These new systems gather and publish information on the prices and quantities of currencies as they are actually traded, thereby revealing details of currency trades that until now the traders have profitably kept to themselves.

The key to the widespread use of computerized foreign-currency trading systems is **liquidity**, as measured by the difference between the rates at which dealers can buy and sell currencies. Liquidity, in turn, requires reaching a critical mass of users. If enough dealers are putting their prices into the system, then users have greater assurance that the system will provide them with the best prices available. Absent this assurance, dealers will avoid the system.

Size

The foreign exchange market is by far the largest financial market in the world. A recent survey of the world's central banks by the Bank for International Settlements placed the average foreign exchange trading volume in 1995 at $1.2 trillion daily, or $300 trillion a year.[2] This figure compares with an average daily trading volume of about $7 billion on the New York Stock Exchange. Indeed, the New York Stock Exchange's biggest day,

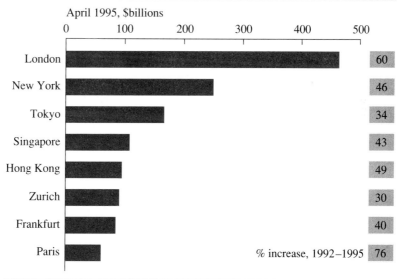

EXHIBIT 5.2A Daily Foreign Exchange Trading Volume by Financial Center

Source: Bank for International Settlements surveys.

---

[2] Survey results appeared in "66th Annual Report," Bank for International Settlements, Basel, June 10, 1996, p. 96.

EXHIBIT 5.2B Daily Global Foreign Exchange Trading Volume

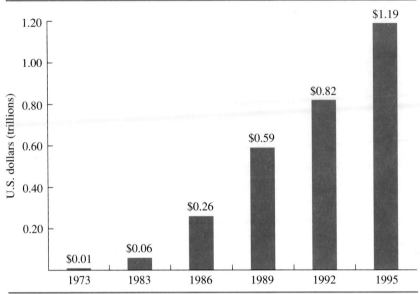

Source: Bank for International Settlements surveys.

Black Monday (October 19, 1987), was only $21 billion, or 4% of the daily foreign exchange volume. As another benchmark, the U.S. gross domestic product was approximately $7.3 trillion in 1995.

According to data from the survey by the Bank for International Settlements mentioned earlier, London is increasing its lead as the world's largest currency trading market, with daily turnover in 1995 estimated at $464 billion, more than that of the next two markets—New York, at $244 billion, and Tokyo, at $161 billion—combined.[3] Exhibit 5.2A shows that the eight biggest financial centers have seen trading volume grow by an average of 47% since the last survey was conducted—slightly faster than the 42% increase between 1989 and 1992—and far outpacing the growth of international trade and the world's output of goods and services. This explosive growth in currency trading—daily volume was estimated at $10 billion in 1973 (see Exhibit 5.2B)—has been attributed to the growing integration of the world's economies and financial markets, as well as a growing desire among companies and financial institutions to manage their currency risk exposure more actively. Dollar-sterling trades used to be the most common, but they have been overtaken by dollar-DM trading.

5.2 THE SPOT MARKET

This section examines the spot market in foreign exchange. It covers spot quotations, transaction costs, and the mechanics of spot transactions.

---

[3] Ibid.

Spot Quotations

Almost all major newspapers print a daily list of exchange rates. For major currencies, up to four different *foreign exchange quotes* (prices) are displayed. One is the *spot price*. The others might include the 30-day, 90-day, and 180-day *forward prices*. These quotes are for trades among dealers in the interbank market. When interbank trades involve dollars (about 60% of such trades do), these rates will be expressed in either **American terms** (numbers of U.S. dollars per unit of foreign currency) or **European terms** (number of foreign currency units per U.S. dollar). In the *Wall Street Journal*, quotes in American and European terms are listed side by side (see Exhibit 5.3). For example, on April 7, 1997,

EXHIBIT 5.3 Foreign Exchange Rate Quotations

EXCHANGE RATES
Monday, April 7, 1997

The New York foreign exchange selling rates below apply to trading among banks in amounts of $1 million and more, as quoted at 4 p.m. Eastern time by Dow Jones and other sources. Retail transactions provide fewer units of foreign currency per dollar.

Country	U.S.$ equiv. Mon	Fri	Currency per U.S.$ Mon	Fri
Argentina (Peso)	1.0012	1.0012	.9988	.9988
Australia (Dollar)	.7764	.7760	1.2880	1.2887
Austria (Schilling)	.08299	.08426	12.050	11.868
Bahrain (Dinar)	2.6525	2.6525	.3770	.3770
Belgium (Franc)	.02832	.02882	35.305	34.698
Brazil (Real)	.9412	.9451	1.0625	1.0581
Britain (Pound)	1.6277	1.6345	.6144	.6118
1-month forward	1.6271	1.6339	.6146	.6120
3-months forward	1.6253	1.6323	.6153	.6126
6-months forward	1.6225	1.6296	.6164	.6137
Canada (Dollar)	.7208	.7183	1.3873	1.3921
1-month forward	.7222	.7197	1.3846	1.3895
3-months forward	.7249	.7222	1.3795	1.3846
6-months forward	.7283	.7254	1.3730	1.3786
Chile (Peso)	.002401	.002400	416.55	416.60
China (Renminbi)	.1201	.1201	8.3266	8.3268
Colombia (Peso)	.0009411	.0009412	1062.58	1062.46
Czech. Rep. (Koruna)	....	....	....	....
Commercial rate	.03445	.03466	29.030	28.848
Denmark (Krone)	.1533	.1557	6.5235	6.4215
Ecuador (Sucre)	....	....	....	....
Floating rate	.0002621	.0002639	3815.00	3789.50
Finland (Markka)	.1958	.1987	5.1063	5.0327
France (Franc)	.1735	.1762	5.7630	5.6755
1-month forward	.1739	.1766	5.7517	5.6638
3-months forward	.1746	.1773	5.7277	5.6407
6-months forward	.1758	.1785	5.6890	5.6025
Germany (Mark)	.5838	.5932	1.7130	1.6859
1-month forward	.5850	.5944	1.7094	1.6824
3-months forward	.5875	.5969	1.7022	1.6752
6-months forward	.5915	.6011	1.6907	1.6637
Greece (Drachma)	.003709	.003750	269.62	266.69
Hong Kong (Dollar)	.1291	.1290	7.7475	7.7490
Hungary (Forint)	.005620	.005667	177.95	176.46
India (Rupee)	.02788	.02787	35.870	35.875
Indonesia (Rupiah)	.0004158	.0004160	2405.00	2404.00
Ireland (Punt)	1.5506	1.5699	.6449	.6370
Israel (Shekel)	.2956	.2967	3.3830	3.3700
Italy (Lira)	.0005928	.0005999	1687.00	1667.00
Japan (Yen)	.007960	.000044	125.63	124.31
1-month forward	.007996	.008079	125.06	123.78
3-months forward	.008067	.008153	123.97	122.66
6-months forward	.008184	.008268	122.19	120.96
Jordan (Dinar)	1.4094	1.4094	.7095	.7095
Kuwait (Dinar)	3.2949	3.3003	.3035	.3030
Lebanon (Pound)	.00066471	.0006471	1545.25	1545.25
Malaysia (Ringgit)	.4003	.4020	2.4982	2.4878
Malta (Lira)	2.6178	2.6420	.3820	.3785
Mexico (Peso)	....	....	....	....
Floating rate	.1267	.1259	7.8950	7.9410
Netherland (Guilder)	.5192	.5273	1.9260	1.8965
New Zealand (Dollar)	.6894	.6909	1.4505	1.4474
Norway (Krone)	.1436	.1461	6.9649	6.8423
Pakistan (Rupee)	.02520	.02520	39.680	39.680
Peru (new Sol)	.3783	.3783	2.6437	2.6437
Philippines (Peso)	.03793	.03792	26.367	36.370
Poland (Zloty)	.3225	.3252	3.1005	3.0755
Portugal (Escudo)	.005824	.005905	171.70	169.35
Russia (Ruble) (a)	.0001744	.0001743	5735.50	5736.00
Saudi Arabia (Riyal)	.2666	.2666	3.7505	3.7505
Singapore (Dollar)	.6942	.6959	1.4405	1.4370
Slovak Rep. (Koruna)	.3080	.3080	32.473	32.473
South Africa (Rand)	.2258	.2266	4.4285	4.4130
South Korea (Won)	.001120	.001120	893.15	892.65
Spain (Peseta)	.006911	.007023	144.70	142.38
Sweden (Krona)	.1298	.1314	7.7025	7.6125
Switzerland (Franc)	.6805	.6932	1.4695	1.4425
1-month forward	.6826	.6955	1.4649	1.4379
3-months forward	.6870	.7000	1.4555	1.4285
6-months forward	.6943	.7074	1.4403	1.4137
Taiwan (Dollar)	.03629	.03630	27.555	27.546
Thailand (Baht)	.03838	.03843	26.057	25.990
Turkey (Lira)	.00000772	.000781129	480.00	128030.00
United Arab (Dirham)	.2723	.2723	3.6720	3.6720
Uruguay (New Peso)	....	....	....	....
Financial	.1101	.1101	9.0850	9.0850
Venezuela (Bolivar)	.002095	.002095	477.37	477.43

SDR	1.3743	1.3834	.7276	.7229
ECU	1.1423	1.1590	....	....

Special Drawing Rights (SDR) are based on exchange rates for the U.S., German, British, French, and Japanese currencies. Source: International Monetary Fund.

European Currency Unit (ECU) is based on a basket of community currencies. a-fixing Moscow Interbank Currency Exchange.

The Wall Street Journal daily foreign exchange data for 1996 and 1997 may be purchased through the Readers' Reference Service (413) 592-3600.

the American quote for the Swiss franc was SFr 1 = $0.6805, and the European quote was $1 = SFr 1.4695. Nowadays, in trades involving dollars, all except U.K. and Irish exchange rates are expressed in European terms.

In their dealings with nonbank customers, banks in most countries use a system of **direct quotation**. A direct exchange rate quote gives the home currency price of a certain quantity of the foreign currency quoted (usually 100 units, but only one unit in the case of the U.S. dollar or the pound sterling). For example, the price of foreign currency is expressed in French francs (FF) in France and in Deutsche marks (DM) in Germany. Thus, in France, the Deutsche mark might be quoted at FF 4, whereas in Germany, the franc would be quoted at DM 0.25.

There are exceptions to this rule, though. Banks in Great Britain quote the value of the pound sterling (£) in terms of the foreign currency—for example, £1 = $1.4420. This method of **indirect quotation** is also used in the United States for domestic purposes and for the Canadian dollar. In their foreign exchange activities abroad, however, U.S. banks adhere to the European method of direct quotation.

American and European terms and direct and indirect quotes are related as follows:

American terms	**European terms**
U.S. dollar price per unit of foreign currency (for example, $0.009251/¥)	Foreign currency units per dollar (for example, ¥108.10/$)
A direct quote in the United States	A direct quote outside the United States
An indirect quote outside the United States	An indirect quote in the United States

Banks do not normally charge a commission on their currency transactions, but they profit from the spread between the buying and selling rates on both spot and forward transactions. Quotes are always given in pairs because a dealer usually does not know whether a prospective customer is in the market to buy or to sell a foreign currency. The first rate is the buy, or bid, price; the second is the sell, or ask, or offer, rate. Suppose the pound sterling is quoted at $1.4419-28. This quote means that banks are willing to buy pounds at $1.4419 and sell them at $1.4428. In practice, because time is money, dealers do not quote the full rate to each other; instead, they quote only the last two digits of the decimal. Thus, sterling would be quoted at 19–28 in the above example. Any dealer who is not sufficiently up-to-date to know the preceding numbers will not remain in business for long.

Note that when American terms are converted to European terms or direct quotations are converted to indirect quotations, bid and ask quotes are reversed; that is, the reciprocal of the American (direct) bid becomes the European (indirect) ask and the reciprocal of the American (direct) ask becomes the European (indirect) bid. So, in the previous example, the reciprocal of the American bid of $1.4419/£ becomes the European ask of £0.6935/$ and the reciprocal of the American ask of $1.4428/£ equals the European bid of £0.6931/$, resulting in a direct quote for the dollar in London of £0.6931-35. Note too that the banks will always buy low and sell high.

TRANSACTION COSTS ■ The **bid-ask spread**—that is, the spread between bid and ask rates for a currency—is based on the breadth and depth of the market for that currency as well as on the currency's volatility. The spread repays traders for the costs they incur in

currency dealing—including earning a profit on the capital tied up in their business—
and compensates them for the risks they bear. It is usually stated as a percentage cost of
transacting in the foreign exchange market, which is computed as follows:

$$\text{Percent spread} = \frac{\text{Ask price} - \text{Bid price}}{\text{Ask price}} \times 100$$

For example, with pound sterling quoted at \$1.4419-28, the percentage spread
equals 0.062%:

$$\text{Percent spread} = \frac{1.4428 - 1.4419}{1.4428} \times 100 = 0.062\%$$

For widely traded currencies, such as the pound, DM, Swiss franc, and yen, the
spread is on the order of 0.05%–0.08%.[4] Less heavily traded currencies, and currencies
having greater volatility, have higher spreads. In response to their higher spreads and
volatility (which increases the opportunity for profit when trading for the bank's own account), the large banks have expanded their trading in emerging market currencies, such
as the Czech koruna, Russian ruble, Turkish lira, and Zambian kwacha. Although these
currencies currently account for less than 5% of the global foreign exchange market, the
forecast is for rapid growth, in line with growing investment in emerging markets.

Mean-forward currency bid-ask spreads are larger than spot spreads, but they are
still small in absolute terms, ranging from 0.09% to 0.15% for actively traded currencies.
There is a growing forward market for emerging currencies, but owing to the thinness of
this market and its lack of liquidity, the bid-ask spreads are much higher.

The quotes found in the financial press are not those that individuals or firms would
get at a local bank. Unless otherwise specified, these quotes are for transactions in the interbank market exceeding \$1 million. (The standard transaction amount in the interbank
market is now about \$10 million.) But competition ensures that individual customers re-
ceive rates that reflect, even if they do not necessarily equal, interbank quotations. For example, a trader may believe that he or she can trade a little more favorably than the market rates indicate—that is, buy from a customer at a slightly lower rate or sell at a
somewhat higher rate than the market rate. Thus, if the current spot rate for the Swiss
franc is \$0.6967-72, the bank may quote a customer a rate of \$0.6964-75. On the other
hand, a bank that is temporarily short in a currency may be willing to pay a slightly more
favorable rate; or if the bank has overbought a currency, it may be willing to sell that currency at a somewhat lower rate.

For these reasons, many corporations will shop around at several banks for quotes
before committing themselves to a transaction. On large transactions, customers also may
get a rate break inasmuch as it ordinarily does not take much more effort to process a
large order than a small order.

The market for traveler's checks and smaller currency exchanges, such as might be
made by a traveler going abroad, is quite separate from the interbank market. The spread

---

[4] Data on mean spot and forward bid-ask spreads appear in Bessembinder, "Bid-Ask Spreads in the Interbank
Foreign Exchange Markets."

EXHIBIT 5.4 Key Currency Cross Rates

KEY CURRENCY CROSS RATES			**Late New York Trading Apr 7, 1997**							
	Dollar	Pound	SFranc	Guilder	Peso	Yen	Lira	D-Mark	FFranc	CdnDlr
Canada	1.3873	2.2581	.94406	.72030	.17572	.01104	.00082	.80987	.24073	....
France	5.7630	9.3804	3.9217	2.9922	.72996	.04587	.00342	3.3643	....	4.1541
Germany	1.7130	2.7883	1.1657	.88941	.21697	.01364	.00102	....	.29724	1.2348
Italy	1687.0	2745.9	1148.0	875.91	213.68	13.428	....	984.82	292.73	1216.0
Japan	125.63	204.49	85.492	65.228	15.913	....	.07447	73.339	21.799	90.557
Mexico	7.8950	12.851	5.3726	4.0992	....	.06284	.00468	4.6089	1.3699	5.6909
Netherlands	1.9260	3.1350	1.3106	....	4.2395	.01533	.00114	1.1243	.33429	1.3883
Switzerland	1.4695	2.3919	....	.76298	.18613	.01170	.00087	.85785	.25499	1.0593
U.K.	.61436	....	.41808	.31898	.07782	.00489	.00036	.35865	.10668	.44285
U.S.	....	1.6277	.68050	.51921	.12666	.00796	.00059	.58377	.17352	.72982

Source: Dow Jones

Source: Wall Street Journal, April 8, 1997, p. C12. Reprinted by permission of the *Wall Street Journal*, © 1997 Dow Jones & Company, Inc. All rights reserved worldwide.

on these smaller exchanges is much wider than that in the interbank market, reflecting the higher average costs banks incur on such transactions. As a result, individuals and firms involved in smaller retail transactions generally pay a higher price when buying, and they receive a lower price when selling foreign exchange than those quoted in newspapers.

CROSS RATES ■ Because most currencies are quoted against the dollar, it may be necessary to work out the **cross rates** for currencies other than the dollar. For example, if the Deutsche mark is selling for $0.60 and the buying rate for the French franc is $0.15, then the DM/FF cross rate is DM 1 = FF 4. A somewhat more complicated cross-rate calcula-

ILLUSTRATION

CALCULATING THE DIRECT QUOTE FOR THE POUND IN FRANKFURT

Suppose sterling is quoted at $1.4419-36, and the Deutsche mark is quoted at $0.6250-67. What is the direct quote for the pound in Frankfurt?

Solution. The bid rate for the pound in Frankfurt can be found by realizing that selling pounds for DM is equivalent to combining two transactions: (1) selling pounds for dollars at the bid rate of $1.4419, and (2) converting those dollars into DM 1.4419/0.6267 = DM 2.3008 per pound at the ask rate of $0.6267. Similarly,

the DM cost of buying one pound (the ask rate) can be found by first buying $1.4436 (the ask rate for £1) with DM and then using those dollars to buy one pound. Buying dollars for DM is equivalent to selling DM for dollars (at the bid rate of $0.6250), therefore, it will take DM 1.4436/0.6250 = DM 2.3098 to acquire the $1.4436 needed to buy one pound. Thus, the direct quotes for the pound in Frankfurt are DM 2.3008-98.

Note that in calculating the cross rates you should always assume that you have to sell a currency at the lower (or bid) rate and buy it at the higher (or ask) rate, giving you the worst possible rate. This method of quotation is how banks make money in foreign exchange.

tion would be the following. Suppose that we have these European quotes for the Japanese yen and the South Korean won:

Japanese yen: ¥135.62/U.S.$
South Korean won: W763.89/U.S.$

In this case, the cross rate of yen per won can be calculated by dividing the rate for the yen by the rate for the won, as follows:

$$\frac{\text{Japanese yen/U.S. dollar}}{\text{Korean won/U.S. dollar}} = \frac{\text{¥135.62/U.S.\$}}{\text{W763.89/U.S.\$}} = \text{¥0.17754/W}$$

Exhibit 5.4 contains cross rates for major currencies on April 7, 1997.

ILLUSTRATION

CALCULATING THE DIRECT QUOTE FOR THE BRAZILIAN REAL IN BANGKOK

Suppose that the Brazilian real is quoted at R 0.9955-1.0076/U.S.$ and the Thai baht is quoted at B 25.2513-3986. What is the direct quote for the real in Bangkok?

Solution. Analogous to the prior example, the direct bid rate for the real in Bangkok can be found by recognizing that selling reals in exchange for baht is equivalent to combining two transactions: (1) selling the real for dollars (which is the same as buying dollars with reals) at the ask rate of R 1.0076/U.S.$ and (2) selling those dollars for baht at the bid rate of B 25.2513/U.S.$. These transactions result in the bid cross rate for the real being the bid rate for the baht divided by the ask rate for the real:

$$\frac{\text{Bid cross rate}}{\text{for the real}} = \frac{\text{Bid rate for Thai baht/U.S.\$}}{\text{Ask rate for Brazilian real/U.S.\$}}$$

$$= \frac{25.2513}{1.0076} = \text{B } 25.0608/R$$

Similarly, the baht cost of buying the real (the ask cross rate) can be found by first buying dollars for Thai baht at the ask rate of B 25.3986/U.S.$ and then selling those dollars to buy Brazilian reals at the ask rate of R 0.9955/U.S.$. Combining these transactions yields the ask cross rate for the real being the ask rate for the baht divided by the bid rate for the real:

$$\frac{\text{Ask cross rate}}{\text{for the real}} = \frac{\text{Ask rate for Thai baht/U.S.\$}}{\text{Bid rate for Brazilian real/U.S.\$}}$$

$$= \frac{25.3986}{0.9955} = \text{B } 25.5134/R$$

Thus, the direct quotes for the real in Bangkok are B 25.0608-5134.

CURRENCY ARBITRAGE ■ Until recently, the pervasive practice among bank dealers was to quote all currencies against the U.S. dollar when trading among themselves. Now, however, about 40% of all currency trades do not involve the dollar, and that percentage is growing.[5] For example, Swiss banks may quote the Deutsche mark against the Swiss franc, and German banks may quote pounds sterling in terms of Deutsche marks.

---

[5] This estimate appeared in the *Wall Street Journal*, March 1, 1991, p. C1.

Exchange traders are continually alert to the possibility of taking advantage, through **currency arbitrage** transactions, of exchange rate inconsistencies in different money centers. These transactions involve buying a currency in one market and selling it in another. Such activities tend to keep exchange rates uniform in the various markets.

Currency arbitrage transactions also explain why such profitable opportunities are fleeting. In the process of taking advantage of an arbitrage opportunity, the buying and selling of currencies tends to move rates in a manner that eliminates the profit opportunity in the future. When profitable arbitrage opportunities disappear, we say that the **no-arbitrage condition** holds. If this condition were violated on an ongoing basis, we would wind up with a money machine, as shown in the following example.

For example, suppose the pound sterling is bid at $1.9809 in New York and the Deutsche mark is offered at $0.6251 in Frankfurt. At the same time, London banks are offering pounds sterling at DM 3.1650. An astute trader would sell dollars for Deutsche marks in Frankfurt, use the Deutsche marks to acquire pounds sterling in London, and sell the pounds in New York.

Specifically, if the trader begins in New York with $1 million, he can acquire DM 1,599,744.04 for $1,000,000 in Frankfurt, sell these Deutsche marks for £505,448.35 in London, and resell the pounds in New York for $1,001,242.64. Thus, a few minutes' work would yield a profit of $1,242.64. In effect, the trader would, by arbitraging through the DM, be able to acquire sterling at $1.9784 in London ($0.6251 × 3.1650) and sell it at $1.9809 in New York. This sequence of transactions, known as **triangular currency arbitrage**, is depicted in Exhibit 5.5.

EXHIBIT 5.5 Triangular Currency Arbitrage

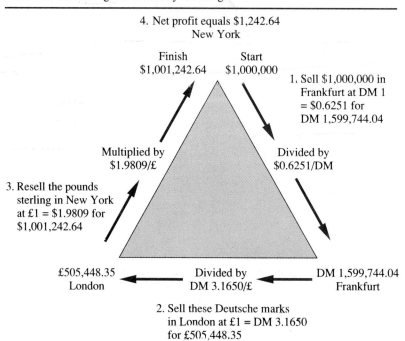

4. Net profit equals $1,242.64
New York

Finish
$1,001,242.64

Start
$1,000,000

1. Sell $1,000,000 in Frankfurt at DM 1 = $0.6251 for DM 1,599,744.04

Multiplied by
$1.9809/£

Divided by
$0.6251/DM

3. Resell the pounds sterling in New York at £1 = $1.9809 for $1,001,242.64

£505,448.35
London

Divided by
DM 3.1650/£

DM 1,599,744.04
Frankfurt

2. Sell these Deutsche marks in London at £1 = DM 3.1650 for £505,448.35

In the preceding example, the arbitrage transactions would tend to cause the Deutsche mark to appreciate vis-à-vis the dollar in Frankfurt and to depreciate against the pound sterling in London; at the same time, sterling would tend to fall in New York against the dollar. Acting simultaneously, these currency changes will quickly eliminate profits from this set of transactions, thereby enforcing the no-arbitrage condition. Otherwise, a money machine would exist, opening up the prospect of unlimited risk-free profits. As we have noted, such profits would quickly attract other traders, whose combined buying and selling activities would bring exchange rates back into equilibrium almost instantaneously.

Opportunities for profitable currency arbitrage have been greatly reduced in recent years, given the extensive network of people—aided by high-speed, computerized information systems—who are continually collecting, comparing, and acting on currency quotes in all financial markets. The practice of quoting rates against the dollar makes currency arbitrage even simpler. The result of this activity is that rates for a specific currency tend to be the same everywhere, with only minimal deviations due to transaction costs.

ILLUSTRATION

CALCULATING THE DIRECT QUOTE FOR THE DEUTSCHE MARK IN NEW YORK

If the direct quote for the dollar is DM 2.5 in Frankfurt, and transaction costs are 0.4%, what are the minimum and maximum possible direct quotes for the DM in New York?

Solution. The object here is to find the *no-arbitrage* range of DM quotes—that is, the widest bid-ask spread within which any potential arbitrage profits are eaten up by transaction costs. It can be found as follows. Begin with an arbitrageur who converts $1 into DM in Frankfurt. The arbitrageur would receive DM 2.5 × 0.996, after paying transaction costs of 0.4%.

Converting these DM into dollars in New York at a direct quote of e, the arbitrageur would keep $2.5 \times 0.996 \times e \times 0.996$. The no-arbitrage condition requires that this quantity must be less than or equal to $1 (otherwise there would be a money machine), or $e \leq 1/[2.5(0.996)^2] = \$0.4032$. Alternatively, an arbitrageur who converted $1 into DM in New York at a rate of e, took those DM to Frankfurt and exchanged them for dollars, would wind up—after paying transaction costs in both New York and Frankfurt—with $(1/e) \times 0.996 \times 1/2.5 \times 0.996$. Because the no-arbitrage condition requires that this quantity must not exceed $1, $(0.996)^2 \times (1/2.5e) \leq 1$, or $e \geq \$0.3968$. Combining these two inequalities yields $\$0.3968 \leq e \leq \$0.4032$.

SETTLEMENT DATE ■ The **value date** for spot transactions, the date on which the monies must be paid to the parties involved, is set as the second working day after the date on which the transaction is concluded. Thus, a spot deal entered into on Thursday in Paris will not be settled until the following Monday (French banks are closed on Saturdays and Sundays). It is possible, although unusual, to get one-day or even same-day value, but the rates will be adjusted to reflect interest differentials on the currencies involved.

EXCHANGE RISK ■ Bankers also act as marketmakers as well as agents, by taking positions in foreign currencies, thereby exposing themselves to **exchange risk**. The immediate adjustment of quotes as traders receive and interpret new political and economic information is the source of both exchange losses and gains by banks active in the foreign exchange market. For instance, suppose a trader quotes a rate of £1:$1.3012 for £500,000, and it is accepted. The bank will receive $650,600 in return for the £500,000. If the bank does not have an offsetting transaction, it may decide within a few minutes to cover its exposed position in the interbank market. If during this brief delay, news of a lower-than-expected British trade deficit reaches the market, the trader may be unable to purchase pounds at a rate lower than $1.3101. Because the bank would have to pay $655,050 to acquire £500,000 at this new rate, the result is a $4,450 ($655,050 − $650,600) exchange loss on a relatively small transaction within just a few minutes. Equally possible, of course, is a gain if the dollar strengthens against the pound.

Clearly, as a trader becomes more and more uncertain about the rate at which she can offset a given currency contract with other dealers or customers, she will demand a greater profit to bear this added risk. This expectation translates into a wider bid-ask spread. For example, during a period of volatility in the exchange rate between the French franc and U.S. dollar, a trader will probably quote a customer a bid for francs that is distinctly lower than the last observed bid in the interbank market; the trader will attempt to reduce the risk of buying francs at a price higher than that at which she can eventually resell them. Similarly, the trader may quote a price for the sale of francs that is above the current asking price.

The Mechanics of Spot Transactions

The simplest way to explain the process of actually settling transactions in the spot market is to work through an example. Suppose a U.S. importer requires HK$1 million to pay his Hong Kong supplier. After receiving and accepting a verbal quote from the trader of a U.S. bank, the importer will be asked to specify two accounts: (1) the account in a U.S. bank that he wants debited for the equivalent dollar amount at the agreed exchange rate—say, U.S.$0.1293 per Hong Kong dollar, and (2) the Hong Kong supplier's account that is to be credited by HK$1 million.

On completion of the verbal agreement, the trader will forward to the settlement section of her bank a dealing slip containing the relevant information. That same day, a *contract note*—which includes the amount of the foreign currency (HK$1 million), the dollar equivalent at the agreed rate ($129,300 = 0.1293 × 1,000,000), and confirmation of the payment instructions—will be sent to the importer. The settlement section will then cable the bank's correspondent (or branch) in Hong Kong, requesting transfer of HK$1 million from its *nostro account*—working balances maintained with the correspondent to facilitate delivery and receipt of currencies—to the account specified by the importer. On the value date, the U.S. bank will debit the importer's account, and the exporter will have his account credited by the Hong Kong correspondent.

At the time of the initial agreement, the trader provides a clerk with the pertinent details of the transaction. The clerk, in turn, constantly updates a *position sheet* that shows the bank's position by currency, as well as by maturities of forward contracts. A number of the major international banks have fully computerized this process to ensure accurate and instantaneous information on individual transactions and on the bank's cumulative currency exposure at any time. The head trader will monitor this information for evidence of possible fraud or excessive exposure in a given currency.

Spot transactions are normally settled (two working days later) so a bank is never certain until one or two days after the deal is concluded whether the payment due the bank has actually been made. To keep this credit risk in bounds, most banks will transact large amounts only with prime names (other banks or corporate customers).

A different type of credit risk is **settlement risk**, also known as **Herstatt risk**. Herstatt risk, named after a German bank that went bankrupt after losing a fortune speculating on foreign currencies, is the risk that a bank will deliver currency on one side of a foreign exchange deal only to find that its counterparty has not sent any money in return. This risk arises because of the way foreign currency transactions are settled. Settlement requires a cash transfer from one bank's account to another at the central banks of the currencies involved. However, because those banks may be in different time zones, there may be a delay. In the case of Herstatt, German regulators closed the bank after it had received Deutsche marks in Frankfurt but before it had delivered dollars to its counterparty banks (because the New York market had not yet opened).

Central banks have been slow to deal with this problem, so some banks have begun to pool their trades in a particular currency, canceling out offsetting ones and settling the balance at the end of the day. In early 1996, 17 of the world's biggest banks went further and announced plans for a global clearing bank that would operate 24 hours a day. If and when this proposal is implemented, banks would trade through the clearing bank, which would settle both sides of foreign exchange trades simultaneously, as long as the banks' accounts had sufficient funds.

5.3 THE FORWARD MARKET

Forward exchange operations carry the same credit risk as spot transactions but for longer periods of time; however, there are significant exchange risks involved.

A **forward contract** between a bank and a customer (which could be another bank) calls for delivery, at a fixed future date, of a specified amount of one currency against dollar payment; the exchange rate is fixed at the time the contract is entered into. Although the Deutsche mark is the most widely traded currency at present, active forward markets exist for the pound sterling, the Canadian dollar, the Japanese yen, and the major Continental currencies—particularly the Swiss franc, French franc, Belgian franc, Italian lira, and Dutch guilder. In general, forward markets for the currencies of less-developed countries (LDCs) are either limited or nonexistent.

In a typical forward transaction, for example, a U.S. company buys textiles from England with payment of £1 million due in 90 days. The importer, thus, is **short** pounds— that is, it owes pounds for future delivery. Suppose the spot price of the pound is $1.71. During the next 90 days, however, the pound might rise against the dollar, raising the dollar cost of the textiles. The importer can guard against this exchange risk by immediately negotiating a 90-day forward contract with a bank at a price of, say, £1 = $1.72. According to the forward contract, in 90 days the bank will give the importer £1 million (which it will use to pay for its textile order), and the importer will give the bank $1.72 million, which is the dollar equivalent of £1 million at the forward rate of $1.72.

In technical terms, the importer is offsetting a short position in pounds by going **long** in the forward market—that is, by buying pounds for future delivery. In effect, the use of the forward contract enables the importer to convert a short underlying position in pounds to a zero net exposed position, with the forward contract receipt of £1,000,000 canceling out the account payable of £1,000,000 and leaving the importer with a net liability of $1,720,000:

Importer's T-Account

Forward contract receipt	£1,000,000	Account payable	£1,000,000
		Forward contract payment	$1,720,000

According to this T-account, the forward contract allows the importer to convert an unknown dollar cost ($1,000,000 \times e_1$, where e_1 is the unknown spot exchange rate—$/£—in 90 days) into a known dollar cost ($1,720,000), thereby eliminating all exchange risk on this transaction.

Exhibit 5.6 plots the importer's dollar cost of the textile shipment with and without the use of a forward contract. It also shows the gain or loss on the forward contract as a function of the contracted forward price and the spot price of the pound when the contract matures.

The gains and losses from long and short forward positions are related to the difference between the contracted forward price and the spot price of the underlying currency at the time the contract matures. In the case of the textile order, the importer is committed to buy pounds at $1.72 apiece. If the spot rate in 90 days is less than $1.72, the importer will suffer an implicit loss on the forward contract because it is buying pounds for more than the prevailing value. However, if the spot rate in 90 days exceeds $1.72, the importer will enjoy an implicit profit because the contract obliges the bank to sell the pounds at a price less than current value.

EXHIBIT 5.6 Hedging a Future Payment with a Forward Contract

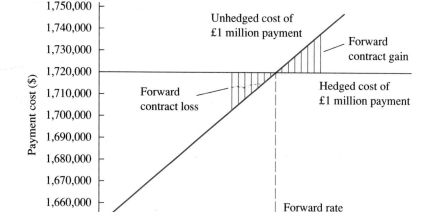

Three points are worth noting. First, the gain or loss on the forward contract is unrelated to the current spot rate of $1.71. Second, the forward contract gain or loss exactly offsets the change in the dollar cost of the textile order that is associated with movements in the pound's value. For example, if the spot price of the pound in 90 days is $1.75, the importer's cost of delivery is $1.75 million. However, the forward contract has a gain of $30,000, or 1,000,000 × (1.75 − 1.72). The net cost of the textile order when covered with a forward contract is $1.72 million, no matter what happens to the spot exchange rate in 90 days. (Chapter 9 elaborates on the use of forward contracts to manage exchange risk.) Third, the forward contract is not an option contract. Both parties must perform the agreed-on behavior, unlike the situation with an option in which the buyer can choose whether to exercise the contract or allow it to expire. The bank must deliver the pounds, and the importer must buy them at the prearranged price. Options are discussed in Chapter 6.

Forward Quotations

Forward rates can be expressed in two ways. Commercial customers are usually quoted the actual price, otherwise known as the outright rate. In the interbank market, however, dealers quote the forward rate only as a discount from, or a premium on, the spot rate. This forward differential is known as the **swap rate**. A foreign currency is at a **forward discount** if the forward rate expressed in dollars is below the spot rate, whereas a **forward premium** exists if the forward rate is above the spot rate. As we see in the next section, the forward premium or discount is closely related to the difference in interest rates on the two currencies.

According to Exhibit 5.3, spot Japanese yen on April 7, 1997, sold at $0.007960, whereas 180-day forward yen were priced at $0.008184. Based on these rates, the swap rate for the 180-day forward yen was quoted as a 224-point premium (0.008184 − 0.007960), where a point, or "pip," refers to the last digit quoted. Similarly, because the 90-day British pound was quoted at $1.6253 while the spot pound was $1.6277, the 90-day forward British pound sold at a 24-point discount. Alternatively, the discount or premium on the foreign currency may be expressed as an annualized percentage deviation from the spot rate using the following formula:

$$\frac{\text{Forward premium}}{\text{or discount}} = \frac{\text{Forward rate} - \text{Spot rate}}{\text{Spot rate}} \times \frac{360}{\text{Forward contract number of days}}$$

where the exchange rate is stated in domestic currency units per unit of foreign currency.

Thus, on April 7, 1997, the 180-day forward Japanese yen was selling at a 5.63% annualized premium:

$$\frac{\text{Forward premium}}{\text{annualized}} = \frac{0.008184 - 0.007960}{0.007960} \times \frac{360}{180} = 0.0563$$

The 90-day British pound was selling at a 0.59% annualized discount:

$$\frac{\text{Forward discount}}{\text{annualized}} = \frac{1.6253 - 1.6277}{1.6277} \times \frac{360}{90} = -0.0059$$

A swap rate can be converted into an outright rate by adding the premium (in points) to, or subtracting the discount (in points) from, the spot rate. Although the swap rates do not carry plus or minus signs, you can determine whether the forward rate is at a discount or a premium using the following rule: When the forward bid in points is smaller than the ask rate in points, the forward rate is at a premium and the points should be added to the spot price to compute the outright quote. Conversely, if the bid in points exceeds the ask in points, the forward rate is at a discount and the points must be subtracted from the spot price to get the outright quotes.[6]

Suppose the following quotes are received for spot, 30-day, 90-day, and 180-day Swiss francs (SFr) and pounds sterling:

Spot		30-day	90-day	180-day
£:$2.0015-30		19-17	26-22	42-35
SFr:$0.6963-68		4-6	9-14	25-38

Bearing in mind the practice of quoting only the last two digits, a dealer would quote sterling at 15-30, 19-17, 26-22, 42-35 and Swiss francs at 63-68, 4-6, 9-14, 25-38.

The outright rates are shown in the following chart.

MATURITY	£			SFR		
	Bid	Ask	Spread (%)	Bid	Ask	Spread (%)
Spot	$2.0015	$2.0030	0.075	$0.6963	$0.6968	0.072
30-day	1.9996	2.0013	0.085	0.6967	0.6974	0.100
90-day	1.9989	2.0008	0.095	0.6972	0.6982	0.143
180-day	1.9973	1.9995	0.110	0.6988	0.7006	0.257

Thus, the Swiss franc is selling at a premium against the dollar and the pound is selling at a discount. Note the slightly wider percentage spread between outright bid and ask on the Swiss franc compared with the spread on the pound. This difference is due to the broader market in pounds. Note too the widening of spreads by maturity for both currencies. This widening is caused by the greater uncertainty surrounding future exchange rates.

EXCHANGE RISK ■ Spreads in the forward market are a function of both the breadth of the market (volume of transactions) in a given currency and the risks associated with forward contracts. The risks, in turn, are based on the variability of future spot rates. Even if the spot market is stable, there is no guarantee that future rates will remain invariant. This uncertainty will be reflected in the forward market. Furthermore, because beliefs

---

[6]This rule is based on two factors: (1) The buying rate, be it for spot or forward delivery, is always less than the selling price; and (2) the forward bid-ask spread always exceeds the spot bid-ask spread. In other words, you can always assume that the bank will be buying low and selling high and that bid-ask spreads widen with the maturity of the contract.

about distant exchange rates are typically less secure than those about nearer-term rates, uncertainty will increase with lengthening maturities of forward contracts. Dealers will quote wider spreads on longer-term forward contracts to compensate themselves for the risk of being unable to reverse their positions profitably. Moreover, the greater unpredictability of future spot rates may reduce the number of market participants. This increased thinness will further widen the bid-ask spread because it magnifies the dealer's risk in taking even a temporary position in the forward market.

CROSS RATES ■ Forward cross rates are figured in much the same way as spot cross rates. For instance, suppose a customer wants to sell 30-day forward Italian lire (Lit) against Dutch guilder (Dfl) delivery. The market rates (expressed in European terms of foreign currency units per dollar) are as follows:

Lit:$ spot	1,890.00–1,892.00
30-day forward	1,894.25–1,897.50
Dfl:$ spot	3.4582–3.4600
30-day forward	3.4530–3.4553

Based on these rates, the forward cross rate for selling lire against (to buy) guilders is found as follows: Forward lire are sold for dollars—that is, dollars are bought at the lira forward selling price of Lit 1,897.50 = $1—and the dollars to be received are simultaneously sold for 30-day forward guilders at a rate of Dfl 3.4530. Thus, Lit 1,897.50 = Dfl 3.4530, or the forward selling price for lire against guilders is Lit 1,897.50/3.4530 = Lit 549.52. Similarly, the forward buying rate for lire against guilders is Lit 1,894.25/3.4553 = Lit 548.22. The spot selling rate is Lit 1,892.0/3.4582 = Lit 547.11. Hence, based on its ask price, the guilder is trading at an annualized 5.29% premium against the lire in the 30-day forward market:

$$\begin{array}{c} \text{Forward premium} \\ \text{annualized} \end{array} = \frac{549.52 - 547.11}{547.11} \times \frac{360}{30} = 5.29\%$$

Forward Contract Maturities

Forward contracts are normally available for 30-day, 60-day, 90-day, 180-day, or 360-day delivery. Banks will also tailor forward contracts for odd maturities (e.g., 77 days) to meet their customers' needs. Longer-term forward contracts can usually be arranged for widely traded currencies, such as the pound sterling, Deutsche mark, or Japanese yen; however, the bid-ask spread tends to widen for longer maturities. As with spot rates, these spreads have widened for almost all currencies since the early 1970s, probably because of the greater turbulence in foreign exchange markets. For widely traded currencies, the 90-day bid-ask spread can vary from 0.1% to 1%.

5.4 INTEREST RATE PARITY THEORY

Spot and forward rates are closely linked to each other and to interest rates in different currencies through the medium of arbitrage. Specifically, the movement of funds between two currencies to take advantage of *interest rate differentials* is a major determinant of

EXHIBIT 5.7 An Example of Interest Rate Parity

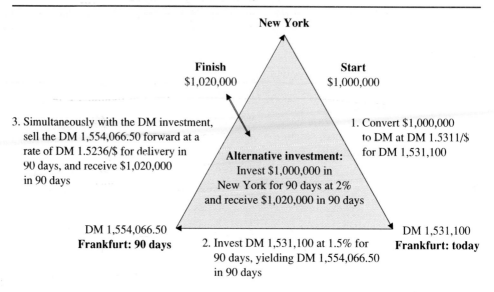

New York

Finish
$1,020,000

Start
$1,000,000

3. Simultaneously with the DM investment, sell the DM 1,554,066.50 forward at a rate of DM 1.5236/$ for delivery in 90 days, and receive $1,020,000 in 90 days

Alternative investment:
Invest $1,000,000 in
New York for 90 days at 2%
and receive $1,020,000 in 90 days

1. Convert $1,000,000 to DM at DM 1.5311/$ for DM 1,531,100

DM 1,554,066.50
Frankfurt: 90 days

2. Invest DM 1,531,100 at 1.5% for 90 days, yielding DM 1,554,066.50 in 90 days

DM 1,531,100
Frankfurt: today

the spread between forward and spot rates. In fact, the forward discount or premium is closely related to the interest differential between the two currencies.

According to *interest rate parity* (IRP) theory, the currency of the country with a lower interest rate should be at a forward premium in terms of the currency of the country with the higher rate. More specifically, in an efficient market with no transaction costs, the interest differential should be (approximately) equal to the forward differential. When this condition is met, the forward rate is said to be at **interest rate parity**, and equilibrium prevails in the money markets.

Interest parity ensures that the return on a hedged (or "covered") foreign investment will just equal the domestic interest rate on investments of identical risk, thereby eliminating the possibility of having a money machine. When this condition holds, the **covered interest differential** — the difference between the domestic interest rate and the hedged foreign rate — is zero. To illustrate this condition, suppose an investor with $1,000,000 to invest for 90 days is trying to decide between investing in U.S. dollars at 8% per annum (2% for 90 days) or in DM at 6% per annum (1.5% for 90 days). The current spot rate is DM 1.5311/$ and the 90-day forward rate is DM 1.5236/$. Exhibit 5.7 shows that regardless of the investor's currency choice, his hedged return will be identical. Specifically, $1,000,000 invested in dollars for 90 days will yield $1,000,000 × 1.02 = $1,020,000. Alternatively, if the investor chooses to invest in DM on a hedged basis, he will

1. Convert the $1,000,000 to DM at the spot rate of DM 1.5311/$. This yields DM 1,531,100 available for investment.

2. Invest the principal of DM 1,531,100 at 1.5% for 90 days. At the end of 90 days, the investor will have DM 1,554,066.50.

3. Simultaneously with the other transactions, sell the DM 1,554,066.50 in principal plus interest forward at a rate of DM 1.5236/$ for delivery in 90 days. This transaction will yield DM 1,554,066.50/1.5236 = $1,020,000 in 90 days.

If the covered interest differential between two money markets is nonzero, there is an arbitrage incentive to move money from one market to the other. This movement of money to take advantage of a covered interest differential is known as **covered interest arbitrage**.

ILLUSTRATION

COVERED INTEREST ARBITRAGE BETWEEN LONDON AND NEW YORK

Suppose the interest rate on pounds sterling is 12% in London, and the interest rate on a comparable dollar investment in New York is 7%. The pound spot rate is $1.75 and the one-year forward rate is $1.68. These rates imply a forward discount on sterling of 4% [(1.68 − 1.75)/ 1.75] and a covered yield on sterling approximately equal to 8% (12% − 4%). Because there is a covered interest differential in favor of London, funds will flow from New York to London.

To illustrate the profits associated with covered interest arbitrage, we will assume that the borrowing and lending rates are identical and the bid-ask spread in the spot and forward markets is zero. Here are the steps the arbitrageur can take to profit from the discrepancy in rates based on a $1 million transaction. Specifically,

as shown in Exhibit 5.8, the arbitrageur will

1. Borrow $1,000,000 in New York at an interest rate of 7%. This means that at the end of one year, the arbitrageur must repay principal plus interest of $1,070,000.
2. Immediately convert the $1,000,000 to pounds at the spot rate of £1 = $1.75. This yields £571,428.57 available for investment.
3. Invest the principal of £571,428.57 in London at 12% for one year. At the end of the year, the arbitrageur will have £640,000.
4. Simultaneously with the other transactions, sell the £640,000 in principal plus interest forward at a rate of £1 = $1.68 for delivery in one year. This transaction will yield $1,075,200 next year.
5. At the end of the year, collect the £640,000, deliver it to the bank's foreign exchange department in return for $1,075,200, and use $1,070,000 to repay the loan. The arbitrageur will earn $5,200 on this set of transactions.

EXHIBIT 5.8 An Example of Covered Interest Arbitrage

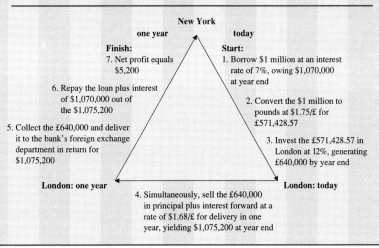

The transactions associated with covered interest arbitrage will affect prices in both the money and foreign exchange markets. In the previous example, as pounds are bought spot and sold forward, boosting the spot rate and lowering the forward rate, the forward discount will tend to widen. Simultaneously, as money flows from New York, interest rates there will tend to increase; at the same time, the inflow of funds to London will depress interest rates there. The process of covered interest arbitrage will continue until interest parity is achieved, unless there is government interference.

If this process is interfered with, covered interest differentials between national money markets will not be arbitraged away. Interference often occurs because many governments regulate and restrict flows of capital across their borders. Moreover, just the risk of controls will be sufficient to yield prolonged deviations from interest rate parity.

The relationship between the spot and forward rates and interest rates in a free market can be shown graphically, as in Exhibit 5.9. Plotted on the vertical axis is the interest differential in favor of the home country. The horizontal axis plots the percentage forward discount (negative) or premium (positive) on the foreign currency relative to the home currency. The interest parity line joins those points for which the forward exchange rate is in equilibrium with the interest differential. For example, if the interest differential in favor of the foreign country is 2%, the currency of that country must be selling at a 2% forward discount for equilibrium to exist.

Point G indicates a situation of disequilibrium. Here, the interest differential is 2%, whereas the forward premium on the foreign currency is 3%. The transfer of funds abroad with exchange risks covered will yield an additional 1% annually. At point H, the forward premium remains at 3%, but the interest differential increases to 4%. Now reversing the flow of funds becomes profitable. The 4% higher interest rate more than makes up for the

EXHIBIT 5.9 Interest Rate Parity Theory

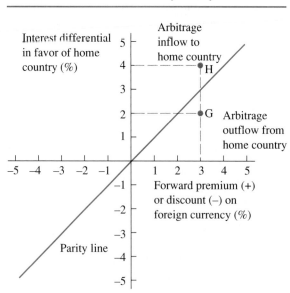

3% loss on the forward exchange transaction, leading to a 1% increase in the interest yield.

In reality, the interest parity line is a band because transaction costs, arising from the spread on spot and forward contracts and brokerage fees on security purchases and sales, cause effective yields to be lower than nominal yields. For example, if transaction costs are 0.75%, a covered yield differential of only 0.5% will not be sufficient to induce a flow of funds. For interest arbitrage to occur, the covered differential must exceed the transaction costs involved.

The covered interest arbitrage relationship can be stated formally. Let e_0 be the current spot rate (dollar value of one unit of foreign currency), and f_1 the end-of-period forward rate. If r_h and r_f are the prevailing interest rates in New York and, say, London, respectively, then one dollar invested in New York will yield $1 + r_h$ at the end of the period. The same dollar invested in London will be worth $(1 + r_f)f_1/e_0$ dollars at maturity. This latter result can be seen as follows: One dollar will convert into $1/e_0$ pounds that, when invested at r_f, will yield $(1 + r_f)/e_0$ pounds at the end of the period. By selling the proceeds forward today, this amount will be worth $(1 + r_f)f_1/e_0$ dollars when the investment matures.

Funds will flow from New York to London if and only if

$$1 + r_h < \frac{(1 + r_f)f_1}{e_0}$$

Conversely, funds will flow from London to New York if and only if

$$1 + r_h > \frac{(1 + r_f)f_1}{e_0}$$

Interest rate parity holds when there are no covered interest arbitrage opportunities. On the basis of the previous discussion, this no-arbitrage condition can be stated as follows:

$$\frac{1 + r_h}{1 + r_f} = \frac{f_1}{e_0} \tag{5.1}$$

ILLUSTRATION

USING INTEREST RATE PARITY TO CALCULATE THE $/¥ FORWARD RATE The interest rate in the United States is 10%; in Japan, the comparable rate is 7%. The spot rate for the yen is $0.003800. If interest rate parity holds, what is the 90-day forward rate?

Solution. According to IRP, the 90-day forward rate on the yen, f_{90}, should be $0.003828:

$$f_{90} = \$0.003800 \times \frac{1 + (0.10/4)}{1 + (0.07/4)} = \$0.003828$$

In other words, the 90-day forward Japanese yen should be selling at an annualized premium of about 2.95% [$4 \times (0.003828 - 0.003800)/0.0038$].

Interest rate parity is often approximated by Equation 5.2[7]:

$$r_h - r_f = \frac{f_1 - e_0}{e_0} \qquad (5.2)$$

In effect, interest rate parity says that *high interest rates on a currency are offset by forward discounts and that low interest rates are offset by forward premiums.*

Transaction costs in the form of bid-ask spreads make the computations more difficult, but the principle is the same: Compute the covered interest differential to see whether there is an arbitrage opportunity.

ILLUSTRATION   COMPUTING THE COVERED INTEREST DIFFERENTIAL WHEN TRANSACTION COSTS EXIST

Suppose the annualized interest rate on 180-day dollar deposits is 6 7/16-5/16%, meaning that dollars can be borrowed at 6 7/16% (the ask rate) and lent at 6 5/16% (the bid rate). At the same time, the annualized interest rate on 180-day Belgian franc deposits is 9 3/8-1/8%. Spot and 180-day forward quotes on Belgian francs are BF 31.5107-46/$ and BF 32.1027-87/$, respectively. Is there an arbitrage opportunity? Compute the profit using BF 10,000,000.

Solution. The only way to determine whether an arbitrage opportunity exists is to examine the two possibilities: Borrow dollars and lend Belgian francs or borrow francs and lend dollars, both on a hedged basis. The key is to ensure that you are using the correct bid or ask interest and exchange rates. In this case, it turns out that there is an arbitrage opportunity from borrowing Belgian francs and lending dollars. The specific steps to implement this arbitrage are as follows:

1. Borrow BF 10,000,000 at the ask rate of 9 3/8% for 180 days. This interest rate translates into a 180-day rate of 0.09375/2 = 4.6875%, requiring repayment of BF 10,468,750 in principal plus interest at the end of 180 days.

2. Immediately convert the BF 10,000,000 to dollars at the spot ask rate of BF 31.5146/$ (the BF cost of buying dollars spot). This yields $317,313.25 ($10,000,000/31.5146) available for investment.

3. Invest the principal of $317,313.25 at 0.063125/2 = 3.15625% for 180 days. In six months, this investment will have grown to $327,328.44 ($317,313.25 × 1.0315625).

4. Simultaneously with the other transactions, sell the $327,328.44 in principal plus interest forward at the bid rate of BF 32.1027 (the rate at which dollars can be converted into BF) for delivery in 180 days. This transaction will yield BF 10,508,126.86 in 180 days.

5. At the end of six months, collect the $327,328.44, deliver it to the bank's foreign exchange department in return for BF 10,508,126.86, and use BF 10,468,750 of the proceeds to repay the loan. The gain on this set of transactions is BF 39,376.86.

---

[7] Subtracting 1 from both sides of Equation 5.1 yields $(f_1 - e_0)/e_0 = (r_h - r_f)/(1 + r_f)$. Equation 5.2 follows if r_f is relatively small.

Empirical Evidence

Interest rate parity is one of the best-documented relationships in international finance. In fact, in the Eurocurrency markets, the forward rate is calculated from the interest differential between the two currencies using the no-arbitrage condition. Deviations from interest parity do occur between national capital markets, however, owing to capital controls (or the threat of them), the imposition of taxes on interest payments to foreigners, and transaction costs. However, as we can see in Exhibit 5.10, these deviations tend to be small and short-lived.

EXHIBIT 5.10 Uncovered and Covered Interest Rate Differentials (U.S. $ Versus Other Currencies)

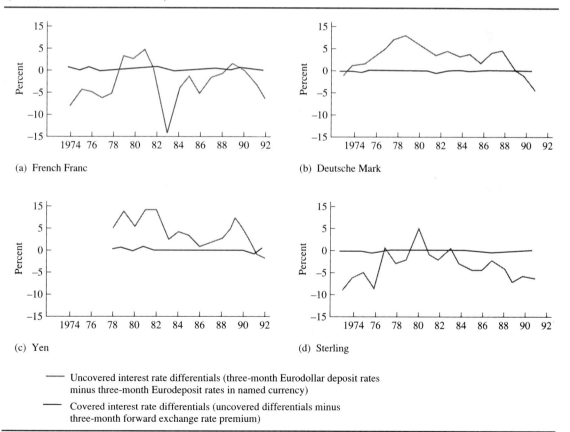

(a) French Franc

(b) Deutsche Mark

(c) Yen

(d) Sterling

——— Uncovered interest rate differentials (three-month Eurodollar deposit rates minus three-month Eurodeposit rates in named currency)

——— Covered interest rate differentials (uncovered differentials minus three-month forward exchange rate premium)

Source: Data from DRI, IMF, and Datastream International reported in "A Survey of the World Economy," *The Economist,* September 19, 1992, p. 23. © 1993 The Economist Newspaper Group, Inc. Used with permission.

⚓ 5.5 SUMMARY AND CONCLUSIONS

In this chapter, we saw that the primary function of the foreign exchange market is to transfer purchasing power denominated in one currency to another and thereby facilitate international trade and investment. The foreign exchange market consists of two tiers: the interbank market, in which major banks trade with each other, and the retail market, in which banks deal with their commercial customers.

In the spot market, currencies are traded for settlement within two business days after the transaction has been concluded. In the forward market, contracts are made to buy or sell currencies for future delivery. Spot and forward quotations are given either in American terms—the dollar price of a foreign currency—or in European terms—the foreign currency price of a dollar. Quotations can also be expressed on a direct basis—the home currency price of another currency—or an indirect basis—the foreign currency price of the home currency.

The major participants in the forward market are categorized as arbitrageurs, traders, hedgers, and speculators. Forward rates can be stated on an outright basis, or as a discount from, or a premium on, the spot rate. This forward differential is known as the swap rate. Because all currencies are quoted against the dollar, the exchange rate between two nondollar currencies—known as a cross rate—must be calculated on the basis of their direct quotes against the dollar.

Finally, we examined the links between the spot and forward markets. We saw that spot and forward rates and interest rates in different currencies are related to each other via covered interest arbitrage, which leads to the interest rate parity condition:

$$\frac{1 + r_h}{1 + r_f} = \frac{f_1}{e}$$

where r_h and r_f are the home and foreign interest rates, respectively, e_0 is the spot value of one unit of foreign currency, and f_1 is the forward rate for delivery of one unit of foreign currency at time 1. In effect, interest rate parity says that the forward premium or discount is approximately equal to the interest differential between the home and foreign currencies.

➤ Questions

1. Answer the following questions using the data in Exhibit 5.3.
   a. How many Swiss francs can you get for one dollar?
   b. How many dollars can you get for one Swiss franc?
   c. What is the three-month forward rate for the Swiss franc?
   d. Is the Swiss franc selling at a forward premium or discount?
   e. What is the 90-day forward discount or premium on the Swiss franc?

2. What risks confront dealers in the foreign exchange market? How can they cope with those risks?

3. Suppose a currency increases in volatility. What is likely to happen to its bid-ask spread? Why?

4. Who are the principal users of the forward market? What are their motives?

5. How does a company pay for the foreign exchange services of a commercial bank?

6. What factors might lead to persistent covered interest arbitrage opportunities among countries?

7. How have forward premiums and discounts relative to the dollar changed over annual intervals during the past five years for the Japanese yen, British pound, French franc, and Deutsche mark? Use beginning-of-year data.

➤ Problems

1. The $:DM exchange rate is DM 1 = $0.35, and the DM:FF exchange rate is FF 1 = DM 0.31. What is the FF:$ exchange rate?

2. Suppose the direct quote for sterling in New York is 1.1110-5.
   a. How much would £500,000 cost in New York?
   b. What is the direct quote for dollars in London?

3. Suppose the quote on pounds is $1.624-31.
   a. If you converted $10,000 to pounds and then back to dollars, how many dollars would you end up with?
   b. Suppose you could buy pounds at the bid rate and sell them at the ask rate. How many dollars would you have to transact in order to earn $1,000 on a roundtrip transaction (buying pounds for dollars and then selling the pounds for dollars)?

4. Using the data in Exhibit 5.3, calculate the 30-day, 90-day, and 180-day forward discounts for the British pound.

5. An investor wishes to buy French francs spot (at $0.1080) and sell French francs forward for 180 days (at $0.1086).
   a. What is the swap rate on French francs?
   b. What is the premium on 180-day French francs?

6. The spot and 90-day forward rates for the pound are $1.1376 and $1.1350, respectively. What is the forward premium or discount on the pound?

7. Suppose Credit Suisse quotes spot and 90-day forward rates of $0.7957-60, 8-13.
   a. What are the outright 90-day forward rates that Credit Suisse is quoting?
   b. What is the forward discount or premium associated with buying 90-day Swiss francs?
   c. Compute the percentage bid-ask spreads on spot and forward Swiss francs.

8. Suppose the spot quote on the Deutsche mark is $0.3302-10, and the spot quote on the French franc is $0.1180-90.
   a. Compute the percentage bid-ask spreads on the DM and franc.
   b. What is the direct spot quote for the franc in Frankfurt?

9. Suppose Dow Chemical receives quotes of $0.009369-71 for the yen and $0.03675-6 for the Taiwan dollar (NT$).
   a. How many U.S. dollars will Dow Chemical receive from the sale of ¥50 million?
   b. What is the U.S. dollar cost to Dow Chemical of buying ¥1 billion?
   c. How many NT$ will Dow Chemical receive for U.S.$500,000?
   d. How many yen will Dow Chemical receive for NT$200 million?
   e. What is the yen cost to Dow Chemical of buying NT$80 million?

10. Suppose you observe the following direct spot quotations in New York and Toronto, respectively: 0.8000-50 and 1.2500-60. What are the arbitrage profits per $1 million?

11. Suppose the DM is quoted at 0.2074-80 in London, and the pound sterling is quoted at 4.7010-32 in Frankfurt.
    a. Is there a profitable arbitrage situation? Describe it.
    b. Compute the percentage bid-ask spreads on the pound and DM.

12. Assuming no transaction costs, suppose £1 = $2.4110 in New York, $1 = FF 3.997 in Paris, and FF 1 = £0.1088 in London. How could you take profitable advantage of these rates?

13. As a foreign exchange trader at Sumitomo Bank, one of your customers would like spot and 30-day forward yen quotes on Australian dollars. Current market rates are

Spot	30-DAY
¥101.37-85/U.S.$1	15-13
A$1.2924-44/U.S.$1	20-26

   a. What bid and ask yen cross rates would you quote on spot Australian dollars?
   b. What outright yen cross rates would you quote on 30-day forward Australian dollars?
   c. What is the forward premium or discount on buying 30-day Australian dollars against yen delivery?

14. Suppose Air France receives the following indirect quotes in New York: FF 5.72-6 and £ 0.63-4. Given these quotes, what range of £/FF bid and ask quotes in Paris will permit arbitrage?

15. Suppose the DM is quoted at $0.5782-92, and the yen is quoted at $0.001760-69.
    a. Given these quotes for the DM and yen, what is the maximum bid-ask spread in the ¥/DM rate for which there is no arbitrage?
    b. What is the maximum bid-ask spread in percentage terms?

16. Assume the pound sterling is worth FF 9.80 in Paris and SFr 5.40 in Zurich.
    a. Show how British arbitrageurs can make profits given that the Swiss franc is worth two French francs. What would be the profit per pound transacted?
    b. What would be the eventual outcome on exchange rates in Paris and Zurich given these arbitrage activities?
    c. Rework Part a, assuming that transaction costs amount to 0.6% of the amount transacted. What would be the profit per pound transacted?
    d. Suppose the Swiss franc is quoted at FF 2 in Zurich. Given a transaction cost of 0.6% of the amount transacted, what are the minimum/maximum French franc prices for the Swiss franc that you would expect to see quoted in Paris?

17. Examine the following currency cross rates. The quotes are expressed as units of the currency represented in the left-hand column per unit of currency shown in the top row (e.g., FF 3.3818-25/DM). Consider the dollar rates to be the quotes off which the cross rates are set. That is, they are the benchmark quotes.

	DM	FF	£
DM	—	0.29570-76	2.4256-67
FF	3.3818-25	—	8.2031-41
£	0.41227-35	0.12381-90	—
¥	78.381-496	23.178-251	190.121-390

	¥	U.S.$
DM	0.01276-78	1.5780-86
FF	0.04315-19	5.3021-33
£	0.00526-29	0.6502-10
¥	—	123.569-707

    a. Do any triangular arbitrage opportunities exist among these currencies? Assume that any deviations from the theoretical cross rates of 5 points or less are due to transaction costs.
    b. How much profit could be made from a $5 million transaction associated with each arbitrage opportunity?

18. Assume that the interest rate is 11% on pounds sterling and 8% on Deutsche marks. If the Deutsche mark is selling at a one-year forward premium of 4% against the pound, is there an arbitrage opportunity? Explain.

19. Suppose the Eurosterling rate is 15%, and the Eurodollar rate is 11.5%. What is the forward premium on the dollar? Explain.

20. If the Swiss franc is $0.68 on the spot market and the 180-day forward rate is $0.70, what is the annualized interest rate in the United States over the next six months? The annualized interest rate in Switzerland is 2%.

21. The interest rate in the United States is 8%; in Japan the comparable rate is 2%. The spot rate for the yen is $0.007692. If interest rate parity holds, what is the 90-day forward rate on the Japanese yen?

22. Suppose the spot rates for the Deutsche mark, pound sterling, and Swiss franc are $0.32, $1.13, and $0.38, respectively. The associated 90-day interest rates (annualized) are 8%, 16%, and 4%; the U.S. 90-day rate (annualized) is 12%. What is the 90-day forward rate on an ACU (ACU 1 = DM 1 + £1 + SFr 1) if interest parity holds?

23. Suppose that three-month interest rates (annualized) in Japan and the United States are 7% and 9%, respectively. If the spot rate is ¥142:$1 and the 90-day forward rate is ¥139:$1,
    a. Where would you invest?
    b. Where would you borrow?
    c. What arbitrage opportunity do these figures present?
    d. Assuming no transaction costs, what would be your arbitrage profit per dollar or dollar-equivalent borrowed?

24. Here are some prices in the international money markets:

$$\text{Spot rate} = \$0.75 : \text{DM}$$
$$\text{Forward rate (one year)} = \$0.77 : \text{DM}$$
$$\text{Interest rate (DM)} = 7\% \text{ per year}$$
$$\text{Interest rate (\$)} = 9\% \text{ per year}$$

    a. Assuming no transaction costs or taxes exist, do covered arbitrage profits exist in the above situation? Describe the flows.
    b. Suppose now that transaction costs in the foreign exchange market equal 0.25% per transaction. Do unexploited covered arbitrage profit opportunities still exist?
    c. Suppose no transaction costs exist. Let the capital gains tax on currency profits equal 25% and the ordinary income tax on interest income equal 50%. In this situation, do covered arbitrage profits exist? How large are they? Describe the transactions required to exploit these profits.

25. On checking the Telerate screen, you see the following exchange rate and interest rate quotes:

    a. Can you find an arbitrage opportunity?
    b. What steps must you take to capitalize on it?
    c. What is the profit per $1,000,000 arbitraged?

CURRENCY	90-DAY INTEREST RATES (ANNUALIZED)	SPOT RATES	90-DAY FORWARD RATES
Dollar	4.99%–5.03%		
Swiss franc	3.14%–3.19%	$0.711-22	$0.726-32

26. On checking the Reuters screen, you see the following exchange rate and interest rate quotes:

    a. Can you find an arbitrage opportunity?
    b. What steps must you take to capitalize on it?
    c. What is the profit per £1,000,000 arbitraged?

CURRENCY	90-DAY INTEREST RATES	SPOT RATES	90-DAY FORWARD RATES
Pound	7 7/16–5/16%	¥159.9696-9912/£	¥145.5731-8692/£
Yen	2 3/8–1/4%		

➤ Bibliography

BESSEMBINDER, HENDRIK. "Bid-Ask Spreads in the Interbank Foreign Exchange Markets." *Journal of Financial Economics*, June 1994, pp. 317–348.

CHRYSTAL, K. ALEC. "A Guide to Foreign Exchange Markets." *Federal Reserve Bank of St. Louis Review*, March 1984, pp. 5–18.

FAMA, EUGENE. "Forward and Spot Exchange Rates." *Journal of Monetary Economics*, v. 14, 1984, pp. 319–338.

GLASSMAN, DEBRA. "Exchange Rate Risk and Transaction Costs: Evidence from Bid-Ask Spreads." *Journal of International Money and Finance*, December 1987, pp. 479–491.

KUBARYCH, ROGER M. *Foreign Exchange Markets in the United States*. New York: Federal Reserve Bank of New York, 1983.

STRONGIN, STEVE. "International Credit Market Connections." *Economic Perspectives*, July/August 1990, pp. 2–10.

CURRENCY FUTURES AND OPTIONS MARKETS

I dipt into the future far as human eye could see, Saw the vision of the world and all the wonder that would be.

Alfred, Lord Tennyson (1842)

$\mathcal{F}$oreign currency futures and options contracts are examples of the new breed of financial instrument known as derivatives. Financial **derivatives** are contracts that derive their value from some underlying asset (such as a stock, bond, or currency), reference rate (such as a 90-day Treasury bill rate), or index (such as the S&P 500 stock index). Popular derivatives include swaps, forwards, futures, and options. The previous chapter discussed forward contracts, and chapter 16 discusses swaps. This chapter describes the nature and valuation of currency futures and options contracts and shows how they can be used to manage foreign exchange risk or take speculative positions on currency movements. It also shows how to read the prices of these contracts as they appear in the financial press.

6.1 FUTURES CONTRACTS

In 1972, the **Chicago Mercantile Exchange** opened its **International Monetary Market** (IMM) division. The IMM provides an outlet for currency speculators and for those looking to reduce their currency risks. Trade takes place in **currency futures**,

EXHIBIT 6.1 Contract Specifications for Foreign Currency Futures

	AUSTRALIAN DOLLAR	BRITISH POUND	CANADIAN DOLLAR	DEUTSCHE MARK	FRENCH FRANC	JAPANESE YEN	SWISS FRANC
Contract Size	A$100,000	£62,500	C$100,000	DM 125,000	FF 500,000	¥12,500,000	SFr 125,000
Symbol	AD	BP	CD	DM	FR	JY	SF
Margin requirements							
Initial	$1,620	$1,418	$675	$1,215	$1,485	$2,025	$1,620
Maintenance	$1,200	$1,050	$500	$900	$1,100	$1,500	$1,200
Minimum price change	0.0001 (1 pt.)	0.0002 (2 pts.)	0.0001 (1 pt.)	0.0001 (1 pt.)	0.00002 (2 pts.)	0.000001 (1 pt.)	0.0001 (1 pt.)
Value of 1 point	$10.00	$6.25	$10.00	$12.50	$10.00	$12.50	$12.50
Months traded	January, March, April, June, July, September October, December, and spot month						
Trading hours	7:20 A.M.–2:00 P.M. (Central Time)						
Last day of trading	The second business day immediately preceding the third Wednesday of the delivery month						

Source: Data collected from Chicago Mercantile Exchange's web site at www.cme.com.

which are contracts for specific quantities of given currencies; the exchange rate is fixed at the time the contract is entered into, and the delivery date is set by the board of directors of the IMM. These contracts are patterned after those for grain and commodity futures contracts, which have been traded on Chicago's exchanges for more than 100 years.

Currency futures contracts currently are available for the Australian dollar, Brazilian real (pronounced rā-äl), British pound, Canadian dollar, Deutsche mark, French franc, Japanese yen, Mexican peso, New Zealand dollar, South African rand, and Swiss franc. The IMM is continually experimenting with new contracts. Those that meet the minimum volume requirements are added and those that do not are dropped. For example, the IMM recently added DM/£ and DM/¥ cross-rate contracts, while dropping contracts in the Dutch guilder and European Currency Unit. The number of contracts outstanding at any one time is called the **open interest**.

Private individuals are encouraged, rather than discouraged, to participate in the market. Contract sizes are standardized by amount of foreign currency—for example, £62,500, C$100,000, SFr 125,000. Exhibit 6.1 shows contract specifications for some of the currencies traded. Leverage is high; margin requirements average less than 2% of the value of the futures contract. The leverage assures that investors' fortunes will be decided by tiny swings in exchange rates.

The contracts have minimum price moves, which generally translate into about $10–$12 per contract. At the same time, most exchanges set daily price limits on their contracts that restrict the maximum daily price move. When these limits are reached, additional margin requirements are imposed and trading may be halted for a short time.

Instead of using the bid-ask spreads found in the interbank market, traders charge commissions. Though commissions will vary, a *round trip*—that is, one buy and one sell—costs as little as $15. This cost works out to less than 0.02% of the value of a sterling contract. The low cost, along with the high degree of leverage, has provided a major inducement for speculators to participate in the market. Other market participants include importers and exporters, companies with foreign currency assets and liabilities, and bankers.

Although volume in the futures market is still small compared with that in the forward market, it can be viewed as an expanding part of a growing foreign exchange market. As we will see shortly, the different segments of this market are linked by arbitrage.

The IMM is still the dominant trader, but other exchanges also trade futures contracts. The most important of these competitors include the London International Financial Futures Exchange (LIFFE), the Chicago Board of Trade (CBOT), the New York Mercantile Exchange, the Singapore International Monetary Exchange (SIMEX), Deutsche Termin Borse (DTB) in Frankfurt, the Hong Kong Futures Exchange (HKFE), the Marché à Termes des Instruments Financiers (MATIF) in Paris, and the Tokyo International Financial Futures Exchange (TIFFE).

A notable feature of the IMM and other futures markets is that deals are struck by brokers face to face on a trading floor rather than over the telephone. There are other, more important distinctions between the futures and forward markets.

Forward Contract versus Futures Contract

One way to understand futures contracts is to compare them with forward contracts. **Futures contracts** are standardized contracts that trade on organized futures markets for specific delivery dates only. In the case of the IMM, the most active currency futures contracts are traded for March, June, September, and December delivery. Contracts expire two business days before the third Wednesday of the delivery month. Contract sizes and maturities are standardized, so all participants in the market are familiar with the types of contracts available, a situation that facilitates trading. Forward contracts, on the other hand, are private deals between two individuals who can sign any type of contract they agree on. For example, two individuals may sign a forward contract for DM 70,000 in 20 months to be paid in Belgian francs. However, IMM contracts trade only in round lots of DM 125,000 priced in U.S. dollars and with a limited range of maturities available. With only a few standardized contracts traded, the trading volume in available contracts is higher, leading to superior liquidity, smaller price fluctuations, and lower transaction costs.

Once a trade is confirmed, the exchange's clearing house—backed by its members' capital—becomes the legal counterparty to both the buyer and the seller of the futures contract. The exchange members, in effect, guarantee both sides of a contract, largely eliminating the default risks of trading. Members of the futures exchange support their guarantee through margin requirements, marking contracts to market daily (explained later), and maintaining a guarantee fund in the event a member defaults. In contrast, a forward contract is a private deal between two parties and is subject to the risk that either side may default on the terms of the agreement.

The contract specifications in Exhibit 6.1 show the margin requirements. The *initial margin* shows how much money must be in the account balance when the contract is entered into. This amount is $1,418 in the case of the pound. A *margin call* is issued if—because of losses on the futures contract—the balance in the account falls below the *mainte-

nance margin, which is $1,050 for the pound. At that time, enough new money must be added to the account balance to bring it up to the maintenance margin. For example, if you start with an initial balance of $1,418 in your account on a pound futures contract and your contract loses, say, $700 in value, you must add $315 ($1,418 − $700 + $332 = $1,050).

The IMM periodically revises its margin requirements in line with changing currency volatilities using a computerized risk-management program call *SPAN*, which stands for Standard Portfolio Analysis of Risk. Note also that the margin requirements set by the IMM are minimums; brokers often require higher margins on more volatile currency contracts.

Profits and losses of futures contracts are paid over every day at the end of trading, a practice called **marking to market**. This daily-settlement feature can best be illustrated with an example. On Tuesday morning, an investor takes a long position in a Swiss franc futures contract that matures on Thursday afternoon. The agreed-on price is $0.75 for SFr 125,000. To begin, the investor must put $1,620 into his initial margin account. At the close of trading on Tuesday, the futures price has risen to $0.755. Because of daily settlement, three things occur. First, the investor receives his cash profit of $625 (125,000 × 0.005). Second, the existing futures contract with a price of $0.75 is canceled. Third, the investor receives a new futures contract with the prevailing price of $0.755. Thus, the value of the futures contracts is set to zero at the end of each trading day.

At Wednesday close, the price has declined to $0.743. The investor must pay the $1,500 loss (125,000 × 0.012) to the other side of the contract and trade in the old contract for a new one with a price of $0.743. At Thursday close, the price drops to $0.74, and the contract matures. The investor pays his $375 loss to the other side and takes delivery of the Swiss francs, paying the prevailing price of $0.74. The investor has had a net loss on the contract of $1,250 ($625 − $1,500 − $375) before paying his commission. Exhibit 6.2 details the daily settlement process.

Daily settlement reduces the default risk of futures contracts relative to forward contracts. Every day, futures investors must pay over any losses or receive any gains from the day's price movements. These gains or losses are generally added to or subtracted

EXHIBIT 6.2 An Example of Daily Settlement with a Futures Contract

TIME	ACTION	CASH FLOW
Tuesday morning	Investor buys SFr futures contract that matures in two days. Price is $0.75.	None
Tuesday close	Futures price rises to $0.755. Position is marked to market.	Investor receives 125,000 × (0.755 − 0.75) = $625.
Wednesday close	Futures price drops to $0.743. Position is marked to market.	Investor pays 125,000 × (0.755 − 0.743) = $1,500.
Thursday close	Futures price drops to $0.74. (1) Contract is marked to market. (2) Investor takes delivery of SFr 125,000.	(1) Investor pays 125,000 × (0.743 − 0.74) = $375. (2) Investor pays 125,000 × 0.74 = $92,500. Net loss on the futures contract = $1,250

EXHIBIT 6.3 Basic Differences Between Forward and Futures Contracts

---

1. **Trading:**
   Forward contracts are traded by telephone or telex.
   Futures contracts are traded in a competitive arena.
2. **Regulation:**
   The forward market is self-regulating.
   The IMM is regulated by the Commodity Futures Trading Commission.
3. **Frequency of Delivery:**
   More than 90% of all forward contracts are settled by actual delivery.
   Less than 1% of the IMM futures contracts are settled by delivery.
4. **Size of Contract:**
   Forward contracts are individually tailored and tend to be much larger than the standardized contracts on the futures market.
   Futures contracts are standardized in terms of currency amount.
5. **Delivery Date:**
   Banks offer forward contracts for delivery on any date.
   IMM futures contracts are available for delivery on only a few specified dates a year.
6. **Settlement:**
   Forward contract settlement occurs on the date agreed on between the bank and the customer.
   Futures contract settlements are made daily via the Exchange's Clearing House; gains on position values may be withdrawn and losses are collected daily. This practice is known as marking to market.
7. **Quotes:**
   Forward prices generally are quoted in European terms (units of local currency per U.S. dollar).
   Futures contracts are quoted in American terms (dollars per one foreign currency unit).
8. **Transaction Costs:**
   Costs of forward contracts are based on bid–ask spread.
   Futures contracts entail brokerage fees for buy and sell orders.
9. **Margins:**
   Margins are not required in the forward market.
   Margins are required of all participants in the futures market.
10. **Credit Risk:**
    The credit risk is borne by each party to a forward contract. Credit limits must therefore be set for each customer.
    The Exchange's Clearing House becomes the opposite side to each futures contract, thereby reducing credit risk substantially.

---

from the investor's margin account. An insolvent investor with an unprofitable position would be forced into default after only one day's trading, rather than being allowed to build up huge losses that lead to one large default at the time the contract matures (as could occur with a forward contract). For example, by deciding to keep his contract in force, rather than closing it out on Wednesday, the investor would have had a margin call for $455 ($1,620 + $625 − $1,500 + $455 = $1,200) at the close of Wednesday trading in order to meet his $1,200 maintenance margin; that is, the investor would have had to add $455 to his account to maintain his futures contract.

Futures contracts can also be closed out with an *offsetting trade*. For example, if a company's long position in DM futures has proved to be profitable, it need not literally take delivery of the DM when the contract matures. Rather, the company can sell futures contracts on a like amount of DM just prior to the maturity of the long position. The two positions cancel on the books of the futures exchange, and the company receives its profit in dollars. Exhibit 6.3 summarizes these and other differences between forward and futures contracts.

ILLUSTRATION

COMPUTING GAINS, LOSSES, AND MAINTENANCE MARGINS ON A FUTURES CONTRACT

On Monday morning, you short one IMM yen futures contract containing ¥12,500,000 at a price of $0.009433. Suppose the broker requires an initial margin of $4,590 and a maintenance margin of $3,400. The settlement prices for Monday through Thursday are $0.009542, $0.009581, $0.009375, and $0.009369, respectively. On Friday, you close out the contract at a price of $0.009394. Calculate the daily cash flows on your account. Describe any margin calls on your account. What is your cash balance with your broker as of the close of business on Friday? Assume that you begin with an initial balance of $4,590 and that your round-trip commission was $27.

Solution

TIME	ACTION	CASH FLOW ON CONTRACT	
Monday morning	Sell one IMM yen futures contract. Price is $0.009433.	None.	
Monday close	Futures price rises to $0.009542. Contract is marked-to-market.	You pay out 12,500,000 × (0.009433 − 0.009542) =	−$1,362.50
Tuesday close	Futures price rises to $0.009581. Contract is marked-to-market.	You pay out an additional 12,500,000 × (0.009542 − 0.009581) =	−$487.50
Wednesday close	Futures price falls to $0.009375. Contract is marked-to-market.	You receive 12,500,000 × (0.009581 − 0.009375) =	+$2,575.00
Thursday close	Futures price falls to $0.009369. Contract is marked-to-market.	You receive an additional 12,500,000 × (0.009375 − 0.009369) =	+$75.00
Friday	You close out your contract at a futures price of $0.009394.	You pay out 12,500,000 × (0.009369 − 0.009394) =	−$312.50
		You pay out a round-trip commission =	−$27.00
		Net gain on the futures contract	$460.50

Your margin calls and cash balances as of the close of each day were as follows:

Monday	With a loss of $1,362.50, your account balance falls to $3,227.50 ($4,590 − $1,362.50). You must add $172.50 ($3,400 − $3,227.50) to your account to restore it to the maintenance margin of $3,400.
Tuesday	With an additional loss of $487.50, your balance falls to $2,912.50 ($3,400 − $487.50). You must add $487.50 to your account to restore it to the maintenance margin of $3,400.
Wednesday	With a gain of $2,575, your balance rises to $5,975.
Thursday	With a gain of $75, your account balance rises further to $6,050.
Friday	With a loss of $312.50, your account balance falls to $5,737.50. After subtracting the round-trip commission of $27, your account balance ends at $5,710.50.

ADVANTAGES AND DISADVANTAGES OF FUTURES CONTRACTS ■ The smaller size of a futures contract and the freedom to liquidate the contract at any time before its maturity in a well-organized futures market differentiate the futures contract from the forward contract. These features of the futures contract attract many users. On the other hand, the limited number of currencies traded, the limited delivery dates, and the rigid contractual amounts of currencies to be delivered are disadvantages of the futures contract to many commercial users. Only by chance will contracts conform exactly to corporate requirements. The contracts are of value mainly to those commercial customers who have a fairly stable and continuous stream of payments or receipts in the traded foreign currencies.

ARBITRAGE BETWEEN THE FUTURES AND FORWARD MARKETS ■ Arbitrageurs play an important role on the IMM. They translate IMM futures rates into interbank forward rates and, by realizing profit opportunities, keep IMM futures rates in line with bank forward rates.

ILLUSTRATION

FORWARD-FUTURES ARBITRAGE

Suppose the interbank forward bid for June 18 on pounds sterling is $1.2927 at the same time that the price of IMM sterling futures for delivery on June 18 is $1.2915. How could the dealer use arbitrage to profit from this situation?

Solution. The dealer would simultaneously buy the June sterling futures contract for $80,718.75 (62,500 × $1.2915) and sell an equivalent amount of sterling forward, worth $80,793.75 (62,500 × $1.2927), for June delivery. Upon settlement, the dealer would earn a profit of $75. Alternatively, if the markets come back together before June 18, the dealer can unwind his position (by simultaneously buying £62,500 forward and selling a futures contract, both for delivery on June 18) and earn the same $75 profit. Although the amount of profit on this transaction is tiny, it becomes $7,500 if 100 futures contracts are traded.

Such arbitrage transactions, as described in the illustration above, will bid up the futures price and bid down the forward price until approximate equality is restored. The word *approximate* is used because there is a difference between the two contracts. Unlike the forward contract, where gains or losses are not realized until maturity, marking to market means that day-to-day futures contract gains (or losses) will have to be invested (or borrowed) at uncertain future interest rates. However, a study of actual rates for the British pound, Canadian dollar, Deutsche mark, Swiss franc, and Japanese yen found that forward and futures prices do not differ significantly.[1]

6.2 CURRENCY OPTIONS

Whatever advantages the forward or the futures contract might hold for its purchaser, the two contracts have a common disadvantage: Although they protect the holder against the risk of adverse movements in exchange rates, they also eliminate the possibility of gaining a windfall profit from favorable movements. This disadvantage was apparently one of the considerations that led some commercial banks to offer **currency options** to their customers. Exchange-traded currency options were first offered in 1983 by the *Philadelphia Stock Exchange (PHLX)*, where they are now traded on the **United Currency Options Market (UCOM)**. Currency options are one of the fastest growing segments of the global foreign exchange market, currently accounting for as much as 7% of daily trading volume.

---

[1] Bradford Cornell and Marc Reinganum, "Forward and Futures Prices: Evidence from the Foreign Exchange Markets," *Journal of Finance*, December 1981, pp. 1035–1045.

In principle, an **option** is a financial instrument that gives the holder the right—but not the obligation—to sell (put) or buy (call) another financial instrument at a set price and expiration date. The seller of the put option or call option must fulfill the contract if the buyer so desires it. Because the option not to buy or sell has value, the buyer must pay the seller of the option some premium for this privilege. As applied to foreign currencies, **call options** give the customer the right to purchase—and **put options** give the right to sell—the contracted currencies at the **expiration date**. Note that because a foreign exchange transaction has two sides, a call (put) option on a foreign currency can be considered a foreign currency put (call) option on the domestic currency. For example, the right to buy DM against dollar payment is equivalent to the right to sell dollars for DM payment. An **American option** can be exercised at any time up to the expiration date; a **European option** can be exercised only at maturity.

An option that would be profitable to exercise at the current exchange rate is said to be **in-the-money**. Conversely, an **out-of-the-money** option is one that would not be profitable to exercise at the current exchange rate. The price at which the option is exercised is called the **exercise price** or **strike price**. An option whose exercise price is the same as the spot exchange rate is termed at-the-money.

EXHIBIT 6.4 Profit from Buying a Call Option for Various Spot Prices at Expiration

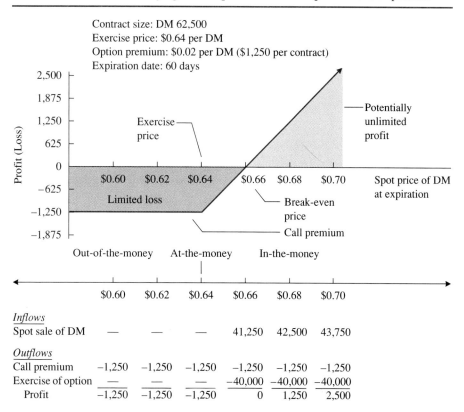

				$0.60	$0.62	$0.64	$0.66	$0.68	$0.70
Inflows									
Spot sale of DM				—	—	—	41,250	42,500	43,750
Outflows									
Call premium				−1,250	−1,250	−1,250	−1,250	−1,250	−1,250
Exercise of option				—	—	—	−40,000	−40,000	−40,000
Profit				−1,250	−1,250	−1,250	0	1,250	2,500

Using Currency Options

To see how currency options might be used, consider a U.S. importer with a DM 62,500 payment to make to a German exporter in 60 days. The importer could purchase a European call option to have the Deutsche marks delivered to him at a specified exchange rate (the strike price) on the due date. Suppose the option premium is $0.02 per DM, and the exercise price is $0.64. The importer has paid $1,250 for a DM 64 call option, which gives him the right to buy DM 62,500 at a price of $0.64 per mark at the end of 60 days. If at the time the importer's payment falls due, the value of the Deutsche mark has risen to, say, $0.70, the option would be in-the-money. In this case, the importer exercises his call option and purchases Deutsche marks for $0.64. The importer would earn a profit of $3,750 (62,500 × 0.06), which would more than cover the $1,250 cost of the option. If the rate has declined below the contracted rate to, say, $0.60, the DM 64 option would be out-of-the-money. Consequently, the importer would let the option expire and purchase the Deutsche marks in the spot market. Despite losing the $1,250 option premium, the importer would still be $1,250 better off than if he had locked in a rate of $0.64 with a forward or futures contract.

Exhibit 6.4 illustrates the importer's gains or losses on the call option. At a spot rate on expiration of $0.64 or lower, the option will not be exercised, resulting in a loss of the

EXHIBIT 6.5 Profit from Selling a Call Option for Various Spot Prices at Expiration

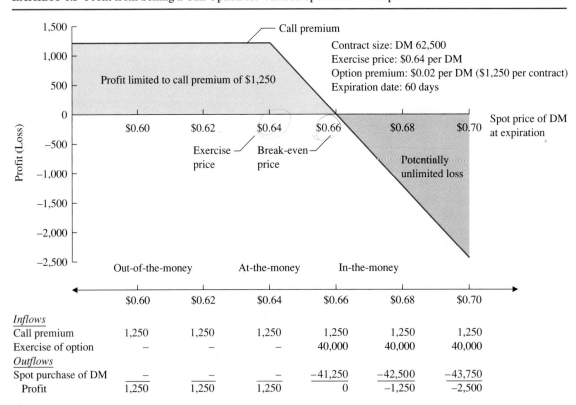

	$0.60	$0.62	$0.64	$0.66	$0.68	$0.70
Inflows						
Call premium	1,250	1,250	1,250	1,250	1,250	1,250
Exercise of option	–	–	–	40,000	40,000	40,000
Outflows						
Spot purchase of DM	–	–	–	–41,250	–42,500	–43,750
Profit	1,250	1,250	1,250	0	–1,250	–2,500

$1,250 option premium (and no spot sale of DM). Between $0.64 and $0.66, the option will be exercised, but the gain is insufficient to cover the premium. The *break-even price*—at which the gain on the option just equals the option premium—is $0.66. Above $0.66 per DM, the option is sufficiently deep in-the-money to cover the option premium and yield a—potentially unlimited—net profit.

Because this is a zero-sum game, the profit from selling a call, shown in Exhibit 6.5, is the mirror image of the profit from buying the call. For example, if the spot rate at expiration is above $0.66/DM, the call option writer is exposed to potentially unlimited losses. Why would an option writer accept such risks? For one thing, the option writer may already be long DM, effectively hedging much of the risk. Alternatively, the writer might be willing to take a risk in the hope of profiting from the option premium because of a belief that the DM will depreciate over the life of the contract. If the spot rate at expiration is $0.64 or less, the option ends out-of-the-money and the call option writer gets to keep the full $1,250 premium. For spot rates between $0.64 and $0.66, the option writer still earns a profit, albeit a diminishing one.

In contrast to the call option, a put option at the same terms (exercise price of $0.64 and put premium of $0.02 per DM) would be in-the-money at a spot price of $0.60 and out-of-the-money at $0.70. Exhibit 6.6 illustrates the profits available on this DM put

EXHIBIT 6.6 Profit from Buying a Put Option for Various Spot Prices at Expiration

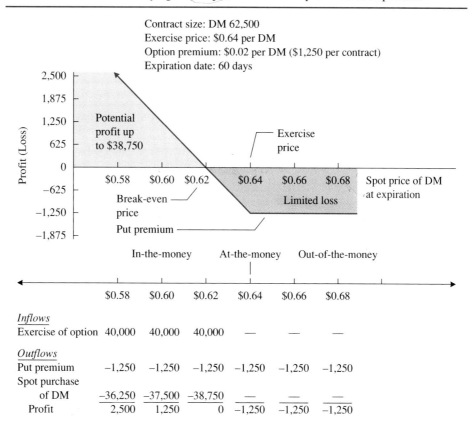

	$0.58	$0.60	$0.62	$0.64	$0.66	$0.68
Inflows						
Exercise of option	40,000	40,000	40,000	—	—	—
Outflows						
Put premium	−1,250	−1,250	−1,250	−1,250	−1,250	−1,250
Spot purchase of DM	−36,250	−37,500	−38,750	—	—	—
Profit	2,500	1,250	0	−1,250	−1,250	−1,250

EXHIBIT 6.7 Profit from Selling a Put Option for Various Spot Prices at Expiration

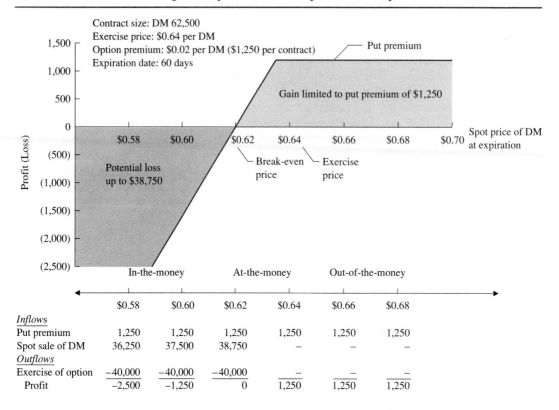

	$0.58	$0.60	$0.62	$0.64	$0.66	$0.68
Inflows						
Put premium	1,250	1,250	1,250	1,250	1,250	1,250
Spot sale of DM	36,250	37,500	38,750	–	–	–
Outflows						
Exercise of option	−40,000	−40,000	−40,000	–	–	–
Profit	−2,500	−1,250	0	1,250	1,250	1,250

option. If the spot price falls to, say, $0.58, the holder of a put option will deliver DM 62,500 worth $36,250 (0.58 × 62,500) and receive $40,000 (0.64 × 62,500). The option holder's profit, net of the $1,250 option premium, is $2,500. As the spot price falls further, the value of the put option rises. At the extreme, if the spot rate falls to zero, the buyer's profit on the contract will reach $38,750 (0.64 × 62,500 − 1,250). Below a spot rate of $0.62, the gain on the put option will more than cover the $1,250 option premium. Between $0.62—the breakeven price for the put option—and $0.64, the holder would exercise the option, but the gain would be less than the option premium. At spot prices above $0.64, the holder would not exercise the option and so would lose the $1,250 premium. Both the put and the call options will be at-the-money if the spot rate in 60 days is $0.64, and the call or put option buyer will lose the $1,250 option premium.

As in the case of the call option, the writer of the put option will have a payoff profile that is the mirror image of that for the buyer. As shown in Exhibit 6.7, if the spot rate at expiration is $0.64 or higher, the option writer gets to keep the full $1,250 premium. As the spot rate falls below $0.64, the option writer earns a decreasing profit down to $0.62. For spot rates below $0.62/DM, the option writer is exposed to increasing losses, up to a maximum potential loss of $38,750. The writer of the put option will accept these risks in the hope of profiting from the put premium. These risks may be minimal if the put option writer is already short DM.

ILLUSTRATION

SPECULATING WITH A JAPANESE YEN CALL OPTION

In March, a speculator who is gambling that the yen will appreciate against the dollar pays $680 to buy a yen June 81 call option. This option gives the speculator the right to buy ¥6,250,000 in June at an exchange rate of ¥1 = $0.0081 (the 81 in the contract de-scription is expressed in hundredths of a cent). By the expiration date in June, the yen spot price has risen to $0.0083. What is the investor's net return on the contract?

Solution. Because the call option is in-the-money by 0.02 cents, the investor will realize a gain of $1,250 ($0.0002 × 6,250,000) on the option contract. This amount less the $680 paid for the option produces a gain on the contract of $570.

Typical users of currency options might be financial firms holding large investments overseas where sizable unrealized gains had occurred because of exchange rate changes and where these gains were thought likely to be partially or fully reversed. Limited use of currency options has also been made by firms that have a foreign currency inflow or out-flow that is possibly forthcoming, but not definitely. In such cases, where future foreign currency cash flows are contingent on an event such as acceptance of a bid, long call or put positions can be safer hedges than either futures or forwards.

For example, assume that a U.S. investor makes a firm bid in pounds sterling to buy a piece of real estate in London. If the firm wishes to hedge the dollar cost of the bid, it can buy pounds forward so that if the pound sterling appreciates, the gain on the forward contract will offset the increased dollar cost of the prospective investment. But if the bid is eventually rejected, and if the pound has fallen in the interim, losses from the forward position will have no offset. If no forward cover is taken and the pound appreciates, the real estate will cost more than expected.

Currency call options can provide a better hedge in such a case. Purchased pound call options would provide protection against a rising pound; and yet, if the bid were re-jected and the pound had fallen, the uncovered hedge loss would be limited to the pre-mium paid for the calls. Note that a U.S. company in the opposite position, such as one bidding on a British project, whose receipt of future pound cash inflows is contingent on acceptance of its bid, would use a long pound put position to provide the safest hedge.

Currency options also can be used by pure speculators, those without an underlying foreign currency transaction to protect against. The presence of speculators in the options markets adds to the breadth and depth of these markets, thereby making them more liquid and lowering transactions costs and risk.

CURRENCY SPREAD ■ A **currency spread** allows speculators to bet on the direction of a currency but at a lower cost than buying a put or a call option alone. It involves buying an option at one strike price and selling a similar option at a different strike price. The currency spread limits the holder's downside risk on the currency bet but at the cost of

EXHIBIT 6.8 Currency Spreads

a. Example of a Bull Spread

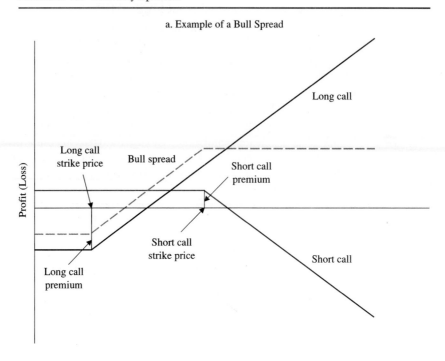

b. Example of a Bear Spread

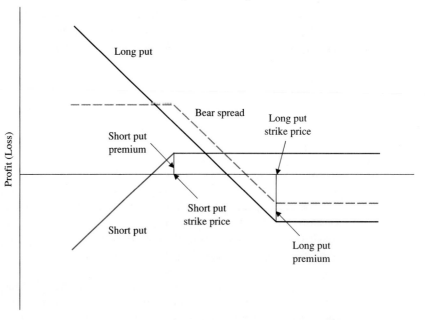

limiting the position's upside potential. As shown in Exhibit 6.8a, a spread designed to bet on a currency's appreciation—also called a **bull spread**—would involve buying a call at one strike price and selling another call at a higher strike price. The net premium paid for this position is positive because the former call will be higher priced than the latter (with a lower strike, the option is less out-of-the-money) but it will be less than the cost of buying the former option alone. At the same time, the upside is limited by the strike price of the latter option. Exhibit 6.8b shows the payoff profile of a currency spread designed to bet on a currency's decline. This spread—also called a **bear spread**—involves buying a put at one strike price and selling another put at a lower strike price.

KNOCKOUT OPTIONS ■ Another way to bet on currency movements at a lower cost than buying a call or a put alone is to use knockout options. A **knockout option** is similar to a standard option except that it is canceled—that is, knocked out—if the exchange rate crosses, even briefly, a pre-defined level called the **outstrike**. If the exchange rate breaches this barrier, the holder cannot exercise this option, even if it ends up in-the-money. Knockout options, also known as *barrier options*, are less expensive than standard currency options precisely because of this risk of early cancellation.

There are different types of knockout options. For example, a *down-and-out call* will have a positive payoff to the option holder if the underlying currency strengthens but is canceled if it weakens sufficiently to hit the outstrike. Conversely, a *down-and-out put* has a positive payoff if the currency weakens but will be canceled if it weakens beyond the outstrike. In addition to lowering cost (albeit at the expense of less protection), down-and-out options are useful when a company believes that if the foreign currency declines below a certain level, it is unlikely to rebound to the point that it will cause the company losses. *Up-and-out options* are canceled if the underlying currency strengthens beyond the outstrike. In contrast to the previous knockout options, *down-and-in* and *up-and-in options* come into existence if and only if the currency crosses a preset barrier. The pricing of these options is extremely complex.

Option Pricing and Valuation

From a theoretical standpoint, the value of an option comprises two components: intrinsic value and time value. The **intrinsic value** of the option is the amount by which the option is in-the-money, or $S - X$, where S is the current spot price and X the exercise price. In other words, the intrinsic value equals the immediate exercise value of the option. Thus, the further into the money an option is, the more valuable it is. An out-of-the-money option has no intrinsic value. For example, the intrinsic value of a call option on Swiss francs with an exercise price of $0.74 and a spot rate of $0.77 would be $0.03 per franc. The intrinsic value of the option for spot rates that are less than the exercise price is zero. Any excess of the option value over its intrinsic value is called the **time value** of the contract. An option will generally sell for at least its intrinsic value. The more out-of-the-money an option is, the lower the option price. These features are shown in Exhibit 6.9.

During the time remaining before an option expires, the exchange rate can move so as to make exercising the option profitable or more profitable. That is, an out-of-the-money option can move into the money, or one already in-the-money can become more so. The chance that an option will become profitable or more profitable is always greater

EXHIBIT 6.9 The Value of a Call Option before Maturity

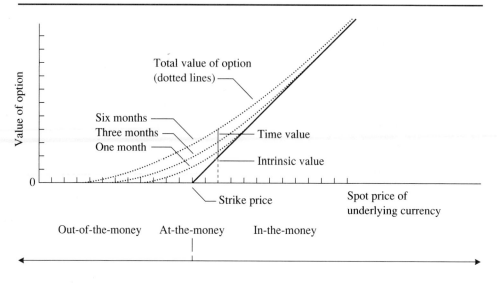

than zero. Consequently, the time value of an option is always positive for an out-of-the-money option and is usually positive for an in-the-money option. Moreover, the more time that remains until an option expires, the higher the time value tends to be. For example, an option with six months remaining until expiration will tend to have a higher price than an option with the same strike price but with only three months until expiration. As the option approaches its maturity, the time value declines to zero.

The value of an American option always exceeds its intrinsic value because the time value is always positive up to the expiration date. For example, if $S > X$, then $C(X) > S - X$, where $C(X)$ is the dollar price of an American call option on one unit of foreign currency. However, the case is more ambiguous for a European option because increasing the time to maturity may not increase its value, given that it can be exercised only on the maturity date.[2] That is, a European currency option may be in-the-money before expiration, yet it may be out-of-the-money by the maturity date.

Before expiration, an out-of-the-money option has only time value, but an in-the-money option has both time value and intrinsic value. At expiration, an option can have only intrinsic value. The time value of a currency option reflects the probability that its intrinsic value will increase before expiration; this probability depends, among other things, on the volatility of the exchange rate. An increase in currency volatility increases the chance of an extremely high or low exchange rate at the time the option expires. The chance of a very high exchange rate benefits the call owner. The chance of a very low exchange rate, however, is irrelevant; the option will be worthless for any exchange rate less than the striking price, whether the exchange rate is "low" or "very low." Inasmuch as the effect of increased volatility is beneficial, the value of the call option is higher. Put options similarly benefit from increased volatility in the exchange rate.

---

[2] For a technical discussion of foreign currency option pricing, see Mark B. Garman and Steven W. Kohlhagen, "Foreign Currency Option Values," *Journal of International Money and Finance*, December 1983, pp. 231–238.

EXHIBIT 6.10 Currency Option Pricing and Valuation		
PRICE EQUALS THE SUM OF	**Intrinsic value:** The amount by which the option is in-the-money	**Time value:** The amount by which the price of the contract exceeds its intrinsic value
CALL OPTION	$= S - X$, where S is the current spot price and X the exercise price Intrinsic value is zero if $S - X < 0$	Positively affected by an increase in: ■ time to expiration (usually) ■ volatility ■ domestic–foreign interest rate differential
PUT OPTION	$= X - S$ Intrinsic value is zero if $X - S < 0$	Positively affected by an increase in: ■ time to expiration (usually) ■ volatility ■ foreign–domestic interest rate differential

Another aspect of time value involves interest rates. In general, options have a present intrinsic value, determined by the exercise price and price of the underlying asset. Because the option is a claim on a specified amount of an asset over a period of time into the future, that claim must have a return in line with market interest rates on comparable instruments. Therefore, a rise in the interest rate will cause call values to rise and put values to fall.

Pricing foreign currency options is more complex because it requires consideration of both domestic and foreign interest rates. A foreign currency is normally at a forward premium or discount vis-à-vis the domestic currency. As we saw in Chapter 5, this premium or discount is determined by relative interest rates. Consequently, for foreign currency options, call values rise and put values fall when the domestic interest rate increases or the foreign interest rate decreases.

The flip side of a more valuable put or call option is a higher option premium. Hence, options become more expensive when exchange rate volatility rises. Similarly, when the domestic-foreign interest differential increases, call options become more expensive and put options less expensive. These elements of option valuation are summarized in Exhibit 6.10.

OPTION PRICING AND ARBITRAGE ■ Option pricing stems from application of the most productive idea in all of finance—arbitrage. The idea underlying arbitrage pricing of a new asset is simple: Create a portfolio of assets with known market prices that exactly duplicates the distribution of payoffs of the new asset. The price of the new asset must equal the cost of purchasing the mimicking portfolio. Otherwise, arbitrageurs would earn riskless profits. This is the technique used by Fischer Black and Myron Scholes in developing the Black-Scholes option pricing model.[3]

In order to develop a closed-form solution for the pricing of a currency option, we must make some assumptions about the statistical properties of the spot and forward

---

[3] Fischer Black and Myron Scholes, "The Pricing of Options and Corporate Liabilities," *Journal of Political Economy*, May–June 1973, pp. 637–659.

exchange rates. Assuming that both these exchange rates are lognormally distributed (i.e., that their natural logarithm follows a normal distribution), one can duplicate the price of a European call option exactly, over a short time interval, with a portfolio of domestic and foreign bonds. This portfolio can be represented as

$$C(t) = aS(t)\,B^*(t,T) + bB(t,T) \qquad (6.1)$$

WHERE

$C(t)$ = call option premium at time t for an option that expires at $t + T$
T = time to expiration of the option, expressed in fractions of a year
$S(t)$ = spot value of the foreign currency at time t
$B^*(t,T)$ = price of a pure discount foreign bond that pays one unit of the foreign currency at $t + T$, or $B^*(t,T) = 1/(1 + r^*T)^4$
$B(t,T)$ = price of a pure discount domestic bond that pays one unit of the domestic currency at $t + T$, or $B(t,T) = 1/(1 + rT)$
r^* = the annualized interest rate on a pure discount foreign bond
r = the annualized interest rate on a pure discount domestic bond
a = amount of the foreign currency bond in the mimicking portfolio
b = amount of the domestic currency bond in the mimicking portfolio

Mark Garman and Stephen Kohlhagen have shown that, given the previously mentioned lognormal distribution assumptions, Equation 6.1 can be expressed as:[5]

$$C(t) = N(d_1) \times S(t) \times B^*(t,T) - N(d_2) \times X \times B(t,T) \qquad (6.2)$$

WHERE

$N(d)$ = the cumulative normal distribution function[6]

$$d_1 = \frac{\ln(SB^*/XB) + 0.5\,\sigma^2 T}{\sigma\sqrt{T}}$$

$$d_2 = \frac{\ln(SB^*/XB) - 0.5\,\sigma^2 T}{\sigma\sqrt{T}} = d_1 - \sigma\sqrt{T}$$

σ = the expected standard deviation of the spot rate, annualized
X = the exercise price on the call option

Equation 6.2 is just the Black-Scholes option pricing formula applied to foreign currency options.

---

[4] The value of a pure discount bond with a continuously compounded interest rate k and maturity T is e^{-kT}. In the examples used in the text, it is assumed that r^* and r are the equivalent interest rates associated with discrete compounding.
[5] Garman and Kohlhagen, "Foreign Currency Option Values."
[6] $N(d)$ is the probability that a random variable that is normally distributed with a mean of zero and a standard deviation of one will have a value less than d.

ILLUSTRATION **P**RICING A SIX-MONTH DM EUROPEAN CALL OPTION

What is the price of a six-month DM European call option having the following characteristics?

S(t) ($/DM)	X ($/DM)	r (6-MONTH)	r* (6-MONTH)	σ (ANNUALIZED)
0.68	0.70	5.8%	6.5%	0.2873

Solution. In order to apply Equation 6.2, we need to estimate $B(t,0.5)$ and $B^*(t,0.5)$ because $T = 0.5$ (6 months equal 0.5 years). Given the annualized interest rates on six-month bonds of 5.8% and 6.5%, the six-month U.S. and German interest rates are 2.9% (5.8/2) and 3.25% (6.5/2), respectively. The associated bond prices are

$$B(t,0.5) = \frac{1}{1.029} = 0.9718$$

$$B^*(t,0.5) = \frac{1}{1.0325} = 0.9685$$

Substituting in the values for B and B^* along with those for S (0.68), X (0.70), and σ (0.2873) in Equation 6.2, we can calculate

$$d_1 = \frac{\ln(SB^*/XB) + 0.5\,\sigma^2 T}{\sigma\sqrt{T}}$$

$$= \frac{\ln(0.68 \times 0.9685/0.7 \times 0.9718) + 0.5(0.2873)^2 0.5}{0.2873\sqrt{0.5}}$$

$$= -0.05786$$

$$d_2 = d_1 - \sigma\sqrt{T}$$

$$= -0.05786 - 0.2873\sqrt{0.5} = 0.26101$$

The easiest way to compute the values of $N(-0.05786)$ and $N(-0.26101)$ is to use a spreadsheet function such as NORMDIST in Excel. This Excel function yields computed values of $N(-0.05786) = 0.47693$ and $N(-0.26101) = 0.39704$.[7] Using Equation 6.2, we can now calculate the value of the six-month DM call option:

$$C(t) = N(d_1) \times S(t) \times B^*(t,T)$$
$$\quad - N(d_2) \times X \times B(t,T)$$
$$= 0.47693 \times 0.68 \times 0.9685$$
$$\quad - 0.39704 \times 0.70 \times 0.9718$$
$$= \$0.04400/DM$$

In other words, the value of the six-month option to acquire DM at an exercise price of $0.70 when the spot rate is $0.68 is 4.400¢/DM. The relatively high volatility of the spot DM has contributed to the significant value of this out-of-the-money call option.

[7]Alternatively, we can use the cumulative normal table in Appendix 6A. However, because this table contains values for d to only two decimal places, we must interpolate to find $N(-0.05786)$ and $N(-0.26101)$. Recognizing that $N(-x) = 1 - N(x)$, we can determine from the table that $N(-0.05) = 0.4801$ and $N(-0.06) = 0.4761$. Similarly, $N(-0.26) = 0.3974$ and $N(-0.27) = 0.3936$. Interpolating between these numbers yields:

$$N(-0.05786) = N(-0.05)\left(\frac{0.06 - 0.05786}{0.06 - 0.05}\right) + N(-0.06)\left(1 - \frac{0.06 - 0.05786}{0.06 - 0.05}\right) = 0.47696$$

$$N(-0.26101) = N(-0.26)\left(\frac{0.27 - 0.26101}{0.27 - 0.26}\right) + N(-0.27)\left(1 - \frac{0.27 - 0.26101}{0.27 - 0.26}\right) = 0.39702$$

These interpolated values are very close to the ones computed using the Excel spreadsheet function.

IMPLIED VOLATILITIES ■ Black-Scholes option prices depend critically on the estimate of volatility (σ) being used. In fact, traders typically use the **implied volatility**—the volatility that, when substituted in Equation 6.2, yields the market price of the option—as an indication of the market's opinion of future exchange rate volatility. Implied volatilities function for options in the same way as yields to maturity do for bonds. They succinctly summarize a great deal of economically relevant information about the price of the asset, and they can be used to compare assets with different contractual terms without having to provide a great deal of detail about the asset.

Indeed, option prices are increasingly being quoted as implied volatilities, which traders by agreement substitute into the Garman-Kohlhagen model (Equation 6.2) to determine the option premium. This is not to say that traders believe that Equation 6.2 and its underlying assumptions are correct. Indeed, they quote different implied volatilities for different strike prices at the same maturity. However, Equation 6.2 by convention is used to map implied volatility quotes to option prices.

SHORTCOMINGS OF THE BLACK-SCHOLES OPTION PRICING MODEL ■ The Black-Scholes model assumes continuous portfolio rebalancing, no transaction costs, stable interest rates, and lognormally distributed and continuously changing exchange rates. Each of these assumptions is violated in periods of currency turmoil, such as occurred during the breakup of the exchange-rate mechanism. With foreign exchange markets shifting dramatically from one moment to the next, continuous portfolio rebalancing turned out to be impossible. And with interest rates being so volatile (for example, overnight interest rates on the Swedish krona jumped from 24% to 600%), the assumption of interest rate stability was violated as well. Moreover, devaluations and revaluations can cause abrupt shifts in exchange rates, contrary to the premise of continuous movements.[8]

A related point is that empirical evidence indicates that there are more extreme exchange rate observations than a lognormal distribution would predict.[9] That is, the distribution of exchange rates is *leptokurtic*, or fat-tailed. Leptokurtosis explains why the typical pattern of implied volatilities is U-shaped (the so-called "volatility smile"). Finally, although prices depend critically on the estimate of volatility used, such estimates may be unreliable. Users can, of course, estimate exchange rate volatility from historical data, but what matters for option pricing is future volatility and this is often difficult to predict because volatility can shift.

Put-Call Option Interest Rate Parity

As we saw in Chapter 5, interest rate parity relates the forward rate differential to the interest differential. Another parity condition relates options prices to the interest differential and, by extension, to the forward differential. We are now going to derive the

---

[8] Other option-pricing models have been developed that allow for discrete jumps in exchange rates. See, for example, David S. Bates, "Jumps and Stochastic Volatility: Exchange Rate Processes Implicit in PHLX Deutschemark Options," Wharton School working paper, 1993. This and other such models are based on the original jump-diffusion model appearing in Robert C. Merton, "Option Pricing When Underlying Stock Returns Are Discontinuous," *Journal of Financial Economics*, January–March 1976, pp. 125–144.

[9] This is primarily a problem for options that mature in one month or less. For options with maturities of three months or more, the lognormal distribution seems to be a good approximation of reality.

relation between put and call option prices, the forward rate, and domestic and foreign interest rates. To do this, we must first define the following parameters:

C = call option premium on a one-period contract

P = put option premium on a one-period contract

X = exercise price on the put and call options (dollars per unit of foreign currency)

Other variables—e_0, e_1, f_1, r_h, and r_f—are as defined earlier.

For illustrative purposes, Germany is taken to be the representative foreign country in the following derivation. In order to price a call option on the DM with a strike price of X in terms of a put option and forward contract, create the following portfolio:

1. Lend $1/(1 + r_f)$ Deutsche marks in Germany. This amount is the present value of DM 1 to be received one period in the future. Hence, in one period, this investment will be worth DM 1, which is equivalent to e_1 dollars.
2. Buy a put option on DM 1 with an exercise price of X.
3. Borrow $X/(1 + r_h)$ dollars. This loan will cost X dollars to repay at the end of the period given an interest rate of r_h.

The payoffs on the portfolio and the call option at expiration depend on the relation between the spot rate at expiration and the exercise price. These payoffs, which are shown pictorially in Exhibit 6.11, are as follows:

	DOLLAR VALUE ON EXPIRATION DATE IF	
SECURITY	$E_1 > X$	$E_1 < X$
I. Portfolio		
1. Lend DM $1/(1 + r_f)$	e_1	e_1
2. Buy a put option on DM 1 with an exercise price of X	0	$X - e_1$
3. Borrow $X/(1 + r_h)$ dollars	$-X$	$-X$
Total	$e_1 - X$	0
II. Buy a DM call option	$e_1 - X$	0

The payoffs on the portfolio and the call option are identical, so both securities must sell for identical prices in the marketplace. Otherwise, a risk-free arbitrage opportunity will exist. Therefore, the dollar price of the call option (which is the call premium, C) must equal the dollar value of the DM loan plus the price of the put option (the put premium, P) less the amount of dollars borrowed. Algebraically, this relation can be expressed as

$$C = \frac{e_0}{1 + r_f} - \frac{X}{1 + r_h} + P \tag{6.3}$$

According to interest rate parity,

$$\frac{e_0}{1 + r_f} = \frac{f_1}{1 + r_h} \tag{6.4}$$

EXHIBIT 6.11 Illustration of Put-Call Option
Interest Rate Parity

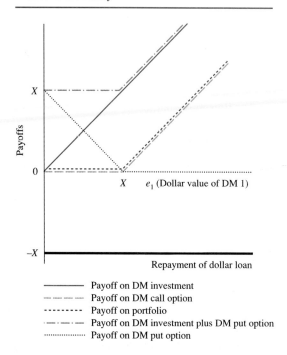

———— Payoff on DM investment
– – – – Payoff on DM call option
- - - - - - - Payoff on portfolio
·—·—·— Payoff on DM investment plus DM put option
·············· Payoff on DM put option

Substituting Equation 6.4 into Equation 6.3 yields a new equation:

$$C = \frac{f_1 - X}{1 + r_h} + P \qquad (6.5)$$

or

$$C - P = \frac{f_1 - X}{1 + r_h} \qquad (6.6)$$

These parity relations say that a long call is equivalent to a long put plus a forward (or futures) contract. The term $f_1 - X$ is discounted because the put and call premia are paid upfront whereas the forward rate and exercise price apply to the expiration date.

ILLUSTRATION

PRICING A DECEMBER DEUTSCHE MARK CALL OPTION
Suppose that the premium on September 15 on a December 15 DM put option is 0.23 cents per DM at a strike price of $0.73. The December 15 forward rate is DM 1 = $0.74 and the quarterly U.S. interest rate is 2.5%. Then, according to Equation 6.6, the December 15 call option should equal

$$C = 0.0023 + \frac{0.74 - 0.73}{1.025} = 0.0121$$

or 1.21 cents per DM.

EXHIBIT 6.12 Declining Exchange Rate Scenario

	JULY 1	SEPTEMBER 1
Spot	$0.6922	$0.6542
September futures	0.6956	0.6558
September 68 put	0.0059	0.0261
September 70 put	0.0144	0.0460

Using Forward or Futures Contracts versus Options Contracts

Suppose that on July 1, an American company makes a sale for which it will receive DM 125,000 on September 1. The firm will want to convert those DM into dollars, so it is exposed to the risk that the mark will fall below its current spot rate of $0.6922 in the meantime. The firm can protect itself against a declining Deutsche mark by selling its expected DM receipts forward (using a futures contract at a futures rate of $0.6956) or by buying a DM put option, both for September delivery.

Exhibit 6.12 shows possible results for each choice, using options with strike prices just above and just below the spot exchange of July 1 ($0.68 and $0.70). The example assumes a DM decline to $0.6542 and the consequent price adjustments of associated futures and options contracts. The put quotes are the option premiums per DM. Thus, the dollar premium associated with a particular quote equals the quote multiplied by the number of DM covered by the put options. For example, the quote of $0.0059 for a September put option with a strike price of 68 (in cents) represents a premium for covering the exporter's DM 125,000 transaction equal to $0.0059 × 125,000 = $737.50.

In the above example, a decision to remain unhedged would yield a loss of 125,000 × (0.6922 − 0.6542), or $4,750. The outcomes of the various hedge possibilities are shown in Exhibit 6.13.

Exhibit 6.13 demonstrates the following differences between the futures and options hedging strategies:

1. The futures hedge offers the closest offset to the loss due to the decline of the Deutsche mark.
2. The purchase of the in-the-money put option (the 70 strike price) offers greater protection (but at a higher premium) than the out-of-the-money put (the 68 strike price).

EXHIBIT 6.13 Hedging Alternatives: Offsetting a $4,750 Loss Due to a Declining DM

Result of Selling Futures

$$(0.6956 - 0.6558) \times 125,000 = \$4,975 \text{ profit}$$

Results of Buying Put Options

$$68 \text{ put: } (0.0261 - 0.0059) \times 125,000 = \$2,525 \text{ profit}$$

$$70 \text{ put: } (0.0460 - 0.0144) \times 125,000 = \$3,950 \text{ profit}$$

EXHIBIT 6.14 Rising Exchange Rate Scenario

	JULY 1	SEPTEMBER 1
Spot	$0.6922	$0.7338
September futures	0.6956	0.7374
September 68 put	0.0059	0.0001
September 70 put	0.0144	0.0001

Note that the September futures price is unequal to the spot rate on September 1 because settlement is not until later in the month. As the DM declines in value, the company would suffer a larger loss on its DM receivables, to be offset by a further increase in the value of the put and futures contracts.

Although the company wants to protect against the possibility of a DM depreciation, what would happen if the DM appreciates? To answer this question—so as to assess fully the options and futures hedge strategies—assume the hypothetical conditions in Exhibit 6.14.

In this scenario, the rise in the DM would increase the value of the unhedged position by $125,000 \times (0.7338 - 0.6922)$, or $5,200. This gain would be offset by losses on the futures or options contracts, as shown in Exhibit 6.15.

We can see that the futures hedge again provides the closest offset. Because these hedges generate losses, however, the company would be better served under this scenario by the smallest offset. With rapidly rising exchange rates, the company would benefit most from hedging with a long put position as opposed to a futures contract; conversely, with rapidly falling exchange rates, the company would benefit most from hedging with a futures contract.

Market Structure

Options are purchased and traded either on an organized exchange (such as the United Currency Options Market of the PHLX) or in the **over-the-counter (OTC) market**. Exchange-traded options or **listed options** are standardized contracts with predetermined

---

EXHIBIT 6.15 Hedging Alternatives: Offsetting a $5,200 Gain Due to a Rising DM

Result of Selling Futures

$$(0.7374 - 0.6956) \times 125,000 = \$5,225 \text{ loss}$$

Results of Buying Put Options

$$68 \text{ put: } (0.0059 - 0.0001) \times 125,000 = \$725 \text{ loss}$$

$$70 \text{ put: } (0.0144 - 0.0001) \times 125,000 = \$1,787.50 \text{ loss}$$

exercise prices, standard maturities (one, three, six, nine, and 12 months), and fixed delivery dates (March, June, September, and December). United Currency Options Market (UCOM) options are available in the ECU and seven currencies—Deutsche mark, pound sterling, French franc, Swiss franc, Japanese yen, Canadian dollar, and Australian dollar—and are traded in standard contracts half the size of the IMM futures contracts (some of which no longer exist). Cross-rate options also are available for the DM/¥ and £/DM. By taking the U.S. dollar out of the equation, cross-rate options allow one to hedge directly the currency risk that arises in dealing with nondollar currencies. Contract specifications are shown in Exhibit 6.16. The PHLX trades both American-style and European-style currency options. It also trades month-end options (listed as EOM, or end of month), which ensure the availability of a short-term (at most, a two- or sometimes three-week) currency option at all times, and long-term options, which extend the available expiration months on PHLX dollar-based and cross-rate contracts—providing for 18- and 24-month European-style options.

In 1994, the PHLX introduced a new option contract, called the **Virtual Currency Option**, which is settled in U.S. dollars rather than in the underlying currency. Virtual Currency Options, also called 3-Ds (dollar-denominated delivery), just formalize the reality that practically no exchange-traded options involve the payment or delivery of the underlying currency. Currently, 3-D options on the Deutsche mark and Japanese yen are available. They are European-style options that have maturities ranging from one week to nine months, and they settle weekly (meaning that each week a new contract is listed). They expire on Monday mornings, so hedging currency risk over a weekend is easy.

The PHLX also offers customized currency options, which allow users to customize various aspects of a currency option, including choice of exercise price, expiration date (up to two years out), and premium quotation as either units of currency or percent of underlying value. Customized currency options can be traded on any combination of the eight currencies (including the ECU) for which standardized options are available, along with the Italian lira, Spanish peseta, and the U.S. dollar.[10] The contract size is the same as that for standardized contracts in the underlying currency.[11] Other organized options exchanges are located in Amsterdam (European Options Exchange), Chicago (Chicago Mercantile Exchange), and Montreal (Montreal Stock Exchange).

Trading volume in PHLX currency options grew dramatically from their introduction in 1983 to 1993, when more than 13 million contracts representing over $600 billion in underlying value were traded. Since their peak in 1993, however, volume on the PHLX has fallen, reaching a low of 2.6 million contracts in 1997. Exhibit 6.17 shows the ups and downs in trading volume on the PHLX since 1983.

Over-the-counter (OTC) options are contracts whose specifications are generally negotiated as to the amount, exercise price and rights, underlying instrument, and expiration. *OTC currency options* are traded by commercial and investment banks in virtually

---

[10] The Australian dollar may be matched only against the U.S. dollar.

[11] Contract sizes for the Italian lira and Spanish peseta, which are available only as customized contracts, are Lit 50 million and Ptas 5 million, respectively. If the U.S. dollar is the underlying currency, the contract size is $50,000.

EXHIBIT 6.16 Philadelphia Stock Exchange Currency Options Specifications

	Australian Dollar	British Pound	Canadian Dollar	Deutsche Mark	European Currency Unit	Swiss Franc	French Franc	Japanese Yen
Symbol								
American style	XAD	XBP	XCD	XDM	ECU	XSF	XFF	XJY
European style	CAD	CBP	CCD	CDM	n.a.	CSF	CFF	CJY
Contract size	A$50,000	£31,250	C$50,000	DM 62,500	ECU 62,500	SFr 62,500	FF 250,000	¥6,250,000
Exercise Price Intervals	1¢	2.5¢	0.5¢	1¢[1]	2¢	1¢[1]	0.25¢	0.01¢[1]
Premium Quotations	Cents per unit	Cents per unit	Cents per unit	Cents per unit	Cents per unit	Cents per unit	Tenths of a cent per unit	Hundredths of a cent per unit
Minimum Price Change	$0.(00)01	$0.(00)01	$0.(00)01	$0.(00)01	$0.(00)01	$0.(00)01	$0.(000)02	$0.(0000)01
Minimum Contract Price Change	$5.00	$3.125	$5.00	$6.25	$6.25	$6.25	$5.00	$6.25
Expiration Months	March, June, September, and December + two near-term months							
Exercise Notice	No automatic exercise of in-the-money options							
Expiration Date	Friday before third Wednesday of the month (Friday is also the last trading day)							
Expiration Settlement Date	Third Wednesday of month							
Daily Price Limits	None							
Issuer & Guarantor	Options Clearing Corporation (OCC)							
Margin for Uncovered Writer	Option premium plus 4% of the underlying contract value less out-of-the-money amount, if any, to a minimum of the option premium plus 3/4% of the underlying contract value. Contract value equal spot price times unit of currency per contract.							
Position & Exercise Limits	100,000 contracts							
Trading Hours	2:30 A.M.–2:30 P.M. Philadelphia time, Monday through Friday[2]							

[1] Half-point strike prices for DM (0.5¢), SFr (0.5¢), and ¥ (0.005¢) in the three near-term months only.

[2] Trading hours for the Canadian dollar are 7:00 A.M.–2:30 P.M. Philadelphia time, Monday through Friday.

Source: Standardized Currency Options Contract Specifications, Philadelphia Stock Exchange, 1995. Used with permission of the Philadelphia Stock Exchange.

all financial centers. OTC activity is concentrated in London and New York and it centers on the major currencies, most often involving U.S. dollars against pounds sterling, Deutsche marks, Swiss francs, Japanese yen, and Canadian dollars. Branches of foreign banks in the major financial centers are generally willing to write options against the currency of their home country. For example, Australian banks in London write options on the Australian dollar. Generally, OTC options are traded in round lots, commonly $5–$10 million in New York and $2–$3 million in London. The average maturity of OTC options ranges from two to six months, and very few options are written for more than one year. American options are most common, but European options are popular in Switzerland and Germany because of familiarity.

The OTC options market consists of two sectors: (1) a *retail market* composed of nonbank customers who purchase from banks what amounts to customized insurance against adverse exchange rate movements and (2) a wholesale market among commercial banks, investment banks, and specialized trading firms; this market may include interbank OTC trading or trading on the organized exchanges. The interbank market in currency options is analogous to the interbank markets in spot and forward exchange. Banks use the wholesale market to hedge, or "reinsure," the risks undertaken in trading with customers and to take speculative positions in options.

Most retail customers for OTC options are either corporations active in international trade or financial institutions with multicurrency asset portfolios. These customers could purchase foreign exchange puts or calls on organized exchanges, but they generally turn to the banks for options in order to find precisely the terms that match their needs. Contracts are generally tailored with regard to amount, strike price, expiration date, and currency.

EXHIBIT 6.17 PHLX Currency Options Trading Volume, 1983–1997*

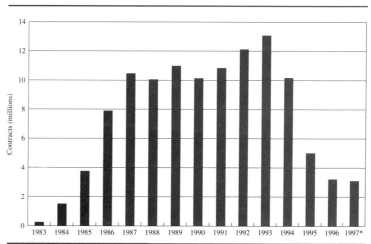

*January–March 1997 annualized.

Source: Philadelphia Stock Exchange, 1997.

The existence of OTC currency options predates exchange-traded options by many years, but trading in OTC options grew rapidly at the same time that PHLX trading began. The acceleration in the growth of options trading in both markets appears to spring from the desire by companies to manage foreign currency risks more effectively and, in particular, from an increased willingness to pay a fee to transfer such risks to another party. Most commentators suggest that corporate demand has increased because the greater volatility of exchange rates has increasingly exposed firms to risks from developments that are difficult to predict and beyond their control.

The growth of listed options, especially for "wholesale" purposes, apparently is putting pressure on the OTC markets for greater standardization in interbank trading. In some instances, OTC foreign currency options are traded for expiration on the third Wednesday of March, June, September, and December, to coincide with expiration dates on the U.S. exchanges.

Although the buyer of an option can lose only the premium paid for the option, the seller's risk of loss is potentially unlimited. Because of this asymmetry between income and risk, few retail customers are willing to write options. For this reason, the market structure is distinctly asymmetrical when compared with the ordinary market for spot and forward foreign exchange, where there is a balance between customers who are purchasing or selling currency and where the interbank market likewise has a reasonable balance.

Futures Options

In January 1984, the IMM introduced a market in options on DM futures contracts. Since then, the **futures option** market has grown to include options on futures in British pounds, Japanese yen, Swiss francs, and Canadian dollars as well. Trading involves purchases and sales of puts and calls on a contract calling for delivery of a standard IMM futures contract in the currency rather than the currency itself. When such a contract is exercised, the holder receives a short or long position in the underlying currency futures contract that is marked-to-market, providing the holder with a cash gain. (If there were a loss on the futures contract, the option would not be exercised.) Specifically,

1. If a call futures option contract is exercised, the holder receives a long position in the underlying futures contract plus an amount of cash equal to the current futures price minus the strike price.

2. If a put futures option is exercised, the holder receives a short position in the futures contract plus an amount of cash equal to the strike price minus the current futures price.

The seller of these options has the opposite position to the holder after exercise: a cash outflow plus a short futures position on a call and a long futures position on a put option.

A futures option contract has this advantage over a futures contract: With a futures contract, the holder must deliver one currency against the other or reverse the contract, regardless of whether this move is profitable. In contrast, with the futures option contract, the holder is protected against an adverse move in the exchange rate but may allow the option to expire unexercised if using the spot market would be more profitable.

EXERCISING A POUND CALL FUTURES OPTION CONTRACT

An investor is holding a pound call futures option contract for June delivery (representing £62,500) at a strike price of $1.5050. The current price of a pound futures contract due in June is $1.5148. What will the investor receive if she exercises her futures option?

Solution. The investor will receive a long position in the June futures contract established at a price of $1.5050, and the option writer has a short position in the same futures contract. These positions are immediately marked to market, triggering a cash payment to the investor from the option writer of 62,500 ($1.5148 − $1.5050) = $612.50. If the investor desires, she can immediately close out her long futures position at no cost, leaving her with the $612.50 payoff.

EXERCISING A SWISS FRANC PUT FUTURES OPTION CONTRACT

An investor is holding one Swiss franc March put futures option contract (representing SFr 125,000) at a strike price of $0.7950. The current price of a Swiss franc futures contract for March delivery is $0.8132. What will the investor receive if she exercises her futures option?

Solution. The investor will receive a short position in the March futures contract established at a price of $0.7950, and the option writer has a long position in the same futures contract. These positions are immediately marked to market and the investor will receive a cash payment from the option writer of 125,000 ($0.8132 − 0.7950) = $2,275. If the investor desires, she can immediately close out her short futures position at no cost, leaving her with the $2,275 payoff.

6.3 READING CURRENCY FUTURES AND OPTIONS PRICES

Futures and exchange-listed options prices appear daily in the financial press. Exhibit 6.18 shows prices for April 7, 1997, as displayed in the *Wall Street Journal* on the following day. Futures prices on the IMM are listed for seven currencies, with one to five contracts quoted for each currency: June, September, and December 1997 for all but the Australian dollar, and March and June 1998 for some currencies. Included are the opening and last settlement (settle) prices, the change from the previous trading day, the range for the day, and the number of contracts outstanding (open interest). For example, the June Deutsche mark futures contract opened at $0.5966 per DM and closed down at $0.5870 per DM. Futures prices are shown in Exhibit 6.18a.

EXHIBIT 6.18 Foreign Currency Futures and Options Quotations

FUTURES

	Open	High	Low	Settle	Change	Lifetime High	Lifetime Low	Open Interest
JAPAN YEN (CME)-12.5 million yen; $ per yen (.00)								
June	.1829	.8151	.8026	.8047	− .0082	.9790	.8026	69,840
Sept	.8220	.8220	.8135	.8158	− .0084	.9790	.8135	943
Dec	.8290	.8290	.8260	.8274	− .0086	.9320	.8260	463

Est vol 23,713; vol Fr 32,092; open int 71,246, + 1,712.

	Open	High	Low	Settle	Change	Lifetime High	Lifetime Low	Open Interest
DEUTSCHEMARK (CME)-125,000 marks; $ per mark								
June	.5966	.5966	.5870	.5876	− .0090	.6947	.5840	62,736
Sept	.5950	.5950	.5910	.5916	− .0090	.6635	.5888	2,533
Dec				.5960	− .0090	.6610	.5933	207

Est vol 23,608; vol Fr 31,844; open int 65,503, + 32.

	Open	High	Low	Settle	Change	Lifetime High	Lifetime Low	Open Interest
CANADIAN DOLLAR (CME)-100,000 dlrs.; $ per Can $								
June	.7219	.7247	.7205	.7244	+ .0030	.7635	.7185	74,644
Sept	.7249	.7284	.7249	.7279	+ .0031	.7662	.7235	4,688
Dec	.7290	.7315	.7290	.7308	+ .0031	.7685	.7270	1,171
Mr98				.7336	+ .0031	.7670	.7324	772
June				.7353	+ .0031	.7470	.7350	102

Est vol 5,778; vol Fr 13,135; open int 81,407, + 1,859.

	Open	High	Low	Settle	Change	Lifetime High	Lifetime Low	Open Interest
BRITISH POUND (CME)-62,500 pds.; $ per pound								
June	1.6328	1.6344	1.5630	1.6268	− .0060	1.6996	1.5520	35,875
Sept	1.6220	1.6262	1.6200	1.6238	− .0064	1.6840	1.5790	841
Dec				1.6208	− .0066	1.6970	1.5800	101

Est vol 9,340; vol Fr 8,826; open int 36,817, −1,508.

	Open	High	Low	Settle	Change	Lifetime High	Lifetime Low	Open Interest
SWISS FRANC (CME)-125,000 francs; $ per franc								
June	.6978	.6987	.6856	.6863	− .0122	.8660	.6767	40,323
Sept	.7025	.7025	.6924	.6933	− .0123	.8082	.6835	2,133
Dec	.7015	.7015	.7000	.7008	− .0124	.7740	.6910	429

Est vol 5,778; vol Fr 18,781; open int 42,935, −381.

	Open	High	Low	Settle	Change	Lifetime High	Lifetime Low	Open Interest
AUSTRALIAN DOLLAR (CME)-100,000 dlrs.; $ per A. $								
June	.7785	.7730	.7753	+ .0003	.8100	.7555		11,772

Est vol 476; vol Fr 2,928; open int 11,787, −366.

	Open	High	Low	Settle	Change	Lifetime High	Lifetime Low	Open Interest
MEXICAN PESO (CME)-500,000 new Mex. peso, $ per MP								
June	.12210	.12260	.12210	.12257	+ .0825	.12330	.10270	16,887
Sept	.11770	.11800	.11760	.11787	+ .0825	.11840	.10250	6,709
Dec	.11300	.11340	.11300	.11340	+ .0850	.11380	.09850	5,333
Mr98	.10910	.10910	.10910	.10930	+ .0850	.11010	.10400	501

Est vol 6,659; vol Fr 8,443; open int 29,528, −424.

(a)

FUTURES OPTIONS

JAPANESE YEN (CME)
12,500,000 yen; cents per 100 yen

Strike Price	Calls-Settle May	Calls-Settle Jun	Calls-Settle Jly	Puts-Settle May	Puts-Settle Jun	Puts-Settle Jly
7950	....	1.92	....	.63	.96	....
8000	1.29	1.63	....	.82	1.16	....
8050	1.00	1.36	....	1.03	1.39	....
8100	.78	1.13	....	1.31	1.65	....
8150	.59	.93	1.72	1.61	1.95	....
8200	.45	.76	1.49	1.97	2.28	....

Est vol 13,777 Fr 16,590 calls 7,380 puts
Op Int Fri 41,734 calls 48,098 puts

DEUTSCHEMARK (CME)
125,000 marks; cents per mark

Strike Price	Calls-Settle May	Calls-Settle Jun	Calls-Settle Jly	Puts-Settle May	Puts-Settle Jun	Puts-Settle Jly
5800	1.13	1.34	....	.37	.59	....
5850	.81	1.06	....	.55	.80	....
5900	.55	.80	....	.79	1.04	....
5950	.37	.60	....	1.11	1.33	....
6000	.25	.44	....	1.48	1.67	....
6050	.16	.32	....	1.89	2.04	....

Est vol 7,731 Fr 11,962 calls 11,721 puts
Op Int Fri 41,018 calls 40,953 puts

CANADIAN DOLLAR (CME)
100,000 Can.$; cents per Can.$

Strike Price	Calls-Settle May	Calls-Settle Jun	Calls-Settle Jly	Puts-Settle May	Puts-Settle Jun	Puts-Settle Jly
7150	....	....	....	.08	.17	....
7200	.63	.75	....	.19	.31	....
7250	.33	.45	....	.39	.51	....
7300	.16	.27	....	.72	.82	....
7350	.07	.16	....	1.13	1.21	....
7400	.03	.10	....	1.58	1.65	....

Est vol 1,082 Fr 3,009 calls 932 puts
Op Int Fri 13,543 calls 6,999 puts

BRITISH POUND (CME)
62,500 pounds; cents per pound

Strike Price	Calls-Settle May	Calls-Settle Jun	Calls-Settle Jly	Puts-Settle May	Puts-Settle Jun	Puts-Settle Jly
16100	3.54	3.08	....	.86	1.42	....
16200	1.92	2.52	....	1.24	1.84	....
16300	1.38	2.00	....	1.70	2.32	....
16400	.96	1.58	....	2.28	2.90	....
16500	.66	1.22	....	2.98	3.52	....
16600	.25	.94	....	3.74	4.24	....

Est vol 6,475 Fr 2,908 calls 4,774 puts
Op Int Fri 27,628 calls 25,969 puts

SWISS FRANC (CME)
125,000 francs; cents per franc

Strike Price	Calls-Settle May	Calls-Settle Jun	Calls-Settle Jly	Puts-Settle May	Puts-Settle Jun	Puts-Settle Jly
6750	....	....	....	.48	.81	....
6800	1.28	1.62	....	.65	1.00	....
6850	1.00	1.36	....	.87	1.23	....
6900	.76	1.12	....	1.13	1.49	....
6950	.58	.92	....	1.45	1.78	....
7000	.43	.75	....	1.79	2.11	....

Est vol 4,007 Fr 4,258 calls 4,516 puts
Op Int Fri 20,065 calls 19,387 puts

BRAZILIAN REAL (CME)
100,00 Braz. reals; $ per reals

Strike Price	Calls-Settle May	Calls-Settle Jun	Calls-Settle Jly	Puts-Settle May	Puts-Settle Jun	Puts-Settle Jly
930	....	....	....	....	0.15	....

Est vol 0 Fr 0 calls 200 puts
Op Int Fri 0 calls 12,681 puts

(b)

OPTIONS
PHILADELPHIA EXCHANGE

		Calls Vol.	Calls Last	Puts Vol.	Puts Last
JYen					**79.63**
6,250,000 Japanese Yen-100thsofacentperunit.					
79	Apr	...	...	27	0.24
79	Jun	25	2.31	3590	0.94
80	Apr	...	...	15	0.66
81	Apr	...	...	2	1.46
81	May	25	0.74	14	1.76
81	Jun	2	1.43	...	...
82	Jun	5	0.84	...	...
83	Jun	50	0.58	152	3.40
85	Jun	180	0.26	...	...
86	Jun	11	0.15	...	...
DMark					**58.48**
62,500 German Mark EOM-European style.					
58	May	...	...	55	0.55
59	Apr	...	...	50	0.74
62,500 German Marks EOM-European style.					
58½	May	...	...	25	0.72
59½	Apr	...	...	25	1.10
Australian Dollar					**77.62**
50,000 Australian Dollars-European Style.					
76	May	...	...	30	0.16
50,000 Australian Dollars-cents per unit.					
78½	May	50	0.41	...	...
50,000 Australian Dollars-cents per unit.					
77	Apr	...	...	50	0.09
78	Apr	...	...	150	0.52
78	May	5	0.64	...	...
British Pound					**162.83**
31,250 British Pounds-European Style.					
163	May	25	1.64	...	...
31,250 British Pounds-cents per unit.					
160	Jun	...	...	1	1.26
165	Jun	1	1.29	...	...
British Pound-GMark					**278.44**
31,250 British Pound-German Mark cross.					
270	May	...	...	16	0.86
276	Apr	64	1.70	...	...

		Calls Vol.	Calls Last	Puts Vol.	Puts Last
			Calls		Puts
		Vol.	Last	Vol.	Last
278	Apr	64	0.80	...	...
Canadian Dollar					**72.10**
50,000 Canadian Dollars-cents per unit.					
72	Jun	20	0.73	...	...
72½	Apr	25	0.07	...	...
ECU					**114.37**
62,500 European Currency Units-cents per unit.					
118	Jun	...	...	12	3.32
French Franc					**173.86**
250,000 French Francs 10ths of a cent per.					
17	Jun	...	...	50	1.12
17½	May	...	...	40	2.52
17½	Jun	...	...	2433	2.90
18	May	...	...	50	6.30
German Mark					**58.48**
62,500 German Marks EOM-cents per unit.					
58½	Apr	...	...	10	0.36
60½	Apr	10	0.12	...	...
62,500 German Marks-European Style.					
58	Apr	...	...	15	0.09
58½	Apr	...	...	20	0.29
59	Apr	15	1.17	15	0.58
59½	Apr	...	...	25	0.73
60	May	24	0.33	...	...
61	May	24	0.14	...	...
62,500 German Marks-cents per unit.					
57	Jun	...	...	10	0.34
57½	Apr	9	1.14	...	...
58	Apr	...	...	420	0.13
58	Jun	...	...	43	0.65
58½	Apr	300	0.33	500	0.13
58½	May	4	0.92	...	...
59	Apr	100	0.15	69	0.60
59	May	...	...	19	0.94
59	Jun	2	1.00	842	1.09
59½	Apr	...	...	10	0.95

		Calls Vol.	Calls Last	Puts Vol.	Puts Last
59½	Jun	...	...	51	1.37
60	May	25	0.38	1	1.25
60	Jun	...	...	221	1.73
61	May	60	0.11	...	...
Japanese Yen					**79.63**
6,250,000 Japanese Yen-100thsofacentperunit.					
79½	Apr	...	...	6	0.40
80½	Apr	2	0.30	4	1.03
80½	Apr	4	0.57	...	...
81½	May	...	...	2	1.90
82½	Jun	2	0.82	...	...
6,250,000 Japanese Yen-European Style.					
79	May	...	...	200	0.76
79½	May	...	...	5	0.87
81	Apr	...	...	100	1.45
81	Jun	4	1.47	...	...
Swiss Franc					**68.17**
62,500 Swiss Francs EOM-cents per unit.					
69	Apr	...	...	2	0.84
62,500 Swiss Francs-European Style.					
68	Apr	10	0.91	...	...
68	May	10	1.59	...	...
68½	May	225	1.19	...	...
69½	May	64	0.65	...	...
70	Apr	50	0.06	...	...
71	Apr	...	...	12	2.60
62,500 Swiss Francs-cents per unit.					
67½	Apr	...	...	10	0.12
68	May	...	...	100	0.50
68	Jun	...	...	50	1.05
68½	Apr	156	0.35	...	...
69	Apr	50	0.18	...	...
69	May	62	0.90	...	...
69	Jun	20	1.24	...	...
69½	Apr	10	0.09	7	0.98
69½	May	...	...	11	1.67
70	Apr	...	...	40	1.01
71	Jun	15	0.60	...	...
Call Vol 11,977				**Open Int ... 154,950**	
Put Vol 100				**Open Int ... 146,634**	

(c)

Source: Wall Street Journal, April 8, 1997, (a), p. C16; (b), p. C17; (c), p. C12. Reprinted with permission of the *Wall Street Journal*, © 1997 Dow Jones & Company, Inc. All rights reserved.

Exhibit 6.18b shows the Chicago Mercantile Exchange (IMM) options on this same futures contract. To interpret the numbers in this column, consider the call options. These are rights to buy the June DM futures contract at specified prices—the strike prices. For example, the call option with a strike price of 5800 means that you can purchase an option to buy a June DM futures contract, up to the June settlement date, for $0.5800 per mark. This option will cost $0.0134 per Deutsche mark, or $1,675, plus brokerage commission, for a DM 125,000 contract. The price is high because the option is in-the-money.

EXHIBIT 6.19 How to Read Futures and Futures Options Quotations

Trading activity can be monitored daily in the business pages of most major newspapers. The following displays are illustrations of the way these prices are shown.

Futures

DEUTSCHE MARK (CME)–125,000 marks; $ per mark

	Open	High	Low	Settle	Chg	Lifetime High	Low	Open Interest
Mar	.5711	.5782	.5703	.5764	+.0053	.6205	.5646	143,059
June	.5680	.5750	.5673	.5632	+.0053	.6162	.5607	8,202
Sept	.5690	.5654	.5690	.5708	+.0053	.6130	.5610	296

Est. vol. 45,259; vol. Wed. 32,307; open int. 151,594 − 2,049

1. Prices represent the open, high, low, and settlement (or closing) price for the previous day.
2. Contract delivery months that are currently traded.
3. Number of contracts traded in the previous two trading sessions.
4. One day's change in the settlement price.
5. The total of the right column, and the change from the prior trading day.
6. The extreme prices recorded for the contract over its trading life.
7. The number of contracts still in effect at the end of the previous day's trading session. Each unit represents a buyer *and* a seller who still have a contract position.

Options on Futures

DEUTSCHE MARK (CME)–125,000 marks; cents per mark

Strike Price	Calls–Settle Feb	Mar	Apr	Puts–Settle Feb	Mar	Apr
5650	1.20	1.37	1.44	0.06	0.24	0.63
5700	0.75	1.04	1.15	0.11	0.40	0.83
5750	0.41	0.75	0.90	0.27	0.61	1.08
5800	0.20	0.52	0.69	0.56	0.88	....
5850	0.09	0.34	0.52	0.95	1.19	....
5900	0.02	0.22	0.39	1.38	1.57	....

Est. vol. 12,585; Wed. vol. 7,875 calls; 9,754 puts
Open interest Wed. 111,163 calls; 74,498 puts

1. Most active strike prices.
2. Expiration months.
3. Closing prices for call options.
4. Closing prices for put options.
5. Volume of options transacted in the previous two trading sessions. Each unit represents both the buyer *and* the seller.
6. The number of options that were still open positions at the end of the previous day's trading session.

Source: An Introduction to Currency Futures and Options, Chicago Mercantile Exchange, 1994, p. 14.

In contrast, the June futures option with a strike price of 6000, which is out-of-the-money, costs only $0.0044 per mark, or $550 for one contract. These option prices indicate that the market expects the dollar price of the Deutsche mark to exceed $0.5800 but not to rise much above $0.6000 by June.

As we have just seen, a futures call option allows you to buy the relevant futures contract, which is settled at maturity. On the other hand, the Philadelphia call options contract is an option to buy foreign exchange spot, which is settled when the call option is exercised; the buyer receives foreign currency immediately.

Price quotes usually reflect this difference. For example, PHLX call options for the June DM (shown in Exhibit 6.18c), with a strike price of $0.5900, are $0.01 per mark (versus $0.0080 for the June futures call option), or $625, plus brokerage fees for one contract of DM 62,500. Brokerage fees here would be about the same as on the IMM: about $16 per transaction round trip per contract. Exhibit 6.19 summarizes how to read price quotations for futures and options on futures using a DM illustration.

❧ 6.4 SUMMARY AND CONCLUSIONS

In this chapter, we examined the currency futures and options markets and looked at some of the institutional characteristics and mechanics of these markets. We saw that currency futures and options offer alternative hedging (and speculative) mechanisms for companies and individuals. Like forward contracts, futures contracts must be settled at maturity. By contrast, currency options give the owner the right but not the obligation to buy (call option) or sell (put option) the contracted currency. An American option can be exercised at any time up to the expiration date; a European option can be exercised only at maturity.

Futures contracts are standardized contracts that trade on organized exchanges. Forward contracts, on the other hand, are custom-tailored contracts, typically entered into between a bank and its customers. Options contracts are sold on both organized exchanges and in the over-the-counter (OTC) market. Like forward contracts, OTC options are contracts whose specifications are generally negotiated as to the terms and conditions between a bank and its customers.

We also derived a put-call parity condition that relates option prices to interest rates and the forward rate. As with interest rate parity, this condition relies on arbitrage for its existence. Formally, this condition is expressed as:

Put-call option interest rate parity

$$C = \frac{e_0}{1 + r_f} - \frac{X}{1 + r_h} + P$$

WHERE

C = call option premium on a one-period contract
P = put option premium on a one-period contract
X = exercise price on the put and call options (HC per unit of foreign currency)

➤ Questions

1. On April 1, the spot price of the British pound was $1.86 and the price of the June futures contract was $1.85. During April the pound appreciated, so that by May 1 it was selling for $1.91. What do you think happened to the price of the June pound futures contract during April? Explain.

2. Suppose that Texas Instruments (TI) must pay a French supplier FF 10 million in 90 days.
   a. Explain how TI can use currency futures to hedge its exchange risk. How many futures contracts will TI need to fully protect itself?
   b. Explain how TI can use currency options to hedge its exchange risk. How many options contracts will TI need to fully protect itself?
   c. Discuss the advantages and disadvantages of using currency futures versus currency options to hedge TI's exchange risk.

3. A forward market already existed, so why was it necessary to establish currency futures and currency options contracts?

4. What are the basic differences between forward and futures contracts? Between futures and options contracts?

5. What is the last day of trading and the settlement day for the IMM Australian dollar futures for September of the current year?

6. Which contract is likely to be more valuable, an American or a European call option? Explain.

7. In Exhibit 6.7, the value of the call option is shown as approaching its intrinsic value as the option goes deeper and deeper in-the-money or further and further out-of-the-money. Explain why this is so.

8. Suppose that Bechtel Group wants to hedge a bid on a Japanese construction project. But because the yen exposure is contingent on acceptance of its bid, Bechtel decides to buy a put option for the ¥15 billion bid amount rather than sell it forward. In order to reduce its hedging cost, however, Bechtel simultaneously sells a call option for ¥15 billion with the same strike price. Bechtel reasons that it wants to protect its downside risk on the contract and is willing to sacrifice the upside potential in order to collect the call premium. Comment on Bechtel's hedging strategy.

9. During September 1992, options on ERM currencies with strike prices outside the ERM bands had positive values. At the same time, actual currency volatility was close to zero.
   a. Is there a paradox here? Explain.
   b. Why might actual currency volatility have been close to zero? What does a zero volatility imply about the value of currency options?
   c. What do the positive values of ERM options outside the bands tell you about the market's perceptions of the possibility of currency devaluations or revaluations?

➤ Problems

1. On Monday morning, an investor takes a long position in a pound futures contract that matures on Wednesday afternoon. The agreed-upon price is $1.78 for £62,500. At the close of trading on Monday, the futures price has risen to $1.79. At Tuesday close, the price rises further to $1.80. At Wednesday close, the price falls to $1.785, and the contract matures. The investor takes delivery of the pounds at the prevailing price of $1.785. Detail the daily settlement process (see Exhibit 6.2). What will be the investor's profit (loss)?

2. On Monday morning, an investor takes a short position in a DM futures contract that matures on Wednesday afternoon. The agreed-upon price is $0.6370 for DM 125,000. At the close of trading on Monday, the futures price has fallen to $0.6315. At Tuesday close, the price falls further to $0.6291. At Wednesday close, the price rises to $0.6420, and the contract matures. The investor delivers the Deutsche marks at the prevailing price of $0.6420. Detail the daily settlement process (see Exhibit 6.2). What will be the investor's profit (loss)?

3. Suppose that the forward ask price for March 20 on DM is $0.7127 at the same time that the price of IMM mark futures for delivery on March 20 is $0.7145. How could an arbitrageur profit from this situation? What will be the arbitrageur's profit per futures contract (contract size is DM 125,000)?

4. On August 6, you go long one IMM yen futures contract at an opening price of $0.00812 with an initial margin of $4,590. The settlement prices for August 6, 7, and 8 are $0.00791, $0.00845, and $0.00894, respectively. On August 9, you close out the contract at a price of $0.00857. Your round-trip commission is $31.48.
   a. Calculate the daily cash flows on your account. Be sure to take into account your required margin and any margin calls.

b. What is your cash balance with your broker on the morning of August 10?

5. Suppose that DEC buys a Swiss franc futures contract (contract size is SFr 125,000) at a price of $0.83. If the spot rate for the Swiss franc at the date of settlement is SFr 1 = $0.8250, what is DEC's gain or loss on this contract?

6. On January 10, Volkswagen agrees to import auto parts worth $7 million from the United States. The parts will be delivered on March 4 and are payable immediately in dollars. VW decides to hedge its dollar position by entering into IMM futures contracts. The spot rate is DM 1.8347/$1, and the March futures price is DM 1.8002.
   a. Calculate the number of futures contracts that VW must buy to offset its dollar exchange risk on the parts contract.
   b. On March 4, the spot rate turns out to be DM 1.7952, while the March futures price is DM 1.7968. Calculate VW's net DM gain or loss on its futures position. Compare this figure with VW's gain or loss on its unhedged position.

7. Citicorp sells a call option on Deutsche marks (contract size is DM 500,000) at a premium of $0.04 per DM. If the exercise price is $0.71 and the spot price of the mark at date of expiration is $0.73, what is Citicorp's profit (loss) on the call option?

8. Ford buys a French franc put option (contract size is FF 250,000) at a premium of $0.01 per franc. If the exercise price is $0.21 and the spot price of the franc at date of expiration is $0.216, what is Ford's profit (loss) on the put option?

9. Suppose you buy three April PHLX DM call options with a 59 strike price at the price shown in Exhibit 6.18c.
   a. What would be your total dollar cost for these calls, ignoring broker fees?
   b. What is the most that you would have paid for the DM 187,500 covered by these calls?
   c. After holding these calls for 60 days, you sell them for 3.8 (¢/DM). What is your net profit on the contracts, assuming that brokerage fees on both entry and exit were $5 per contract and that your opportunity cost was 8% per annum on the money tied up in the premium?

10. A trader executes a "bear spread" on the Japanese yen consisting of a long PHLX 103 March put and a short PHLX 101 March put.
   a. If the price of the 103 put is 2.81 (100ths of ¢/¥), while the price of the 101 put is 1.6 (100ths of ¢/¥), what is the net cost of the bear spread?
   b. What is the maximum amount the trader can make on the bear spread in the event the yen depreciates against the dollar?

c. Redo your answers to parts a and b, assuming the trader executes a "bull spread" consisting of a long PHLX 101 March call priced at 1.96 (100ths of ¢/¥) and a short PHLX 99 March call priced at 3.21(100ths of ¢/¥). What is the trader's maximum profit? Maximum loss?

11. Apex Corporation must pay its Japanese supplier ¥125 million in three months. It is thinking of buying 20 yen call options (contract size is ¥6.25 million) at a strike price of $0.00800 in order to protect against the risk of a rising yen. The premium is 0.015 cents per yen. Alternatively, Apex could buy 10 three-month yen futures contracts (contract size is ¥12.5 million) at a price of $0.007940/¥. The current spot rate is ¥1 = $0.007823. Apex's treasurer believes that the most likely value for the yen in 90 days is $0.007900, but the yen could go as high as $0.008400 or as low as $0.007500.
   a. Diagram Apex's gains and losses on the call option position and the futures position within its range of expected prices (see Exhibit 6.4). Ignore transaction costs and margins.
   b. Calculate what Apex would gain or lose on the option and futures positions if the yen settled at its most likely value.
   c. What is Apex's breakeven future spot price on the option contract? On the futures contract?
   d. Calculate and diagram the corresponding profit and loss and breakeven positions on the futures and options contracts for the sellers of these contracts.

12. Biogen expects to receive royalty payments totaling £1.25 million next month. It is interested in protecting these receipts against a drop in the value of the pound. It can sell 30-day pound futures at a price of $1.6513/£ or it can buy pound put options with a strike price of $1.6612 at a premium of 2.0 cents per pound. The spot price of the pound is currently $1.6560, and the pound is expected to trade in the range of $1.6250 to $1.70100. Biogen's treasurer believes that the most likely price of the pound in 30 days will be $1.6400.
   a. How many futures contracts will Biogen need to protect its receipts? How many options contracts?
   b. Diagram Biogen's profit and loss on the put option position and the futures position within its range of expected exchange rates (see Exhibit 6.6). Ignore transaction costs and margins.
   c. Calculate what Biogen would gain or lose on the option and futures positions within the range of expected future exchange rates and if the pound settled at its most likely value.

d. What is Biogen's breakeven future spot price on the option contract? On the futures contract?

e. Calculate and diagram the corresponding profit and loss and breakeven positions on the futures and options contracts for those who took the other side of these contracts.

13. Assume that the spot price of the British pound is $1.55, the 30-day annualized sterling interest rate is 10%, the 30-day annualized U.S. interest rate is 8.5%, and the annualized standard deviation of the dollar : pound exchange rate is 17%. Calculate the value of a 30-day PHLX call option on the pound at a strike price of $1.57.

14. Suppose the spot price of the yen is $0.0109, the 3-month annualized yen interest rate is 3%, the 3-month annualized dollar rate is 6%, and the annualized standard deviation of the dollar : yen exchange rate is 13.5%. What is the value of a three-

month PHLX call option on the Japanese yen at a strike price of $0.0099/¥?

15. Suppose that the premium on March 20 on a June 20 yen put option is 0.0514 cents per yen at a strike price of $0.0077. The forward rate for June 20 is ¥1 = $0.00787 and the quarterly U.S. interest rate is 2%. If put-call parity holds, what is the current price of a June 20 PHLX yen call option with an exercise price of $0.0077?

16. On June 25, the call premium on a December 25 PHLX contract is 6.65 cents per pound at a strike price of $1.81. The 180-day (annualized) interest rate is 7.5% in London and 4.75% in New York. If the current spot rate is £1 = $1.8470 and put-call parity holds, what is the put premium on a December 25 PHLX pound contract with an exercise price of $1.81?

➤ Bibliography

BATES, DAVID S. "Jumps and Stochastic Volatility: Exchange Rate Processes Implicit in PHLX Deutschemark Options," Wharton School working paper, 1993.

BLACK, FISCHER, and MYRON SCHOLES. "The Pricing of Options and Corporate Liabilities." *Journal of Political Economy*, May–June 1973, pp. 637–659.

BODURTHA, JAMES, N., JR., and GEORGES R. COURTADON. "Tests of an American Option Pricing Model on the Foreign Currency Options Market." *Journal of Financial and Quantitative Analysis*, June 1987, pp. 153–167.

Chicago Mercantile Exchange. *Using Currency Futures and Options*. Chicago: CME, 1987.

CORNELL, BRADFORD, and MARC REINGANUM. "Forward and Futures Prices: Evidence from the Foreign

Exchange Markets." *Journal of Finance*, December 1991, pp. 1035–1045.

GARMAN, MARK B., and STEVEN W. KOHLHAGEN. "Foreign Currency Option Values." *Journal of International Money and Finance*, December 1983, pp. 231–237.

JORION, PHILLIPE. "On Jump Processes in the Foreign Exchange and Stock Markets." *Review of Financial Studies* 1, no. 4 (1988): 427–445.

SHASTRI, KULEEP, and KULPATRA WETHYAVIVORN. "The Valuation of Currency Options for Alternate Stochastic Processes." *Journal of Financial Research*, Winter 1987, pp. 283–293.

PARITY CONDITIONS IN INTERNATIONAL FINANCE AND CURRENCY FORECASTING

It is not for its own sake that men desire money, but for the sake of what they can purchase with it.

Adam Smith (1776)

On the basis of the flows of goods and capital discussed in Chapter 2, this chapter presents a simple, yet elegant, set of equilibrium relationships that should apply to product prices, interest rates, and spot and forward exchange rates if markets are not impeded. These relationships, or *parity conditions*, provide the foundation for much of the remainder of this text; they should be clearly understood before you proceed further. The final section of this chapter examines the usefulness of a number of models and methodologies in profitably forecasting currency changes under both fixed-rate and floating-rate systems.

7.1 ARBITRAGE AND THE LAW OF ONE PRICE

In competitive markets, characterized by numerous buyers and sellers having low-cost access to information, exchange-adjusted prices of identical tradable goods and financial assets must be within transaction costs of equality worldwide. This idea, referred to as the

law of one price, is enforced by international arbitrageurs who follow the profit-guaranteeing dictum of "buy low, sell high" and prevent all but trivial deviations from equality. Similarly, in the absence of market imperfections, risk-adjusted expected returns on financial assets in different markets should be equal.

Five key theoretical economic relationships, which are depicted in Exhibit 7.1 (we have already encountered interest rate parity in Chapter 5), result from these arbitrage activities:

■ Purchasing power parity (PPP)
■ Fisher effect (FE)
■ International Fisher effect (IFE)
■ Interest rate parity (IRP)
■ Forward rates as unbiased predictors of future spot rates (UFR)

The framework of Exhibit 7.1 emphasizes the links among prices, spot exchange rates, interest rates, and forward exchange rates. According to the diagram, if inflation in, say, France is expected to exceed inflation in the United States by 3% for the coming year, then the French franc should decline in value by about 3% relative to the dollar. By

EXHIBIT 7.1 Five Key Theoretical Relationships Among Spot Rates, Forward Rates, Inflation Rates, and Interest Rates

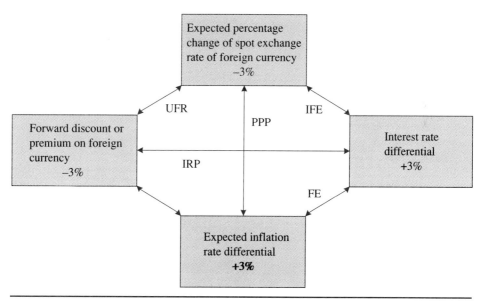

Note: UFR = Forward rates as unbiased predictors of future spot rates
　　　 PPP = Purchasing power parity
　　　 IFE = International Fisher effect
　　　 FE = Fisher effect
　　　 IRP = Interest rate parity

the same token, the one-year forward French franc should sell at a 3% discount relative to the U.S. dollar. Similarly, one-year interest rates in France should be about 3% higher than one-year interest rates on securities of comparable risk in the United States.

The common denominator of these parity conditions is the adjustment of the various rates and prices to inflation. According to modern monetary theory, inflation is the logical outcome of an expansion of the money supply in excess of real output growth. Although this view of the origin of inflation is not universally subscribed to, it has a solid microeconomic foundation. In particular, it is a basic precept of price theory that as the supply of one commodity increases relative to supplies of all other commodities, the price of the first commodity must decline relative to the prices of other commodities. Thus, for example, a bumper crop of corn should cause corn's value in exchange—its exchange rate—to decline. Similarly, as the supply of money increases relative to the supply of goods and services, the purchasing power of money—the exchange rate between money and goods—must decline.

The mechanism that brings this adjustment about is simple and direct. Suppose, for example, that the supply of U.S. dollars exceeds the amount that individuals desire to hold. In order to reduce their excess holdings of money, individuals increase their spending on goods, services, and securities, causing U.S. prices to rise.

ILLUSTRATION **B**OLIVIA ENDS ITS HYPERINFLATION

In the spring of 1985, Bolivia's inflation rate was running at 25,000% a year, one of the highest rates in history. At the time, Bolivian government revenues covered less than 15% of its spending, with most of the rest being paid for by printing new pesos. Inflation threatened the very fabric of society. Prices changed by the minute, and people literally carried money around in suitcases. Currency, which was printed abroad, was the third-largest import in 1984. The two-inch stack of money needed to buy a chocolate bar far outweighed the candy. The government eventually solved the stacks-of-money problem by issuing 1-million-, 2-million-, and 10-million-peso notes. But its failure to solve the inflation problem led to its replacement by a new government that announced an anti-inflation program on August 29, 1985. The program had two basic thrusts: Cut spending, and shut down the printing presses. To cut spending, the new government adopted the simple rule that it would not spend more than it received. Each day the finance minister signed checks only up to the value of the revenues the treasury had received that day. By October, the monthly inflation rate had fallen to zero, from more than 60% in August (see Exhibit 7.2). Economists consider this performance to be a singular verification of basic monetary theory.

A further link in the chain relating money supply growth, inflation, interest rates, and exchange rates is the notion that money is neutral. That is, money should have no impact on real variables. Thus, for example, a 10% increase in the supply of money relative to the demand for money should cause prices to rise by 10%. This view has important implications for international finance. Specifically, although a change in the quantity of money will affect prices and exchange rates, this change should not affect the rate at

EXHIBIT 7.2 Bolivia Ends Its Hyperinflation in 1985 by Shutting Down the Printing Presses

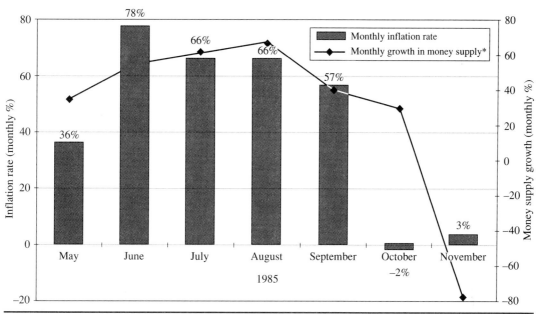

* Money supply data are not seasonally adjusted.

Source: International Financial Statistics, International Monetary Fund, various editions.

which domestic goods are exchanged for foreign goods or the rate at which goods today are exchanged for goods in the future. These ideas are formalized as purchasing power parity and the Fisher effect, respectively. We will examine them here briefly and then in greater detail in the next two sections.

The international analogue to inflation is home-currency depreciation relative to foreign currencies. The analogy derives from the observation that inflation involves a change in the exchange rate between the home currency and domestic goods, whereas home currency depreciation—a decline in the foreign currency value of the home currency—results in a change in the exchange rate between the home currency and foreign goods.

That inflation and currency depreciation are related is no accident. Excess money-supply growth, through its impact on the rate of aggregate spending, affects the demand for goods produced abroad as well as goods produced domestically. In turn, the domestic demand for foreign currencies changes, and, consequently, the foreign exchange value of the domestic currency changes. Thus, the rate of domestic inflation and changes in the exchange rate are jointly determined by the rate of domestic money growth relative to the growth of the amount that people—domestic and foreign—want to hold.

If international arbitrage enforces the law of one price, then the exchange rate between the home currency and domestic goods must equal the exchange rate between the home currency and foreign goods. In other words, a unit of home currency (HC) should have the same purchasing power worldwide. Thus, if a dollar buys a pound of bread in the United States, it should also buy a pound of bread in Great Britain. For this to happen, the foreign exchange rate must change by (approximately) the difference between the domestic and foreign rates of inflation. This relationship is called **purchasing power parity** (PPP).

Similarly, the **nominal interest rate**, the price quoted on lending and borrowing transactions, determines the exchange rate between current and future dollars (or any other currency). For example, an interest rate of 10% on a one-year loan means that one dollar today is being exchanged for 1.1 dollars a year from now. But what really matters according to the **Fisher effect** is the exchange rate between current and future purchasing power, as measured by the real interest rate. Simply put, the lender is concerned with how many more goods can be obtained in the future by forgoing consumption today, whereas the borrower wants to know how much future consumption must be sacrificed to obtain more goods today. This condition is the case regardless of whether the borrower and lender are located in the same or different countries. As a result, if the exchange rate between current and future goods—the **real interest rate**—varies from one country to the next, arbitrage between domestic and foreign capital markets, in the form of international capital flows, should occur. These flows will tend to equalize real interest rates across countries. By looking more closely at these and related parity conditions, we can see how they can be formalized and used for management purposes.

7.2 PURCHASING POWER PARITY

Purchasing power parity (PPP) was first stated in a rigorous manner by the Swedish economist Gustav Cassel in 1918. He used it as the basis for recommending a new set of official exchange rates at the end of World War I that would allow for the resumption of normal trade relations.[1] Since then, PPP has been widely used by central banks as a guide to establishing new par values for their currencies when the old ones were clearly in disequilibrium. From a management standpoint, purchasing power parity is often used to forecast future exchange rates, for purposes ranging from deciding on the currency denomination of long-term debt issues to determining in which countries to build plants.

In its *absolute* version, purchasing power parity states that price levels should be equal worldwide when expressed in a common currency. In other words, a unit of home currency (HC) should have the same purchasing power around the world. This theory is just an application of the law of one price to national price levels rather than to individual prices. (That is, it rests on the assumption that free trade will equalize the price of any good in all countries; otherwise, arbitrage opportunities would exist.) However, absolute PPP ignores the effects on free trade of transportation costs, tariffs, quotas and other restrictions, and product differentiation.

The *relative* version of purchasing power parity, which is used more commonly now, states that the exchange rate between the home currency and any foreign currency will adjust to reflect changes in the price levels of the two countries. For example, if inflation is 5% in the United States and 1% in Japan, then the dollar value of the Japanese yen must rise by about 4% to equalize the dollar price of goods in the two countries.

---

[1] Gustav Cassel, "Abnormal Deviations in International Exchanges," *Economic Journal*, December 1918, pp. 413–415.

Formally, if i_h and i_f are the periodic price level increases (rates of inflation) for the home country and the foreign country, respectively; e_0 is the dollar (HC) value of one unit of foreign currency at the beginning of the period; and e_t is the spot exchange rate in period t, then

$$\frac{e_t}{e_0} = \frac{(1 + i_h)^t}{(1 + i_f)^t} \tag{7.1}$$

If Equation 7.1 holds, then

$$e_t = e_0 \times \frac{(1 + i_h)^t}{(1 + i_f)^t} \tag{7.2}$$

The value of e_t appearing in Equation 7.2 is known as the PPP rate. For example, if the United States and Switzerland are running annual inflation rates of 5% and 3%, respectively, and the spot rate is SFr 1 = \$0.75, then according to Equation 7.2 the PPP rate for the Swiss franc in three years should be

$$e_3 = 0.75 \left(\frac{1.05}{1.03} \right)^3 = \$0.7945$$

If purchasing power parity is expected to hold, then \$0.7945/SFr is the best prediction for the franc spot rate in three years. The one-period version of Equation 7.2 is commonly used. It is

$$e_1 = e_0 \times \frac{1 + i_h}{1 + i_f} \tag{7.3}$$

ILLUSTRATION

CALCULATING THE PPP RATE FOR THE DM

Suppose the current U.S. price level is at 112 and the German price level is at 107, relative to base price levels of 100. If the initial value of the Deutsche mark was \$0.48, then according to PPP, the dollar value of the DM should have risen to approximately \$0.5024 [0.48 × (112/107)], an appreciation of 4.67%. On the other hand, if the German price level now equals 119, then the Deutsche mark should have depreciated by about 5.88%, to \$0.4518 [0.48 × 112/119)].

Purchasing power parity is often represented by the following approximation of Equation 7.3[2]:

$$\frac{e_1 - e_0}{e_0} = i_h - i_f \tag{7.4}$$

---

[2] Dividing both sides of Equation 7.3 by e_0 and then subtracting 1 from both sides yields

$$\frac{e_1 - e_0}{e_0} = \frac{i_h - i_f}{1 + i_f}$$

Equation 7.4 follows if i_f is relatively small.

EXHIBIT 7.3 Purchasing Power Parity

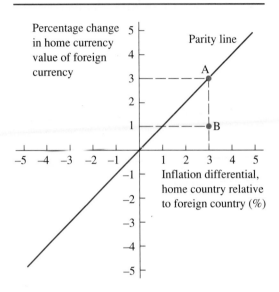

That is, the exchange rate change during a period should equal the inflation differential for that same time period. In effect, PPP says that *currencies with high rates of inflation should devalue relative to currencies with lower rates of inflation.*

Equation 7.4 is illustrated in Exhibit 7.3. The vertical axis measures the percentage currency change, and the horizontal axis shows the inflation differential. Equilibrium is reached on the parity line, which contains all those points at which these two differentials are equal. At point A, for example, the 3% inflation differential is just offset by the 3% appreciation of the foreign currency relative to the home currency. Point B, on the other hand, depicts a situation of disequilibrium, where the inflation differential of 3% is greater than the appreciation of 1% in the HC value of the foreign currency.

The Lesson of Purchasing Power Parity

Purchasing power parity bears an important message: Just as the price of goods in one year cannot be meaningfully compared with the price of goods in another year without adjusting for interim inflation, so exchange rate changes may indicate nothing more than the reality that countries have different inflation rates. In fact, according to PPP, exchange rate movements should just cancel out changes in the foreign price level relative to the domestic price level. These offsetting movements should have no effects on the relative competitive positions of domestic firms and their foreign competitors. Thus, changes in the **nominal exchange rate**—that is, the actual exchange rate—may be of little significance in determining the true effects of currency changes on a firm and a nation. In terms of currency changes affecting relative competitiveness, therefore, the focus must be not on nominal exchange rate changes but instead on changes in the real purchasing power of one currency relative to another. Here we consider the concept of the real exchange rate.

The **real exchange rate** is the nominal exchange rate adjusted for changes in the relative purchasing power of each currency since some base period. In technical terms, the real exchange rate at time t (dollars or HC per unit of foreign currency), e'_t, relative to the base period (specified as time 0) is defined as

$$e'_t = e_t \frac{P_f}{P_h} \tag{7.5}$$

WHERE

P_f = the foreign price level
P_h = the home price level at time t
(both indexed to 100 at time 0).

By indexing these price levels to 100 as of the base period, their ratio reflects the change in the relative purchasing power of these currencies since time 0. Note that increases in the foreign price level and foreign-currency depreciation have offsetting effects on the real exchange rate. Similarly, home price-level increases and foreign-currency appreciation offset each other.

An alternative—and equivalent—way to represent the real exchange rate is to directly reflect the change in relative purchasing powers of these currencies by adjusting the nominal exchange rate for inflation in both countries since time 0, as follows:

$$e'_t = e_t \frac{(1 + i_f)^t}{(1 + i_h)^t} \tag{7.6}$$

WHERE the various parameters are the same as those defined previously.

If changes in the nominal exchange rate are fully offset by changes in the relative price levels between the two countries, then the real exchange rate remains unchanged (note that the real exchange rate in the base period is just the nominal rate e_0). Specifically, if PPP holds, then we can substitute the value of e_t from Equation 7.2 into Equation 7.6. Making this substitution yields $e'_t = e_0$; that is, the real exchange rate remains constant at e_0. Alternatively, a change in the real exchange rate is equivalent to a deviation from PPP.

The distinction between the nominal exchange rate and the real exchange rate has important implications for foreign exchange risk measurement and management. As we will see in Chapter 10, if the real exchange rate remains constant (i.e., if purchasing power parity holds), currency gains or losses from nominal exchange rate changes will generally be offset over time by the effects of differences in relative rates of inflation, thereby reducing the net impact of nominal devaluations and revaluations. Deviations from purchasing power parity, however, will lead to real exchange gains and losses. In the case of Japanese exporters, the real appreciation of the yen forced them to cut costs and develop new products less subject to pricing pressures. We will discuss their responses in more detail in Chapter 10.

ILLUSTRATION CALCULATING THE REAL EXCHANGE RATE FOR THE JAPANESE YEN

Between 1980 and 1995, the ¥/$ exchange rate moved from ¥226.63/$ to ¥93.96. During this same 15-year period, the consumer price index (CPI) in Japan rose from 91.0 to 119.2, and the U.S. CPI rose from 82.4 to 152.4.

a. If PPP had held over this period, what would the ¥/$ exchange rate have been in 1995?

Solution. According to Equation 7.2, in 1995, the ¥/$ exchange rate should have been ¥160.51/$:

$$\text{¥/\$ PPP rate} = 226.63 \times \frac{(119.2/91.0)}{(152.4/82.4)} = \text{¥}160.51/\$$$

In working this problem, note that Equation 7.2 was inverted because we are expressing the exchange rate in direct terms rather than indirect terms. Note too that the ratio of CPIs is equal to the cumulative price level increase. Comparing the PPP rate of ¥160.51/$ to the actual rate of ¥93.96/$, we can see that the yen has appreciated more than PPP would suggest.

b. What happened to the real value of the yen in terms of dollars during this period?

Solution. To estimate the real value of the yen, we convert the yen from European to American terms and apply Equation 7.5 (using the yen quoted in European terms would yield us the real value of the U.S. dollar in terms of yen):[3]

$$e_t' = e_t \frac{P_f}{P_h} = \frac{1}{93.96} \times \frac{(119.2/91.0)}{(152.4/82.4)}$$

$$= \$0.007538/\text{¥}$$

To interpret this real exchange rate and see how it changed since 1980, we compare it to the real exchange rate in 1980, which just equals the nominal rate (quoted in American terms) at that time of 1/226.63 = $0.004412/¥ (because the real and nominal rates are equal in the base period). This comparison reveals that during the 15-year period 1980–1995 the yen appreciated in real terms by (0.007538 − 0.004412)/0.004412 = 71%. This dramatic appreciation in the inflation-adjusted value of the Japanese yen put enormous competitive pressure on Japanese exporters as the dollar prices of their goods rose far more than the U.S. rate of inflation would justify.

[3]Dividing both current price levels by their base levels effectively indexes each to 100 as of the base period.

Expected Inflation and Exchange Rate Changes

Changes in expected, as well as actual, inflation will cause exchange rate changes. An increase in a currency's expected rate of inflation, all other being things equal, makes that currency more expensive to hold over time (because its value is being eroded at a faster rate) and less in demand at the same price. Consequently, the value of higher-inflation currencies will tend to be depressed relative to the value of lower-inflation currencies, other things being equal.

The Monetary Approach

More recently, purchasing power parity has been reformulated into the *monetary approach* to exchange rate determination. It is based on the quantity theory of money:

$$\frac{M}{P} = \frac{y}{v} \tag{7.7}$$

WHERE

M = the national money supply
P = the general price level
y = real GNP
v = the velocity of money

We can rewrite Equation 7.7 in terms of growth rates to give the determinants of domestic inflation:

$$i_h = \mu_h - g_{yh} + g_{vh} \tag{7.8}$$

WHERE

i_h = the domestic inflation rate
μ_h = the rate of domestic money supply expansion
g_{yh} = the growth in real domestic GNP
g_{vh} = the change in the velocity of the domestic money supply

For example, if U.S. money supply growth is forecast at 5%, real GNP is expected to grow at 2%, and the velocity of money is expected to fall by 0.5%, then Equation 7.8 predicts that the U.S. inflation rate will be 5% − 2% + (−0.5%) = 2.5%.

A similar equation will hold for the predicted foreign rate of inflation. Combining these two equations along with purchasing power parity leads to the following predicted exchange rate change:

$$\frac{e_1 - e_0}{e_0} = i_h - i_f = (\mu_h - \mu_f) - (g_{yh} - g_{yf}) + (g_{vh} - g_{vf}) \tag{7.9}$$

where the subscript *f* refers to the corresponding rates for the foreign country

Empirical Evidence

The strictest version of purchasing power parity — that all goods and financial assets obey the law of one price — is demonstrably false. The risks and costs of shipping goods internationally, as well as government-erected barriers to trade and capital flows, are at times high enough to cause exchange-adjusted prices to systematically differ between countries. On the other hand, there is clearly a relationship between relative inflation rates and changes in exchange rates. This relationship is shown in Exhibit 7.4, which compares the relative change in the purchasing power of 22 currencies (as measured by their relative inflation rates) with the relative change in the exchange rates for those currencies for the period 1982 through 1988. As expected, those currencies with the largest relative decline (gain) in purchasing power saw the sharpest erosion (appreciation) in their foreign exchange values.

EXHIBIT 7.4 Purchasing Power Parity: Empirical Data, 1982–1988

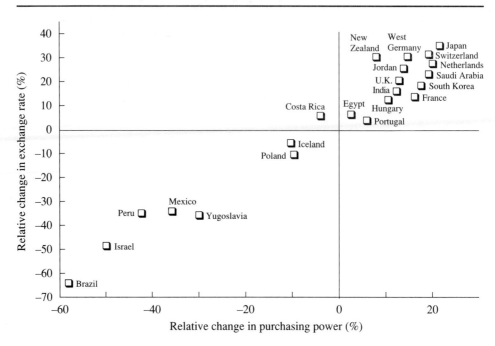

The general conclusion from empirical studies of PPP is that the theory holds up well in the long run, but not as well over shorter time periods.[4] The difference between the short-run and long-run effects can be seen in Exhibit 7.5, which compares the actual dollar exchange rate for 12 countries with their PPP rates. Despite substantial short-run deviations from purchasing power parity, currencies have a distinct tendency to move toward their PPP-predicted rates. Another way to view this evidence is that, despite fluctuations, the real exchange rate tends to revert back to its predicted value of e_0. That is, if $e_t' > e_0$, then the real exchange rate should fall over time towards e_0, whereas if $e_t' < e_0$, the real exchange rate should rise over time towards e_0. Additional support for the existence of mean-reverting behavior of real exchange rates is provided by data spanning two centuries on the dollar-sterling and French franc-sterling real exchange rates.[5] Mean reversion has important implications for currency risk management, which will be explored in Chapter 11.

A common explanation for the failure of PPP to hold is that goods prices are sticky, leading to short-term violations of the law of one price. Adjustment to PPP eventually occurs, but it does so with a lag. An alternative explanation for the failure of most tests to support PPP in the short run is that these tests ignore the problems caused by the combi-

---

[4] Perhaps the best known of these studies is Henry J. Gailliot, "Purchasing Power Parity as an Explanation of Long-Term Changes in Exchange Rates," *Journal of Money, Credit, and Banking*, August 1971, pp. 348–357.

[5] See James R. Lothian and Mark P. Taylor, "Real Exchange Rate Behavior: The Recent Float from the Perspective of the Past Two Centuries," *Journal of Political Economy*, June 1996.

EXHIBIT 7.5 Purchasing Power Parity and Actual Exchange Rates

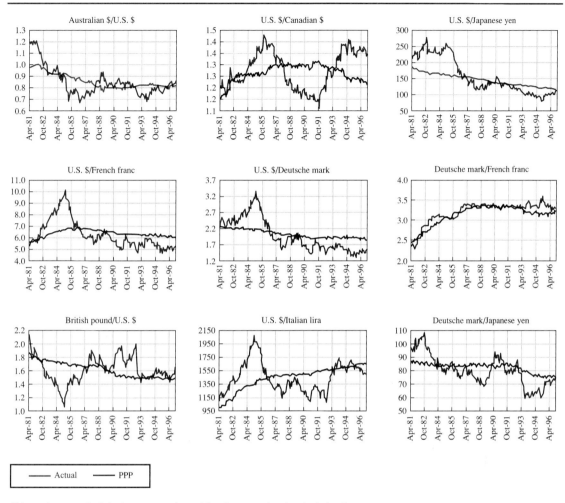

If the quoting convention is foreign currency units per dollar, the currency is undervalued when the actual rate is above the PPP rate. The opposite is true if the quoting convention is units of dollar per foreign currency unit.

Source: "Currency Review," Bank of America Foreign Currency Research; Winter 1996/1997, pp. 42–45.

nation of differently constructed price indices, relative price changes, and nontraded goods and services.

One problem arises because the price indices used to measure inflation vary substantially between countries as to the goods and services included in the "market basket" and the weighting formula used. Thus, changes in the relative prices of various goods and services will cause differently constructed indices to deviate from each other, falsely signaling deviations from PPP. Measured deviations from PPP tend to be far smaller when using the same weights than when using different weights in calculating the U.S. and foreign price indices.

In addition, relative price changes could lead to changes in the equilibrium exchange rate, even in the absence of changes in the general level of prices. For example, an increase in the relative price of oil will lead to an increase in the exchange rates of oil-exporting countries, even if other prices adjust so as to keep all price levels constant. In general, a relative price change that increases a nation's wealth will also increase the value of its currency. Similarly, an increase in the real interest rate in one nation relative to real interest rates elsewhere will induce flows of foreign capital to the first nation. The result will be an increase in the real value of that nation's currency.

Finally, price indices heavily weighted with nontraded goods and services will provide misleading information about a nation's international competitiveness. Over the longer term, increases in the price of medical care or the cost of education will affect the cost of producing traded goods. But, in the short run, such price changes will have little effect on the exchange rate.

Exhibit 7.6 shows the often substantial gaps that can arise between nontraded goods and services between countries. You cannot easily substitute a Hong Kong Big Mac

EXHIBIT 7.6 Comparison of Prices of Nontraded Goods and Services

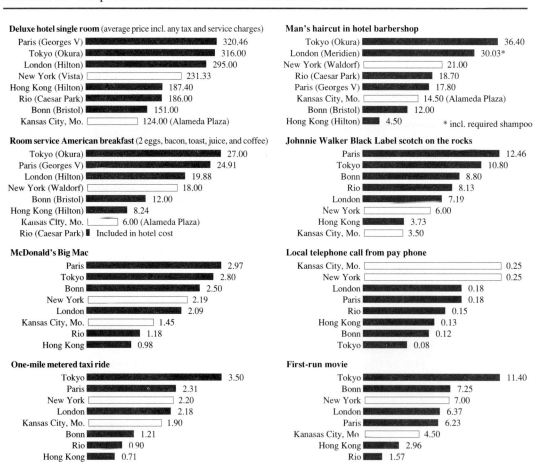

Deluxe hotel single room (average price incl. any tax and service charges)
- Paris (Georges V) 320.46
- Tokyo (Okura) 316.00
- London (Hilton) 295.00
- New York (Vista) 231.33
- Hong Kong (Hilton) 187.40
- Rio (Caesar Park) 186.00
- Bonn (Bristol) 151.00
- Kansas City, Mo. 124.00 (Alameda Plaza)

Room service American breakfast (2 eggs, bacon, toast, juice, and coffee)
- Tokyo (Okura) 27.00
- Paris (Georges V) 24.91
- London (Hilton) 19.88
- New York (Waldorf) 18.00
- Bonn (Bristol) 12.00
- Hong Kong (Hilton) 8.24
- Kansas City, Mo. 6.00 (Alameda Plaza)
- Rio (Caesar Park) Included in hotel cost

McDonald's Big Mac
- Paris 2.97
- Tokyo 2.80
- Bonn 2.50
- New York 2.19
- London 2.09
- Kansas City, Mo. 1.45
- Rio 1.18
- Hong Kong 0.98

One-mile metered taxi ride
- Tokyo 3.50
- Paris 2.31
- New York 2.20
- London 2.18
- Kansas City, Mo. 1.90
- Bonn 1.21
- Rio 0.90
- Hong Kong 0.71

Man's haircut in hotel barbershop
- Tokyo (Okura) 36.40
- London (Meridien) 30.03*
- New York (Waldorf) 21.00
- Rio (Caesar Park) 18.70
- Paris (Georges V) 17.80
- Kansas City, Mo. 14.50 (Alameda Plaza)
- Bonn (Bristol) 12.00
- Hong Kong (Hilton) 4.50

* incl. required shampoo

Johnnie Walker Black Label scotch on the rocks
- Paris 12.46
- Tokyo 10.80
- Bonn 8.80
- Rio 8.13
- London 7.19
- New York 6.00
- Hong Kong 3.73
- Kansas City, Mo. 3.50

Local telephone call from pay phone
- Kansas City, Mo. 0.25
- New York 0.25
- London 0.18
- Paris 0.18
- Rio 0.15
- Hong Kong 0.13
- Bonn 0.12
- Tokyo 0.08

First-run movie
- Tokyo 11.40
- Bonn 7.25
- New York 7.00
- London 6.37
- Paris 6.23
- Kanasas City, Mo 4.50
- Hong Kong 2.96
- Rio 1.57

costing $0.98 for a Paris Big Mac at $2.97. Similarly, if you are in Tokyo and you do not like the $316 price of a room at the Hotel Okura, you cannot easily substitute a $124 room at the Alameda Plaza hotel in Kansas City. Because PPP is driven by arbitrage, including such prices in estimating PPP will not help to determine whether exchange rates are in equilibrium.

Despite the problems caused by relative price changes, most tests of relative PPP as a long-term theory of exchange rate determination seem to support its validity. The reason is that over long periods of time with a moderate inflation differential, the general trend in the price level ratio will tend to dominate the effects of relative price changes. This factor also explains why standard tests of PPP support it even in the short run during periods of hyperinflation: With high inflation, changes in the general level of prices quickly swamp the effects of relative price changes.

In summary, despite often lengthy departures from PPP, there is a clear correspondence between relative inflation rates and changes in the nominal exchange rate. However, for reasons that have nothing necessarily to do with market disequilibrium, the correspondence is not perfect.

7.3 THE FISHER EFFECT

The interest rates that are quoted in the financial press are nominal rates. That is, they are expressed as the rate of exchange between current and future dollars. For example, a nominal interest rate of 8% on a one-year loan means that $1.08 must be repaid in one year for $1.00 loaned today. But what really matters to both parties to a loan agreement is the real interest rate, the rate at which current goods are being converted into future goods.

Looked at one way, the real rate of interest is the net increase in wealth that people expect to achieve when they save and invest their current income. Alternatively, it can be viewed as the added future consumption promised by a corporate borrower to a lender in return for the latter's deferring current consumption. From the company's standpoint, this exchange is worthwhile as long as it can find suitably productive investments.

However, because virtually all financial contracts are stated in nominal terms, the real interest rate must be adjusted to reflect expected inflation. The **Fisher effect** (FE) states that the nominal interest rate r is made up of two components: (1) a real required rate of return a and (2) an inflation premium equal to the expected amount of inflation i. Formally, the Fisher effect is

$$1 + \text{Nominal rate} = (1 + \text{Real rate})(1 + \text{Expected inflation rate}) \qquad (7.10)$$

$$1 + r = (1 + a)(1 + i)$$

or

$$r = a + i + ai$$

Equation 7.10 is often approximated by the equation $r = a + i$.

The Fisher equation says, for example, that if the required real return is 3% and expected inflation is 10%, then the nominal interest rate will be about 13% (13.3%, to be exact). The logic behind this result is that $1 next year will have the purchasing power of $0.90 in terms of today's dollars. Thus, the borrower must pay the lender $0.103 to compensate for the erosion in the purchasing power of the $1.03 in principal and interest payments, in addition to the $0.03 necessary to provide a 3% real return.

ILLUSTRATION

BRAZILIANS SHUN NEGATIVE REAL INTEREST RATES ON SAVINGS

In 1981, the Brazilian government spent $10 million on an advertising campaign to help boost national savings, which dropped sharply in 1980. According to the *Wall Street Journal* (January 12, 1981, p. 23), the decline in savings occurred, "because the pre-fixed rates on savings deposits and treasury bills for 1980 were far below the rate of inflation, currently 110%." Clearly, the Brazilians were not interested in investing money at interest rates less than the inflation rate.

The generalized version of the Fisher effect asserts that real returns are equalized across countries through arbitrage—that is, $a_h = a_f$, where the subscripts h and f refer to home and foreign real rates. If expected real returns were higher in one currency than another, capital would flow from the second to the first currency. This process of arbitrage would continue, in the absence of government intervention, until expected real returns were equalized.

In equilibrium, then, with no government interference, it should follow that the nominal interest rate differential will approximately equal the anticipated inflation rate differential, or

$$\frac{1 + r_h}{1 + r_f} = \frac{1 + i_h}{1 + i_f} \qquad (7.11)$$

where r_h and r_f are the nominal home and foreign currency interest rates, respectively.

If r_f and i_f are relatively small, then this exact relationship can be approximated by Equation 7.12[6]:

$$r_h - r_f = i_h - i_f \qquad (7.12)$$

In effect, the generalized version of the Fisher effect says that *currencies with high rates of inflation should bear higher interest rates than currencies with lower rates of inflation.*

For example, if inflation rates in the United States and the United Kingdom are 4% and 7%, respectively, the Fisher effect says that nominal interest rates should be about 3% higher in the United Kingdom than in the United States. A graph of Equation 7.12 is

---

[6]Equation 7.11 can be converted into Equation 7.12 by subtracting 1 from both sides and assuming that r_f and i_f are relatively small.

EXHIBIT 7.7 The Fisher Effect

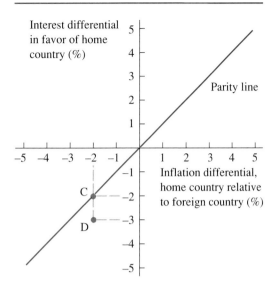

shown in Exhibit 7.7. The horizontal axis shows the expected difference in inflation rates between the home country and the foreign country, and the vertical axis shows the interest differential between the two countries for the same time period. The parity line shows all points for which $r_h - r_f = i_h - i_f$.

Point C, for example, is a position of equilibrium because the 2% higher rate of inflation in the foreign country ($i_h - i_f = -2\%$) is just offset by the 2% lower HC interest rate ($r_h - r_f = -2\%$). At point D, however, where the real rate of return in the home country is 1% lower than in the foreign country (an inflation differential of 2% versus an interest differential of 3%), funds should flow from the home country to the foreign country to take advantage of the real differential. This flow will continue until expected real returns are again equal.

Empirical Evidence

Exhibit 7.8 illustrates the relationship between interest rates and inflation rates for 22 countries as of April 1996. It is evident from the graph that nations with higher inflation rates generally have higher interest rates. Thus, the empirical evidence is consistent with the hypothesis that most of the variation in nominal interest rates across countries can be attributed to differences in inflationary expectations.

The proposition that expected real returns are equal between countries cannot be tested directly. However, many observers believe it unlikely that significant real interest differentials could long survive in the increasingly internationalized capital markets. Most market participants agree that arbitrage, via the huge pool of liquid capital that operates in international markets these days, is forcing pre-tax real interest rates to converge across all the major nations.

EXHIBIT 7.8 Fisher Effect: Nominal Interest Rate Versus Inflation Rate for 22 Developed and Developing Countries as of April 1996

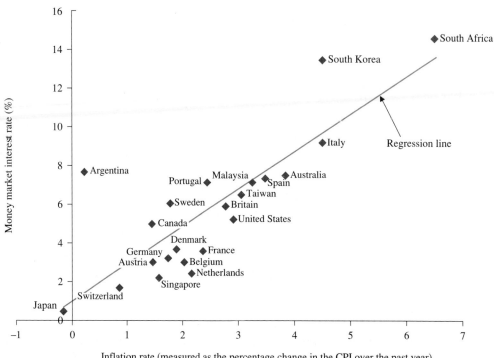

Inflation rate (measured as the percentage change in the CPI over the past year)

   To the extent that arbitrage is permitted to operate unhindered, capital markets are integrated worldwide. **Capital market integration** means that real interest rates are determined by the global supply and global demand for funds. This is in contrast to **capital market segmentation**, where real interest rates are determined by local credit conditions. The difference between capital market segmentation and capital market integration is depicted in Exhibit 7.9. With a segmented capital market, the real interest rate in the United States, a_{us}, is based on the national demand D_{us} and national supply S_{us} of credit. Conversely, the real rate in the rest of the world, a_{rw}, is based on the rest-of-world supply S_{rw} and demand D_{rw}. In this example, the U.S. real rate is higher than the real rate outside the United States, or $a_{us} > a_{rw}$.

   Once the U.S. market opens up, the U.S. real interest rate falls (and the rest-of-world rate rises) to the new world rate a_w, which is determined by the world supply S_w ($S_{us} + S_{rw}$) and world demand D_w ($D_{us} + D_{rw}$) for credit. The mechanism whereby equilibrium is brought about is a capital inflow to the United States. It is this same capital flow that drives up the real interest rate outside the United States.[7]

---

[7] The net gain from the transfer of capital equals the higher returns on the capital imported to the United States less the lower returns foregone in the rest of the world. Returns on capital must be higher in the United States prior to the capital inflow because the demand for capital depends on the expected return on capital. Thus, a higher real interest rate indicates a higher real return on capital.

EXHIBIT 7.9 The Distinction Between Capital Market Integration and
Capital Market Segmentation

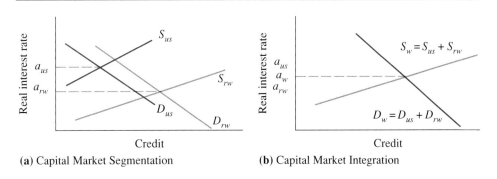

Credit   Credit
(a) Capital Market Segmentation   (b) Capital Market Integration

As shown by Exhibit 7.9, in an integrated capital market, the domestic real interest rate depends on what is happening outside as well as inside the United States. For example, a rise in the demand for capital by German companies to finance investments in Eastern Europe will raise the real interest rate in the United States as well as in Germany. Similarly, a rise in the U.S. savings rate, other things being equal, will lower the real cost of capital both in the United States and in the rest of the world. Conversely, a fall in U.S. inflation will lower the nominal U.S. interest rate (the Fisher effect), while leaving unchanged real interest rates worldwide.

Capital market integration has homogenized markets around the world, eroding much—although not all—of the real interest rate differentials between comparable domestic and offshore securities, and strengthening the link between assets that are denominated in different currencies but carry similar credit risks.[8] To the extent that real interest differentials do exist, they must be due to either currency risk or some form of political risk.

A real interest rate differential could exist without being arbitraged away if investors strongly preferred to hold domestic assets in order to avoid currency risk, even if the expected real return on foreign assets were higher. The evidence on this point is somewhat mixed. The data indicate a tendency toward convergence in real interest rates internationally, indicating that arbitrage does occur, but real rates still appear to differ from each other.[9] Moreover, the estimated currency risk premium appears to be highly variable and unpredictable, leading to extended periods of apparent differences in real interest rates between nations.[10]

---

[8] An offshore security is one denominated in the home currency but issued abroad. They are generally referred to as Eurosecurities.

[9] See, for example, Frederick S. Mishkin, "Are Real Interest Rates Equal across Countries? An International Investigation of Parity Conditions," *Journal of Finance*, December 1984, pp. 1345–1357. He finds that, although capital markets may be integrated, real interest rates appear to differ across countries because of currency risk. His findings are consistent with those of Baghar Modjtahedi, "Dynamics of Real Interest Rate Differentials: An Empirical Investigation," *European Economic Review* 32, no. 6 (1988): 1191–1211.

[10] Adrian Throop, "International Financial Market Integration and Linkages of National Interest Rates," *Federal Reserve Bank of San Francisco Economic Review*, no. 3 (1994): 3–18, found that exchange risk caused persistent real interest rate differentials among developed nations for the years 1981–1993.

EXHIBIT 7.10 Real Interest Rate Versus Nominal Interest Rate for 22 Developed and Developing Nations as of April 1996

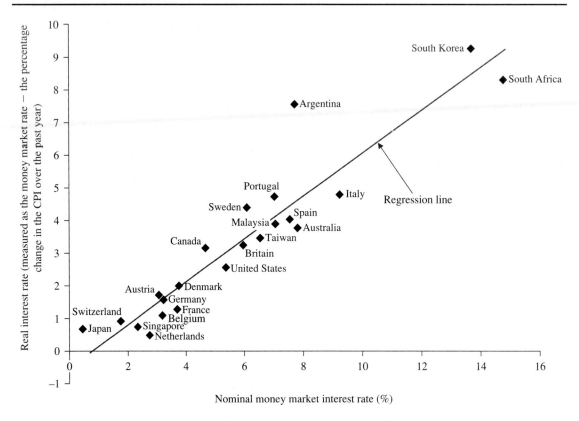

These differences are displayed in Exhibit 7.10, which compares real interest rates (measured as the nominal interest rate minus the past year's inflation rate as a surrogate for the expected inflation rate) as of April 1996 versus nominal rates for the same 22 countries shown in Exhibit 7.8. According to this exhibit, countries with higher nominal interest rates (implying higher expected inflation and greater currency risk) tend to have higher real interest rates, resulting in large real-rate differentials among some countries.

In addition to currency and inflation risk, real interest rate differentials in a closely integrated world economy can stem from countries pursuing sharply differing tax policies or imposing regulatory barriers to the free flow of capital.

In many developing countries, however, currency controls and other government policies impose political risk on foreign investors. In effect, political risk can drive a wedge between the returns available to domestic investors and those available to foreign investors. For example, if political risk in Brazil causes foreign investors to demand a 7% higher interest rate than they demand elsewhere, then foreign investors would consider a 10% expected real return in Brazil to be equivalent to a 3% expected real return in the United States. Hence, real interest rates in developing countries can exceed those in developed countries without presenting attractive arbitrage opportunities to foreign

 FRANCE SEGMENTS ITS CAPITAL MARKET
Throughout the European Monetary System's September 1992 crisis, the French franc managed to stay within the ERM. Exhibit 7.11 suggests why. It plots two interest-rate differentials: (1) the gap between three-month domestic money-market rates and the corresponding Eurofranc rate, which is the rate on francs deposited in London in the Eurocurrency market (to be discussed in Chapter 15); and (2) the gap between the domestic money-market rate and the bank prime rate.

France supposedly has ended capital market controls, so arbitrage should ensure that the first of these interest differentials (shown by the black line) should be approximately zero—as it was until the crisis that began in mid-Septem-

ber. Once trouble began, however, the Euromarket rate exceeded the domestic rate by a big margin for almost two weeks, indicating that the French government was impeding the flow of capital out of France.

The other series sheds more light on what was going on during this period. Until pressures began building in the spring, the prime lending rate slightly exceeded the money-market rate—as you might expect, because money-market rates help determine the banks' cost of funds. In May, however, the money-market rates rose above the prime lending rate, widening to more than four percentage points at the height of the crisis.

The obvious conclusion is that the French government was using high money-market rates to defend the franc, while forcing banks to lend money at a loss in order to avoid the adverse impact of high interest rates on the French economy.

EXHIBIT 7.11 France Segements Its Money Market to Defend the Franc

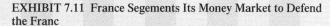
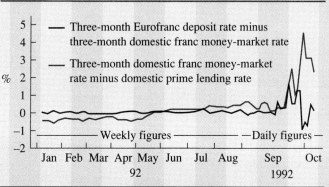

Source: Datastream, in *The Economist*, October 10, 1992, p. 97. © 1992 The Economist Newspaper Group, Inc. Reprinted with permission.

investors. The combination of a relative shortage of capital and high political risk in most developing countries is likely to cause real interest rates in these countries to exceed real interest rates in the developed countries. Indeed, the countries in Exhibit 7.10 with the highest real rates of interest are all developing countries.

Investors' tolerance of economic mismanagement in developed nations also has fallen dramatically, as financial deregulation, abolition of foreign exchange controls, and the process of global portfolio diversification have swollen the volume of international capital flows. With modern technology enabling investors to move capital from one market to another at low cost and almost instantaneously, the pressure on central banks to seem to "do the right thing" is intense. Conversely, those nations who must attract a disproportionate amount of global capital to finance their national debts and who have no credible policies to deal with their problems in a noninflationary way will be forced to pay a rising risk premium. Canada provides a good example of both these trends.

ILLUSTRATION CANADA'S HIGH REAL INTEREST RATE COMES DOWN

In early 1995, the Canadian dollar slipped to an 8½ year low against the U.S. dollar. At the same time, with Canada's inflation rate under 1% and its 10-year government bonds yielding 9.3% (about 1.5 percentage points more than 10-year U.S. Treasury bonds), Canada had the highest real long-term interest rates in the world. The weak Canadian dollar and high real interest rates stemmed from the same source—a lack of confidence in Canada's longer-term inflation prospects.

Canada had a large current-account deficit, driven by large budget deficits, political uncertainty, and other structural problems that led to investor worries that the current low rate of inflation was only temporary. The persistently high budget deficits, in turn, reflect big spending on generous social-welfare programs and overly rigid labor markets, along with a lack of political will to attack these problems. At the same time, investors feared that the government would rely more on tax increases than on spending cuts to reduce the deficit. Further increases in the already high Canadian tax rates would likely drive more of the economy underground and aggravate capital flight. Investors were concerned that if higher tax rates did not reduce the deficit, and the government would not cut spending, Canada might be tempted at some point to monetize its deficits, reigniting inflation.

Adding fuel to these fears was the resignation of John Crow, the highly respected head of the Bank of Canada, Canada's central bank, and a strong advocate of price stability. Some analysts contended he was forced out by government officials who opposed his tough low-inflation targets. His successor as head of the Bank of Canada followed a relatively lax monetary policy. Investors responded to these worries by driving down the value of the Canadian dollar and by demanding higher interest rates. In the background was the ever-present fear that Quebec separatists would manage to secede from Canada.

By late 1995, Quebec's separatists had lost their referendum for independence, the federal and provincial governments began slashing their budget deficits and planned even bigger cuts in the future, and Canada's largest province, Ontario, announced large tax cuts as well. Perceived political risk declined and investors began focusing on Canada's low inflation rate. As a result of these favorable trends, the Canadian dollar strengthened and, in early 1996, short-term Canadian interest rates fell below U.S. rates, after having stayed above U.S. rates for more than a decade. But the yield on 10-year Canadian government bonds stayed about 1 percentage point above that on U.S. Treasuries. With continued low inflation, however, by late 1997, Canada paid about 0.5 percentage points *less* for 10-year money than the United States.

Before we move to the next parity condition, a caveat is in order. We must keep in mind that there are numerous interest differentials just as there are many different interest rates in a market. The rate on bank deposits, for instance, will not be identical to that on Treasury bills. In computation of an interest differential, therefore, the securities on which this differential is based must be of identical risk characteristics save for currency risk. Otherwise, there is the danger of comparing apples with oranges (or at least temple oranges with navel oranges).

ADDING UP CAPITAL MARKETS INTERNATIONALLY ■ Central to understanding how we can add yen and DM and dollar capital markets together is to recognize that money is only a veil: All financial transactions, no matter how complex, ultimately involve exchanges of goods today for goods in the future. As we saw in Chapter 4, you supply credit (capital) when you consume less than you produce; you demand credit when you consume more than you produce. Thus, the supply of credit can be thought of as the excess supply of goods and the demand for credit as the excess demand for goods. When we add up the capital markets around the world, we are adding up the excess demands for goods and the excess supplies of goods. A car is still a car, whether it is valued in yen or dollars.

7.4 THE INTERNATIONAL FISHER EFFECT

The key to understanding the impact of relative changes in nominal interest rates among countries on the foreign exchange value of a nation's currency is to recall the implications of PPP and the generalized Fisher effect. PPP implies that exchange rates will move to offset changes in inflation rate differentials. Thus, a rise in the U.S. inflation rate relative to those of other countries will be associated with a fall in the dollar's value. It will also be associated with a rise in the U.S. interest rate relative to foreign interest rates. Combine these two conditions and the result is the **international Fisher effect** (IFE):

$$\frac{(1 + r_h)^t}{(1 + r_f)^t} = \frac{\overline{e_t}}{e_0} \qquad (7.13)$$

where $\overline{e_t}$ is the expected exchange rate in period t. The single-period analogue to Equation 7.13 is

$$\frac{1 + r_h}{1 + r_f} = \frac{\overline{e_t}}{e_0} \qquad (7.14)$$

According to Equation 7.14, the expected return from investing at home, $1 + r_h$, should equal the expected HC return from investing abroad, $(1 + r_f)e_1/e_0$. As discussed in the previous section, however, despite the intuitive appeal of equal expected returns, domestic and foreign expected returns might not equilibrate if the element of currency risk restrained the process of international arbitrage.

ILLUSTRATION

USING THE IFE TO FORECAST U.S.$ AND SFR RATES
In July, the one-year interest rate is 4% on Swiss francs and 13% on U.S. dollars.
a. If the current exchange rate is SFr 1 = $0.63, what is the expected future exchange rate in one year?
Solution. According to the international Fisher effect, the spot exchange rate expected in one year equals 0.63 × 1.13/1.04 = $0.6845.

b. If a change in expectations regarding future U.S. inflation causes the expected future spot rate to rise to $0.70, what should happen to the U.S. interest rate?
Solution. If r_{us} is the unknown U.S. interest rate, and the Swiss interest rate stayed at 4% (because there has been no change in expectations of Swiss inflation), then according to the international Fisher effect, $0.70/0.63 = (1 + r_{us})/1.04$, or $r_{us} = 15.56\%$.

If r_f is relatively small, Equation 7.15 provides a reasonable approximation to the international Fisher effect:[11]

$$r_h - r_f = \frac{\overline{e}_1 - e_0}{e_0} \tag{7.15}$$

In effect, the IFE says that *currencies with low interest rates are expected to appreciate relative to currencies with high interest rates.*

A graph of Equation 7.15 is shown in Exhibit 7.12. The vertical axis shows the expected change in the home-currency value of the foreign currency, and the horizontal axis shows the interest differential between the two countries for the same time period. The parity line shows all points for which $r_h - r_f = (\overline{e}_1 - e_0)/e_0$.

Point E is a position of equilibrium because it lies on the parity line, with the 4% interest differential in favor of the home country just offset by the anticipated 4% appreciation in the HC value of the foreign currency. Point F, however, illustrates a situation of disequilibrium. If the foreign currency is expected to appreciate by 3% in terms of the HC, but the interest differential in favor of the home country is only 2%, then funds would flow from the home to the foreign country to take advantage of the higher exchange-adjusted returns there. This capital flow will continue until exchange-adjusted returns are equal in the two nations.

Essentially what the IFE says is that arbitrage between financial markets—in the form of international capital flows—should ensure that the interest differential between any two countries is an *unbiased predictor* of the future change in the spot rate of exchange. This condition does not mean, however, that the interest differential is an especially accurate predictor; it just means that prediction errors tend to cancel out over time. Moreover, an implicit assumption that underlies IFE is that investors view foreign and

---

[11] Subtracting 1 from both sides of Equation 7.14 yields

$$\frac{(\overline{e}_1 - e_0)}{e_0} = \frac{r_h - r_f}{1 + r_f}$$

Equation 7.15 follows if i_f is relatively small.

EXHIBIT 7.12 International Fisher Effect

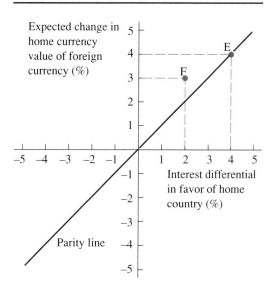

domestic assets as perfect substitutes. To the extent that this condition is violated (see the discussion on the Fisher effect) and investors require a risk premium (in the form of a higher expected real return) to hold foreign assets, IFE will not hold exactly.

Empirical Evidence

As predicted, there is a clear tendency for currencies with high interest rates (for example, Mexico and Brazil) to depreciate and those with low interest rates (for example, Japan and Switzerland) to appreciate. This tendency is shown in Exhibit 7.13, which graphs the nominal interest differential (relative to the U.S. interest rate) against exchange rate changes (relative to the U.S. dollar) for 21 currencies during the period 1982 to 1988. The ability of interest differentials to anticipate currency changes is also supported by several empirical studies that indicate the long-run tendency for these differentials to offset exchange rate changes.[12] The international Fisher effect also appears to hold even in the short run in the case of nations facing very rapid rates of inflation. Thus, at any given time, currencies bearing higher nominal interest rates can reasonably be expected to depreciate relative to currencies bearing lower interest rates.

Despite this apparently convincing evidence for the international Fisher effect, a large body of empirical evidence now indicates that the IFE does not hold up very well in the short run for nations with low to moderate rates of inflation.[13] One possible explana-

---

[12] See, for example, Ian H. Giddy and Gunter Dufey, "The Random Behavior of Flexible Exchange Rates," *Journal of International Business Studies*, Spring 1975, pp. 1–32; and Robert A. Aliber and Clyde P. Stickney, "Accounting Measures of Foreign Exchange Exposure: The Long and Short of It," *The Accounting Review*, January 1975, pp. 44–57.

[13] Much of this research is summarized in Kenneth A. Froot, "Short Rates and Expected Asset Returns," NBER working paper No. 3247, January 1990.

EXHIBIT 7.13 International Fisher Effect: Empirical Data, 1982–1988

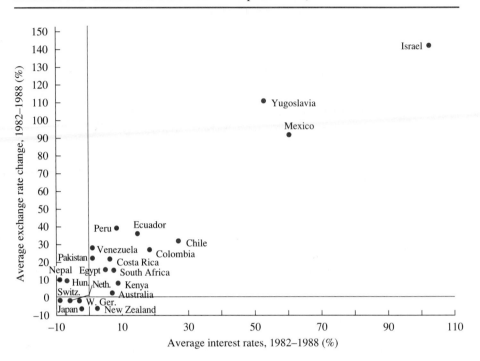

tion for this result relies on the existence of a time-varying exchange risk premium. However, this explanation for the failure of the IFE to hold in the short run has been challenged by empirical evidence indicating that the currency risk premium, to the extent it exists, is very small.[14]

A more plausible explanation relies on the nature of the Fisher effect. According to the Fisher effect, changes in the nominal interest differential can be due to changes in either the real interest differential or relative inflationary expectations. These two possibilities have opposite effects on currency values. For example, suppose that the nominal interest differential widens in favor of the United States. If this spread is due to a rise in the real interest rate in the United States relative to that of other countries, the value of the dollar will rise. Alternatively, if the change in the nominal interest differential is caused by an increase in inflationary expectations for the United States, the dollar's value will drop.

The key to understanding short-run changes in the value of the dollar or other currency, then, is to distinguish changes in nominal interest rate differentials that are caused by changes in real interest rate differentials from those caused by changes in relative inflation expectations. Historically, changes in the nominal interest differential have been dominated, at times, by changes in the real interest differential; at other times, they have been dominated by changes in relative inflation expectations. Consequently,

---

[14] See, for example, Kenneth A. Froot and Jeffrey A. Frankel, "Forward Discount Bias: Is It an Exchange Risk Premium?" *Quarterly Journal of Economics*, February 1989, pp. 139–161.

there is no stable, predictable relationship between changes in the nominal interest differential and exchange rate changes.

7.5 THE RELATIONSHIP BETWEEN THE FORWARD RATE AND THE FUTURE SPOT RATE

Our current understanding of the workings of the foreign exchange market suggests that, in the absence of government intervention in the market, both the spot rate and the forward rate are influenced heavily by current expectations of future events; and the two rates move in tandem, with the link between them based on interest differentials. New information, such as a change in interest rate differentials, is reflected almost immediately in both spot and forward rates.

Suppose a depreciation of pounds sterling is anticipated. Recipients of sterling will begin selling sterling forward, and sterling-area dollar earners will slow their sales of dollars in the forward market. These actions will tend to depress the price of forward sterling. At the same time, banks will probably try to even out their long (net purchaser) positions in forward sterling by selling sterling spot. In addition, sterling-area recipients of dollars will tend to delay converting dollars into sterling, and earners of sterling will speed up their collection and conversion of sterling. In this way, pressure from the forward market is transmitted to the spot market, and vice versa.

Ignoring risk for the moment, equilibrium is achieved only when the forward differential equals the expected change in the exchange rate. At this point, there is no longer any incentive to buy or sell the currency forward. This condition is illustrated in Exhibit 7.14. The vertical axis measures the expected change in the home currency value of the

EXHIBIT 7.14 Relationship Between the Forward Rate and the Future Spot Rate

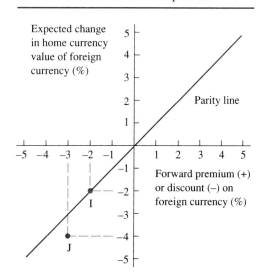

foreign currency, and the horizontal axis shows the forward discount or premium on the foreign currency. Parity prevails at point I, for example, where the expected foreign currency depreciation of 2% is just matched by the 2% forward discount on the foreign currency. Point J, however, is a position of disequilibrium because the expected 4% depreciation of the foreign currency exceeds the 3% forward discount on the foreign currency. We would, therefore, expect to see speculators selling the foreign currency forward for home currency, taking a 3% discount in the expectation of covering their commitment with 4% fewer units of HC.

A formal statement of the **unbiased nature of the forward rate** (UFR) is that the forward rate should reflect the expected future spot rate on the date of settlement of the forward contract:

$$f_t = \overline{e_t} \qquad\qquad (7.16)$$

where $\overline{e_t}$ is the expected future exchange rate at time t (units of home currency per unit of foreign currency) and f_t is the forward rate for settlement at time t.

USING UFR TO FORECAST THE FUTURE $/DM SPOT RATE
If the 90-day forward rate is DM 1 = $0.5987, what is the expected value of the DM in 90 days?

Solution. Arbitrage should ensure that the market expects the spot value of the DM in 90 days to be about $0.5987.

Equation 7.16 can be transformed into the one reflected in the parity line appearing in Exhibit 7.14, which is that the forward differential equals the expected change in the exchange rate, by subtracting 1 (e_0/e_0), from both sides, where e_0 is the current spot rate (HC per unit of foreign currency):[15]

$$\frac{f_1 - e_0}{e_0} = \frac{\overline{e_1} - e_0}{e_0} \qquad\qquad (7.17)$$

It should be noted that market efficiency requires that people process information and form reasonable expectations; it does not require that $f_1 = \overline{e_1}$. Market efficiency allows for the possibility that risk-averse investors will demand a risk premium on forward contracts, much the same as they demand compensation for bearing the risk of

---

[15] Note that this condition can be derived through a combination of the international Fisher effect and interest parity theory. Specifically, interest rate parity says that the interest differential equals the forward differential, whereas the IFE says that the interest differential equals the expected change in the spot rate. Things equal to the same thing are equal to each other, so the forward differential will equal the expected exchange rate change if both interest rate parity and the IFE hold.

investing in stocks. In this case, the forward rate will not reflect exclusively the expectation of the future spot rate.

The principal argument against the existence of a risk premium is that currency risk is largely diversifiable. If foreign exchange risk can be diversified away, no risk premium need be paid for holding a forward contract; the forward rate and expected future spot rate will be approximately equal. Ultimately, therefore, the unbiased nature of forward rates is an empirical, and not a theoretical, issue.

Empirical Evidence

A number of studies have examined the relation between forward rates and future spot rates.[16] Of course, it would be unrealistic to expect a perfect correlation between forward and future spot rates because the future spot rate will be influenced by events, such as an oil crisis, that can be forecast only imperfectly, if at all.

Nonetheless, the general conclusion from early studies was that forward rates are unbiased predictors of future spot rates. More recent studies, using more powerful econometric techniques, argue that the forward rate is a biased predictor, probably because of a risk premium.[17] However, the premium appears to change signs—being positive at some times and negative at other times—and averages near zero. This result, which casts doubt on the risk premium story, should not be surprising given that testing the unbiased nature of the forward rate is equivalent to testing the international Fisher effect (assuming covered interest parity holds).

In effect, we wind up with the same conclusions: Over time, currencies bearing a forward discount (higher interest rate) depreciate relative to currencies with a forward premium (lower interest rate). That is, on average, the forward rate is unbiased. On the other hand, at any point in time, the forward rate appears to be a biased predictor of the future spot rate. More specifically, the evidence indicates that one can profit on average by buying currencies selling at a forward discount (that is, the currency whose interest rate is relatively high) and selling currencies trading at a forward premium (that is, the currency whose interest rate is relatively low). Nonetheless, research also suggests that this evidence of forward market inefficiency may be difficult to profit from on a risk-adjusted basis. One reason is the existence of what is known as the peso problem.

The **peso problem** refers to the possibility that during the time period studied investors anticipated significant events that did not materialize, thereby invalidating statistical inferences based on data drawn from that period. The term derives from the experience of Mexico during the period 1955–1975. During this entire 21-year period, the peso was fixed at a rate of $0.125 yet continually sold at a forward discount because investors anticipated a large peso devaluation. This devaluation eventually occurred in 1976, thereby validating the prediction imbedded in the forward rate (and relative interest rates). However, had someone limited their analysis on the relation between forward and future

---

[16] See, for example, Giddy and Dufey, "The Random Behavior of Flexible Exchange Rates"; and Bradford Cornell, "Spot Rates, Forward Rates, and Market Efficiency," *Journal of Financial Economics*, January 1977, pp. 55–65.

[17] See, for example, Lars P. Hansen and Robert J. Hodrick, "Forward Rates as Optimal Predictions of Future Spot Rates," *Journal of Political Economy*, October 1980, pp. 829–853.

spot rates to data drawn only from the 1955–1975 period, they would have falsely concluded that the forward rate was a biased predictor of the future spot rate.

In their comprehensive survey of the research on bias in forward rates, Froot and Thaler conclude, "Whether or not there is really money to be made based on the apparent inefficiency of foreign exchange markets, it is worth emphasizing that the risk-return tradeoff for a single currency is not very attractive . . . Although much of the risk in these [single currency] strategies may be diversifiable in principle, more complex, diversified strategies may be much more costly, unreliable, or difficult to execute."[18] This evidence of bias suggests that the selective use of forward contracts—sell forward if the currency is at a forward premium and buy it forward if it is selling at a discount—may increase expected profits but at the expense of higher risk.

7.6 INFLATION RISK AND ITS IMPACT ON FINANCIAL MARKETS

We have seen from the Fisher effect that both borrowers and lenders factor expected inflation into the nominal interest rate. The problem, of course, is that actual inflation could turn out to be higher or lower than expected. This possibility introduces the element of **inflation risk**, by which is meant the divergence between actual and expected inflation.

It is easy to see why inflation risk can have such a devastating impact on bond prices. Bonds promise investors fixed cash payments until maturity. Even if those payments are guaranteed, as in the case of default-free U.S. Treasury bonds, investors face the risk of random changes in the dollar's purchasing power. If actual inflation could vary between 5% and 15% annually, then the real interest rate associated with a 13% nominal rate could vary between 8% (13 − 5) and −2% (13 − 15). What matters to people is the real value—not the quantity—of the money they will receive, therefore a high and variable rate of inflation will result in lenders demanding a premium to bear inflation risk.

Borrowers also face inflation risk. Assume that a firm issues a 20-year bond priced to yield 15%. If this nominal rate is based on a 12% expected rate of inflation, then the firm's expected real cost of debt will be about 3%. But suppose the rate of inflation averages 2% over the next 20 years. Instead of a 3% real cost of debt, the firm must pay a real interest rate of 13%. Thus, borrowers will also demand to be compensated for bearing inflation risk. Of course, if inflation turns out to be higher than expected, the borrower will gain by having a lower real cost of funds than anticipated. Inflation is a zero-sum game, so the lender will lose exactly what the borrower gains. When inflation is lower than expected, however, the lender will profit at the borrower's expense. Thus, the presence of uncertain inflation introduces an element of risk into financial contracts even when, as with government debt, default risk is absent. Under conditions of high and variable inflation, therefore, we would expect to find an inflation risk premium, p, added to the basic Fisher equation. The modified Fisher equation would then be

$$r = a + i + ai + p \tag{7.18}$$

---

[18] Kenneth A. Froot and Richard H. Thaler, "Anomalies: Foreign Exchange," *Journal of Economic Perspectives*, Summer 1990, pp. 179–192.

Yet a basic problem remains. Although inflation risk on a financial contract stated in nominal terms affects both borrower and lender, the two cannot be compensated simultaneously for bearing this risk. Therefore, lender and borrower will decide under some circumstances to shun fixed-rate debt contracts altogether.

Inflation and Bond Price Fluctuations

The effect of inflation on bond prices is comparable to that of interest rate changes because both affect the real value of cash flows to be received in the future. For example, the real, or inflation-adjusted, value of a dollar to be received in one year when inflation is 5% per annum equals 1/1.05 or $0.9524. That same dollar to be received two years hence has a real value of $1/(1.05)^2 = \$0.9070$. An increase in the inflation rate to 8% per annum will change the real values of the dollars received in the first and second years to $0.9259 and $0.8573, respectively. The 3% increase in the rate of inflation reduces the real value of the dollar received in year 1 by $0.0265, and the real value of the dollar received in year 2 drops by $0.0497.

This example illustrates a more general phenomenon: The longer the maturity of a bond, the greater the impact on the present value of that bond associated with a given change in the rate of inflation. In effect, a change in the inflation rate is equivalent to a change in the rate at which future cash flows are discounted back to the present. Thus, inflation risk is most devastating on long-term, fixed-rate bonds.

Responses to Inflation Risk

The problem of inflation risk increases with the maturity of a bond. Therefore, the presence of volatile inflation will make corporate borrowers less willing to issue, and investors less willing to buy, long-term, fixed-rate debt. The result will be a decline in the use of long-term, fixed-rate financing and an increased reliance on debt with shorter maturities, floating-rate bonds, and indexed bonds.

SHORTER MATURITIES ■ With shorter maturities, investors and borrowers lock themselves in for shorter periods of time. They reduce their exposure to inflation risk because the divergence between actual and expected inflation decreases as the period of time over which inflation is measured decreases. When the security matures, the loan can be "rolled over" or borrowed again at a new rate that reflects revised expectations of inflation.

FLOATING-RATE BONDS ■ The problem with short maturities for the corporate borrower, however, is that there is no guarantee that additional funds will be available at maturity. A floating-rate bond solves this problem. The funds are automatically rolled over every three to six months or so, at an adjusted interest rate. The new rate is typically set at a fixed margin above a mutually agreed-upon interest rate "index" such as the London interbank offer rate (LIBOR) for Eurodollar deposits (see Chapter 15), the corresponding Treasury bill rate, or the prime rate. For example, if the floating rate is set at prime plus 2, then a prime rate of 8.5% will yield a loan rate of 10.5%.

INDEXED BONDS ■ The real interest rate on a floating-rate bond can still change if real interest rates in the market change. This problem can be alleviated by issuing indexed bonds that pay interest tied to the inflation rate. For example, the British government has sold several billion pounds of indexed bonds since 1981. The way indexation works is that the interest rate is set equal to, say, 3% plus an adjustment for the amount of inflation during the past year. If inflation was 17%, the interest rate will be $3 + 17 = 20\%$. In this way, although the nominal rate of interest will fluctuate with inflation, the real rate of interest will be fixed at 3%. Both borrower and lender are protected against inflation risk.

The international evidence clearly supports these conjectures. In countries that historically have had high inflation rates—such as Argentina, Brazil, Israel, and Mexico—long-term fixed-rate financing has disappeared. Instead, long-term financing is done with floating-rate bonds or indexed debt. Similarly, in the United States during the late 1970s and early 1980s, when inflation risk was at its peak, 30-year conventional fixed-rate mortgages were largely replaced with so-called adjustable-rate mortgages. The interest rate on this type of mortgage is adjusted periodically in line with the changing short-term cost of funds. To the extent that short-term interest rates track actual inflation rates closely—a reasonable premise according to the available evidence—an adjustable-rate mortgage protects both borrower and lender from inflation risk. As inflation subsided in the United States, long-term fixed-rate financing reappeared. One would expect a similar outcome in Argentina and other countries that now have inflation under control.

7.7 CURRENCY FORECASTING

Forecasting exchange rates has become an occupational hazard for financial executives of multinational corporations. The potential for periodic—and unpredictable—government intervention makes currency forecasting all the more difficult. But this difficulty has not dampened the enthusiasm for currency forecasts or the willingness of economists and others to supply them. Unfortunately, though, enthusiasm and willingness are not sufficient conditions for success.

Requirements for Successful Currency Forecasting

Currency forecasting can lead to consistent profits only if the forecaster meets at least one of the following four criteria.[19] He or she

- Has exclusive use of a superior forecasting model
- Has consistent access to information before other investors
- Exploits small, temporary deviations from equilibrium
- Can predict the nature of government intervention in the foreign exchange market

The first two conditions are self-correcting. Successful forecasting breeds imitators, and the second situation is unlikely to last long in the highly informed world of international finance. The third situation describes how foreign exchange traders actually earn their living and also why deviations from equilibrium are not likely to last long. The fourth situation is the one worth searching out. Countries that insist on managing their

---

[19] These criteria were suggested by Giddy and Dufey, "The Random Behavior of Flexible Exchange Rates."

EXHIBIT 7.15 Forecasting in a Fixed-Rate System

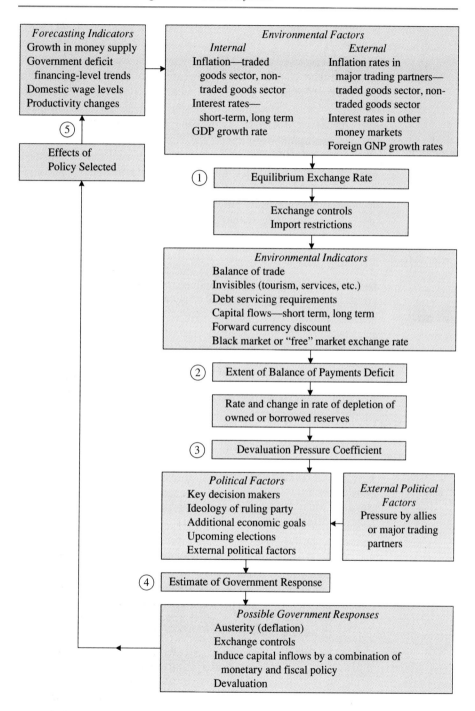

Forecasting Indicators
Growth in money supply
Government deficit
 financing-level trends
Domestic wage levels
Productivity changes

⑤

Effects of
Policy Selected

Environmental Factors
Internal *External*
Inflation—traded   Inflation rates in
 goods sector, non-   major trading partners—
 traded goods sector   traded goods sector, non-
Interest rates—   traded goods sector
 short-term, long term   Interest rates in other
GDP growth rate   money markets
    Foreign GNP growth rates

① Equilibrium Exchange Rate

Exchange controls
Import restrictions

Environmental Indicators
Balance of trade
Invisibles (tourism, services, etc.)
Debt servicing requirements
Capital flows—short term, long term
Forward currency discount
Black market or "free" market exchange rate

② Extent of Balance of Payments Deficit

Rate and change in rate of depletion of
owned or borrowed reserves

③ Devaluation Pressure Coefficient

Political Factors
Key decision makers
Ideology of ruling party
Additional economic goals
Upcoming elections
External political factors

*External Political
Factors*
Pressure by allies
 or major trading
 partners

④ Estimate of Government Response

Possible Government Responses
Austerity (deflation)
Exchange controls
Induce capital inflows by a combination of
 monetary and fiscal policy
Devaluation

exchange rates, and are willing to take losses to achieve their target rates, present specu-lators with potentially profitable opportunities. Simply put, consistently profitable predictions are possible in the long run only if it is not necessary to outguess the market to win.

As a general rule, in a fixed-rate system, the forecaster must focus on the govern-mental decision-making structure because the decision to devalue or revalue at a given time is clearly political. During the Bretton Woods system, for example, many speculators did quite well by "stepping into the shoes of the key decision makers" to forecast their likely behavior. The basic forecasting methodology in a fixed-rate system, therefore, involves first ascertaining the pressure on a currency to devalue or revalue and then deter-mining how long the nation's political leaders can, and will, persist with this particular level of disequilibrium. Exhibit 7.15 depicts a five-step procedure for performing this analysis. In the case of a floating-rate system, where government intervention is sporadic or nonexistent, currency prognosticators have the choice of using either market- or model-based forecasts, neither of which guarantees success.

Market-Based Forecasts

So far, we have identified several equilibrium relationships that should exist between exchange rates and interest rates. The empirical evidence on these relationships implies that, in general, the financial markets of developed countries efficiently incorporate expected currency changes in the cost of money and forward exchange. This means that currency forecasts can be obtained by extracting the predictions already embodied in interest and forward rates.

FORWARD RATES ■*Market-based forecasts* of exchange rate changes can be derived most simply from current forward rates. Specifically, f_1—the forward rate for one period from now—will usually suffice for an unbiased estimate of the spot rate as of that date. In other words, f_1 should equal $\overline{e}_1$, where $\overline{e}_1$ is the expected future spot rate.

INTEREST RATES ■Although forward rates provide simple and easy-to-use currency fore-casts, their forecasting horizon is limited to about one year because of the general absence of longer-term forward contracts. Interest rate differentials can be used to supply exchange rate predictions beyond one year. For example, suppose five-year interest rates on dollars and Deutsche marks are 12% and 8%, respectively. If the current spot rate for the DM is $0.40 and the (unknown) value of the DM in five years is e_5, then $1.00 invested today in Deutsche marks will be worth $(1.08)^5 e_5/0.4$ dollars at the end of five years; if invested in the dollar security, it will be worth $(1.12)^5$ in five years. The market's forecast of e_5 can be found by assuming that investors demand equal returns on dollar and DM securities, or

$$\frac{(1.08)^5 e_5}{0.4} = (1.12)^5$$

Thus, the five-year DM spot rate implied by the relative interest rates is $e_5 = \$0.4798$ $(0.40 \times 1.12^5/1.08^5)$.

Model-Based Forecasts

The two principal model-based approaches to currency prediction are known as fundamental analysis and technical analysis. Each approach has its advocates and detractors.

FUNDAMENTAL ANALYSIS ■ **Fundamental analysis** is the most common approach to forecasting future exchange rates. It relies on painstaking examination of the macroeconomic variables and policies that are likely to influence a currency's prospects. The variables examined include relative inflation and interest rates, national income growth, and changes in money supplies. The interpretation of these variables and their implications for future exchange rates depend on the analyst's model of exchange rate determination.

The simplest form of fundamental analysis involves the use of PPP. We have previously seen the value of PPP in explaining exchange rate changes. Its application in currency forecasting is straightforward.

ILLUSTRATION USING PPP TO FORECAST THE PESETA'S FUTURE SPOT RATE

The U.S. inflation rate is expected to average about 4% annually, and the Spanish rate of inflation is expected to average about 9% annually. If the current spot rate for the peseta is $0.008, what is the expected spot rate in two years?

Solution. According to PPP (Equation 7.2), the expected spot rate for the peseta in two years is $0.008 \times (1.04/1.09)^2 = \$0.00728$.

Most analysts use more complicated forecasting models whose analysis usually centers on how the different macroeconomic variables are likely to affect the demand and supply for a given foreign currency. The currency's future value is then determined by estimating the exchange rate at which supply just equals demand—when any current-account imbalance is just matched by a net capital flow.

Forecasting based on fundamental analysis has inherent difficulties. First, you must be able to select the right fundamentals; then you must be able to forecast them—itself a problematic task (think about forecasting interest rates); finally, your forecasts of the fundamentals must differ from those of the market. Otherwise, the exchange rate will have already discounted the anticipated change in the fundamentals. Another difficulty that forecasters face is the variability in the lag between when changes in fundamentals are forecast to occur and when they actually affect the exchange rate.

Despite these difficulties, Robert Cumby developed a sophisticated regression model—incorporating forward premia along with real variables such as relative inflation rates and current-account balances—that yielded predictable return differentials (between investing in uncovered foreign deposits and domestic deposits) on the order of 10% to 30% per annum.[20]

---

[20]Robert Cumby, "Is It Risk? Deviations from Uncovered Interest Parity," *Journal of Monetary Economics*, September 1988, pp. 279–300.

TECHNICAL ANALYSIS ■ **Technical analysis** is the antithesis of fundamental analysis in that it focuses exclusively on past price and volume movements—while totally ignoring economic and political factors—to forecast currency winners and losers. Success depends on whether technical analysts can discover price patterns that repeat themselves and are, therefore, useful for forecasting.

There are two primary methods of technical analysis: *charting* and *trend analysis*. Chartists examine bar charts or use more sophisticated computer-based extrapolation techniques to find recurring price patterns. They then issue buy or sell recommendations if prices diverge from their past pattern. Trend-following systems seek to identify price trends via various mathematical computations.

Model Evaluation

The possibility that either fundamental or technical analysis can be used to profitably forecast exchange rates is inconsistent with the efficient market hypothesis, which says that current exchange rates reflect all publicly available information. Because markets are forward-looking, exchange rates will fluctuate randomly as market participants assess and then react to new information, much as security and commodity prices in other asset markets respond to news. Thus, exchange rate movements are unpredictable; otherwise, it would be possible to earn arbitrage profits. Such profits could not persist in a market—such as the foreign exchange market—that is characterized by free entry and exit and an almost unlimited amount of money, time, and energy that participants are willing to commit in pursuit of profit opportunities.

In addition to the theoretical doubts surrounding forecasting models, a variety of statistical and technical assumptions underlying these models have been called into question as well. For all practical purposes, though, the quality of a currency forecasting model must be viewed in relative terms. That is, a model can be said to be "good" if it is better than alternative means of forecasting currency values. Ultimately, a currency forecasting model is "good" only to the extent that its predictions will lead to better decisions.

Certainly interest differentials and/or forward rates provide low-cost alternative forecasts of future exchange rates. At a minimum, any currency forecasting model should be able to consistently outperform the market's estimates of currency changes. In other words, one relevant question is whether *profitable* decisions can be made in the forward and/or money markets by using any of these models.

Currency forecasters charge for their services, so researchers periodically evaluate the performance of these services to determine whether the forecasts are worth their cost. The evaluation criteria generally fall into two categories: accuracy and correctness. The accuracy measure focuses on the deviations between the actual and the forecasted rates, and the correctness measure examines whether or not the forecast predicts the right direction of the change in exchange rates.

An accurate forecast may not be correct in predicting the direction of change, and a correct forecast may not be very accurate. The two criteria are sometimes in conflict. Which of these two criteria should be used in evaluation depends on how the forecasts are to be used.

An analysis of forecasting errors—the difference between the forecast and actual exchange rate—will tell us little about the profit-making potential of econometric forecasts. Instead, we need to link these forecasts to actual decisions and then calculate the resulting profits or losses. For example, if the forecasts are to be used to decide whether or not to hedge with forward contracts, the relative predictive abilities of the forecasting services can be evaluated by using the following decision rule:

If $f_1 > \overline{e}_1$, sell forward

If $f_1 < \overline{e}_1$, buy forward

where f_1 is the forward rate and $\overline{e}_1$ is the forecasted spot rate at the forward contract's settlement date. In other words, if the forecasted rate is below the forward rate, the currency should be sold forward; if the forecasted rate is above the forward rate, the currency should be bought forward.

Where e_1 is the *actual* spot rate being forecasted, the percentage profit (loss) realized from this strategy equals $100[(f_1 - e_1)/e_1]$ when $f_1 > e_1$, and equals $100[(e_1 - f_1)/e_1]$ when $f_1 < e_1$.

ILLUSTRATION

THE DISTINCTION BETWEEN AN ACCURATE FORECAST AND A PROFITABLE FORECAST Suppose that the ¥/$ spot rate is currently ¥110/$. A 90-day forecast puts the exchange rate at ¥102/$; the 90-day forward rate is ¥109/$. According to our decision rule, we should buy the yen forward. If we buy $1 million worth of yen forward and the actual rate turns out to be ¥108/$, then our decision will yield a profit of $9,259 [(109,000,000 − 108,000,000)/108]. In contrast, if the forecasted value of the yen had been ¥111/$, we would have sold yen forward and lost $9,259. Thus, an accurate forecast, off by less than 3% (3/108), leads to a loss; and a less accurate forecast, off by almost 6% (6/108), leads to a profitable decision.

When deciding on a new investment or planning a revised pricing strategy, however, the most critical attribute of a forecasting model is its accuracy. In the latter case, the second forecast would be judged superior.

Despite the theoretical skepticism over successful currency forecasting, a study of 14 forecast advisory services by Richard Levich indicates that the profits associated with using several of these forecasts seem too good to be explained by chance.[21] Of course, if the forward rate contains a risk premium, these returns would have to be adjusted for the risks borne by speculators. It is also questionable whether currency forecasters would continue selling their information rather than acting on it themselves if they truly believed it could yield excess risk-adjusted returns. That being said, it is hard to attribute expected return differentials of up to 30% annually (Cumby's results) to currency risk when the estimated equity risk premium on the U.S. stock exchange is only about 8% for a riskier investment.

---

[21] Richard M. Levich, "The Use and Analysis of Foreign Exchange Forecasts: Current Issues and Evidence," paper presented at the Euromoney Treasury Consultancy Program, New York, September 4–5, 1980.

Of course, if you take a particular data sample and run every possible regression, you are likely to find some apparently profitable forecasting model. But that does not mean it is a reliable guide to the future. To control for this tendency to "data mine," you must do *out-of-sample forecasting*. That is, you must see if your model forecasts well enough to be profitable in time periods not included in the original data sample. Hence, the profitable findings of Cumby and others may stem from the fact that their results are based on the in-sample performance of their regressions. That is, they used the same data sample both to estimate their model and to check its forecasting ability. Indeed, Richard Meese and Kenneth Rogoff concluded that sophisticated models of exchange rate determination make poor forecasts.[22] Their conclusion is similar to that of Jeffrey Frankel, who —after reviewing the research on currency forecasting—stated that

> the proportion of exchange rate changes that are forecastable in any manner—by the forward discount, interest rate differential, survey data, or models based on macroeconomic fundamentals—appears to be not just low, but almost zero.[23]

Frankel's judgment is consistent with the existence of an efficient market in which excess risk-adjusted returns have a half-life measured in minutes, if not seconds.

Forecasting Controlled Exchange Rates

A major problem in currency forecasting is that the widespread existence of exchange controls, as well as restrictions on imports and capital flows, often masks the true pressures on a currency to devalue. In such situations, forward markets and capital markets are invariably nonexistent or subject to such stringent controls that interest and forward differentials are of little practical use in providing market-based forecasts of exchange rate changes. An alternative forecasting approach in such a controlled environment is to use *black-market exchange rates* as useful indicators of devaluation pressure on the nation's currency.

Black-market exchange rates for a number of countries are regularly reported in *Pick's Currency Yearbook and Reports*. These black markets for foreign exchange are likely to appear whenever exchange controls cause a divergence between the equilibrium exchange rate and the official, or *controlled*, exchange rate. Those potential buyers of foreign exchange without access to the central banks will have an incentive to find another market in which to buy foreign currency. Similarly, sellers of foreign exchange will prefer to sell their holdings at the higher black-market rate.

The black-market rate depends on the difference between the official and equilibrium exchange rates, as well as on the expected penalties for illegal transactions. The existence of these penalties and the fact that some transactions do go through at the official rate mean that the black-market rate is not influenced by exactly the same set of supply-and-demand forces that influences the free-market rate. Therefore, the black-market rate in itself cannot be regarded as indicative of the true equilibrium rate that would prevail in the absence of

---

[22] Richard A. Meese and Kenneth Rogoff, "Empirical Exchange Rate Models of the Seventies: Do They Fit Out of Sample?" *Journal of International Economics* 14, no. 1/2 (1983): 3–24.

[23] Jeffrey Frankel, "Flexible Exchange Rates: Experience Versus Theory," *Journal of Portfolio Management*, Winter 1989, pp. 45–54.

controls. Economists normally assume that for an overvalued currency, the hypothetical equilibrium rate lies somewhere between the official rate and the black-market rate.

The usefulness of the black-market rate is that it is a good indicator of where the official rate is likely to go if the monetary authorities give in to market pressure. Although the official rate can be expected to move toward the black-market rate, we should not expect to see it coincide with that rate because of the bias induced by government sanctions. The black-market rate seems to be most accurate in forecasting the official rate one month ahead and is progressively less accurate as a forecaster of the future official rate for longer time periods.[24]

Exhibit 7.16a graphs monthly movements in the official and black-market rates for the Brazilian cruzeiro from December 1968 to March 1977. Exhibit 7.16b graphs

EXHIBIT 7.16 Black Market Versus Official Exchange Rates: Brazilian Cruzeiro

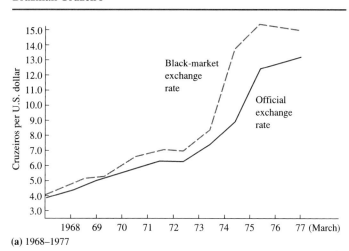

(a) 1968–1977

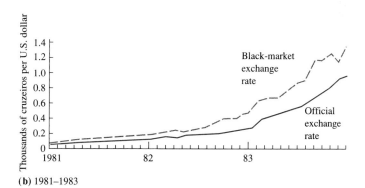

(b) 1981–1983

[24] See, for example, Ian Giddy, "Black Market Exchange Rates as a Forecasting Tool," working paper, Columbia University, May 1978.

monthly movements in the two rates from 1981 to 1983. The data clearly show that the black-market cruzeiro rate was invariably above the official rate and was a useful indicator of future devaluations.[25]

❧ 7.8 SUMMARY AND CONCLUSIONS

In this chapter, we examined four relationships, or parity conditions, that should apply to spot rates, inflation rates, and interest rates in different currencies: purchasing power parity (PPP), the Fisher effect (FE), the international Fisher effect (IFE), and the forward rate as an unbiased forecast of the future spot rate (UFR). (We examined a fifth parity condition, interest rate parity or IRP in Chapter 5.) These parity conditions follow from the law of one price, the notion that in the absence of market imperfections, arbitrage ensures that exchange-adjusted prices of identical traded goods and financial assets are within transaction costs worldwide.

The technical description of these four equilibrium relationships is summarized as follows:

■ Purchasing power parity

$$\frac{e_t}{e_0} = \frac{(1 + i_h)^t}{(1 + i_f)^t}$$

WHERE

e_t = the home currency value of the foreign currency at time t
e_0 = the home currency value of the foreign currency at time 0
i_h = the periodic domestic inflation rate
i_f = the periodic foreign inflation rate

■ Fisher effect

$$1 + r = (1 + a)(1 + i)$$

WHERE

r = the nominal rate of interest
a = the real rate of interest
i = the rate of expected inflation

■ Generalized version of Fisher effect

$$\frac{(1 + r_h)^t}{(1 + r_f)^t} = \frac{(1 + i_h)^t}{(1 + i_f)^t}$$

WHERE

r_h = the periodic home currency interest rate
r_f = the periodic foreign currency interest rate

---

[25] The cruzeiro has since been replaced by the *real*.

■ International Fisher effect

$$\frac{(1 + r_h)^t}{(1 + r_f)^t} = \frac{\overline{e}_t}{e_0}$$

WHERE

$\overline{e}_t$ = the expected home currency value of the foreign currency at time t

   ■ Forward rate as an unbiased predictor of the future spot rate

$$f_t = \overline{e}_t$$

WHERE

$\overline{e}_t$ = the expected home currency value of the foreign currency at time t
f_t = the forward rate for delivery of one unit of foreign currency at time t

Despite the mathematical precision with which these parity conditions are expressed, they are only approximations to reality. A variety of factors can lead to significant and prolonged deviations from parity. For example, both currency risk and inflation risk may cause real interest rates to differ across countries. Similarly, various shocks can cause the real exchange rate—defined as the nominal, or actual, exchange rate adjusted for changes in the relative purchasing power of each currency since some base period—to change over time. Moreover, the short-run relation between changes in the nominal interest differential and changes in the exchange rate is not so easily determined. The lack of definiteness in this relation stems from the differing effects on exchange rates of purely nominal interest rate changes and real interest rate changes.

We examined the concept of the real exchange rate in more detail as well. The real exchange rate, e_t', incorporates both the nominal exchange rate between two currencies and the inflation rates in both countries. It is defined as follows:

   ■ Real exchange rate

$$e_t' = e_t \frac{P_f}{P_h}$$

WHERE

P_f = the foreign price level at time t indexed to 100 at time 0
P_h = the home price level at time t indexed to 100 at time 0

We also analyzed a series of forecasting models that purport to outperform the market's own forecasts of future exchange rates as embodied in interest and forward differentials. We concluded that the foreign exchange market is no different from any other financial market in its susceptibility to profitable predictions.

Those who have inside information about events that will affect the value of a currency or a security should benefit handsomely. Those who do not have this access will have to trust either to luck or to the existence of a market imperfection, such as government intervention, to assure themselves of above-average, risk-adjusted profits. In fact, given the widespread availability of information and the many knowledgeable participants in the foreign exchange market, only the latter situation—government manipulation of exchange rates—holds the promise of superior risk-adjusted returns from currency forecasting. When governments spend money to control exchange rates, this money flows into the hands of private participants who bet against the government. The trick is to predict government actions.

➤Questions

1. What are some reasons for deviations from purchasing power parity?

2. Under what circumstances can purchasing power parity be applied?

3. One proposal to stabilize the international monetary system involves setting exchange rates at their purchasing power parity rates. Once exchange rates were correctly aligned (according to PPP), each nation would adjust its monetary policy so as to maintain them. What problems might arise from using the PPP rate as a guide to the equilibrium exchange rate?

4. Suppose the dollar/rupiah rate is fixed but Indonesian prices are rising faster than U.S. prices. Is the Indonesian rupiah appreciating or depreciating in real terms?

5. If the dollar is appreciating against the Polish zloty in nominal terms but depreciating against the zloty in real terms, what do we know about Polish and U.S. inflation rates?

6. Suppose the nominal peso/dollar exchange rate is fixed. If the inflation rates in Mexico and the United States are constant (but not necessarily equal in both countries), will the real value of the peso/dollar exchange rate also be constant over time?

7. If the average rate of inflation in the world rises from 5% to 7%, what will be the likely effect on the U.S. dollar's forward premium or discount relative to foreign currencies?

8. Comment on the following statement. "It makes sense to borrow during times of high inflation because you can repay the loan in cheaper dollars."

9. Which is likely to be higher, a 150% ruble return in Russia or a 15% dollar return in the United States?

10. The interest rate in England is 12%; in Switzerland it is 5%. What are possible reasons for this interest rate differential? What is the most likely reason?

11. Over the period 1982–1988, Peru and Chile stand out as countries whose interest rates are not consistent with their inflation experience. Specifically, Peru's inflation and interest rates averaged about 125% and 8%, respectively, over this period, whereas Chile's inflation and interest rates averaged about 22% and 38%, respectively.
    a. How would you characterize the real interest rates of Peru and Chile (e.g., close to zero, highly positive, highly negative)?

    b. What might account for Peru's low interest rate relative to its high inflation rate? What are the likely consequences of this low interest rate?
    c. What might account for Chile's high interest rate relative to its inflation rate? What are the likely consequences of this high interest rate?
    d. In Exhibit 7.13, Peru is shown as having a small interest differential and yet a large average exchange rate change. How would you reconcile this experience with the international Fisher effect and with your answer to part b?

12. A number of countries (e.g., Pakistan, Hungary, Venezuela) are shown in Exhibit 7.13 as having a small or negative interest rate differential and a large average annual depreciation against the dollar. How would you explain these data? Can you reconcile these data with the international Fisher effect?

13. The empirical evidence shows that there is no consistent relationship between the spot exchange rate and the nominal interest rate differential. Why might this be?

14. During 1988, the U.S. prime rate—the rate of interest banks charge on loans to their best customers—stood at 9.5%. Japan's prime rate, meanwhile, was about 3.5%. Pointing to that discrepancy, a number of commentators argued that the cost of capital must come down for U.S. business to remain competitive with Japanese companies. What additional information would you need to properly assess this claim? Why might interest rates be lower in Japan than in the United States?

15. In the late 1960s, Firestone Tire decided that Swiss francs at 2% were cheaper than U.S. dollars at 8% and borrowed about SFr 500 million. Comment on this choice.

16. Comment on the following quote from a story in the *Wall Street Journal* (August 27, 1984, p. 6) that discusses the improving outlook for Britain's economy: "Recovery here will probably last longer than in the U.S. because there isn't a huge budget deficit to pressure interest rates higher."

17. In early 1989, Japanese interest rates were about four percentage points below U.S. rates. The wide difference between Japanese and U.S. interest rates prompted some U.S. real estate developers to borrow in yen to finance their projects. Comment on this strategy.

18. In early 1990, Japanese and German interest rates rose while U.S. rates fell. At the same time, the yen

and DM fell against the U.S. dollar. What might explain the divergent trends in interest rates?

19. In late December 1990, one-year German Treasury bills yielded 9.1%, whereas one-year U.S. Treasury bills yielded 6.9%. At the same time, the inflation rate during 1990 was 6.3% in the United States, double the German rate of 3.1%.
    a. Are these inflation and interest rates consistent with the Fisher effect?
    b. What might explain this difference in interest rates between the United States and Germany?

20. The spot rate on the Deutsche mark is $0.63, and the 180-day forward rate is $0.64. What are possible reasons for the difference between the two rates?

21. Comment on the following headline that appeared in the *Wall Street Journal* (December 19, 1990, p. C10): "Dollar Falls Across the Board as Fed Cuts Discount Rate to 6.5% From 7%." The discount rate is the interest rate the Fed charges member banks for loans.

22. In late 1990, the U.S. government announced that it might try to reduce the budget deficit by imposing a 0.5% transfer tax on all sales and purchases of securities in the United States, with the exception of Treasury securities. It projected that the tax would raise $10 billion annually in federal revenues—an amount arrived at by multiplying 0.5% by the value of the $2 trillion trading on the New York Stock Exchange each year.
    a. What are the likely consequences of this tax? Consider its effects on trading volume in the United States and stock and bond prices.
    b. Why does the U.S. government plan to exclude its securities from this tax?
    c. Critically assess the government's estimates of the revenue it will raise from this tax.

23. It has been argued that the U.S. government's economic policies, particularly as they affect the U.S. budget deficit, are severely constrained by the world's financial markets. Do you agree or disagree? Discuss.

24. In 1991, the U.S. government imposed a stiff import tariff on the active-matrix LCD screens that now appear in next-generation laptop computers.
    a. Assess the likely consequences of the import duty for U.S. laptop computer manufacturers.
    b. How are these manufacturers likely to react to this import duty?

25. "High real interest rates can be a cause for celebration, not alarm." Discuss.

26. In an integrated world capital market, will higher interest rates in, say, Japan mean higher interest rates in, say, the United States?

27. German government bonds, or Bunds, currently are paying higher interest rates than comparable U.S. Treasury bonds. Suppose the Bundesbank eases the money supply to drive down interest rates. How is an American investor in Bunds likely to fare?

28. In France in 1994, short-term interest rates and bond yields remained higher than in Germany, despite a better outlook for inflation in France. Does this situation indicate a violation of the Fisher effect? Explain.

29. On February 15, 1993, President Clinton previewed his State of the Union message to Congress in a toughly worded talk on television about how the growing federal budget deficit made tax increases necessary. Financial markets reacted by pushing bond prices up and pummeling stock prices. President Clinton said that the rise in Treasury bond prices was a "very positive" response to his televised speech the night before. How would you interpret the reaction of the financial markets to President Clinton's speech?

30. At the same time that it was talking down the dollar, the Clinton administration was talking about the need for low interest rates to stimulate economic growth. Comment.

31. In 1993 and early 1994, Turkish banks borrowed abroad at relatively low interest rates to fund their lending at home. The banks earned high profits because rampant inflation in Turkey forced up domestic interest rates. At the same time, Turkey's central bank was intervening in the foreign exchange market to maintain the value of the Turkish lira. Comment on the Turkish banks' funding strategy.

32. One idea to curb potentially destabilizing international movements of capital has been devised by James Tobin, a Nobel Prize-winning economist. He proposes putting a small tax on foreign exchange transactions. He claims that his "Tobin tax" would make short-term speculation more costly while having little effect on long-term investment.
    a. Why would the Tobin tax have a disproportionate impact on short-term investments?
    b. Is the Tobin tax likely to accomplish its objective? Explain.

Problems

1. From base price levels of 100 in 1987, West German and U.S. price levels in 1988 stood at 102 and 106, respectively. If the 1987 $:DM exchange rate was $0.54, what should the exchange rate be in 1988? In fact, the exchange rate in 1988 was DM 1 = $0.56. What might account for the discrepancy? (Price levels were measured using the consumer price index.)

2. Two countries, the United States and England, produce only one good, wheat. Suppose the price of wheat in the United States is $3.25 and in England it is £1.35.
   a. According to the law of one price, what should the $:£ spot exchange rate be?
   b. Suppose the price of wheat over the next year is expected to rise to $3.50 in the United States and to £1.60 in England. What should the one-year $:£ forward rate be?
   c. If the U.S. government imposes a tariff of $0.50 per bushel on wheat imported from England, what is the maximum possible change in the spot exchange rate that could occur?

3. In early 1996, the short-term interest rate in France was 3.7%, and forecast French inflation was 1.8%. At the same time, the short-term German interest rate was 2.6% and forecast German inflation was 1.6%.
   a. Based on these figures, what were the real interest rates in France and Germany?
   b. To what would you attribute any discrepancy in real rates between France and Germany?

4. In July, the one-year interest rate is 12% on British pounds and 9% on U.S. dollars.
   a. If the current exchange rate is $1.63:£1, what is the expected future exchange rate in one year?
   b. Suppose a change in expectations regarding future U.S. inflation causes the expected future spot rate to decline to $1.52:£1. What should happen to the U.S. interest rate?

5. If expected inflation is 100% and the real required return is 5%, what will the nominal interest rate be according to the Fisher effect?

6. Suppose that in Japan the interest rate is 8% and inflation is expected to be 3%. Meanwhile, the expected inflation rate in France is 12%, and the English interest rate is 14%. To the nearest whole number, what is the best estimate of the one-year forward exchange premium (discount) at which the pound will be selling relative to the French franc?

7. Chase Econometrics has just published projected inflation rates for the United States and Germany for the next five years. U.S. inflation is expected to be 10% per year, and German inflation is expected to be 4% per year.
   a. If the current exchange rate is $0.65/DM, forecast the exchange rates for the next five years.
   b. Suppose that U.S. inflation over the next five years turns out to average 3.2%, German inflation averages 1.5%, and the exchange rate in five years is $0.79/DM. What has happened to the real value of the DM over this five-year period?

8. During 1995, the Mexican peso exchange rate rose from Mex$5.33/U.S.$ to Mex$7.64/U.S.$. At the same time, U.S. inflation was approximately 3%, in contrast to Mexican inflation of about 48.7%.
   a. By how much did the nominal value of the peso change during 1995?
   b. By how much did the real value of the peso change over this period?

9. The inflation rate in Great Britain is expected to be 4% per year, and the inflation rate in France is expected to be 6% per year. If the current spot rate is £1 = FF 12.50, what is the expected spot rate in two years?

10. If the $:¥ spot rate is $1 = ¥218 and interest rates in Tokyo and New York are 6% and 12%, respectively, what is the expected $:¥ exchange rate one year hence?

11. Suppose three-year deposit rates on Eurodollars and Eurofrancs (Swiss) are 12% and 7%, respectively. If the current spot rate for the Swiss franc is $0.3985, what is the spot rate implied by these interest rates for the franc three years from now?

12. Suppose that on January 1, the cost of borrowing French francs for the year is 18%. During the year, U.S. inflation is 5%, and French inflation is 9%. At the same time, the exchange rate changes from FF 1 = $0.15 on January 1 to FF 1 = $0.10 on December 31. What was the real U.S. dollar cost of borrowing francs for the year?

13. In late 1990, following Britain's entry into the exchange-rate mechanism of the European Monetary System, 10-year British Treasury bonds yielded 11.5%, and the German equivalent offered a yield of just 9%. Under terms of its entry, Britain established a central rate against the DM of DM 2.95 and pledged to maintain this rate within a band of plus and minus 6%.
   a. By how much would sterling have to fall against the DM over a 10-year period for the German bond to offer a higher overall return than the British one? Assume that the Trea-

suries are zero-coupon bonds with no interest paid until maturity.

b. How does the exchange rate established in Part a compare with the lower limit that the British government is pledged to maintain for sterling against the DM?

c. What accounts for the difference between the two rates? Does this difference violate the international Fisher effect?

14. Assume that the interest rate is 16% on pounds sterling and 7% on Deutsche marks. At the same time, inflation is running at an annual rate of 3% in Germany and 9% in England.

a. If the Deutsche mark is selling at a one-year forward premium of 10% against the pound, is there an arbitrage opportunity? Explain.

b. What is the real interest rate in Germany? In England?

c. Suppose that during the year the exchange rate

changes from DM 2.7:£1 to DM 2.65:£1. What are the real costs to a German company of borrowing pounds? Contrast this cost to its real cost of borrowing DM.

d. What are the real costs to a British firm of borrowing DM? Contrast this cost to its real cost of borrowing pounds.

15. Suppose today's exchange rate is $0.62/DM. The 6-month interest rates on dollars and DM are 6% and 3%, respectively. The 6-month forward rate is $0.6185. A foreign exchange advisory service has predicted that the DM will appreciate to $0.64 within six months.

a. How would you use forward contracts to profit in the above situation?

b. How would you use money market instruments (borrowing and lending) to profit?

c. Which alternatives (forward contracts or money market instruments) would you prefer? Why?

➤ Bibliography

CORNELL, BRADFORD. "Relative Price Changes and Deviations from Purchasing Power Parity." *Journal of Banking and Finance*, 3 (1979): 263–279.

DUFEY, GUNTER, and IAN H. GIDDY. "Forecasting Exchange Rates in a Floating World." *Euromoney*, November 1975, pp. 28–35.

——. *The International Money Market.* Englewood Cliffs, N.J.: Prentice Hall, 1978.

GAILLIOT, HENRY J. "Purchasing Power Parity as an Explanation of Long-Term Changes in Exchange Rates." *Journal of Money, Credit, and Banking*, August 1971, pp. 348–357.

GIDDY, IAN H. "An Integrated Theory of Exchange Rate Equilibrium." *Journal of Financial and Quantitative Analysis*, December 1976, pp. 883–892.

GIDDY, IAN H., and GUNTER DUFEY. "The Random Behavior of Flexible Exchange Rates," *Journal of International Business Studies*, Spring 1975, pp. 1–32.

LEVICH, RICHARD M. "Analyzing the Accuracy of Foreign Exchange Advisory Services: Theory and

Evidence." In *Exchange Risk and Exposure*, edited by Richard Levich and Clas Wihlborg. Lexington, Mass: D.C. Heath, 1980.

OFFICER, LAWRENCE H. "The Purchasing-Power-Parity Theory of Exchange Rates: A Review Article." *IMF Staff Papers*, March 1976, pp. 1–60.

ROLL, RICHARD. "Violations of the Law of One Price and Their Implications for Differentially-Denominated Assets." In *International Finance and Trade*, vol. 1, edited by Marshall Sarnat and George Szego. Cambridge, Mass.: Ballinger, 1979.

SHAPIRO, ALAN C. "What Does Purchasing Power Parity Mean?" *Journal of International Money and Finance*, December 1983, pp. 295–318.

STRONGIN, STEVE. "International Credit Market Connections." *Economic Perspectives*, July/August 1990, pp. 2–10.

TREUHERZ, ROLF M. "Forecasting Foreign Exchange Rates in Inflationary Economies." *Financial Executive*, February 1969, pp. 57–60.

Case Studies

CASE I.1

OIL LEVIES: THE ECONOMIC IMPLICATIONS

BACKGROUND

The combination of weakening oil prices and the failure of Congress to deal with the budget deficit by cutting spending led some to see the possibility of achieving two objectives at once: (1) protecting U.S. oil producers from "cheap" foreign competition and (2) reducing the budget deficit. The solution was an oil-import fee or tariff. A tax on imported crude and refined products that matched a world oil-price decline, for example, would leave oil and refined-product prices in the United States unchanged. Thus, it was argued, such a tax would have little effect on U.S. economic activity. It merely represents a transfer of funds from foreign oil producers to the U.S. Treasury. Moreover, it would provide some price relief to struggling U.S. refineries and encourage the production of U.S. oil. Finally, at the current level of imports, a $5/barrel tariff on foreign crude oil and a separate tariff of $10/barrel-equivalent on refined products would raise more than $11.5 billion a year.

QUESTIONS

1. Suppose the tariff were levied solely on imported crude. In an integrated world economy, who will bear the burden of the import tariff? Who will benefit? Why? What will be the longer-term consequences?
2. If a $10/barrel tariff were levied on imported refined products (but no tariff were levied on crude oil), who would bear the burden of such a tariff? Who would benefit? Why? What would be the longer-term consequences?
3. What would be the economic consequences of the combined $5/barrel tariff on imported crude and a $10/barrel tariff on refined oil products? How would these tariffs affect domestic consumers, oil producers, refiners, companies competing against imports, and exporters?
4. How would these proposed import levies affect foreign suppliers to the United States of crude oil and refined products?
5. During the 1970s, price controls on crude oil—but not on refined products—were in effect in the United States. On the basis of your previous analysis, what differences would you expect to see between heating oil and gasoline prices in New York and in Rotterdam (the major refining center in northwestern Europe)?

CASE I.2

PRESIDENT CARTER LECTURES THE FOREIGN EXCHANGE MARKETS

At a press conference in March 1978, President Jimmy Carter—responding to a falling dollar—lectured the international financial markets as follows:

*I've spent a lot of time studying about the American dollar, its value in international monetary markets, the causes of its recent deterioration as it relates to other major currencies. I can say with complete assurance that the basic principles of monetary values are not being adequately addressed on the current international monetary market.**

* "President Canute," *Wall Street Journal*, March 8, 1978, p. 16.

President Carter then offered three reasons why the dollar should improve: (1) the "rapidly increasing" attractiveness of investment in the U.S. economy due to high nominal interest rates, (2) an end to growth in oil imports, and (3) a decline in the real growth of the U.S. economy relative to the rest of the world's economic growth.

QUESTIONS

1. How were financial markets likely to respond to President Carter's lecture? Explain.

2. At the time President Carter made his remarks, the inflation rate was running at about 10% annually and accelerating as the Federal Reserve continued to pump up the money supply to finance the growing government budget deficit. Meanwhile, the interest rate on long-term Treasury bonds had risen to about 8.5%. Was President Carter correct in his assessment of the positive effects on the dollar of the higher interest rates? Explain. Note that during 1977, the movement of private capital had switched to an outflow of $6.6 billion in the second half of the year, from an inflow of $2.9 billion in the first half.

3. Comment on the consequences of a reduction in U.S. oil imports for the value of the U.S. dollar. Next, consider that President Carter's energy policy involved heavily taxing U.S. oil production, imposing price controls on domestically produced crude oil and gasoline, and providing rebates to users of heating oil. How was this energy policy likely to affect the value of the dollar? Explain.

4. What were the likely consequences of the slowdown in U.S. economic growth for the value of the U.S. dollar? For the U.S. trade balance?

5. If President Carter had listened to the financial markets instead of trying to lecture them, what might he have heard? That is, what were the markets trying to tell him about his policies?

CASE I.3

RESCUING THE INDONESIAN RUPIAH WITH A CURRENCY BOARD

On January 9, 1998, the Indonesian rupiah hit an historic low of 11,000 after a record low of 9,900 the previous day (see Exhibit I 3.1). It was down from 2,450 in July 1997 before the Asian currency crisis began. Jakarta's Stock Exchange Index plunged 12%. As news of Indonesia's troubles spread, stock markets around the world were battered; the Dow Jones Industrial Average fell more than 240 points to 7560.74. The rupiah's plunge threw hundreds of banks and companies that had borrowed heavily abroad into bankruptcy (collectively, Indonesian businesses are estimated to be saddled with $74 billion in foreign debt).

President Bill Clinton spoke to President Suharto and urged him to implement tough economic reforms to stop the deepening crisis from getting out of control. A senior Indonesian official said that Suharto agreed with Clinton's assessment and promised to implement serious reforms. A senior U.S. economic delegation accompanied by the IMF's top two officials was sent to Jakarta. Investors took a dismal view of Indonesia's prospects; from July 1997 when the Asian crisis began through the end of January 1998, the Jakarta Stock Exchange lost more than 84% of its value in dollar terms (see Exhibit I 3.2).

The drop in the Indonesian stock market and the rupiah began the week before, when Suharto announced a draft budget that contained wildly unrealistic projections (GDP growth of 4%, inflation remaining at 9%, the rupiah doubling in value from its current level, and a balanced budget based on projections of a 10% increase in non-oil tax receipts) and reneged on taking any of

EXHIBIT I 3.1 The Rupiah Rides a Rollercoaster

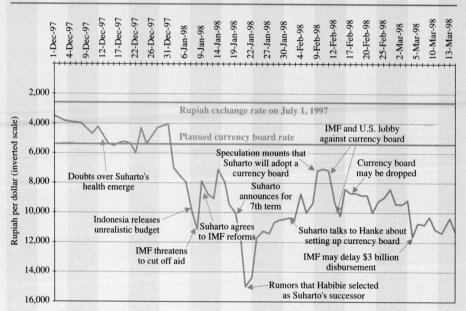

EXHIBIT I 3.2 Indonesia's Stock Market Index in U.S. Dollars: January 1995–March 1998

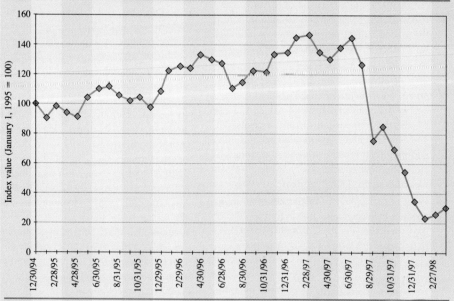

Source: Morgan Stanley Capital International.

the tough structural reform measures it agreed to in October 1997 in return for a $43 billion IMF-led rescue package. Specifically, Suharto had promised to restructure the banking system, maintain a tight monetary policy, raise sales taxes, cut food, fuel, and electricity subsidies, end government funding of several huge, money-losing investments (including projects to design and manufacture a national car and a national airplane), and disband domestic monopolies, cartels, and special entitlements that had enriched his children and cronies.

Most analysts viewed the budget as an attempt to maintain the status quo at a time that Indonesia was on the brink of disaster (private economists forecast a decline in GDP of 15% and inflation of 55%). They had hoped for austerity, a blueprint for the handling of cash-strapped and insolvent companies and banks, and repeals on foreign-ownership limits on property and financial institutions. Instead, Mr. Suharto pledged a 32% increase in government spending, including a 13 trillion rupiah increase in subsidies for fuel and food. The budget also allocated funds for the development by Suharto's eldest daughter of a new power plant on Java, even though Java already had an electricity surplus.

In response, the IMF threatened to pull the plug on its $43 billion bailout program for Indonesia. IMF officials said they could understand Indonesia's desire to place a high priority on social and humanitarian concerns in its budget but they criticized the government for doing so without making a good-faith effort to implement the agreed-on structural economic reforms. Under the IMF-led rescue plan, Indonesia was required to achieve a budget surplus of 1% of GDP. However, IMF officials said they were willing to renegotiate the fiscal targets provided Indonesia implemented such reforms. The heightened tensions sent the rupiah plummeting and led to a sharp rise in food and other prices (see Exhibit I 3.3).

EXHIBIT I 3.3 Inflation Accelerates in Indonesia

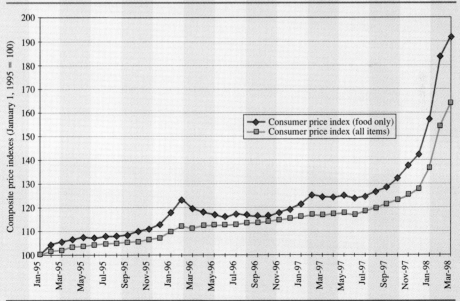

Source: Central Bank of Indonesia.

Citizens began hoarding food and fears of unemployment and social unrest were spreading. The threat of political instability was not taken lightly by its neighbors and others. Indonesia, with 200 million people, is the fourth-most-populous nation in the world. However, the income distribution is skewed, with the 4% of the population that is ethnic Chinese controlling about 60% of the nation's wealth. In the 1960s, Indonesia was torn by bloody ethnic riots that led to a change of power, when Suharto deposed his predecessor, President Sukarno. An estimated 500,000 Indonesians died in those riots, many of them ethnic Chinese.

The long economic boom since then, with GDP growing at an average rate of 7% annually (see Exhibit I 3.4 for growth rates since 1980), helped salve the ethnic antagonisms, but they always lie just beneath the surface. So it was predictable that with the economic hardships anti-Chinese riots became more common.

Political stability also was threatened by Indonesia's endemic corruption and the autocratic regime's resulting lack of popular support. Not to put too fine a point on it, President Suharto's regime was a kleptocracy, run for the financial benefit of his family and friends. They exploited to the full the opportunities for corruption and profit that their connections gave them. The scale of corruption was breathtaking; Suharto and his six children had an estimated net worth of $40 billion. Since Indonesia's economic meltdown began in 1997, President Suharto's actions appeared designed to protect his family's financial interests and preserve his power rather than to promote the public good. With 7% real economic growth, enough prosperity reached the streets to keep the populace quiescent, if not happy. Authoritarianism, nepotism, and outright corruption were tolerated as long as Suharto delivered the goods. But there was no reservoir of popular support to carry President Suharto and his regime through the

EXHIBIT I 3.4 Growth in Indonesia Real Gross Domestic Product: 1980–1996

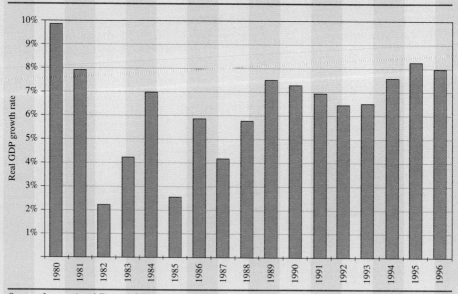

Source: *International Financial Statistics*, various editions.

price increases and other painful reform measures mandated by the IMF and necessitated by the current crisis. Moreover, there was no obvious successor to the 76-year-old Suharto, whose health is a question mark.

On January 15, 1998, Suharto agreed to the sweeping economic reforms he had reneged on the week before. In return, the IMF agreed to begin disbursing funds to Indonesia. The rupiah and Jakarta stock exchange staged a modest recovery.

Immediately after making these promises, Indonesia began pursuing contradictory policies that led to a sharp selloff in the rupiah. The central bank agreed to compensate depositors in 16 closed banks, while printing rupiah (ultimately more than 100 trillion rupiah) to keep the remaining 220 afloat. Rumors also began circulating that Suharto intended to appoint Bucharuddin Habibie, the minister of research and technology, as his next vice president and likely successor. These rumors drove the rupiah down on January 22nd, at one point to 17,000 to the dollar, largely because Mr. Habibie was the man behind many of Indonesia's controversial spending programs—programs (such as the attempt to build a national aircraft industry from scratch) whose financing President Suharto had just agreed to cut.

Most controversially from the IMF's standpoint, President Suharto began to flirt with the idea of establishing a currency board that would tie the value of the rupiah to the dollar. He was introduced to the idea of a currency board by Steven Hanke, an American economist who pointed to the experiences of Hong Kong, Argentina, and other countries as demonstrating that such a system would stabilize the currency and bring down soaring interest rates. By pegging the rupiah at a rate of 5,500 to the dollar, about twice its current value, President Suharto and his advisers became convinced that Indonesia could stop the rupiah's slide, rein in soaring prices, and restore the confidence of local and foreign investors—all without resorting to the IMF's bitter economic medicine. As word of the currency board spread, the rupiah soared in value.

U.S., IMF, and EU economic officials, however, contended that a currency board is a quick fix that won't work. They argued that it could lead to disastrously high interest rates, further troubles for the Indonesian banking system, and increased social unrest. In addition, many currency traders believed that a currency board would soon fail, taking much of the banking system with it. The government was guaranteeing all bank deposits, which it could not do under a currency board (because it could issue rupiah only if there were dollars to back them). Banks might suffer a run as depositors rushed to convert rupiah into dollars. Critics also complained that Suharto was only interested in a currency board because raising the rupiah's value would rescue his associates who have dollar-denominated debt.

The IMF responded by delaying a disbursement scheduled for March 15th. It then promised flexibility, especially on food and electricity subsidies. Most observers felt that the IMF had no choice: Between a poor harvest, massive unemployment, and soaring prices, there was a real fear of a humanitarian, social, and political disaster.

During February, food riots and looting erupted in dozens of towns across Indonesia. Most of the violence was directed at shops owned by the ethnic Chinese. Students and other political protesters staged peaceful demonstrations against the government. Worse, Suharto appeared to have lost the support of many of Indonesia's middle class, who tolerated him and his family as long as he delivered economic growth and a rising standard of living.

Only the Indonesian armed forces appeared to stand in the way of anarchy. Its half million men were supposed to protect the state against internal threats as well as external ones. In practice the army had done the president's bidding. In return, its officers profited from the numerous opportunities for corruption that their pervasive presence in the country's administration gave them. As long as Suharto retained the loyalty and support of the army, his grip on power was secure. However, if the social unrest got out of hand, the soldiers might decide that killing hundreds or even thousands of their countrymen to perpetuate his reign was not worth the personal cost. Such bloodshed would also make the army unpopular for years to come and threaten its privileged role in Indonesian society.

To President Suharto and some of his economic advisers, the idea of a currency board promised a way out of the crisis without making the fundamental changes demanded by the IMF. They estimated that there were enough foreign exchange reserves to immediately restore the value of the currency to about 5,500 rupiah to the dollar. This jump in the rupiah's value would reassure investors that their Indonesian investments would retain their value. It also would reduce the inflation that was eating away at the purchasing power of the average Indonesian's wages and would make it easier for Indonesian companies and banks to service their foreign debts (overseas borrowing gave them capital at about five percentage points less than at home). With the time bought with a currency board, Indonesia's natural economic strengths could reassert themselves. It produces all its own oil and exports billions of dollars' worth, and its manufacturing industry exports goods worth more than double its energy exports.

Skeptics pointed out that what worried investors most was not inflation eating away at their wealth but that many of the country's businesses and financial institutions could go bankrupt, or even that the country could descend into the chaos of the 1960s. At the same time, committing Indonesia's reserves to a currency board meant they would be unavailable to pay for imports or debt service. The debt service alone was enormous, given the $140 billion that Indonesia's public and private entities had borrowed abroad, much of it short term (for example, Indonesian companies had $43.2 billion in foreign debt due within one year). In addition, critics claimed that a rupiah fixed at 5,500 to the dollar would give the nation's wealthy elite, including President Suharto's children and associates, a chance to trade their rupiah for dollars and deposit them overseas. Indonesia's dollar reserves would disappear, interest rates would skyrocket, and the economy would be battered even more. Supporters, however, dismissed these concerns, claiming that money would flow into Indonesia, not out, thanks to new international confidence in the currency and country.

Another option being discussed—possibly in conjunction with a currency board—was a debt moratorium. A moratorium would presumably

support the rupiah because debtors no longer would have to buy dollars to service their foreign debts.

QUESTIONS

1. What monetary policies could Suharto follow that would restore the rupiah's pre-crisis value? What problems would those policies face?
2. What were the costs and benefits of an Indonesian debt moratorium?
3. How did undermining the social contract between Suharto and the middle class affect the value of the rupiah?
4. How did a weak banking system affect the prospects for a currency board?
5. Should the IMF have withheld disbursements if Indonesia did not honor its commitments? What were the pros and cons?
6. Did the IMF's prescription for Indonesia make sense? Explain.
7. Why had Suharto found it so hard to implement the IMF's provisions?
8. What suggestions do you have for stemming the rupiah's slide and strengthening its value?
9. Did cuts in fuel and food subsidies make sense?
10. Should Indonesia have established a currency board? What considerations would you weigh in that decision? If you decide against a currency board, what alternative would you suggest?

CASE I.4

BRAZIL FIGHTS A *REAL* BATTLE

In 1994, with inflation running at close to 50% *monthly* and Brazil's economy near collapse, Fernando Cardoso was elected president. A former socialist mugged by reality, Cardoso opened the economy, adopted a number of free market policies, stabilized the currency, and brought inflation down to just 7% annually by 1997. Now the Asian financial turmoil is threatening President Cardoso's proudest achievement and the cornerstone of his popularity: a stable Brazilian *real*. Fallout from the Asian crisis has been more acute in Brazil than elsewhere in Latin America, partly because speculators believe that the *real* is overvalued. They also see parallels to Asia in Brazil's large trade deficit financed by foreign capital inflows. The pressure on the *real* has been ratcheted up because Brazilian financial players are unusually aggressive and sophisticated, traits that are a legacy of the past three decades when rampant inflation and abrupt policy shifts forced traders to be brutally pragmatic and move quickly to survive.

The speculators may have met their match in President Cardoso, however. Cardoso is unlikely to give up easily because he has staked Brazil's fortunes, and his own, on the *Real Plan,* which he had introduced in 1993 while he was finance minister. Its linchpin is a crawling peg designed to end the cycle of runaway inflation followed by massive devaluations that has given Brazil 30 years of economic chaos and eroded living standards for much of the population. By permitting only a tightly controlled and limited 0.6% depreciation of the *real* against the dollar each month (see Exhibit I 4.1), Cardoso has managed to halt Brazil's hyperinflation and bring a new confidence that is spurring foreign investment (see Exhibit I 4.2 for statistics on the Brazilian economy). However, Brazil's current account deficit is mounting, its foreign exchange reserves are falling, and financial markets are getting nervous that the *real* will suffer the same fate as that of the Asian currencies. The loss of confidence in the *real* is showing up in the form of capital flight, with $10 billion fleeing Brazil in the last two weeks of October 1997 alone.

In addition, a fractious Congress has kept President Cardoso from carrying out the budgetary reforms that would allow him to control the fiscal deficits that have been the engine of hyperinflation and thereby institutionalize Brazil's

EXHIBIT I 4.1 Brazil's Crawling Peg Leads to a Gradual Devaluation of the Real Against the U.S. Dollar

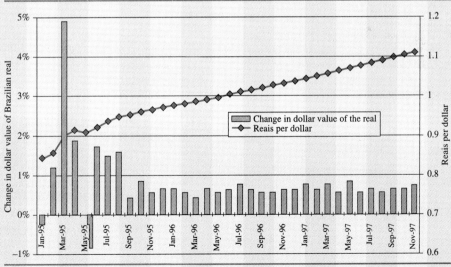

Source: International Financial Statistics, various editions.

EXHIBIT I 4.2 Key Brazilian Economic Statistics
(In U.S. $ Billions Unless Otherwise Indicated)

	1990	1991	1992	1993	1994	1995	1996	1997
GROSS DOMESTIC PRODUCT	$445.9	$386.2	$387.3	$429.7	$543.1	$705.4	$775.4	$803.0
CURRENT-ACCOUNT BALANCE	−$3.8	−$1.5	$6.1	$0.0	−$1.2	−$18.1	−$24.3	−$33.8
CURRENT-ACCOUNT BALANCE (% OF GDP)	−0.9%	−0.4%	1.6%	0.0%	−0.2%	−2.6%	−3.1%	−4.2%
BUDGET DEFICIT (% OF GDP)[1]	−1.4%	−1.4%	2.2%	−0.2%	−1.3%	4.8%	3.9%	5.9%
FOREIGN INVESTMENT[2]	$0.8	$3.9	$16.4	$15.8	$23.8	$25.8	$34.3	$55.1
PORTFOLIO	$0.5	$3.8	$14.5	$15.0	$21.6	$22.6	$24.7	$37.2
DIRECT	$0.3	$0.1	$1.9	$0.9	$2.2	$3.3	$9.6	$17.9
RESERVES EXCLUDING GOLD	$7.4	$8.0	$22.5	$30.6	$37.1	$49.7	$58.3	$51.4
EXCHANGE RATE (END OF PERIOD, REAIS/$)[3]	0.000064	0.0004	0.005	0.119	0.846	0.973	1.039	1.116
INFLATION (CONSUMER PRICE INDEX)	1657.7%	493.8%	1156.0%	2828.7%	992.0%	25.9%	11.3%	7.2%
U.S. INFLATION RATE (CPI)	5.4%	4.2%	3.2%	3.0%	2.6%	2.8%	3.0%	2.3%

[1] A negative sign indicates a budget surplus.
[2] Other components of foreign investment are relatively minor.
[3] Exchange rate for the *real* prior to 1994 reflects the effects of two earlier currency replacements.
Source: Central Bank of Brazil Bulletin, April 1998 and *Economic Report of the President,* February 1998.

new-found macroeconomic stability. These reforms, which include revamping the tax, civil service, and social security systems, and privatizing key industries from telecom to mining, would allow Brazil to consolidate the economic gains it has already made, give investors greater confidence in the economy's future, and speed up Brazil's growth rate. As of now, Brazil has a bloated public sector (government workers cannot be dismissed, regardless of performance), large budget deficits (5.9% of GDP in 1997), protected industries, high tax rates with low tax collections, and one of the most unequal distributions of income and wealth in the world. One legacy of Brazil's large budget deficits is that the country must now roll over about $20 billion a month in government debt.

Nonetheless, until the Asian crisis hit, investors had responded to currency stabilization and the free-market reforms already begun by pushing up the Brazilian bolsa (stock market) by about 158% from January 1994, when the *Real Plan* took hold, to October 1997 (see Exhibit I 4.3). However, contagion from the Asian crisis has recently pummeled the bolsa, sending it down 29% from August through October.

Although it now appears that Cardoso will have to wait until his second term in office to carry out most of the proposed reforms, he is nonetheless reputed to be preparing a new push on his long-stalled reform program. One of the stalled reforms is a bill to privatize the Brazilian social security system. This system, which pays out benefits to some workers in their 40s, is

EXHIBIT I 4.3 Brazilian Bolsa in U.S. Dollars: January 1995–November 1997

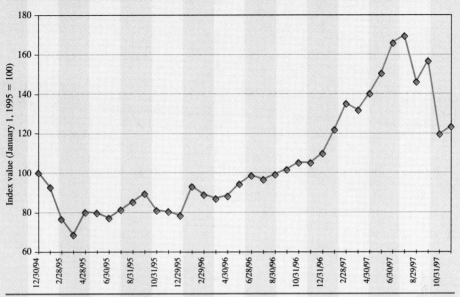

Source: Morgan Stanley Capital International.

widely viewed as contributing to a Brazilian savings rate of just 16% of GDP (to put this number into perspective, Asian countries average closer to 30%). Until the presidential election in October 1998, Cardoso expects to maintain a tight-money policy that will support the *real* and, it is hoped, prevent the trade deficit from spiraling out of control. In addition to raising interest rates, the government's main weapon to stanch Brazil's fiscal and trade deficits is a newly-invigorated privatization program. In the first half of 1997, a large share of the $7.4 billion in foreign direct investment came in the form of purchases of state-owned businesses. Altogether, Brazil expects to take in an estimated $17 billion from privatizations in 1997 and at least $22 billion in 1998.

Fortunately, the government is making enough progress toward economic reform to continue to entice foreign investors, who appear mesmerized by Brazil's enormous potential. With a population of 160 million and an estimated GDP of about $800 billion that has grown at an average rate of more than 4% annually since 1993 (see Exhibit I 4.4), Brazil is by far the largest market in Latin America (representing about 45% of total Latin American GDP) and provides a launching pad for investments throughout the region. Moreover, it is rapidly industrializing, as evidenced by the fact that industrial products now account for 75% of Brazil's exports. Historically, coffee made up 80% of exports; now it accounts for only about 5%.

Rather than devaluing the currency, the traditional response by Brazilian governments faced with speculative attacks, Brazil's central bank—the Banco Central do Brasil—has defended the *Real Plan* by doubling the basic interest rate to 43% (see Exhibit I 4.5) and spending the nation's dollar reserves to buy up excess reais (plural of *real*). In early November 1997, Cardoso also managed to push through Congress some budgetary reforms along with $18 billion in spending cuts

EXHIBIT I 4.4 Brazil's Economic Growth Rate: 1990–1997

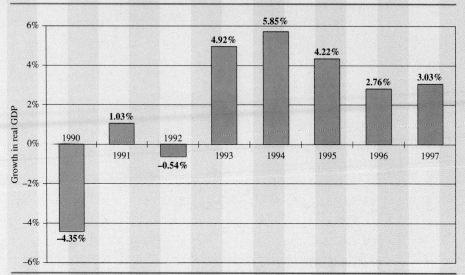

Source: Central Bank of Brazil Bulletin, April 1998.

EXHIBIT I 4.5 Brazilian Interest Rates

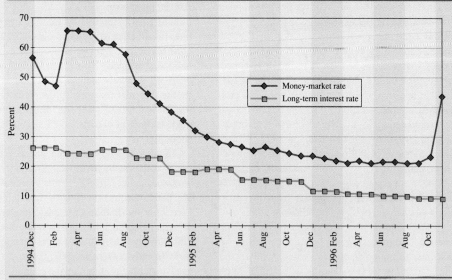

Source: Central Bank of Brazil Bulletin, April 1998.

and tax increases, equivalent to more than 2% of GDP. The *real's* defense is not cost-free, however. The high real interest rates are expected to slow GDP growth in 1998 to 1.5%, from 3% in 1997. They are also pushing up the unemployment rate and worsening the credit quality of Brazil's banks. Cardoso's economic reforms have been attacked as benefiting the wealthy and coming at the expense of the poor, who are the first to lose their jobs. These charges are worrisome in a presidential-election year. Some observers have argued that by increasing the pace of the devaluations, Brazil could afford to lower interest rates and boost growth and jobs—and at the same time improve Cardoso's prospects in next October's presidential elections. However, many analysts believe that any significant deviation from the *Real Plan,* especially before an election, would look like politics as usual and cast doubt on the sustainability of all economic reforms. Backsliding on currency stability is also likely to be unpopular. The *Real Plan* has made politicians and the populace aware of the strong link between devaluation and inflation. Thus, any policy changes that threaten the low inflation wrought by currency stability is likely to be politically risky. However, the recent austerity measures are also unpopular and are eroding Cardoso's approval ratings. With a recession looming as Brazil enters into an election year, Cardoso's supporters fear that rising joblessness will cut deeply into his electoral chances.

Banks and other financial institutions have found that speculating against the *real* is an expensive and difficult proposition because the Banco do Brasil maintains a wide range of controls on the foreign exchange market. One of the most onerous and costly requires that buyers of dollars deposit a large portion of them in special, low-interest accounts with the central bank. In times of intense speculative pressure, the central bank has cut the interest rate paid on these accounts to zero. Thus investors who borrowed reais and converted them to dollars would start losing money the moment they deposited them

with the central bank, unless the *real* fell enough to offset the loss of interest.

QUESTIONS

1. How does Brazil hope to control its current-account deficit through a tight monetary policy? What alternatives are available to control Brazil's current-account deficit?

2. How will Brazil's tight money policy affect its fiscal deficit? How will it affect Brazil's real (inflation-adjusted) interest rates, both short-term and long-term rates?

3. Why have Brazil's interest rates generally fallen in recent years?

4. How would reform and privatization of the social security system improve Brazil's savings rate? What would be the likely consequences of this improvement for Brazil's current-account balance and the *real's* value? Explain.

5. What are the costs and benefits of using currency controls to defend the *real?*

6. Why might speculators view the *real* as being overvalued? Based on the data in the case, what is your best estimate as to the *real's* degree of overvaluation?

7. What are the tradeoffs that President Cardoso must consider in deciding whether to accelerate the *real's* depreciation?

8. Could Brazil have avoided the recessionary impacts of its monetary policy if it had devalued the *real* instead?

9. What would a Brazilian devaluation do to the currencies and economies of Argentina and Chile, its neighbors and largest trading partners?

10. What is the link between Brazil's budget deficits and its historical hyperinflation?

11. What mix of fiscal and monetary policy would you recommend to President Cardoso? Should he devalue or defend the *real?*

PART TWO

FOREIGN EXCHANGE RISK MANAGEMENT

MEASURING ACCOUNTING EXPOSURE

The stream of time sweeps away errors, and leaves the truth for the inheritance of humanity.

George Brandes

The general concept of *exposure* refers to the degree to which a company is affected by exchange rate changes. This impact can be measured in several ways. As so often happens, economists tend to favor one approach to measuring foreign exchange exposure, while accountants favor an alternative approach. This chapter focuses on accounting exposure and presents alternative methods that accountants have developed for measuring that exposure. The chapter also discusses the differences between accounting requirements and economic reality, making recommendations to accountants and financial executives on how to adjust reporting standards in order to reconcile those differences. The next chapter discusses how companies can manage their accounting exposure.

8.1 ALTERNATIVE MEASURES OF FOREIGN EXCHANGE EXPOSURE

The three basic types of exposure are accounting exposure, transaction exposure, and operating exposure. Transaction exposure and operating exposure combine to form **265**

EXHIBIT 8.1 Comparison of Accounting, Transaction, and Translation Exposure

ACCOUNTING EXPOSURE	OPERATING EXPOSURE
Changes in the book value of balance sheet assets and liabilities and income statement items that are caused by an exchange rate change. The resulting exchange gains and losses are determined by accounting rules and are paper only. The measurement of accounting exposure is retrospective in nature as it is based on activities that occurred in the past.	Changes in the amount of future operating cash flows caused by an exchange rate change. The resulting exchange gains or losses are determined by changes in the firm's future competitive position and are real. The measurement of operating exposure is prospective in nature as it is based on future activities.
Impacts: Balance sheet assets and liabilities and income statement items that already exist.	*Impacts:* Revenues and costs associated with future sales.

Exchange rate
change occurs

Impacts: Contracts already entered into, but . . .   to be settled at a later date.

TRANSACTION EXPOSURE

Changes in the value of outstanding foreign currency-denominated contracts (i.e., contracts that give rise to future foreign currency cash flows) that are brought about by an exchange rate change. The resulting exchange gains and losses are determined by the nature of the contracts already entered into and are real. The measurement of transaction exposure mixes the retrospective and prospective because it is based on activities that occurred in the past but will be settled in the future.

---

economic exposure. Exhibit 8.1 illustrates and contrasts accounting, transaction, and operating exposure.

Accounting Exposure

Accounting exposure, also known as **translation exposure**, arises from the need, for purposes of reporting and consolidation, to convert the financial statements of foreign operations from the local currencies (LC) involved to the home currency (HC). If exchange rates have changed since the previous reporting period, this *translation*, or restatement, of those assets, liabilities, revenues, expenses, gains, and losses that are denominated in foreign currencies will result in foreign exchange gains or losses. The possible extent of these gains or losses is measured by the translation exposure figures. The rules that govern translation are devised by an accounting association such as the **Financial Accounting Standards Board** in the United States, the parent firm's government, or the firm itself. This chapter will focus on **Statements of Financial Accounting Standards No. 8** and **No. 52**—the past (No. 8) and present (No. 52) currency translation methods prescribed by the Financial Accounting Standards Board (FASB).

Transaction Exposure

Transaction exposure results from transactions that give rise to known, contractually-binding future foreign currency-denominated cash inflows or outflows. As exchange rates change between now and when these transactions settle, so does the value of their associated foreign currency cash flows, leading to currency gains and losses. Examples of transaction exposure for a U.S. company would be the account receivable associated with a sale denominated in French francs or the obligation to repay a Japanese yen debt. Although transaction exposure is rightly part of economic exposure, it is usually lumped under accounting exposure.

Operating Exposure

Operating exposure measures the extent to which currency fluctuations can alter a company's future operating cash flows, that is, its future revenues and costs. Any company whose revenues or costs are affected by currency changes has operating exposure, even it is a purely domestic corporation and has all its cash flows denominated in home currency.

The two cash flow exposures—operating exposure and transaction exposure—combine to equal a company's economic exposure. In technical terms, **economic exposure** is the extent to which the value of the firm—as measured by the present value of its expected cash flows—will change when exchange rates change.

8.2 ALTERNATIVE CURRENCY TRANSLATION METHODS

Companies with international operations will have foreign currency-denominated assets and liabilities, revenues, and expenses. However, because home-country investors and the entire financial community are interested in home currency values, the foreign currency balance-sheet accounts and income statement must be assigned HC values. In particular, the financial statements of an MNC's overseas subsidiaries must be translated from local currency to home currency before consolidation with the parent's financial statements.

If currency values change, foreign exchange translation gains or losses may result. Assets and liabilities that are translated at the current (postchange) exchange rate are considered to be exposed; those translated at a historical (prechange) exchange rate will maintain their historic HC values and, hence, are regarded as not exposed. **Translation exposure** is simply the difference between exposed assets and exposed liabilities. The controversies among accountants center on which assets and liabilities are exposed and on when accounting-derived foreign exchange gains and losses should be recognized (reported on the income statement). A crucial point to realize in putting these controversies in perspective is that such gains or losses are of an accounting nature—that is, no cash flows are necessarily involved.

Four principal translation methods are available: the current/noncurrent method, the monetary/nonmonetary method, the temporal method, and the current rate method. In practice, there are also variations of each method.

Current/Noncurrent Method

At one time, the **current/noncurrent method**, whose underlying theoretical basis is maturity, was used by almost all U.S. multinationals. With this method, all the foreign subsidiary's current assets and liabilities are translated into home currency at the **current exchange rate**. Each noncurrent asset or liability is translated at its **historical exchange rate**; that is, at the rate in effect at the time the asset was acquired or the liability incurred. Hence, a foreign subsidiary with positive local currency working capital will give rise to a translation loss (gain) from a devaluation (revaluation) with the current/noncurrent method, and vice versa if working capital is negative.

The income statement is translated at the average exchange rate of the period, except for those revenues and expense items associated with noncurrent assets or liabilities. The latter items, such as depreciation expense, are translated at the same rates as the corresponding balance sheet items. Thus, it is possible to see different revenue and expense items with similar maturities being translated at different rates.

Monetary/Nonmonetary Method

The **monetary/nonmonetary method** differentiates between *monetary* assets and liabilities — that is, those items that represent a claim to receive, or an obligation to pay, a fixed amount of foreign currency units — and *nonmonetary*, or physical, assets and liabilities. Monetary items (for example, cash, accounts payable and receivable, and long-term debt) are translated at the current rate; nonmonetary items (for example, inventory, fixed assets, and long-term investments) are translated at historical rates.

Income statement items are translated at the average exchange rate during the period, except for revenue and expense items related to nonmonetary assets and liabilities. The latter items, primarily depreciation expense and cost of goods sold, are translated at the same rate as the corresponding balance sheet items. As a result, the cost of goods sold may be translated at a rate different from that used to translate sales.

Temporal Method

The **temporal method** appears to be a modified version of the monetary/nonmonetary method. The only difference is that under the monetary/nonmonetary method, inventory is always translated at the historical rate. Under the temporal method, inventory is normally translated at the historical rate, but it can be translated at the current rate if the inventory is shown on the balance sheet at market values. Despite the similarities, however, the theoretical bases of the two methods are different. The choice of exchange rate for translation is based on the type of asset or liability in the monetary/nonmonetary method; in the temporal method, it is based on the underlying approach to evaluating cost (historical versus market). Under a historical cost accounting system, as the United States now has, most accounting theoreticians probably would argue that the temporal method is the appropriate method for translation.

Income statement items normally are translated at an average rate for the reporting period. However, cost of goods sold and depreciation and amortization charges related to balance sheet items carried at past prices are translated at historical rates.

Current Rate Method

The **current rate method** is the simplest; all balance sheet and income items are translated at the current rate. This method is widely employed by British companies. Under this method, if a firm's foreign currency-denominated assets exceed its foreign currency-denominated liabilities, a devaluation must result in a loss and a revaluation, in a gain. One variation is to translate all assets and liabilities except net fixed assets at the current rate.

Exhibit 8.2 applies the four methods to a hypothetical balance sheet that is affected by both a 25% devaluation and a revaluation of 37.5%. Depending on the method chosen, the translation results for the LC devaluation can range from a loss of $205,000 to a gain of $215,000; LC revaluation results can vary from a gain of $615,000 to a loss of $645,000. The assets and liabilities that are considered exposed under each method are the ones that change in dollar value. Note that the translation gains or losses for each method show up as the change in the equity account. For example, the LC devaluation combined with the current rate method results in a $205,000 reduction in the equity account ($1,025,000 − $820,000), which equals the translation loss for this method. Another way to calculate this loss is to take the net LC translation exposure, which equals exposed assets minus exposed liabilities (for the current rate method this figure is LC 4,100,000, which, not coincidentally, equals its equity value) and multiply it by the $0.05 ($0.25 − $0.20) change in the exchange rate. This calculation yields a translation loss of $205,000 ($0.05 × 4,100,000), the same as calculated in Exhibit 8.2. Another way to calculate this loss is to multiply the net dollar translation exposure by the fractional change in the exchange rate, or $1,025,000 × 0.05/0.25 = $205,000. Either approach gives the correct answer.

8.3 STATEMENT OF FINANCIAL ACCOUNTING STANDARDS NO. 8

Such a wide variation in results as those of Exhibit 8.2 led the Financial Accounting Standards Board to issue a new ruling: **Statement of Financial Accounting Standards No. 8** (FASB-8). **FASB-8** established uniform standards for the translation into dollars of foreign currency-denominated financial statements and transactions of U.S.-based multinational companies.

FASB-8, which was based on the temporal method, became effective on January 1, 1976. Its principal virtue was its consistency with generally accepted accounting practice that requires balance sheet items to be valued (translated) according to their underlying measurement basis (that is, current or historical).

Almost immediately upon its adoption, controversy ensued over FASB-8. A major source of corporate dissatisfaction with FASB-8 was the ruling that all reserves for currency losses be disallowed. Before FASB-8, many companies established a reserve and were able to defer unrealized translation gains and losses by adding them to, or charging them against, the reserve. In that way, corporations generally were able to cushion the impact of sharp changes in currency values on reported earnings. With FASB-8, however, fluctuating values of pesos, pounds, marks, Canadian dollars, Australian dollars, and

EXHIBIT 8.2 Financial Statement Impact of Translation Alternatives (U.S. $ Thousands)

	Local Currency	U.S. Dollars Prior to Exchange Rate Change (LC 4 = $1)	After Devaluation of Local Currency (LC 5 = $1)			
			Monetary/ Non- Monetary	Temporal	Current/ Non- Current	Current Rates for All Assets and Liabilities
Assets						
Current assets						
Cash, marketable securities, and receivables	LC 2,600	$ 650	$ 520	$ 520	$ 520	$ 520
Inventory (at market)	3,600	900	900	720	720	720
Prepaid expenses	200	50	50	50	40	40
Total current assets	6,400	1,600	1,470	1,290	1,280	1,280
Fixed assets less accumulated depreciation	3,600	900	900	900	900	720
Goodwill	1,000	250	250	250	250	200
Total assets	LC 11,000	$2,750	$2,620	$2,440	$2,430	$2,200
Liabilities						
Current liabilities	3,400	850	680	680	680	680
Long-term debt	3,000	750	600	600	750	600
Deferred income taxes	500	125	100	100	125	100
Total liabilities	6,900	1,725	1,380	1,380	1,555	1,380
Capital stock	1,500	375	375	375	375	375
Retained earnings	2,600	650	865	685	500	445
Total equity	4,100	1,025	1,240	1,060	875	820
Total liabilities plus equity	LC 11,000	$2,750	$2,620	$2,440	$2,430	$2,200
Translation Gain (Loss)	—	—	$ 215	$ 35	$ (150)	$ (205)

other foreign currencies often had far more impact on profit-and-loss statements than did the sales and profit margins of multinational manufacturers' product lines.

The experience of Sony, the Japanese electronics producer, which follows U.S. accounting rules because its shares are traded on the New York Stock Exchange, illustrates the impact of FASB-8. In 1979, Sony's consolidated second-quarter earnings were reduced by a foreign exchange loss of $49.3 million, whereas in 1978, the second-quarter net was restated to include a $19.3 million gain on currency conversion—a $68.6 million swing in pre-tax net income that might never be realized. The result was a 49% slump in Sony's second-quarter consolidated earnings, despite a near tripling in operating earnings from a year earlier. For the first half of 1979, Sony reported that net earnings had declined by 36% from the year before, despite a 98% increase in operating earnings. Sony's plight was caused almost entirely by an $85.8 million earnings movement resulting from a 1979

			CURRENT RATES FOR
MONETARY/ NON- MONETARY	TEMPORAL	CURRENT/ NON- CURRENT	ALL ASSETS AND LIABILITIES

AFTER REVALUATION OF
LOCAL CURRENCY (LC 2.5 = $1)

MONETARY/ NON- MONETARY	TEMPORAL	CURRENT/ NON- CURRENT	CURRENT RATES FOR ALL ASSETS AND LIABILITIES
$1,040	$1,040	$1,040	$1,040
900	1,440	1,440	1,440
50	50	80	80
1,990	2,530	2,560	2,560
900	900	900	1,440
250	250	250	400
$3,140	$3,680	$3,710	$4,400
1,360	1,360	1,360	1,360
1,200	1,200	750	1,200
200	200	125	200
2,760	2,760	2,235	2,760
375	375	375	375
5	545	1,100	1,265
380	920	1,475	1,640
$3,140	$3,680	$3,710	$4,400
$ (645)	$ (105)	$ 450	$ 615

first-half foreign exchange loss of $59.4 million, compared to a gain of $26.4 million a year earlier.

8.4 STATEMENT OF FINANCIAL ACCOUNTING STANDARDS NO. 52

In 1981, widespread dissatisfaction by corporate executives over FASB-8 led to a new translation standard: **Statement of Financial Accounting Standards No. 52** (FASB-52). According to **FASB-52**, firms must use the current rate method to translate foreign currency-denominated assets and liabilities into dollars. All foreign currency revenue and expense items on the income statement must be translated at either the exchange rate in effect on the date these items are recognized or at an appropriately weighted average exchange rate for the period. The most important aspect of the new standard is that, un-

EXHIBIT 8.3 Factors Indicating the Appropriate Functional Currency

FOREIGN UNIT'S	LOCAL CURRENCY INDICATORS	DOLLAR INDICATOR
Cash Flows	Primarily in the local currency; do not directly affect parent company cash flows	Direct impact on parent company; cash flow available for remittance
Sales Prices	Not responsive to exchange rate changes in the short run; determined more by local conditions	Determined more by world-wide competition; affected in the short run by exchange rate changes
Sales Market	Active local market for entity's products	Products sold primarily in the United States; sales contracts denominated in dollars
Expenses	Labor, materials, and other costs denominated primarily in local currency	Inputs primarily from sources in the United States or otherwise denominated in dollars
Financing	Primarily in local currency; operations generate sufficient funds to service these debts	Primarily from the parent company or otherwise denominated in dollars; operations don't generate sufficient dollars to service its dollar debts
Intercompany Transactions	Few intracorporate transactions; little connection between local and parent operations	High volume of intracorporate transactions; extensive interrelationship between local and parent operations

like the case with FASB-8, most FASB-52 translation gains and losses bypass the income statement and are accumulated in a separate equity account on the parent's balance sheet. This account is usually called something like "cumulative translation adjustment."

FASB-52 differentiates for the first time between the functional currency and the reporting currency. An affiliate's **functional currency** is the currency of the primary economic environment in which the affiliate generates and expends cash. If the enterprise's operations are relatively self-contained and integrated within a particular country, the functional currency would generally be the currency of that country. An example of this case would be an English affiliate that both manufactures and sells most of its output in England. Alternatively, if the foreign affiliate's operations are a direct and integral component or extension of the parent company's operations, the functional currency would be the U.S. dollar. An example would be a Hong Kong assembly plant for radios that sources the components in the United States and sells the assembled radios in the United States. It is also possible that the functional currency is neither the local currency nor the dollar but, rather, is a third currency. However, in the remainder of this chapter, we will assume that if the functional currency is not the local currency, then it is the U.S. dollar.

Guidelines for selecting the appropriate functional currency are presented in Exhibit 8.3. There is sufficient ambiguity to give companies some leeway in selecting the functional currency. However, in the case of a **hyperinflationary country**—defined as one that has cumulative inflation of approximately 100% or more over a three-year period—the functional currency must be the dollar.

Companies usually will explain in the notes to their annual report how they accounted for foreign currency translation. A typical statement is that found in Dow Chemical's 1989 Annual Report:

The U.S. dollar has been used as the functional currency throughout the world except for operations in Germany and Japan, for which local currencies have been used. Effective December 31, 1988, the functional currency of the subsidiary companies in six European countries was changed to the local currency. These countries are Belgium, France, Italy, the Netherlands, Spain, and U.K. The impact on the balance sheet is noted in the Consolidated Statement of Stockholders' Equity. This change did not impact income in 1988.

Where the U.S. dollar is used as the functional currency, foreign currency gains and losses are reflected in income currently. Translation gains and losses of those operations that use local currencies as the functional currency, and the effects of exchange rate changes on transactions designated as hedges of net foreign investments, are included as a separate component of stockholders' equity.

The **reporting currency** is the currency in which the parent firm prepares its own financial statements; that is, U.S. dollars for a U.S. firm. FASB-52 requires that the financial statements of a foreign unit first be stated in the functional currency, using generally accepted accounting principles of the United States. At each balance sheet date, any assets and liabilities denominated in a currency other than the functional currency of the recording entity must be adjusted to reflect the current exchange rate on that date. Transaction gains and losses that result from adjusting assets and liabilities denominated in a currency other than the functional currency, or from settling such items, generally must appear on the foreign unit's income statement. The only exceptions to the general requirement to include transaction gains and losses in income as they arise are listed as follows:

1. Gains and losses attributable to a foreign currency transaction that is designated as an economic hedge of a net investment in a foreign entity must be included in the separate component of shareholders' equity in which adjustments arising from translating foreign currency financial statements are accumulated. An example of such a transaction would be a Deutsche mark borrowing by a U.S. parent. The transaction would be designated as a hedge of the parent's net investment in its German subsidiary. See, for example, the statement in Dow's 1989 Annual Report.

2. Gains and losses attributable to intercompany foreign currency transactions that are of a long-term investment nature must be included in the separate component of shareholders' equity. The parties to the transaction in this case are accounted for by the equity method in the reporting entity's financial statements.

3. Gains and losses attributable to foreign currency transactions that hedge identifiable foreign currency commitments are to be deferred and included in the measurement of the basis of the related foreign transactions.

The requirements regarding translation of transactions apply both to transactions entered into by a U.S. company and denominated in a currency other than the U.S. dollar and to transactions entered into by a foreign affiliate of a U.S. company and denominated in a currency other than its functional currency. Thus, for example, if a German subsidiary of a U.S. company owed $180,000 and the DM declined from $0.60 to $0.50, the Deutsche mark amount of the liability would increase from DM 300,000 (180,000/0.60) to DM 360,000 (180,000/0.50), for a loss of DM 60,000. If the subsidiary's functional currency is the Deutsche mark, the DM 60,000 loss must be translated into dollars at the average exchange rate for the period (say $0.55), and the resulting amount ($33,000) must be included as a transaction loss in the U.S. company's consolidated statement of income. This loss results even though the liability is denominated in the parent company's reporting currency because the subsidiary's functional currency is the Deutsche mark, and its financial statements must be measured in terms of that currency. Under FASB-8, there was no gain or loss included in consolidated net income on debt denominated in the parent company's reporting currency. Similarly, under FASB-52, if the subsidiary's functional currency is the U.S. dollar, no gain or loss will arise on the $180,000 liability.

After all financial statements have been converted into the functional currency, the functional currency statements are then translated into dollars, with translation gains and losses flowing directly into the parent's foreign exchange equity account.

If the functional currency is the dollar, the unit's local currency financial statements must be remeasured in dollars. The objective of the remeasurement process is to produce the same results that would have been reported if the accounting records had been kept in dollars rather than the local currency. Translation of the local currency accounts into dollars takes place according to the temporal method previously required by FASB-8; thus, the resulting translation gains and losses *must* be included in the income statement.

A large majority of firms have opted for the local currency as the functional currency for most of their subsidiaries. The major exceptions are those subsidiaries operating in Latin American and other highly inflationary countries, which must use the dollar as their functional currency.

Application of FASB No. 52

Sterling Ltd., the British subsidiary of a U.S. company, started business and acquired fixed assets at the beginning of a year when the exchange rate for the pound sterling was £1 = $1.50. The average exchange rate for the period was $1.40, the rate at the end of the period was $1.30, and the historical rate for inventory was $1.45. Refer to Exhibits 8.4 and 8.5 for the discussion that follows.

EXHIBIT 8.4 Translation of Sterling Ltd.'s Income Statements Under FASB-52 (Millions)

| | | FUNCTIONAL CURRENCY | | | |
| | | POUND STERLING | | U.S. DOLLAR | |
	POUND STERLING	RATES USED	U.S. DOLLARS	RATES USED	U.S. DOLLARS
Revenue	£120	$1.40	$168	$1.40	$168
Cost of goods sold	(50)	1.40	(70)	1.45	(73)
Depreciation	(20)	1.40	(28)	1.50	(30)
Other expenses, net	(10)	1.40	(14)	1.40	(14)
Foreign exchange gain					108
Income before taxes	40		56		159
Income taxes	(20)	1.40	(28)		(28)
Net income	£ 20		$ 28		$131
Ratios					
Net income to revenue	0.17		0.17		0.78
Gross profit to revenue	0.58		0.58		0.57
Debt to equity	7.33		7.33		4.07

EXHIBIT 8.5 Translation of Sterling Ltd.'s Balance Sheets Under
FASB-52 (Millions)

| | | FUNCTIONAL CURRENCY | | | |
| | | POUND STERLING | | U.S. DOLLAR | |
	POUND STERLING	RATES USED	U.S. DOLLARS	RATES USED	U.S. DOLLARS
Assets					
Cash	£ 100	$1.30	$ 130	$1.30	$ 130
Receivables	200	1.30	260	1.30	260
Inventory	300	1.30	390	1.45	435
Fixed assets, net	400	1.30	520	1.50	600
Total assets	£1,000		$1,300		$1,425
Liabilities					
Current liabilities	180	1.30	234	1.30	234
Long-term debt	700	1.30	910	1.30	910
Stockholders' equity					
Common stock	100	1.50	150	1.50	150
Retained earnings	20		28		131
Cumulative translation adjustment			(22)		
Total liabilities plus equity	£1,000		$1,300		$1,425

During the year, Sterling Ltd. has income after tax of £20 million, which goes into retained earnings—that is, no dividends are paid. Thus, retained earnings rise from 0 to £20 million. Exhibit 8.4 shows how the income statement would be translated into dollars under two alternatives: (1) The functional currency is the pound sterling, and (2) the functional currency is the U.S. dollar. The second alternative yields results similar to those under FASB-8.

If the functional currency is the pound sterling, Sterling Ltd. will have a translation loss of $22 million, which bypasses the income statement (because the functional currency is identical to the local currency) and appears on the balance sheet as a separate item called *cumulative translation adjustment* under the stockholder's equity account. The translation loss is calculated as the number that reconciles the equity account with the remaining translated accounts to balance assets with liabilities and equity. Exhibit 8.5 shows the balance sheet translations for Sterling Ltd. under the two alternative functional currencies.

Similarly, if the dollar is the functional currency, the foreign exchange translation gain of $108 million, which appears on Sterling Ltd.'s income statement (because the functional currency differs from the local currency), is calculated as the difference between translated income before currency gains ($23 million) and the retained earnings figure ($131 million). This amount just balances Sterling Ltd.'s books.

Two comments are appropriate here.

1. Fluctuations in reported earnings in the example above are reduced significantly under FASB-52 when the local currency is the functional currency, as compared with the case when the U.S. dollar is the functional currency. Using the U.S. dollar as the functional currency is similar to the situation that prevailed when FASB-8 was in effect.

2. Key financial ratios and relationships—such as net income-to-revenue, gross profit, and debt-to-equity—are the same when translated into dollars under FASB-52, using the local currency as the functional currency, as they are in the local currency financial statements. These ratios and relationships were significantly different under FASB-8, represented here by using the dollar as the functional currency. The ratios appear at the bottom of Exhibit 8.4.

8.5 TRANSACTION EXPOSURE

Companies often include transaction exposure as part of their accounting exposure, although as a cash flow exposure it is rightly part of a company's economic exposure. As we have already seen, **transaction exposure** stems from the possibility of incurring future exchange gains or losses on transactions already entered into and denominated in a foreign currency. For example, when IBM sells a mainframe computer to Royal Dutch Shell in England, it typically will not be paid until a later date. If that sale is priced in pounds, IBM has a pound transaction exposure.

A company's transaction exposure is measured currency by currency and equals the difference between contractually fixed future cash inflows and outflows in each currency. Some of these unsettled transactions, including foreign currency-denominated debt and accounts receivable, are already listed on the firm's balance sheet. However, other obligations, such as contracts for future sales or purchases, are not.

ILLUSTRATION

COMPUTING TRANSACTION EXPOSURE

Suppose Boeing Airlines sells five 747s to Garuda, the Indonesian airline, in rupiahs. The rupiah price is Rp 140 billion. To help reduce the impact on Indonesia's balance of payments, Boeing agrees to buy parts from various Indonesian companies worth Rp 55 billion.

a. If the spot rate is $0.004/Rp, what is Boeing's net rupiah transaction exposure?
Solution. Boeing's net rupiah exposure equals its projected rupiah inflows minus its projected rupiah outflows, or Rp 140 billion − Rp 55 billion = Rp 85 billion. Converted into dollars at the spot rate of $0.004/Rp, Boeing's transaction exposure equals $340 million.

b. If the rupiah depreciates to $0.0035/Rp, what is Boeing's transaction loss?
Solution. Boeing will lose an amount equal to its rupiah exposure multiplied by the change in the exchange rate, or 85,000,000,000(0.004 − 0.0035) = $42.5 million. This loss can also be determined by multiplying Boeing's exposure in dollar terms, by the fractional change in the exchange rate, or 340,000,000(0.0005/0.004) = $42.5 million.

Although translation and transaction exposures overlap, they are not synonymous. Some items included in translation exposure, such as inventories and fixed assets, are excluded from transaction exposure, whereas other items included in transaction exposure, such as contracts for future sales or purchases, are not included in translation exposure. Thus, it is possible for transaction exposure in a currency to be positive and translation exposure in that same currency to be negative and vice versa. For example, Sterling Ltd. has translation exposure of £120,000 (£1,000,000 − £880,000). At the same time, ignoring off-balance sheet items, it has transaction exposure of −£580,000 (by eliminating £700,000 in inventory and fixed assets from our calculations).

8.6 ACCOUNTING PRACTICE AND ECONOMIC REALITY

Many multinationals have responded to increased currency volatility by devoting more resources to the management of **foreign exchange risk**. In order to develop an effective strategy for managing currency risk, management must first determine what is at risk. This determination requires an appropriate definition of foreign exchange risk. However, there is a major discrepancy between accounting practice and economic reality in terms of measuring exposure.

Accounting measures of exposure focus on the effect of currency changes on previous decisions of the firm, as reflected in the book values of assets acquired and liabilities incurred in the past. However, book values (which represent historical cost) and market values (which reflect future cash flows) of assets and liabilities typically differ. Therefore, *retrospective* accounting techniques, no matter how refined, cannot truly account for the economic (that is, cash flow) effects of a devaluation or revaluation on the value of a firm because these effects are primarily *prospective* in nature.

Since the change in accounting net worth produced by a movement in exchange rates often bears little relationship to the change in the market value of the firm, information derived from a historical-cost accounting system can provide a misleading picture of a firm's true economic exposure. As we saw earlier, **economic exposure** equals the company's cash flow exposure; that is, it measures the extent to which an exchange rate change will change the value of the company through its impact on the present value of the company's future cash flows. Although all items on a firm's balance sheet represent future cash flows, not all future flows appear there. Moreover, these items are not adjusted to reflect the distorting effects of inflation and relative price changes on their associated future cash flows.

The definition of exposure based on market value assumes that management's goal is to maximize the value of the firm. Whether management actually behaves in this fashion has been vigorously debated. Some managers undoubtedly will prefer to pursue other objectives. Nevertheless, the assumption that management seeks to maximize (risk-adjusted) cash flow remains standard in much of the finance literature. Moreover, the principle of maximizing stockholder wealth provides a rational guide to financial decision making.

Our evolving understanding of what exchange risk is, however, is often at odds with current management practice. Many top managers seem to be preoccupied with potential accounting-based currency gains or losses; perhaps these managers believe that the stock market evaluates a firm on the basis of its reported earnings or changes in accounting net worth, regardless of the underlying cash flows. Or perhaps their compensation is tied to earnings instead of to market value. For whatever reason, failure to distinguish between the accounting description of foreign exchange risk and the business reality of the effects of currency movements can cause corporate executives to make serious errors of judgment.

Recommendations for International Financial Executives

Most chief executives want to generate the smooth pattern of year-to-year earnings gains so cherished by security analysts. That is probably why empirical research by academicians, as well as statements by practitioners, show such a strong relationship between accounting translation methods and corporate financial policies that are designed to manage currency risk. However, because the real effect of currency changes is on a firm's future cash flows, it is obvious that information based on retrospective accounting techniques may bear no relationship to a firm's actual operating results. Furthermore, basing management decisions on these flawed accounting data can lead to financial policies that will adversely affect the real economic growth of foreign operations.

The myopia of acting on the basis of **balance-sheet exposure** rather than economic impact has been scathingly portrayed by Gunter Dufey.[1] In Dufey's example, the French subsidiary of an American multinational corporation was instructed to reduce its working-capital balances in light of a forecasted French franc devaluation. To do so would have forced it to curtail its operations; however, the French subsidiary was selling all of its output to other subsidiaries located in Germany and Belgium. Because the dollar value of its output would remain constant while franc costs expressed in dollars would decline, a 10% franc devaluation was expected to increase the French subsidiary's dollar profitability by more than 25%. The French manager, therefore, argued (correctly) that the plant should begin expanding its operations, rather than contracting them, to take advantage of the anticipated devaluation.

To be sure, the distortions associated with accounting measures of exposure do not mean that accounting statements are irrelevant; clearly the statements serve a useful purpose and are necessary for consolidating the results of a worldwide network of operating units. The danger is that the results will be misinterpreted—not by financial executives, but by stockholders, bankers, security analysts, and the board of directors. In fact, financial managers generally are aware of the misleading nature of many of these results. Despite that knowledge, however, most financial executives undertake cosmetic exchange risk management actions because they worry that others will not understand the real, as opposed to the accounting, effects of currency changes. Clearly, if the capital markets did not rationally price corporate securities, managers would be hard-pressed to design a foreign exchange strategy that could be expected to maximize the firm's value.

Is there a solution to this dichotomy between accounting and economic reality? Fortunately, the problem may be more apparent than real. A large body of research on financial markets suggests that investors are relatively sophisticated in responding to publicly available information; they appear able to understand detailed financial statements and properly interpret various accounting conventions behind corporate balance sheets and income statements.

Although the view of efficient capital markets in which stock prices correctly reflect all available information is not universally held, there is a good deal of empirical evidence that investors can effectively discriminate between accounting gimmickry and economic reality. In particular, when accounting numbers diverge significantly from cash

---

[1] Gunter Dufey, "Corporate Finance and Exchange Rate Variations," *Financial Management*, Summer 1978, pp. 51–57.

flows, changes in security prices generally reflect changes in cash flows rather than reported earnings. Consider, as an example, changes in accounting practices for reporting (but not tax) purposes, such as switching depreciation or inventory valuation methods, that affect reported earnings but not cash flows. The changes do not appear to have any discernible, statistically significant effect on security prices.[2]

The implication of these results for multinational firms is that as long as there is complete disclosure, it probably does not matter which translation method is used. In an efficient market, translation gains or losses will be placed in a proper perspective by investors and, therefore, should not affect an MNC's stock price. To help the market correctly interpret the translation outcomes, though, companies should clearly and openly disclose which translation methods they use. Furthermore, nothing prevents management from including a note in the financial statement explaining its view of the economic consequences of exchange rate changes.

Despite its inconclusive nature, the debate over the adoption by the Financial Accounting Standards Board of new currency translation methods has helped increase Wall Street's insight into the effects of currency changes on foreign operations. The debate has focused attention on the all-important distinction between the accounting and the cash-flow approaches to measuring exposure.

Against that background, some multinational firms are now taking a longer-term look at their degree of exchange risk. This look involves an examination of the risk due to the potential impact of uncertain exchange rate changes on future cash flows. Chapter 10 looks more closely at what constitutes this real exposure.

❧ 8.7 SUMMARY AND CONCLUSIONS

In this chapter, we examined the concept of exposure to exchange rate changes from the perspective of the accountant. The accountant's concern is the appropriate way to translate foreign currency-denominated items on financial statements to their home currency values. If currency values change, translation gains or losses may result. We surveyed the four principal translation methods available: the current/noncurrent method, the monetary/nonmonetary method, the temporal method, and the current rate method. In addition, we analyzed the past and present translation methods mandated by the Financial Accounting Standards Board, FASB-8 and FASB-52, respectively.

Regardless of the translation method selected, measuring accounting exposure is conceptually the same. It involves determining which foreign currency-denominated assets and liabilities will be translated at the current (postchange) exchange rate and which will be translated at the historical (prechange) exchange rate. The former items are considered to be exposed, whereas the latter items are regarded as not exposed. Translation exposure is simply the difference between exposed assets and exposed liabilities.

By far the most important feature of the accounting definition of exposure is the exclusive focus on the balance sheet effects of currency changes. We saw that this focus is misplaced because it has led the accounting profession to ignore the more important effect that these changes may have on future cash flows.

---

[2] See, for example, Robert S. Kaplan and Richard Roll, "Investor Evaluation of Accounting Information: Some Empirical Evidence," *Journal of Business*, April 1972, pp. 225–257.

Questions

1. What is translation exposure? Transaction exposure?
2. What are the basic translation methods? How do they differ?
3. Why was FASB-8 so widely criticized? How did the Financial Accounting Standards Board respond to this criticism?
4. What factors affect a company's translation exposure? What can the company do to affect its degree of translation exposure?

Problems

1. Rolls-Royce, the British jet engine manufacturer, sells engines to U.S. airlines and buys parts from U.S. companies. Suppose it has accounts receivable of $1.5 billion and accounts payable of $740 million. It also has borrowed $600 million. The current spot rate is $1.5128/£.
   a. What is Rolls-Royce's dollar transaction exposure in dollar terms? in pound terms?
   b. Suppose the pound appreciates to $1.7642/£. What is Rolls-Royce's gain or loss, in pound terms, on its dollar transaction exposure?

2. Suppose that at the start and at the end of the year, Bell U.K., the British subsidiary of Bell U.S., has current assets of £1 million, fixed assets of £2 million, and current liabilities of £1 million. Bell has no long-term liabilities.
   a. What is Bell U.K.'s translation exposure under the current/noncurrent, monetary/nonmonetary, temporal, and current rate methods?
   b. Assuming the pound is the functional currency, if the pound depreciates during the year from $1.50 to $1.30, what will be the FASB-52 translation gain (loss) to be included in the equity account of Bell's U.S. parent?
   c. Redo part (b) assuming the dollar is the functional currency. Included in current assets is inventory of £0.5 million. The historical exchange rates for inventory and fixed assets are $1.45 and $1.65, respectively. If the dollar is the functional currency, where does Bell U.K.'s translation gain or loss show up on Bell U.S.'s financial statements?

3. Paragon U.S.'s Japanese subsidiary, Paragon Japan, has exposed assets of ¥8 billion and exposed liabilities of ¥6 billion. During the year, the yen appreciates from ¥125/$ to ¥95/$.
   a. What is Paragon Japan's net translation exposure at the beginning of the year in yen? in dollars?
   b. What is Paragon Japan's translation gain or loss from the change in the yen's value?

   c. At the start of the next year, Paragon Japan adds exposed assets of ¥1.5 billion and exposed liabilities of ¥2 billion. During the year, the yen depreciates from ¥95/$ to ¥130/$. What is Paragon Japan's translation gain or loss for this year? What is its total translation gain or loss for the two years?

4. Suppose that on January 1, American Golf's French subsidiary, Golf du France, had a balance sheet that showed current assets of FF 1 million; current liabilities of FF 300,000; total assets of FF 2.5 million; and total liabilities of FF 900,000. On December 31, Golf du France's balance sheet in francs was unchanged from the figures given above, but the franc had declined in value from $0.1270 at the start of the year to $0.1180 at the end of the year. Under FASB-52, what is the translation amount to be shown on American Golf's equity account for the year if the franc is the functional currency? How would your answer change if the dollar were the functional currency?

5. Halon France, the French subsidiary of a U.S. company, Halon, Inc., has the following balance sheet:

ASSETS (FF THOUSANDS)	
Cash, marketable securities	FF 7,000
Accounts receivable	18,000
Inventory	31,000
Net fixed assets	63,000
	FF 119,000

LIABILITIES (FF THOUSANDS)	
Accounts payable	FF 14,000
Short-term debt	8,000
Long-term debt	45,000
Equity	52,000
	FF 119,000

a. At the current spot rate of $0.21/FF, calculate Halon France's accounting exposure under the current/noncurrent, monetary/nonmonetary, temporal, and current rate methods.
b. Suppose the French franc depreciates to $0.17. Produce balance sheets for Halon France at the new exchange rate under each of the four alternative translation methods.
c. Calculate the translation gains or losses associated with the FF depreciation for each of the four methods. Relate these gains and losses to the exposure calculations performed in part (a) combined with the exchange rate change. Where would these translation gains or losses show up in the balance sheets prepared for part (b)?
6. Zapata Auto Parts, the Mexican affiliate of American Diversified, Inc., had the balance sheet on January 1, 1992:

The exchange rate on January 1, 1992, was Ps 8,000 = $1.
a. What is Zapata's FASB-52 peso translation exposure on January 1, 1992?

ASSETS (Ps MILLIONS)	
Cash, marketable securities	Ps 1,000
Accounts receivable	50,000
Inventory	32,000
Net fixed assets	111,000
Total assets	Ps 194,000

LIABILITIES (Ps MILLIONS)	
Current liabilities	Ps 47,000
Long-term debt	12,000
Equity	135,000
Liabilities plus equity	Ps 194,000

b. Suppose the exchange rate on December 31, 1992, is Ps 12,000. What will be Zapata's translation loss for the year?
c. Zapata can borrow an additional Ps 15,000. What will happen to its translation exposure if it uses the funds to pay a dividend to its parent? If it uses the funds to increase its cash position?

➤Bibliography

Accounting for the Translation of Foreign Currency Transactions and Foreign Currency Financial Statements, Statement of Financial Accounting Standards No. 8. Stamford, Conn.: Financial Accounting Standards Board, October 1975.

DUFEY, GUNTER. "Corporate Finance and Exchange Rate Variations," *Financial Management,* Summer 1978, pp. 51–57.

DUKES, ROLAND. *An Empirical Investigation of the Effects of Statement of Financial Accounting Standards No. 8 on Security Return Behavior.* Stamford, Conn.: Financial Accounting Standards Board, 1978.

EVANS, THOMAS G., WILLIAM R. FOLKS JR., and

MICHAEL JILLING. *The Impact of Statement of Financial Accounting Standards No. 8 on the Foreign Exchange Risk Management Practices of American Multinationals.* Stamford, Conn.: Financial Accounting Standards Board, November 1978.

GIDDY, IAN H. "What Is FAS No. 8's Effect on the Market's Valuation of Corporate Stock Prices?" *Business International Money Report,* May 26, 1978, p. 165.

Statement of Financial Accounting Standards No. 52. Stamford, Conn.: Financial Accounting Standards Board, December 1981.

MANAGING ACCOUNTING EXPOSURE

Unfortunately the values of today's currencies oscillate wildly—to the despair of international companies that want to plan ahead.

The Economist, May 28, 1988, p. 81

The pressure to monitor and manage foreign currency risks has led many companies to develop sophisticated computer-based systems to keep track of their foreign exchange exposure and aid in managing that exposure. This chapter deals with the management of accounting exposure, including both translation and transaction exposure. Management of accounting exposure centers around the concept of hedging. **Hedging** a particular currency exposure means establishing an offsetting currency position such that whatever is lost or gained on the original currency exposure is exactly offset by a corresponding foreign exchange gain or loss on the currency hedge. Regardless of what happens to the future exchange rate, therefore, hedging locks in a dollar (home currency) value for the currency exposure. In this way, hedging can protect a firm from unforeseen currency movements.

A variety of hedging techniques are available, but before a firm uses them it must decide on which exposures to manage. Once the firm has determined the exposure position it intends to manage, how should it manage that position? How much of that position should it hedge, and which exposure-reducing technique(s) should it employ? In addition, how should exchange rate considerations be incorporated into operating decisions that will affect the firm's exchange risk posture? This chapter deals with these and other issues.

9.1 DESIGNING A HEDGING STRATEGY

An essential ingredient in any successful hedging program is for the firm to specify an operational set of goals for those involved in exchange risk management. Failure to do so can lead to possibly conflicting and costly actions on the part of employees. Although many firms do have objectives, their goals are often sufficiently vague and simplistic (for example, "eliminate all exposure" or "minimize reported foreign exchange losses") as to provide little realistic guidance to managers. For example, should an employee told to eliminate all exposure do so by using forward contracts, currency options, or by borrowing in the local currency? And, if hedging is not possible in a particular currency, should sales in that currency be foregone even if it means losing potential profits? The latter policy is likely to present a manager with the dilemma of choosing between the goals of increased profits and reduced exchange losses. Moreover, as we saw in the previous chapter, reducing translation exposure could lead to an increase in transaction exposure and vice versa. What tradeoffs, if any, should a manager be willing to make between these two types of exposure?

These and similar questions demonstrate the need for a coherent and effective strategy. The following elements are suggested for an effective exposure-management strategy:[1]

1. Determine the types of exposure to be monitored.
2. Formulate corporate objectives and give guidance in resolving potential conflicts in objectives.
3. Ensure that these corporate objectives are consistent with maximizing shareholder value and can be implemented.
4. Clearly specify who is responsible for which exposures, and detail the criteria by which each manager is to be judged.
5. Make explicit any constraints on the use of exposure-management techniques, such as limitations on entering into forward contracts.
6. Identify the channels by which exchange rate considerations are incorporated into operating decisions that will affect the firm's exchange risk posture.
7. Develop a system for monitoring and evaluating exchange risk management activities.

Objectives

The usefulness of a particular hedging strategy depends on both *acceptability* and *quality*. Acceptability refers to approval by those in the organization who will implement the strategy, and quality refers to the ability to provide better decisions. To be acceptable, a hedging strategy must be consistent with top management's values and overall corporate objectives. In turn, these values and objectives are strongly motivated by

---

[1] Most of these elements are suggested in Thomas G. Evans and William R. Folks, Jr., "Defining Objectives for Exposure Management," *Business International Money Report*, February 2, 1979, pp. 37–39.

management's beliefs about financial markets and how its performance will be evaluated. The quality, or value to the shareholders, of a particular hedging strategy is, therefore, related to the congruence between those perceptions and the realities of the business environment.

The most frequently occurring objectives, explicit and implicit, in management behavior include the following:[2]

1. *Minimize translation exposure.* This common goal necessitates a complete focus on protecting foreign currency-denominated assets and liabilities from changes in value due to exchange rate fluctuations. Given that translation and transaction exposures are not synonymous, reducing the former could cause an increase in the latter (and vice versa).

2. *Minimize quarter-to-quarter (or year-to-year) earnings fluctuations owing to exchange rate changes.* This goal requires a firm to consider both its translation exposure and its transaction exposure.

3. *Minimize transaction exposure.* This objective involves managing a subset of the firm's true cash flow exposure.

4. *Minimize economic exposure.* To achieve this goal, a firm must ignore accounting earnings and concentrate on reducing cash flow fluctuations stemming from currency fluctuations.

5. *Minimize foreign exchange risk management costs.* This goal requires a firm to balance off the benefits of hedging with its costs. It also assumes risk neutrality.

6. *Avoid surprises.* This objective involves preventing large foreign exchange losses.

The most appropriate way to rank these objectives is on their consistency with the overarching goal of maximizing shareholder value. To establish what hedging can do to further this goal, we return to our discussion of total risk in Chapter 1. In that discussion, we saw that total risk tends to adversely affect a firm's value by leading to lower sales and higher costs. Consequently, actions taken by a firm that decrease its total risk will improve its sales and cost outlooks, thereby increasing its expected cash flows.

Reducing total risk can also ensure that a firm will not run out of cash to fund its planned investment program. Otherwise, potentially profitable investment opportunities may be passed up because of corporate reluctance to tap the financial markets when internally generated cash is insufficient.[3]

There are other explanations for hedging as well, all of which relate to the idea that there is likely to be an inverse relation between total risk and shareholder value.[4] Given these considerations, the view taken here is that the basic purpose of hedging is to reduce

---

[2] See, for example, David B. Zenoff, "Applying Management Principles to Foreign Exchange Exposure," *Euromoney*, September 1978, pp. 123–130.

[3] This explanation appears in Kenneth Froot, David Scharfstein, and Jeremy Stein, "A Framework for Risk Management," *Harvard Business Review*, November 1994, pp. 91–102. The reluctance to raise additional external capital may stem from the problem of information asymmetry—this problem arises when one party to a transaction knows something relevant to the transaction that the other party does not know—which could lead investors to impose higher costs on the company seeking capital.

[4] For a good summary of these other rationales for corporate hedging, see Matthew Bishop, "A Survey of Corporate Risk Management," *The Economist*, February 10, 1996, special section.

exchange risk, where exchange risk is defined as that element of cash-flow variability attributable to currency fluctuations. This is Objective 4.

To the extent that earnings fluctuations or large losses can adversely affect the company's perceptions in the minds of potential investors, customers, employees, and so on, there may be reason to also pay attention to Objectives 2 and 6.[5] However, despite these potential benefits, there are likely to be few, if any, advantages to devoting substantial resources to managing earnings fluctuations or accounting exposure more generally (Objectives 1 and 3). To begin, trying to manage accounting exposure is inconsistent with a large body of empirical evidence that investors have the uncanny ability to peer beyond the ephemeral and concentrate on the firm's true cash-flow generating ability. In addition, whereas balance sheet gains and losses can be dampened by hedging, operating earnings will also fluctuate in line with the combined and offsetting effects of currency changes and inflation. Further, hedging costs themselves will vary unpredictably from one period to the next, leading to unpredictable earnings changes. Thus, it is impossible for firms to protect themselves from earnings fluctuations due to exchange rate changes except in the very short run.

Given the questionable benefits of managing accounting exposure, the emphasis in this text is on managing economic exposure. However, this chapter will describe the techniques used to manage transaction and translation exposure because many of these techniques are equally applicable to hedging cash flows.

In operational terms, hedging to reduce the variance of cash flows translates into the following exposure management goal: *to arrange a firm's financial affairs in such a way that however the exchange rate may move in the future, the effects on dollar returns are minimized.* This objective is not universally subscribed to, however. Instead, many firms follow a selective hedging policy designed to protect against anticipated currency movements. A selective hedging policy is especially prevalent among those firms that organize their treasury departments as profit centers. In such firms, the desire to reduce the expected costs of hedging (Objective 5)—and thereby increase profits—often leads to taking higher risks by hedging only when a currency change is expected and going unhedged otherwise.

However, if financial markets are efficient, firms cannot hedge against *expected* exchange rate changes. Interest rates, forward rates, and sales-contract prices should already reflect currency changes that are anticipated, thereby offsetting the loss-reducing benefits of hedging with higher costs. In the case of Mexico, for instance, the one-year forward discount in the futures market was close to 100% just before the peso was floated in 1982. The unavoidable conclusion is that a firm can protect itself only against *unexpected* currency changes.

Moreover, there is always the possibility of bad timing. For example, big Japanese exporters such as Toyota and Honda have incurred billions of dollars in foreign exchange losses. One reason for these losses is that Japanese companies often try to predict where the dollar is going and hedge (or not hedge) accordingly. At the beginning of 1994, many thought that the dollar would continue to strengthen and thus failed to hedge their exposure. When the dollar plummeted instead, they lost billions.

---

[5] Fluctuating earnings could also boost a company's taxes by causing it to alternate between high and low tax brackets (see Rene Stulz, "Rethinking Risk Management," Ohio State University working paper).

Costs and Benefits of Standard Hedging Techniques

Standard techniques for responding to anticipated currency changes are summarized in Exhibit 9.1. Such techniques, however, are vastly overrated in terms of their ability to minimize hedging costs. If a devaluation is unlikely, they are costly and inefficient ways of doing business. If a devaluation is expected, the cost of using the techniques (like the cost of local borrowing) rises to reflect the anticipated devaluation. Just before the August 1982 peso devaluation, for example, every company in Mexico was trying to delay peso payments. Of course, this technique cannot produce a net gain because one company's payable is another company's receivable. As another example, if one company wants peso trade credit, another must offer it. Assuming that both the borrower and the lender are rational, a deal will not be struck until the interest cost rises to reflect the expected decline in the peso.

EXHIBIT 9.1 Basic Hedging Techniques

DEPRECIATION	APPRECIATION
■ Sell local currency forward	■ Buy local currency forward
■ Reduce levels of local currency cash and marketable securities	■ Increase levels of local currency cash and marketable securities
■ Tighten credit (reduce local currency receivables)	■ Relax local currency credit terms
■ Delay collection of hard-currency receivables	■ Speed up collection of soft-currency receivables
■ Increase imports of hard-currency goods	■ Reduce imports of soft-currency goods
■ Borrow locally	■ Reduce local borrowing
■ Delay payment of accounts payable	■ Speed up payment of accounts payable
■ Speed up dividend and fee remittances to parent and other subsidiaries	■ Delay dividend and fee remittances to parent and other subsidiaries
■ Speed up payment of intersubsidiary accounts payable	■ Delay payment of intersubsidiary accounts payable
■ Delay collection of intersubsidiary accounts receivable	■ Speed up collection of intersubsidiary accounts receivable
■ Invoice exports in foreign currency and imports in local currency	■ Invoice exports in local currency and imports in foreign currency

Even shifting funds from one country to another is not a costless means of hedging. The net effect of speeding up remittances while delaying receipt of intercompany receivables is to force a subsidiary in a devaluation-prone country to increase its local currency (LC) borrowings to finance the additional working capital requirements. The net cost of shifting funds, therefore, is the cost of the LC loan minus the profit generated from the use of the funds—for example, prepaying a hard-currency loan—with both adjusted for expected exchange rate changes. As mentioned previously, loans in local currencies subject to devaluation fears carry higher interest rates that are likely to offset any gains from LC devaluation.

Reducing the level of cash holdings to lower exposure can adversely affect a subsidiary's operations, whereas selling LC-denominated marketable securities can entail an opportunity cost (the lower interest rate on hard-currency securities). A firm with excess cash or marketable securities should reduce its holdings regardless of whether a devaluation is anticipated. After cash balances are at the minimum level, however, any further reductions will involve real costs that must be weighed against the expected benefits.

Invoicing exports in the foreign currency and imports in the local currency may cause the loss of valuable sales or may reduce a firm's ability to extract concessions on import prices. Similarly, tightening credit may reduce profits more than costs.

In summary, hedging exchange risk costs money and should be scrutinized like any other purchase of insurance. The costs of these hedging techniques are summarized in Exhibit 9.2.

A company can benefit from the preceding techniques only to the extent that it can forecast future exchange rates more accurately than the general market. For example, if the company has a foreign currency cash inflow, it would hedge only if the forward rate exceeds its estimate of the future spot rate. Conversely, with a foreign currency cash

---

EXHIBIT 9.2 Cost of the Basic Hedging Techniques

DEPRECIATION	COSTS
■ Sell local currency forward	■ Transaction costs; difference between forward and future spot rates
■ Reduce levels of local currency cash and marketable securities	■ Operational problems; opportunity cost (loss of higher interest rates on LC securities)
■ Tighten credit (reduce local currency receivables)	■ Lost sales and profits
■ Delay collection of hard-currency receivables	■ Cost of financing additional receivables
■ Increase imports of hard-currency goods	■ Financing and holding costs
■ Borrow locally	■ Higher interest rates
■ Delay payment of accounts payable	■ Harm to credit reputation
■ Speed up dividend and fee remittances to parent and other subsidiaries	■ Borrowing cost if funds not available or loss of higher interest rates if LC securities must be sold
■ Speed up payment of intersubsidiary accounts payable	■ Opportunity cost of money
■ Delay collection of intersubsidiary accounts receivable	■ Opportunity cost of money
■ Invoice exports in foreign currency and imports in local currency	■ Lost export sales or lower price; premium price for imports

outflow, it would hedge only if the forward rate was below its estimated future spot rate. In this way, it would apparently be following the profit-guaranteeing dictum of buy low-sell high. The key word, however, is *apparently*. Attempting to profit from foreign exchange forecasting is speculating rather than hedging. The hedger is well-advised to assume that the market knows as much as he or she does. Those who feel that they have superior information may choose to speculate, but this activity should not be confused with hedging.

ILLUSTRATION **S**ELECTIVE **S**HEDGING

In March, Multinational Industries, Inc. (MII), assessed the September spot rate for sterling at the following rates:

$1.30/£ with probability 0.15
$1.35/£ with probability 0.20
$1.40/£ with probability 0.25
$1.45/£ with probability 0.20
$1.50/£ with probability 0.20

a. What is the expected spot rate for September?
Solution. The expected future spot rate is 1.30(.15) + 1.35(.2) + 1.40(.25) + 1.45(.20) + 1.50(.20) = $1.405.

b. If the six-month forward rate is $1.40, should the firm sell forward its £500,000 pound receivables due in September?
Solution. If MII sells its pound proceeds forward, it will lock in a value of $700,000

(1.40 × 500,000). Alternatively, if it decides to wait until September and sell its pound proceeds in the spot market, it expects to receive $702,500 ($1.405 × 500,000). Based on these figures, if MII wants to maximize expected profits, it should retain its pound receivables and sell the proceeds in the spot market upon receipt.

c. What factors are likely to affect Multinational Industries' hedging decision?
Solution. Risk aversion could lead MII to sell its receivables forward to hedge their dollar value. However, if MII has pound liabilities, they could provide a natural hedge. Exposure netting would then reduce or eliminate the amount necessary to hedge. The existence of a cheaper hedging alternative, such as borrowing pounds and converting them to dollars for the duration of the receivables, would also make undesirable the use of a forward contract. This latter situation assumes that interest rate parity is violated. The tax treatment of foreign exchange gains and losses on forward contracts could also affect the hedging decision.

Under some circumstances, it is possible for a company to benefit at the expense of the local government without speculating. Such a circumstance would involve the judicious use of market imperfections or existing tax asymmetries or both. In the case of an overvalued currency, such as the Mexican peso in 1982, if exchange controls are not imposed to prevent capital outflows and if hard currency can be acquired at the official exchange rate, then money can be moved out of the country via intercompany payments. For instance, a subsidiary can speed payments of intercompany accounts payable, make immediate purchases from other subsidiaries, or speed remittances to the parent. Unfortunately, governments are not unaware of these tactics. During a currency crisis, when hard currency is scarce, the local government can be expected to block such transfers or at least make them more expensive.

Another often-cited reason for market imperfection is that individual investors may not have equal access to capital markets. For example, because forward exchange markets exist only for the major currencies, hedging often requires local borrowing in heavily regulated capital markets. As a legal citizen of many nations, the MNC normally has greater access to these markets.

Similarly, if forward contract losses are treated as a cost of doing business, whereas gains are taxed at a lower capital gains rate, the firm can engage in tax arbitrage. In the absence of financial market imperfections or tax asymmetries, however, the net expected value of hedging over time should be zero. Despite the questionable value to shareholders of hedging balance sheet exposure or even transaction exposure, however, managers often try to reduce these exposures because they are evaluated, at least in part, on translation or transaction gains or losses.

Centralization versus Decentralization

In the area of foreign exchange risk management, there are good arguments both for and against centralization. Favoring centralization is the reasonable assumption that local treasurers want to optimize their own financial and exposure positions, regardless of the overall corporate situation. An example is a multibillion-dollar U.S. consumer-goods firm that gives its affiliates a free hand in deciding on their hedging policies. The firm's local treasurers ignore the possibilities available to the corporation to trade off positive and negative currency exposure positions by consolidating exposure worldwide. If subsidiary A sells to subsidiary B in sterling, then from the corporate perspective, these sterling exposures net out on a consolidated translation basis (but only before tax). If A or B or both hedge their sterling positions, however, unnecessary hedging takes place or a zero sterling exposure turns into a positive or negative position. Furthermore, in their dealings with external customers, some affiliates may wind up with a positive exposure and others with a negative exposure in the same currency. Through lack of knowledge or incentive, individual subsidiaries may undertake hedging actions that increase rather than decrease overall corporate exposure in a given currency.

A further benefit of centralized exposure management is the ability to take advantage, through exposure netting, of the portfolio effect discussed previously. Thus, centralization of exchange risk management should reduce the amount of hedging required to achieve a given level of safety.

After the company has decided on the maximum currency exposure it is willing to tolerate, it can then select the cheapest option(s) worldwide to hedge its remaining exposure. Tax effects can be crucial at this stage, in computing both the amounts to hedge and the costs involved, but only headquarters will have the required global perspective. Centralized management also is needed to take advantage of the before-tax hedging cost variations that are likely to exist among subsidiaries because of market imperfections.

All these arguments for centralization of currency risk management are powerful. Against the benefits must be weighed the loss of local knowledge and the lack of incentive for local managers to take advantage of particular situations that only they may be familiar with. Companies that decentralize the hedging decision may allow local units to manage their own exposures by engaging in forward contracts with a central unit at negotiated rates. The central unit, in turn, may or may not lay off these contracts in the marketplace.

℥ *Risk Management*

A number of highly publicized cases of derivatives-related losses have highlighted the potential dangers in the use of derivatives such as futures and options. Although these losses did not involve the use of currency derivatives, several lessons for risk management can be drawn from these cases, which include the bankruptcies of Orange County and Barings PLC and the huge losses taken at Metallgesellschaft, Kidder Peabody, Sumitomo, Union Bank of Switzerland, and Procter & Gamble. The most important lesson to be learned is that risk management failures have their origins in inadequate systems and controls rather than from any risk inherent in the use of derivatives themselves. In every case of large losses, senior management did not understand fully the activities of those taking positions in derivatives and failed to monitor and supervise their activities adequately. Some specific lessons include the following.

First, segregate the duties of those trading derivatives from those supposed to monitor them. For example, Nicholas Leesing, the trader who sank Barings, was in charge of trading and also kept his own books. When he took losses, he covered them up and doubled his bets. Similarly, the manager responsible for the profits generated by trading derivatives at UBS also oversaw the risks of his position. No one else at the bank was allowed to examine the risks his department was taking. And a rogue trader at Sumitomo, who lost $1.8 billion, oversaw the accounts that kept track of his dealing. These conflicts of interest are a recipe for disaster. Second, derivatives positions should be limited to prevent the possibility of catastrophic losses, and they should be marked to market every day to avoid the possibility of losses going unrecognized and being allowed to accumulate. As in the cases of Barings and Sumitomo, traders who can roll over their positions at nonmarket prices tend to make bigger and riskier bets to recoup their losses.

Third, compensation arrangements should be designed to shift more of the risk onto the shoulders of those taking the risks. For example, deferring part of traders' salaries until their derivatives positions actually pay off would make them more cognizant of the risks they are taking. Fourth, pay attention to warning signs. For example, Barings was slow to respond to an audit showing significant discrepancies in Leeson's accounts. Similarly, Kidder Peabody's executives ignored a trader who was generating record profits while supposedly engaged in risk-free arbitrage. A related lesson is that there's no free lunch. Traders and others delivering high profits deserve special scrutiny by independent auditors. The auditors must pay particular attention to the valuation of exotic derivatives—specialized contracts not actively traded. Given the lack of ready market prices for exotics, it is easy for traders to overvalue their positions in exotics without independent oversight. Finally, those who value reward above risk will likely wind up with risk at the expense of reward.

9.2 MANAGING TRANSACTION EXPOSURE

A transaction exposure arises whenever a company is committed to a foreign currency-denominated transaction. Since the transaction will result in a future foreign currency cash inflow or outflow, any change in the exchange rate between the time the transaction is entered into and the time it is settled in cash will lead to a change in the dollar (HC) amount of the cash inflow or outflow. Protective measures to guard against transaction exposure involve entering into foreign currency transactions whose cash flows exactly offset the cash flows of the transaction exposure.

These protective measures include using forward contracts, price adjustment clauses, currency options, and borrowing or lending in the foreign currency. Alternatively, the company could try to invoice all transactions in dollars and to avoid transaction exposure entirely. However, eliminating transaction exposure does not eliminate all foreign exchange risk. The firm still is subject to exchange risk on its future revenues and costs—its operating cash flows.

We will now look at the various techniques for managing transaction exposure by examining the case of General Electric's Deutsche mark exposure. Suppose that on January 1, GE is awarded a contract to supply turbine blades to Lufthansa, the German airline. On December 31, GE will receive payment of DM 25 million for these blades. The most direct way for GE to hedge this receivable is to sell a DM 25 million forward contract for delivery in one year. Alternatively, it can use a money market hedge, which would involve borrowing DM 25 million for one year, converting it into dollars, and investing the proceeds in a security that matures on December 31. As we will see, if interest rate parity holds, the two methods will yield the same results. GE can also manage its transaction exposure through risk shifting, risk sharing, exposure netting, and currency options.

Forward Market Hedge

In a **forward market hedge**, a company that is long a foreign currency will sell the foreign currency forward, whereas a company that is short a foreign currency will buy the currency forward. In this way, the company can fix the dollar value of future foreign currency cash flow. For example, by selling forward the proceeds from its sale of turbine blades, GE can effectively transform the currency denomination of its DM 25 million receivable from Deutsche marks to dollars, thereby eliminating all currency risk on the sale. For example, suppose the current spot price for the Deutsche mark is $0.40/DM, and the one-year forward rate is $0.3828/DM. Then, a forward sale of DM 25 million for delivery in one year will yield GE $9.57 million on December 31. Exhibit 9.3 shows the cash-flow consequences of combining the forward sale with the Deutsche mark receivable, given three possible exchange rate scenarios.

Regardless of what happens to the future spot rate, Exhibit 9.3 demonstrates that GE still gets to collect $9.57 million on its turbine sale. Any exchange gain or loss on the forward contract will be offset by a corresponding exchange loss or gain on the receiv-

EXHIBIT 9.3 Possible Outcomes of Forward Market Hedge as of December 31					
SPOT EXCHANGE RATE	VALUE OF ORIGINAL RECEIVABLE (1)	+	GAIN (LOSS) ON FORWARD CONTRACT (2)	=	TOTAL CASH FLOW (3)
DM 1 = $0.40	$10,000,000		($430,000)		$9,570,000
DM 1 = $0.3828	9,570,000		0		$9,570,000
DM 1 = $0.36	9,000,000		570,000		$9,570,000

able. The effects of this transaction also can be seen with the following simple T-account describing GE's position as of December 31:

DECEMBER 31: GE T-ACCOUNT (MILLIONS)			
Account receivable	DM 25	Forward contract payment	DM 25
Forward contract receipt	$9.57		

Without hedging, GE will have a DM 25 million asset whose value will fluctuate with the exchange rate. The forward contract creates an equal DM liability, offset by an asset worth $9.57 million dollars. The DM asset and liability cancel each other out, and GE is left with a $9.57 million asset.

This example illustrates another point as well: *Hedging with forward contracts eliminates the downside risk but at the expense of foregoing the upside potential.*

THE TRUE COST OF HEDGING ■ Exhibit 9.3 also shows that the true cost of hedging cannot be calculated in advance because it depends on the future spot rate, which is unknown at the time the forward contract is entered into. In the example above, the actual cost of hedging can vary from +$430,000 to −$570,000; a plus (+) represents a cost, and a minus (−) represents a negative cost or a gain. In percentage terms, the cost varies from −5.7% to +4.3%.

This example points out the distinction between the traditional method of calculating the cost of a forward contract and the correct method, which measures its opportunity cost. Specifically, the cost of a forward contract is usually measured as its forward discount or premium:

$$\frac{f_1 - e_0}{e_0}$$

where e_0 is the current spot rate (dollar price) of the foreign currency and f_1 is the forward rate. In GE's case, this cost would equal 4.3%.

However, this approach is wrong because the relevant comparison must be between the dollars per unit of foreign currency received with hedging, f_1, and the dollars received in the absence of hedging, e_1, where e_1 is the future (unknown) spot rate on the date of settlement. That is, the real cost of hedging is an *opportunity cost*. In particular, if the forward contract had not been entered into, the future value of each unit of foreign currency would have been e_1 dollars. Thus, the true dollar cost of the forward contract per dollar's worth of foreign currency sold forward equals

$$\frac{f_1 - e_1}{e_0}$$

The expected cost (value) of a forward contract depends on whether or not a risk premium or other source of bias exists. Absent such bias, the expected cost of hedging via a forward contract will be zero. Otherwise, there would be an arbitrage opportunity.

Suppose, for example, that management at General Electric believes that despite a one-year forward rate of $0.3828, the Deutsche mark will actually be worth about $0.3910 on December 31. Then GE could profit by buying (rather than selling) DM forward for one year at $0.3828 and, on December 31, completing the contract by selling DM in the spot market at $0.3910. If GE is correct, it will earn $0.0082 (0.3910 − 0.3828) per Deutsche mark sold forward. On a DM 25 million forward contract, this profit would amount to $205,000—a substantial reward for a few minutes of work.

The prospect of such rewards would not go unrecognized for long, which explains why, on average, the forward rate appears to be unbiased. Therefore, unless GE or any other company has some special information about the future spot rate that it has good reason to believe is not adequately reflected in the forward rate, it should accept the forward rate's predictive validity as a working hypothesis and avoid speculative activities. After the fact, of course, the actual cost of a forward contract will turn out to be positive or negative (unless the future spot rate equals the forward rate), but the sign cannot be predicted in advance.

On the other hand, the evidence presented in Chapter 7 points to the possibility of bias in the forward rate at any point in time. The nature of this apparent bias suggests that the selective use of forward contracts in hedging may reduce expected hedging costs; but beware of the peso problem—the possibility that historical returns may be unrepresentative of future returns. The specific cost-minimizing selective hedging policy to take advantage of this bias would depend on whether you are trying to hedge a long or a short position in a currency and is as follows:

■ If you are long a currency, hedge (by selling forward) if the currency is at a forward premium; if the currency is at a forward discount, do not hedge.

■ If you are short a currency, hedge (by buying forward) if the currency is selling at a forward discount; if the currency is at a forward premium, do not hedge.

As discussed in Chapter 7, this selective hedging policy does not come free; it may reduce expected costs but at the expense of higher risk. Absent other considerations, therefore, the impact on shareholder wealth of selective hedging via forward contracts should be minimal, with any expected gains likely to be offset by higher risk.

Money Market Hedge

An alternative to a forward market hedge is to use a money market hedge. A **money market hedge** involves simultaneous borrowing and lending activities in two different currencies to lock in the dollar value of a future foreign currency cash flow. For example, suppose Deutsche mark and U.S. dollar interest rates are 15% and 10%, respectively. Using a money market hedge, General Electric will borrow DM (25/1.15) million = DM 21.74 million for one year, convert it into $8.7 million in the spot market, and invest the $8.7 million for one year. On December 31, GE will receive 1.10 × $8.7 million = $9.57 million from its dollar investment. GE then will use the proceeds of its DM receivable, collectible on that date, to repay the 1.15 × DM 21.74 million = DM 25 million it owes in principal and interest. As Exhibit 9.4 shows, the exchange gain or loss on

EXHIBIT 9.4 Possible Outcomes of Money Market Hedge on December 31

SPOT EXCHANGE RATE	VALUE OF ORIGINAL RECEIVABLE (1)	+	GAIN (LOSS) ON MONEY MARKET HEDGE (2)	=	TOTAL CASH FLOW (3)
DM 1 = $0.40	$10,000,000		($430,000)		$9,570,000
DM 1 = $0.3828	9,570,000		0		$9,570,000
DM 1 = $0.36	9,000,000		570,000		$9,570,000

the borrowing and lending transactions exactly offsets the dollar loss or gain on GE's DM receivable.

The gain or loss on the money market hedge can be calculated simply by subtracting the cost of repaying the DM debt from the dollar value of the investment. For example, in the case of an end-of-year spot rate of $0.40, the DM 25 million in principal and interest will cost $10 million to repay. The return on the dollar investment is only $9.57 million, leaving a loss on the money market hedge of $430,000.

We can also view the effects of this transaction with the simple T-account used earlier:

DECEMBER 31: GE T-ACCOUNT (MILLIONS)

Account receivable	DM 25	Loan repayment (including interest)	DM 25
Investment return (including interest)	$9.57		

As with the forward contract, the DM asset and liability (the loan repayment) cancel each other out, and GE is left with a $9.57 million asset (its investment).

The equality of the net cash flows from the forward market and money market hedges is not coincidental. The interest rates and forward and spot rates were selected so that interest rate parity holds. In effect, the simultaneous borrowing and lending transactions associated with a money market hedge enable GE to create a "home-made" forward contract. The effective rate on this forward contract will equal the actual forward rate if interest rate parity holds. Otherwise, a covered interest arbitrage opportunity would exist.

In reality, of course, there are transaction costs associated with hedging: the bid-ask spread on the forward contract and the difference between borrowing and lending rates. These transaction costs must be factored in when comparing a forward contract hedge with a money market hedge. The key to making these comparisons, as shown in Chapter 5, is to ensure that the correct bid and ask and borrowing and lending rates are used.

Risk Shifting

General Electric can avoid its transaction exposure altogether if Lufthansa allows it to price the sale of turbine blades in dollars. Dollar invoicing, however, does not eliminate currency risk; it simply shifts that risk from GE to Lufthansa, which now has dollar ex-

---

ILLUSTRATION

COMPARING HEDGING ALTERNATIVES WHEN THERE ARE TRANSACTION COSTS

PepsiCo would like to hedge its C$40 million payable to Alcan, a Canadian aluminum producer, which is due in 90 days. Suppose it faces the following exchange and interest rates.

Spot rate:	$0.7307–12/C$
Forward rate (90 days):	$0.7320–41/C$
Canadian dollar 90-day interest rate (annualized):	4.71%–4.64%
U.S. dollar 90-day interest rate (annualized):	5.50%–5.35%

Which hedging alternative would you recommend? The first interest rate is the borrowing rate and the second one is the lending rate.
Solution. The hedged cost of the payable using the forward market is U.S.$29,364,000 (0.7341 × 40,000,000), remembering that PepsiCo must buy forward Canadian dollars at the ask rate. Alternatively, PepsiCo could use a money market hedge. This hedge would entail the following steps:

1. Borrow U.S. dollars at 5.50% annualized for 90 days (the borrowing rate). The actual interest rate for 90 days will be 1.375% (5.50% × 90/360).

2. Convert the U.S. dollars into Canadian dollars at $0.7312 (the ask rate).

3. Invest the Canadian dollars for 90 days at 4.63% annualized for 90 days (the lending rate) and use the loan proceeds to pay Alcan. The actual interest rate for 90 days will be 1.16% (4.64% × 90/360).

Since PepsiCo needs C$40 million in 90 days and will earn interest equal to 1.16%, it must invest the present value of this sum or C$39,541,321 (40,000,000/1.0116). This sum is equivalent to U.S.$28,912,614 converted at the spot ask rate (39,541,321 × 0.7312). At a 90-day borrowing rate of 1.375%, PepsiCo must pay back principal plus interest in 90 days of U.S.$29,310,162 (28,912,614 × 1.01375). Thus, the hedged cost of the payable using the money market hedge is $29,310,162.

Comparing the two hedged costs, we see that by using the money market hedge instead of the forward market hedge, PepsiCo will save $53,838 (29,364,000 − 29,310,162). Other things being equal, therefore, this is the recommended hedge for PepsiCo.

---

posure. Lufthansa may or may not be better able, or more willing, to bear it. Despite the fact that this form of *risk shifting* is a zero-sum game, it is common in international business. Firms typically attempt to invoice exports in strong currencies and imports in weak currencies.

Is it possible to gain from risk shifting? Not if one is dealing with informed customers or suppliers. To see why, consider the GE-Lufthansa deal. If Lufthansa is willing to be invoiced in dollars for the turbine blades, that must be because Lufthansa calculates that its Deutsche mark equivalent cost will be no higher than the DM 25 million price it was originally prepared to pay. Since Lufthansa does not have to pay for the turbine blades until December 31, its cost will be based on the spot price of the dollars as of that date. By buying dollars forward at the one-year forward rate of $0.3828/DM, Lufthansa can convert a dollar price of P into a DM cost of $P/0.3828$. Thus, the maximum dollar price P_M that Lufthansa should be willing to pay for the turbine blades is the solution to

$$\frac{P_M}{0.3828} = \text{DM 25 million}$$

or

$$P_M = \$9.57 \text{ million}$$

Considering that GE can guarantee itself $9.57 million by pricing in Deutsche marks and selling the resulting DM 25 million forward, it will certainly not accept a lower dollar price. The bottom line is that both Lufthansa and General Electric will be indifferent between a U.S. dollar price and a Deutsche mark price only if the two prices are equal at the forward exchange rate. Therefore, because the Deutsche mark price arrived at through arm's-length negotiations is DM 25 million, the dollar price that is equally acceptable to Lufthansa and GE can only be $9.57 million. Otherwise, one or both of the parties involved in the negotiations has ignored the possibility of currency changes. Such naiveté is unlikely to exist for long in the highly competitive world of international business.

Pricing Decisions

Notwithstanding the view expressed above, top management sometimes has failed to take anticipated exchange rate changes into account when making operating decisions, leaving financial management with the essentially impossible task, through purely financial operations, of recovering a loss already incurred at the time of the initial transaction. To illustrate this type of error, suppose that GE has priced Lufthansa's order of turbine blades at $10 million and then, because Lufthansa demands to be quoted a price in Deutsche marks, converts the dollar price to a Deutsche mark quote of DM 25 million, using the spot rate of $0.40/DM.

In reality, the quote is worth only $9.57 million—even though it is booked at $10 million— because that is the risk-free price that GE can guarantee for itself by using the forward market. If GE management wanted to sell the blades for $10 million, it should have set a Deutsche mark price equal to DM 10,000,000/0.3828 = DM 26.12 million. Thus, GE lost $430,000 the moment it signed the contract (assuming that Lufthansa would have agreed to the higher price rather than turn to another supplier). This loss is not an exchange loss; it is a loss due to management inattentiveness.

The general rule on credit sales overseas is to convert between the foreign currency price and the dollar price by using the forward rate, not the spot rate. If the dollar price is high enough, the exporter should follow through with the sale. Similarly, if the dollar price on a foreign currency-denominated import is low enough, the importer should follow through on the purchase. All this rule does is to recognize that a Deutsche mark (or any other foreign currency) tomorrow is not the same as a Deutsche mark today. This rule is the international analogue to the insight that a dollar tomorrow is not the same as a dollar today. In the case of a sequence of payments to be received at several points in time, the foreign currency price should be a weighted average of the forward rates for delivery on those dates.

ILLUSTRATION

WEYERHAEUSER QUOTES A FRENCH FRANC PRICE FOR ITS LUMBER

Weyerhaeuser is asked to quote a price in French francs for lumber sales to a French company. The lumber will be shipped and paid for in four equal, quarterly installments. Weyerhaeuser requires a minimum price of $1 million to accept this contract. If P_F is the French franc price contracted for, then Weyerhaeuser will receive $0.25P_F$ every three months, beginning 90 days from now. Suppose the spot and forward rates for the French franc are as follows:

SPOT	90-DAY	180-DAY	270-DAY	360-DAY
$0.1772	$0.1767	$0.1761	$0.1758	$0.1751

On the basis of these forward rates, the certainty-equivalent dollar value of this franc revenue is $0.25P_F(0.1767 + 0.1761 + 0.1758 + 0.1751)$, or $0.25P_F(0.7037) = \$0.1759P_F$. In order for Weyerhaeuser to realize $1 million from this sale, the minimum French franc price must be the solution to

$$\$0.1759\, P_F = \$1,000,000$$

or

$$P_F = FF\ 5,685,048$$

At any lower franc price, Weyerhaeuser cannot be assured of receiving the $1 million it demands for this sale. Note that the spot rate did not enter into any of these calculations.

Exposure Netting

Exposure netting involves offsetting exposures in one currency with exposures in the same or another currency, where exchange rates are expected to move in such a way that losses (gains) on the first exposed position should be offset by gains (losses) on the second currency exposure. This portfolio approach to hedging recognizes that the total variability or risk of a currency exposure portfolio should be less than the sum of the individual variabilities of each currency exposure considered in isolation. The assumption underlying exposure netting is that the net gain or loss on the entire currency exposure portfolio is what matters, rather than the gain or loss on any individual monetary unit.

It is easy to see, for example, that a DM 1 million receivable and DM 1 million payable cancel each other out, with no net (before-tax) exposure. It may be less obvious that such exposure netting can also be accomplished by using positions in different currencies. However, companies practice multicurrency exposure netting all the time. According to one executive: "We might be willing to tolerate a short position in Swiss francs if we had a long position in Deutsche marks. They're fellow travelers. We look at our exposures as a portfolio."[6] In practice, exposure netting involves one of three possibilities:

1. A firm can offset a long position in a currency with a short position in that same currency.

---

[6]"How Corporations Are Playing the Currency Game," *Institutional Investor*, May 1976, p. 31.

2. If the exchange rate movements of two currencies are positively correlated (for example, the Swiss franc and Deutsche mark), then the firm can offset a long position in one currency with a short position in the other.

3. If the currency movements are negatively correlated, then short (or long) positions can be used to offset each other.

USING EXPOSURE NETTING TO MANAGE TRANSACTION EXPOSURE

Suppose that Apex Computers has the following transaction exposures:

APEX T-ACCOUNT (MILLIONS)

Marketable securities	DM 3.5	Accounts payable	Mex $15.4
Accounts receivable	SFr 6.2	Bank loan	SFr 14.8
		Tax liability	DM 1.3

On a net basis, before taking currency correlations into account, Apex's transaction exposures—now converted into dollar terms—is:

APEX T-ACCOUNT (MILLIONS)

Deutsche mark (2.2)	$1.2	Swiss franc (8.6)	$5.1
		Mexican peso	$1.8

Given the historical positive correlation between the DM and Swiss franc, Apex decides to net out its DM long position from its franc short position, leaving it with a net short position in the Swiss franc of $3.9 million ($1.2 million − $5.1 million). Lastly, Apex takes into account the historical negative correlation between the Mexican peso and the Swiss franc and offsets these two short positions. The result is a net short position in Swiss francs of $2.1 million ($3.9 million − $1.8 million). By hedging only this residual transaction exposure, Apex can dramatically reduce the volume of its hedging transactions. Of course, the latter exposure netting—offsetting DM, Swiss franc, and Mexican peso exposures with each other—depends on the strength of the correlations among these currencies. Specifically, Apex's offsetting its exposures on a dollar-for-dollar basis will be fully effective and appropriate only if the correlations are +1 for the DM/SFr currency pair and −1 for the SFr/Mex$ currency pair.

Currency Risk Sharing

In addition to, or instead of, a traditional hedge, General Electric and Lufthansa can agree to share the currency risks associated with their turbine blade contract. **Currency risk sharing** can be implemented by developing a customized hedge contract embedded in the underlying trade transaction. This hedge contract typically takes the form of a *price adjustment clause*, whereby a base price is adjusted to reflect certain exchange rate changes. For example, the base price could be set at DM 25 million, but the parties would share the currency risk beyond a neutral zone. The *neutral zone* represents the currency range in which risk is not shared.

Suppose the neutral zone is specified as a band of exchange rates: $0.39−0.41/DM, with a base rate of $0.40/DM. This means that the exchange rate can fall as far as $0.39/DM or rise as high as $0.41/DM without reopening the contract. Within the neutral

EXHIBIT 9.5 Currency Risk Sharing: GE and Lufthansa

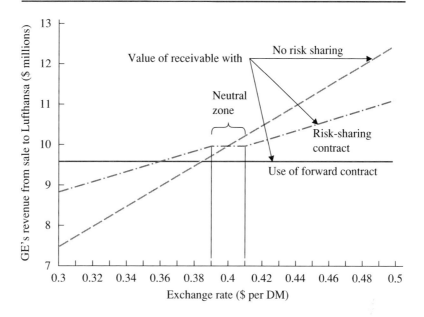

zone, Lufthansa must pay GE the dollar equivalent of DM 25 million at the base rate of $0.40, or $10 million. Thus, Lufthansa's cost can vary from DM 24.39 million to DM 25.64 million (10,000,000/0.41 to 10,000,000/0.39). However, if the DM depreciates from $0.40 to, say, $0.35, the actual rate will have moved $0.04 beyond the lower boundary of the neutral zone ($0.39/DM). This amount is shared equally. Thus, the exchange rate actually used in settling the transaction is $0.38/DM ($0.40 − .04/2). The new price of the turbine blades becomes DM 25,000,000 × 0.38, or $9.5 million. Lufthansa's cost rises to DM 27.14 million (9,500,000/0.35). In the absence of a risk-sharing agreement, the contract value to GE would have been $8.75 million. Of course, if the Deutsche mark appreciates beyond the upper bound to, say, $0.45, GE does not get the full benefit of the DM's rise in value. Instead, the new contract exchange rate becomes $0.42 (0.40 + 0.04/2). GE collects DM 25,000,000 × 0.42, or $10.5 million, and Lufthansa pays a price of DM 23.33 million (10,500,000/0.45).

Exhibit 9.5 compares the currency risk protection features of the currency risk-sharing arrangement with that of a traditional forward contract (at a forward rate of $0.3828) and a no-hedge alternative. Within the neutral zone, the dollar value of GE's contract under the risk-sharing agreement stays at $10 million. This situation is equivalent to Lufthansa selling GE a forward contract at the current spot rate of $0.40. Beyond the neutral zone, the contract's dollar value rises or falls only half as much under the risk-sharing agreement as under the no-hedge alternative. The value of the hedged contract remains the same, regardless of the exchange rate.

EXHIBIT 9.6 Currency Range Forward: GE and Lufthansa

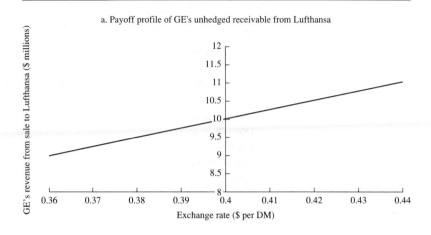

a. Payoff profile of GE's unhedged receivable from Lufthansa

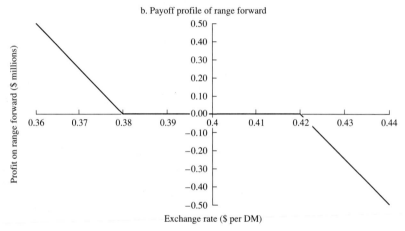

b. Payoff profile of range forward

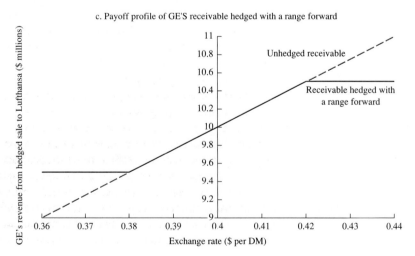

c. Payoff profile of GE'S receivable hedged with a range forward

Currency Collars

Suppose that GE is prepared to take some but not all of the risk associated with its DM receivable. In this case, it could buy a **currency collar**, which is a contract that provides protection against currency moves outside an agreed-upon range. For example, suppose that GE is willing to accept variations in the value of its DM receivable associated with fluctuations in the DM in the range of $0.38 to $0.42. Beyond that point, however, it wants protection. With a currency collar, also known as a **range forward**, GE will convert its DM receivable at the following range forward rate, RF, which depends on the actual future spot rate, e_1:

If $e_1 < \$0.38$, then RF = \$0.38
If $\$0.38 \le e_1 \le \$0.42$, then RF = e_1
If $e_1 > \$0.42$, then RF = \$0.42

In effect, GE is agreeing to convert its DM proceeds at the future spot rate if that rate falls within the range $0.38–$0.42 and at the boundary rates beyond that range. Specifically, if the future spot rate exceeds $0.42, then it will convert the DM proceeds at $0.42, giving the bank a profit on the range forward. Alternatively, if the future spot rate falls below $0.38, then GE will convert the proceeds at $0.38 and the bank suffers a loss.

Exhibit 9.6 shows that with the range forward, GE has effectively collared its exchange risk (hence the term, currency collar). Exhibit 9.6a shows the payoff profile of the DM receivable. Exhibit 9.6b shows the payoff profile for the currency collar. Exhibit 9.6c shows the payoff profile for GE's receivable hedged with the collar. With the collar, GE is guaranteed a minimum cash flow of 25,000,000 × $0.38, or $9,500,000. Its maximum cash flow with the collar is $10,500,000, which it receives for any exchange rate beyond $0.42. For exchange rates within the range, it receives 25,000,000 × (actual spot rate).

Why would GE accept a contract that limits its upside potential? In order to lower its cost of hedging its downside risk. The cost saving can be seen by recognizing that a currency collar can be created by simultaneously buying an out-of-the-money put option and selling an out-of-the-money call option of the same size. In effect, the purchase of the put option is financed by the sale of the call option. By selling off the upside potential with the call option, GE can reduce the cost of hedging its downside risk with the put option. The payoff profile of the combined put purchase and call sale, also known as a **cylinder**, is shown in Exhibit 9.7. By adjusting the strike prices such that the put premium just equals the call premium, you can always create a cylinder with a zero net cost, in which case you have a range forward. In this exhibit, it is assumed that the put premium at a strike price of $0.39 just equals the call premium at a strike price of $0.41.

Cross-Hedging

Hedging with futures is very similar to hedging with forward contracts. However, a firm that wants to manage its exchange risk with futures may find that the exact futures contract it requires is unavailable. In this case, it may be able to **cross-hedge** its exposure by using futures contracts on another currency that is correlated with the one of interest.

The idea behind cross-hedging is as follows: If we cannot find a futures/forward contract on the currency in which we have an exposure, we will hedge our exposure via a

EXHIBIT 9.7 Use of a Currency Cylinder to Hedge GE's Receivable

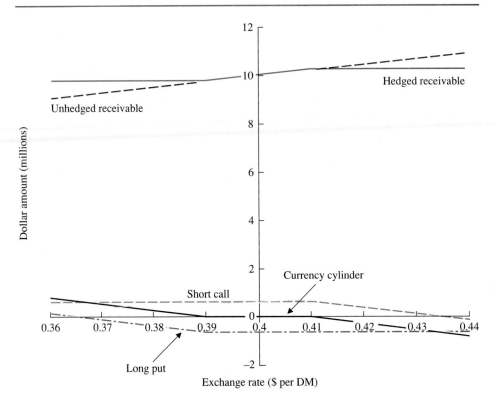

futures/forward contract on a related currency. Lacking a model or theory to tell us the exact relationship between the exchange rates of the two related currencies, we estimate the relationship by examining the historical association between these rates. The resulting regression coefficient tells us the sign and approximate size of the futures/forward position we should take in the related currency. However, the cross-hedge is only as good as the stability and economic significance of the correlation between the two currencies. A key output of the regression equation, such as the one between the Belgian franc and DM, is the R^2, which measures the fraction of variation in the exposed currency that is explained by variation in the hedging currency. In general, the greater the R^2 of the regression of one exchange rate on the other, the better the cross-hedge will be.

Foreign Currency Options

Thus far, we have examined how firms can hedge known foreign currency transaction exposures. Yet, in many circumstances, the firm is uncertain whether the hedged foreign currency cash inflow or outflow will materialize. For example, the previous assumption was that GE learned on January 1 that it had won a contract to supply turbine blades to Lufthansa. But suppose that, although GE's bid on the contract was submitted on January

Hedging a Belgian Franc Exposure Using Deutsche Mark Futures

An exporter with a receivable denominated in Belgian francs will not find Belgian franc futures available. Although an exact matching futures contract is unavailable, the firm may be able to find something that comes close. The exporter can cross-hedge his Belgian franc position with German mark futures, as the dollar values of those currencies tend to move in unison.

Suppose it is October 15 and our exporter expects to collect a BF 5 million receivable on December 15. The exporter can always sell the Belgian francs on the spot market at that time but is concerned about a possible fall in the franc's value between now and then. The exporter's treasurer has copied the spot prices of the Belgian franc and Deutsche mark from the *Wall Street Journal* every day for the past three months and has estimated the following regression relationship using this information:

$$\Delta BF/\$ = .8(\Delta DM/\$)$$

where $\Delta = e_t - e_{t-1}$ and e_t is the spot rate for day t (that is, Δ is the change in the exchange rate). In addition, the R^2 of the regression is 0.91, meaning that 91% of the variation in the Belgian franc is explained by movements in the DM. With an R^2 this high, the exporter can confidently use DM futures contracts to cross-hedge the Belgian franc.

According to this relationship, a 1¢ change in the value of the DM leads to a 0.8¢ change in the value of the BF. To cross-hedge the forthcoming receipt of BF, 0.8 units of DM futures must be sold for every unit of BF to be sold on December 15. With a Belgian franc exposure of BF 5 million, the exporter must sell DM futures contracts in the amount of DM 4 million ($0.8 \times 5,000,000$). With a DM futures contract size of DM 125,000, this DM amount translates into 32 contracts (4,000,000/125,000). The example illustrates the idea that the DM futures can be used to effectively offset the risk posed by the BF receivable.

1, the announcement of the winning bid would not be until April 1. During the three-month period from January 1 to April 1, GE does not know if it will receive a payment of DM 25 million on December 31. This uncertainty has important consequences for the appropriate hedging strategy.

GE would like to guarantee that the exchange rate does not move against it between the time it bids and the time it gets paid, should it win the contract. The danger of not hedging is that its bid will be selected and the Deutsche mark will decline in value, possibly wiping out GE's anticipated profit margin. For example, if the forward rate on April 1 for delivery December 31 falls to DM 1 = $0.36, the value of the contract will drop from $9.57 million to $9 million, for a loss in value of $570,000.

The apparent solution is for GE to sell the anticipated DM 25 million receivable forward on January 1. However, if GE does that and loses the bid on the contract, it still has to sell the currency—which it will have to get by buying on the open market, perhaps at a big loss. For example, suppose the forward rate on April 1 for December 31 delivery has risen to $0.4008. To eliminate all currency risk on its original forward contract, GE would have to buy DM 25 million forward at a price of $0.4008. The result would be a loss of $450,000 [(0.3828 − 0.4008) × 25,000,000] on the forward contract entered into on January 1 at a rate of $0.3828.

Until recently, GE, or any company that bid on a contract denominated in a foreign currency and was not assured of success, would be unable to resolve its foreign exchange risk dilemma. The advent of **currency options** has changed all that. Specifically, the solution to managing its currency risk in this case is for GE, at the time of its bid, to purchase an option to sell DM 25 million on December 31. For example, suppose that on January 1, GE can buy for $100,000 the right to sell Citibank DM 25 million on December 31 at a price of $0.3828/DM. If it enters into this put option contract with Citibank, GE will guarantee itself a minimum price ($9.57 million) should its bid be selected, while simultaneously ensuring that if it lost the bid, its loss would be limited to the price paid for the option contract (the premium of $100,000). Should the spot price of the Deutsche mark on December 31 exceed $0.3828, GE would let its option contract expire unexercised and convert the DM 25 million at the prevailing spot rate.

Instead of a straight put option, GE could use a futures put option. This would entail GE buying a put option on a December futures contract with the option expiring in April. If the put were in-the-money on April 1, GE would exercise it and receive a short position in a DM futures contract plus a cash amount equal to the strike price minus the December futures price as of April 1. Assuming it had won the bid, GE would hold on to the December futures contract. If it had lost the bid, GE would pocket the cash and immediately close out its short futures position at no cost.

As we saw in Chapter 6, two types of options are available to manage exchange risk. A **currency put option**, such as the one appropriate to GE's situation, gives the buyer the right, but not the obligation, to sell a specified number of foreign currency units to the option seller at a fixed dollar price, up to the option's expiration date. Alternatively, a **currency call option** is the right, but not the obligation, to buy the foreign currency at a specified dollar price, up to the expiration date.

A call option is valuable, for example, when a firm has offered to buy a foreign asset, such as another firm, at a fixed foreign currency price but is uncertain whether its bid will be accepted. By buying a call option on the foreign currency, the firm can lock in a maximum dollar price for its tender offer, while limiting its downside risk to the call premium in the event its bid is rejected.

Currency options are a valuable risk-management tool in other situations as well. Conventional transaction-exposure management says you wait until your sales are booked or your orders placed before hedging them. If a company does that, however, it faces potential losses from exchange rate movements because the foreign currency price does not necessarily adjust right away to changes in the value of the dollar. As a matter of policy, to avoid confusing customers and salespeople, most companies do not change their price list every time the exchange rate changes. Unless and until the foreign currency price changes, the unhedged company may suffer a decrease in its profit margin. Because of the uncertainty of anticipated sales or purchases, however, forward contracts are an imperfect tool to hedge the exposure.

For example, a company that commits to a foreign currency price list for, say, three months has a foreign currency exposure that depends on the unknown volume of sales at those prices during this period. Thus, the company does not know what volume of forward contracts to enter into to protect its profit margin on these sales. For the price of the premium, currency put options allow the company to insure its profit margin against adverse movements in the foreign currency while guaranteeing fixed prices to foreign customers. Without options, the firm might be forced to raise its foreign currency prices sooner than the competitive situation warranted.

ILLUSTRATION

HEWLETT-PACKARD USES CURRENCY OPTIONS TO PROTECT ITS PROFIT MARGINS

Hewlett-Packard (H-P), the California-based computer firm, uses currency options to protect its dollar profit margins on products built in the United States but sold in Europe. The firm needs to be able to lower local currency (LC) prices if the dollar weakens and hold LC prices steady for about three months (the price adjustment period) if the dollar strengthens.

Suppose H-P sells anticipated DM sales forward at DM 2.5/$ to lock in a dollar value for those sales. If one month later the dollar weakens to DM 2/$, H-P faces tremendous competitive pressure to lower its Deutsche mark prices. H-P would be locked into a loss on the forward contracts that would not be offset by a gain on its sales because it had to cut DM prices. With DM put options, H-P would just let them expire, and it would lose only the put premium. Conversely, options help H-P delay LC price increases when the dollar strengthens until it can raise them without suffering a competitive disadvantage. The reduced profit margin on local sales is offset by the gain on the put option.

Currency options also can be used to hedge exposure to shifts in a competitor's currency. Companies competing with firms from other nations may find their products at a price disadvantage if a major competitor's currency weakens, allowing the competitor to reduce its prices. Thus, the company will be exposed to fluctuations in the competitor's currency even if it has no sales in that currency. For example, a Swiss engine manufacturer selling in Germany will be placed at a competitive disadvantage if dollar depreciation allows its principal competitor, located in the United States, to sell at a lower price in Germany. Purchasing out-of-the-money put options on the dollar and selling them for a profit if they move into the money (which will happen if the dollar depreciates enough) will allow the Swiss firm to partly compensate for its lost competitiveness. The exposure is not contractually set, so forward contracts again are not as useful as options in this situation.

The ideal use of forward contracts is when the exposure has a straight risk-reward profile: Forward contract gains or losses are exactly offset by losses or gains on the underlying transaction. If the transaction exposure is uncertain, however, because the volume or the foreign currency prices of the items being bought or sold are unknown, a forward contract will not match it. By contrast, currency options are a good hedging tool in situations in which the quantity of foreign exchange to be received or paid out is uncertain.

How Cadbury Schweppes Uses Currency Options

Cadbury Schweppes, the British candy manufacturer, uses currency options to hedge uncertain payables. The price of its key product input, cocoa, is quoted in sterling but is really a dollar-based product. That is, as the value of the dollar changes, the sterling price of cocoa changes as well. The objective of the company's foreign exchange strategy is to eliminate the currency element in the decision to purchase the commodity, thus leaving the company's buyers able to concentrate on fundamentals. However, this task is complicated by the fact that the company's projections of its future purchases are highly uncertain.

As a result, Cadbury Schweppes has turned to currency options. After netting its total exposure, the company covers with forward contracts a base number of exposed, known payables. It covers the remaining—uncertain—portion with options. The options act as an insurance policy.

A company could, of course, use currency options to hedge its exposure in lieu of forward contracts. However, each type of hedging instrument is more advantageous in some situations and it makes sense to match the instrument to the specific situation. The general rules to follow when choosing between currency options and forward contracts for hedging purposes are summarized as follows:

1. When the quantity of a foreign currency cash outflow is known, buy the currency forward; when the quantity is unknown, buy a call option on the currency.
2. When the quantity of a foreign currency cash inflow is known, sell the currency forward; when the quantity is unknown, buy a put option on the currency.
3. When the quantity of a foreign currency cash flow is partially known and partially uncertain, use a forward contract to hedge the known portion and an option to hedge the maximum value of the uncertain remainder.[7]

These rules presume that the financial manager's objective is to reduce risk and not to speculate on the direction or volatility of future currency movements. They also presume that both forward and options contracts are fairly priced. In an efficient market, the expected value or cost of either of these contracts should be zero. Any other result would introduce the possibility of arbitrage profits. The presence of such profits would attract arbitrageurs as surely as bees are attracted to honey. Their subsequent attempts to profit from inappropriate prices would return these prices to their equilibrium values.

A Hedging Caveat

It is worth noting that at times it may be better not to hedge at all. Hedging can lock in a company's dollar cost, but doing so may place the hedged company at a competitive disadvantage and increase its risk. For example, suppose that Trader Joe buys 4,000 bottles

[7] For elaboration, see Ian H. Giddy, "The Foreign Exchange Option as a Hedging Tool," *Midland Corporate Finance Journal*, Fall 1983, pp. 32–42.

of French champagne to be delivered and paid for in 90 days. The price is FF 50 per bottle. At the current spot rate of FF 1 = $0.19, the price is equivalent to $9.50. If the 90-day forward rate is $0.1850, Trader Joe can lock in a price of $9.25 per bottle.

But suppose that, after Trader Joe buys French francs forward to pay for its purchase, the franc depreciates to $0.16 and the price of French champagne remains at FF 50 per bottle. Trader Joe will now be facing competition from other wine importers whose cost per bottle of French champagne is only $8.00–$1.25 below Trader Joe's cost. This competitive pressure will drive down the price at which Trader Joe can sell its French champagne. Thus, if it hedges its future purchases of French wine, Trader Joe's dollar profit margin will be hurt by a French franc depreciation. Of course, it will also benefit from an appreciation of the French franc. The important point, though, is that hedging will increase the variability – and the risk – of Trader Joe's profit margin. Hedging will fix Trader Joe's dollar cost, but its dollar price will vary in line with the dollar value of the French franc. If Trader Joe does not hedge, its dollar cost and dollar revenue will fluctuate in unison, preserving a relatively constant dollar margin.

ILLUSTRATION   **S**HOWA **S**HELL **M**ISCALCULATES
In 1992, Royal Dutch/Shell discovered that its Japanese affiliate, Showa Shell, had lost almost $1.6 billion on more than $6 billion of forward contracts. Showa Shell, an oil refiner, purchased the forward contracts to hedge its dollar expenses for oil by locking in its yen costs. When the dollar's value fell by almost a third against the yen, Showa Shell had locked in a large loss on every dollar bought forward. Ordinarily, this loss on the forward contracts should have been offset by the higher profits it earned on sales of the refined products made out of the lower-cost (in yen terms) oil. Unfortunately for Showa Shell, competitive pressures drove down the yen price of refined products in line with the fall in the yen cost of oil.

9.3 MANAGING TRANSLATION EXPOSURE

Firms have three available methods for managing their translation exposure: (1) adjusting fund flows, (2) entering into forward contracts, and (3) exposure netting. The basic hedging strategy for reducing translation exposure shown in Exhibit 9.8 uses these methods. Essentially, the strategy involves increasing **hard-currency** (likely to appreciate) assets and decreasing **soft-currency** (likely to depreciate) assets, while simultaneously decreasing hard-currency liabilities and increasing soft-currency liabilities. For example, if a devaluation appears likely, the basic hedging strategy would be executed as follows: Reduce the level of cash, tighten credit terms to decrease accounts receivable, increase LC borrowing, delay accounts payable, and sell the weak currency forward.

Despite their prevalence among firms, however, these hedging activities are not automatically valuable. If the market already recognizes the likelihood of currency appreciation or depreciation, this recognition will be reflected in the costs of the various

EXHIBIT 9.8 Basic Strategy for Hedging Translation Exposure		
	ASSETS	LIABILITIES
Hard currencies (Likely to appreciate)	Increase	Decrease
Soft currencies (Likely to depreciate)	Decrease	Increase

hedging techniques. Only if the firm's anticipations differ from the market's and are also superior to the market's can hedging lead to reduced costs. Otherwise, the principal value of hedging would be to protect a firm from unforeseen currency fluctuations.

Funds Adjustment

Most techniques for hedging an impending local currency (LC) devaluation reduce LC assets or increase LC liabilities, thereby generating LC cash. If accounting exposure is to be reduced, these funds must be converted into hard-currency assets. For example, a company will reduce its translation loss if, before an LC devaluation, it converts some of its LC cash holdings to the home currency. This conversion can be accomplished, either directly or indirectly, by means of funds adjustment techniques.

Funds adjustment involves altering either the amounts or the currencies (or both) of the planned cash flows of the parent or its subsidiaries to reduce the firm's local currency accounting exposure. If an LC devaluation is anticipated, direct funds-adjustment methods include pricing exports in hard currencies and imports in the local currency, investing in hard-currency securities, and replacing hard-currency borrowings with local currency loans. The indirect methods, which will be elaborated on in Chapter 14, include adjusting transfer prices on the sale of goods between affiliates; speeding up the payment of dividends, fees, and royalties; and adjusting the leads and lags of intersubsidiary accounts. The last method, which is the one most frequently used by multinationals, involves speeding up the payment of intersubsidiary accounts payable and delaying the collection of intersubsidiary accounts receivable. These hedging procedures for devaluations would be reversed for revaluations (see Exhibit 9.1).

Some of these techniques or tools may require considerable lead time and—as is the case with a transfer price—once they are introduced, they cannot easily be changed. In addition, techniques such as transfer price, fee and royalty, and dividend flow adjustments fall into the realm of corporate policy and are not usually under the treasurer's control, although this situation may be changing. It is, therefore, incumbent on the treasurer to educate other decision makers about the impact of these tools on the costs and management of corporate exposure.

Although entering forward contracts is the most popular coverage technique, leading and lagging of payables and receivables is almost as important. For those countries in which a formal market in LC forward contracts does not exist, leading and lagging and LC borrowing are the most important techniques. The bulk of international business, however, is conducted in those few currencies for which forward markets do exist.

Forward contracts can reduce a firm's translation exposure by creating an offsetting asset or liability in the foreign currency. For example, suppose that IBM U.K. has translation exposure of £40 million (that is, sterling assets exceed sterling liabilities by that amount). IBM U.K. can eliminate its entire translation exposure by selling £40 million forward. Any loss (gain) on its translation exposure will then be offset by a corresponding gain (loss) on its forward contract. Note, however, that the gain (or loss) on the forward contract is of a cash-flow nature and is netted against an unrealized translation loss (or gain).

Selecting convenient (less-risky) currencies for invoicing exports and imports and adjusting transfer prices are two techniques that are less frequently used, perhaps because of constraints on the use of those techniques. It is often difficult, for instance, to make a customer or supplier accept billing in a particular currency.

Exposure netting is an additional exchange-management technique that is available to multinational firms with positions in more than one foreign currency or with offsetting positions in the same currency. As defined earlier, this technique involves offsetting exposures in one currency with exposures in the same or another currency such that gains and losses on the two currency positions will offset each other.

Evaluating Alternative Hedging Mechanisms

Ordinarily, the selection of a funds-adjustment strategy cannot proceed by evaluating each possible technique separately without risking suboptimization; for example, whether a firm chooses to borrow locally is not independent of its decision to use or not use those funds to import additional hard-currency inventory. However, where the level of forward contracts that the financial manager can enter into is unrestricted, the following two-stage methodology allows the optimal level of forward transactions to be determined apart from the selection of what funds-adjustment techniques to use.[8] Moreover, this methodology is valid regardless of the manager's (or firm's) attitude toward risk.

Stage 1: Compute the profit associated with each funds-adjustment technique on a covered, after-tax basis. Transactions that are profitable on a covered basis ought to be undertaken regardless of whether they increase or decrease the firm's accounting exposure. However, such activities should not be termed *hedging*; rather, they involve the use of *arbitrage* to exploit market distortions.

Stage 2: Any unwanted exposure resulting from the first stage can be corrected in the forward market. Stage 2 is the selection of an optimal level of forward transactions based on the firm's initial exposure, adjusted for the impact on exposure of decisions made in Stage 1. Where the forward market is nonexistent, or where access to it is limited, the firm must determine both what techniques to use and what their appropriate levels are. In the latter case, a comparison of the net cost of a funds-adjustment technique with the anticipated currency depreciation will indicate whether the hedging transaction is profitable on an expected-value basis.

---

[8] This methodology is presented in William R. Folks, Jr., "Decision Analysis for Exchange Risk Management," *Financial Management*, Winter 1972, pp. 101–112.

9.4 ILLUSTRATION: MANAGING TRANSACTION EXPOSURE FOR THE TORONTO BLUE JAYS

During the first half of the 1985 baseball season, the Toronto Blue Jays had the best won-loss record in the major leagues. Yet their profits at the gate did not match their performance at the plate. Attendance was up, and so were ticket prices, but the Blue Jays budgeted for a loss of more than C$2 million in 1985. The reason: The Blue Jays get most of their revenue in the form of Canadian dollars (C$) but pay most of their bills in U.S. dollars.

Projected 1985 expenses included about $19 million in U.S. dollars and the equivalent of only about $4.5 million in Canadian currency. Projected revenues of roughly $21 million were almost all in Canadian dollars except for income from a U.S. television package and 20% of gate receipts from the Jays' games in U.S. ballparks. As a result of this imbalance of currency inflows and outflows, it was estimated that each $0.01 drop of the Canadian dollar against its U.S. counterpart cost the Jays about C$135,000 in 1985.

Although major-league teams usually lose money, it was believed that the Jays' ecstatic fans would have made the team profitable in 1985 if it were not for the currency problem. The magnitude of this problem is indicated by the changed fortune of the Canadian dollar. When the Toronto franchise was created in 1976, the Canadian currency was worth $1.04; at midseason 1985, it was trading at about $0.73.

The biggest expense in 1985 was U.S.$10 million for players' salaries. (All major-league ballplayers are paid in U.S. dollars; none of the Jays' players are Canadians anyway.) By 1989, that figure had reached about U.S.$18 million. At an exchange rate of $0.80, a 1 cent drop in the Canadian dollar added C$285,000 to annual salary costs.

Like other businesses with foreign exchange problems, the Blue Jays and their fellow sufferers, the Montreal Expos, make forward purchases of U.S. dollars to protect against swings in exchange rates. Late in 1984, for example, the Jays contracted to buy about 60% of the team's projected 1985 U.S. currency needs at about 75 cents per Canadian dollar. The profit on this position enabled the team to offset most of the losses on its U.S. dollar outflows.

❧ 9.5 SUMMARY AND CONCLUSIONS

Hedging exchange risk is a complicated and difficult task. As a first step, the firm must specify an operational set of goals for those involved in exchange risk management. Failure to do so can lead to possibly conflicting and costly actions on the part of employees. We saw that the hedging objective that is most consistent with the overarching objective of maximizing shareholder value is to reduce exchange risk, where exchange risk is defined as that element of cash-flow variability attributable to currency fluctuations. This objective translates into the following exposure management goal: to arrange a firm's financial affairs in such a way that however the exchange rate may move in the future, the effects on dollar returns are minimized.

We saw that firms normally cope with anticipated currency changes by engaging in forward contracts, borrowing locally, and adjusting their pricing and credit policies. However, there is reason to question the value of much of this activity. In fact, we have seen that, in normal circumstances, hedging cannot provide protection against expected exchange rate changes.

A number of empirical studies indicate that, on average, the forward rate appears to be an unbiased estimate of the future spot rate. On the other hand, the evidence also points to the possibility of bias in the forward rate at any point in time. However, trying to take advantage of this apparent bias via selective hedging is likely to expose the company to increased risk. Furthermore, according to the international Fisher effect, in the absence of government controls, interest rate differentials among countries should equal anticipated currency devaluations or revaluations. Empirical research substantiates that over time, gains or losses on debt in hard currencies tend to be offset by low interest rates; in soft currencies, they will be offset by higher interest rates unless, of course, there are barriers that preclude equalization of real interest rates. Again, to the extent that bias exists in the interest rate differential—owing to a risk premium or other factor—the risk associated with selective hedging is likely to offset any expected gains.

In fact, no other results would be consistent with the existence of a well-informed market with numerous participants—as is represented by the international financial community. Persistent differences between forward and future spot rates, for instance, would provide profitable opportunities for speculators. However, the very act of buying or selling forward to take advantage of these differences would tend to bring about equality between hedging costs and expected currency changes (taking risk into account).

The other hedging methods, which involve factoring anticipated exchange rate changes into pricing and credit decisions, can be profitable only at the expense of others. Thus, to consistently gain by these trade-term adjustments, it is necessary to deal continuously with less-knowledgeable people. Certainly, though, a policy predicated on the continued existence of naive firms is unlikely to be viable for very long in the highly competitive and well-informed world of international business. The real value to a firm of factoring currency change expectations into its pricing and credit decisions is to prevent others from profiting at its expense.

The basic value of hedging, therefore, is to protect a company against unexpected exchange rate changes; however, by definition, these changes are unpredictable and, consequently, impossible to profit from. Of course, to the extent that a government does not permit interest or forward rates to fully adjust to market expectations, a firm with access to these financial instruments can expect, on average, to gain from currency changes. Nevertheless, the very nature of these imperfections severely restricts a company's ability to engage in such profitable financial operations.

➢ Questions

1. A U.S. firm has fully hedged its sterling receivables and has bought credit insurance to cover the risk of default. Has this firm eliminated all risk on these receivables? Explain.

2. What is the basic translation hedging strategy? How does it work?

3. What alternative hedging transactions are available to a company seeking to hedge the translation exposure of its German subsidiary?

4. Referring to Question 3, how would the appropriate hedge change if the German affiliate's functional currency is the U.S. dollar?

5. Multinational firms can always reduce the foreign exchange risk faced by their foreign affiliates by borrowing in the local currency. True or false? Why?

6. Can hedging provide protection against expected exchange rate changes? Explain.

7. What is the domestic counterpart to exchange risk? Explain.

8. If a currency that a company is long in threatens to weaken, many companies will sell that currency forward. Comment on this policy.

9. In order to eliminate all risk on its exports to Japan, a company decides to hedge both its actual and anticipated sales there. To what risk is the company exposing itself? How could this risk be managed?

10. Instead of its previous policy of always hedging its foreign currency receivables, Sun Microsystems has decided to hedge only when it believes the dollar

will strengthen. Otherwise, it will go uncovered. Comment on this new policy.

11. Studies have shown that in trade dealings between nations that have high and volatile inflation rates, most export prices are quoted in dollars. What might account for this finding?

12. Your bank is working with an American client who wishes to hedge its long exposure in the Malaysian ringgit. Suppose it is possible to invest in ringgit but not borrow in that currency. However, you can both borrow and lend in U.S. dollars.
    a. Assuming there is no forward market in ringgit, can you create a homemade forward contract that would allow your client to hedge its ringgit exposure?
    b. Several of your Malaysian clients are interested in selling their U.S. dollar export earnings forward for ringgit. Can you accommodate them by creating a forward contract?

13. Eastman Kodak gives its traders bonuses if their selective hedging strategies are less expensive than the cost of hedging all their transaction exposure on a continuous basis. What problems can you foresee from this bonus plan?

14. Many managers prefer to use options to hedge their exposure because it allows them the possibility of capitalizing on favorable movements in the exchange rate. In contrast, a company using forward contracts avoids the downside but also loses the upside potential as well. Comment on this strategy.

15. In January 1988, Arco bought a 24.3% stake in the British oil firm Britoil PLC. It intended to buy a further $1 billion worth of Britoil stock if Britoil was agreeable. However, Arco was uncertain whether Britoil, which had expressed a strong desire to remain independent, would accept its bid. To guard against the possibility of a pound appreciation in the interim, Arco decided to convert $1 billion into pounds and place them on deposit in London, pending the outcome of its discussions with Britoil's management. What exchange risk did Arco face, and did it choose the best way to protect itself from that risk?

16. Sumitomo Chemical of Japan has one week in which to negotiate a contract to supply products to a U.S. company at a dollar price that will remain fixed for one year. What advice would you give Sumitomo?

17. Kemp & Beatley, Inc., is a New York importer of table linens and accessories. It hedges all its import orders using forward contracts. Does Kemp & Beatley face any exchange risk? Explain.

18. U.S. Farm-Raised Fish Trading Co., a catfish concern in Jackson, Mississippi, tells its Japanese customers that it wants to be paid in dollars. According to its director of export marketing, this simple strategy eliminates all its currency risk. Is he right? Why?

19. The Montreal Expos are a major-league baseball team located in Montreal, Canada. What currency risk is faced by the Expos, and how can this exchange risk be managed?

20. General Electric recently had to put together a $50 million bid, denominated in Swiss francs, to upgrade a Swiss power plant. If it won, GE expected to pay subcontractors and suppliers in five currencies. The payment schedule for the contract stretched over a five-year period.
    a. How should General Electric establish the Swiss franc price of its $50 million bid?
    b. What exposure does GE face on this bid? How can it hedge that exposure?

21. Because Liz Claiborne contracts out much of its production to foreign manufacturers, the company faces currency risk.
    a. What currency risk does Liz Claiborne face?
    b. How might Liz Claiborne go about hedging its currency risk?
    c. What danger does it face from locking in currency rates today?

22. Dell Computer produces its machines in Asia with components largely imported from the United States and sells its products in various Asian nations in local currencies.
    a. What is the likely impact on Dell's Asian profits of a strengthened dollar? Explain.
    b. What hedging technique(s) can Dell employ to lock in a desired currency conversion rate for its Asian sales during the next year?
    c. Suppose Dell wishes to lock in a specific conversion rate but does not want to foreclose the possibility of profiting from future currency moves. What hedging technique would be most likely to achieve this objective?
    d. What are the limits of Dell's hedging approach?

Problems

1. Walt Disney expects to receive a Mex$16 million theatrical fee from Mexico in 90 days. The current spot rate is $0.1321/Mex$ and the 90-day forward rate is $0.1242/Mex$.

a. What is Disney's peso transaction exposure associated with this fee?

b. If the spot rate expected in 90 days is $0.1305, what is the expected U.S. dollar value of the fee?

c. What is the hedged dollar value of the fee?

2. An importer has a payment of £8 million due in 90 days.

   a. If the 90-day pound forward rate is $1.4201, what is the hedged cost of making that payment?

   b. If the spot rate expected in 90 days is $1.4050, what is the expected cost of payment?

   c. What factors will influence the hedging decision?

3. A foreign exchange trader assesses the French franc exchange rate three months hence as follows:

   $0.11 with probability 0.25

   $0.13 with probability 0.50

   $0.15 with probability 0.25

   The 90-day forward rate is $0.12.

   a. Will the trader buy or sell French francs forward against the dollar if she is concerned solely with expected values? In what volume?

   b. In reality, what is likely to limit the trader's speculative activities?

   c. Suppose the trader revises her probability assessment as follows:

   $0.09 with probability 0.33

   $0.13 with probability 0.33

   $0.17 with probability 0.33

   If the forward rate remains at $0.12, will this new assessment affect the trader's decision? Explain.

4. International Worldwide would like to execute a money market hedge to cover a ¥250,000,000 shipment from Japan of sound systems it will receive in six months. The current exchange rate for yen is ¥124/$.

   a. How would International structure the hedge? What would it do to hedge the Japanese yen it must pay in six months? The annual yen interest rate is 4%.

   b. The yen may rise to as much as ¥140/$ or fall to ¥115/$. What will the total dollar cash flow be in six months in either case?

5. An investment manager hedges a portfolio of Bunds (German government bonds) with a six-month forward contract. The current spot rate is DM 1.64/$ and the 180-day forward rate is DM 1.61/$. At the end of the six-month period, the Bunds have risen in value by 3.75% (in DM terms), and the spot rate is now DM 1.46/$.

   a. If the Bunds earn interest at the annual rate of 5%, paid semiannually, what is the investment manager's total dollar return on the hedged Bunds?

   b. What would the return on the Bunds have been without hedging?

   c. What was the true cost of the forward contract?

6. A French corporate treasurer expects to receive a DM 11 million payment in 90 days from a German customer. The current spot rate is DM 0.29870/FF, and the 90-day forward rate is DM 0.29631/FF. In addition, the annualized three-month EuroDM and Eurofranc (French) rates are 9.8% and 12.3%, respectively.

   a. What is the hedged value of the DM receivable using the forward contract?

   b. Describe how the French treasurer could use a money market hedge to lock in the franc value of the DM receivable. What is the hedged value of the DM receivable? What is the effective forward rate that the treasurer can obtain using this money market hedge?

   c. Given your answers in parts a and b, is there an arbitrage opportunity? How could the treasurer take advantage of it?

   d. At what 90-day forward rate would interest rate parity hold?

7. Magnetronics, Inc., a U.S. company, owes its Taiwanese supplier NT$205 million in three months. The company wishes to hedge its NT$ payable. The current spot rate is NT$1 = U.S.$0.03987, and the three-month forward rate is NT$1 = U.S.$0.04051. Magnetronics can also borrow or lend U.S. dollars at an annualized interest rate of 12% and Taiwanese dollars at an annualized interest rate of 8%.

   a. What is the U.S. dollar accounting entry for this payable?

   b. What is the minimum U.S. dollar cost that Magnetronics can lock in for this payable? Describe the procedure it would use to get this price.

8. Cooper Inc., a U.S. firm, has just invested £500,000 in a note that will come due in 90 days and is yielding 9.5% annualized. The current spot value of the pound is $1.5612 and the 90-day forward rate is $1.5467.

   a. What is the hedged dollar value of this note at maturity?

   b. What is the annualized dollar yield on the hedged note?

   c. Cooper anticipates that the value of the pound in 90 days will be $1.5550. Should it hedge? Why or why not?

   d. Suppose that Cooper has a payable of £980,000 coming due in 180 days. Should this affect its decision of whether to hedge its sterling note? How and why?

9. Plantronics owes SKr 50 million, due in one year, for some electrical equipment it recently bought from ABB Asea Brown Boveri. At the current spot rate of $0.1480/SKr, this payable is $7.4 million. It wishes to hedge this payable but is undecided about how to do it. The one-year forward rate is currently $0.1436. Plantronics' treasurer notes that the company has $10 million in a marketable U.S. dollar CD yielding 7% per annum. At the same time, SE Banken in Stockholm is offering a one-year time deposit rate of 10.5%.
   a. What is the low-cost hedging alternative for Plantronics? What is the cost?
   b. Suppose interest rate parity held. What would the one-year forward rate be?

10. DKNY owes Ptas 70 million in 30 days for a recent shipment of Spanish textiles. It faces the following interest and exchange rates:

Spot rate:	Ptas 130/$
Forward rate (30 days)	Ptas 131/$
30-day put option on dollars at Ptas 131/$	1% premium
30-day call option on dollars at Ptas 129/$	3% premium
U.S. dollar 30-day interest rate (annualized):	7.5%
Peseta 30-day interest rate (annualized):	15%

   a. What is the hedged cost of DKNY's payable using a forward market hedge?
   b. What is the hedged cost of DKNY's payable using a money market hedge?
   c. What is the hedged cost of DKNY's payable using a put option?
   d. At what exchange rate is the cost of the put option just equal to the cost of the forward market hedge? to the cost of the money market hedge?
   e. How can DKNY construct a currency collar? What is the net premium paid for the currency collar? Using this currency collar, what is the net dollar cost of the payable if the spot rate in 30 days is Ptas 128/$? Ptas 131/$? Ptas 134/$?
   f. What is the preferred alternative?
   g. Suppose that DKNY expects the 30-day spot rate to be Ptas 134/$. Should it hedge this payable? What other factors should go into DKNY's hedging decision?

11. Dow Chemical has sold SFr 25 million in chemicals to Ciba-Geigy. Payment is due in 180 days. Suppose Dow faces the following interest and exchange rates:

Spot rate:	$0.7957/SFr
180-day forward rate	$0.8095/SFr
180-day U.S. dollar interest rate (annualized)	5.25%
180-day Swiss franc interest rate (annualized)	1.90%
180-day call option at $0.80/SFr	2% premium
180-day put option at $0.80/SFr	1% premium

   a. What is the hedged value of Dow's receivable using the forward market hedge? the money market hedge?
   b. What alternatives are available to Dow to use currency options to hedge its receivable? Which option hedging strategy would you recommend?
   c. Which of the hedging alternatives analyzed in parts (a) and (b) would you recommend to Dow? Why?

12. American Airlines is trying to decide how to go about hedging DM 70 million in ticket sales receivable in 180 days. Suppose it faces the following exchange and interest rates.

Spot rate:	$0.6433–42/DM
Forward rate (180 days):	$0.6578–99/DM
DM 180-day interest rate (annualized):	4.01%–3.97%
U.S. dollar 180-day interest rate (annualized):	8.01%–7.98%

   a. What is the hedged value of American's ticket sales using a forward market hedge?
   b. What is the hedged value of American's ticket sales using a money market hedge? Assume the first interest rate is the rate at which money can be borrowed and the second one the rate at which it can be lent.
   c. Which hedge is less expensive?
   d. Is there an arbitrage opportunity here?
   e. Suppose the expected spot rate in 180 days is $0.67/DM, with a most likely range of $0.64–$0.70/DM. Should American hedge? What factors should enter into its decision?

13.* Metallgesellschaft, a leading German metal processor, has scheduled a supply of 20,000 metric tons of copper for October 1. On April 1,

---

copper is quoted on the London Metals Exchange at £562 per metric ton for immediate delivery and £605 per metric ton for delivery on October 1. Monthly storage costs are £10 for a metric ton in London and DM 30 in Hamburg, payable on the first day of storage. Exchange rate quotations are as follows: The pound is worth DM 3.61 on April 1 and is selling at a 6.3% annualized forward discount. The opportunity cost of capital for Metalgesselschaft is estimated at 8% annually, and the pound sterling is expected to depreciate at a yearly rate of 6.3% throughout the next 12 months. Compute the Deutsche mark cost for Metalgesselschaft on April 1 of the following options:

a. Buy 20,000 metric tons of copper on April 1 and store it in London until October 1.
b. Buy a forward contract of 20,000 metric tons on April 1, for delivery in six months. Cover sterling debt by purchasing forward pounds sterling on April 1.
c. Buy 20,000 metric tons of copper on October 1.
d. Can you identify other options available to Metalge·selschaft? Which one would you recommend?

14. Cosmo, a Japanese exporter, wishes to hedge its $15 million in dollar receivables coming due in 60 days. In order to reduce its net cost of hedging to zero, however, Cosmo sells a 60-day dollar call option for $20 million with a strike price of ¥98/$ and uses the premium of $314,000 to buy a 60-day $15 million put option at a strike price of ¥90/$.
a. Graph the payoff on Cosmo's hedged position over the range ¥80/$–¥110/$. What risk is Cosmo subjecting itself to with this option hedge?
b. What is the net yen value of Cosmo's option hedged position at the following future spot rates: ¥85/$, ¥95/$, and ¥105/$?
c. As an alternative to using options, Cosmo could have hedged with a 60-day forward contract at a price of ¥97/$. What would be the yen value of Cosmo's hedged receivable if it had used a forward contract to hedge?
d. At what exchange rate would the hedged value of Cosmo's dollar receivables be the same whether it used the option hedge or forward hedge?

15. Madison Inc. imports olive oil from Greek firms and the invoices are always denominated in drachma (Dr). It currently has a payable in the amount of Dr 250 million that it would like to hedge. Unfortunately, there are no drachma futures contracts available and Madison is having difficulty arranging a drachma forward contract. Its treasurer, who recently received her MBA, suggests using Italian lira to cross-hedge the drachma exposure. She recently ran the following regression of the change in the exchange rate for the drachma against the change in the lira exchange rate:

$$\Delta Dr/\$ = 1.6(\Delta Lit/\$)$$

a. Although there is no lira futures contract, there is an active market in forward lira. To cross-hedge Madison's drachma exposure, should the treasurer buy or sell lira forward?
b. What is the risk-minimizing amount of lira that the treasurer would have to buy or sell forward to hedge Madison's drachma exposure?

➤ Bibliography

BISHOP, MATTHEW. "A Survey of Corporate Risk Management." *The Economist*, February 10, 1996, special section.

CORNELL, BRADFORD, and ALAN C. SHAPIRO. "Managing Foreign Exchange Risks." *Midland Corporate Finance Journal*, Fall 1983, pp. 16–31.

DUFEY, GUNTER, and SAM L. SRINIVASULU. "The Case for Corporate Management of Foreign Exchange Risk." *Financial Management*, Summer 1984, pp. 54–62.

EVANS, THOMAS G., and WILLIAM R. FOLKS, JR., Most of these elements are suggested in Thomas G. Evans and William R. Folks, Jr., "Defining Objectives for Exposure Management," *Business International Money Report*, February 2, 1979, pp. 37–39.

GIDDY, IAN H. "The Foreign Exchange Option as a Hedging Tool." *Midland Corporate Finance Journal*, Fall 1983, pp. 32–42.

GOELTZ, RICHARD K. "Managing Liquid Funds on an International Scope." New York: Joseph E. Seagram and Sons, 1971.

SHAPIRO, ALAN C., and DAVID P. RUTENBERG. "Managing Exchange Risks in a Floating World." *Financial Management*, Summer 1976, pp. 48–58.

SRINIVASULU, SAM, and EDWARD MASSURA. "Sharing Currency Risks in Long-Term Contracts." *Business International Money Reports*, February 23, 1987, pp. 57–59.

MEASURING ECONOMIC EXPOSURE

Let's face it. If you've got 75% of your assets in the U.S. and 50% of your sales outside it, and the dollar's strong, you've got problems.

Donald V. Fites
Executive Vice President
Caterpillar Inc.

*C*hapter 8 focused on the accounting effects of currency changes. As we saw in that chapter, the adoption of FASB-52 has helped to moderate the wild swings in the translated earnings of overseas subsidiaries. Nevertheless, the problem of coping with volatile currencies remains essentially unchanged. Fluctuations in exchange rates will continue to have "real" effects on the cash profitability of foreign subsidiaries—complicating overseas selling, pricing, buying, and plant-location decisions.

This chapter develops an appropriate definition of foreign exchange risk. It discusses the nature and origins of exchange risk and presents a theory of the *economic*, as distinguished from the accounting, consequences of currency changes on a firm's value. The chapter also illustrates how economic exposure can be measured and provides an operational measure of exchange risk.

10.1 FOREIGN EXCHANGE RISK AND ECONOMIC EXPOSURE

The most important aspect of foreign exchange risk management is to incorporate currency change expectations into *all* basic corporate decisions. In performing this task, the firm must know what is at risk. However, there is a major discrepancy between accounting practice and economic reality in terms of measuring *exposure*, which is the degree to which a company is affected by exchange rate changes.

As we saw in Chapter 8, those who use an accounting definition of exposure—whether FASB-8, FASB-52, or some other method—divide the balance sheet's assets and liabilities into those accounts that will be affected by exchange rate changes and those that will not. In contrast, economic theory focuses on the impact of an exchange rate change on future cash flows; that is, **economic exposure** is based on the extent to which the value of the firm—as measured by the present value of its expected future cash flows—will change when exchange rates change.

Specifically, if *PV* is the present value of a firm, then that firm is exposed to currency risk if $\Delta PV/\Delta e$ is not equal to zero, where ΔPV is the change in the firm's present value associated with an exchange rate change, Δe. **Exchange risk**, in turn, is defined as the variability in the firm's value that is caused by uncertain exchange rate changes. Thus, exchange risk is viewed as the possibility that currency fluctuations can alter the expected amounts or variability of the firm's future cash flows.

Economic exposure can be separated into two components: transaction exposure and operating exposure. We saw that **transaction exposure** stems from exchange gains or losses on foreign currency-denominated contractual obligations. Although transaction exposure is often included under accounting exposure, as it was in Chapter 8, it is more properly a cash-flow exposure and, hence, part of economic exposure. However, even if the company prices all contracts in dollars or otherwise hedges its transaction exposure, the residual exposure—longer-term operating exposure—still remains.

Operating exposure arises because currency fluctuations can alter a company's future revenues and costs—that is, its operating cash flows. Consequently, measuring a firm's operating exposure requires a longer-term perspective, viewing the firm as an ongoing concern with operations whose cost and price competitiveness could be affected by exchange rate changes.

Thus, the firm faces operating exposure the moment it invests in servicing a market subject to foreign competition or in sourcing goods or inputs abroad. This investment includes new-product development, a distribution network, foreign supply contracts, or production facilities. Transaction exposure arises later on, and only if the company's commitments lead it to engage in foreign currency-denominated sales or purchases. Exhibit 10.1 shows the time pattern of economic exposure.

The measurement of economic exposure is made especially difficult because it is impossible to assess the effects of an exchange rate change without simultaneously considering the impact on cash flows of the underlying relative rates of inflation associated with each currency. Reconsidering the concept of the real exchange rate will help clarify the discussion of exposure. As presented in Chapter 7, the **real exchange rate** is defined as the nominal exchange rate (for example, the number of dollars per franc) adjusted for

EXHIBIT 10.1 The Time Pattern of Economic Exposure

NONCONTRACTUAL	QUASI-CONTRACTUAL	CONTRACTUAL	
Investment in new product development, distribution facilities, brand name, marketing, foreign production capacity, foreign supplier relationships	Quote foreign currency price, receive a foreign currency price quote	Ship product/bill customers in foreign currency, receive bill for supplies in foreign currency	Collect foreign currency receivables, pay foreign currency liabilities

changes in the relative purchasing power of each currency since some base period. Specifically,

$$e'_t = e_t \times \frac{(1 + i_{f,t})}{(1 + i_{h,t})} \tag{10.1}$$

WHERE

e'_t = the real exchange rate (home currency per one unit of foreign currency) at time t
e_t = the nominal exchange rate (home currency per one unit of foreign currency) at time t
$i_{f,t}$ = the amount of foreign inflation between times 0 and t
$i_{h,t}$ = the amount of domestic inflation between times 0 and t

Given that the base period nominal rate, e_0, is also the real base period exchange rate, the change in the real exchange rate can be computed as follows:

$$\frac{e'_t - e_0}{e_0} \tag{10.2}$$

For example, suppose the Italian lira has devalued by 5% during the year. At the same time, Italian and U.S. inflation rates were 3% and 2%, respectively. Then, according to Equation 10.1, if e_0 is the exchange rate (dollar value of the lira) at the beginning of the year, the real exchange rate for the lira at year's end is

$$0.95e_0 \times \frac{1.03}{1.02} = 0.96e_0$$

Applying Equation 10.2, we can see that the real value of the lira has declined by 4% during the year:

$$\frac{0.96e_0 - e_0}{e_0} = -4\%$$

In effect the lira's 5% nominal devaluation more than offset the 1% inflation differential between Italy and the United States, leading to a 4% decline in the real value of the lira.

Importance of the Real Exchange Rate

The distinction between the nominal exchange rate and the real exchange rate is important because of their vastly different implications for exchange risk. A dramatic change in the nominal exchange rate accompanied by an equal change in the price level should have no effects on the relative competitive positions of domestic firms and their foreign competitors and, therefore, will not alter real cash flows. Alternatively, if the real exchange rate changes, it will cause relative price changes—changes in the ratio of domestic goods' prices to prices of foreign goods. In terms of currency changes affecting relative competitiveness, therefore, the focus must be not on nominal exchange rate changes, but instead on changes in the purchasing power of one currency relative to another.

Inflation and Exchange Risk

Let us begin by holding relative prices constant and looking only at the effects of general inflation. This condition means that if the inflation rate is 10%, the price of every good in the economy rises by 10%. In addition, we will initially assume that all goods are traded in a competitive world market without transaction costs, tariffs, or taxes of any kind. Given these conditions, economic theory tells us that the law of one price must prevail. That is, the price of any good, measured in a common currency, must be equal in all countries.

If the law of one price holds and if there is no variation in the relative prices of goods or services, then the rate of change in the exchange rate must equal the difference between the inflation rates in the two countries. The implications of a constant real exchange rate—that is, purchasing power parity (PPP) holds—are worth exploring further. To begin, purchasing power parity does not imply that exchange rate changes will necessarily be small or easy to forecast. If a country has high and unpredictable inflation (for example, Russia), then the country's exchange rate will also fluctuate randomly.

Nonetheless, without relative price changes, a multinational company faces no real operating exchange risk. As long as the firm avoids contracts fixed in foreign currency terms, its foreign cash flows will vary with the foreign rate of inflation. Because the exchange rate also depends on the difference between the foreign and the domestic rates of inflation, the movement of the exchange rate exactly cancels the change in the foreign price level, leaving real dollar cash flows unaffected.

Of course, the above conclusion does not hold if the firm enters into contracts fixed in terms of the foreign currency. Examples of such contracts are debt with fixed interest rates, long-term leases, labor contracts, and rent. However, if the real exchange rate remains constant, the risk introduced by entering into fixed contracts is not exchange risk; it is inflation risk. For instance, a Mexican firm with fixed-rate debt in pesos faces the same risk as the subsidiary of an American firm with peso debt. If the rate of inflation declines, the real interest cost of the debt rises, and the real cash flow of both companies falls. The solution to the problem of inflation risk is to avoid writing contracts fixed in nominal terms in countries with unpredictable inflation. If the contracts are indexed and if the real exchange rate remains constant, exchange risk is eliminated.

CALCULATING THE EFFECTS OF EXCHANGE RATE CHANGES AND INFLATION ON APEX SPAIN Apex Spain, the Spanish subsidiary of Apex Company, produces and sells medical imaging devices in Spain. At the current peseta exchange rate of Ptas. 1 = $0.01, the devices cost Ptas. 40,000 ($400) to produce and sell for Ptas. 100,000 ($1000). The profit margin of Ptas. 60,000 provides a dollar margin of $600. Suppose that Spanish inflation during the year is 20%, and the U.S. inflation rate is zero. All prices and costs are assumed to move in line with inflation. If we assume that purchasing power parity holds, the peseta will devalue to $0.0083 [0.01 × (1/1.2)]. The real value of the peseta stays at $0.01 [0.0083 × (1.2/1.0)], so Apex Spain's dollar profit margin will remain at $600. These effects are shown in Exhibit 10.2.

EXHIBIT 10.2 The Effects of Nominal Exchange Rate Changes and Inflation on Apex Spain

PRICE LEVEL	SPAIN	UNITED STATES
Beginning of year	100	100
End of year	120	100

EXCHANGE RATE	BEGINNING OF YEAR	END OF YEAR
Nominal rate	Ptas. 1 = $0.01	Ptas. 1 = $0.0083
Real Rate	Ptas. 1 = $0.01	Ptas. 1 = $0.0083 × 1.2/1 = $0.01

	BEGINNING OF YEAR		END OF YEAR	
PROFIT IMPACT	PESETAS	U.S. DOLLARS	PESETAS	U.S. DOLLARS
Price*	Ptas. 100,000	$1,000	Ptas. 120,000	$1,000
Cost of production*	40,000	400	48,000	400
Profit margin	Ptas. 60,000	$ 600	Ptas. 72,000	$ 600

*Peseta prices and costs are assumed to increase at the 20% rate of Spanish inflation.

Real Exchange Rate Changes and Exchange Risk

In general, a decline in the real value of a nation's currency makes its exports and import-competing goods more competitive. Conversely, an appreciating currency hurts the nation's exporters and those producers competing with imports.

During the late 1970s, for example, worldwide demand for Swiss franc-denominated assets caused the Swiss franc to appreciate in real terms. As a result, Swiss watchmakers were squeezed. Because of competition from Japanese companies, Swiss firms could not significantly raise the dollar price of watches sold in the United States. Yet, at the same time, the *dollar* cost of Swiss labor was rising because the franc was appreciating against the dollar.

American companies faced similar problems when the real value of the dollar began rising against other currencies during the early 1980s. U.S. exporters found themselves with the Hobson's choice of either keeping dollar prices constant and losing sales volume (because foreign currency prices rose in line with the appreciating dollar), or setting prices in the foreign currency to maintain market share, with a corresponding erosion in dollar revenues and profit margins. At the same time, the dollar cost of American labor remained the same or rose in line with U.S. inflation. The combination of lower dollar revenues and unchanged or higher dollar costs resulted in severe hardship for those U.S. companies selling abroad. Similarly, U.S. manufacturers competing domestically with imports whose dollar prices were declining saw both their profit margins and sales volumes reduced. Now the shoe is on the other foot, however, as Japanese firms attempt to cope with a yen that appreciated by more than 150% in real terms between 1985 and 1995.

ILLUSTRATION

YEN APPRECIATION HARMS JAPANESE TV PRODUCERS

For most of 1985, the yen traded at about ¥240 = $1. By 1995, the yen's value had risen to about ¥90 = $1, without a commensurate increase in U.S. inflation. This rise had a highly negative impact on Japanese television manufacturers. If it cost, say, ¥100,000 to build a color TV in Japan, ship it to the United States, and earn a normal profit, that TV could be sold in 1985 for about $417 (100,000/240). However, in 1995, the price would have had to be about $1,111 (100,000/90) for Japanese firms to break even, presenting them with the following dilemma. Because other U.S. prices had not risen much, as Japanese firms raised their dollar price to compensate for yen appreciation, Americans would buy fewer Japanese color TVs, and yen revenues would fall. If Japanese TV producers decided to keep their price constant at $417 to preserve market share in the United States, they would have to cut their yen price to about ¥37,530 (417 × 90). In general, whether they held the line on yen prices or cut them, real yen appreciation was bad news for Japanese TV manufacturers. Subsequent yen depreciation has eased the pressure on Japanese companies.

Alternatively, Industrias Penoles, the Mexican firm that is the world's largest refiner of newly mined silver, increased its dollar profits by more than 200% after the real devaluation of the Mexican peso relative to the dollar in 1982. Similarly, when the peso plunged in 1995, the company saw its profits rise again. The reason for the firm's success is that its costs, which are in pesos, declined in dollar terms, and the dollar value of its revenues, which are derived from exports, held steady.

In summary, the economic impact of a currency change on a firm depends on whether the exchange rate change is fully offset by the difference in inflation rates or whether (because of price controls, a shift in monetary policy, or some other reason) the real exchange rate and, hence, relative prices, change. It is these relative price changes that ultimately determine a firm's long-run exposure.

A less obvious point is that a firm may face more exchange risk if nominal exchange rates do *not* change. Consider, for example, a Brazilian shoe manufacturer producing for export to the United States and Europe. If the Brazilian *real*'s exchange rate remains fixed in the face of Brazil's typically high rate of inflation, then both the *real*'s real exchange rate and the manufacturer's dollar costs of production will rise.

EXHIBIT 10.3 The Effects of Real Exchange Rate Changes on the Brazilian Shoe Manufacturer

PRICE LEVEL	BRAZIL	UNITED STATES
Beginning of year	100	100
End of year	200	100

SCENARIO 1	BEGINNING OF YEAR	END OF YEAR
Nominal Exchange Rate	R1 = $0.02	R1 = $0.02
Real Exchange Rate	R1 = $0.02	R1 = $0.02 × 2/1 = $0.04

	BEGINNING OF YEAR		END OF YEAR	
PROFIT IMPACT	REAIS*	U.S. DOLLARS	REAIS	U.S. DOLLARS
Price	500	10.00	500	10.00
Cost of production	200	4.00	400	8.00
Profit margin	300	6.00	100	2.00

SCENARIO 2	BEGINNING OF YEAR	END OF YEAR
Nominal Exchange Rate	R1 = $0.02	R1 = $0.01
Real Exchange Rate	R1 = $0.02	R1 = $0.01 × 2/1 = $0.02

	BEGINNING OF YEAR		END OF YEAR	
PROFIT IMPACT	REAIS	U.S. DOLLARS	REAIS	U.S. DOLLARS
Price	500	10.00	1,000	10.00
Cost of production	200	4.00	400	4.00
Profit margin	300	6.00	600	2.00

*Reais is the plural of real.

Therefore, unless the *real* devalues, the Brazilian exporter will be placed at a competitive disadvantage vis-à-vis producers located in countries with less rapidly rising costs, such as Taiwan and South Korea.

Suppose, for example, that the Brazilian firm sells its shoes in the U.S. market for $10. Its profit margin is $6, or R300, because the shoes cost $4 to produce at the current exchange rate of R1 = $0.02. If Brazilian inflation is 100% but the nominal exchange rate remains constant, it will cost the manufacturer $8 to produce these same shoes by the end of the year. Assuming no U.S. inflation, the firm's profit margin will drop to $2. The basic problem is the 100% real appreciation of the *real* (0.02 × 2/1). This situation is shown in Exhibit 10.3 as Scenario 1.

In order to preserve its dollar profit margin (but not its inflation-adjusted *real* margin), the firm will have to raise its price to $14. (Why?[1]) But if it does that, it will be

---

[1] If the price is raised to $14, the profit margin is $6 ($14 − $8). However, at an exchange rate of R1 = $0.02, the real margin is still R300. With 100% inflation, the inflation-adjusted value of this margin is equivalent to only half of today's margin of R300 (R2 at year's end has the purchasing power of R1 today).

CHILE MISMANAGES ITS EXCHANGE RATE

A particularly dramatic illustration of the unfortunate effects of a fixed nominal exchange rate combined with high domestic inflation is provided by Chile. As part of its plan to bring down the rate of Chilean inflation, the government fixed the exchange rate in the middle of 1979 at 39 pesos to the U.S. dollar. Over the next 2½ years, the Chilean price level rose 60%, but U.S. prices rose by only about 30%. Thus, by early 1982, the Chilean peso had appreciated in real terms by approximately 23% $(1.6/1.3 - 1)$ against the U.S. dollar. These data are summarized in Exhibit 10.4.

An 18% "corrective" devaluation was en-

EXHIBIT 10.4 Nominal and Real Exchange Rates for Chile, 1979–1982

PRICE LEVEL		CHILE	UNITED STATES
1979		100	100
1982		160	130
Nominal Exchange Rate	1979	Ps. 1 = $0.02564	
	1982	Ps. 1 = $0.02564	
Real Exchange Rate	1979	Ps. 1 = $0.02564	
	1982	Ps. 1 = $0.02564 \times \dfrac{1.60}{1.30}$	
		= $0.03156	
Increase in Real Value of the Chilean Peso		$\dfrac{0.03156 - 0.02564}{0.02564} = 23.1\%$	

Result: Economic Devastation
—Loss of export markets
—Loss of domestic markets to imports
—Massive unemployment
—Numerous bankruptcies
—Numerous bank failures

acted in June 1982. Overall, the peso fell 90% over the next 12 months. However, the artificially high peso had already done its double damage to the Chilean economy: It made Chile's manufactured products more expensive abroad, pricing many of them out of international trade; and it made imports cheaper, undercutting Chilean domestic industries. The effects of the overvalued peso were devastating. Banks became insolvent, factories and copper smelters were thrown into bankruptcy, copper mines were closed, construction projects were shut down, and farms were put on the auction block. Unemployment approached 25%, and some areas of Chile resembled industrial graveyards.

The implosion of the Chilean peso did have a silver lining: Chilean companies became dynamic exporters, which today sell chopsticks and salmon to Japan, wine to Europe, and machinery to the United States. It also sped the acceptance of free market economic policies, which have given Chile one of the strongest growth rates in the world.

placed at a competitive disadvantage. By contrast, Scenario 2 shows that if the *real* devalues by 50%, to $0.01, the real exchange rate will remain constant at $0.02 ($.01 × 2/1), the Brazilian firm's competitive situation will be unchanged, and its profit margin will stay at $6. Its inflation-adjusted *real* profit margin also remains the same. Note that with

100% inflation, today's R300 profit margin must rise to R600 by year's end (which it does) to stay constant in inflation-adjusted *real* terms.

The Chilean example illustrates a critical point: *An increase in the real value of a currency acts as a tax on exports and a subsidy on imports.* Hence, firms that export or that compete with imports are hurt by an appreciating home currency. Conversely, such firms benefit from home-currency depreciation. These general principles identify a company's economic exposure.

10.2 THE ECONOMIC CONSEQUENCES OF EXCHANGE RATE CHANGES

We now examine more closely the specifics of a firm's economic exposure. Solely for the purpose of exposition, the discussion of exposure is divided into its component parts: transaction exposure and real operating exposure.

Transaction Exposure

Transaction exposure arises out of the various types of transactions that require settlement in a foreign currency. Examples are cross-border trade, borrowing and lending in foreign currencies, and the local purchasing and sales activities of foreign subsidiaries. Strictly speaking, of course, the items already on a firm's balance sheet, such as loans and receivables, capture some of these transactions. However, a detailed transaction exposure report must contain a number of off-balance-sheet items as well, including future sales and purchases, lease payments, forward contracts, loan repayments, and other contractual or anticipated foreign currency receipts and disbursements.

In terms of measuring economic exposure, though, a transaction exposure report, no matter how detailed, has a fundamental flaw: the assumption that local currency cost and revenue streams remain constant after an exchange rate change.

That assumption does not permit an evaluation of the typical adjustments that consumers and firms can be expected to undertake under conditions of currency change. Hence, attempting to measure the likely exchange gain or loss by simply multiplying the projected predevaluation (prerevaluation) local currency cash flows by the forecast devaluation (revaluation) percentage will lead to misleading results. Given the close relationship between nominal exchange rate changes and inflation as expressed in purchasing power parity, measuring exposure to a currency change without reference to the accompanying inflation is also a misguided task.

We will now examine more closely the typical demand and cost effects that result from a real exchange rate change and how these effects combine to determine a firm's true operating exposure. In general, an appreciating real exchange rate can be expected to have the opposite effects. The dollar is assumed to be the home currency (HC).

Operating Exposure

A real exchange rate change affects a number of aspects of the firm's operations. With respect to dollar (HC) appreciation, the key issue for a domestic firm is its degree of *pricing flexibility*—that is, can the firm maintain its dollar margins both at home and abroad?

Can the company maintain its dollar price on domestic sales in the face of lower-priced foreign imports? In the case of foreign sales, can the firm raise its foreign currency selling price sufficiently to preserve its dollar profit margin?

The answers to these questions depend largely on the **price elasticity of demand**. The less price elastic the demand, the more price flexibility a company will have to respond to exchange rate changes. Price elasticity, in turn, depends on the degree of competition and the location of key competitors. The more differentiated (distinct) a company's products are, the less competition it will face and the greater its ability will be to maintain its domestic currency prices both at home and abroad. Examples here are IBM and Daimler-Chrysler (producer of Mercedes Benz cars). Similarly, if most competitors are based in the home country, then all will face the same change in their cost structure from home currency appreciation, and all can raise their foreign currency prices without putting any of them at a competitive disadvantage relative to their domestic competitors. Examples of this situation are in the precision instrumentation and high-end telecommunications industries, in which virtually all the important players are U.S.-based companies.

Conversely, the less differentiated a company's products are and the more internationally diversified its competitors (for example, the low-priced end of the auto industry), the greater the price elasticity of demand for its products will be and the less pricing flexibility it will have. These companies face the greatest amount of exchange risk.

ILLUSTRATION

PRODUCT DIFFERENTIATION AND SUSCEPTIBILITY TO EXCHANGE RISK OF THE U.S. APPAREL AND TEXTILE INDUSTRIES

The U.S. textile and apparel industries are highly competitive, with each composed of many small manufacturers. In addition, nearly every country has a textile industry, and apparel industries are also common to most countries.

Despite these similarities, the textile industry exists in a more competitive environment than the apparel industry because textile products are more standardized than apparel products. Buyers of textiles can easily switch from a firm that sells a standard good at a higher price to one that sells virtually the same good at a lower price. Because they are more differentiated, the products of competing apparel firms are viewed as more distinct and are less sensitive than textile goods to changes in prices. Thus, even though both textile and apparel firms operate in highly competitive industries, apparel firms—with their greater degree of pricing flexibility—are less subject to exchange risk than are textile firms.

Another important determinant of a company's susceptibility to exchange risk is its ability to shift production and the sourcing of inputs among countries. The greater a company's flexibility to substitute between home-country and foreign-country inputs or production, the less exchange risk the company will face. Other things being equal, firms with worldwide production systems can cope with currency changes by increasing production in a nation whose currency has undergone a real devaluation and decreasing production in a nation whose currency has revalued in real terms.

With respect to a multinational corporation's foreign operations, the determinants of their economic exposure will be similar to those just mentioned. A foreign subsidiary selling goods or services in its local market generally will be unable to raise its local currency (LC) selling price to the full extent of an LC devaluation, causing it to register a decline in its postdevaluation dollar revenues. However, because an LC devaluation will also reduce import competition, the more import competition the subsidiary was facing before the devaluation, the smaller its dollar revenue decline will be.

The harmful effects of LC devaluation will be mitigated somewhat because the devaluation should lower the subsidiary's dollar production costs, particularly those attributable to local inputs. However, the higher the import content of local inputs, the less dollar production costs will decline. Inputs used in the export or import-competing sectors will decline less in dollar price than other domestic inputs.

Suppose, for example, that IBM Germany currently is earning DM 300 million. This figure translates into $200 million at the current exchange rate of DM 1.5/$1. If the dollar rises to DM 2, other things being equal, IBM Germany's profit will translate into $150 million. This estimate, which entails a $50 million drop in profit, assumes that IBM Germany's DM revenues and costs will remain constant regardless of the exchange rate. However, to the extent that IBM Germany is buying U.S. components at set dollar prices, its costs in DM will rise and its profit loss will exceed $50 million. Of course, if IBM Germany can replace U.S. components with German components or if it can raise its DM prices without suffering much loss in sales volume, it can offset some of the erosion in its dollar profits. This example points out that the impact of a rise in the value of the dollar depends on two effects: a translation effect and an adjustment effect. The *translation effect* involves converting current foreign earnings into dollars at the new exchange rate. The *adjustment effect* includes factors such as the nature of the company's cost structure and the extent to which it can raise its prices and alter its cost structure.

An MNC using its foreign subsidiary as an export platform will benefit from an LC devaluation because its export revenues should stay about the same, whereas its dollar costs will decline. The net result will be a jump in dollar profits for the exporter. For example, Seagate Technology, the world's largest disk-drive maker, produces many of its drives in Thailand and Malaysia for sale worldwide. Hence, the steep currency declines in those countries meant lower dollar manufacturing costs for Seagate and an improved competitive position.

Local currency movements also can affect the dollar value of the depreciation tax shield. The cash flow associated with the tax write-off of depreciable assets can have a substantial present value, particularly for a capital-intensive corporation. Unless indexation of fixed assets is permitted (as in Argentina, Brazil, and Israel), the dollar value of the local currency-denominated tax shield will unambiguously decline by the percentage of nominal devaluation.

The major conclusion is that the sector of the economy in which a firm operates (export, import-competing, or purely domestic), the sources of the firm's inputs (imports, domestic traded or nontraded goods), and fluctuations in the real exchange rate are far more important in delineating the firm's true economic exposure than is any accounting definition. The economic effects are summarized in Exhibit 10.5.

EXHIBIT 10.5 Characteristic Economic Effects of Exchange Rate Changes on Multinational Corporations

CASH-FLOW CATEGORIES	RELEVANT ECONOMIC FACTORS	DEVALUATION IMPACT	REVALUATION IMPACT
		Parent-Currency Revenue Impact	**Parent-Currency Revenue Impact**
Revenue			
Export sales	Price-sensitive demand	Increase (++)	Decrease (− −)
	Price-insensitive demand	Slight increase (+)	Slight decrease (−)
Local sales	Weak prior import competition	Sharp decline (− −)	Increase (++)
	Strong prior import competition	Decrease (−) (less than devaluation %)	Slight increase (+)
		Parent-Currency Cost Impact	**Parent-Currency Cost Impact**
Costs			
Domestic inputs	Low import content	Decrease (− −)	Increase (++)
	High import content/ inputs used in export or import-competing sectors	Slight decrease (−)	Slight increase (+)
Imported inputs	Small local market	Remain the same (0)	Remain the same (0)
	Large local market	Slight decrease (−)	Slight increase (+)
		Cash-Flow Impact	**Cash-Flow Impact**
Depreciation			
Fixed assets	No asset valuation adjustment	Decrease by devaluation % (− −)	Increase by revaluation % (++)
	Asset valuation adjustment	Decrease (−)	Increase (+)

Note: To interpret the above chart, and taking the impact of a devaluation on local demand as an example, it is assumed that if import competition is weak, local prices will climb slightly, if at all; in such a case, there would be a sharp contraction in parent-company revenue. If imports generate strong competition, local-currency prices are expected to increase, although not to the full extent of the devaluation; in this instance, only a moderate decline in parent-company revenue would be registered.

Source: Alan C. Shapiro, "Developing a Profitable Exposure Management System," reprinted from p. 188 of the June 17, 1977, issue of *Business International Money Report,* with the permission of the Economist Intelligence Unit, NA, Incorporated.

ILLUSTRATION

NIKE'S GAINS FROM THE INDONESIAN RUPIAH'S DEPRECIATION ARE LIMITED

In the six-month period beginning July 1997, the Indonesian rupiah plummeted more than 70%. The rupiah decline helped lower costs for Nike, which manufactures some of its shoes in Indonesia for sale elsewhere in the world. However, a Nike spokesperson said that prices on U.S. retail shelves would not change much because 65% of the materials of shoes made in Indonesia are imported. We can see why the impact on the retail shoe would be much less than expected with a simple example.

Suppose that the shoes Nike makes in Indonesia sell at retail for $85. They cost Nike $20 to manufacture and are sold to retailers for $45 (the extra $25 covers Nike's overhead expenses plus profit). The retailers then mark the shoes up another $40 to cover their expenses and profit. Assuming the absence of any inflation after devaluation, the impact of a 70% rupiah devaluation will be to cut Nike's manufacturing costs by 70% of the 35% value added in Indonesia, or $4.90 ($0.70 \times 0.35 \times \$20$). To the extent Indonesian prices begin rising (which they did), the gains from rupiah devaluation will be even smaller. In any event, even ignoring inflation, shoe prices would fall by no more than about $5, less than 6% of the original $85 retail price.

A surprising implication of this analysis is that domestic facilities that supply foreign markets normally entail much greater exchange risk than do foreign facilities that supply local markets. The explanation is that material and labor used in a domestic plant are paid for in the home currency, whereas the products are sold in a foreign currency. For example, take a Japanese company, such as Nissan Motors, that builds a plant to produce cars for export, primarily to the United States. The company will incur an exchange risk from the point at which it invests in facilities to supply a foreign market (the United States) because its yen expenses will be matched with dollar revenues rather than yen revenues. The point seems obvious; however, all too frequently, firms neglect those effects when analyzing a proposed foreign investment.

Similarly, a firm (or its affiliate) producing solely for the domestic market and using only domestic sources of inputs can be strongly affected by currency changes, even though its accounting exposure is zero. Consider, for example, American Motors (now a Chrysler

---

ILLUSTRATION

A TALE OF TWO THAI COMPANIES

As seen in Exhibit 10.6, the 40% loss in the value of the Thai baht in the second half of 1997 had very different effects on two Thai companies, Delta Electronics and Thai Petrochemical In-dustries. Delta makes electronic parts for export with a combination of local labor and locally-sourced raw materials, such as wire, plastic, steel, and glass. With most of its costs sourced in baht, Delta's dollar costs have declined in line with baht devaluation. With about 98% of its sales in dollars, Delta's dollar profits have risen, although by less than might be expected given its cost and revenue structure. The reason for the lower-than-expected increase in dollar profits: declining dollar revenues, owing to stiff competition from companies in South Korea and other Asian countries whose currencies have also fallen against the dollar. The net result of Delta's higher dollar earnings and the steep decline in the dollar cost of servicing Delta's debt, most of which is in baht, has led to stock price appreciation in both baht and dollar terms.

Baht devaluation has been an ill wind for Thai Petrochemical Industries (TPI), however. Most of its debt is in dollars, its raw material is oil priced in dollars, and it sells 70% of its plastics output domestically in baht. Its exports are facing competitive pressures from similarly stricken Asian firms trying to export their way out of trouble. The result has been huge losses (in late 1997, TPI reported the largest loss for a listed firm in Thai history) and a stock price sinking with the baht.

EXHIBIT 10.6 Good, Bad and Ugly: Share Prices and the Thai baht

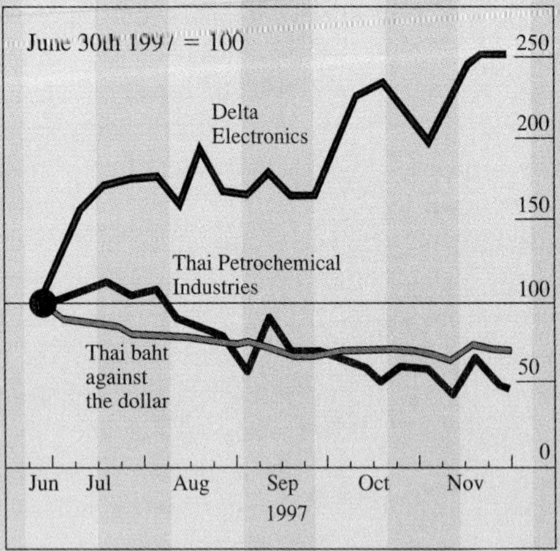

Source: *The Economist*, November 29th, 1997, p. 67.

subsidiary), which produces and sells cars only in the United States and uses only U.S. labor and materials. Because it buys and sells only in dollars, by U.S. accounting standards it has no balance-sheet exposure. However, its principal emphasis has been on the compact, economy-minded end of the auto market—the segment most subject to competition from less-expensive Korean, Japanese, Italian, and German imports. Dollar devaluations have certainly enhanced American Motors' competitive position or, at least, have slowed down its erosion, enabling the company to enjoy higher dollar profits than it would have in the absence of these currency changes. Appreciation of the dollar has had opposite effects.

10.3 IDENTIFYING ECONOMIC EXPOSURE

At this point it makes sense to illustrate some of the concepts just discussed by examining several firms to see in what ways they may be susceptible to exchange risk. The companies are Aspen Skiing Company, Petróleos Mexicanos, and Toyota Motor Company.

Aspen Skiing Company

Aspen Skiing Company owns and operates ski resorts in the Colorado Rockies, catering primarily to Americans. It buys all its supplies in dollars and uses only American labor and materials. All guests pay in dollars. Because it buys and sells only in dollars, by U.S. standards it has no accounting exposure. Yet, Aspen Skiing Company does face economic exposure because changes in the value of the dollar affect its competitive position. For example, the strong dollar in the early 1980s adversely affected the company because it led to bargains abroad that offered stiff competition for domestic resorts, including the Rocky Mountain ski areas.

Despite record snowfalls in the Rockies during the early 1980s, many Americans decided to ski in the European Alps instead. Although airfare to the Alps cost much more than a flight to Colorado, the difference between expenses on the ground made a European ski holiday less expensive. For example, in January 1984, American Express offered a basic one-week ski package in Aspen for $439 per person, including double-occupancy lodging, lift ticket, and free rental car or bus transfer from Denver.[2] Throw in round-trip airfare between New York and Denver of $300 and the trip's cost totaled $739.

At the same time, skiers could spend a week in Chamonix in the French Alps for $234, including lodging, lift ticket, breakfast and a bus transfer from Geneva, Switzerland. Adding in round-trip airfare from New York of $579 brought the trip's cost to $813. The Alpine vacation became less expensive than the one in the Rockies when the cost of meals was included: an estimated $50 a day in Aspen versus $30 a day in Chamonix.

In effect, Aspen Skiing Company is operating in a global market for skiing or, more broadly, vacation services. As the dollar appreciates in real terms, both foreigners and Americans find less-expensive skiing and vacation alternatives outside the United States. In addition, even if California and other West Coast skiers find that high transportation costs continue to make it more expensive to ski in Europe than in the Rockies, they are not restricted to the American Rockies. They have the choice of skiing in the Canadian Rockies, where the skiing is fine and their dollars go further.

---

[2] Report in the *Wall Street Journal*, January 17, 1984, p. 1.

Conversely, a depreciating dollar makes Aspen Skiing Company more competitive and should increase its revenues and profits. In either event, the use of American products and labor means that its costs will not be significantly affected by exchange rate fluctuations.

Petróleos Mexicanos

Petróleos Mexicanos, or Pemex, is the Mexican national oil company. It is the largest company in Mexico and ranked number 40 in the 1992 *Fortune* directory of the biggest non-U.S. industrial companies. Most of its sales are overseas. Suppose Pemex borrows U.S. dollars. If the peso devalues, is Pemex a better or worse credit risk?

The instinctive response of most people is that peso devaluation makes Pemex a poorer credit risk. This response is wrong. Consider Pemex's revenues. Assume that it exports all its oil. Oil is priced in dollars, so Pemex's dollar revenues will remain the same after peso devaluation. Its dollar costs, however, will change. Most of its operating costs are denominated in pesos. These costs include labor, local supplies, services, and materials. Although the peso amount of these costs may go up somewhat, they will not rise to the extent of the devaluation of the peso. Hence, the dollar amount of peso costs will decline. Pemex also uses a variety of sophisticated equipment and services to aid in oil exploration, drilling, and production. Because these inputs are generally from foreign sources, their dollar costs are likely to be unaffected by peso devaluation. Inasmuch as some costs will fall in dollar terms and other costs will stay the same, the overall effect of peso devaluation is a decline in Pemex's dollar costs.

Its dollar revenue will stay the same and the dollar amount of its costs will fall, so the net effect on Pemex of a peso devaluation is to increase its dollar cash flow. Hence, it becomes a better credit risk in terms of its ability to service dollar debt.

Might this conclusion be reversed if it turns out that Pemex sells much of its oil domestically? Surprisingly, the answer is no if we add the further condition that the Mexican government does not impose oil price controls. Suppose the price of oil is $20 a barrel. If the initial peso exchange rate is Ps 1 = $0.04, that means that the price of oil in Mexico must be Ps 500 a barrel. Otherwise, there would be an arbitrage opportunity because oil transportation costs are a small fraction of the price of oil. If the peso now devalues to $0.02, the price of oil must rise to Ps 1,000. Consider what would happen if the price stayed at Ps 500. The dollar equivalent price would now be $10. But why would Pemex sell oil in Mexico for $10 a barrel when it could sell the same oil outside Mexico for $20 a barrel? It would not do so unless there were price controls in Mexico and the government allocated a certain amount of oil to the Mexican market at this price. Hence, in the absence of government intervention, the peso price of oil must rise to Ps 1,000, and Pemex's dollars profits will rise whether it exports all or part of its oil.

This situation points out the important distinction between the currency of denomination and the currency of determination. The **currency of denomination** is the currency in which contracts are stated. For example, oil prices in Mexico are stated in pesos. However, although the currency of denomination for oil sales in Mexico is the peso, the peso price itself is determined by the dollar price of oil; that is, as the peso:dollar exchange rate changes, the peso price of oil changes to equate the dollar equivalent price of oil in Mexico with the dollar price of oil in the world market. Thus, the **currency of determination** for Pemex's domestic oil sales is the U.S. dollar.

Toyota Motor Company

Toyota is the largest Japanese auto company and the fourth largest non-U.S. industrial firm in the world. More than half of its sales are overseas, primarily in the United States. If the yen appreciates, Toyota has the choice of keeping its yen price constant or its dollar price constant. If Toyota holds its yen price constant, the dollar price of its auto exports will rise and sales volume will decline. On the other hand, if Toyota decides to maintain its U.S. market share, it must hold its dollar price constant. In either case, its yen revenues will fall.

Even if Toyota decides to focus on the Japanese market, it will face the *flow-back effect*, as previously exported products flow back into the home market. Flow-back occurs because other Japanese firms find that a high yen makes it difficult to export their cars, so they emphasize Japanese sales as well. The result is increased domestic competition and lower profit margins on domestic sales.

Toyota's yen production costs also will be affected by yen appreciation. Steel, copper, aluminum, oil (from which plastics are made), and other materials that go into making a car are all imported. As the yen appreciates, the yen cost of these imported materials will decline. Yen costs of labor and domestic services, products, and equipment will likely stay the same. The net effect of lower yen costs for some inputs and constant yen costs for other inputs is a reduction in overall yen costs of production.

The net effect on profits of lower yen revenues and lower yen costs is an empirical question. This question can be answered by examining the profit consequences of yen appreciation. Here, the answer is unambiguous: Yen appreciation hurts Toyota; the reduction in its revenues more than offsets the reduction in its costs.

These three examples illustrate a progression of ideas. Aspen Skiing Company's revenues were affected by exchange rate changes, but its costs were largely unaffected. By contrast, Pemex's costs, but not its revenues, were affected by exchange rate changes. Toyota had both its costs and its revenues affected by exchange rate changes. The process of examining these companies includes a systematic approach to identifying a company's exposure to exchange risk. Exhibit 10.7 summarizes this approach by presenting a series of questions that underlie the analysis of economic exposure.

EXHIBIT 10.7 Key Questions to Ask That Help Identify Exchange Risk

1. Where is the company selling?
   Domestic versus foreign sales breakdown
2. Who are the company's key competitors?
   Domestic versus foreign companies
3. How sensitive is demand to price?
   Price-sensitive demand versus price-insensitive demand
4. Where is the company producing?
   Domestic production versus foreign production
5. Where are the company's inputs coming from?
   Domestic inputs versus foreign inputs
6. How are the company's inputs or outputs priced?
   Priced in a world market or in a domestic market; the currency of determination as opposed to the currency of denomination

---

10.4 CALCULATING ECONOMIC EXPOSURE

We will now work through a hypothetical, though comprehensive, example illustrating all the various aspects of exposure that have been discussed so far. This example emphasizes the quantitative, rather than qualitative, determination of economic exposure. It shows how critical the underlying assumptions are.

Spectrum Manufacturing AB is the wholly owned Swedish affiliate of a U.S. multi-national industrial plastics firm. It manufactures patented sheet plastic in Sweden, with 60% of its output currently being sold in Sweden and the remaining 40% exported to other European countries. Spectrum uses only Swedish labor in its manufacturing process, but it uses both local and foreign sources of raw material. The effective Swedish tax rate on corporate profits is 40%, and the annual depreciation charge on plant and equipment, in Swedish kronor (SKr), is SKr 900,000. In addition, Spectrum AB has outstanding SKr 3 million in debt, with interest payable at 10% annually.

Exhibit 10.8 presents Spectrum's projected sales, costs, after-tax income, and cash flow for the coming year, based on the current exchange rate of SKr 4 = $1. All sales are invoiced in kronor (singular, krona).

Spectrum's Accounting Exposure

Exhibit 10.9 shows Spectrum's balance sheet before and after an exchange rate change. To contrast the economic and accounting approaches to measuring exposure, assume that the Swedish krona devalued by 20%, from SKr 4 = $1 to SKr 5 = $1. The third column of Exhibit 10.9 shows that under the current rate method mandated by FASB-52, Spectrum will have a translation loss of $685,000. Use of the monetary/nonmonetary method leads to a much smaller reported loss of $50,000.

EXHIBIT 10.8 Summary of Projected Operations for Spectrum Manufacturing AB: Base Case

	UNITS (HUNDRED THOUSANDS)	UNIT PRICE (SKR)	TOTAL	
Domestic sales	6	20	(SKr) 12,000,000	
Export sales	4	20	8,000,000	
Total revenue				20,000,000
Total operating expenditures				10,800,000
Overhead expenses				3,500,000
Interest on Krona debt @ 10%				300,000
Depreciation				900,000
Net profit before tax				(SKr) 4,500,000
Income tax @ 40%				1,800,000
Profit after tax				(SKr) 2,700,000
Add back depreciation				900,000
Net cash flow in kronor				(SKr) 3,600,000
Net cash flow in dollars (SKr 4 = $1)				$900,000

EXHIBIT 10.9 Impact of Krona Devaluation on Spectrum AB's Financial Statement under FASB-52

	KRONOR	U.S. DOLLARS BEFORE KRONA DEVALUATION (SKR 4 = $1)	U.S. DOLLARS AFTER KRONA DEVALUATION (SKR 5 = $1) Current Rate	Monetary/ Nonmonetary
Assets				
Cash	SKr 1,000,000	$ 250,000	$ 200,000	$ 200,000
Accounts				
Receivable	5,000,000	1,250,000	1,000,000	1,000,000
Inventory	2,700,000	675,000	540,000	675,000
Net fixed				
assets	10,000,000	2,500,000	2,000,000	2,500,000
Total assets	SKr 18,700,000	$4,675,000	$3,740,000	$4,375,000
Liabilities				
Accounts				
payable	2,000,000	500,000	400,000	400,000
Long-term				
debt	3,000,000	750,000	600,000	600,000
Equity	13,700,000	3,425,000	2,740,000	3,375,000
Total	SKr 18,700,000	$4,675,000	$3,740,000	$4,375,000
liabilities				
plus equity				
Translation gain			$ (685,000)	$ (50,000)
(loss)				

FASB-52 = Statement of Financial Accounting Standards no. 52.

Spectrum's Economic Exposure

On the basis of current information, it is impossible to determine just what the economic impact of the krona devaluation will be. Therefore, three different scenarios have been constructed, with varying degrees of plausibility, and Spectrum's economic exposure has been calculated under each scenario. The three scenarios are

1. All variables remain the same.
2. Krona sales prices and all costs rise; volume remains the same.
3. There are partial increases in prices, costs, and volume.

Scenario 1: All variables remain the same. If all prices remain the same (in kronor) and sales volume does not change, then Spectrum's krona cash flow will stay at SKr 3,600,000. At the new exchange rate, this amount will equal $720,000 (3,600,000/5). Then the net loss in dollar operating cash flow in year one can be calculated as follows:

First-year cash flow (SKr 4 = $1)	$900,000
First-year cash flow (SKr 5 = $1)	720,000
Net loss from devaluation	$180,000

Moreover, this loss will continue until relative prices adjust. Part of this loss, however, will be offset by the $150,000 gain that will be realized when the SKr 3 million loan is repaid (3 million $\times$ 0.05).[3] If a three-year adjustment process is assumed and the krona loan will be repaid at the end of year 3, then the present value of the economic loss from operations associated with the krona devaluation, using a 15% discount rate, equals $312,420:

YEAR	POSTDEVALUATION CASH FLOW (1)	−	PREDEVALUATION CASH FLOW (2)	=	CHANGE IN CASH FLOW (3)	×	15% PRESENT VALUE FACTOR (4)	=	PRESENT VALUE (5)
1	$720,000		$900,000		−$180,000		0.870		−$156,600
2	720,000		900,000		− 180,000		0.756		136,080
3	870,000*		900,000		− 30,000		0.658		19,740

This loss is primarily due to the inability to raise the sales price. The resulting constant krona profit margin translates into a 20% reduction in dollar profits. The economic loss of $312,420 contrasts with the accounting recognition of a $685,000 foreign exchange loss. In reality, of course, the prices, costs, volume, and input mix are unlikely to remain fixed. The discussion will now focus on the economic effects of some of these potential adjustments.

EXHIBIT 10.10 Summary of Projected Operations for Spectrum Manufacturing AB: Scenario 2

	UNITS (HUNDRED THOUSANDS)	UNIT PRICE (SKR)	TOTAL
Domestic sales	6	25	SKr 15,000,000
Export sales	4	25	10,000,000
Total revenue			25,000,000
Total operating expenditures			13,500,000
Overhead expenses			4,375,000
Interest on krona debt @ 10%			300,000
Depreciation			900,000
Net profit before tax			SKr 5,925,000
Income tax @ 40%			2,370,000
Profit after tax			SKr 3,555,000
Add back depreciation			900,000
Net cash flow in kronor			SKr 4,455,000
Net cash flow in dollars (SKr 5 = $1)			$891,000

---

[3] No Swedish taxes will be owed on this gain because SEK 3 million were borrowed and SEK 3 million were repaid. These tax effects are elaborated on in Chapters 13 and 18.

Scenario 2: Krona sales prices and all costs rise; volume remains the same. It is assumed here that all costs and prices increase in proportion to the krona devaluation, but unit volume remains the same. However, the operating cash flow in kronor does not rise to the same extent because depreciation, which is based on historical cost, remains at SKr 900,000. As a potential offset, interest payments also hold steady at SKr 300,000. Working through the numbers in Exhibit 10.10 gives us an operating cash flow of $891,000.

The $9,000 reduction in cash flow equals the decreased dollar value of the SKr 900,000 depreciation tax shield less the decreased dollar cost of paying the SKr 300,000 in interest. Before devaluation, the tax shield was worth (900,000 × 0.4)/4 dollars, or $90,000. After devaluation, the dollar value of the tax shield declines to (900,000 × 0.4)/5 dollars = $72,000, or a loss of $18,000 in cash flow. Similarly, the dollar cost of paying SEK 300,000 in interest declines by $15,000 to $60,000 (from $75,000). After tax, this decrease in interest expense equals $9,000. Adding the two figures (−$18,000 + $9,000) yields a net loss of $9,000 annually in operating cash flow.

The net economic gain over the coming three years, relative to predevaluation expectations, is $78,220.

YEAR	POSTDEVALUATION CASH FLOW (1)	−	PREDEVALUATION CASH FLOW (2)	=	CHANGE IN CASH FLOW (3)	×	15% PRESENT VALUE FACTOR (4)	=	PRESENT VALUE (5)
1	$ 891,000		$900,000		− $9,000		0.870		−$ 7,830
2	891,000		900,000		− 9,000		0.756		− 6,804
3	1,041,000*		900,000		+141,000		0.658		92,778
							Net Gain		$78,144

*Includes a gain of $150,000 on loan repayment.

Most of this gain in economic value comes from the gain on repayment of the krona loan.

Scenario 3: Partial increases in prices, costs, and volume. In the most realistic situation, all variables will adjust somewhat. It is assumed here that the sales price at home rises by 10% to SEK 22 and the export price is raised to SKr 24—still providing a competitive advantage in dollar terms over foreign products. The result is a 20% increase in domestic sales and a 15% increase in export sales.

Local input prices are assumed to go up, but the dollar price of imported material stays at its predevaluation level. As a result of the change in relative cost, some substitutions are made between domestic and imported goods. The result is an increase in SKr unit cost of approximately 17%. Overhead expenses rise by only 10% because some components of this account, such as rent and local taxes, are fixed in value.

The net result of all these adjustments is an operating cash flow of $1,010,800, which is a gain of $110,800 over the predevaluation level of $900,000. This scenario is shown in Exhibit 10.11.

Over the next three years, cash flows and the firm's economic value will change as follows:

YEAR	POSTDEVALUATION CASH FLOW (1)	−	PREDEVALUATION CASH FLOW (2)	=	CHANGE IN CASH FLOW (3)	×	15% PRESENT VALUE FACTOR (4)	=	PRESENT VALUE (5)
1	$1,010,800		$900,000		$110,800		0.870		$ 96,396
2	1,010,800		900,000		110,800		0.756		83,765
3	1,160,800*		900,000		260,800		0.658		171,606
							Net Gain		$351,767

*Includes a gain of $150,000 on loan repayment.

Thus, under this scenario, the economic value of the firm will increase by $351,767. This gain reflects the increase in operating cash flow combined with the gain on loan repayment.

CASE ANALYSIS ■ The three preceding scenarios demonstrate the sensitivity of a firm's economic exposure to assumptions concerning its price elasticity of demand, its ability to adjust its mix of inputs as relative costs change, its pricing flexibility, subsequent local inflation, and its use of local currency financing. Perhaps most important of all, this example makes clear the lack of any necessary relationship between accounting-derived measures of exchange gains or losses and the true impact of currency changes on a firm's economic value. The economic effects of this devaluation under the three alternative scenarios are summarized in Exhibit 10.12.

EXHIBIT 10.11 Summary of Projected Operations for Spectrum Manufacturing AB: Scenario 3

	UNITS (HUNDRED THOUSANDS)	UNIT PRICE (SKR)	TOTAL	
Domestic sales	7.2	22	SKr 15,840,000	
Export sales	4.6	24	11,040,000	
Total revenue				26,880,000
Total operating expenditures				14,906,000
Overhead expenses				3,850,000
Interest on krona debt @ 10%				300,000
Depreciation				900,000
Net profit before tax				SKr 6,924,000
Income tax @ 40%				2,769,000
Profit after tax				SKr 4,154,000
Add back depreciation				900,000
Net cash flow in kronor				SKr 5,054,000
Net cash flow in dollars (SKr 5 = $1)				$1,010,800

EXHIBIT 10.12 Summary of Economic Exposure Impact of Krona Devaluation on Spectrum Manufacturing AB

	FORECAST CHANGE IN CASH FLOWS		
YEAR	SCENARIO 1	SCENARIO 2	SCENARIO 3
1	−$180,000	−$ 9,000	$110,800
2	−$180,000	− 9,000	110,800
3	− 30,000*	141,000*	260,800*
Change in present value (15% discount factor)	−$312,420	$ 78,114	$351,767

*Includes a gain of $150,000 on loan repayment.

10.5 AN OPERATIONAL MEASURE OF EXCHANGE RISK

The preceding example demonstrates that determining a firm's true economic exposure is a daunting task, requiring a singular ability to forecast the amounts and exchange rate sensitivities of future cash flows. Most firms that follow the economic approach to managing exposure, therefore, must settle for a measure of their economic exposure and the resulting exchange risk that often is supported by nothing more substantial than intuition.

This section presents a workable approach for determining a firm's true economic exposure and susceptibility to exchange risk. The approach avoids the problem of using seat-of-the-pants estimates in performing the necessary calculations.[4] The technique is straightforward to apply, and it requires only historical data from the firm's actual operations or, in the case of a *de novo* venture, data from a comparable business.

This approach is based on the following operational definition of the exchange risk faced by a parent or one of its foreign affiliates: *A company faces exchange risk to the extent that variations in the dollar value of the unit's cash flows are correlated with variations in the nominal exchange rate.* This correlation is precisely what a *regression analysis* seeks to establish. A simple and straightforward way to implement this definition, therefore, is to regress the changes in actual cash flows from past periods, converted into their dollar values, on changes in the average exchange rate during the corresponding period. Specifically, this involves running the following regression[5]:

$$\Delta CF_t = a + \beta \Delta EXCH_t + u_t \tag{10.3}$$

---

[4] This section is based on C. Kent Garner and Alan C. Shapiro, "A Practical Method of Assessing Foreign Exchange Risk," *Midland Corporate Finance Journal*, Fall 1984, pp. 6–17.

[5] The application of the regression approach to measuring exposure to currency risk is illustrated in Garner and Shapiro, "A Practical Method of Assessing Foreign Exchange Risk," and in Michael Adler and Bernard Dumas, "Exposure to Currency Risk: Definition and Measurement," *Financial Management*, Summer 1984, pp. 41–50. We use changes, rather than levels, of the variables in the regression because the variables are nonstationary.

WHERE

ΔCF_t $= CF_t - CF_{t-1}$, and CF_t equals the dollar value of total affiliate (parent) cash flows in period t

$\Delta EXCH_t$ $= EXCH_t - EXCH_{t-1}$, and $EXCH_t$ equals the average nominal exchange rate (dollar value of one unit of the foreign currency) during period t

u = a random error term with mean 0

The output from such a regression includes three key parameters: (1) the foreign exchange beta (β) coefficient, which measures the sensitivity of dollar cash flows to exchange rate changes; (2) the t-statistic, which measures the statistical significance of the beta coefficient; and (3) the R^2, which measures the fraction of cash flow variability explained by variation in the exchange rate. The higher the beta coefficient, the greater the impact of a given exchange rate change on the dollar value of cash flows. Conversely, the lower the beta coefficient, the less exposed the firm is to exchange rate changes. A larger t-statistic means a higher level of confidence in the value of the beta coefficient.

However, even if a firm has a large and statistically significant beta coefficient and thus faces real exchange risk, this situation does not necessarily mean that currency fluctuations are an important determinant of overall firm risk. What really matters is the percentage of total corporate cash-flow variability that is due to these currency fluctuations. Thus, the most important parameter, in terms of its impact on the firm's exposure management policy, is the regression's R^2. For example, if exchange rate changes explain only 1% of total cash-flow variability, the firm should not devote much in the way of resources to foreign exchange risk management, even if the beta coefficient is large and statistically significant.

Limitations

The validity of this method is clearly dependent on the sensitivity of future cash flows to exchange rate changes being similar to their historical sensitivity. In the absence of additional information, this assumption seems to be reasonable. However, the firm may have reason to modify the implementation of this method. For example, the nominal foreign currency tax shield provided by a foreign affiliate's depreciation is fully exposed to the effects of currency fluctuations. If the amount of depreciation in the future is expected to differ significantly from its historical values, then the depreciation tax shield should be removed from the cash flows used in the regression analysis and treated separately. Similarly, if the firm has recently entered into a large purchase or sales contract fixed in terms of the foreign currency, it might decide to consider the resulting transaction exposure apart from its operating exposure.

10.6 ILLUSTRATION: LAKER AIRWAYS

The crash of Sir Freddie Laker's Skytrain had little to do with the failure of its navigational equipment or its landing gear; indeed, it can largely be attributed to misguided management decisions. Laker's management erred in selecting the financing mode for the acquisition of the aircraft fleet that would accommodate the booming transatlantic business spearheaded by Sir Freddie's sound concept of a "no-frill, low-fare, stand-by" air travel package.

In 1981, Laker was a highly leveraged firm with a debt of more than $400 million. The debt resulted from the mortgage financing provided by the U.S. Eximbank and the U.S. aircraft manufacturer McDonnell Douglas. As most major airlines do, Laker Airways incurred three major categories of cost: (1) fuel, typically paid for in U.S. dollars (even though the United Kingdom is more than self-sufficient in oil); (2) operating costs incurred in sterling (administrative expenses and salaries), but with a nonnegligible dollar cost component (advertising and booking in the United States); and (3) financing costs from the purchase of U.S.-made aircraft, denominated in dollars. Revenues accruing from the sale of transatlantic airfare were about evenly divided between sterling and dollars. The dollar fares, however, were based on the assumption of a rate of $2.25 to the pound. The imbalance in the currency denomination of cash flows (dollar-denominated cash outflows far exceeding dollar-denominated cash inflows) left Laker vulnerable to a sterling depreciation below the budgeted exchange rate of £1 = $2.25. Indeed, the dramatic plunge of the exchange rate to £1 = $1.60 over the 1981–1982 period brought Laker Airways to default.

Could Laker have hedged its "natural" dollar liability exposure? The first option of indexing the sale of sterling airfare to the day-to-day exchange rate was not a viable alternative. Advertisements, based on a set sterling fare, would have had to be revised almost daily and would have discouraged the "price-elastic, budget-conscious" clientele of the company. Another possibility would have been for Laker to direct more of its marketing efforts toward American travelers, thereby giving it a more diversified demand structure. When the pound devalued against the dollar, fewer British tourists would vacation in the United States, but more Americans would travel to Britain. Laker also could have financed the acquisition of DC 10 aircraft in sterling rather than in dollars, thereby more closely matching its pound outflows with its pound inflows. This example points out that the currency denomination of debt financing can ill afford to be determined apart from the currency risk faced by the firm's total business portfolio.

❧ 10.7 SUMMARY AND CONCLUSIONS

In this chapter, we examined the concept of exposure to exchange rate changes from the perspective of the economist. We saw that the focus of the accounting profession on the balance sheet impact of currency changes has led accountants to ignore the more important effect that these changes may have on future cash flows. Moreover, it is now apparent that currency risk and inflation risk are intertwined—that through the theory of purchasing power parity, these risks are, to a large extent, offsetting. Hence, for firms incurring costs and selling products in foreign countries, the net effect of currency appreciations and depreciations may be less important in the long run.

One implication of this close association between inflation and currency fluctuations is that to measure exposure properly, we must focus on inflation-adjusted, or real, exchange rates instead of on nominal, or actual, exchange rates. Therefore, economic exposure has been defined as the extent to which the value of a firm is affected by currency fluctuations, inclusive of price-level changes. Thus, any accounting measure that focuses on the firm's past activities and decisions, as reflected in its current balance-sheet accounts, is likely to be misleading.

Although exchange risk is conceptually easy to identify, it is difficult in practice to determine what the actual economic impact of a currency change will be. For a given firm, this impact depends on a great number of variables including the location of its major markets and competitors, supply and demand elasticities, substitutability of inputs, and offsetting inflation. We did see a technique that avoids many of these problems by using regression analysis to determine an operation's exposure to exchange risk. However, its applicability is limited by the assumption that the past is representative of the future.

➤ Questions

1. Please answer the following questions.
   a. Define exposure, differentiating between accounting and economic exposure. What role does inflation play?
   b. Describe at least three circumstances under which economic exposure is likely to exist.
   c. Of what relevance are the international Fisher effect and purchasing power parity to your answers to parts a and b?
   d. What is exchange risk, as distinct from exposure?

2. Under what circumstances might multinational firms be less subject to exchange risk than purely domestic firms in the same industry?

3. Suppliers of the equipment used to make semiconductors, such as Applied Materials and LAM Research, who produce in the United States but are heavily dependent on sales to Asia, saw their share prices plummet in the wake of the Asian financial crisis. Explain.

4. Malaysian palm oil producers export more than 90% of their product for sale in dollars. Virtually all their costs, however, are in Malaysian ringgit.
   a. How would the 30% fall in the value of the ringgit during 1997 affect the ringgit profitability of these producers? Explain.
   b. How would the ringgit's depreciation affect the dollar profits of these producers? Explain.

5. Should Laker have financed its purchase of DC 10 aircraft by borrowing sterling from a British bank rather than using the dollar-denominated financing supplied by McDonnell Douglas and the Eximbank? Explain.

6. Many business people and the business press are convinced that a devalued dollar offers a significant advantage to foreign bidders for American companies and real estate. Comment on this position.

7. The sharp decline of the U.S. dollar between 1985 and 1995 significantly improved the profitability of U.S. firms both at home and abroad.
   a. In what sense is this profit improvement false prosperity?
   b. How would you incorporate the decline in the dollar in evaluating management performance? In making investment decisions?

8. E & J Gallo is the largest vintner in the United States. It gets its grapes in California (some of which it grows itself) and sells its wines throughout the United States. Does Gallo face currency risk? Why and how?

9. Chrysler Motors exports vans to Europe in competition with the Japanese. Similarly, Digital Equipment exports computers to Europe. However, all of Digital's biggest competitors are American companies—IBM, Hewlett-Packard, and Tandem. Assuming all else is equal, which of these companies—Chrysler or Digital—is likely to benefit more from a weak dollar? Explain.

10. Saint-Gobain, a French firm, and Pilkington PLC, a British firm, are arch rivals in the European flat-glass business. After Britain's exit from the exchange-rate mechanism in September 1992, the pound fell by 15% against the franc.
    a. What was the likely impact on Saint-Gobain's profitability of the pound devaluation?
    b. What was the likely impact on Pilkington's profitability of the pound devaluation?

11. In 1994, the Singapore dollar rose by 9% in real terms against the U.S. dollar. What was the likely impact of the strong Singapore dollar on U.S. electronics manufacturers using Singapore as an export platform? Consider the following facts. On average, materials and components—85% of which are purchased abroad—account for about 60% of product costs. Labor accounts for an additional 15%; other operating costs account for the remaining 25%.

12. Bakrie, an Indonesian conglomerate, is assessing the likely consequences of the rupiah's precipitous decline on its different businesses. These businesses

include a telecommunications company that is building a network (using mostly imported equipment) throughout Jakarta to offer wireless service to its residents; a company that sells pipe to the Western firms exploiting Indonesia's oil and gas fields; and a big agricultural business (54% of its revenues are in dollars, compared with 40% of its costs) that owns rubber and palm plantations feeding a large refining and distribution operation.

   a. Assess the likely impact of the rupiah's depreciation on Bakrie's three different businesses.

   b. Which of Bakrie's businesses will be most hurt by the rupiah's fall? Will any of these businesses actually benefit from rupiah depreciation?

   c. Bakrie has about $1 billion in foreign debt. Will this debt increase or decrease its currency exposure? Explain.

13. Midwestern Bank has lent $10 million to finance an equipment sale to Thailand by Lasertech, a major exporter located in Michigan. Both the loan and the sale are priced in U.S. dollars.

   a. Is Midwestern's loan to Lasertech exposed to exchange risk? Explain.

   b. Suppose Midwestern has lent money to Lasertech secured by the general credit of the company. Are these loans exposed to exchange risk? Explain.

⮞ Problems

Problems 1 and 2 are based on the Spectrum Manufacturing AB case (scenarios 1, 2, and 3) presented in the chapter. Calculate Spectrum's economic exposure under the following new scenarios:

1. *Scenario 4: Sales and import prices rise; domestic materials substituted for imported materials; other variables remain the same.*

   a. Spectrum is able to raise the krona price of its sheet plastic to SKr 25 to exactly offset the effect of the devaluation.

   b. Because of domestic materials substitutions, krona operating expenditures rise by only 4% relative to the base case.

   c. Physical sales volume stays at its predevaluation level.

2. *Scenario 5: Volume and import prices rise; other variables remain the same.*

   a. The krona sales price is held constant at SKr 20.

   b. Unit sales volume rises by 50%, both domestically and abroad, owing to the lower dollar price.

   c. Because krona costs of local labor and materials stay the same, krona unit operating expenditures rise by only 5.6%.

   d. The firm's various overhead expenses do not change.

3. Hilton International is considering investing in a new Swiss hotel. The required initial investment is $1.5 million (or SFr 2.38 million at the current exchange rate of $0.63 = SFr 1). Profits for the first 10 years will be reinvested, at which time Hilton will sell out to its partner. Based on projected earnings, Hilton's share of this hotel will be worth SFr 3.88 million in 10 years.

   a. What factors are relevant in evaluating this investment?

   b. How will fluctuations in the value of the Swiss franc affect this investment?

   c. How would you forecast the $:SFr exchange rate 10 years ahead?

4. A proposed foreign investment involves a plant whose entire output of 1 million units per annum is to be exported. With a selling price of $10 per unit, the yearly revenue from this investment equals $10 million. At the present rate of exchange, dollar costs of local production equal $6 per unit. A 10% devaluation is expected to lower unit costs by $0.30, and a 15% devaluation will reduce these costs by an additional $0.15. Suppose a devaluation of either 10% or 15% is likely, with respective probabilities of 0.4 and 0.2 (the probability of no currency change is 0.4). Depreciation at the current exchange rate equals $1 million annually, and the local tax rate is 40%.

   a. What will annual dollar cash flows be if no devaluation occurs?

   b. Given the currency scenario described, what is the expected value of annual after-tax dollar cash flows assuming no repatriation of profits to the United States?

5. On January 1, the U.S. dollar:Japanese yen exchange rate is $1 = ¥250. During the year, U.S. inflation is 4% and Japanese inflation is 2%. On December 31, the exchange rate is $1 = ¥235. What are the likely competitive effects of this exchange rate change on Caterpillar Inc., the American earth-moving equipment manufacturer, whose toughest competitor is Japan's Komatsu?

6. Mucho Macho is the leading beer in Patagonia, with a 65% share of the market. Because of trade barriers, it faces essentially no import competition.

Exports account for less than 2% of sales. Although some of its raw material is bought overseas, the large majority of the value added is provided by locally supplied goods and services. Over the past five years, Patagonian prices have risen by 300%, and U.S. prices have risen by about 10%. During this time period, the value of the Patagonian peso has dropped from P 1 = $1.00 to P 1 = $0.50.

   a. What has happened to the real value of the peso over the past five years? Has it gone up or down? A little or a lot?

   b. What has the high inflation over the past five years likely done to Mucho Macho's peso profits? Has it moved profits up or down? A lot or a little? Explain.

   c. Based on your answer to part a, what has been the likely effect of the change in the peso's real value on Mucho Macho's peso profits converted into dollars? Have dollar-equivalent profits gone up or down? A lot or a little? Explain.

   d. Mucho Macho has applied for a dollar loan to finance its expansion. Were you to look solely at its past financial statements in judging its creditworthiness, what would be your likely response to Mucho Macho's dollar loan request?

   e. What foreign exchange risk would such a dollar loan face? Explain.

7. In 1990, General Electric acquired Tungsram Ltd., a Hungarian light bulb manufacturer. Hungary's inflation rate was 28% in 1990 and 35% in 1991, while the forint (Hungary's currency) was devalued 5% and 15%, respectively, during those years. Corresponding inflation for the U.S. was 6.1% in 1990 and 3.1% in 1991.

   a. What happened to the competitiveness of GE's Hungarian operations during 1990 and 1991? Explain.

   b. In early 1992, GE announced that it would cut back its capital investment in Tungsram. What might have been the purpose of GE's publicly announced cutback?

8. You are asked to lend money for a major commercial real estate development in the town of Calexico, which is on the California side of the Mexican border. There is some talk about a further devaluation of the Mexican peso. What information do you need to assess the creditworthiness of this project?

9. About two-thirds of all California almonds are exported. The ups and downs of the U.S. dollar, therefore, cause headaches for almond growers. To avoid these problems, a grower decides to concentrate on domestic sales. Does that grower bear exchange risk? Why and how?

10. Aldridge Washmon Co. is one of the largest distrib- utors of heavy farming equipment in Brownsville, Texas, located on the border with Mexico. The time is late 1981. Sales have increased dramatically over the past two years, and Aldridge is requesting an expansion of its credit line. What information would you as a banker need before you make a decision?

11. Assess the likely consequences of a declining dollar on Fluor Corporation, the international construction-engineering contractor based in Irvine, California. Most of Fluor's value added involves project design and management; most of its costs are for U.S. labor in design, engineering, and construction-management services.

12. The European chemical industry pays for an estimated 79% of its oil-based feedstock in dollars. Thus, its costs are declining sharply because of the drop in the price of oil combined with the sharp decline in the value of the dollar. What is the likely impact of the dollar's decline on the European chemical industry's profits? Will it now be more competitive relative to the American chemical industry?

13. Cooper Industries is a maker of compressors, pneumatic tools, and electrical equipment. It does not face much foreign competition in the United States, and exports account for only 7% of its sales. Does it face exchange risk?

14. The Edmonton Oilers (Canada) of the National Hockey League are two-time defending Stanley Cup champions. (The Stanley Cup playoff is hockey's equivalent of football's Super Bowl or baseball's World Series.) As is true of all NHL teams, most of the Oilers' players are Canadian. How are the Oilers affected by changes in the Canadian-dollar·U.S.-dollar exchange rate?

15. South Korean companies such as Goldstar, Samsung, and Daewoo have captured more than 10% of the U.S. color television market with their small, low-priced sets. They are also becoming more significant exporters of videocassette recorders and small microwave ovens. What currency risk do these firms face?

16. A common complaint leveled against the Japanese government is that it deliberately holds down the value of the yen to boost exports of Japanese products. American steelmakers have been particularly vocal in their complaints. As a remedy, steelmakers in 1985 asked President Reagan to curtail Japanese steel imports further and to impose a 25% tariff to offset what they describe as the "artificial" undervaluation of the yen. Does Nippon Steel profit from a weak yen? What are the likely consequences of the recent appreciation of the yen? Here are some facts. Imports of U.S. raw materials priced in dol-

lars account for about one-third of Nippon's costs, and exports to the United States generate about 4% to 5% of its revenues. Nippon Steel currently is exporting as much steel as it can to the United States under existing quota restrictions. What additional information do you need to fully assess the impact of currency changes on Nippon Steel?

17. Monsanto Co., the St. Louis chemical firm, is a major seller of herbicides. Its two brand-name herbicides, Roundup and Lasso, have a large share of the U.S. and foreign markets. Its major competitors are other U.S. chemical companies. How are sales and profits of these products, as well as Monsanto's other chemicals, likely to be affected by changes in the value of the dollar?

18. Black & Decker Manufacturing Co., of Towson, Maryland, has roughly 45% of its assets and 40% of its sales overseas. How does a soaring dollar affect its profitability, both at home and abroad?

19. The shipbuilding industry is facing a worldwide capacity surplus. Although Japan currently controls about 50% of the world market, it is facing severe competition from the South Koreans. Japanese shipyards are extraordinarily productive, but at current price levels they were just about breaking even with an exchange rate of ¥240 = $1. What are the likely effects on Japanese shipbuilders of a yen apprecia-

tion to ¥85 = $1? Assume the South Korean won has maintained its dollar value.

20. Nissho Iwai American Corporation is the American arm of a large Japanese trading company that deals in everything from steel to tuna fish. Assess the credit-risk implications for Nissho Iwai of a 30% rise in the dollar value of the yen.

21. Thomasville Plastics Corp. has contracted to buy $1.1 million worth of Japanese plastic-injection molding machines. The contract price is set in dollars. Does Thomasville bear any currency risk associated with this purchase? Explain.

22. Over the past year, China has experienced an inflation rate of about 22%, in contrast to U.S. inflation of about 3%. At the same time, the exchange rate has gone from Y8.7/U.S.$1 to Y8.3/U.S.$1.
    a. What has happened to the real value of the yuan over the past year? Has it gone up or down? A little or a lot?
    b. What are the likely effects of the change in the yuan's real value on the dollar profits of a company like Procter & Gamble that sells almost exclusively in the local market?
    c. What are the likely effects of the change in the yuan's real value on the dollar profits of a textile manufacturer that exports most of its output to the United States?

➤ Bibliography

ADLER, MICHAEL, and BERNARD DUMAS. "Exposure to Currency Risk: Definition and Measurement." *Financial Management*, Summer 1984, pp. 41–50.

ANKROM, ROBERT K. "Top Level Approach to the Foreign Exchange Problem." *Harvard Business Review*, July–August 1974, pp. 79–90.

CORNELL, BRADFORD, and ALAN C. SHAPIRO. "Managing Foreign Exchange Risks." *Midland Corporate Finance Journal*, Fall 1983, pp. 16–31.

EAKER, MARK R. "The Numeraire Problem and Foreign Exchange Risk." *Journal of Finance*, May 1981, pp. 419–426.

GARNER, C. KENT, and ALAN C. SHAPIRO. "A Practical Method of Assessing Foreign Exchange Risk." *Midland Corporate Finance Journal*, Fall 1984, pp. 6–17.

GIDDY, IAN H. "Exchange Risk: Whose View?" *Financial Management*, Summer 1977, pp. 23–33.

LESSARD, DONALD R., and JOHN B. LIGHTSTONE. "Volatile Exchange Rates Can Put Operations at Risk." *Harvard Business Review*, July–August 1986, pp. 107–114.

SHAPIRO, ALAN C. "Exchange Rate Changes, Inflation, and the Value of the Multinational Corporation." *Journal of Finance*, May 1975, pp. 485–502.

MANAGING ECONOMIC EXPOSURE

We concluded there's enough risk in the car business without trying to get rich in foreign exchange; so our policy is essentially defensive.

Executive at Ford Motor Co.

The explosive rise of the dollar's value that began in 1981 cut deeply into the ability of American manufacturers to export their products, even as it gave a welcome boost to the U.S. sales of foreign companies. While still paying attention to the short-term balance sheet effects of the strong dollar, most firms responded to its economic consequences by making the longer-term operating adjustments described in the headline above. These firms were then well-positioned to take advantage of the stunning reversal of the dollar's fortunes that began in 1985.

Based on these and similar experiences, this chapter discusses the basic considerations that go into the design of a strategy for managing foreign exchange risk, along with the marketing, production, and financial management strategies that are appropriate for coping with the economic consequences of exchange rate changes. Its basic message is straightforward: Because currency risk affects all facets of a company's operations, it should not be the concern of financial managers alone. Operating managers, in particular, should develop marketing and production initiatives that help to ensure profitability over the long run. They should also devise anticipatory or proactive, rather than reactive, strategic alternatives in order to gain competitive leverage internationally.

The focus on the real (economic) effects of currency changes and how to cope with the associated risks suggests that a sensible strategy for exchange risk management is one that is designed to protect the dollar (HC) earning power of the company as a whole. Although firms can easily hedge exposures based on projected foreign-currency cash flows, *competitive exposures*—those arising from competition with firms based in other currencies—are longer-term, harder to quantify, and cannot be dealt with solely through financial hedging techniques. Rather, they require making the longer-term operating adjustments that are described in this chapter.

11.1 AN OVERVIEW OF OPERATING EXPOSURE MANAGEMENT

In order for a currency depreciation or appreciation to significantly affect a firm's value, it must lead to changes in the relative prices either of the firm's inputs or of the products bought or sold in various countries. The impact of these currency-induced relative price changes on corporate revenues and costs depends on the extent of the firm's commitment to international business, its competitive environment, and its degree of operational flexibility.

To the extent that exchange rate changes do bring about relative price changes, the firm's competitive situation will be altered. As a result, management may wish to adjust its production process or its marketing mix to accommodate the new set of relative prices.

By making the necessary marketing and production revisions, companies can either counteract the harmful effects of, or capitalize on the opportunities presented by, a currency appreciation or depreciation. We turn again here to the concept of the real exchange rate (discussed in Chapter 7 and elaborated on in Chapter 10). The distinction between the nominal and the real exchange rates has important implications for those marketing and production decisions that bear on exchange risk. As we saw in Chapter 10, nominal currency changes that are fully offset by differential inflation do not entail a material degree of real exchange risk for a firm unless that firm has major contractual agreements in the foreign currency. A real exchange rate change, however, can strongly affect the competitive positions of local firms and their foreign competitors.

The following are some of the proactive marketing and production strategies that a firm can pursue in response to anticipated or actual real exchange rate changes.

MARKETING INITIATIVES	PRODUCTION INITIATIVES
Marketing selection	Product sourcing
Product strategy	Input mix
Pricing strategy	Plant location
Promotional strategy	Raising productivity

The appropriate response to an anticipated or actual real exchange rate change depends crucially on the length of time that real change is expected to persist. For example, after a real home currency appreciation, the exporter has to decide whether, and how much, to raise its foreign currency prices. If the change is expected to be temporary and if regaining market share will be expensive, the exporter will probably prefer to maintain its foreign currency prices at existing levels. Although this response will mean a temporary reduction in unit profitability, the alternative—raising prices now and reducing them later when the real exchange rate declines—could be even more costly. A longer-lasting change in the real exchange rate, however, will probably lead the firm to raise its foreign currency prices, at the expense of losing some export sales. Assuming a still more permanent shift, management might choose to build production facilities overseas. Alternatively, if the cost of regaining market share is sufficiently great, the firm can hold foreign currency prices constant and count on shifting production overseas to preserve its longer-term profitability.

These considerations are illustrated by the experience of Millipore Corp., a Bedford, Massachusetts, maker of industrial equipment for domestic and foreign companies. Confronted by the strong dollar, Millipore decided to cut prices in dollar terms on some products and hold back increases on others to maintain its market share against foreign competition. Although earnings were damaged in the short run, Millipore's management decided that it would be more expensive in the long run to regain market share. To cope with the pressure on its profit margins, Millipore opened a plant in Japan and expanded another one in France.

In general, real exchange rate movements that narrow the gap between the current rate and the equilibrium rate are likely to be longer lasting than are those that widen the gap. Neither, however, will be permanent. Rather, in a world without an anchor for currency values, there will be a sequence of equilibrium rates, each of which has its own implications for marketing and production strategies.

11.2 MARKETING MANAGEMENT OF EXCHANGE RISK

The design of a firm's marketing strategy under conditions of home currency (HC) fluctuation presents considerable opportunity for gaining competitive leverage. Thus one of the international marketing manager's tasks should be to identify the likely effects of a currency change and then act on them by adjusting pricing and product policies.

Market Selection

Major strategic considerations for an exporter are the markets in which to sell—that is, *market selection*—and the relative marketing support to devote to each market. As a result of the strong dollar, for example, some discouraged U.S. firms pulled out of markets that foreign competition made unprofitable. From the perspective of foreign companies, however, the strong U.S. dollar was a golden opportunity to gain market share at the expense of their U.S. rivals. Japanese and European companies also used their dollar cost advantage to carve out market share against American competitors in third markets. The subsequent drop in the dollar helped U.S. firms turn the tables on their foreign competitors, both at home and abroad.

It is also necessary to consider the issue of *market segmentation* within individual countries. A firm that sells differentiated products to more affluent customers may not be harmed as much by a foreign currency devaluation as will a mass marketer. On the other hand, after a depreciation of the home currency, a firm that sells primarily to upper-income groups may find it is now able to penetrate mass markets abroad.

ILLUSTRATION U.S. TEXTILE MILLS COPE WITH CURRENCY RISK

Thanks to mammoth modernization efforts, the productivity of U.S. textile mills is now the highest in the world. Nevertheless, although American textile manufacturers are currently on an equal footing with those of Italy and Japan, their costs are 30% to 40% higher for products such as lightweight shirting than costs in the Latin American and East Asian newly industrialized countries. Thus, the industry is concentrating on sophisticated materials such as industrial fabrics and on goods such as sheets and towels that require little direct labor.

American producers are also competing by developing a service edge in the domestic market. For example, the industry developed a computerized inventory management and ordering program, called Quick Response, that provides close coordination between textile mills, apparel manufacturers, and retailers. The system cuts in half the time between a fabric order and delivery of the garment to a retailer and gives all the parties better information for planning, thereby placing foreign manufacturers at a competitive disadvantage. In response, Japanese companies—and even some Korean ones—are looking to set up U.S. factories.

Market selection and market segmentation provide the basic parameters within which a company may adjust its marketing mix over time. In the short term, however, neither of these two basic strategic choices can be altered in reaction to actual or anticipated currency changes. Instead, the firm must select certain tactical responses such as adjustments of pricing, promotional, and credit policies. In the long run, if the real exchange rate change persists, the firm will have to revise its marketing strategy.

Pricing Strategy

Two key issues that must be addressed when developing a *pricing strategy* in the face of currency volatility are whether to emphasize market share or profit margin and how frequently to adjust prices.

MARKET SHARE VERSUS PROFIT MARGIN ■ In the wake of a rising dollar, a U.S. firm selling overseas or competing at home against foreign imports faces a Hobson's choice: Does it keep its dollar price constant to preserve its profit margin and, thereby, lose sales volume, or does it cut its dollar price to maintain market share and, thereby, suffer a reduced profit margin? Conversely, does the firm use a weaker dollar to regain ground lost to foreign competitors, or does it use the weak dollar to raise prices and recoup losses incurred from the strong dollar?

To begin the analysis, a firm selling overseas should follow the standard economic proposition of setting the price that maximizes dollar profits (by equating marginal revenues and marginal costs). In making this determination, however, firms should translate profits using the forward exchange rate that reflects the true expected dollar value of the receipts upon collection.

After appreciation of the dollar, which is equivalent to a foreign currency (FC) depreciation, a firm selling overseas should consider opportunities to increase the FC prices of its products. The problem, of course, is that local producers now will have a competitive cost advantage, limiting an exporter's ability to recoup dollar profits by raising FC selling prices.

At best, therefore, an exporter will be able to raise its product prices by the extent of the FC devaluation. For example, suppose Avon is selling cosmetics in England priced at £2.00 when the exchange rate is £1 = $1.80. This gives Avon revenue of $3.60 per unit. If the pound devalues to £1 = $1.50, Avon's revenue will fall to $3.00, unless it can raise its selling price to £2.40 (2.40 × 1.5 = 3.60). At worst, in an extremely competitive situation, the exporter will be forced to absorb a reduction in home currency revenues equal to the percentage decline in the value of the local currency. For example, if Avon cannot raise its pound price, its new dollar price of $3.00 represents a 16.7% decline in revenue, the same percentage decline as the drop in the value of the pound [(1.80 − 1.50)/1.80].

In the most likely case, foreign currency prices can be raised somewhat, and the exporter will make up the difference through a lower profit margin on its foreign sales. This is the strategy followed by the Japanese auto industry. Exhibit 11.1a shows that Japanese car prices in the United States rose by much less than the value of the yen from 1985 to 1994, even as the automakers upgraded their cars. Nonetheless, the data also show that real Japanese car prices climbed over this 10-year period, hurting Japanese market share. In response, the Japanese companies became even more aggressive. For example, Exhibit 11.1b shows that even as the dollar fell dramatically against the yen over the period 1991–1994 (the 1992 model actually comes out in 1991, and so on), Nissan actually lowered the price of its Maxima sedan by almost $500. Yen depreciation since 1995 facilitated the efforts of Japanese automakers to regain market share by cutting their dollar prices. For example, the 1997 Lexus ES300, a low-end luxury sedan imported from Japan, was priced at $30,395, $2,500 lower than the 1996 model it replaced.

Under conditions of dollar depreciation, it follows that U.S. exports will gain a competitive price advantage on the world market. An exporter now has the option of increasing unit profitability—that is, *price skimming*—or expanding its market share—*penetration pricing*. Continuing with the previous example of Avon, suppose the exchange rate rises to £1 = $2. At a price of £2.00, Avon's revenues will rise to $4 per unit, from $3.60. Avon has the felicitous choice of holding its pound price constant and boosting its dollar profit margin or holding its dollar price constant (by cutting its pound price to £1.80 per unit) and boosting its sales volume.

The pricing decision is influenced by such factors as whether this change is likely to persist, economies of scale, the cost structure of expanding output, consumer price sensitivity, and the likelihood of attracting competition if high unit profitability is obvious.

The greater the **price elasticity of demand**—the change in demand for a given change in price—the greater the incentive to hold down price and thereby expand sales and revenues. Similarly, if significant **economies of scale** exist, it generally will be worthwhile to hold down price, expand demand, and thereby lower unit production costs. This

EXHIBIT 11.1a Japanese Automakers Kept Their Prices Down: 1985–1994

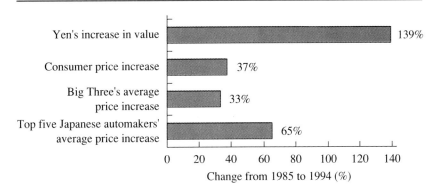

Yen's increase in value — 139%
Consumer price increase — 37%
Big Three's average price increase — 33%
Top five Japanese automakers' average price increase — 65%

Change from 1985 to 1994 (%)

EXHIBIT 11.1b Nissan Holds Down the Maxima's Price as the Yen Rises

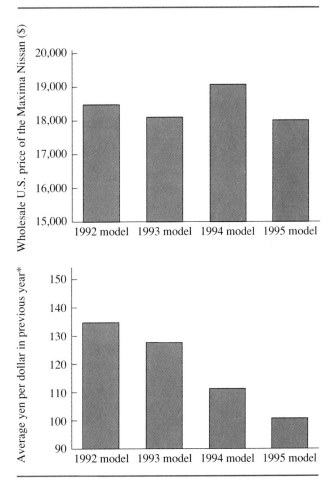

Wholesale U.S. price of the Maxima Nissan ($)

1992 model   1993 model   1994 model   1995 model

Average yen per dollar in previous year*

1992 model   1993 model   1994 model   1995 model

* The 199x model came out in 199x-1, which is why the exchange rate depicted is for the year prior to the model year.

Source: International Monetary Fund, *Los Angeles Times*, August 8, 1994, p. D1.

is the strategy followed by Japanese (but usually not American) manufacturers. The reverse is true if economies of scale are nonexistent or if price elasticity is low.

ILLUSTRATION

A.T. CROSS MARKS DOWN ITS PEN PRICES

In February 1993, after a 10% decline in the dollar (from ¥123 to ¥111), A.T. Cross cut the yen prices of its pens by 20%. For example, a 10-carat gold Cross pen was marked down to ¥8,000 from ¥10,000. Suppose that the manufacturing and shipping costs of this Cross pen were $25 and distribution costs were ¥2,000, giving Cross a pre-exchange rate change contribution margin of $40 (¥10,000/123 − $25 − ¥2,000/123). Assuming these costs stay the same, how much added volume must Cross generate in order to maintain its prechange dollar profits on this pen?

Solution: The contribution margin at the new price and exchange rate equals $29 (¥8,000/111 − $25 − ¥2,000/111). In order to maintain dollar profits on this pen at its previous level, unit sales must rise by 38% (40/29). A sales increase of this magnitude implies a price elasticity of demand of 1.9 (38/20).

Historically, many of the exports of U.S. multinationals appear to have fit the latter category (low price elasticity of demand) because they were technologically innovative or differentiated, often without close substitutes. Thus, there was a pronounced tendency for U.S. firms not to decrease prices after the real dollar devaluations of the 1970s.

Similarly, after dollar appreciation in the early 1980s, European and Japanese automakers were able to keep the dollar prices up on their car exports to the United States. European cars figure largely in the luxury car market, which is fairly insensitive to price swings. Import quotas enabled the Japanese car companies to avoid the price cutting they find necessary in their highly competitive home market. This factor, combined with the strong dollar, resulted in the major Japanese automakers' earning about 80% of their 1984 worldwide profits in the United States.

By the late 1980s, however, the United States was no longer the across-the-board leader in the development and application of new technology to manufactured goods. As the U.S. technological edge has eroded, more American companies have faced competition from companies in other industrial nations. Absent other strategies to reduce costs or the price sensitivity of demand, these American firms have become more subject to exchange risk.

In general, firms dealing in commodities, such as table wines, face more exchange risk than those selling differentiated products, such as Mercedes Benzes. For example, when French winemakers refused to cut their franc prices as the dollar declined, French table wine exports to the United States dropped about 27% in 1987.

In deciding whether to raise prices after a foreign currency devaluation, companies must consider not just sales that will be lost today but also the likelihood of losing future sales as well. The reason is that once they build market share, foreign firms are unlikely to pull back. For example, foreign capital goods manufacturers used the period when they had a price advantage to build strong U.S. distribution and service networks. U.S. firms that had not previously bought foreign-made equipment became loyal customers. When the dollar fell, foreign firms opened U.S. plants to supply their distribution systems and hold on to their customers.

The same is true in many other markets as well: *A customer who is lost may be lost forever*. Americans who discover California wines may not switch back to French wines even after a franc devaluation. Similarly, in the auto business, a customer who is satisfied with a foreign model may stick with that brand for a long time.

FREQUENCY OF PRICE ADJUSTMENTS ■ Firms in international competition differ in their ability and willingness to adjust prices in response to exchange rate changes. Some firms constantly adjust their prices for exchange rate changes. Other companies, however, feel that stable prices are a key ingredient in maintaining their customer base. For example, customers who have invested in machinery or other assets that operate best with supplies from the selling firm may value a contract that is fixed in both price and quantity. Similarly, the company may try to shield risk-averse customers by offering them prices fixed in their local currency for a certain period of time.

It is important also not to neglect the effect of frequent price changes on the exporter's distributors, who must constantly adjust their margins to conform to the prices they pay. A number of firms now have different list prices for domestic and foreign customers in order to shield their foreign customers—especially those who sell through catalogues—from continual revisions of overseas prices.

Turning now to domestic pricing after devaluation, a domestic firm facing strong import competition may have much greater latitude in pricing. It then has the choice of potentially raising prices consistent with import price increases or of holding prices constant in order to improve market share. Again, the strategy depends on such variables as economies of scale and *consumer price sensitivity*.

In early 1978, for instance, General Motors and Ford took advantage of price increases on competitive foreign autos to raise prices on their Chevette and Pinto models. The prices of those small cars had previously been held down, and even reduced, in an attempt to combat the growing market share of German and Japanese imports. However, the declining value of the U.S. dollar relative to the Deutsche mark and yen led the German and Japanese automakers to raise their dollar prices. The price increases by the U.S. manufacturers, which were less than the sharp rise in import prices, improved profit margins and kept U.S. cars competitive with their foreign rivals.

Promotional Strategy

Promotional strategy should similarly take into account anticipated exchange rate changes. A key issue in any marketing program is the size of the promotional budget for advertising, personal selling, and merchandising. Promotional decisions should explicitly build in exchange rates, especially in allocating budgets among countries. The appreciation of the U.S. dollar in the early 1980s illustrates these promotional considerations. European countries wooed U.S. skiers from the Rocky Mountains with campaigns capitalizing on their lower costs and comparable Alpine skiing.

A firm exporting its products after a domestic devaluation may well find that the return per dollar expenditure on advertising or selling is increased because of the product's improved price positioning. In the wake of the falling dollar, *Business Week* reported that "food, beverage, and tobacco companies are boosting their promotional spending overseas to capitalize on their newfound pricing advantage."[1] A foreign currency devaluation, on the other hand, is likely to reduce the return on marketing expenditures and may require a more fundamental shift in the firm's product policy.

One should also consider the interactions between pricing and promotional strategy. For example, after the sharp drop in the value of the won in 1997, Korean automakers decided to generally hold the line on prices in the U.S. market. Among other things, they feared that aggressive pricing would reinforce the undesirable image that they produce cheap and shoddy goods. Instead, they decided to fatten their profit margins and use the additional funds for advertising and other forms of image-building, adding more features, and financing cheaper leases (which is a form of price cutting).

Product Strategy

Companies often respond to exchange risk by altering their *product strategy*, which deals with such areas as new-product introduction, product-line decisions, and product innovation. One way to cope with exchange rate fluctuations is to change the timing of the introduction of new products. For example, because of the competitive price advantage, the period after a home currency depreciation may be the ideal time to develop a brand franchise. Societe Claude Havrey, a French maker of women's clothes, began its U.S. sales push in 1984, after a significant strengthening of the dollar against the French franc. According to the export sales manager, "If the dollar were weak, we might have waited a while before starting. You have to choose the right time to start—the hard part is implanting yourself in the foreign market."[2] The strong dollar enabled the firm to price its clothes competitively in the United States.

Exchange rate fluctuations also affect product-line decisions. After home currency devaluation, a firm will potentially be able to expand its product line and cover a wider spectrum of consumers both at home and abroad. Conversely, after appreciation of the home currency, a firm may have to reorient its product line and target it to a higher-income, more quality-conscious, less price-sensitive constituency. Volkswagen, for example, achieved its export prominence on the basis of low-priced, stripped-down, low-maintenance cars. The appreciation of the Deutsche mark in the early 1970s, however, effectively ended VW's ability to compete primarily on the basis of price. The company lost more than $310 million in 1974 alone attempting to maintain its market share by lowering DM prices. To compete in the long run, Volkswagen was forced to revise its product line and sell relatively high-priced cars to middle-income consumers, from an extended product line, on the basis of quality and styling rather than cost.

The equivalent strategy for firms selling to the industrial, rather than consumer, market and confronting a strong home currency is *product innovation*, financed by an expanded research and development (R&D) budget. Kollmorgen Corp., a Connecticut-based electronic components company, responded to the strong dollar in part by increasing its R&D budget by 40%. According to Kollmorgen's chairman, Robert Swiggett, "We're not counting on being able to increase foreign sales very substantially unless we can keep in-

---

[1] "Weak Dollar, Strong Profits," *Business Week*, July 11, 1994, p. 39.
[2] *Wall Street Journal*, January 18, 1984, p. 16.

troducing new product lines. That Bunsen burner is burning an awful lot brighter these days."[3] For example, in order to stay competitive in Asia after the steep decline in Asian currencies in 1997, Minneapolis-based Medtronic designed a simple $1,000 pacemaker to replace its more sophisticated $5,000 U.S.-made pacemakers.

11.3 PRODUCTION MANAGEMENT OF EXCHANGE RISK

The adjustments discussed so far involve attempts to alter the dollar value of foreign currency revenues. However, sometimes exchange rates move so much that pricing or other marketing strategies cannot save the product. This was the case for U.S. firms in the early 1980s and for Japanese firms in the late 1980s. Firms facing this situation must either drop uncompetitive products or cut their costs.

Product sourcing and *plant location* are the principal variables that companies manipulate to manage competitive risks that cannot be dealt with through marketing changes alone. Consider, for example, the possible responses of U.S. firms to a strong dollar. The basic strategy would involve shifting the firm's manufacturing base overseas, but this can be accomplished in more than one way.

Input Mix

Outright additions to facilities overseas naturally accomplish a manufacturing shift. A more flexible solution is to change the input mix by purchasing more components overseas. After the rise of the dollar in the early 1980s, most U.S. companies increased their global sourcing. For example, Medtronic cut costs by producing its $1,000 pacemakers in a new plant in China. Similarly, Caterpillar responded to the soaring U.S. dollar and a tenacious competitor, Japan's Komatsu, by "shopping the world" for components. More than 50% of the pistons that the company uses in the United States now come from abroad, mainly from a Brazilian company. Some work previously done by Caterpillar's Milwaukee plant was moved to a subsidiary in Mexico. Caterpillar also stopped most U.S. production of lift trucks and began importing a new line—complete with Cat's yellow paint and logo—from South Korea's Daewoo. The steep decline in the value of the won has cut the dollar cost of manufacturing in Korea still further.

For a firm already manufacturing overseas, the cost savings associated with using a higher proportion of foreign produced goods and services, after dollar appreciation, will depend on subsequent foreign price behavior. Goods and services with a low import content or with little involvement in international trade will exhibit greater dollar price decreases than those used in international trade or with a high import content.

For the longer term, when increasing production capacity, the firm should consider the option of designing new facilities that provide added flexibility in making substitutions among various sources of goods. Maxwell House, for instance, can blend the same coffee whether using coffee beans from Brazil, the Ivory Coast, or other producers. The

---

[3] *Wall Street Journal*, August 1, 1984, p. 16.

JAPANESE AUTOMAKERS OUTSOURCE TO COPE WITH A STRONG YEN

Japanese automakers have protected themselves against a strong yen by purchasing a significant percentage of intermediate components from independent suppliers. This practice, called *outsourcing*, gives them the flexibility to shift purchases of intermediate inputs toward suppliers with costs least affected by exchange rate changes. Some of these inputs come from South Korea and Taiwan, nations whose currencies have been closely linked to the U.S. dollar. Thus, even if such intermediate goods are not priced in dollars, their yen-equivalent prices tend to decline with the dollar and, thereby, lessen the impact of a falling dollar on the cost of Japanese cars sold in the United States.

Outsourcing in countries whose currencies are linked to the currency of the export market also creates competitive pressures on domestic suppliers of the same intermediate goods. To cope in such an environment, domestic suppliers must themselves have flexible arrangements with their own input suppliers. In many cases, these smaller firms can survive because they have greater ability to recontract their costs than do the larger firms specializing in assembly and distribution. When the suppliers are faced with the reality of an exchange rate change that reduces the competitive price of their outputs, they are able to recontract with their own inputs (typically by lowering wages) to reduce costs sufficiently to remain economically viable.

advantages of being able to respond to relative price differences among domestic and imported inputs must be weighed, of course, against the extra design and construction costs.

Shifting Production Among Plants

Multinational firms with worldwide production systems can allocate production among their several plants in line with the changing dollar costs of production, increasing production in a nation whose currency has devalued and decreasing production in a country where there has been a revaluation. Contrary to conventional wisdom, therefore, multinational firms may well be subject to less exchange risk than an exporter, given the MNC's greater ability to adjust its production (and marketing) operations on a global basis, in line with changing relative production costs.

A good example of this flexibility is provided by the former Westinghouse Electric Corp. (now CBS) of Pittsburgh, Pennsylvania. Westinghouse's businesses (now largely sold off to other companies) quote their customers prices from numerous foreign affiliates: gas turbines from Canada, generators from Spain, circuit breakers and robotics from Britain, and electrical equipment from Brazil. Its sourcing decisions take into account both the effect of currency values and subsidized export financing available from foreign governments.

Of course, the theoretical ability to shift production is more limited in reality. The limitations depend on many factors, not the least of which is the power of the local labor unions involved. However, the innovative nature of the typical MNC means a continued generation of new products. The sourcing of those new products among the firm's various plants can certainly be done with an eye to the costs involved.

A strategy of *production shifting* presupposes that the MNC already has created a portfolio of plants worldwide. For example, as part of its global sourcing strategy, Caterpillar now has dual sources, domestic and foreign, for some products. These sources allow Caterpillar to "load" the plant that offers the best economies of production, given exchange rates at any moment, but multiple plants also create manufacturing redundancies and impede cost cutting.

The cost of multiple sourcing is especially great where there are economies of scale that would ordinarily dictate the establishment of only one or two plants to service the global market. However, most firms have found that in a world of uncertainty, significant benefits may be derived from production diversification. In effect, having redundant capacity is the equivalent of buying an option to execute volume shifts fairly easily. As in the case of currency options, the value of such a real option increases with the volatility of the exchange rate. Hence, despite the higher unit costs associated with smaller plants and excess capacity, currency risk may provide one more reason for the use of multiple production facilities. Indeed, 63% of foreign exchange managers surveyed cited having locations "to increase flexibility by shifting plant loading when exchange rates changed" as a factor in international siting.[4]

The case of the auto industry illustrates the potential value of maintaining a globally balanced distribution of production facilities in the face of fluctuating exchange rates. For Japanese and Swedish auto manufacturers, which historically located all their production facilities domestically, it has been feast or famine. When the home currency appreciates, as in the 1970s or the late 1980s and early 1990s, the firms' exports suffer from a lack of cost competitiveness. On the other hand, a real depreciation of the home currency, as in the early 1980s, is a time of high profits.

By contrast, Ford and General Motors, with their worldwide manufacturing facilities, have substantial leeway in reallocating various stages of production among their several plants in line with relative production and transportation costs. For example, Ford can shift production among the United States, Spain, Germany, Great Britain, Brazil, and Mexico.

Plant Location

A firm without foreign facilities that is exporting to a competitive market whose currency has devalued may find that sourcing components abroad is insufficient to maintain unit profitability. Despite its previous hesitancy, the firm may have to locate new plants abroad. For example, the economic response by the Japanese to the strong yen is well under way. Many Japanese companies have built new plants in the United States as opposed to expanding plants in Japan. In 1992, for example, transplants accounted for 45% of Japanese automakers' U.S. sales. This figure was expected to rise to 60% by 1998. Similarly, German automakers such as BMW and Mercedes-Benz have built plants in the United States, in part to shield themselves from currency fluctuations.

Third-country plant locations are also a viable alternative in many cases, depending especially on the labor intensity of production or the projections for further monetary

---

[4]Donald B. Lessard, "Survey on Corporate Responses to Volatile Exchange Rates," MIT Sloan School of Management working paper, 1990.

realignments. Many Japanese firms, for example, have shifted production offshore—to Taiwan, South Korea, Singapore, and other developing nations, as well as to the United States—first to cope with the high yen and then to take advantage of falling Asian currencies. Japanese automakers have been particularly aggressive in making these shifts. It is estimated that by the year 2000, they will produce 42% of their cars overseas, up from just 8% in 1985.

Before making such a major commitment of its resources, management should attempt to assess the length of time a particular country will retain its cost advantage. If the devaluation was due to inflationary conditions that are expected to persist, a country's apparent cost advantage may soon reverse itself. In Mexico, for example, the wholesale price index rose 18% relative to U.S. prices between January 1969 and May 1976. This rise led to a 20% peso devaluation in September 1976. Within one month, though, the Mexican government allowed organized labor to raise its wages by 35%–40%. As a result, the devaluation's effectiveness was nullified, and the government was forced to devalue the peso again in less than two months. Once again, however, the Mexican government fixed the nominal value of the peso while inflation persisted at a high level.

Yet, shifting production abroad when the home currency rises is not always the best approach. Toyota Motors, for example, decided it was cheaper to produce parts in high volume in efficient Japanese factories that are close to domestic suppliers and to assembly plants. Producing at home also improves coordination between design and manufacturing and avoids problems of quality control. For firms that rely heavily on such coordination and closeness to suppliers, raising domestic productivity is preferable to producing abroad. Nonetheless, after a certain point, currency considerations may outweigh the benefits of home country production. For example, Toyota now has three assembly lines in Canada and the United States and recently doubled the capacity—to 400,000 vehicles annually—at its Georgetown, Kentucky facility. It has also stepped up its parts purchases in the United States to an estimated $6.5 billion annually in 1996.

Raising Productivity

Many U.S. companies assaulted by foreign competition made prodigious efforts to improve their productivity—closing inefficient plants, automating heavily, and negotiating wage and benefit cutbacks and work-rule concessions with unions. Many also began programs to heighten productivity and improve product quality through employee motivation. These cost cuts now stand U.S. firms in good stead as the firms attempt to use the weaker dollar to gain back market share lost to foreign competitors.

Others, most notably the steel and auto industries, have successfully sought government import restrictions. But import quotas illustrate a dilemma facing U.S. industry. Bicycle manufacturers, for example, sought government restrictions on imports of finished bicycles; at the same time, they tried to save money by importing more parts and materials from foreign suppliers.

Another way to improve productivity and lower one's cost structure is by revising product offerings. This is the route now being taken by the Japanese. Despite their vaunted super-lean production systems, many Japanese firms, in an attempt to gain market share, have created too much product variety and offered too many options to cus-

---

THE AMERICAN PAPER AND PULP INDUSTRY RESTRUCTURES

The restructuring of the American paper and pulp industry is a classic case. It enjoys comparative advantages galore. With a $110 billion domestic market, the industry can achieve tremendous economies of scale. Then there are the fast-sprouting southern pines, which mature in half the time it takes northern types.

But plenty of other countries were giving the United States a hard time even before the dollar rose. The Finns have offset their high costs by pioneering new technologies. The Brazilians, with abundant and cheap power supplies, are the low-cost pulp producers.

The phenomenal restructuring wave that swept America, swallowing giants such as Scott Paper, Crown Zellerbach, St. Regis, and Diamond International, has rationalized product lines, held down compensation, and retired old machinery. Today the industry boasts some of the most modern facilities, probably the highest productivity, and quite possibly the fiercest cost-cutting management in the world. These strengths, and the dollar's fall, helped manufacturers regain overseas markets.

---

tomers. The result is that parts makers and assembly plants have to accommodate very small and very rare orders too frequently. This variety requires too much design work, too much capital investment for small-volume parts, too many parts inventories, and constant equipment setups and changeovers. By slashing variety to the 20% or so of models and product variations that account for 80% of sales and profits, and reducing unique parts by 30% to 50% for new models, Japanese companies are finding that they can dramatically reduce costs without sacrificing much in the way of market share.

NISSAN REVERSES COURSE

In 1993, Nissan renounced a decade-long quest to build cars in ever more sizes, colors, and functions to cater to every conceivable consumer whim. That effort spun out of control. For the 1993 model lineup alone, Nissan offered 437 different kinds of dashboard meters, 110 types of radiators, 1,200 types of floor carpets, and more than 300 varieties of ashtrays. Its Laurel model alone had 87 variations of steering wheel and 62 varieties of electrical harnesses (which link up electrical components in a car). To assemble these vehicles, Nissan used more than 6,000 different fasteners.

The payoff from this product proliferation was pathetic. Nissan engineers discovered that 70 of the 87 types of steering wheel accounted for just 5% of Laurel's sales. Overall, 50% of Nissan's model variations contributed only about 5% of total sales.

Nissan has ordered its designers to reduce the number of unique parts in its vehicles by 40%. Model variations, which had ballooned to more than 2,200, are to be rolled back 50%. The goal of this reformation is to reduce annual production costs by at least ¥200 billion ($2 billion at an exchange rate of ¥100/$1). These production cost savings are what Nissan is counting on to maintain its profitability even as it cuts its prices in the United States to remain competitive in the face of a surging yen.

Planning for Exchange Rate Changes

The marketing and production strategies advocated thus far assume knowledge of exchange rate changes. Even if currency changes are unpredictable, however, contingency plans can be made. This planning involves developing several plausible currency scenarios (see Chapter 10), analyzing the effects of each scenario on the firm's competitive position, and deciding on strategies to deal with these possibilities.

When a currency change actually occurs, the firm is able to quickly adjust its marketing and production strategies in line with the plan. Given the substantial costs of gathering and processing information, a firm should focus on scenarios that have a high probability of occurrence and that also would have a strong impact on the firm.

ILLUSTRATION

KODAK PLANS FOR CURRENCY CHANGES

Historically, Eastman Kodak focused its exchange risk management efforts on hedging near-term transactions. It now looks at exchange rate movements from a strategic perspective. Kodak's moment of truth came in the early 1980s when the strong dollar enabled overseas rivals such as Fuji Photo Film of Japan to cut prices and make significant inroads into its market share. This episode convinced Kodak that it had been defining its currency risk too narrowly. It appointed a new foreign exchange planning director, David Fiedler, at the end of 1985. According to Mr. Fiedler, "We were finding a lot of things that didn't fit our definition [of exposure] very well, and yet would have a real economic impact on the corporation."[5] To make sure such risks no longer go unrecognized, Mr. Fiedler now spends about 25% of his time briefing Kodak's operating managers on foreign exchange planning, advising them on everything from sourcing alternatives to market pricing. Kodak's new approach figured in a 1988 decision against putting in a factory in Mexico. Kodak decided to locate the plant elsewhere because of its assessment of the peso's relative strength. In the past, currency risk would have been ignored in such project assessments. According to Kodak's chief financial officer, before their reassessment of the company's foreign exchange risk management policy, its financial officers "would do essentially nothing to assess the possible exchange impact until it got to the point of signing contracts for equipment."[6]

The ability to plan for volatile exchange rates has fundamental implications for exchange risk management because there is no longer such a thing as the "natural" or "equilibrium" rate. Rather, there is a sequence of equilibrium rates, each of which has its own implications for corporate strategy. Success in such an environment—where change is the only constant—depends on a company's ability to react to change within a shorter time horizon than ever before. To cope, companies must develop competitive options—such as outsourcing, flexible manufacturing systems, a global network of production facilities, and shorter product cycles.

---

[5] Quoted in Christopher J. Chipello, "The Market Watcher," *Wall Street Journal*, September 23, 1988, p. 14.
[6] Quoted in Chipello, "The Market Watcher."

In a volatile world, these investments in flexibility are likely to yield high returns. For example, flexible manufacturing systems permit faster production response times to shifting market demand. Similarly, foreign facilities, even if they are uneconomical at the moment, can pay off by enabling companies to shift production in response to changing exchange rates or other relative cost shocks.

The greatest boost to competitiveness comes from compressing the time it takes to bring new and improved products to market. The edge a company gets from shorter **product cycles** is dramatic: Not only can it charge a premium price for its exclusive products, but it can also incorporate more up-to-date technology in its goods and respond faster to emerging market niches and changes in taste. Often held up as the ideal is the speedy way that retailers such as The Limited operate (see Chapter 1, Exhibit 1.2): If they see that an item is catching on with the public, they can have it manufactured in quantity and on their shelves within perhaps three weeks.

This speedy delivery is harder for a company that produces, say, automobiles or heavy machinery. Apparently, however, radical improvements in new-product delivery time are within reach even in those industries. In response to *endaka*—the soaring yen—which made their old products less competitive, Japanese automakers made a frantic effort to reduce—from four years to less than two years—the time between the initial design and the actual production of a new car. With better planning and more competitive options, corporations can now change their strategies substantially before the impact of any currency change can make itself felt.

As a result, the adjustment period after a large exchange rate change has been compressed dramatically. The 100% appreciation of the Japanese yen against the dollar from

---

ILLUSTRATION

TOSHIBA COPES WITH A RISING YEN BY CUTTING COSTS

By 1988, Toshiba's cost cutting reduced its cost-to-sales ratio to where it was before the yen began rising. The company shifted production of low-tech products to developing nations and moved domestic production to high-value-added products. At a VCR plant outside of Tokyo, it halved the number of assembly-line workers by minimizing inventories and simplifying operations. Other cost-reducing international activities included production of color picture tubes with Westinghouse in the United States, photocopier production in a joint venture with Rhone-Poulenc in France, assembly of videocassette recorders in Tennessee, production of similar VCRs in Germany, and establishment of a new plant in California for assembling and testing telephones and medical electronics equipment. Overall, Toshiba is estimated to have saved ¥115 billion—¥53 billion by redesigning products, ¥47 billion in parts cutbacks and lower raw material costs, and ¥15 billion in greater operating efficiency. Similarly, by 1989, with the dollar around ¥125, Fujitsu Fanuc, a robot maker, had streamlined itself so thoroughly and differentiated its products so effectively that it estimated it could break even with only a fifth of its plant in use and a dollar down to ¥70. The Japanese auto industry has seen its efforts pay off as well. It is estimated that a fall in the yen in 1996, from ¥95/$1 to ¥105/$1, translated into more than $6 billion in additional operating income for Japanese automakers.

1985 to 1988, for example, sparked some changes in Japanese corporate strategy that are likely to be long-lasting: increased production in the United States and East Asia to cope with the high yen and to protect their foreign markets from any trade backlash; purchasing more parts overseas to take advantage of lower costs; upscaling to reduce the price sensitivity of their products and broaden their markets; massive cost-reduction programs in their Japanese plants, with a long-term impact on production technology; and an increase in joint ventures between competitors.

When the yen was soaring, the aim among Japanese exporters was to prepare themselves to compete at 90 yen to the dollar. Some were even more ambitious. Toyota restructured itself in an attempt to remain competitive even with the yen at 80 to the dollar. Its breakeven point in 1994 was estimated at about ¥105/$1. If the Japanese are prepared to compete at 90 yen, the corollary is that American companies must be ready to compete at ¥140/$1, a level reached by the yen in 1998.

11.4 FINANCIAL MANAGEMENT OF EXCHANGE RISK

The one attribute that all the strategic marketing and production adjustments have in common is that they take time to accomplish in a cost-effective manner. The role of financial management in this process is to structure the firm's liabilities in such a way that during the time the strategic operational adjustments are under way, the reduction in asset earnings is matched by a corresponding decrease in the cost of servicing these liabilities.

One possibility is to finance the portion of a firm's assets used to create export profits so that any shortfall in operating cash flows due to an exchange rate change is offset by a reduction in debt-servicing expenses. For example, a firm that has developed a sizable export market should hold a portion of its liabilities in that country's currency. The portion to be held in the foreign currency depends on the size of the loss in profitability associated with a given currency change. No more definite recommendations are possible because the currency effects will vary from one company to another.

Volkswagen is a case in point. To hedge its operating exposure, VW should have used dollar financing in proportion to its net dollar cash flow from U.S. sales. This strategy would have cushioned the impact of the DM revaluation that almost brought VW to its knees. For the longer term, though, VW could manage its competitive exposure only by developing new products with lower price elasticities of demand and by establishing production facilities in lower-cost nations.

The implementation of a hedging policy is likely to be quite difficult in practice, if only because the specific cash-flow effects of a given currency change are hard to predict. Trained personnel are required to implement and monitor an active hedging program. Consequently, hedging should be undertaken only when the effects of anticipated exchange rate changes are expected to be significant.

A highly simplified example can illustrate the application of the financing rule developed previously; namely, that the liability structure of the combined MNC—parent and subsidiaries—should be set up in such a way that any change in the inflow on assets due to a currency change should be matched by a corresponding change in the outflow on the liabilities used to fund those assets. Consider the effect of a local currency change on

EXHIBIT 11.2 Statement of Projected Cash Flow

	UNITS (HUNDRED THOUSANDS)	UNIT PRICE (LC)	TOTAL	
LC 1 = $0.25				
Domestic sales	4	20	8,000,000	
Export sales	4	20	8,000,000	
Total revenue				16,000,000
Local labor (hours)	8	10	8,000,000	
Local material	8	3	2,400,000	
Imported material	6	4	2,400,000	
Total expenditures				12,800,000
Net cash flow from operations in LC				LC 3,200,000
Net cash flow from operations in US$				$800,000
LC 1 = $0.20				
Domestic sales	3	24	7,200,000	
Export sales	5	24	12,000,000	
Total revenue				19,200,000
Local labor (hours)	8	12	9,600,000	
Local material	10	3.5	3,500,000	
Imported material	4.5	5	2,250,000	
Total expenditures				15,350,000
Net cash flow from operations in LC				LC 3,850,000
Net cash flow from operations in US$				$770,000

the subsidiary depicted in Exhibit 11.2. In the absence of any exchange rate changes, the subsidiary is forecast to have an operating profit of $800,000. If a predicted 20% devaluation of the local currency from LC 1 = $0.25 to LC 1 = $0.20 occurs, the subsidiary's LC profitability is expected to rise to LC 3.85 million from LC 3.2 million because of price increases. However, that LC profit rise still entails a loss of $30,000, despite a reduction in the dollar cost of production.

Suppose the subsidiary requires assets equaling LC 20 million, or $5 million at the current exchange rate. It can finance these assets by borrowing dollars at 8% and converting them into their local currency equivalent, or it can use LC funds at 10%. How can the parent structure its subsidiary's financing in such a way that a 20% devaluation will reduce the cost of servicing the subsidiary's liabilities by $30,000 and thus balance operating losses with a decrease in cash outflows?

Actually, a simple procedure is readily available. If S is the dollar outflow on local debt service, then it is necessary that $0.2S$, the dollar gain on devaluation, equal $30,000, the operating loss on devaluation. Hence, $S = \$150,000$, or LC 600,000 at the current exchange rate. At a local currency interest rate of 10%, that debt-service amount corresponds to local currency debt of LC 6 million. The remaining LC 14 million can be provided by borrowing $3.5 million. Exhibit 11.3 illustrates the offsetting cash effects associated with such a financial structure.

This example would certainly become more complex if the effects of taxes, depreciation, and working capital were included. Although the execution becomes more difficult, a rough equivalence between operating losses (gains) and debt-service gains (losses) can still be achieved as long as all cash flows are accounted for. The inclusion of other foreign

EXHIBIT 11.3 Effect of Financial Structure on Net Cash Flow

	LC 1 = $0.25		LC 1 = $0.20	
	LOCAL CURRENCY	DOLLARS	LOCAL CURRENCY	DOLLARS
Operating cash flows	LC 3,200,000	$800,000	LC 3,850,000	$770,000
Debt service requirements:				
Local currency debt	600,000	150,000	600,000	120,000
Dollar debt	1,120,000	280,000	1,400,000	280,000
Total debt service outflow	1,720,000	430,000	2,000,000	400,000
Net cash flow	LC 1,480,000	$370,000	LC 1,850,000	$370,000

operations just requires the aggregation of the cash-flow effects over all affiliates because the corporation's total exchange risk is based on the sum of the changes of the profit contributions of each individual unit.

As mentioned earlier, this approach concentrates exclusively on risk reduction rather than on cost reduction. Where financial market imperfections are significant, a firm might consider exposing itself to more exchange risk in order to lower its expected financing costs. Regardless of how sophisticated a company's currency risk management program is, however, it can only do so much. A variety of evidence indicates that currency risk accounts for no more than 10% of cash flow volatility at most companies

---

ILLUSTRATION

SOUTH KOREAN COMPANIES AND BANKS MISMATCH THEIR CURRENCIES

An important contributing factor to the magnitude of the South Korean won's collapse was the currency mismatch faced by Korean companies and banks. Specifically, Korean banks lent huge amounts of won to the Korean *chaebol*, or conglomerates. The banks, in turn, financed their loans by borrowing dollars, yen, and other foreign currencies. The chaebol also borrowed large amounts of foreign currencies and invested the proceeds in giant industrial projects both at home and abroad. Considering how highly leveraged the chaebol were already, with debt-to-equity ratios on the order of 10:1, everything had to go right in order for them to be able to service their debts. When the won lost 40% of its value against the dollar during 1997, the chaebol had difficulty servicing their debts and many of them became insolvent. To the extent the Korean banks continued to receive won interest and debt repayments from the chaebol, won devaluation meant that the banks' won cash flows were insufficient to service their foreign debts. Similarly, although the chaebols' overseas projects were expected to generate foreign exchange to service their dollar debts, these projects turned out to be ill-conceived money losers. With both banks and chaebol scrambling to come up with dollars to service their foreign debts, the won was put under additional pressure and fell further, exacerbating the problems faced by Korean borrowers. In the last three months of 1997, 8 out of the 30 largest chaebol went bankrupt.

ILLUSTRATION

AVON IS CALLING IN ASIA

The currency turmoil in Asia in 1997 was unexpected but Avon Products was prepared to deal with it. An examination of what Avon did before the crisis and how it responded afterward illustrates many of the principles of managing operating exposure. It also provides insights into the role of financial officers as key members of strategic management teams.

Avon has a long history of international operations. As a general rule, Avon tries to hedge its currency risk by buying almost all its raw materials and making nearly all its products in the markets in which they are sold. For example, Avon Asia-Pacific has factories making cosmetics in its largest markets—China, Indonesia, the Philippines, and Japan—and contracts out production in six other Asian countries. It further hedged its currency risk by financing its local operations with local currency loans. Altogether, the ten Asian countries in which Avon operated accounted for $751 million of its $4.8 billion in revenue in 1996.

When the crisis began in Thailand in July 1997, Avon's executives did not anticipate that Thailand's problems would spread but as a precaution decided to further reduce currency risk by having the Asian units remit earnings weekly instead of monthly. By late August, however, the currency markets got nervous after the remarks of Malaysia's Prime Minister Mahathir Mohamad, who complained that Asia's economic crisis was provoked by an international cabal of Jewish financiers intent on derailing the region's growth. The head of Avon's Asia-Pacific region, Jose Ferriera Jr., also considered the possibility that other Asian countries would have to allow their currencies to depreciate to maintain their export competitiveness. In response, Avon decided to sell about $50 million worth of five Asian currencies forward against the dollar for periods of up to 15 months.

Having done what it could financially, Avon then turned to its operating strategy. Anticipating tough times ahead, Avon Asia-Pacific decided to redirect its marketing budget to hire more salespeople in Asia to bring in more customers rather than offering incentives to the existing sales force to get their current, cash-strapped customers to spend more money. Mr. Ferriera also urged his country managers to step up their purchase of local materials wherever possible and not allow local vendors to pass on all of their cost increases. At the same time, Avon began planning to compete more aggressively against disadvantaged competitors who have to import their products and raw materials. Finally, Avon began to analyze the incremental profits it could realize by using its Asian factories to supply more of the noncosmetic products sold in the United States. Avon Asia-Pacific was helped by a team of Latin American executives who traveled to Asia to share their experiences of how they had managed to cope in similar circumstances of currency turmoil in their countries.

In all of these deliberations and decisions, Avon Treasurer Dennis Ling was a full and active participant. For example, Mr. Ling helped the head of Avon's jewelry business renegotiate the terms of its contract with a Korean company that supplies jewelry for sale in the United States. The result was a substantial price discount based on the won's steep decline against the dollar. According to Mr. Ling, "Part of my job is to help our managers of operations understand and take advantage of the impact of currencies on their business."[7]

---

[7]Fred R. Bleakely, "How U.S. Firm Copes with Asian Crisis," *Wall Street Journal*, December 26, 1997, p. A2.

and often substantially less.[8] Even eliminating currency risk, therefore, is unlikely to reduce overall cash flow volatility by more than 10%. The import of this discussion is that currency risk management is most likely to be of value in the event of a catastrophic currency failure, like that of the Mexican peso and the East Asian currencies.

❧ 11.5 SUMMARY AND CONCLUSIONS

We saw in this chapter that currency risk affects all facets of a company's operations; therefore, it should not be the concern of financial managers alone. Operating managers, in particular, should develop marketing and production initiatives that help to ensure profitability over the long run. They should also devise anticipatory or proactive, rather than reactive, strategic alternatives in order to gain competitive leverage internationally.

The key to effective exposure management is to integrate currency considerations into the general management process. One approach used by a number of MNCs to develop the necessary coordination among executives responsible for different aspects of exchange risk management is to establish a committee for managing foreign currency exposure. Besides financial executives, such committees should—and often do—include the senior officers of the company such as the vice president-international, top marketing and production executives, the director of corporate planning, and the chief executive officer. This arrangement is desirable because top executives are exposed to the problems of exchange risk management, so they can incorporate currency expectations into their own decisions.

In this kind of integrated exchange risk program, the role of the financial executive is four-fold: (1) to provide local operating management with forecasts of inflation and exchange rates, (2) to identify and highlight the risks of competitive exposure, (3) to structure evaluation criteria so that operating managers are not rewarded or penalized for the effects of unanticipated currency changes, and (4) to estimate and hedge whatever operating exposure remains after the appropriate marketing and production strategies have been put in place.

➤ Questions

1. Why should managers focus on marketing and production strategies to cope with foreign exchange risk?

2. What marketing and production techniques can firms initiate to cope with exchange risk?

3. What is the role of finance in protecting against exchange risk?

4. Comment on the following statement: "The sharp appreciation of the U.S. dollar during the early 1980s might have been the best thing that ever happened to American industry."

5. In what sense is the boost in profits of American companies due to a falling dollar artificial?

6. How does a shorter product cycle time help companies reduce the exchange risk they face?

7. Why do exchange rate changes bring feast or famine for Volvo, but neither feast nor famine for Ford? Consider the distribution and concentration of their production facilities worldwide.

8. In order to cut costs when the dollar was at its peak, Caterpillar shifted production of small construction equipment overseas. By contrast, Caterpillar's main competitors in that area, Deere & Co. and J.I. Case, make most of their small construction equipment in the United States. What are the most likely competitive consequences of this restructuring?

---

[8]See, for example, Thomas E. Copeland and Yash Joshi, "Why Derivatives Don't Reduce Currency Risk," *The McKinsey Quarterly*, 1996, Number 1, pp. 66–79.

9. When the dollar was strong, and it could no longer earn a reasonable profit margin on European sales, Osmose International gave up its government permits to sell its chemical wood preservatives in much of Europe. "Why pay for a permit when you can't sell anything there anyway?" explained its president.
   a. What response do you have for the president of Osmose?
   b. How might you go about assessing the trade-offs involved for Osmose?

10. In order to avoid speculation, Honda hedges only the sales it has clinched, not the ones it expects. Comment on Honda's currency risk strategy.

11. A U.S. company needs to borrow $100 million for a period of seven years. It can issue dollar debt at 7% or yen debt at 3%.
    a. Suppose the company is a multinational firm with sales in the United States and inputs purchased in Japan. How should this affect its financing choice?
    b. Suppose the company is a multinational firm with sales in Japan and inputs that are primarily determined in dollars. How should this affect its financing choice?

➤ Problems

1.* Nissan produces a car that sells in Japan for ¥1.8 million. On September 1, the beginning of the model year, the exchange rate is ¥150:$1. Consequently, Nissan sets the U.S. sticker price at $12,000. By October 1, the exchange rate has dropped to ¥125:$1. Nissan is upset because it now receives only $12,000 × 125 = ¥1.5 million per sale.
    a. What scenarios are consistent with the U.S. dollar's depreciation?
    b. What alternatives are open to Nissan to improve its situation?
    c. How should Nissan respond in this situation?
    d. Suppose that on November 1, the U.S. Federal Reserve intervenes to rescue the dollar, and the exchange rate adjusts to ¥220:$1 by the next July. What problems and/or opportunities does this situation present for Nissan and for General Motors?

2.* Middle American Corporation (MAC) produces a line of corn silk cosmetics. All of the inputs are purchased domestically and processed at the factory in Des Moines, Iowa. Sales are only in the United States, primarily west of the Mississippi.
    a. Is there any sense in which MAC is exposed to the risk of foreign exchange rate changes that affect large multinational firms? If yes, how could MAC protect itself from these risks?
    b. If MAC opens a sales office in Paris, will this move increase its exposure to exchange rate risks? Explain.

3.* Gizmo, U.S.A. is investigating medium-term financing of $10 million in order to build an addition to its factory in Toledo, Ohio. Gizmo's bank has suggested the following alternatives:

TYPE OF LOAN	RATE (%)
Three-year U.S. dollar loan	14
Three-year Deutsche mark loan	8
Three-year Swiss franc loan	4

   a. What information does Gizmo require to decide among the three alternatives?
   b. Suppose the factory will be built in Geneva, Switzerland, rather than Toledo. How does this affect your answer in part a?

4. Chemex, a U.S. maker of specialty chemicals, exports 40% of its $600 million in annual sales: 5% goes to Canada and 7% each to Japan, Britain, Germany, France, and Italy. It incurs all its costs in U.S. dollars, and most of its export sales are priced in the local currency.
   a. How is Chemex affected by exchange rate changes?
   b. Distinguish between Chemex's transaction exposure and its operating exposure.
   c. How can Chemex protect itself against transaction exposure?
   d. What financial, marketing, and production techniques can Chemex use to protect itself against operating exposure?
   e. Can Chemex eliminate its operating exposure by hedging its position every time it makes a foreign sale or by pricing all foreign sales in dollars? Why or why not?

---

*Problems 1–3 contributed by Richard M. Levich.

5. In September 1992, Dow Chemical reacted to the currency chaos in Europe by switching to DM pricing for all its products in Europe. The purpose, said a Dow executive, was to shift currency risk from Dow to its European customers. Moreover, said the Dow executive, the policy was fairer: By setting the same DM price throughout Europe, Dow's new policy would nullify any advantage that a Dow customer in one country might have over competitors in another country on the basis of currency swings.
   a. What is Dow really trying to accomplish with its new pricing policy?
   b. What is the likelihood that this new policy will reduce Dow's currency risk?
   c. How are Dow's customers likely to respond to this new policy?

6. Boeing Commercial Airplane Co. manufactures all its planes in the United States and prices them in dollars, even the 50% of its sales destined for overseas markets. Assess Boeing's currency risk. How can it cope with this risk?

7. Fire King International, an Indiana manufacturer of fire-resistant filing cabinets and disk-storage units, has sought to protect itself from currency risk by pricing its export sales in dollars and holding firm on price. What currency risk does Fire King face from a rising dollar? How can Fire King manage that risk?

8. Cost Plus Imports is a West Coast chain specializing in low-cost imported goods, principally from Japan. It has to put out its semiannual catalogue with prices that are good for six months. Advise Cost Plus Imports on how it can protect itself against currency risk.

9. Matsushita exports about half of its TV set production to the United States under its Panasonic, Quasar, and Technics brand names. It prices its products in yen. Suppose the yen moves from ¥130 = $1 to ¥110 = $1. What currency risk is Matsushita facing? How can it cope with this currency risk?

10. During 1993, the Japanese yen appreciated by 11% against the dollar. In response to the lower cost of the main imported ingredients—beef, cheese, potatoes, and wheat for burger buns—MacDonald's Japanese affiliate reduced the price on certain set menus. For example, a cheeseburger, soda, and small order of french fries were marked down to ¥410 from ¥530. Suppose the higher yen lowered the cost of ingredients for this meal by ¥30.
   a. How much of a volume increase is necessary to justify the price cut from 530 to 410 yen? Assume that the previous profit margin (contribution to overhead) for this meal was ¥220. What is

the implied price elasticity of demand associated with this necessary rise in demand?
   b. Suppose sales volume of this meal rises by 60%. What will be the percentage change in MacDonald's dollar profit from this meal?
   c. What other reasons might MacDonald's have had for cutting price besides raising its profits?

11. Lyle Shipping, a British company, has chartered out ships at fixed-U.S.-dollar freight rates. How can Lyle use financing to hedge against its exposure? How will your recommendation affect Lyle's translation exposure? Lyle uses the current rate method to translate foreign currency assets and liabilities. However, the charters are off-balance-sheet items.

12. Di Giorgio International, a subsidiary of California-based Di Giorgio Corp., processes fruit juices and packages condiments in Turnhout, Belgium. It buys Brazilian orange concentrate in dollars, German apples in marks, Italian peaches in lire, and cartons in Dutch guilders. At the same time, it exports 85% of its production. Assess Di Giorgio International's currency risk and determine how it can structure its financing to reduce this risk.

13. In 1985, Japan Airlines (JAL) bought $3 billion of foreign exchange contracts at ¥180/$1 over 11 years to hedge its purchases of U.S. aircraft. By 1994, with the yen at about ¥100/$1, JAL had incurred more than $1 billion in cumulative foreign exchange losses on that deal.
   a. What was the economic rationale behind JAL's hedges?
   b. Did JAL's forward contracts constitute an economic hedge? That is, is it likely that JAL's losses on its forward contracts were offset by currency gains on its operations?

14. Texas Instruments (TI) manufactures integrated circuits and memory chips that it sells around the world. It has major markets in Europe. TI's primary competitors are Japanese companies.
   a. What factors will influence TI's exposure to movements in the dollar value of European currencies?
   b. Does TI's European business have yen exposure? Explain.
   c. How can TI use financing to reduce its yen exposure, to the extent that this exposure exists?

15. South Korea's Korean Air Lines (KAL) is the world's 12th largest passenger airline and its second-largest cargo carrier. It has borrowed $5 billion (much of it denominated in dollars) to finance its fleet of planes.
   a. In what ways is KAL affected by depreciation of the won against the dollar?

b. How can KAL use financing to reduce its currency risk?

c. KAL argues that its jet fleet naturally hedges its currency exposure. Do you agree or disagree? Explain.

d. At the end of 1997, KAL decided to sell off its older planes, use the proceeds to pay down some of its debt, and replace the sold planes with aircraft it leases through a subsidiary in Ireland. Will this strategy lower KAL's high debt ratio?

16. In 1990, a Japanese investor paid $100 million for an office building in downtown Los Angeles. At the time, the exchange rate was ¥145/$. When the investor went to sell the building five years later, in early 1995, the exchange rate was ¥85/$ and the building's value had collapsed to $50 million.

a. What exchange risk did the Japanese investor face at the time of his purchase?

b. How could the investor have hedged his risk?

c. Suppose the investor financed the building with a 10% downpayment in yen and a 90% dollar loan accumulating interest at the rate of 8% per annum. This is a zero-coupon loan, so the interest on it (along with the principal) is not due and payable until the building is sold. How much has the investor lost in yen terms? In dollar terms?

d. Suppose the investor financed the building with a 10% downpayment in yen and a 90% yen loan accumulating interest at the rate of 3% per annum. This is a zero-coupon loan, so the interest on it (along with the principal) is not due and payable until the building is sold. How much has the investor lost in yen terms? In dollar terms?

➤ Bibliography

DUFEY, GUNTER. "Corporate Financial Policies and Floating Exchange Rates." Address presented at the meeting of the International Fiscal Association in Rome, October 14, 1974.

LESSARD, DONALD R., and JOHN B. LIGHTSTONE. "Volatile Exchange Rates Can Put Operations at Risk." *Harvard Business Review*, July–August 1986, pp. 107–114.

SHAPIRO, ALAN C., and THOMAS S. ROBERTSON. "Managing Foreign Exchange Risks: The Role of Marketing Strategy." Working paper, The Wharton School, University of Pennsylvania, 1976.

Case Studies

CASE II.1

BRITISH MATERIALS CORPORATION*

In January 1981, Vulkan Inc., a U.S. firm relatively new to international business, acquired British Materials Corp., or BMC, an English firm. BMC operated two detinning plants in England, one in Manchester and the other in Birmingham, and a scrap collection depot just outside of London.

Detinning involves the separation and recovery of tin and detinned steel from tinplate scrap. The principal sources of tinplate scrap are the waste cuttings and stampings from the manufacture of articles made from tinplate trimmings and rejects from steel companies that manufacture tinplate. Both the steel and the tin recovered in this process are high-quality, high-purity premium metals.

BMC was the only detinning company operating in the United Kingdom and had established clear domination of the industrial tinplate scrap market. At the time of its acquisition, approximately 80% of BMC's scrap supply was provided by 39 tinplate fabricators, the largest of which provided nearly half of BMC's scrap. BMC did not buy the scrap supplied to it by these firms. Rather, it had signed contracts with them to process their scrap for a fee. These contracts all had similar provisions. They were cost-plus, and they prescribed a profit to BMC equal to 15% of the prices BMC received for the detinned steel and the recovered tin.

Costs covered by the contracts included all variable costs as well as an agreed-upon amount for fixed costs excluding depreciation and financing charges. The management of Vulkan felt that the fixed-cost recovery provisions were adequate

to cover projected out-of-pocket fixed costs. The remaining 20% of BMC's tinplate scrap requirement was met through open market purchases.

Detinned steel recovered by BMC was sold primarily to British Steel, with the remainder exported to companies in Western Europe. During 1974–1976, only 2% of BMC's detinned steel sales revenue arose from foreign sales, whereas 33% of its 1980 sales revenue came from export sales. Most of the tin recovered is sold to various firms in the market areas surrounding the detinning plants. These firms convert the tin into inorganic tin chemicals consumed by the glass, plating, and chemical industries.

The acquisition of BMC was effected through Vulkan's newly formed United Kingdom subsidiary, Vulkan U.K. or VUK, which purchased all of the outstanding common and preference shares of BMC. Subsequently, BMC and its primary subsidiaries were liquidated into VUK. As it considered the alternatives for funding this acquisition, a paramount concern of Vulkan was the possible foreign exchange exposure associated with the sterling revenues and costs generated by VUK. On the basis of 1980's proportions of pound sterling- and U.S. dollar-denominated sales, and assuming that sterling prices were invariant to exchange rate changes, Vulkan tested the sensitivity of VUK's income and debt-service capacity to likely changes in the dollar-sterling exchange rate. These analyses tended to indicate that dollar-denominated earnings and cash flows were sensitive to exchange rate fluctuations. In addition, Thomas Alan, Vulkan's financial vice-president, consulted with several investment and commercial bankers. A typical opinion is the one from Diane Ronningen, the partner in charge of international finance at the investment banking firm of Ronningen and Simnowitz (see Exhibit II 1.1).

To minimize the economic gains and losses on its investment in BMC resulting from fluctua-

---

Source: Copyright © 1985 by Alan C. Shapiro.

EXHIBIT II 1.1 Opinion on BMC's Exchange Risk

January 5, 1981

Mr. Thomas Alan
Vice President-Finance
Vulkan, Inc.
30 Golden Triangle
Pittsburgh, Pennsylvania 15217

Dear Tom:

Following our recent conversations, I am writing to give you our thoughts on the appropriate currency Vulkan should use for financing the acquisition of British Materials Corp. (BMC). You have asked specifically that we review alternatives in pounds sterling, U.S. dollars, Deutsche marks, and Swiss francs.

We believe that financing the acquisition of BMC with sterling or a sterling equivalent makes the most financial and business sense. It is sterling revenues and income which BMC generates in its daily operations and sterling which Vulkan would then have available to service any debt used for the acquisition. If BMC were a substantial exporter or competed in the United Kingdom against firms which set their prices on a dollar base (e.g., the U.K. computer industry, North Sea oil, etc.) the appropriate currency might be dollars. Since this is not the case, a financing in dollars places an unnecessary foreign exchange exposure burden on Vulkan. Vulkan's primary business is not currency speculation. Since neither you nor we know the future movements of the sterling exchange rate over the next few years and since sterling has been one of the most volatile and least predictable currencies in the world recently, incurring such an exchange risk would, in our opinion, be ill-advised.

Borrowing on the Deutsche mark or Swiss franc markets on an unhedged basis to fund the acquisition makes even less sense for Vulkan since you have no natural exposure in either of these currencies. On a hedged basis, the cost would theoretically be similar to those for the dollar borrowing alternative.

I hope this letter clarifies our recommendations. Please don't hesitate to call if you have questions. Best regards.

Sincerely,
Diane M. Ronningen
Senior Partner
Ronningen & Simnowitz

tions in the rate of exchange between the U.S. dollar and the pound sterling, Vulkan concluded that the acquisition should be funded entirely in pounds sterling. This decision was based on the following factors:

- All BMC's assets would be denominated in pounds sterling
- The high probability that most, if not all, of BMC's future revenues and costs would be denominated in pounds sterling or would be determined on a pound sterling-equivalent basis
- Vulkan's projected income and debt-service sensitivity analyses

- U.K. and U.S. tax laws and U.K. corporate law
- The advice of Vulkan's investment and commercial banks

Accordingly, in January 1981, Vulkan and VUK borrowed £2,355,000 and £1,137,000, respectively, for 10 years on a floating-rate basis (LIBOR plus a margin) to fund part of the purchase of all the outstanding common and preference shares of BMC. The balance of the purchase price was funded by VUK's borrowing under a sterling overdraft facility and its issuance of short-term sterling notes. VUK's obligations were not

EXHIBIT II 1.2 Operating Data for BMC, 1972–1983

Year: Quarter	Exchange Rate[1]	£Cash Flow (BIT)[2]	£Cash Flow (BDIT)[3]	Home Price[4]	Export Price[5]	Average Price[6]
72:1	2.599	—	—	—	—	—
72:2	2.599	—	—	—	—	—
72:3	2.445	—	—	—	—	—
72:4	2.364	—	—	—	—	—
73:1	2.420	—	—	—	—	—
73:2	2.530	—	—	—	—	—
73:3	2.480	—	—	—	—	—
73:4	2.379	—	—	—	—	—
74:1	2.279	127.000	51.000	21.190	25.860	21.930
74:2	2.397	186.000	142.000	28.020	36.280	28.920
74:3	2.350	−11.000	−57.000	34.040	—	34.040
74:4	2.330	220.000	171.000	39.120	51.500	39.570
75:1	2.391	325.000	280.000	40.740	—	40.740
75:2	2.325	392.000	345.000	36.720	38.400	36.750
75:3	2.129	235.000	175.000	35.160	29.990	35.030
75:4	2.043	354.000	305.000	33.610	30.140	33.430
76:1	2.000	32.000	−10.000	39.010	—	39.010
76:2	1.807	693.000	648.000	48.010	—	48.010
76:3	1.767	416.000	363.000	40.490	—	40.490
76:4	1.651	207.000	154.000	39.300	—	39.300
77:1	1.714	65.000	40.000	36.290	—	36.290
77:2	1.719	−54.000	−146.000	35.050	28.180	33.350
77:3	1.735	417.000	365.000	32.250	26.410	30.410
77:4	1.815	688.000	638.000	29.600	22.460	27.040
78:1	1.927	53.000	2.000	29.240	21.900	28.020
78:2	1.835	597.000	539.000	33.360	30.610	32.830
78:3	1.932	401.000	342.000	38.830	35.760	38.250
78:4	1.984	800.000	728.000	45.230	42.210	44.690
79:1	2.016	−35.000	−94.000	57.800	65.380	58.350
79:2	2.080	616.000	553.000	59.580	53.420	58.530
79:3	2.232	760.000	693.000	60.310	47.400	58.420
79:4	2.159	829.000	760.000	51.890	47.390	50.450
80:1	2.254	186.000	109.000	53.750	51.410	52.770
80:2	2.285	379.000	299.000	46.870	46.360	46.640
80:3	2.381	120.000	27.000	35.290	36.190	35.870
80:4	2.386	−141.000	−246.000	30.910	31.280	31.120
81:1	2.310	838.000	803.000	34.620	33.310	34.180
81:2	2.081	—	—	35.130	39.270	35.990
81:3	1.837	332.000	274.000	35.920	38.930	36.600
81:4	1.884	545.000	477.000	39.720	35.530	39.210
82:1	1.847	552.000	496.000	47.880	40.320	46.150
82:2	1.780	177.000	116.000	42.260	45.620	42.920
82:3	1.725	5.000	−60.000	41.740	44.720	42.360
82:4	1.650	370.000	297.000	35.310	38.880	36.180
83:1	1.534	−57.000	−171.000	36.140	37.920	36.500

[1] Average spot exchange rate during the quarter (U.S. dollars/British pounds).
[2] Cash flow equals income before interest and taxes plus depreciation plus or minus changes in working capital.
[3] Same as in note 2 but without depreciation.
[4] Average sales price in U.K. in £/ton.
[5] Average export sales price in £/ton.
[6] Volume weighted, average total sales price in £/ton.

guaranteed by Vulkan. On the date of these borrowings, the exchange rate was $2.4060:£1.00.

It should be emphasized that Vulkan decided to finance its acquisition of BMC with sterling debt to hedge against the effects of unanticipated exchange rate changes, not to profit from the possibility that sterling would devalue by more than the amount already reflected in the sterling:dollar interest rate differential. Pursuing the latter objective would have constituted currency speculation, not hedging. And it was an article of faith among Vulkan's management that its comparative advantage lay in production and marketing, not in currency speculation.

During April 1983, the average U.S. dollar:pound sterling exchange rate was $1.5362. On the basis of quarterly exchange rates between 1981:1 and 1983:1, the nominal or actual, sterling depreciation against the dollar was 33.6%. In real or inflation-adjusted terms, using the implicit price indices in both countries to measure inflation, sterling depreciated 31.0%. This significant and rapid depreciation of the pound sterling in both nominal and real terms raised the question: Had the sterling borrowing to finance the acquisition of BMC provided an effective hedge of the economic foreign exchange exposure believed to be inherent in its operations? Vulkan's management accordingly decided to reexamine its original conclusion that the acquisition of BMC created a "long" pound sterling exposure.

Although Ms. Ronningen's reasoning still seemed persuasive, Mr. Alan decided to call in an independent consultant, Robert Daniels, for a second opinion of the advisability of funding VUK with pound debt. Mr. Daniels, who is noted for his expertise in the area of currency risk management, requested all available data on BMC's past operations.

Thomas Alan managed to assemble operating data for BMC from the first quarter of 1974 through the first quarter of 1983. Due to unusual transactions that occurred during the second quarter of 1981, he decided to exclude these data. In addition, Mr. Alan included the average exchange rate (dollars:pound), as well as some price data on detinned steel, for each quarter. These data are contained in Exhibit II 1.2.

Now it was up to Mr. Daniels to interpret these data and come to some conclusion concerning the extent to which VUK was subject to exchange risk. His opinion would have a major impact on whether Vulkan would maintain its pound sterling debt or refund this debt and replace it with dollar financing.

QUESTIONS

1. Is VUK subject to exchange risk? How, if at all, do your analysis and conclusions differ from those of Ms. Ronningen?

2. Should Vulkan refund the pound debt it used for the acquisition of BMC and replaced it with dollar financing? Why or why not? What criteria are you using to reach your decision?

3. Suppose it is concluded that VUK is not subject to exchange risk. Should Vulkan repay its pound debts? Should VUK repay its pound loans and replace them with dollar financing? Consider the tax consequences of replacing the pound debt with dollar financing in both the United Kingdom and the United States.

4. Does Vulkan's foreign exchange risk management objective make sense? From what perspective?

CASE II.2

EUCLIDES ENGINEERING LTD.*

The submission of a bid to the Mexican government's agency in charge of the rural electrification project had been most disappointing. In November 1984, Sam Finkel, manager-finance of the Power Systems Management Division at Euclides Engineering, was notified

---

*Source: Copyright © 1985 by Laurent L. Jacque. Revised by Alan C. Shapiro

EXHIBIT II 2.1 WEFA Exchange Rate Forecasts (DM per U.S.$, End of Period)

YEAR	JAN.	FEB.	MAR.	APR.	MAY	JUNE
1985	3.18	3.16	3.14	3.11	3.09	3.07
1986	2.87	2.83	2.80	2.77	2.74	2.71
1987	2.55	2.54	2.53	2.51	2.49	2.47

YEAR	JULY	AUG.	SEPT.	OCT.	NOV.	DEC.
1985	3.05	3.02	2.99	2.96	2.93	2.90
1986	2.68	2.66	2.64	2.61	2.59	2.57
1987	2.46	2.44	2.43	2.41	2.39	2.38

that Euclides had been underbid to the tune of $13 million by the Swiss-West German consortium Brown-Boveri & Siemens.

Euclides had entered the bidding contest for the installation of five high-voltage transmission units near Monterrey, Mexico's second-largest industrial center. The bid submitted in March 1984 was in the amount of $67 million to be paid in three equal installments on July 1, 1986; December 31, 1986; and July 1, 1987, with installation to be completed in the last six months of 1985. At-

tached with the reply from the Mexican government was a photocopy of the two bids, which were virtually identical from the standpoint of technical specifications but which varied in terms of payment.

- Brown-Boveri & Siemens: equivalent of $54 million (denominated in Deutsche marks at the rate of DM 3.14:$1). Same payment schedule as Euclides, but in three equal installments of Deutsche marks.
- Euclides Engineering: $67 million

EXHIBIT II 2.2 Exchange Rate Quotations (DM per U.S.$)

WEFA FORECAST FOR THE DEUTSCHE MARK, DECEMBER 3, 1984

	1 Month	3 Months	6 Months	12 Months
Forecast	3.1563	3.1611	3.0899	2.9285
Forward	3.1045	3.0863	3.0550	2.9875

FOREIGN CURRENCY OPTIONS (PHILADELPHIA EXCHANGE), 1985

	PREMIUM ON CALL CONTRACT		PREMIUM ON PUT CONTRACT	
STRIKE PRICE	March	June	March	June
0.31	1.20	2.25	0.33	0.55
0.32	1.10	1.65	0.67	0.92
0.33	0.62	1.19	1.19	
0.34	0.33	0.79		
0.35	0.20	0.52		

Note: Strike prices are expressed in U.S. dollars per DM and premiums in cents per DM.

A second round of bidding was to be held on December 10, with the winner to be announced on December 20. Sam Finkel was concerned that a strong dollar had just about closed his export market, where Euclides used to be price competitive even when lavish export credits were offered by its foreign competitors. Sam felt that in spite of his new financial responsibilities, his background and the last 15 years of his career as a civil engineer with Euclides did not quite equip him with the creative financing skills that could close the seemingly unbridgeable gap between the two bids. Fortunately, Sam felt he could depend on his newly hired assistant, Gerardo Wehmann, a Mexican national with graduate education in electrical engineering from Stanford and an MBA in international business from the Wharton School.

Gerardo, who had gone over the files, felt that the exchange rate consideration had much to do with Euclide's problem. He decided to study the situation further. To begin, he examined the Deutsche mark exchange rate forecasts put out by Wharton Econometric Forecasting Associates (WEFA). He also studied the forward rates and the rates on several options contracts as of December 3. These data are contained in Exhibits II 2.1 and II 2.2, respectively.

QUESTIONS

1. In view of the relative values of the U.S. dollar and the Deutsche mark, how can you explain the discrepancy between the U.S. and the Swiss-West German bids?

2. Can Euclides match the Swiss-West German bid without changing its dollar price? How?

3. If you were to advise the Mexican government on how to compare bids denominated in different currencies, what would your advice be?

CASE II.3

ROLLS-ROYCE LIMITED

Rolls-Royce Limited, the British aeroengine manufacturer, suffered a loss of £58 million in 1979 on worldwide sales of £848 million. The company's annual report for 1979 (page 4) blamed the loss on the dramatic revaluation of the pound sterling against the dollar, from £1 = $1.71 in early 1977 to £1 = $2.12 by the end of 1979.

The most important reason for the loss was the effect of the continued weakness of the U.S. dollar against sterling. The large civil engines that Rolls-Royce produces are supplied to American air frames. Because of U.S. dominance in civil aviation, both as producer and customer, these engines are usually priced in U.S. dollars and escalated accordingly to U.S. indices. . . .

A closer look at Rolls-Royce's competitive position in the global market for jet engines reveals the sources of its dollar exposure. For the previous several years Rolls-Royce's export sales had accounted for a stable 40% of total sales and had been directed at the U.S. market. This market is dominated by two U.S. competitors, Pratt and Whitney Aircraft Group (United Technologies) and General Electric's aerospace division. As the clients of its mainstay engine, the RB 211, were U.S. aircraft manufacturers (Boeing's 747SP and 747,200 and Lockheed's L1011), Rolls-Royce had little choice in the currency denomination of its export sales but to use the dollar.

Indeed, Rolls-Royce won some huge engine contracts in 1978 and 1979 that were fixed in dollar terms. Rolls-Royce's operating costs, on the other hand, were almost exclusively incurred in sterling (wages, components, and debt servicing). These contracts were mostly pegged to an exchange rate of about $1.80 for the pound, and Rolls-Royce officials, in fact, expected the pound to fall further to $1.65. Hence, they didn't cover their dollar exposures. If the officials were correct, and the dollar strengthened, Rolls-Royce would enjoy windfall profits. When the dollar weakened instead, the combined effect of fixed dollar revenues and sterling costs resulted in foreign exchange losses in 1979 on its U.S. engine contracts that were estimated by *The Wall Street Journal* (March 11, 1980, p. 6) to be equivalent to as much as $200 million.

Moreover, according to that same *Wall Street Journal* article, "the more engines produced and sold under the previously negotiated contracts, the greater Rolls-Royce's losses will be."

QUESTIONS

1. Describe the factors you would need to know to assess the economic impact on Rolls-Royce of the change in the dollar:sterling exchange rate. Does inflation affect Rolls-Royce's exposure?

2. Given these factors, how would you calculate Rolls-Royce's economic exposure?

3. Suppose Rolls-Royce had hedged its dollar contracts. Would it now be facing any economic exposure? How about inflation risk?

4. What alternative financial management strategies might Rolls-Royce have followed that would have reduced or eliminated its economic exposure on the U.S. engine contracts?

5. What nonfinancial tactics might Rolls-Royce now initiate to reduce its exposure on the remaining engines to be supplied under the contracts? On future business (e.g., diversification of export sales)?

6. What additional information would you require to ascertain the validity of the statement that "the more engines produced and sold under the previously negotiated contracts, the greater Rolls-Royce's losses will be"?

CASE II.4

THE MEXICAN PESO

The basic purpose of this case is to have you conduct an in-depth analysis of government macroeconomic policies on firms and banks doing business with Mexico. The vehicle being used is the Mexican peso. See Exhibit II 4.1 for statistics related to exchange rates and price indexes in the United States and Mexico for the period 1976–1997. Using these data, please address the following questions.

QUESTIONS

1. What are the causes of the continuing devaluation of the peso since August 1976? Analyze both the immediate causes (e.g., balance-of-payments deficits) and longer-term, more fundamental causes (e.g., inflation, the political and economic environment). Concentrate especially on the 1982 and 1994–1995 devaluations of the peso.

EXHIBIT II 4.1 The Mexican Peso's Key Statistics: 1975–1997

| Year:Quarter | NOMINAL EXCHANGE RATES (PERIOD AVERAGE) | | | |
	Peso:Dollar	Dollar:Peso	CPI Mexico	CPI United States
75:1	12.5	$0.08000	100.0	100.0
75:2	12.5	$0.08000	103.2	101.9
75:3	12.5	$0.08000	107.0	104.1
75:4	12.5	$0.08000	110.4	105.8
76:1	12.5	$0.08000	115.2	106.8
76:2	12.5	$0.08000	118.2	108.2
76:3	19.9	$0.05038	122.0	109.8
76:4	20.0	$0.05013	137.3	111.0
77:1	22.7	$0.04407	149.1	113.0
77:2	23.0	$0.04348	155.7	115.5
77:3	22.7	$0.04407	162.4	117.1
77:4	22.7	$0.04399	168.6	118.5

EXHIBIT II 4.1 Continued

NOMINAL EXCHANGE RATES (PERIOD AVERAGE)

Year:Quarter	Peso:Dollar	Dollar:Peso	CPI Mexico	CPI United States
78:1	22.7	$0.04398	177.7	120.4
78:2	22.8	$0.04384	183.9	123.6
78:3	22.7	$0.04399	191.4	126.5
78:4	22.7	$0.04401	197.6	129.0
79:1	22.8	$0.04380	209.5	132.3
79:2	22.8	$0.04378	217.0	136.8
79:3	22.8	$0.04390	225.2	141.3
79:4	22.8	$0.04386	235.6	145.4
80:1	22.9	$0.04376	256.8	151.1
80:2	22.9	$0.04361	271.5	156.6
80:3	23.1	$0.04337	289.3	159.4
80:4	23.3	$0.04301	303.9	163.7
81:1	23.8	$0.04209	328.5	168.0
81:2	24.4	$0.04103	348.4	171.9
81:3	25.2	$0.03968	365.9	176.9
81:4	26.3	$0.03804	389.4	179.3
82:1	45.5	$0.02198	435.0	180.9
82:2	48.0	$0.02082	501.5	183.5
82:3	50.0	$0.02000	606.1	187.0
82:4	96.5	$0.01036	730.8	187.5
83:1	102.0	$0.00980	926.2	187.3
83:2	120.0	$0.00833	1,076.9	189.7
83:3	132.0	$0.00758	1,217.2	191.9
83:4	143.9	$0.00695	1,369.6	193.6
84:1	155.8	$0.00642	1,601.9	195.7
84:2	167.6	$0.00597	1,807.3	197.9
84:3	179.6	$0.00557	1,987.7	200.1
84:4	192.6	$0.00519	2,196.5	201.5
85:1	208.9	$0.00479	2,552.7	202.8
85:2	228.0	$0.00439	2,800.9	205.3
85:3	305.1	$0.00328	3,097.0	206.7
85:4	371.7	$0.00269	3,527.5	212.6
86:1	473.6	$0.00211	4,254.7	213.1
86:2	575.4	$0.00174	4,957.8	212.6
86:3	752.0	$0.00133	5,930.4	214.3
86:4	923.5	$0.00108	7,164.7	215.4
87:1	1126.0	$0.00089	8,909.4	217.8
87:2	1353.7	$0.00074	11,120.8	220.7
87:3	1570.8	$0.00064	13,889.6	223.1
87:4	2209.7	$0.00045	15,228.2	225.0
88:1	2281.0	$0.00044	29,116.0	229.4
88:2	2281.0	$0.00044	32,453.0	232.2
88:3	2281.0	$0.00044	33,924.0	235.3
88:4	2281.0	$0.00044	34,968.1	237.7
89:1	2369.0	$0.00042	36,943.5	240.5
89:2	2460.0	$0.00041	38,435.6	244.3
89:3	2551.0	$0.00039	39,687.9	246.4
89:4	2641.0	$0.00038	41,501.0	248.7
90:1	2733.0	$0.00037	45,621.2	253.0
90:2	2817.0	$0.00035	48,104.5	255.5
90:3	2890.6	$0.00035	50,789.0	260.0
90:4	2945.4	$0.00034	53,783.8	264.3

EXHIBIT II 4.1 Continued

NOMINAL EXCHANGE RATES (PERIOD AVERAGE)

Year:Quarter	Peso:Dollar	Dollar:Peso	CPI Mexico	CPI United States
91:1	2981.0	$0.00034	57,724.0	266.4
91:2	3018.2	$0.00033	59,806.8	267.9
91:3	3055.8	$0.00033	61,443.3	270.0
91:4	3071.0	$0.00033	64,270.0	272.1
92:1	3083.5	$0.00032	67,741.4	274.0
92:2	3122.3	$0.00032	69,576.3	276.3
92:3	3116.3	$0.00032	70,964.8	278.1
92:4	3115.4	$0.00032	72,750.1	280.1
93:1	3097.6	$0.00032	75,130.5	282.5
93:2	3121.2	$0.00032	76,519.0	284.5
93:3	3117.8	$0.00032	77,758.8	285.3
93:4	3105.9	$0.00032	79,048.2	288.5
94:1	3359.8	$0.00030	80,585.5	290.4
94:2	3391.8	$0.00029	81,825.3	292.5
94:3	3404.0	$0.00029	83,015.4	295.4
94:4	5325.0	$0.00019	84,552.8	296.3
95:1	6817.5	$0.00015	92,685.7	299.8
95:2	6309.2	$0.00016	109,447.5	302.4
95:3	6419.5	$0.00016	117,630.0	303.7
95:4	7642.5	$0.00013	125,713.4	305.3
96:1	7547.9	$0.00013	137,218.5	307.9
96:2	7610.8	$0.00013	146,789.6	311.0
96:3	7537.4	$0.00013	153,533.9	313.1
96:4	7850.9	$0.00013	161,071.8	314.7
97:1	7890.5	$0.00013	172,180.2	315.7
97:2	7958.0	$0.00013	178,031.9	316.6
97:3	7819.9	$0.00013	182,991.0	317.6
97:4	8136.0	$0.00012	188,793.2	318.5

2. What role did oil price changes play in Mexico's difficulties?

3. What indicators of peso devaluation prior to 1982 and 1994 were there?

4. What were the likely effects of the peso devaluation between 1976 and January 1982 on
   a. Mexican companies?
   b. Foreign firms operating in Mexico?
   c. U.S. companies in border towns catering to Mexicans?

5. Redo question 4, focusing on the effects of peso devaluation subsequent to February 1982 and prior to December 1994.

6. In August 1982, the Mexican government devalued the peso, froze all dollar accounts in Mexican banks, and imposed currency controls. What are the government's objectives? How did these actions affect the black market value of the peso? Why?

7. How did the Mexican government's expropriation of Mexico City real estate, following the September 1985 earthquake, affect the value of the peso and why?

8. Consider the trust factor with respect to Mexican policies. What have been the probable effects of trust or its lack on investment in Mexico, Mexican citizens' investment choices, and the peso's value?

9. Are dollar loans to the Mexican government and Mexican companies exposed to exchange risk? Explain.

10. How did Mexico's economic policies contribute to its debt crisis? How have subsequent government policies affected Mexico's financial health?

Case II.5

LINK TECHNOLOGIES*

Link Technologies, a small firm located in San Jose, California, is currently engaged in the development, manufacture, and sale of high-speed fax modems for use in personal computers. The company was created in 1980 by Mr. James Lee, a researcher who was employed at IBM's T.J. Watson research center, and two former graduate school classmates from Cornell University.

At the time, the personal computer business was in its infancy, and the company sought to fill a niche by providing communication networks to link mainframe computers. Although the company grew rapidly, Mr. Lee felt that the real opportunities for growth lay in personal computers. In the mid-1980s, he decided to make a major switch to the production of high-speed modems. Link Technologies' new products were extraordinarily well received in the marketplace.

Many viewed the company's culture as an essential element in its rapid growth. Mr. Lee emphasized the need for high quality and techno-logical superiority. To ensure that his company was on the cutting edge, he provided strong financial incentives for production managers to develop new or more-efficient products. Further, he contained costs by decentralization, requiring that each branch of the firm operate as an independent profit center. As a result, Link Technologies was rated by a prominent business periodical as one of the 10 best-managed small firms in the United States.

Mr. Lee and his team had developed a new method of transmitting data reliably at high speed, and this technological breakthrough, together with the proprietary communications software that came bundled with the product, led to strong demand for the company's main product, the PCI 2000 modem. Indeed, the company's modems were recognized to be of the highest quality, comparable with the products of much larger competitors such as Intel, Motorola, and U.S. Robotics. The company was especially successful in marketing its products abroad because the proprietary communications software included with each modem was extremely easy to use. Further the company ensured that detailed manuals were provided in the local language, a strategy that paid off handsomely in some countries, such as Finland and Turkey. Indeed, by 1994, almost two-thirds of the company's revenues came from sales abroad. Exhibit II 5.1 provides details of the company's revenues in domestic and foreign markets from 1990 to 1994.

EXHIBIT II 5.1 Foreign and Domestic Operations

YEAR	DOMESTIC REVENUES (%)	FOREIGN REVENUES (%)	NET INCOME ($ MILLIONS)
1989	66	34	5.89
1990	51	49	6.03
1991	38	62	6.49
1993	32	68	7.14

*Copyright Ananth Madhavan, 1994. This case was developed for teaching purposes at the Marshall School of Business at the University of Southern California; please do not reproduce or use without permission.

The rapid expansion into foreign markets presented a new problem for Mr. Lee, who was both the chief executive officer and the largest shareholder in the firm. The initial capital for the company came from the savings of the three founders and a substantial investment from a silent partner. Before going public in 1989, the company had sought to conserve cash by compensating key employees partially with shares in the fledgling enterprise. The rapid expansion in production facilities was financed primarily through intermediate-term loans. However, the company eventually had to sell new common stock to the public to raise the necessary capital. Even so, Mr. Lee himself still retained almost 32% of the outstanding shares, and this represented almost all of his personal wealth. Other founders and their families also had substantial holdings. Mr. Lee, although well aware of the potential benefits of diversification, was unwilling to reduce his holdings in the company. He still felt extremely confident about the company's future prospects and its ability to create new products, and he enjoyed the status and benefits associated with controlling the firm. But Mr. Lee was concerned about the risks to his personal wealth arising from foreign currency fluctuations. In some countries, especially some of the smaller nations where Link Technologies had high market share, currency fluctuations could be quite severe, and in one case had had a substantial impact on profits.

In November 1990, Mr. Lee asked the firm's chief financial officer, Mr. Stanford Brown, to explore various alternatives to reduce the volatility in sales revenue arising from foreign currency movements. Mr. Brown readily admitted that he was not at ease discussing the most recent approaches to risk reduction or hedging. He had received his MBA from Harvard in the 1960s and had spent most of his career working for a company that had little international exposure. Moreover, he was not familiar with derivatives such as currency options, which until recently were not widely traded. However, Mr. Brown had recently hired an assistant, Mr. Dan Pross, who had some knowledge of hedging and derivatives. As a student at UCLA, Mr. Pross had traded various types of derivatives for his own portfolio and was familiar with how they were traded. Although Mr. Pross did not have a finance background, he was, in Mr. Brown's opinion, extremely intelligent and highly capable. Mr. Brown suggested that Mr. Pross make a presentation to the senior management on the use of derivatives to reduce risk.

The presentation was scheduled for the first week of December 1990. Mr. Pross outlined the use of various derivatives, noting that they differed widely in their ability to reduce risk. If the company was, say, placing a large bid to buy a building abroad, one might prefer to use foreign currency options to hedge the currency risk in the event the deal fell through. He argued, however, that foreign currency futures were best suited to hedge the fluctuations in revenues arising from currency movements. Mr. Pross proposed a plan to hedge currency risk using futures which he termed the *derivatives plan*. Mr. Lee and the other senior managers (many of whom also held substantial blocks of the company's stock) were impressed by the presentation. They enthusiastically approved the plan submitted by Mr. Pross, subject to supervision by Mr. Brown. Mr. Lee, however, cautioned both men that he would be monitoring the results of the so-called derivatives plan very closely.

The derivatives plan was implemented in February 1991. In the first week of May 1991, Mr. Lee called both Mr. Brown and Mr. Pross to his office. He had been looking over the figures for the derivatives plan, which was managed by Mr. Brown's treasury group. He noted with concern that in the first three months since its inception, the total losses from derivatives trading amounted to almost half a million dollars. At this rate, he observed, the operation was going to have a significant negative impact on the foreign operations results. Mr. Lee emphasized that he expected the treasury group to "pull its own weight," reminding both men that their bonuses would reflect their success in this regard. He felt that the problem lay in the use of futures contracts rather than options contracts. With call options, the downside risk was known; you could not lose more than what you paid for the call. This was not the case with futures contracts. He suggested that they use options rather than futures contracts in the future.

Mr. Lee felt that his "heart-to-heart" talk had had an immediate impact on the derivatives operation, staunching the large losses associated with its inception. Indeed, the program yielded a net profit over the entire calendar year, and it continued to be a profit center for the company in 1992, 1993, and most of 1994.

In August 1994, Mr. Lee first heard of the case of a huge German metals conglomerate, Metallgesellschaft (MG). MG's U.S. energy unit had suffered a $1.6 billion loss in oil futures trading, almost resulting in the company's bankruptcy. Ironically, the trading was supposed to be part of a risk reduction strategy. He had also heard through his contacts in the computer industry that Dell Computer Corporation had lost over $30 million in derivatives trading so far that year. Mr. Lee decided that it would be an opportune time to review the operations of the derivatives program at Link Technologies and to obtain a better under-

standing of the risks associated with derivatives trading. He asked Mr. Pross to provide him with detailed information on the derivatives operation to be reviewed by him and other senior managers later that month.

The information provided by Mr. Pross is contained in Exhibit II 5.2. For each month from February 1991 to June 1994 (the last date for which data were available), the exhibit shows the unit shipments abroad, the revenue remitted from foreign operations (converted to dollars), and the revenue from the derivatives program (in dollars). Mr. Pross noted that the amounts to be shipped to foreign customers (usually stores specializing in the sale of personal computers, not individual customers) were usually received the previous month. He pointed out that these figures were "lined up right"; that is, the revenue in February 1991 actually corresponded to the sale of those units listed in the February column for unit shipments. From early 1991 to mid-1994, with

EXHIBIT II 5.2 Revenue from Foreign Operations and Currency Hedging, February 1991 to June 1994

Month	Unit Shipments Abroad	Revenue from Foreign Operations ($)	Revenue from Currency Hedging ($)	Month	Unit Shipments Abroad	Revenue from Foreign Operations ($)	Revenue from Currency Hedging ($)
Feb. 1991	15,760	3,743,262	(330,059)	Nov. 1992	20,373	4,394,101	17,158
Mar. 1991	18,021	3,899,820	(27,008)	Dec. 1992	14,603	3,046,495	(138,065)
Apr. 1991	16,303	3,635,379	(129,874)				
May 1991	15,927	3,509,714	24,964	Jan. 1993	15,707	3,465,664	349,908
June 1991	19,232	4,219,470	373,566	Feb. 1993	11,249	2,361,102	(139,123)
July 1991	19,855	4,188,058	102,208	Mar. 1993	13,001	2,781,478	(120,682)
Aug. 1991	17,491	3,728,835	(30,059)	Apr. 1993	20,779	4,512,325	31,901
Sept. 1991	19,616	4,130,203	22,794	May 1993	14,611	3,096,276	27,576
Oct. 1991	19,061	4,020,544	(107,065)	June 1993	15,354	3,203,224	(84,124)
Nov. 1991	18,300	3,940,144	37,984	July 1993	16,721	3,566,408	2,383
Dec. 1991	14,102	3,272,762	217,586	Aug. 1993	12,703	2,887,545	110,910
				Sept. 1993	18,456	3,815,000	(126,504)
Jan. 1992	13,234	2,768,417	(65,804)	Oct. 1993	19,384	4,126,439	(1,071)
Feb. 1992	14,860	3,153,290	(19,984)	Nov. 1993	17,882	3,872,616	(24,388)
Mar. 1992	11,835	2,610,311	31,731	Dec. 1993	20,021	4,446,400	(650)
Apr. 1992	13,241	2,760,794	(93,494)				
May 1992	16,488	3,557,012	10,867	Jan. 1994	17,931	3,889,644	(73,377)
June 1992	17,384	3,761,208	(28,820)	Feb. 1994	14,607	3,147,837	285,960
July 1992	18,722	4,084,854	18,950	Mar. 1994	16,009	3,588,134	73,804
Aug. 1992	18,408	3,988,505	119,363	Apr. 1994	11,192	2,415,616	135,004
Sept. 1992	14,095	3,116,734	5,008	May 1994	13,808	3,094,086	199,594
Oct. 1992	16,123	3,380,041	(2,903)	June 1994	13,358	2,784,303	(65,520)

most major markets in recession, sales were relatively stable at about 16,000 units per month, but they were somewhat volatile. Facing potentially severe competition, the company had elected to maintain the price of the company's mainstay, the PCI 2000 modem, at the equivalent of its U.S. dollar price of $215. In most of the foreign markets where Link Technologies was strong, the dollar fluctuated in value, but it did not seem consistently to appreciate or depreciate, so the $215 figure could be taken as a reasonable approximation of the expected dollar revenue per unit shipped. Since the modems were manufactured in the United States and used few foreign components, currency fluctuations had little impact on product costs, which were roughly $185 per unit.

Mr. Pross used Exhibit II 5.2 to bolster his argument that the derivatives program was a positive contribution to the company. He noted that from May 1991 onward (i.e., excluding the initial three months), the program had been a source of profits for the company. Further, the variance of the currency hedging revenues, in proportion to the variance of the revenue from foreign operations, was not significant. Thus, it seemed unlikely that the derivatives program could produce the kinds of losses reported in the financial press. Finally, from May 1991 onward the program relied more heavily on options as opposed to futures contracts, again limiting possible risks. He argued that, if anything, the derivatives program should be expanded in its scale of operations. Mr Brown strongly seconded these arguments.

A dissenting opinion was offered by Ms. Anne Cohen, the chief operations officer. She initially had been in favor of the derivatives program but had reversed her opinion over time. Although she did not provide specific facts to support her position, she argued forcefully that the derivatives program had exposed the company to new risks and was not being correctly implemented. She recommended that the program be radically al-

tered or be scrapped altogether if it could not be done right. Another executive also voiced the opinion that the risks from the derivatives operations were too difficult to assess and should be eliminated, noting that the currency risk could be eliminated by simply pricing in U.S. dollars.

As the discussion continued without any resolution, Mr. Lee was forced to adjourn the meeting. He decided that the evaluation of the derivatives program was more complicated than he had thought and that further analysis was clearly warranted.

QUESTIONS

1. Is the derivatives program at Link Technologies reducing risk? Are Ms. Cohen's arguments correct, or is the program performing as expected?

2. What impact, if any, does the corporate culture have on the implementation of the derivatives program at Link Technologies? How should the program's success or failure be judged?

3. What types of derivatives would you use if you were putting together a hedging program for the company?

4. Carefully evaluate the suggestion that pricing in dollars would eliminate the foreign currency risk facing the company.

CASE II.6

THE CASE OF THE DEPRECIATING INDIAN RUPEE*

It is October 21, 1995 and Mr. Hemang Patel, the Finance Manager of Surat Emerald Exports (SEE), is not a happy man. His concerns stem from the ongoing depreciation of the Indian rupee. The rupee's recent slide began in September and

---

*Prepared by Ganesh Kumar Nidugala, Associate Professor, T. A. Pai Management Institute, Manipal, India and Alan C. Shapiro, Professor of Banking and Finance, Marshall School of Business, University of Southern California. This case was prepared while Professor Nidugala was a Visiting Scholar, Department of Economics, University of Southern California.

shows no sign of abating. During the past month, the rupee's cumulative depreciation has been close to 10% (see Exhibits II 6.1 and II 6.2). As SEE is a company with a large volume of export-import transactions relative to its total sales, its bottom line is significantly affected by movements in exchange rates.

The foreign exchange market was very steady during the period from May 1993 to July 1995, with the rupee: U.S. dollar exchange rate fluctuating in the narrow range of Rs. 31.3-Rs. 31.4 per U.S. dollar (see Exhibits II 6.3 and II 6.4). Hence, many market observers were taken by surprise by the sudden and steep depreciation of the rupee during September–October, 1995.

BACKGROUND OF SEE

SEE was founded by Hemang Patel's father, Mr. Kirit Patel, in 1975 to produce and export gems and jewelry at Surat, India. It is a 100% family-owned enterprise. SEE imports raw (unprocessed) precious stones and adds value to them through cutting and polishing and by turning them into attractive pieces of jewelry. Import content in the finished product is almost 50%. All imports and exports of SEE are invoiced in U.S. dollars. The company's projected foreign exchange inflows and outflows during the next six months are given in Exhibit II 6.5.

FOREIGN EXCHANGE RISK MANAGEMENT PRACTICE AT SEE

SEE's policy has been to routinely cover its entire foreign exchange exposure (arising out of both imports and exports) in the forward market. This policy meant that once the exchange rates were locked in, further currency fluctuations would not affect the bottom line of the company in rupees. Mr. Kirit Patel was of the opinion that being a small manufacturing firm with little expertise in the foreign exchange markets, the financial objective should be "risk minimization."

EXHIBIT II 6.1 Daily Movements in the Exchange Rate During September–October 1995

DATE	RUPEES/U.S. DOLLAR	DATE	RUPEES/U.S. DOLLAR
July 14, 1995	31.3700	Sept. 22	33.8400
Aug. 1, 1995	31.4100	Sept. 25	33.9100
Aug. 14, 1995	31.4800	Sept. 26	33.9700
Sept. 1, 1995	31.9400	Sept. 27	33.9500
Sept. 4	31.9400	Sept. 28	34.0000
Sept. 5	32.1300	Sept. 29	34.0100
Sept. 6	32.4200	Oct. 4, 1995	33.9000
Sept. 7	32.2100	Oct. 5	33.8300
Sept. 8	32.2500	Oct. 6	33.9200
Sept. 11	32.4100	Oct. 9	33.9100
Sept. 12	32.6200	Oct. 11	33.8700
Sept. 13	32.8800	Oct. 12	33.9100
Sept. 14	33.1600	Oct. 13	34.0100
Sept. 15	34.1500	Oct. 16	34.7200
Sept. 18	33.8900	Oct. 17	34.6200
Sept. 19	34.0200	Oct. 18	34.8100
Sept. 20	33.8600	Oct. 19	34.8300
Sept. 21	33.9400	Oct. 20	35.5000

EXHIBIT II 6.2 Daily Spot and Forward Rate During October 1995

Date	Bid	Bid	Bid	Bid	Ask	Ask	Ask	Ask
Oct. 20	Spot Rs./U.S.$	1-Month Forward	3-Month Forward	6-Month Forward	Spot Rs./U.S.$	1-Month Forward	3-Month Forward	6-Month Forward
Oct. 4	33.66	33.91	34.28	34.90	33.99	34.26	34.63	35.25
Oct. 5	33.72	33.97	34.34	34.96	34.06	34.33	34.7	35.32
Oct. 6	33.7	33.95	34.32	34.94	34.04	34.31	34.68	35.3
Oct. 9	33.67	33.92	34.29	34.91	34.01	34.28	34.65	35.27
Oct. 10	33.7	33.95	34.32	35.02	34.04	34.31	34.68	35.38
Oct. 11	33.66	33.94	34.35	35.05	34.0	34.3	34.71	35.41
Oct. 12	33.75	34.03	34.44	35.14	34.09	34.39	34.80	35.50
Oct. 13	33.34	34.22	34.63	35.33	34.28	34.58	34.99	35.69
Oct. 16	34.33	34.61	35.02	35.38	34.67	34.87	35.3	36.21
Oct. 17	35.54	34.82	35.23	36.09	34.88	35.18	35.59	36.45
Oct. 18	34.75	35.06	35.48	36.17	35.1	35.43	35.85	36.54
Oct. 19	34.78	35.14	35.61	36.32	35.12	35.5	35.97	36.68
Oct. 20	35.32	35.67	36.15	36.91	35.66	36.05	36.53	37.20

Note: The figures here may not exactly match the spot rates in Exhibit II 6.1 because of the differences in bid and ask rates and differences in the sources of data.

THE NEW EXPERTISE

Mr. Hemang Patel joined his father in May 1995 after completing his bachelor's degree in economics and receiving an MBA in finance, both from a well-known university in California. After his return to India he was given charge of handling finances of the company, though on crucial matters the final decision still rested with Mr. Kirit Patel. Mr. Hemang Patel disagreed with the policy of 100% coverage of SEE's foreign exchange expo-

sures. He felt that by following a policy of 100% foreign exchange exposure coverage, SEE was missing an opportunity to profit from exchange rate movements.

Hemang Patel felt that during the past several years there were some occasions when SEE should have covered only its imports and left its exports unhedged in order to exploit the difference between the forward and spot rates. For example, when the rupee was devalued by almost 22% in July 1991, the company lost the opportu-

EXHIBIT II 6.3 Foreign Exchange Rates (1980–81 to 1994–95)

Year	Rs./U.S. Dollar	Year	Rs./U.S. Dollar
1980–81	7.91	1988–89	14.48
1981–82	8.97	1989–90	16.65
1982–83	9.67	1990–91	17.94
1983–84	10.34	1991–92	24.47
1984–85	11.89	1992–93	28.96
1985–86	12.24	1993–94	31.37
1986–87	12.78	1994–95	31.39
1987–88	12.97		

Note: The years indicated in Exhibit II 6.3 are Indian financial years spanning from April to March. The exchange rates refer to the average spot rates during the respective years.

EXHIBIT II 6.4 Monthly Movements in the Exchange Rate During 1995

MONTH	EXCHANGE RATE (RS./U.S.$)
January 1995	31.374
February	31.378
March	31.650
April	31.414
May	31.416
June	31.402
July	31.379
August	31.590
September	33.263

EXHIBIT II 6.5 SEE's Exports and Imports (Projected, in '000 Dollars)

MONTH	EXPORTS	IMPORTS
December 1995	200	100
January 1996	50	100
February	300	100
March	500	300
April	400	100
May	100	100

nity to sell its export earnings at the higher (depreciated) rates. SEE had entered into forward sales contracts for exports at much lower rates. However on the import side, he felt it was a good strategy to buy dollars forward because it allowed the company to pick up dollars at lower rates than it could realize in the spot market. On the whole, because of this devaluation, the company realized an opportunity loss since the net foreign exchange earnings of the company was positive (that is, exports were greater than imports). Even after July 1991, the company faced similar situations but with a lower opportunity cost because the amount of rupee depreciation was much smaller. The company continued with the policy of 100% coverage of exports. Now, SEE is faced with the same situation: The rupee is rapidly depreciating and Mr. Hemang Patel is obviously a worried man because of this development.

Although on past occasions Hemang Patel has suggested to his father that they might increase their profits by taking calculated risks, Mr. Kirit Patel always argued that it was better for the company to follow a risk-free strategy of covering all foreign exchange exposure. However, Hemang could not have asked for a better opportunity than now to push for his ideas. He is faced with a real-life decision problem. He must decide what course of action the company should take with regard to SEE's foreign exchange exposures for the next six months. Hemang learned in school

that the financial objective of a company is to maximize value to the shareholders. Here he is faced with the problem of value maximization under conditions of uncertainty.

HEMANG'S PREPARATION

Hemang starts to collect the necessary material for his decision. Hemang's decision whether to cover both exports and imports, only one of them, or none of them depends on his view of what will happen to the exchange rate from now on. He obviously cannot depend on what the newspapers say since they (Exhibit II 6.6) point in all directions as to the future course of the rupee: possible depreciation, appreciation, or no change in the rate from the current levels. Hemang remembers the inputs he received from various courses on international finance at his school in the United States and decides to put this skill to the task at hand. He collects information on the Indian exchange rate regime and foreign exchange market, and gathers data on crucial macroeconomic variables which he thinks could influence the exchange rate. This information is presented in Exhibits II 6.7 through II 6.12. Since Mr. Kirit Patel is still calling the shots with respect to crucial financial matters, Hemang has to convince his father with logical explanations for his decision.

EXHIBIT II 6.6 Newspaper Reports on the Fall of the Rupee

The popular press cited the following reasons for the declining rupee: government repayments of IMF loans; massive outflows from the foreign currency deposits owned by non-resident Indians; heavy buying of foreign exchange by leading oil companies; fewer security issues being sold in the international capital markets because of slump in the stock market. During the financial year 1995–1996, the estimated net outflow on the IMF account is estimated to be close to $1.7 billion. These outflows are repayment of the structural adjustment loans borrowed during the economic crisis of the financial year 1990–1991. Further, one of the leading private agencies that publishes data relating to Indian economy provided the following estimates for some balance-of-payments items for the financial year 1995–1996:

Export Growth (%)	20.0
Import Growth (%)	25.0
Current account/GDP ratio (%)	− 2.0
Eurocurrency issues	U.S. $200 million
IMF payment	U.S. $1.7 billion

All these figures pointed towards a worsening of the balance of payments in relation to 1994–1995 (refer to Exhibit II 6.9 for balance-of-payments statistics for 1994–1995). The inflation rate was expected to average 8.5% for the year but was expected to fall by the end of year to an annualized rate below 6%.

The Central Bank (Reserve Bank of India, the RBI) had intervened in the foreign exchange market by selling nearly $50 million to acquire rupees at a rate of Rs.34.9/U.S.$ on October 16, 1995. But the RBI was fighting a losing battle against the army of traders. Despite the intervention, the rupee kept falling.

Future Course of the Rupee.   Reports in leading finance dailies regarding the future course of the rupee are very confusing. Some market observers believe that the rupee has already reached its "true value" and it is just speculation which is driving the market. They believe that it will stabilize around the current levels of Rs.35–36/U.S.$. They also argue that the RBI will not allow the rupee to fall below Rs.36/dollar level. They cite the RBI intervention of selling dollars in the market on October 16 at lower levels of around Rs.34.90/U.S.$ as an indication of what can be expected of the central bank. Others in the market believe that the RBI will not be able to arrest the present depreciation. They pointed out that the central bank's reserve of only $20 billion would not last for long at an intervention rate of $50–100 million a day. According to them, the rupee-dollar rate might reach Rs.40/$ in the near future. There are some others who believe that the rupee has already depreciated too much. They point out that over the past 5 months the exchange rate has moved from Rs.31.53 to Rs.36, a level at which they believe it has overshot its equilibrium level. They predict that the rate will settle down to Rs.33 per dollar very soon.

The newspapers also observed that since 1950 the rupee had dropped in value so steeply on only two previous occasions. The first time was in 1966 when the rupee was devalued from Rs.4.76/$ to Rs.7.50/$. The second was in July 1991 when the rupee was devalued by about 22%. On both occasions the exchange rate regime was fixed.

EXHIBIT II 6.7 India's Exchange Rate Regime

From the 1970s to March 1, 1992, India was under a fixed exchange rate regime. The exchange rate was determined by the central bank, the Reserve Bank of India (RBI). The rate was adjusted in relation to a basket of currencies. Normally the rates were continuously adjusted by small margins (see Exhibit II 6.4). However, in July 1991, there was a steep devaluation of the rupee, by about 22% vis-à-vis dollar. The devalued rupee stood at Rs.26.10–26.20 per dollar immediately following this devaluation. In March 1992, the rupee was partially floated and a dual exchange rate system came into existence. Under this regime, there were two rates: an official rate (40% of all exports had to be surrendered at this rate) and a market rate (most imports, and 60% of exports were to take place at this rate). During March 1992–Feb. 1993, the average official rate was about Rs.26.20/$, whereas market rate averaged about Rs.30/$. On March 1, 1993, the dual exchange rates were unified into one market exchange rate. Since then, all transactions take place at exchange rates determined by the forces of demand and supply.

However, these forces include the central bank, which periodically intervenes to regulate the market. Thus, the current system is a managed float. For example, following the full float of the rupee in March 1993, by mid–1993 there was a surge in foreign capital flows (owing in large part to the adoption of several measures designed to encourage foreign capital). The RBI ended up buying a large portion of foreign exchange inflow, which resulted in an accumulation of foreign exchange reserves (see Exhibit II 6.11). This also led to an increase in the money supply (the RBI did not sterilize the impact of its foreign exchange market operations); consequently, the inflation rate jumped (see Exhibit II 6.12). The objective of buying up dollars was to prevent a steep appreciation of the rupee, which could hurt exports. Now that the rupee is depreciating steeply, the RBI is selling foreign exchange.

EXHIBIT II 6.8 The Foreign Exchange Market in India

The foreign exchange market in India is still in its early stages of development. Financial services such as options and futures are yet to be introduced. There are no futures exchanges. The spot market is reasonably well developed. There is an active inter-bank market and retail market. Individual investors still cannot buy and sell currencies for speculative purposes. Foreign currencies can be bought and sold by individuals and corporations in the spot and forward markets only if they have a genuine underlying transaction. Genuine underlying transactions would include export receivables, import payables, interest payables, and interest receivables. Individuals are not allowed to speculate in the market. However, the players in the inter-bank market are allowed to take speculative positions in the spot and forward markets. But the RBI could impose limits on the net position the traders could take in foreign currencies.

The forward premium/discounts on foreign currencies are determined by the forward demand and supply of currencies. Unlike in developed-country markets, the forward premium is not determined by interest rate differentials. This is because the flow of capital is highly restricted and arbitrage opportunities, if any, cannot be exploited. The exporters (importers) may prefer to book their receipts (payments) forward or wait to transact in the spot market when the actual payment/receipts happen. When there is an expectation of currency depreciation, importers rush to cover their imports whereas exporters prefer to stay away from the forward market. This makes the discount on the domestic currency rise. Thus, the forward rate may not reflect the interest rate differential, particularly when the spot market is in turmoil.

Both the spot and forward markets exist only within India. The Reserve Bank prohibits any international speculative access to the rupee.

EXHIBIT II 6.9 India's Balance of Payments (1990–91 to 1994–95)

	U.S. $ Millions				
	1990–91	1991–92	1992–93	1993–94	1994–95
1. Exports	18,477	18,266	18,869	22,700	26,857
2. Imports	27,917	21,064	23,237	23,985	31,672
3. Trade Balance	− 9,437	− 2,798	− 4,368	− 1,285	− 4,815
4. Invisibles (net)	− 243	1,620	842	970	2,191
NonFactor Services	979	1,207	1,128	777	− 494
Investment Income	− 3,752	− 3,830	− 3,422	− 4,002	− 3,905
Pvt. Transfers	2,069	3,873	2,773	3,825	6,200
Official Grants	461	370	363	370	390
5. Current a/c Balance	− 9,680	− 1,178	− 3,526	− 315	− 2,624
6. External Assistance	2,210	3,037	1,859	1,700	1,250
7. Commercial Borrowings (net)	2,249	1,256	− 358	1,252	1,029
8. IMF (net)	1,214	786	1,288	191	− 1,146
9. Nonresident Indian Deposits	1,536	290	2,001	940	847
10. Rupee Debt Service	− 1,193	− 1,240	− 878	− 745	1,050
11. Foreign Investment	68	154	585	4,110	4,895
i. Foreign Direct Investment	68	154	344	620	1,314
ii. Portfolio Investment by Foreign Institutional Investors	0	0	1	1,665	1,503
iii. Euro Issues by Indian Companies	0	0	240	1,460	2,078
12. Other Flows (net)	2,318	271	− 243	1,735	1,556
13. Capital Account (net)	8,402	4,754	4,524	9,183	7,381
14. Reserves (− indicates increase)	1,278	− 3,756	− 728	− 8,868	− 4,757

Important Ratios					
Current a/c − GDP (%)	− 3.2	− 0.4	− 1.8	− 0.1	− 0.8
External Debt Service − Exports (%)	35.3	30.2	28.6	25.2	26.6
External Debt − GDP ratio (%)	30.4	41.0	39.8	35.9	32.7

Note: The years indicated above are Indian Financial years spanning from April to March.

EXHIBIT II 6.10 Major Economic Indicators of the Indian Economy 1992–1995

Indicator	1992–93	1993–94	1994–95
Real GDP growth	5.1	5.0	6.3
Inflation (CPI based)	6.1	9.9	9.7
Growth of M3 (broad money)	15.7	18.4	22.3
Growth of Imports (in $)	12.7	6.5	22.9
Growth of Exports (in $)	3.8	20.0	18.4
Foreign Currency Assets (billion U.S. $)	6.43	15.07	20.81

EXHIBIT II 6.11 Indian Foreign Exchange Reserves

MONTH	FOREIGN EXCHANGE RESERVES (U.S.$ MILLIONS)
March 1991	2,236
March 1992	5,631
March 1993	6,434
June 1993	6,553
Sept. 1993	7,629
Dec. 1993	9,807
March 1994	15,068
June 1994	16,372
Sept. 1994	18,856
Dec. 1994	19,386
March 1995	20,809
June 1995	19,601
Sept. 1995	19,064

EXHIBIT II 6.12 Consumer Price Indices and Prime Lending Rates in India and the United States

YEAR/MONTH	CONSUMER PRICE INDEX IN INDIA	CONSUMER PRICE INDEX IN UNITED STATES	PRIME LENDING RATE IN INDIA	PRIME LENDING RATE IN UNITED STATES
1990	100.0	100.0	16.5	10.01
1991	113.9	104.2	17.88	8.46
1992	127.3	107.4	18.92	6.25
1993	135.4	110.6	16.25	6.00
1994	149.2	113.4	16.00	7.14
1993 June	134.2	110.6	16.25	6.00
1994 December	155.2	114.6	16.00	8.50
1995 January	155.2	115.0	16.00	8.50
1995 February	156.2	115.5	16.00	9.00
1995 March	157.3	115.9	16.00	9.00
1995 April	158.4	116.3	16.00	9.00
1995 May	161.1	116.5	16.00	9.00
1995 June	164.3	116.7	16.00	9.00
1995 July	168.1	116.7	16.00	8.80
1995 August	169.1	117.0	16.00	8.75
1995 September	170.2	117.3	16.00	8.75

Note: Figures for the CPI and the prime lending rate for both India and the United States are period averages.

QUESTIONS

1. Could the depreciation of the rupee during September–October 1995 have been predicted?

2. What is your prediction as to the future course of the rupee?

3. What should SEE do now?

NOTES

1. In July 1991, there was a steep devaluation of the rupee by about 22% vis-à-vis the dollar. The devalued rupee stood at about Rs.26.10–26.20 per dollar in the spot market immediately after the devaluation.

2. The rupee was partially floated in March 1992. This dual exchange rate regime lasted for one year (1992–93). The exchange rate shown in Exhibit II 6.3 for 1992–93 is the average of the market rate and the official rate. The market rate was around Rs.30 and the official rate was maintained at about Rs.26 during 1992–93. Immediately before the full float of the rupee in March 1993, the market exchange rate was close to Rs.31 and the official exchange rate was around Rs.26.20 per U.S. dollar.

3. In March 1993, the dual exchange rate was abolished and a unified, market-determined exchange rate came into force (see Exhibit II 6.7 for details).

MULTINATIONAL
WORKING
CAPITAL
MANAGEMENT

FINANCING FOREIGN TRADE

The development of a new product is a three-step process: first, a U.S. firm announces an invention; second, the Russians claim they made the same discovery 20 years ago; third, the Japanese start exporting it.

Anonymous

$\mathcal{M}$ost multinational corporations are heavily involved in foreign trade in addition to their other international activities. The financing of trade-related working capital requires large amounts of money, as well as financial services such as letters of credit and acceptances. It is vital, therefore, that the multinational financial executive have knowledge of the institutions and documentary procedures that have evolved over the centuries to facilitate the international movement of goods. Much of the material in this chapter is descriptive in nature, but interspersed throughout will be discussions of the role of these special financial techniques and their associated advantages and disadvantages.

The main purpose of this chapter is to describe and analyze the various payment terms possible in international trade, along with the necessary documentation associated with each procedure. It also examines the different methods and sources of export financing and credit insurance that are available from the public sector. The final section discusses the rise of countertrade, a sophisticated word for barter.

391

12.1 PAYMENTS TERMS IN INTERNATIONAL TRADE

Every shipment abroad requires some kind of financing while in transit. The exporter also needs financing to buy or manufacture its goods. Similarly, the importer has to carry these goods in inventory until the goods are sold. Then, it must finance its customers' receivables.

A financially strong exporter can finance the entire trade cycle out of its own funds by extending credit until the importer has converted these goods into cash. Alternatively, the importer can finance the entire cycle by paying cash in advance. Usually, however, some in-between approach is chosen, involving a combination of financing by the exporter, the importer, and one or more financial intermediaries.

The five principal means of payment in international trade, ranked in terms of increasing risk to the exporter, are

- Cash in advance
- Letter of credit
- Draft
- Consignment
- Open account

As a general rule, the greater the protection afforded the exporter, the less convenient are the payment terms for the buyer (importer). Some of these methods, however, are designed to protect both parties against commercial and/or political risks. It is up to the exporter when choosing among these payment methods to weigh the benefits in risk reduction against the cost of lost sales. The five basic means of payment are discussed in the following paragraphs.

Cash in Advance

Cash in advance affords the exporter the greatest protection because payment is received either before shipment or upon arrival of the goods. This method also allows the exporter to avoid tying up its own funds. Although less common than in the past, cash payment upon presentation of documents is still widespread.

Cash terms are used where there is political instability in the importing country or where the buyer's credit is doubtful. Political crises or exchange controls in the purchaser's country may cause payment delays or even prevent fund transfers, leading to a demand for cash in advance. In addition, where goods are made to order, prepayment is usually demanded, both to finance production and to reduce marketing risks.

Letter of Credit

Importers often will balk at paying cash in advance and will demand credit terms instead. When credit is extended, the **letter of credit** (L/C) offers the exporter the greatest degree of safety.

If the importer is not well known to the exporter or if exchange restrictions exist or are possible in the importer's country, the exporter selling on credit may wish to have the importer's promise of payment backed by a foreign or domestic bank. On the other hand, the importer may not wish to pay the exporter until it is reasonably certain that the merchandise has been shipped in good condition. A letter of credit satisfies both of these conditions.

In essence, the letter of credit is a letter addressed to the seller, written and signed by a bank acting on behalf of the buyer. In the letter, the bank promises it will honor drafts drawn on itself if the seller conforms to the specific conditions set forth in the L/C. (The draft, which is a written order to pay, is discussed in the next part of this section.) Through an L/C, the bank substitutes its own commitment to pay for that of its customer (the importer). The letter of credit, therefore, becomes a financial contract between the issuing bank and a designated beneficiary that is separate from the commercial transaction.

The advantages to the exporter are as follows:

1. Most important, an L/C eliminates credit risk if the bank that opens it is of undoubted standing. Therefore, the firm need check only on the credit reputation of the issuing bank.
2. An L/C also reduces the danger that payment will be delayed or withheld owing to exchange controls or other political acts. Countries generally permit local banks to honor their letters of credit. Failure to honor them could severely damage the country's credit standing and credibility.
3. An L/C reduces uncertainty. The exporter knows all the requirements for payment because they are clearly stipulated on the L/C.
4. The L/C can also guard against preshipment risks. The exporter who manufactures under contract a specialized piece of equipment runs the risk of contract cancellation before shipment. Opening a letter of credit will provide protection during the manufacturing phase.
5. Last, and certainly not least, the L/C facilitates financing because it ensures the exporter a ready buyer for its product. It also becomes especially easy to create a banker's acceptance—a draft accepted by a bank.

Most advantages of an L/C are realized by the seller; nevertheless, there are some advantages to the buyer as well.

1. Because payment is only in compliance with the L/C's stipulated conditions, the importer is able to ascertain that the merchandise is actually shipped on, or before, a certain date by requiring an on-board bill of lading. The importer also can require an inspection certificate.
2. Any documents required are carefully inspected by clerks with years of experience. Moreover, the bank bears responsibility for any oversight.
3. An L/C is about as good as cash in advance, so the importer usually can command more advantageous credit terms and/or prices.
4. Some exporters will sell only on a letter of credit. Willingness to provide one expands a firm's sources of supply.
5. L/C financing may be cheaper than the alternatives. There is no tie-up of cash if the L/C substitutes for cash in advance.
6. If prepayment is required, the importer is better off depositing its money with a bank than with the seller because it is then easier to recover the deposit if the seller is unable or unwilling to make a proper shipment.

The mechanics of letter-of-credit financing are quite simple, as illustrated by the case of U.S.A. Importers, Inc., of Los Angeles. The company is buying spare auto parts worth $38,000 from Japan Exporters, Inc., of Tokyo, Japan. U.S.A. Importers applies for,

EXHIBIT 12.1 Letter of Credit

IRREVOCABLE
COMMERCIAL
LETTER OF
CREDIT

Since 1852

WELLS FARGO BANK,N.A.

☐ 475 SANSOME STREET, SAN FRANCISCO, CALIFORNIA 94111
☐ 770 WILSHIRE BLVD. LOS ANGELES, CALIFORNIA 90017

INTERNATIONAL DIVISION   COMMERCIAL L/C DEPARTMENT   CABLE ADDRESS WELLS

OUR LETTER
OF CREDIT NO. XYZ9000   AMOUNT US$38,000   DATE MAY 6, 19XX
THIS NUMBER MUST BE MENTIONED
ON ALL DRAFTS AND CORRESPONDENCE

. JAPAN EXPORTERS INC.   . BANK OF TOKYO
. TOKYO, JAPAN   . TOKYO, JAPAN
.   .
.   .

GENTLEMEN:

   BY ORDER OF   U.S.A. IMPORTERS INC.

AND FOR ACCOUNT OF   SAME

WE HEREBY AUTHORIZE YOU TO DRAW ON OURSELVES

UP TO AN AGGREGATE AMOUNT OF THIRTY EIGHT THOUSAND AND NO/100 U.S. DOLLARS

AVAILABLE BY YOUR DRAFTS AT ON OURSELVES, IN DUPLICATE, AT 90 DAYS SIGHT
ACCOMPANIED BY
SIGNED INVOICE IN TRIPLICATE
PACKING LIST IN DUPLICATE
FULL SET OF CLEAN OCEAN BILLS OF LADING, MADE OUT TO ORDER OF SHIPPER,
   BLANK ENDORSED, MARKED FREIGHT PREPAID AND NOTIFY: U.S.A. IMPORTERS,
   INC., LOS ANGELES, DATED ON BOARD NOT LATER THAN MAY 30, 19XX.
INSURANCE POLICY/CERTIFICATE IN DUPLICATE FOR 110% OF INVOICE VALUE,
   COVERING ALL RISKS.

COVERING: SHIPMENT OF AUTOMOBILE SPARE PARTS, AS PER BUYER'S ORDER NO.
   900 DATED MARCH 15, 19XX FROM ANY JAPANESE PORT C.I.F.
   LOS ANGELES, CALIFORNIA
PARTIAL SHIPMENTS ARE PERMITTED.
TRANSHIPMENT IS NOT PERMITTED.
DOCUMENTS MUST BE PRESENTED WITHIN 7 DAYS AFTER THE BOARD DATE OF
   THE BILLS OF LADING, BUT IN ANY EVENT NOT LATER THAN JUNE 6, 19XX.

SPECIMEN

DRAFTS MUST BE DRAWN AND NEGOTIATED NOT LATER THAN   JUNE 6, 19XX
ALL DRAFTS DRAWN UNDER THIS CREDIT MUST BEAR ITS DATE AND NUMBER AND THE AMOUNTS
MUST BE ENDORSED ON THE REVERSE SIDE OF THIS LETTER OF CREDIT BY THE NEGOTIATING BANK.
WE HEREBY AGREE WITH THE DRAWERS, ENDORSERS, AND BONA FIDE HOLDERS OF ALL DRAFTS
DRAWN UNDER AND IN COMPLIANCE WITH THE TERMS OF THIS CREDIT, THAT SUCH DRAFTS WILL
BE DULY HONORED UPON PRESENTATION TO THE DRAWEE.
THIS CREDIT IS SUBJECT TO THE UNIFORM CUSTOMS AND PRACTICE FOR DOCUMENTARY CREDITS
(1974 REVISION). INTERNATIONAL CHAMBER OF COMMERCE PUBLICATION NO. 290.

SPECIMEN

AUTHORIZED SIGNATURE

Used by permission of Wells Fargo Bank, N.A.

EXHIBIT 12.2 Relationships Among the Three Parties to a Letter of Credit

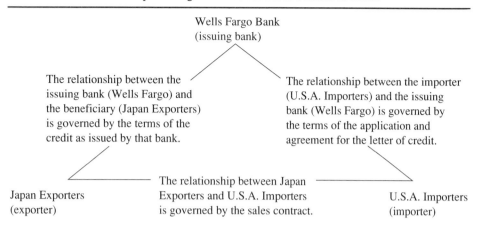

Wells Fargo Bank
(issuing bank)

The relationship between the issuing bank (Wells Fargo) and the beneficiary (Japan Exporters) is governed by the terms of the credit as issued by that bank.

The relationship between the importer (U.S.A. Importers) and the issuing bank (Wells Fargo) is governed by the terms of the application and agreement for the letter of credit.

Japan Exporters
(exporter)

The relationship between Japan Exporters and U.S.A. Importers is governed by the sales contract.

U.S.A. Importers
(importer)

and receives, a letter of credit for $38,000 from its bank, Wells Fargo. The actual L/C is shown in Exhibit 12.1. Exhibit 12.2, in turn, shows the relationships between the three parties to the letter of credit.

After Japan Exporters has shipped the goods, it draws a draft against the issuing bank (Wells Fargo) and presents it, along with the required documents, to its own bank, the Bank of Tokyo. The Bank of Tokyo, in turn, forwards the bank draft and attached documents to Wells Fargo; Wells Fargo pays the draft upon receiving evidence that all conditions set forth in the L/C have been met. Exhibit 12.3 details the sequence of steps in the L/C transaction.

Most L/Cs issued in connection with commercial transactions are *documentary*— that is, the seller must submit, together with the draft, any necessary invoices and the like. The documents required from Japan Exporters are listed on the face of the letter of credit in Exhibit 12.1 following the words "accompanied by." A *nondocumentary*, or *clean, L/C* is normally used in other than commercial transactions.

The letter of credit can be revocable or irrevocable. A revocable L/C is a means of arranging payment, but it does not carry a guarantee. It can be revoked, without notice, at any time up to the time a draft is presented to the issuing bank. An **irrevocable L/C**, on the other hand, cannot be revoked without the specific permission of all parties concerned, including the exporter. Most credits between unrelated parties are irrevocable; otherwise, the advantage of commitment to pay is lost. In the case of Japan Exporters, the L/C is irrevocable.

Although the essential character of a letter of credit—the substitution of the bank's name for the merchant's—is absent with a revocable credit, this type of L/C is useful in some respects. Just the fact that a bank is willing to open a letter of credit for the importer gives an indication of the customer's creditworthiness. Thus, it is safer than sending goods on a collection basis, where payment is made by a draft only after the goods have been shipped. Of equal, if not greater, importance is the probability that imports covered by letters of credit will be given priority in the allocation of foreign exchange should currency controls be imposed.

EXHIBIT 12.3 Example of Letter of Credit Financing of U.S. Imports

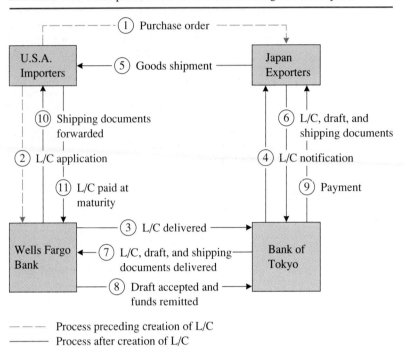

A letter of credit also can be confirmed or unconfirmed. A *confirmed L/C* is an L/C issued by one bank and confirmed by another, obligating both banks to honor any drafts drawn in compliance. An unconfirmed L/C is the obligation of only the issuing bank.

An exporter will prefer an irrevocable letter of credit by the importer's bank with confirmation by a domestic bank. In this way, the exporter need look no further than a bank in its own country for compliance with terms of the letter of credit. For example, if the Bank of Tokyo had confirmed the letter of credit issued by Wells Fargo, and Wells Fargo, for whatever reason, failed to honor its irrevocable L/C, Japan Exporters could collect $38,000 from the Bank of Tokyo, assuming that Japan Exporters met all the necessary conditions. This arrangement serves two purposes. Most exporters are not in a position to evaluate or deal with a foreign bank directly should difficulties arise. Domestic confirmation avoids this problem. In addition, should the foreign bank be unable to fulfill its commitment to pay, whether because of foreign exchange controls or political directives, that is of no concern to the exporter. The domestic confirming bank still must honor all drafts in full.

Thus, the three main types of L/C, in order of safety for the exporter, are (1) the irrevocable, confirmed L/C; (2) the irrevocable, unconfirmed L/C; and (3) the revocable L/C. Selecting the type of L/C to use depends on an evaluation of the risks associated with the transaction and the relative costs involved. One of the costs is the possibility of lost sales if the importer can get better credit terms elsewhere.

An exporter who acts as an intermediary may have to provide some assurance to its supplier that the supplier will be paid. It can provide this assurance by transferring or assigning the proceeds of the letter of credit opened in its name to the manufacturer.

A *transferable L/C* is one under which the beneficiary has the right to instruct the paying bank to make the credit available to one or more secondary beneficiaries. No L/C is transferable unless specifically authorized in the credit; moreover, it can be transferred only once. The stipulated documents are transferred along with the L/C.

An *assignment*, in contrast to a transfer, assigns part or all of the proceeds to another party but does not transfer to the party the required documents. This provision is not as safe to the assignee as a transfer because the assignee does not have control of the required merchandise and documentation.

Draft

Commonly used in international trade, a **draft** is an unconditional order in writing—usually signed by the exporter (seller) and addressed to the importer (buyer) or the importer's agent—ordering the importer to pay on demand, or at a fixed or determinable future date, the amount specified on its face. Such an instrument, also known as a *bill of exchange*, serves three important functions:

■ It provides written evidence, in clear and simple terms, of a financial obligation.
■ It enables both parties to potentially reduce their costs of financing.
■ It provides a negotiable and unconditional instrument. (That is, payment must be made to any holder in due course despite any disputes over the underlying commercial transaction.)

Using a draft also enables an exporter to employ its bank as a collection agent. The bank forwards the draft or bill of exchange to the foreign buyer (either directly or through a branch or correspondent bank), collects on the draft, and then remits the proceeds to the exporter. The bank has all the necessary documents for control of the merchandise and turns them over to the importer only when the draft has been paid or accepted in accordance with the exporter's instructions.

The conditions for a draft to be negotiable under the U.S. Uniform Commercial Code are that it must be

■ In writing
■ Signed by the issuer (drawer)
■ An unconditional order to pay
■ A certain sum of money
■ Payable on demand or at a definite future time
■ Payable to order of bearer

There are usually three parties to a draft. The party who signs and sends the draft to the second party is called the *drawer*; payment is made to the third party, the *payee*. Normally, the drawer and payee are the same person. The party to whom the draft is addressed is the *drawee*, who may be either the buyer or, if a letter of credit was used, the buyer's bank. In the case of a confirmed L/C, the drawee would be the confirming bank.

EXHIBIT 12.4 Time Draft

TOKYO, JAPAN	MAY 26	, 19 XX	No. 712	

AT NINETY DAYS _____ SIGHT OF THIS **ORIGINAL** OF EXCHANGE (DUPLICATE UNPAID)

PAY TO THE ORDER OF ___ BANK OF TOKYO _____ U.S. $ __ 38,000.00 ___

THE SUM OF ___ THIRTY EIGHT THOUSAND AND NO/100 * U.S. Dollars

DRAWN UNDER LETTER OF CREDIT NO.	DATED	ISSUED BY
X Y Z 9000	MAY 6, 19XX	WELLS FARGO BANK

To   WELLS FARGO BANK   *SPECIMEN*

770 WILSHIRE BLVD.

LOS ANGELES, CALIFORNIA   JAPAN EXPORTERS INC.

Used by permission of Wells Fargo Bank.

In the previous example, Japan Exporters is the drawer, and the Bank of Tokyo is the payee. The drawee is Wells Fargo under the terms of the L/C. This information is included in the draft shown in Exhibit 12.4.

Drafts may be either sight or time drafts. **Sight drafts** must be paid on presentation or else dishonored. **Time drafts** are payable at some specified future date and as such become a useful financing device. The maturity of a time draft is known as its *usance* or *tenor*. As mentioned earlier, for a draft to qualify as a negotiable instrument, the date of payment must be determinable. For example, a time draft payable "upon delivery of goods" is not specific enough, given the vagaries of ocean freight; the vague date of payment will likely nullify its negotiability. As shown in Exhibit 12.4, the draft drawn under the letter of credit by Japan Exporters is a time draft with a tenor of 90 days, indicated by the words "at ninety days sight." Thus, the draft will mature on August 24, 90 days after it was drawn (May 26).

A time draft becomes an **acceptance** after being accepted by the drawee. Accepting a draft means writing *accepted* across its face, followed by an authorized person's signature and the date. The party accepting a draft incurs the obligation to pay it at maturity. A draft accepted by a bank becomes a **banker's acceptance**; one drawn on and accepted by a commercial enterprise is termed a **trade acceptance**. Exhibit 12.5 is the time draft in Exhibit 12.4 after being accepted by Wells Fargo.

The exporter can hold the acceptance or sell it at a discount from face value to its bank, to some other bank, or to an acceptance dealer. The discount normally is less than the prevailing prime rate for bank loans. These acceptances enjoy a wide market and are an important tool in the financing of international trade. They are discussed in more detail in the next section. An acceptance can be transferred from one holder to another simply by endorsement.

Drafts can be clean or documentary. A **clean draft**, one unaccompanied by any other papers, normally is used only for nontrade remittances. Its primary purpose is to put pressure on a recalcitrant debtor that must pay or accept the draft or else face damage to its credit reputation.

EXHIBIT 12.5 Banker's Acceptance

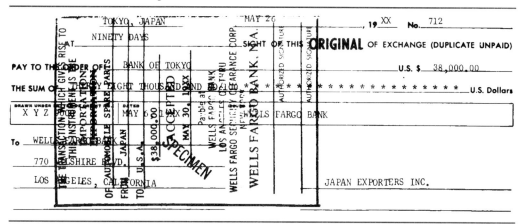

Used by permission of Wells Fargo Bank.

Most drafts used in international trade are documentary. A **documentary draft**, which can be either sight or time, is accompanied by documents that are to be delivered to the drawee on payment or acceptance of the draft. Typically, these documents include the bill of lading in negotiable form, the commercial invoice, the consular invoice where required, and an insurance certificate. The bill of lading in negotiable form is the most important document because it gives its holder the right to control the goods covered. A documentary sight draft also is known as a D/P (documents against payment) draft; if documents are delivered on acceptance, it is a D/A draft.

There are two significant aspects to shipping goods under documentary time drafts for acceptance. First, the exporter is extending credit to the importer for the usance of the draft. Second, the exporter is relinquishing control of the goods in return for a signature on the acceptance to assure it of payment.

It is important to bear in mind that sight drafts are not always paid at presentation, nor are time drafts always paid at maturity. Firms can get bank statistics on the promptness of sight and time draft payments, by country, from bank publications such as Chase Manhattan's *Collection Experience* bulletin.

Unless a bank has accepted a draft, the exporter ultimately must look to the importer for payment. Thus, use of a sight or accepted time draft is warranted only when the exporter has faith in the importer's financial strength and integrity.

Consignment

Goods sent on **consignment** are only shipped, but not sold, to the importer. The exporter (consignor) retains title to the goods until the importer (consignee) has sold them to a third party. This arrangement is normally made only with a related company because of the large risks involved. There is little evidence of the buyer's obligation to pay, and should the buyer default, it will prove difficult to collect.

The seller must carefully consider the credit risks involved and also the availability of foreign exchange in the importer's country. Imports covered by documentary drafts receive priority to foreign exchange over imports shipped on consignment.

Open Account

Open account selling is shipping goods first and billing the importer later. The credit terms are arranged between the buyer and the seller, but the seller has little evidence of the importer's obligation to pay a certain amount at a certain date. Sales on open account, therefore, are made only to a foreign affiliate or to a customer with which the exporter has a long history of favorable business dealings. However, open account sales have greatly expanded because of the major increase in international trade, the improvement in credit information about importers, and the greater familiarity with exporting in general. The benefits include greater flexibility (no specific payment dates are set) and involve lower costs, including fewer bank charges than with other methods of payment. As with shipping on consignment, the possibility of currency controls is an important factor because of the low priority in allocating foreign exchange normally accorded this type of transaction.

Exhibit 12.6 summarizes some of the advantages and disadvantages associated with the various means of arranging payment in international trade.

Banks and Trade Financing

Historically, banks have been involved in only a single step in international trade transactions such as providing a loan or a letter of credit. However, as financing has become an integral part of many trade transactions, U.S. banks—especially major money-center banks—have evolved as well. They have gone from financing individual trade deals to providing comprehensive solutions to trade needs. Such comprehensive services include combining bank lending with subsidized funds from government export agencies, interna-

EXHIBIT 12.6 International Methods of Payment: Advantages and Disadvantages (Ranked by Risk)

METHOD	RISK*	CHIEF ADVANTAGE	CHIEF DISADVANTAGE
Cash in advance	L	No credit extension required	Can limit sales potential, disturb some potential customers
Sight draft	M/L	Retains control and title; ensures payment before goods are delivered	If customer does not or cannot accept goods, goods remain at port of entry and no payment is due

Letters of credit
  Irrevocable
  Revocable
	
	M
	M/H
		Banks accept responsibility to pay; payment upon presentation of papers; costs go to buyer	If revocable, terms can change during contract work

Time draft	M/H	Lowers customer resistance by allowing extended payment after receipt of goods	Same as sight draft, plus goods are delivered before payment is due or received
Consignment sales	M/H	Facilitates delivery; lowers customer resistance	Capital tied up until sales; must establish distributor's creditworthiness; need political risk insurance in some countries; increased risk from currency controls
Open account	H	Simplified procedure; no customer resistance	High risk; seller must finance production; increased risk from currency controls

*L: low risk; M: medium risk; H: high risk.

tional leasing, and other nonbank financing sources, along with political and economic risk insurance.

Collecting Overdue Accounts

Typically, 1%–3% of a company's export sales go uncollected. Small businesses, however, take more risks than do large ones, often selling on terms other than a confirmed letter of credit. One reason is that they are eager to develop a new market opportunity; another reason is that they are not as well versed in the mechanics of foreign sales. Thus, their percentage of uncollected export sales may be higher than that of large companies.

Once an account becomes delinquent, sellers have three options: (1) They can try to collect the account themselves; (2) they can hire an attorney who is experienced in international law; or (3) they can engage the services of a collection agency.

The first step is for sellers to attempt to recover the money themselves. Turning the bill over to a collection agency or a lawyer too quickly will hurt the customer relationship. However, after several telephone calls, telexes, and/or personal visits, the firm must decide whether to write the account off or pursue it further.

The cost of hiring a high-priced U.S. lawyer, who then contacts an expensive foreign lawyer, is a deterrent to following the second option for receivables of less than $100,000. With such a relatively small amount, a collection agency usually would be more appropriate. Unlike lawyers, who charge by the hour for their services, regardless of the amount recovered, collection agencies work on a percentage basis. A typical fee is 20%–25% of the amount collected, but if the claim is more than $25,000 or so, the agency often will negotiate a more favorable rate.

Even with professional help, there are no guarantees of collecting on foreign receivables. This reality puts a premium on checking a customer's credit before filling an order, but getting credit information on specific foreign firms is often difficult.

One good source of credit information is the U.S. Department of Commerce's International Trade Administration (ITA). Its *World Data Trade Reports* covers nearly 200,000

EXHIBIT 12.7 Your Check Is Not in the Mail Yet: Length of Time Required for U.S. Companies to Collect on the Average Bill from Concerns in Selected Foreign Countries in the Second Quarter of 1992

COUNTRY	NUMBER OF DAYS
Iran	337
Kenya	129
Argentina	123
Brazil	119
Italy	90
India	80
Mexico	74
Taiwan	73
United Kingdom	70
Japan	58
Germany	54

Source: Michael Selz, "Small Firms Hit Foreign Obstacles in Billing Overseas," *Wall Street Journal,* December 18, 1992, p. B2. Reprinted by permission of the *Wall Street Journal,* © Dow Jones & Company, Inc. 1992. All rights reserved worldwide.

foreign establishments and can be obtained from district offices of ITA for $75. Other places to check on the creditworthiness of foreign companies and governments are export management companies and the international departments of commercial banks. Also, Dun & Bradstreet International publishes *Principal International Businesses*, a book with information on about 50,000 foreign enterprises in 133 countries.

The National Association of Credit Management collects data on how much time it takes to collect on the average bill from importers in various foreign countries. Exhibit 12.7, which contains these data for the second quarter of 1992, shows the wide variation in collection times by country.

12.2 DOCUMENTS IN INTERNATIONAL TRADE

The most important supporting document required in commercial bank financing of exports is the bill of lading. Of secondary importance are the commercial invoice, consular invoice, and insurance certificate.

Bill of Lading

Of the shipping documents, the **bill of lading** (B/L) is the most important. It serves three main and separate functions:

1. It is a contract between the carrier and shipper (exporter) in which the former agrees to carry the goods from port of shipment to port of destination.
2. It is the shipper's receipt for the goods.
3. The *negotiable B/L*, its most common form, is a document that establishes control over the goods.

A bill of lading can be either a straight or an order B/L. A *straight B/L* consigns the goods to a specific party, normally the importer, and is not negotiable. Title cannot be transferred to a third party merely by endorsement and delivery; therefore, a straight B/L is not good collateral and is used only when no financing is involved.

Most trade transactions do involve financing, which requires transfer of title, so the vast majority of bills of lading are order B/Ls. Under an *order B/L,* the goods are consigned to the order of a named party, usually the exporter. In this way, the exporter retains title to the merchandise until it endorses the B/L on the reverse side. The exporter's representative may endorse to a specific party or endorse it in blank by simply signing his or her name. The shipper delivers the cargo in the port of destination to the bearer of the endorsed order B/L, who must surrender it.

An order B/L represents goods in transit that are probably readily marketable and fully insured, so this document is generally considered to be good collateral by banks. It is required under L/C financing and for discounting of drafts.

Bills of lading also can be classified in several other ways. An *on-board* B/L certifies that the goods have actually been placed on board the vessel. By contrast, a *received-for-shipment* B/L merely acknowledges that the carrier has received the goods for shipment. It does not state that the ship is in port or that space is available. The cargo can, therefore, sit on the dock for weeks, or even months, before it is shipped. When goods are seasonal or perishable, therefore, the received-for-shipment B/L is never satisfactory to either the shipper or the importer. A received-for-shipment B/L can easily be converted into an on-board B/L by stamping it "on-board" and supplying the name of the vessel, the date, and the signature of the captain or the captain's representative.

A *clean B/L* indicates that the goods were received in apparently good condition. However, the carrier is not obligated to check beyond the external visual appearance of the boxes. If boxes are damaged or in poor condition, this observation is noted on the B/L, which then becomes a foul B/L. It is important that the exporter get a clean B/L—that is, one with no such notation—because foul B/Ls generally are not acceptable under a letter of credit.

Commercial Invoice

A **commercial invoice** contains an authoritative description of the merchandise shipped, including full details on quality, grades, price per unit, and total value. It also contains the names and addresses of the exporter and importer, the number of packages, any distinguishing external marks, the payment terms, other expenses such as transportation and insurance charges, any fees collectible from the importer, the name of the vessel, the ports of departure and destination, and any required export or import permit numbers.

Insurance

All cargoes going abroad are insured. Most of the insurance contracts used today are under an *open*, or *floating*, *policy*. This policy automatically covers all shipments made by the exporter, thereby eliminating the necessity of arranging individual insurance for each shipment. To evidence insurance for a shipment under an open policy, the exporter makes out an insurance certificate on forms supplied by the insurance company. This certificate contains information on the goods shipped. All entries must conform exactly with the information on the B/L, on the commercial invoice, and, where required, on the consular invoice.

Consular Invoice

Exports to many countries require a special **consular invoice**. This invoice, which varies in its details and information requirements from nation to nation, is presented to the local consul in exchange for a visa. The form must be filled out very carefully, for even trivial inaccuracies can lead to substantial fines and delays in customs clearance. The consular invoice does not convey any title to the goods being shipped and is not negotiable.

12.3 FINANCING TECHNIQUES IN INTERNATIONAL TRADE

In addition to straight bank financing, there are several other techniques available for trade financing: bankers' acceptances, discounting, factoring, and forfaiting.

Bankers' Acceptances

Bankers' acceptances have played an important role in financing international trade for many centuries. As we saw in the previous section, a banker's acceptance is a time draft drawn on a bank. By "accepting" the draft, the bank makes an unconditional promise to pay the holder of the draft a stated amount on a specified day. Thus, the bank effectively substitutes its own credit for that of a borrower, and, in the process, it creates a negotiable instrument that may be freely traded.

CREATING AN ACCEPTANCE ■ A typical acceptance transaction is shown in Exhibit 12.8. An importer of goods seeks credit to finance its purchase until the goods can be resold. If

EXHIBIT 12.8 Example of Banker's Acceptance Financing of U.S. Imports: Created, Discounted, Sold, and Paid at Maturity

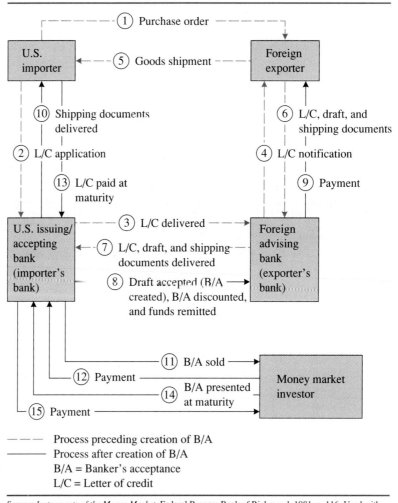

— — —   Process preceding creation of B/A
————   Process after creation of B/A
B/A = Banker's acceptance
L/C = Letter of credit

Source: Instruments of the Money Market. Federal Reserve Bank of Richmond, 1981, p. 116. Used with permission.

EXHIBIT 12.9 Bankers' Acceptances Outstanding by Transaction Type, Quarterly Averages of Month-End Figures

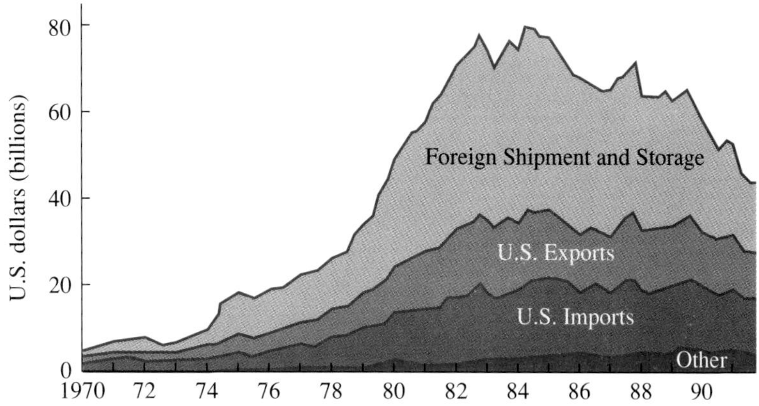

Source: Data from the Board of Governors' *Annual Statistical Digest* and *Federal Reserve Bulletin*, in Robert K. La Roche, "Bankers' Acceptances," Federal Reserve Bank of Richmond *Quarterly Review*, Winter 1993, p. 77.

the importer does not have a close relationship with and cannot obtain financing from the exporter it is dealing with, it may request acceptance financing from its bank. Under an acceptance agreement, the importer will have its bank issue a letter of credit on its behalf, authorizing the foreign exporter to draw a time draft on the bank in payment for the goods. On the basis of this authorization, the exporter ships the goods on an order B/L made out to itself and presents a time draft and the endorsed shipping documents to its bank. The foreign bank then forwards the draft and the appropriate shipping documents to the importer's bank; the importer's bank accepts the draft and, by so doing, creates a banker's acceptance. The exporter discounts the draft with the accepting bank and receives payment for the shipment. The shipping documents are delivered to the importer, and the importer now may claim the shipment. The accepting bank may either buy (discount) the B/A and hold it in its own portfolio or sell (rediscount) the B/A in the money market. In Exhibit 12.8, the bank sells the acceptance in the money market.

Acceptances also are created to finance the shipment of goods within the United States and to finance the storage of goods in the United States and abroad. As shown in Exhibit 12.9, domestic shipment and storage acceptances, which are included under "Other," have been a small part of the market in recent years. On average during 1991, they accounted for 10% of acceptances outstanding. Acceptances arising from imports into the United States accounted for 28%, those arising from U.S. exports accounted for 24%, and those arising from storage of goods or shipment of goods between foreign countries accounted for 38% of acceptances created in 1991.

TERMS OF ACCEPTANCE FINANCING ■ Typical maturities on bankers' acceptances are 30 days, 90 days, and 180 days, with the average being 90 days. Maturities can be tailored, though, to cover the entire period needed to ship and dispose of the goods financed.

For an investor, a banker's acceptance is a close substitute for other bank liabilities, such as certificates of deposit (CDs). Consequently, bankers' acceptances trade at rates very close to those on CDs. Market yields, however, do not give a complete picture of the costs of acceptance financing to the borrower because the accepting bank levies a fee, or commission, for accepting the draft. The fee varies depending on the maturity of the draft as well as the creditworthiness of the borrower, but it averages less than 1% per annum. The bank also receives a fee if a letter of credit is involved. In addition, the bank may hope to realize a profit on the difference between the price at which it purchases and the price at which it resells the acceptance.

On the maturity date of the acceptance, the accepting bank is required to pay the current holder the amount stated on the draft. The holder of a bank acceptance has recourse for the full amount of the draft from the last endorser in the event of the importer's unwillingness or inability to pay at maturity. The authenticity of an accepted draft is separated from the underlying commercial transaction and may not be dishonored for reason of a dispute between the exporter and importer. This factor, of course, significantly enhances its marketability and reduces its riskiness.

In recent years, the demand for acceptance financing has fallen off. One factor has been the increased availability of funding from nonbank investors in the U.S. commercial paper market. **Commercial paper** (CP) is a short-term, unsecured promissory note that is generally sold by large corporations on a discount basis to institutional investors and other corporations. Prime commercial paper generally trades at rates near those on acceptance liabilities of prime banks. For firms with access to this market, the overall cost—including placement fees charged by dealers and fees for back-up line of credit—is usually below the all-in cost of acceptance financing. The fall-off in acceptance financing in international trade is shown in Exhibit 12.10. Its share of U.S. international-trade

EXHIBIT 12.10 Shares of U.S. and Non-U.S. International Trade Financed by Bankers' Acceptances

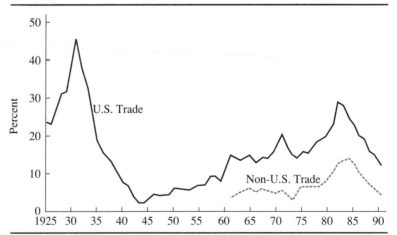

Source: Acceptance data are from the Board of Governors' *Banking and Monetary Statistics, Annual Statistical Digest,* and *Federal Reserve Bulletin.* U.S. imports and exports are from the U.S. Bureau of the Census's *Historical Statistics, Statistical Abstract,* and FT 900 release. Trade data for the rest of the world are from the International Monetary Fund's *International Financial Statistics.* Shares financed are based on average maturity of 90 days. In Robert K. La Roche, "Bankers' Acceptances," Federal Reserve Bank of Richmond *Quarterly Review,* Winter 1993, p. 81

financing fell from more than 25% in the early 1980s to about 10% by 1990. The changes in the use of acceptance financing over time are closely tied to their favorable (or unfavorable) treatment by the Federal Reserve.

EVALUATING THE COST OF ACCEPTANCE FINANCING ■ Suppose that the discount rate on a $1 million acceptance for 90 days is 9.8% per annum and the acceptance fee is 2% per annum. If the exporter chooses to hold the acceptance, then in 90 days it will receive the face amount less the acceptance fee:

Face amount of acceptance	$1,000,000
Less: 2% per annum commission for 90 days	−5,000
Amount received by exporter in 90 days	$ 995,000

Alternatively, the exporter can sell the acceptance at a 2.45% discount (9.8%/4) and receive $970,500 immediately:

Face amount of acceptance	$1,000,000
Less: 2% per annum commission for 90 days	−5,000
Less: 9.8% per annum discount for 90 days	−24,500
Amount received by exporter immediately	$ 970,500

Whether the exporter should discount the acceptance or wait depends on the opportunity cost of money. Suppose that the exporter's opportunity cost of money is 10.2%. Then the present value of holding onto the acceptance is $995,000/[1 + (.102/4)], or $970,258. In this case, the exporter would come out ahead by selling the acceptance.

Discounting

Even if a trade draft is not accepted by a bank, the exporter still can convert the trade draft into cash by means of **discounting**. The exporter places the draft with a bank or other financial institution and, in turn, receives the face value of the draft less interest and commissions. By insuring the draft against both commercial and political risks, the exporter often will pay a lower interest rate. If losses covered by the insurer do occur, the insuring agency will reimburse the exporter or any institution to which the exporter transfers the draft.

The discount rate for trade paper is often lower than interest rates on overdrafts, bank loans, and other forms of local funding. This lower rate is usually a result of export promotion policies that lead to direct or indirect subsidies of rates on export paper.

Discounting may be done with or without *recourse*. With recourse, the bank can collect from the exporter if the importer fails to pay the bill when due. The bank bears the collection risk if the draft is sold without recourse.

Factoring

Firms with a substantial export business and companies too small to afford a foreign credit and collections department can turn to a **factor**. Factors buy a company's receivables at a discount, thereby accelerating their conversion into cash. Most **factoring** is done on a **nonrecourse** basis, which means that the factor assumes all the credit and political risks except for those involving disputes between the transacting parties. In order to avoid being stuck with only receivables of risky customers (with good credits not being factored), factors usually insist on handling most or all sales. This selection bias is not an issue in factoring *with recourse*, where the exporter assumes all risks.

Factoring is becoming increasingly popular as a trade financing vehicle. The value of world exports financed through factoring now exceeds $10 billion. By using a factor, a firm can ensure that its terms are in accord with local practice and are competitive. For instance, customers can be offered payment on open account rather than being asked for a letter of credit or stiffer credit requirements. If the margin on its factored sales is not sufficiently profitable, then the firm can bear the credit risks itself or forgo that business. Even if an exporter chooses not to discount its foreign receivables with a factor, it still can use the factor's extensive credit information files to ascertain the creditworthiness of prospective customers.

An exporter that has established an ongoing relationship with a factor will submit new orders directly to the factor. After evaluating the creditworthiness of the new claim, the factor will make a recourse/nonrecourse decision within two days to two weeks, depending on the availability of information.

Although the factors may consider their fees to be nominal considering the services provided, they are not cheap. Export factoring fees are determined on an individual company basis and are related to the annual turnover (usually a minimum of $500,000 to $1 million is necessary), the average invoice size (smaller invoices are more expensive because of the fixed information-gathering costs), the creditability of the claims, and the terms of sale. In general, these fees run from 1.75% to 2% of sales.

EVALUATING THE COST OF FACTORING ■ Suppose that a factor will buy an exporter's receivables at a 2.5% per *month* discount. In addition, the factor will charge an extra 1.75% fee for nonrecourse financing. If the exporter decides to factor $1 million in 90-day receivables without recourse, then it will receive $907,500 today:

Face amount of receivable	$1,000,000
Less: 1.75% nonrecourse fee	−17,500
Less: 2.5% monthly factoring fee for three months	−75,000
Amount received by exporter	$ 907,500

On an annualized basis, factoring is costing this company 41.34%:

$$\text{Annual percentage rate (APR)} = \frac{17{,}500 + 75{,}000}{1{,}000{,}000 - 17{,}500 - 75{,}000} \times \frac{365}{90}$$

$$= 41.34\%$$

Despite these high costs, factoring can be quite worthwhile to many firms for one principal reason: The cost of bearing the credit risk associated with a given receivable can be substantially lower to a factor than to the selling firm. First, the factor's greater credit information makes it more knowledgeable about the actual, as opposed to the perceived, risks involved and, thereby, reduces its required risk premium. Second, by holding a well-diversified portfolio of receivables, the factor can eliminate some of the risks associated with individual receivables.

In general, factoring is most useful for (1) the occasional exporter and (2) the exporter having a geographically diverse portfolio of accounts receivable. In both cases, it would be organizationally difficult and expensive to internalize the accounts-receivable collection process. Such companies generally would be small or else would be involved on a limited scale in foreign markets.

Forfaiting

The specialized factoring technique known as forfaiting is sometimes used in the case of extreme credit risk. **Forfaiting** is the discounting—at a fixed rate without recourse—of medium-term export receivables denominated in fully convertible currencies (U.S. dollar, Swiss franc, Deutsche mark). This technique is usually used in the case of capital-goods exports with a five-year maturity and repayment in semiannual installments. The discount is set at a fixed rate: about 1.25% above the local cost of funds.

Forfaiting is especially popular in Western Europe (primarily in Switzerland and Austria), and many forfaiting houses are subsidiaries of major international banks, such as Credit Suisse. These houses also provide help with administrative and collection problems.

12.4 GOVERNMENT SOURCES OF EXPORT FINANCING AND CREDIT INSURANCE

In the race for export orders, particularly for capital equipment and other "big-ticket" items requiring long repayment arrangements, most governments of developed countries have attempted to provide their domestic exporters with a competitive edge in the form of low-cost export financing and concessionary rates on political and economic risk insurance. Nearly every developed nation has its own export-import agency for development and trade financing.

Export Financing

Procedures for extending credit vary greatly among agencies. Many agencies offer funds in advance of the actual export contract, whereas private sources extend financing only after the sale has been made. Some programs extend credit only to the supplier—called *supplier credits*—to pass on to the importer; others grant credit directly to the buyer—called *buyer credits*—who then pays the supplier. The difference is that in the first arrangement, the supplier bears the credit risk, whereas in the latter case, the government is the risk bearer. Of course, the government often provides credit insurance in conjunction with supplier credits.

EXPORT-IMPORT BANK ■ The **Export-Import Bank (Eximbank)** is the only U.S. government agency dedicated solely to financing and facilitating U.S. exports. Eximbank loans provide competitive, fixed-rate financing for U.S. export sales facing foreign competition backed with subsidized official financing. Evidence of foreign competition is not required for exports produced by small businesses where the loan amount is $2.5 million or less. Eximbank also provides guarantees of loans made by others. The loan and guarantee programs cover up to 85% of the U.S. export value and have repayment terms of one year or more.

Eximbank operations generally conform to five basic principles:

1. Loans are made for the specific purpose of financing U.S. exports of goods and services. If a U.S. export item contains foreign-made components, Eximbank will cover up to 100% of the U.S. content of exports provided that the total amount financed or guaranteed does not exceed 85% of the total contract price of the item and that the total U.S. content accounts for at least half of the contract price.

2. Eximbank will not provide financing unless private capital is unavailable in the amounts required. It supplements, rather than competes with private capital.

3. Loans must have reasonable assurance of repayment and must be for projects that have a favorable impact on the country's economic and social well-being. The host government must be aware of, and not object to, the project.

4. Fees and premiums charged for guarantees and insurance are based on the risks covered.

5. In authorizing loans and other financial assistance, Eximbank is obliged to take into account any adverse effects on the U.S. economy or balance of payments that might occur.

In recent years, the Eximbank has become very aggressive in financing U.S. trade with China. Indeed, China—already the bank's biggest lending market in Asia—will likely become its biggest lending market in the world before the turn of the century.

The interest rates on Eximbank's loans are based on an international arrangement among the 29 members of the Organization for Economic Cooperation and Development (OECD). The purpose of the arrangement, which sets minimum rates that an official export finance agency must charge on export credits, is to limit the interest subsidies used by many industrial countries to gain competitive advantage vis-à-vis other nations. The OECD minimum rates are based on the weighted average interest rate on government bond issues denominated in the U.S. dollar, German mark, British pound, French franc, and Japanese yen. In this way, rates on export credits are brought closer to market interest rates.

Eximbank extends direct loans to foreign buyers of U.S. exports and intermediary loans to financial institutions that extend loans to the foreign buyers. Both direct and intermediary loans are provided when U.S. exporters face officially subsidized foreign competition.

Eximbank's medium-term loans to intermediaries (where the loan amount is $10 million or less and the term is seven years or less) are structured as "standby" loan commitments. The intermediary may request disbursement by Eximbank at any time during the term of the underlying debt obligation.

Eximbank guarantees provide repayment protection for private-sector loans to creditworthy buyers of exported U.S. goods and services. The guarantees are available alone or may be combined with an intermediary loan. Most guarantees provide comprehensive coverage of both political and commercial risks. Eximbank also will guarantee payments on cross-border or international leases.

Exporters also may have access to an Eximbank program that guarantees export-related working-capital loans to creditworthy small and medium-sized businesses. All

Eximbank guarantees carry the full faith and credit of the U.S. government, so loans provided under these guarantee programs are made at close to the risk-free interest rate. In effect, low-cost guarantees are another form of government-subsidized export financing.

Repayment terms vary with the project and type of equipment purchased. For capital goods, long-term credits are normally provided for a period of five years to ten years. Loans for projects and large product acquisitions are eligible for longer terms, whereas lower-unit-value items receive shorter terms. Loan amortization is made in semiannual installments, beginning six months after delivery of the exported equipment.

Another program run by Eximbank provides a *preliminary commitment* that outlines the amount, terms, and conditions of financing it will extend to importers of U.S. goods and services. This commitment gives U.S. firms a competitive advantage in bidding on foreign projects because it enables the firms to offer financing along with their equipment proposals. Preliminary commitments are issued without cost (there is a $100 processing fee) or obligation to applicants.

Eximbank charges a front-end exposure fee, assessed on each disbursement of a loan by Eximbank or the guaranteed or intermediary lender. Exposure fees, which are adjusted periodically, vary according to the term of the loan, the classification of the borrower or guarantor, and the borrower's country. For exposure-fee purposes, Eximbank classifies countries in five country categories according to risk. Under each country category, there are three borrower/guarantor classifications:

- ■ Class I: Sovereign borrowers or guarantors, or for political-risks-only coverage
- ■ Class II: Creditworthy nonsovereign public institutions or banks, or highly creditworthy private buyers
- ■ Class III: Other creditworthy private buyers

In recent years, Eximbank has become more aggressive in fighting perceived abuses by foreign export-credit agencies. One area that Eximbank has targeted is foreign *mixed-credit financing* — the practice of tying grants and low-interest loans to the acceptance of specific commercial contracts. For years, U.S. capital-equipment exporters, engineering firms, and high-tech producers have lost overseas bids to foreign firms backed by government mixed credits. Eximbank now offers its own mixed credits. It will even offer mixed-credit deals before the fact. Any deal that has a chance of attracting a foreign mixed-credit bid is considered. However, an Eximbank spokesperson noted, "This is not an export promotion. We are not out there to match every mixed credit. We're out to end mixed credits, and will only offer one if it helps us to make a specific negotiating point."[1]

PRIVATE EXPORT FUNDING CORPORATION ■ The **Private Export Funding Corporation** (PEFCO) was created in 1970 by the *Bankers' Association for Foreign Trade* to mobilize private capital for financing the export of big-ticket items by U.S. firms. It purchases the medium- to long-term debt obligations of importers of U.S. products at fixed interest rates. PEFCO finances its portfolio of foreign importer loans through the sale of its own securities. Eximbank fully guarantees repayment of all PEFCO foreign obligations.

---

[1]"How U.S. Firms Benefit from Eximbank's War on Foreign Mixed Credits," *Business International Money Report*, February 10, 1986, p. 41.

PEFCO normally extends its credits jointly with one or more commercial banks and Eximbank. The maturity of the importers' notes purchased by PEFCO varies from 2.5 years to 12 years; the banks take the short-term maturity and Eximbank takes the long-term portion of a PEFCO loan. Much of this money goes to finance purchases of U.S.-manufactured jet aircraft and related equipment such as jet engines.

TRENDS ■ There are several trends in public-source export financing, including:

1. *A shift from supplier to buyer credits.*   Many capital goods exports that cannot be financed under the traditional medium-term supplier credits become feasible under buyer credits, where the payment period can be stretched up to 20 years.

2. *A growing emphasis on acting as catalysts to attract private capital.*   This action includes participating with private sources, either as a member of a financial consortium or as a partner with an individual private investor, in supplying export credits.

3. *Public agencies as a source of refinancing.*   Public agencies are becoming an important source for refinancing loans made by bankers and private financiers. Refinancing enables a private creditor to discount its export loans with the government.

4. *Attempts to limit competition among agencies.*   The virtual export-credit war among governments has led to several attempts to agree upon and coordinate financing terms. These attempts, however, have been honored more in the breach than in the observance.

Export-Credit Insurance

Export financing covered by government credit insurance, also known as *export-credit insurance*, provides protection against losses from political and commercial risks. It serves as collateral for the credit and is often indispensable in making the sale. The insurance does not usually provide an ironclad guarantee against all risks, however. Having this insurance results in lowering the cost of borrowing from private institutions because the government agency is bearing those risks set forth in the insurance policy. The financing is nonrecourse to the extent that risks and losses are covered. Often, however, the insurer requires additional security in the form of a guarantee by a foreign local bank or a certificate from the foreign central bank that foreign exchange is available for repayment of the interest and principal.

The purpose of export-credit insurance is to encourage a nation's export sales by protecting domestic exporters against nonpayment by importers. The existence of medium- and long-term credit insurance policies makes banks more willing to provide nonrecourse financing of big-ticket items that require lengthy repayment maturities, provided the goods in question have been delivered and accepted.

FOREIGN CREDIT INSURANCE ASSOCIATION ■ In the United States, the export-credit insurance program is administered by the **Foreign Credit Insurance Association** (FCIA). The FCIA is a cooperative effort of Eximbank and a group of approximately 50 of the leading marine, casualty, and property insurance companies. FCIA insurance offers protection from political and commercial risks to U.S. exporters: The private insurers cover commercial risks, and the Eximbank covers political risks. The exporter (or the financial institution providing the loan) must self-insure that portion not covered by the FCIA.

Short-term insurance is available for export credits up to 180 days (360 days for bulk agricultural commodities and consumer durables) from the date of shipment. Coverage is of two types: comprehensive (90%–100% of political and 90%–95% of commercial risks) and political only (90%–100% coverage). Coinsurance is required presumably because of the element of moral hazard: the possibility that exporters might take unreasonable risks knowing that they would still be paid in full.

Rather than sell insurance on a case-by-case basis, the FCIA approves discretionary limits within which each exporter can approve its own credits. Insurance rates are based on the terms of sale, type of buyer, and the country of destination and can vary from a low of 0.1% to a high of 2%. The greater the loss experience associated with the particular exporter and the countries and customers it deals with, the higher the insurance premium charged. The FCIA also offers preshipment insurance up to 180 days from the time of sale.

Medium-term insurance, guaranteed by Eximbank and covering big-ticket items sold on credit, usually from 181 days to five years, is available on a case-by-case basis. As with short-term coverage, the exporter must reside in, and ship from, the United States. However, the FCIA will provide medium-term coverage for that portion only of the value added that originated in the United States. As before, the rates depend on the terms of sale and the destination.

Under the FCIA lease insurance program, lessors of U.S. equipment and related services can cover both the stream of lease payments and the fair market value of products leased outside the United States. The FCIA charges a risk-based premium that is determined by country, lease term, and the type of lease.

Taking Advantage of Government-Subsidized Export Financing

Government-subsidized export-credit programs often can be employed advantageously by multinationals. The use will depend on whether the firm is seeking to export or import goods or services, but the basic strategy remains the same: Shop around among the various export-credit agencies for the best possible financing arrangement.

EXPORT FINANCING STRATEGY ◼Massey-Ferguson (now Varity Corp.), the multinational Canadian farm-equipment manufacturer, illustrates how MNCs are able to generate business for their foreign subsidiaries at minimum expense and risk by playing off national export-credit programs against each other.

The key to this *export financing strategy* is to view the foreign countries in which the MNC has plants not only as markets but also as potential sources of financing for exports to third countries. For example, in early 1978, Massey-Ferguson was looking to ship 7,200 tractors (worth $53 million) to Turkey, but it was unwilling to assume the risk of

ILLUSTRATION

TEXAS INSTRUMENTS SEARCHES FOR LOW-COST CAPITAL

Texas Instruments (TI) is seeking to finance an aggressive capital spending program through a series of joint ventures and other cooperative arrangements with foreign governments and corporations. In Italy, TI received a package of development grants and low-cost loans from the government that offset more than half of TI's investment in a state-of-the-art semiconductor plant there—an investment expected to total more than $1 billion over many years. TI was able to negotiate the incentive package because the Italian government was seeking to improve its technological infrastructure in the area selected by TI for the new plant.

In Taiwan, TI and a Taiwanese customer, Acer Computer Company, established a joint venture in which Acer's majority stake is financed with Taiwanese equity capital that would be unavailable to a U.S. company acting alone. In Japan, TI entered into a joint venture with Kobe Steel, a company seeking diversification. Here, too, TI relies on its foreign partner to supply a majority of the equity. In both Asian joint ventures, however, TI has an option to convert its initial minority stake into a majority holding.

currency inconvertibility.[2] Turkey at that time already owed $2 billion to foreign creditors, and it was uncertain whether it would be able to come up with dollars to pay off its debts (especially because its reserves were at about zero).

Massey solved this problem by manufacturing the tractors at its Brazilian subsidiary, Massey-Ferguson of Brazil, and selling them to Brazil's Interbras—the trading-company arm of Petrobras, the Brazilian national oil corporation. Interbras, in turn, arranged to sell the tractors to Turkey and pay Massey in cruzeiros. The cruzeiro financing for Interbras came from Cacex, the Banco do Brasil department that is in charge of foreign trade. Cacex underwrote all the political, commercial, and exchange risks as part of the Brazilian government's intense export promotion drive. Before choosing Brazil as a supply point, Massey made a point of shopping around to get the best export-credit deal available.

IMPORT FINANCING STRATEGY ■ Firms engaged in projects that have sizable import requirements may be able to finance these purchases on attractive terms. A number of countries, including the United States, make credit available to foreign purchasers at low (below-market) interest rates and with long repayment periods. These loans are almost always tied to procurement in the agency's country; thus, the firm must compile a list of goods and services required for the project and relate them to potential sources by country. Where there is overlap among the potential suppliers, the purchasing firm may have leverage to extract more-favorable financing terms from the various export-credit agencies involved. This strategy is illustrated by the hypothetical example of a copper mining venture in Exhibit 12.11.

---

[2] "Massey-Ferguson's No-Risk Tractor Deal," *Business International Money Report*, February 3, 1978, pp. 35–36.

EXHIBIT 12.11 Alternative Sources of Procurement: Hypothetical Copper Mine (U.S. $ Millions)

ITEM	TOTAL PROJECT	UNITED STATES	FRANCE	GERMANY	JAPAN	UNITED KINGDOM	SWEDEN	ITALY
Mine Equipment								
Shovels	$12	$12	$ 8	$12	$12	$12	$10	—
Trucks	20	20	—	20	20	10	20	12
Other	8	8	5	3	6	8	—	4
Mine Facilities								
Shops	7	7	7	7	3	7	5	6
Offices	3	3	3	3	2	3	3	2
Preparation Plant								
Crushers	11	11	8	11	11	11	—	—
Loading	15	15	10	10	15	12	15	7
Environmental	13	13	5	8	5	10	7	5
Terminal								
Ore handling	13	13	10	13	13	13	8	9
Shiploader	6	6	6	6	6	6	2	4
Bulk Commodities								
Steel	20	20	20	20	20	15	8	20
Electrical	17	17	12	14	10	15	5	8
Mechanical	15	15	8	—	12	10	6	—
Total potential foreign purchases	$160	$160	$102	$127	$135	$132	$89	$77

Perhaps the best-known application of this *import financing strategy* in recent years is the financing of the Soviet gas pipeline to Western Europe. The former Soviet Union played off European and Japanese suppliers and export financing agencies against each other and managed to get extraordinarily favorable credit and pricing terms.

12.5 COUNTERTRADE

In recent years, more and more multinationals have had to resort to **countertrade** to sell overseas: purchasing local products to offset the exports of their own products to that market. Countertrade transactions often can be complex and cumbersome. They may involve two-way or three-way transactions, especially where a company is forced to accept unrelated goods for resale by outsiders.

If swapping goods for goods sounds less efficient than using cash or credit, that's because it is less efficient, but it is preferable to having no sales in a given market. More firms are finding it increasingly difficult to conduct business without being prepared to countertrade. Although precise numbers are impossible to come by, this transaction is growing in importance. One estimate placed countertrade volume at from 20% to 30% of all international trade.[3]

When a company exports to a nation requiring countertrade, it must take back goods that the country cannot (or will not try to) sell in international markets. To unload these goods, the company usually has to cut the prices at which the goods are nominally

---

[3] Thomas R. Hofstedt, "An Overview of Countertrade," July 29, 1987, unpublished.

valued in the barter arrangement. Recognizing this necessity, the firm typically will pad the price of the goods it sells to its countertrade customer. When a German machine-tool maker sells to Romania, for instance, it might raise prices by 20%. Then, when it unloads the blouses it gets in return, the premium covers the reduction in price.

Usually, an exporting company wants to avoid the trouble of marketing those blouses, so it hands over the 10% premium to a countertrade specialist. This middleman splits the premium with a blouse buyer, keeping perhaps 2% and passing the remaining 8% along in price cuts. The result: Romania pays above the market for imports, making international trade less attractive than it should be, and dumps its own goods through backdoor price shaving. In the long run, however, the practice is self-defeating. Having failed to set up continuing relationships with customers, Romania never learns what the market really wants—what style blouses, for instance—or how it might improve its competitiveness.

Countertrade takes several specific forms.

1. *Barter* is a direct exchange of goods between two parties without the use of money. For example, Iran might swap oil for guns.
2. *Counterpurchase*, also known as parallel barter, is the sale and purchase of goods that are unrelated to each other. For example, Pepsico sold soft drinks to the former Soviet Union for vodka.
3. *Buyback* is repayment of the original purchase price through the sale of a related product. For example, Western European countries delivered various pipeline materials to the Soviet Union for construction of a gas pipeline from Siberian gas fields and in return agreed to purchase 28 billion cubic meters of gas per year.

The unanswered question in countertrade is: Why go through such a convoluted sales process? Why not sell the goods directly at their market price (which is what ultimately happens anyway) using experts to handle the marketing? One argument is that countertrade enables members of cartels such as OPEC to undercut an agreed-upon price without formally doing so. Another argument is that countertrade keeps bureaucrats busy in centrally planned economies. Countertrade also may reduce the risk faced by a country that contracts for a new manufacturing facility. If the contractor's payment is received in the form of goods supplied by the facility, the contractor has an added incentive to do quality work and to ensure that the plant's technology is suitable for the skill levels of the available workers.

Regardless of its reason for being, countertrading is replete with problems for the firms involved. First, the goods that can be taken in countertrade are usually undesirable. Those that could be readily converted into cash have already been converted, so, although a firm shipping computers to Brazil might prefer to take coffee beans in return, the only goods available might be Brazilian shoes. Second, the trading details are difficult to work out. (How many tons of naphtha is a pile of shoddy Eastern European goods worth?) The inevitable result is a very high ratio of talk to action, with only a small percentage of deals that are talked about getting done. And lost deals cost money.

Until recently, most countertrade centered on the government foreign-trade organizations (FTOs) of Eastern European countries. In order to sell a machine or an entire plant to an FTO, a Western firm might be required to take at least some of its pay in goods (for example, tomatoes, linen, and machine parts). Sometimes these deals will stretch over several years. Centered in Vienna, the countertrade experts in this business use their contacts with Eastern European officials and their knowledge of available products to earn their keep. However, the restructuring currently taking place in Eastern Europe has eliminated most FTOs and thereby reduced the scale and profitability of the countertrade business.

The loss of Eastern European business, however, has been more than offset by the explosive growth in countertrade with Third World countries. The basis for the new wave of countertrade is the cutting off of bank credit to developing nations. Third World countertrade involves more commodities and fewer hard-to-sell manufactured goods. A typical deal, arranged by Sears World Trade, involved bartering U.S. breeding swine for Dominican sugar. Another countertrader swapped BMWs for Ecuadoran tuna fish. The dissolution of the former Soviet Union also has created new opportunities for countertrade as the following convoluted example illustrates.

ILLUSTRATION

MARC RICH & CO. RECREATES THE FORMER SOVIET UNION'S SUPPLY SYSTEM

In 1992, Marc Rich & Co., a giant Swiss-based commodity trader, engineered a remarkable $100 million deal involving enterprises and governments in five newly independent countries. This deal essentially pieced together shattered supply links that Moscow controlled when the U.S.S.R. was one giant planned economy.

Here's how the deal worked. Marc Rich bought 70,000 tons of raw sugar from Brazil on the open market. The sugar was shipped to Ukraine, where, through a "tolling contract," it was processed at a local refinery. After paying the sugar refinery with part of the sugar, Marc Rich sent 30,000 tons of refined sugar east to several huge Siberian oil refineries, which need sugar to supply their vast work forces.

Strapped for hard currency, the oil refineries paid instead with oil products, much of it low-grade A-76 gasoline, which has few export markets. But one market is Mongolia, with which Marc Rich has long traded. The company shipped 130,000 tons of oil products there; in payment, the Mongolians turned over 35,000 tons of copper concentrate. The company sent most of that back across the border to Kazakhstan, where it was refined into copper metal. Then the metal was shipped westward to a Baltic seaport and out to the world market where, several months after the deal began, Marc Rich earned a hard currency profit.

In an effort to make it easier for Third World countries to buy their products, big manufacturers—including General Motors, General Electric, and Caterpillar—have set up countertrading subsidiaries. Having sold auto and truck parts to Mexico, for example, GM's countertrade subsidiary, Motors Trading Corp., arranged tour groups to the country and imported Mexican slippers and gloves. Similarly, arms manufacturers selling to developing countries are often forced to accept local products in return—for example, Iraqi oil for French Exocet missiles or Peruvian anchovies for Spanish Piranha patrol boats.

Authorities in countertrading countries are concerned that goods taken in countertrade will cannibalize their existing cash markets. Proving that countertrade goods go to new markets is difficult enough in the area of manufactured goods; it's impossible for commodities, whose ultimate use cannot be identified with its source. For example, some Indonesian rubber taken in countertrade inevitably will displace rubber that Indonesia sells for cash.

Interest in countertrade and its variations appears to be growing, even among developed countries, despite the obvious difficulties it presents to the firms and countries involved. For example, in order for McDonnell Douglas to sell F-15s to the Japanese air force, it had to offset the cost to Japan in currency and jobs by agreeing to teach Japanese manufacturers to make military aircraft.

The growth in countertrade is reflected in the scramble for experienced specialists. It has been said that a good countertrader combines the avarice and opportunism of a commodities trader, the inventiveness and political sensitivity of a crooked bureaucrat, and the technical knowledge of a machine-tool salesperson.

12.6 SUMMARY AND CONCLUSIONS

In this chapter, we examined different financing arrangements and documents involved in international trade. The most important documents encountered in bank-related financing are the draft, which is a written order to pay; the letter of credit, which is a bank guarantee of payment provided that certain stipulated conditions are met; and the bill of lading, the document covering actual shipment of the merchandise by a common carrier and title. Documents of lesser importance include commercial and consular invoices and the insurance certificate.

These instruments serve four primary functions. They

- Reduce both buyer and seller risk
- Define who bears those risks that remain
- Facilitate the transfer of risk to a third party
- Facilitate financing

Each instrument evolved over time as a rational response to the additional risks in international trade posed by greater distances, the lack of familiarity between exporters and importers, the possibility of government imposition of exchange controls, and the greater costs involved in bringing suit against a party domiciled in another nation. Were it not for the latter two factors and publicly financed export-promotion programs, we might expect that, with the passage of time, the financial arrangements in international trade would differ little from those encountered in purely domestic commercial transactions.

We also examined some of the government-sponsored export-financing programs and credit-insurance programs. The number of these institutions and their operating scope have grown steadily, in line with national export drives. From the standpoint of international financial managers, the most significant difference between public and private sources of financing is that public lending agencies offer their funds and credit insurance at lower-than-normal commercial rates. The multinational firm can take advantage of these subsidized rates by structuring its marketing and production programs in accord with the different national financial programs.

❧ Questions

1. What are the basic problems arising in international trade financing and how do the main financing instruments help solve those problems?

2. The different forms of export financing distribute risks differently between the exporter and the importer. Analyze the distribution of risk in the following export-financing instruments.
   a. Confirmed, revocable letter of credit
   b. Confirmed, irrevocable letter of credit
   c. Open account credit
   d. Time draft, D/A
   e. Cash with order
   f. Cash in advance
   g. Consignment
   h. Sight draft

3. Describe the different steps and documents involved in exporting motors from Kansas to Hong Kong using a confirmed letter of credit, with payment terms of 90 days sight. What alternatives are available to the exporter to finance this shipment?

4. Explain the advantages and disadvantages of each of the following forms of export financing.
   a. Bankers' acceptances
   b. Discounting
   c. Factoring
   d. Forfaiting

5. In order to "meet the competition" from its counterparts overseas, Eximbank will mechanically match the terms of a loan provided by a rival export-financing agency—including the interest rate—when it finances U.S. exports.
   a. What problems might arise from this rule of matching nominal interest rates?
   b. As of January 15, 1988, the minimum interest rate on government-supplied export credits to rich countries was set at a flat rate of 10.4% for all nations providing such credits. What prob-

lems might arise with this rule? Comment on which governments would push for such a rule. Which would be against it?
   c. How should minimum interest rates on export credits be set so as to ensure comparability across countries?
   d. Suppose that instead of subsidizing interest rates, governments turn to export insurance subsidies. Is this move an improvement vis-à-vis export-credit subsidies? Explain.

6. One of the purposes of Eximbank is to absorb credit risks on export sales that the private sector will not accept. Comment on this purpose.

7. Comment on the following statement: "Eximbank does not compete with private financial institutions. It offers assistance only in cases in which the export-credit transaction would not take place without its help. Eximbank does not offer direct-loan assistance to foreign buyers when private institutions will provide comparable financing on reasonable terms."

8. These questions relate to the Foreign Credit Insurance Association.
   a. Describe the different risks covered by FCIA. Why does the FCIA require coinsurance?
   b. What factors affect the insurance premium charged by the FCIA?
   c. Describe the basic features of a typical FCIA short-term policy.
   d. Describe the basic features of a typical FCIA medium-term policy.

9. Low-cost export financing is often a bad sign. Explain.

10. What is countertrade? Why is it termed a sophisticated form of barter?

11. What are the potential advantages and disadvantages of countertrade for the parties involved?

❧ Problems

1. Texas Computers (TC) recently has begun selling overseas. It currently has 30 foreign orders outstanding, with the typical order averaging $2,500. TC is considering the following three alternatives to protect itself against credit risk on these foreign sales:

   ■ *Request a letter of credit from each customer.* The cost to the customer would be $75 plus 0.25% of the invoice amount. To remain competitive, TC would have to absorb the cost of the letter of credit.

   ■ *Factor the receivables.* The factor would charge a nonrecourse fee of 1.6%.

   ■ *Buy FCIA insurance.* The FCIA would charge a 1% insurance premium.

   a. Which of these alternatives would you recommend to Texas Computers? Why?
   b. Suppose that TC's average order size rose to $250,000. How would that affect your decision?

2. L.A. Cellular has received an order for phone switches from Singapore. The switches will be

exported under the terms of a letter of credit issued by Sumitomo Bank on behalf of Singapore Telecommunications. Under the terms of the L/C, the face value of the export order, $12 million, will be paid six months after Sumitomo accepts a draft drawn by L.A. Cellular. The current discount rate on six-month acceptances is 8.5% per annum, and the acceptance fee is 1.25% per annum. In addition, there is a flat commission, equal to 0.5% of the face amount of the accepted draft, that must be paid if it is sold.

a. How much cash will L.A. Cellular receive if it holds the acceptance until maturity?

b. How much cash will it receive if it sells the acceptance at once?

c. Suppose L.A. Cellular's opportunity cost of funds is 8.75% per annum. If it wishes to maximize the present value of its acceptance, should it discount the acceptance?

3. Suppose Minnesota Machines (MM) is trying to price an export order from Russia. Payment is due nine months after shipping. Given the risks involved, MM would like to factor its receivable without recourse. The factor will charge a monthly discount of 2% plus a fee equal to 1.5% of the face value of the receivable for the nonrecourse financing.

a. If Minnesota Machines desires revenue of $2.5 million from the sale, after paying all factoring charges, what is the minimum acceptable price it should charge?

b. Alternatively, CountyBank has offered to discount the receivable, but with recourse, at an annual rate of 14% plus a 1% fee. What price will net MM the $2.5 million it desires to clear from the sale?

c. On the basis of your answers to parts a and b, should Minnesota Machines discount or factor its Russian receivables? MM is competing against Nippon Machines for the order, so the higher MM's price, the lower the probability that its bid will be accepted. What other considerations should influence MM's decision?

d. What other alternatives might be available to MM to finance its sale to Russia?

▶ Bibliography

Business International Corporation. *Financing Foreign Operations*. New York: BIC, various issues.

Chase World Information Corporation. *Methods of Export Financing*, 2nd ed. New York: Chase World Information Corporation, 1976.

SCHNEIDER, GERHARD W.   *Export-Import Financing*. New York: The Ronald Press, 1974.

CURRENT ASSET MANAGEMENT AND SHORT-TERM FINANCING

A penny saved is a penny earned.

Benjamin Franklin

$\mathcal{T}$he management of **working capital** in the multinational corporation is similar to its domestic counterpart. Both are concerned with selecting that combination of **current assets**—cash, marketable securities, accounts receivable, and inventory—and **current liabilities**—short-term funds to finance those current assets—that will maximize the value of the firm. The essential differences between domestic and international working-capital management include the impact of currency fluctuations, potential exchange controls, and multiple tax jurisdictions on these decisions, in addition to the wider range of short-term financing and investment options available.

Chapter 14 discusses the mechanisms by which the multinational firm can shift liquid assets among its various affiliates; it also examines the tax and other consequences of these maneuvers. This chapter deals with the management of working-capital items available to each affiliate. The focus is on international cash, accounts receivable, and inventory management and short-term financing.

13.1 INTERNATIONAL CASH MANAGEMENT

International money managers attempt to attain on a worldwide basis the traditional domestic objectives of cash management: (1) bringing the company's cash resources under control as quickly and efficiently as possible, and (2) achieving the optimum conservation and utilization of these funds. Accomplishing the first goal requires establishing accurate, timely forecasting and reporting systems, improving cash collections and disbursements, and decreasing the cost of moving funds among affiliates. The second objective is achieved by minimizing the required level of cash balances, making money available when and where it is needed, and increasing the risk-adjusted return on those funds that can be invested.

This section is divided into seven key areas of *international cash management*: (1) organization, (2) collection and disbursement of funds, (3) netting of interaffiliate payments, (4) investment of excess funds, (5) establishment of an optimal level of worldwide corporate cash balances, (6) cash planning and budgeting, and (7) bank relations.

Organization

When compared with a system of autonomous operating units, a fully centralized international cash management program offers a number of advantages:

1. The corporation is able to operate with a smaller amount of cash; pools of excess liquidity are absorbed and eliminated; each operation will maintain transactions balances only and not hold speculative or precautionary ones.
2. By reducing total assets, profitability is enhanced and financing costs are reduced.
3. The headquarters staff, with its purview of all corporate activity, can recognize problems and opportunities that an individual unit might not perceive.
4. All decisions can be made using the overall corporate benefit as the criterion.
5. By increasing the volume of foreign exchange and other transactions done through headquarters, firms encourage banks to provide better foreign exchange quotes and better service.
6. Greater expertise in cash and portfolio management exists if one group is responsible for these activities.
7. Less can be lost in the event of an expropriation or currency controls restricting the transfer of funds because the corporation's total assets at risk in a foreign country can be reduced.

The forgoing benefits have long been understood by many experienced multinational firms. Today the combination of volatile currency and interest rate fluctuations, questions of capital availability, increasingly complex organizations and operating arrangements, and a growing emphasis on profitability virtually mandates a highly centralized international cash management system. There is also a trend to place much greater responsibility in corporate headquarters.

Centralization does not necessarily imply control by corporate headquarters of all facets of cash management. Instead, a concentration of decision making at a sufficiently high level within the corporation is required so that all pertinent information is readily available and can be used to optimize the firm's position.

Collection and Disbursement of Funds

Accelerating collections both within a foreign country and across borders is a key element of international cash management. Material potential benefits exist because long delays often are encountered in collecting receivables, particularly on export sales, and in transferring funds among affiliates and corporate headquarters. Allowing for mail time and bank processing, delays of eight to ten business days are common from the moment an importer pays an invoice to the time when the exporter is credited with **good funds**—that is, when the funds are available for use. Given high interest rates, wide fluctuations in the foreign exchange markets, and the periodic imposition of credit restrictions that have characterized financial markets in recent years, cash in transit has become more expensive and more exposed to risk.

With increasing frequency, corporate management is participating in the establishment of an affiliate's credit policy and the monitoring of collection performance. The principal goals of this intervention are to minimize *float*—that is, the transit time of payments—to reduce the investment in accounts receivable and to lower banking fees and other transaction costs. By converting receivables into cash as rapidly as possible, a company can increase its portfolio or reduce its borrowing and thereby earn a higher investment return or save interest expense.

Considering either national or international collections, accelerating the receipt of funds usually involves the following: (1) defining and analyzing the different available payment channels, (2) selecting the most efficient method, which can vary by country and by customer, and (3) giving specific instructions regarding procedures to the firm's customers and banks.

In addressing the first two points, the full costs of using the various methods must be determined, and the inherent delay of each must be calculated. There are two main sources of delay in the collections process: the time between the dates of payment and of receipt and the time for the payment to clear through the banking system. Inasmuch as banks will be as "inefficient" as possible to increase their float, understanding the subtleties of domestic and international money transfers is requisite if a firm is to reduce the time that funds are held and extract the maximum value from its banking relationships. Exhibit 13.1 lists the different methods multinationals use to expedite their collection of receivables.

With respect to payment instructions to customers and banks, the use of *cable remittances* is a crucial means for companies to minimize delays in receipt of payments and

---

EXHIBIT 13.1 How Multinationals Expedite Their Collection of Receivables

PROCEDURES FOR EXPEDITING RECEIPT OF PAYMENT	PROCEDURES FOR EXPEDITING CONVERSION OF PAYMENTS INTO CASH
Cable remittances	Cable remittances
Mobilization centers	Establishing accounts in customers' banks
Lock boxes	Negotiations with banks on value-dating
Electronic fund transfers	

in conversion of payments into cash, especially in Europe, because European banks tend to defer the value of good funds when the payment is made by check or draft.

In the case of international cash movements, having all affiliates transfer funds by telex enables the corporation to better plan because the vagaries of mail time are eliminated. Third parties, too, will be asked to use wire transfers. For most amounts, the fees required for telex are less than the savings generated by putting the money to use more quickly.

One of the cash manager's biggest problems is that bank-to-bank wire transfers do not always operate with great efficiency or reliability. Delays, crediting the wrong account, availability of funds, and many other operational problems are common. One solution to these problems is to be found in the *SWIFT* network (Society for Worldwide Interbank Financial Telecommunications), first mentioned in Chapter 5. SWIFT has standardized international message formats and employs a dedicated computer network to support funds-transfer messages.

The SWIFT network connects more than 2,700 banks in North America, Western Europe, and the Far East and processes more than two million transactions a day. Its mission is to quickly transmit standard forms to allow its member banks to automatically process data by computer. All types of customer and bank transfers are transmitted, as well as foreign exchange deals, bank account statements, and administrative messages. To use SWIFT, the corporate client must deal with domestic banks that are subscribers and with foreign banks that are highly automated.

To cope with some of the transmittal delays associated with checks or drafts, customers are instructed to remit to "mobilization" points that are centrally located in regions with large sales volumes. These funds are managed centrally or are transmitted to the selling subsidiary. For example, European customers may be told to make all payments to Switzerland, where the corporation maintains a staff specializing in cash and portfolio management and collections.

Sometimes customers are asked to pay directly into a designated account at a branch of the bank that is mobilizing the MNC's funds internationally. This method is particularly useful when banks have large branch networks. Another technique used is to have customers remit funds to a designated **lock box**, which is a postal box in the company's name. One or more times daily, a local bank opens the mail received at the lock box and deposits any checks immediately.

Multinational banks now provide firms with rapid transfers of their funds among branches in different countries, generally giving their customers **same-day value**—that is, funds are credited that same day. Rapid transfers also can be accomplished through a bank's correspondent network, although it becomes somewhat more difficult to arrange same-day value for funds.

Chief financial officers are increasingly relying on computers and worldwide telecommunications networks to help manage the company cash portfolio. Many multinational firms will not deal with a bank that does not have a leading-edge electronic banking system.

At the heart of today's high-tech corporate treasuries are the **treasury workstation** software packages that many big banks sell as supplements to their cash management systems. Linking the company with its bank and branch offices, the workstations let treasury personnel compute a company's worldwide cash position on a "real-time" basis, meaning that the second a transaction is made in, say, Rio de Janeiro, it is electronically recorded in Tokyo as well. This simultaneous record keeping lets companies keep their funds active at all times. Treasury personnel can use their workstation to initiate fund transfers from units with surplus cash to those units that require funds, thereby reducing the level of bank borrowings.

ILLUSTRATION

INTERNATIONAL CASH MANAGEMENT AT NATIONAL SEMICONDUCTOR

After computerizing its cash management system, National Semiconductor was able to save significant interest expenses by quickly transferring money from locations with surplus cash to those needing money. In a typical transaction, the company shifted a surplus $500,000 from its Japanese account to its Philippine operations—avoiding the need to borrow the half-million dollars and saving several thousand dollars in interest expense. Before computerization, it would have taken five or six days to discover the surplus.

Management of *disbursements* is a delicate balancing act: holding onto funds versus staying on good terms with suppliers. It requires a detailed knowledge of individual country and supplier nuances, as well as the myriad payment instruments and banking services available around the world. Exhibit 13.2 presents some questions that corporate treasurers should address in reviewing their disbursement policies.

Payments Netting in International Cash Management

Many multinational corporations are now in the process of rationalizing their production on a global basis. This process involves a highly coordinated international interchange of materials, parts, subassemblies, and finished products among the various units of the MNC, with many affiliates both buying from, and selling to, each other.

The importance of these physical flows to the international financial executive is that they are accompanied by a heavy volume of *interaffiliate fund flows*. Of particular importance is the fact that there is a measurable cost associated with these cross-border

---

EXHIBIT 13.2 Reviewing Disbursements: Auditing Payment Instruments

1. What payment instruments are you using to pay suppliers, employees, and government entities (e.g., checks, drafts, wire transfers, direct deposits)?
2. What are the total disbursements made through each of these instruments annually?
3. What is the mail and clearing float for these instruments in each country?
4. What techniques, such as remote disbursement, are being used to prolong the payment cycle?
5. How long does it take suppliers to process the various instruments and present them for payment?
6. What are the bank charges and internal processing costs for each instrument?
7. Are banking services such as controlled disbursement and zero-balance accounts used where available?

fund transfers, including the cost of purchasing foreign exchange (the foreign exchange spread), the opportunity cost of float (time in transit), and other transaction costs, such as cable charges. These transaction costs are estimated to vary from 0.25% to 1.5% of the volume transferred. Thus, there is a clear incentive to minimize the total volume of inter-company fund flows. This can be achieved by payments netting.

BILATERAL AND MULTILATERAL NETTING ■The idea behind a *payments netting* system is very simple: Payments among affiliates go back and forth, whereas only a netted amount need be transferred. Suppose, for example, the German subsidiary of an MNC sells goods worth $1 million to its Italian affiliate that in turn sells goods worth $2 million to the German unit. The combined flows total $3 million. On a net basis, however, the German unit necd remit only $1 million to the Italian unit. This type of *bilateral netting* is valuable, though, only if subsidiaries sell back and forth to each other.

Bilateral netting would be of less use where there is a more complex structure of in-ternal sales, such as in the situation depicted in Exhibit 13.3a, which presents the pay-ment flows (converted first into a common currency, assumed here to be the dollar) that take place among four European affiliates, located in France, Belgium, Sweden, and the Netherlands. On a multilateral basis, however, there is greater scope for reducing cross-border fund transfers, by netting out each affiliate's inflows against its outflows.

Since a large percentage of multinational transactions are internal, leading to a relatively large volume of interaffiliate payments—the payoff from *multilateral netting* can be large, relative to the costs of such a system. Many companies find they can elimi-nate 50% or more of their intercompany transactions through multilateral netting, with annual savings in foreign exchange transactions costs and bank transfer charges that

EXHIBIT 13.3a Multilateral Netting

PAYMENT FLOWS BEFORE MULTILATERAL NETTING

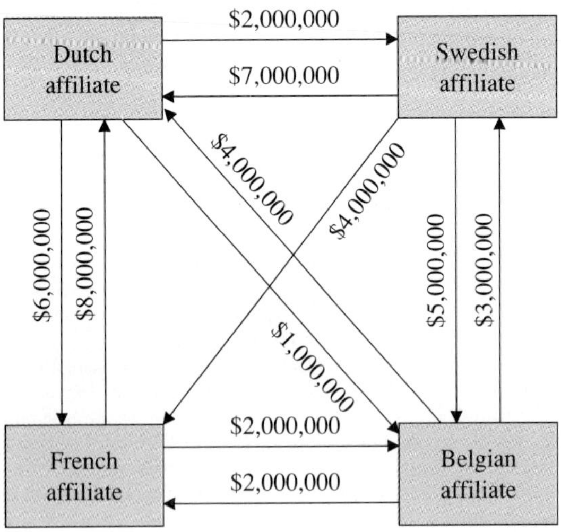

EXHIBIT 13.3b Intercompany Payments Matrix (U.S.$ Millions)

Receiving Affiliates	PAYING AFFILIATES				Net Total Receipts	Net Receipt (Payment)
	Netherlands	France	Sweden	Belgium		
Netherlands	—	8	7	4	19	10
France	6	—	4	2	12	2
Sweden	2	0	—	3	5	(11)
Belgium	1	2	5	—	8	(1)
Total Payments	9	10	16	9	44	

average between 0.5% and 1.5% per dollar netted. For example, SmithKline Beckman estimates that it saves $300,000 annually in foreign exchange transactions costs and bank-transfer charges by using a multilateral netting system.[1] Similarly, Baxter International estimates it saves $200,000 per year by eliminating approximately 60% of its inter-company transactions through netting.[2]

INFORMATION REQUIREMENTS ■ Essential to any netting scheme is a centralized control point that can collect and record detailed information on the intracorporate accounts of each participating affiliate at specified time intervals. The control point, called a *netting*

EXHIBIT 13.3c Payment Flows After Multilateral Netting

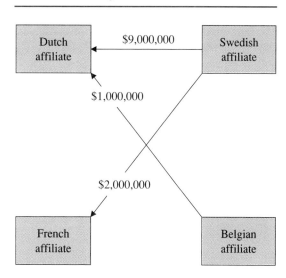

[1] "How Centralized Systems Benefit Managerial Control: SmithKline Beckman," *Business International Money Report*, June 23, 1986, p. 198.
[2] Business International Corporation, *Solving International Financial and Currency Problems* (New York: BIC, 1976), p. 29.

center, is a subsidiary company set up in a location with minimal exchange controls for trade transactions.

The netting center will use a matrix of payables and receivables to determine the net payer or creditor position of each affiliate at the date of clearing. An example of such a matrix is provided in Exhibit 13.3b, which takes the payment flows from Exhibit 13.3a and shows the amounts due to and from each of the affiliated companies. Note that in an intercompany system, the payables will always equal the receivables on both a gross basis and a net basis. Typically, the impact of currency changes on the amounts scheduled for transfer is minimized by fixing the exchange rate at which these transactions occur during the week that netting takes place.

Without netting, the total payments in the system would equal $44 million and the number of separate payments made would be 11. Multilateral netting will pare these transfers to $12 million, a net reduction of 73%, and the number of payments can be reduced to three, a net reduction of 73%, as well. One possible set of payments is shown in Exhibit 13.3c. Assuming foreign exchange and bank-transfer charges of 0.75%, this company will save $240,000 through netting (0.0075 × $32 million).

Notice that alternative sets of multilateral payments were also possible in this example. For example, the Swedish unit could have paid $11 million to the Dutch unit, with the Dutch and Belgian units then sending $1 million each to the French unit. The choice of which affiliate(s) each payer pays depends on the relative costs of transferring funds between each pair of affiliates. The per-unit costs of sending funds between two affiliates can vary significantly from month to month because one subsidiary may receive payment from a third party in a currency that is needed by the other subsidiary. Using this currency for payment can eliminate one or more foreign exchange conversions. This conclusion implies that the cost of sending funds from Germany to France, for example, can differ greatly from the cost of moving money from France to Germany.

For example, Volvo has a policy of transferring a currency, without conversion, to a unit needing that currency to pay a creditor.[3] To see how this policy works, suppose that Volvo Sweden buys automotive components from a German manufacturer and Volvo Belgium purchases automotive kits from Volvo Sweden. At the same time, a German dealer buys automobiles from Volvo Belgium and pays in Deutsche marks. Volvo Belgium then will use these DM to pay Volvo Sweden, which in turn will use them to pay its German creditor.

FOREIGN EXCHANGE CONTROLS ■ Before implementing a payments netting system, a company needs to know whether any restrictions on netting exist. Firms sometimes may be barred from netting or be required to obtain permission from the local monetary authorities.

ANALYSIS ■ The higher the volume of intercompany transactions and the more back-and-forth selling that takes place, the more worthwhile netting is likely to be. A useful approach to evaluating a netting system would be to establish the direct cost savings of the netting system and then use this figure as a benchmark against which to measure the

---

[3] Business International Corporation, *Solving International Financial and Currency Problems*, p. 32.

costs of implementation and operation. These set-up costs have been estimated at less than $20,000.[4]

An additional benefit from running a netting system is the tighter control that management can exert over corporate fund flows. The same information required to operate a netting system also will enable an MNC to shift funds in response to expectations of currency movements, changing interest differentials, and tax differentials.

ILLUSTRATION

COST/BENEFIT ANALYSIS OF AN INTERNATIONAL CASH MANAGEMENT SYSTEM

Although company A already operates a multilateral netting system, it commissioned a study to show where additional improvements in cash management could be made.[5] The firm proposed to establish a finance company (FINCO) in Europe. FINCO's primary function would be to act as a collecting and paying agent for divisions of company A that export to third parties. All receivables would be gathered into the international branch network of bank X. Each branch would handle receivables denominated in the currency of its country of domicile.

These branch accounts would be monitored by both FINCO and the exporting unit via the bank's electronic reporting facility.

Intercompany payments and third-party collection payments from FINCO to each exporter would be included in the existing multilateral netting system, which would be administered by FINCO. Payments for imports from third-party suppliers also would be included. Finally, the netting system would be expanded to include intercompany payments from operations in the United States, Canada, and one additional European country.

The feasibility study examined six basic savings components and two cost components. The realizable, annualized savings are summarized below:

SAVINGS COMPONENT	COST SAVINGS
1. Optimized multilateral netting	$ 29,000
2. Reduced remittance-processing time by customer and remitting bank	26,000
3. Reduction in cross-border transfer float by collecting currencies in their home country	46,000
4. Reduction in cross-border transfer commissions and charges by collecting currencies in their home country	41,000
5. Use of incoming foreign currencies to source outgoing foreign payments in the same currencies	16,000
6. Use of interest-bearing accounts	8,000
Total estimated annual savings	$166,000

COST COMPONENT	COST
1. Computer time-sharing charges for accessing Bank X's system	$ 17,000
2. Communications charges for additional cross-border funds transfers	13,000
Total estimated annual costs	$ 30,000
Total net savings	$136,000

---

[4] Business International Corporation, "The State of the Art," in *New Techniques in International Exposure and Cash Management*, vol. 1 (New York: BIC, 1977), p. 244.

[5] This illustration appears in "Cost/Benefit Analysis of One Company's Cash Management System," *Business International Money Report*, April 14, 1986, pp. 119–120.

Management of the Short-Term Investment Portfolio

A major task of international cash management is to determine the levels and currency denominations of the multinational group's investment in cash balances and money market instruments. Firms with seasonal or cyclical cash flows have special problems, such as spacing investment maturities to coincide with projected needs. To manage this investment properly requires (1) a forecast of future cash needs based on the company's current budget and past experience and (2) an estimate of a minimum cash position for the coming period. These projections should take into account the effects of inflation and anticipated currency changes on future cash flows.

Successful management of an MNC's required cash balances and of any excess funds generated by the firm and its affiliates depends largely on the astute selection of appropriate short-term money market instruments. Rewarding opportunities exist in many countries, but the choice of an investment medium depends on government regulations,

EXHIBIT 13.4 Key Money Market Instruments

INSTRUMENT	BORROWER	MATURITIES	COMMENTS
Treasury bills (T-bills)	Central governments of many countries	Up to 1 year	Safest and most liquid short-term investment
Federal funds (U.S.)	U.S. commercial banks temporarily short of legal reserve requirements	Overnight to 3 days	Suitable for very short-term investment of large amounts ($1 million or more)
Government agency notes (U.S.)	Issued by U.S. government agencies such as Federal National Mortgage Association	30 days to 270 days	Similar to local authority notes in the United Kingdom. Both offer slightly higher yields than T-bills.
Demand deposits	Commercial banks	On demand	Governments sometimes impose restrictions on interest rates banks can offer (as in the U.S.).
Time deposits	Commercial banks	Negotiable but advance notice usually required	Governments sometimes regulate interest rates and/or maturities.
Deposits with nonbank financial institutions	Nonbank financial institutions	Negotiable	Usually offer higher yields than banks do
Certificates of deposit (CDs)	Commercial banks	Negotiable but normally 30, 60, or 90 days	Negotiable papers representing a term bank deposit; more liquid than straight deposits because they can be sold
Bankers' acceptances	Bills of exchange guaranteed by a commercial bank	Up to 180 days	Highest-quality investment next to T-bills
Commercial paper (also known as trade paper or, in the United Kingdom, fine trade bills)	Large corporations with high credit ratings	30 days to 270 days	Negotiable, unsecured promissory notes; available in all major money markets
Temporary corporate loans	Corporations	Negotiable	Usually offer higher returns than those available from financial institutions but are not liquid because they must be held to maturity

the structure of the market, and the tax laws, all of which vary widely. Available money instruments differ among the major markets, and at times, foreign firms are denied access to existing investment opportunities. Only a few markets, such as the broad and diversified U.S. market and the Eurocurrency markets, are truly free and international. Capsule summaries of key money market instruments are provided in Exhibit 13.4.

Once corporate headquarters has fully identified the present and future needs of its affiliates, it must then decide on a policy for managing its liquid assets worldwide. This policy must recognize that the value of shifting funds across national borders to earn the highest possible risk-adjusted return depends not only on the risk-adjusted yield differential, but also on the transaction costs involved. In fact, the basic reason for holding cash in several currencies simultaneously is the existence of currency conversion costs. If these costs are zero and government regulations permit, all cash balances should be held in the currency having the highest effective risk-adjusted return net of withdrawal costs.

Given that transaction costs do exist, the appropriate currency denomination mix of an MNC's investment in money and near-money assets is probably more a function of the currencies in which it has actual and projected inflows and outflows than of effective yield differentials or government regulations. The reason why is simple: Despite government controls, it would be highly unusual to see an annualized risk-adjusted interest differential of even 2%. Although seemingly large, a 2% annual differential yields only an additional 0.167% for a 30-day investment or 0.5% extra for a 90-day investment. Such small differentials can easily be offset by foreign exchange transaction costs. Thus, even large annualized risk-adjusted interest spreads may not justify shifting funds for short-term placements.

PORTFOLIO GUIDELINES ■ Commonsense guidelines for globally managing the marketable securities portfolio are as follows:

1. Diversify the instruments in the portfolio to maximize the yield for a given level of risk. Don't invest only in government securities. Eurodollar and other instruments may be nearly as safe.
2. Review the portfolio daily to decide which securities should be liquidated and what new investments should be made.
3. In revising the portfolio, make sure that the incremental interest earned more than compensates for such added costs as clerical work, the income lost between investments, fixed charges such as the foreign exchange spread, and commissions on the sale and purchase of securities.
4. If rapid conversion to cash is an important consideration, then carefully evaluate the security's marketability (liquidity). Ready markets exist for some securities, but not for others.
5. Tailor the maturity of the investment to the firm's projected cash needs, or be sure a secondary market for the investment with high liquidity exists.
6. Carefully consider opportunities for covered or uncovered interest arbitrage.

Optimal Worldwide Cash Levels

Centralized cash management typically involves the transfer of an affiliate's cash in excess of minimal operating requirements into a centrally managed account, or *cash pool*. Some firms have established a special corporate entity that collects and disburses funds through a single bank account.

With cash pooling, each affiliate need hold locally only the minimum cash balance required for transactions purposes. All precautionary balances are held by the parent or in the pool. As long as the demands for cash by the various units are reasonably independent of each other, centralized cash management can provide an equivalent degree of protection with a lower level of cash reserves.

Another benefit from pooling is that either less borrowing need be done or more excess funds are available for investment where returns will be maximized. Consequently, interest expenses are reduced or investment income is increased. In addition, the larger the pool of funds, the more worthwhile it becomes for a firm to invest in cash management expertise. Furthermore, pooling permits exposure arising from holding foreign currency cash balances to be centrally managed.

EVALUATION AND CONTROL ■ Taking over control of an affiliate's cash reserves can create motivational problems for local managers unless some adjustments are made to the way in which these managers are evaluated. One possible approach is to relieve local managers of profit responsibility for their excess funds. The problem with this solution is that it provides no incentive for local managers to take advantage of specific opportunities of which only they may be aware.

An alternative approach is to present local managers with interest rates for borrowing or lending funds to the pool that reflect the opportunity cost of money to the parent corporation. In setting these *internal interest rates* (IIRs), the corporate treasurer, in effect, is acting as a bank, offering to borrow or lend currencies at given rates. By examining these IIRs, local treasurers will be more aware of the opportunity cost of their idle cash balances, as well as having an added incentive to act on this information. In many instances, they will prefer to transfer at least part of their cash balances (where permitted) to a central pool in order to earn a greater return. To make pooling work, managers must have access to the central pool whenever they require money.

ILLUSTRATION

AN ITALIAN CASH MANAGEMENT SYSTEM
An Italian firm has created a centralized cash management system for its 140 operating units within Italy. At the center is a holding company that manages banking relations, borrowings, and investments.

In the words of the firm's treasurer, "We put ourselves in front of the companies as a real bank and say, 'If you have a surplus to place, I will pay you the best rates.' If the company finds something better than that, they are free to place the funds outside the group. But this doesn't happen very often."[6] In this way, the company avoids being overdrawn with one bank while investing with another.

---

[6]"Central Cash Management Step by Step: The European Approach," *Business International Money Report*, October 19, 1984, p. 331.

Cash Planning and Budgeting

The key to the successful global coordination of a firm's cash and marketable securities is a good reporting system. Cash receipts must be reported and forecast in a comprehensive, accurate, and timely manner. If the headquarters staff is to fully and economically use the company's worldwide cash resources, they must know the financial positions of affiliates, the forecast cash needs or surpluses, the anticipated cash inflows and outflows, local and international money market conditions, and likely currency movements.

As a result of rapid and pronounced changes in the international monetary arena, the need for more frequent reports has become acute. Firms that had been content to receive information quarterly now require monthly, weekly, or even daily data. Key figures often are transmitted by telex or fax machine instead of by mail.

MULTINATIONAL CASH MOBILIZATION ■ A **multinational cash mobilization** system is designed to optimize the use of funds by tracking current and near-term cash positions. The information gathered can be used to aid a multilateral netting system, to increase the operational efficiency of a centralized cash pool, and to determine more effective short-term borrowing and investment policies.

The operation of a multinational cash mobilization system is illustrated here with a simple example centered around a firm's four European affiliates. Assume that the European headquarters maintains a regional cash pool in London for its operating units located in England, France, Germany, and Italy. Each day, at the close of banking hours, every affiliate reports to London its current cash balances in *cleared funds*—that is, its cash accounts net of all receipts and disbursements that have cleared during the day. All balances are reported in a common currency, which is assumed here to be the U.S. dollar, with local currencies translated at rates designated by the manager of the central pool.

One report format is presented in Exhibit 13.5. It contains the end-of-day balance as well as a revised five-day forecast. According to the report for July 12, the Italian affiliate has a cash balance of $400,000. This balance means the affiliate could have disbursed an additional $400,000 that day without creating a cash deficit or having to use its overdraft facilities. The French affiliate, on the other hand, has a negative cash balance of $150,000, which it is presumably covering with an overdraft. Alternatively, it might have borrowed funds from the pool to cover this deficit. The British and German subsidiaries report cash surpluses of $100,000 and $350,000, respectively.

The manager of the central pool then can assemble these individual reports into a more usable form, such as that depicted in Exhibit 13.6. This report shows the cash balance for each affiliate, its required minimum operating cash balance, and the resultant cash surplus or deficit for each affiliate individually and for the region as a whole. According to the report, both the German and Italian affiliates ended the day with funds in excess of their operating needs, whereas the English unit wound up with $25,000 less than it normally requires in operating funds (even though it had $100,000 in cash). The French affiliate was short $250,000, including its operating deficit and minimum required balances. For the European region as a whole, however, there was excess cash of $75,000.

The information contained in these reports can be used to decide how to cover any deficits and where to invest temporary surplus funds. Netting also can be facilitated by breaking down each affiliate's aggregate inflows and outflows into their individual cur-

EXHIBIT 13.5 Daily Cash Reports of European Central Cash Pool (U.S. $ Thousands)

	DATE: JULY 12, 200X AFFILIATE: FRANCE CASH POSITION: −150 FIVE-DAY FORECAST:				DATE: JULY 12, 200X AFFILIATE: GERMANY CASH POSITION: +350 FIVE-DAY FORECAST:		
Day	Deposit	Disburse	Net	Day	Deposit	Disburse	Net
1	400	200	+200	1	430	50	+380
2	125	225	−100	2	360	760	−400
3	300	700	−400	3	500	370	+130
4	275	275	0	4	750	230	+520
5	250	100	150	5	450	120	+330
		Net for period	−150			Net for period	+960

	DATE: JULY 12, 200X AFFILIATE: ITALY CASH POSITION: +400 FIVE-DAY FORECAST:				DATE: JULY 12, 200X AFFILIATE: ENGLAND CASH POSITION: +100 FIVE-DAY FORECAST:		
Day	Deposit	Disburse	Net	Day	Deposit	Disburse	Net
1	240	340	−100	1	100	50	+ 50
2	400	275	+125	2	260	110	+150
3	480	205	+275	3	150	350	−200
4	90	240	−150	4	300	50	+250
5	300	245	+ 55	5	200	300	−100
		Net for period	+205			Net for period	+150

rency components. This breakdown will aid in deciding what netting operations to perform and in which currencies.

The cash forecasts contained in the daily reports can aid in determining when to transfer funds to or from the central pool and the maturities of any borrowings or investments. For example, although the Italian subsidiary currently has $250,000 in excess funds, it projects a deficit tomorrow of $100,000. One possible strategy is to have the Italian unit remit $250,000 to the pool today and, in turn, have the pool return $100,000 tomorrow to cover the projected deficit. However, unless interest differentials are large

EXHIBIT 13.6 Aggregate Cash Position of European Central Cash Pool (U.S. $ Thousands)

	DAILY CASH POSITION, JULY 12, 200X		
Affiliate	Closing Balance	Minimum Required	Cash Balance Surplus (Deficit)
France	−150	100	−250
Germany	+350	250	100
Italy	+400	150	250
England	+100	125	− 25
Regional surplus (deficit)			+ 75

EXHIBIT 13.7 Five-Day Cash Forecast of European Central Cash Pool
(U.S. $ Thousands)

Affiliate	+1	+2	+3	+4	+5	Five-Day Total
			DAYS FROM JULY 12, 200X			
France	+200	−100	−400	0	+150	− 150
Germany	+380	−400	+130	+520	+330	+ 960
Italy	−100	+125	+275	−150	+ 55	+ 205
England	+ 50	+150	−200	+250	−100	+ 150
Forecast regional surplus (deficit) by day	+530	−225	−195	+620	+435	+1,165

and/or transaction costs are minimal, it may be preferable to instruct the Italian unit to remit only $150,000 to the pool and invest the remaining $100,000 overnight in Italy.

Similarly, the five-day forecast shown in Exhibit 13.7, based on the data provided in Exhibit 13.6, indicates that the $75,000 European regional surplus generated today can be invested for at least two days before it is required (because of the cash deficit forecasted two days from today).

The cash mobilization system illustrated here has been greatly simplified in order to bring out some key details. In reality, such a system should include longer-term forecasts of cash flows broken down by currency, forecasts of intercompany transactions (for netting purposes), and interest rates paid by the pool (for decentralized decision making).

Bank Relations

Good bank relations are central to a company's international cash management effort. Although some companies may be quite pleased with their banks' services, others may not even realize that they are being poorly served by their banks. Poor cash management services mean lost interest revenues, overpriced services, and inappropriate or redundant services. Many firms that have conducted a bank relations audit find that they are dealing with too many banks. Here are some considerations involved in auditing the company's banks.

Some common problems in bank relations are

1. *Too many relations*: Using too many banks can be expensive. It also invariably generates idle balances, higher compensating balances, more check-clearing float, suboptimal rates on foreign exchange and loans, a heavier administrative workload, and diminished control over every aspect of banking relations.

2. *High banking costs*: To keep a lid on bank expenses, treasury management must carefully track not only the direct costs of banking services—including rates, spreads, and commissions—but also the indirect costs rising from check float, *value-dating*—that is, when value is given for funds—and compensating balances. This monitoring is especially important in developing countries of Latin America and Asia. In these countries, compensating balance requirements—the fraction of an outstanding loan balance required to held on deposit in a noninterest-bearing account—may range as high as

30%–35%, and check-clearing times may drag on for days or even weeks. It also pays off in such European countries as Italy, where banks enjoy value-dating periods of as long as 20–25 days.

3. *Inadequate reporting*: Banks often do not provide immediate information on collections and account balances. This delay can cause excessive amounts of idle cash and prolonged float. To avoid such problems, firms should instruct their banks to provide daily balance information and to clearly distinguish between *ledger* and *collected balances*—that is, posted totals versus immediately available funds.

4. *Excessive clearing delays*: In many countries, bank float can rob firms of funds availability. In such nations as Mexico, Spain, Italy, and Indonesia, checks drawn on banks located in remote areas can take weeks to clear to headquarters accounts in the capital city. Fortunately, firms that negotiate for better float times often meet with success. Whatever method is used to reduce clearing time, it is crucial that companies constantly check up on their banks to ensure that funds are credited to accounts as expected.

Negotiating better service is easier if the company is a valued customer. Demonstrating that it is a valuable customer requires the firm to have ongoing discussions with its bankers to determine the precise value of each type of banking activity and the value of the business it generates for each bank. Armed with this information, the firm should make up a monthly report that details the value of its banking business. By compiling this report, the company knows precisely how much business it is giving to each bank it uses. With such information in hand, the firm can negotiate better terms and better service from its banks.

> ### ILLUSTRATION
> ### How Morton Thiokol Manages Its Bank Relations
>
> Morton Thiokol, a Chicago-based manufacturer with international sales of about $300 million, centralizes its banking policy for three main reasons: Cash management is already centralized; small local staffs may not have time to devote to bank relations; and overseas staffs often need the extra guidance of centralized bank relations. Morton Thiokol is committed to trimming its overseas banking relations to cut costs and streamline cash management. A key factor in maintaining relations with a bank is a bank's willingness to provide the firm with needed services at reasonable prices. Although Morton Thiokol usually tries to reduce the number of banks with which it maintains relations, it will sometimes add banks to increase competition and thereby improve its chances of getting quality services and reasonable prices.

13.2 ACCOUNTS RECEIVABLE MANAGEMENT

Firms grant trade credit to customers, both domestically and internationally, because they expect the investment in receivables to be profitable, either by expanding sales volume or by retaining sales that otherwise would be lost to competitors. Some companies also earn a profit on the financing charges they levy on credit sales.

The need to scrutinize *credit terms* is particularly important in countries experiencing rapid rates of inflation. The incentive for customers to defer payment, liquidating their debts with less valuable money in the future, is great. Furthermore, credit standards abroad are often more relaxed than in the home market, especially in countries lacking alternative sources of credit for small customers. To remain competitive, MNCs may feel compelled to loosen their own credit standards. Finally, the compensation system in many companies tends to reward higher sales more than it penalizes an increased investment in accounts receivable. Local managers frequently have an incentive to expand sales even if the MNC overall does not benefit.

The effort to better manage receivables overseas will not get far if finance and marketing don't coordinate their efforts. In many companies, finance and marketing work at cross purposes. Marketing thinks about selling, and finance thinks about speeding up cash flows. One way to ease the tensions between finance and marketing is to educate the sales force on how credit and collection affect company profits. Another way is to tie bonuses for salespeople to *collected* sales or to adjust sales bonuses for the interest cost of credit sales. Forcing managers to bear the opportunity cost of working capital ensures that their credit, inventory, and other working-capital decisions will be more economical.

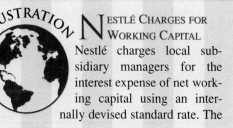

NESTLÉ CHARGES FOR WORKING CAPITAL
Nestlé charges local subsidiary managers for the interest expense of net working capital using an internally devised standard rate. The inclusion of this finance charge encourages country managers to keep a tight rein on accounts receivable and inventory because the lower the net working capital, the lower the theoretical interest charge, and the higher their profits.

Credit Extension

Two key credit decisions to be made by a firm selling abroad are the amount of credit to extend and the currency in which credit sales are to be billed. Nothing need be added here to Chapter 9's discussion (see Section 9.2, Managing Transaction Exposure) of the latter decision except to note that competitors will often resolve the currency-of-denomination issue.

The easier the credit terms are, the more sales are likely to be made. Generosity is not always the best policy. Balanced against higher revenues must be the risk of default, increased interest expense on the larger investment in receivables, and the deterioration (through currency devaluation) of the dollar value of accounts receivable denominated in the buyer's currency. These additional costs may be partly offset if liberalized credit terms enhance a firm's ability to raise its prices.

The bias of most personnel evaluation systems is in favor of higher revenues, but another factor often tends to increase accounts receivable in foreign countries. An uneconomic expansion of local sales may occur if managers are credited with dollar sales when accounts receivable are denominated in the local currency. Sales managers should be charged for the expected depreciation in the value of local currency accounts receivable. For instance, if the current exchange rate is LC 1 = $0.10, but the expected exchange rate 90 days hence (or the three-month forward rate) is $0.09, managers providing three-month credit terms should be credited with only $0.90 for each dollar in sales booked at the current spot rate.

The following five-step approach enables a firm to compare the expected benefits and costs associated with extending credit internationally:

1. Calculate the current cost of extending credit.
2. Calculate the cost of extending credit under the revised credit policy.
3. Using the information from steps 1 and 2, calculate incremental credit costs under the revised credit policy.
4. Ignoring credit costs, calculate incremental profits under the new credit policy.
5. If, and only if, incremental profits exceed incremental credit costs, select the new credit policy.

ILLUSTRATION

EVALUATING CREDIT EXTENSION OVERSEAS

Suppose a subsidiary in France currently has annual sales of $1 million with 90-day credit terms. It is believed that sales will increase by 6%, or $60,000, if terms are extended to 120 days. Of these additional sales, the cost of goods sold is $35,000. Monthly credit expenses are 1% in financing charges. In addition, the French franc is expected to depreciate an average of 0.5% every 30 days.

If we ignore currency changes for the moment, but consider financing costs, the value today of $1 of receivables to be collected at the end of 90 days is approximately $0.97. When the 1.5% (3 × 0.5%) expected French franc depreciation over the 90-day period is taken into account, this value declines to 0.97(1 − 0.015),

or $0.955, implying a 4.5% cost of carrying French franc receivables for three months. Similarly, $1 of receivables collected 120 days from now is worth (1 − 4 × 0.01)(1 − 0.02) today, or $0.941. Then the incremental cost of carrying French franc receivables for the fourth month equals 0.955 − 0.941 dollars, or 1.4%.

Applying the five-step evaluation approach and using the information generated above yields current 90-day credit costs of $1,000,000 × 0.045 = $45,000. Lengthening terms to 120 days will raise this cost to $1,000,000 × 0.059 = $59,000. The cost of carrying for 120 days the incremental sales of $60,000 is $60,000 × 0.059 = $3,540. Thus, incremental credit costs under the new policy equal $59,000 + $3,540 − $45,000 = $17,540. Since this amount is less than the incremental profit of $25,000 (60,000 − 35,000), it is worthwhile providing a fourth month of credit.

13.3 INVENTORY MANAGEMENT

Inventory in the form of raw materials, work in process, or finished goods is held (1) to facilitate the production process by both ensuring that supplies are at hand when needed and allowing a more even rate of production, and (2) to make certain that goods are available for delivery at the time of sale.

Although, conceptually, the inventory management problems faced by multinational firms are not unique, they may be exaggerated in the case of foreign operations. For instance, MNCs typically find it more difficult to control their inventory and realize inventory turnover objectives in their overseas operations than in domestic ones. There are a variety of reasons: long and variable transit times if ocean transportation is used, lengthy customs proceedings, dock strikes, import controls, high duties, supply disruption, and anticipated changes in currency values.

Production Location and Inventory Control

Many U.S. companies have eschewed domestic manufacturing for offshore production to take advantage of low-wage labor and a grab bag of tax holidays, low-interest loans, and other government largess. However, a number of firms have found that low manufacturing cost is not everything. Aside from the strategic advantages associated with U.S. production, such as maintaining close contact with domestic customers, onshore manufacturing allows for a more efficient use of capital. In particular, because of the delays in inter-

---

ILLUSTRATION

CYPRESS SEMICONDUCTOR DECIDES TO STAY ONSHORE

The added inventory expenses that foreign manufacture would entail is an important reason that Cypress Semiconductor decided to manufacture integrated circuits in San Jose, California, instead of going abroad. Cypress makes relatively expensive circuits (they average around $8 apiece), so time-consuming international shipments would have tied up the company's capital in a very expensive way. Even though offshore production would save about $0.032 per chip in labor costs, the company estimated that the labor saving would be more than offset by combined shipping and customs duties of $0.025 and an additional $0.16 in the capital cost of holding inventory.

According to Cypress Chairman L. J.

Sevin, "Some people just look at the labor rates, but it's inventory cost that matters. It's simply cheaper to sell a part in one week than in five or six. You have to figure out what you could have done with the inventory or the money you could have made simply by pulling the interest on the dollars you have tied up in the part."[7]

The estimate of $0.16 in carrying cost can be backed out as follows: As the preceding quotation indicates, parts manufactured abroad were expected to spend an extra five weeks or so in transit. This means that parts manufactured abroad would spend five more weeks in work-in-process inventory than would parts manufactured domestically. Assuming an opportunity cost of 20% (not an unreasonable number considering the volatility of the semiconductor market) and an average cost per chip of $8 yields the following added inventory-related interest expense associated with overseas production:

$$\text{Added interest expense} = \frac{\text{Opportunity cost}}{\text{of funds}} \times \frac{\text{Added time}}{\text{in transit}} \times \frac{\text{Cost per}}{\text{part}}$$

$$= 0.20 \times 5/52 \times \$8 = \$0.154$$

---

[7] Joel Kotkin, "The Case for Manufacturing in America," *Inc.*, March 1985, p. 54.

national shipment of goods and potential supply disruptions, firms producing abroad typically hold larger work-in-process and finished goods inventories than do domestic firms. The result is higher inventory-carrying costs.

Advance Inventory Purchases

In many developing countries, forward contracts for foreign currency are limited in availability or are nonexistent. In addition, restrictions often preclude free remittances, making it difficult, if not impossible, to convert excess funds into a hard currency. One means of hedging is to engage in anticipatory purchases of goods, especially imported items. The trade-off involves owning goods for which local currency prices may be increased, thereby maintaining the dollar value of the asset even if devaluation occurs, versus forgoing the return on local money market investments.

Inventory Stockpiling

Because of long delivery lead times, the often limited availability of transport for economically sized shipments, and currency restrictions, the problem of supply failure is of particular importance for any firm that is dependent on foreign sources. These conditions may make the knowledge and execution of an optimal stocking policy, under a threat of a disruption to supply, more critical in the MNC than in the firm that purchases domestically.

The traditional response to such risks has been advance purchases. Holding large amounts of inventory can be quite expensive, though. In fact, the high cost of *inventory stockpiling*—including financing, insurance, storage, and obsolescence—has led many companies to identify low inventories with effective management. In contrast, production and sales managers typically desire a relatively large inventory, particularly when a cutoff in supply is anticipated. One way to get managers to take into account the trade-offs involved—the costs of stockpiling versus the costs of shortages—is to adjust the profit performances of those managers who are receiving the benefits of additional inventory on hand to reflect the added costs of stockpiling.

It is obvious that as the probability of disruption increases or as holding costs go down, more inventory should be ordered. Similarly, if the cost of a stock-out rises or if future supplies are expected to be more expensive, it will pay to stockpile additional inventory. Conversely, if these parameters move in the opposite direction, less inventory should be stockpiled.

13.4 SHORT-TERM FINANCING

Financing the working capital requirements of a multinational corporation's foreign affiliates poses a complex decision problem. This complexity stems from the large number of financing options available to the subsidiary of an MNC. Subsidiaries have access to funds from sister affiliates and the parent, as well as from external sources. This section is concerned with the following four aspects of developing a short-term overseas financing strategy: (1) identifying the key factors, (2) formulating and evaluating objectives, (3) describing available short-term borrowing options, and (4) developing a methodology for calculating and comparing the effective dollar costs of these alternatives.

Key Factors in Short-Term Financing Strategy

Expected costs and risks, the basic determinants of any funding strategy, are strongly influenced in an international context by six key factors.

1. If forward contracts are unavailable, the crucial issue is whether differences in nominal interest rates among currencies are matched by anticipated exchange rate changes. For example, is the difference between an 8% dollar interest rate and a 3% Swiss franc interest rate due solely to expectations that the dollar will devalue by 5% relative to the franc? The key issue here, in other words, is whether there are deviations from the international Fisher effect. If deviations do exist, then expected dollar borrowing costs will vary by currency, leading to a decision problem. Trade-offs must be made between the expected borrowing costs and the exchange risks associated with each financing option.

2. The element of exchange risk is the second key factor. Many firms borrow locally to provide an offsetting liability for their exposed local currency assets. On the other hand, borrowing a foreign currency in which the firm has no exposure will increase its exchange risk. That is, the risks associated with borrowing in a specific currency are related to the firm's degree of exposure in that currency.

3. The third essential element is the firm's degree of risk aversion. The more risk averse the firm (or its management) is, the higher the price it should be willing to pay to reduce its currency exposure. Risk aversion affects the company's risk-cost trade-off and consequently, in the absence of forward contracts, influences the selection of currencies it will use to finance its foreign operations.

4. If forward contracts are available, however, currency risk should not be a factor in the firm's borrowing strategy. Instead, relative borrowing costs, calculated on a covered basis, become the sole determinant of which currencies to borrow in. The key issue here is whether the nominal interest differential equals the forward differential—that is, whether interest rate parity holds. If it does hold, then the currency denomination of the firm's debt is irrelevant. Covered costs can differ among currencies because of government capital controls or the threat of such controls. Because of this added element of risk, the annualized forward discount or premium may not offset the difference between the interest rate on the LC loan versus the dollar loan—that is, interest rate parity will not hold.

5. Even if interest rate parity does hold before tax, the currency denomination of corporate borrowings does matter where tax asymmetries are present. These tax asymmetries are based on the differential treatment of foreign exchange gains and losses on either forward contracts or loan repayments. For example, English firms or affiliates have a disincentive to borrow in strong currencies because Inland Revenue, the British tax agency, taxes exchange gains on foreign currency borrowings but disallows the deductibility of exchange losses on the same loans. An opposite incentive (to borrow in stronger currencies) is created in countries such as Australia that may permit exchange gains on forward contracts to be taxed at a lower rate than the rate at which forward contract losses are deductible. In such a case, even if interest parity holds before tax, after-tax forward contract gains may be greater than after-tax interest costs. Such tax asymmetries lead to possibilities of borrowing arbitrage, even if interest rate parity holds before tax. The essential point is that, in comparing relative borrowing costs, firms must compute these costs on an after-tax covered basis.

6. A final factor that may enter into the borrowing decision is **political risk**. Even if local financing is not the minimum cost option, multinationals often will still try to maximize their local borrowings if they believe that expropriation or exchange controls are serious possibilities. If either event occurs, an MNC has fewer assets at risk if it has used local, rather than external, financing.

Short-Term Financing Objectives

Four possible objectives can guide a firm in deciding where and in which currencies to borrow.

1. *Minimize expected cost*: By ignoring risk, this objective reduces information requirements, allows borrowing options to be evaluated on an individual basis without considering the correlation between loan cash flows and operating cash flows, and lends itself readily to breakeven analysis (see Calculating the Dollar Costs of Alternative Financing Options, below).

2. *Minimize risk without regard to cost*: A firm that followed this advice to its logical conclusion would dispose of all its assets and invest the proceeds in government securities. In other words, this objective is impractical and contrary to shareholder interests.

3. *Trade off expected cost and systematic risk*: The advantage of this objective is that, like the first objective, it allows a company to evaluate different loans without considering the relationship between loan cash flows and operating cash flows from operations. Moreover, it is consistent with shareholder preferences as described by the capital asset pricing model. In practical terms, however, there is probably little difference between expected borrowing costs adjusted for systematic risk and expected borrowing costs without that adjustment. The reason for this lack of difference is that the correlation between currency fluctuations and a well-diversified portfolio of risky assets is likely to be quite small.

4. *Trade off expected cost and total risk*: The theoretical rationale for this approach was described in Chapter 1. Basically, it relies on the existence of potentially substantial costs of financial distress. On a more practical level, management generally prefers greater stability of cash flows (regardless of investor preferences). Management typically will self-insure against most losses but might decide to use the financial markets to hedge against the risk of large losses. To implement this approach, it is necessary to take into account the covariances between operating and financing cash flows. This approach (trading off expected cost and total risk) is valid only where forward contracts are unavailable. Otherwise, selecting the lowest-cost borrowing option, calculated on a covered after-tax basis, is the only justifiable objective.[8]

---

[8]These possible objectives are suggested by Donald R. Lessard, "Currency and Tax Considerations in International Financing," Teaching Note No. 3, Massachusetts Institute of Technology, Spring 1979.

Short-Term Financing Options

Firms typically prefer to finance the temporary component of current assets with short-term funds. The three principal short-term financing options that may be available to an MNC include: (1) the intercompany loan, (2) the local currency loan, and (3) Euronotes and Euro-commercial paper.

Intercompany Financing

A frequent means of affiliate financing is to have either the parent company or sister affiliate provide an **intercompany loan**. At times, however, these loans may be limited in amount or duration by official exchange controls. In addition, interest rates on intercompany loans are frequently required to fall within set limits. The relevant parameters in establishing the cost of such a loan include the lender's opportunity cost of funds, the interest rate set, tax rates and regulations, the currency of denomination of the loan, and expected exchange rate movements over the term of the loan.

Local Currency Financing

Like most domestic firms, affiliates of multinational corporations generally attempt to finance their working capital requirements locally, for both convenience and exposure management purposes. All industrial nations and most LDCs have well-developed commercial banking systems, so firms desiring local financing generally turn there first. The major forms of bank financing include overdrafts, discounting, and term loans. Non-bank sources of funds include commercial paper and factoring (see Chapter 12, Financing Techniques in International Trade).

BANK LOANS ■Loans from commercial banks are the dominant form of short-term interest-bearing financing used around the world. These loans are described as *self-liquidating* because they are usually used to finance temporary increases in accounts receivable and inventory. These increases in working capital soon are converted into cash, which is used to repay the loan.

Short-term bank credits are typically unsecured. The borrower signs a note evidencing its obligation to repay the loan when it is due, along with accrued interest. Most notes are payable in 90 days; the loans must, therefore, be repaid or renewed every 90 days. The need to periodically roll over bank loans gives a bank substantial control over the use of its funds, reducing the need to impose severe restrictions on the firm. To further ensure that short-term credits are not being used for permanent financing, a bank will usually insert a **cleanup clause** requiring the company to be completely out of debt to the bank for a period of at least 30 days during the year.

Forms of Bank Credit. Bank credit provides a highly flexible form of financing because it is readily expandable and, therefore, serves as a financial reserve. Whenever the firm needs extra short-term funds that cannot be met by trade credit, it is likely to turn first to bank credit. Unsecured bank loans may be extended under a line of credit, under a revolving credit arrangement, or on a transaction basis. Bank loans can be originated in either the domestic or the Eurodollar market.

1. *Term loans*: **Term loans** are straight loans, often unsecured, that are made for a fixed period of time, usually 90 days. They are attractive because they give corporate treasurers complete control over the timing of repayments. A term loan typically is made for a specific purpose with specific conditions and is repaid in a single lump sum. The loan provisions are contained in the promissory note that is signed by the customer. This type of loan is used most often by borrowers who have an infrequent need for bank credit.

2. *Line of credit*: Arranging separate loans for frequent borrowers is a relatively expensive means of doing business. One way to reduce these transaction costs is to use a **line of credit**. This informal agreement permits the company to borrow up to a stated maximum amount from the bank. The firm can draw down its line of credit when it requires funds and pay back the loan balance when it has excess cash. Although the bank is not legally obligated to honor the line-of-credit agreement, it almost always does unless it or the firm encounters financial difficulties. A line of credit is usually good for one year, with renewals renegotiated every year.

3. *Overdrafts*: In countries other than the United States, banks tend to lend through overdrafts. An **overdraft** is simply a line of credit against which drafts (checks) can be drawn (written) up to a specified maximum amount. These overdraft lines often are extended and expanded year after year, thus providing, in effect, a form of medium-term financing. The borrower pays interest on the debit balance only.

4. *Revolving credit agreement*: A **revolving credit agreement** is similar to a line of credit except that now the bank (or syndicate of banks) is *legally committed* to extend credit up to the stated maximum. The firm pays interest on its outstanding borrowings plus a commitment fee, ranging between 0.125% and 0.5% per annum, on the *unused* portion of the credit line. Revolving credit agreements are usually renegotiated every two or three years.

The danger that short-term credits are being used to fund long-term requirements is particularly acute with a revolving credit line that is continuously renewed. Inserting an out-of-debt period under a cleanup clause validates the temporary need for funds.

5. *Discounting*: The *discounting of trade bills* is the preferred short-term financing technique in many European countries—especially in France, Italy, Belgium, and, to a lesser extent, Germany. It is also widespread in Latin America, particularly in Argentina, Brazil, and Mexico. These bills often can be rediscounted with the central bank.

Discounting usually results from the following set of transactions. A manufacturer selling goods to a retailer on credit draws a bill on the buyer, payable in, say, 30 days. The buyer endorses (accepts) the bill or gets his or her bank to accept it, at which point it becomes a **banker's acceptance**. The manufacturer then takes the bill to his or her bank, and the bank accepts it for a fee if the buyer's bank has not already accepted it. The bill is then sold at a discount to the manufacturer's bank or to a money market dealer. The rate of interest varies with the term of the bill and the general level of local money market interest rates.

The popularity of discounting in European countries stems from the fact that according to European commercial law, which is based on the Code Napoleon, the claim of the bill holder is independent of the claim represented by the underlying transaction. (For example, the bill holder must be paid even if the buyer objects to the quality of the merchandise.) This right makes the bill easily negotiable and enhances its liquidity (or tradability), thereby lowering the cost of discounting relative to other forms of credit.

Interest Rates on Bank Loans. The interest rate on bank loans is based on personal negotiation between the banker and the borrower. The loan rate charged to a specific customer reflects that customer's creditworthiness, previous relationship with the bank, the maturity of the loan, and other factors. Ultimately, of course, bank interest rates are based on the same factors as the interest rates on the financial securities issued by a borrower: the risk-free return, which reflects the time value of money, plus a risk premium based on the borrower's credit risk. However, there are certain bank-loan pricing conventions that you should be familiar with.

Interest on a loan can be paid at maturity or in advance. Each payment method gives a different effective interest rate, even if the quoted rate is the same. The **effective interest rate** is defined as follows:

$$\text{Effective interest rate} = \frac{\text{Annual interest paid}}{\text{Funds received}}$$

Suppose you borrow $10,000 for one year at 11% interest. If the interest is paid at maturity, you owe the lender $11,100 at the end of the year. This payment method yields an effective interest rate of 11%, the same as the stated interest rate:

$$\text{Effective interest rate when interest is paid at maturity} = \frac{\$1,100}{\$10,000} = 11\%$$

If the loan is quoted on a **discount basis**, the bank deducts the interest in advance. On the $10,000 loan, you will receive only $8,900 and must repay $10,000 in one year. The effective rate of interest exceeds 11% because you are paying interest on $10,000, but have the use of only $8,900:

$$\text{Effective interest rate on discounted loan} = \frac{\$1,100}{\$8,900} = 12.4\%$$

An extreme illustration of the difference in the effective interest rate between paying interest at maturity and paying interest in advance is provided by the Mexican banking system. In 1985, the nominal interest rate on a peso bank loan was 70%, about 15 percentage points higher than the inflation rate. But high as it was, the nominal figure did not tell the whole story. By collecting interest in advance, Mexican banks boosted the effective rate dramatically. Consider, for example, the cost of a Ps 10,000 loan. By collecting interest of 70%, or Ps 7,000, in advance, the bank actually loaned out only Ps 3,000 and received Ps 10,000 at maturity. The effective interest rate on the loan was 233%:

$$\text{Effective interest rate on Mexican loan} = \frac{\text{Ps } 7,000}{\text{Ps } 3,000} = 233\%$$

Compensating Balances. Many banks require borrowers to hold from 10% to 20% of their outstanding loan balance on deposit in a noninterest-bearing account. These **compensating balance** requirements raise the effective cost of a bank credit because not all of the loan is available to the firm:

$$\begin{array}{c}\text{Effective interest}\\\text{rate with compensating}\\\text{balance requirement}\end{array} = \frac{\text{Annual interest paid}}{\text{Usable funds}}$$

Usable funds equal the net amount of the loan less the compensating balance requirement.

Returning to the previous example, suppose you borrow $10,000 at 11% interest paid at maturity, and the compensating balance requirement is 15%, or $1,500. Thus, the $10,000 loan provides only $8,500 in usable funds for an effective interest rate of 12.9%:

$$\begin{array}{c}\text{Effective interest when}\\\text{interest is paid at maturity}\end{array} = \frac{\$1,100}{\$8,500} = 12.9\%$$

If the interest is prepaid, the amount of usable funds declines by a further $1,100—that is, to $7,400—and the effective interest rate rises to 14.9%:

$$\begin{array}{c}\text{Effective interest rate}\\\text{on discounted loan}\end{array} = \frac{\$1,100}{\$7,400} = 14.9\%$$

In both instances, the compensating balance requirement raises the effective interest rate above the stated interest rate. This higher rate is the case even if the bank pays interest on the compensating balance deposit because the loan rate invariably exceeds the deposit rate.

COMMERCIAL PAPER ■ One alternative to borrowing short term from a bank is to issue commercial paper. As defined in Chapter 12, **commercial paper** (CP) is a short-term unsecured promissory note that is generally sold by large corporations on a discount basis to institutional investors and to other corporations. Because commercial paper is unsecured and bears only the name of the issuer, the market has generally been dominated by the largest, most creditworthy companies.

Commercial paper is one of the most-favored short-term nonbank financing methods for MNCs, but CP markets are not all alike. Perhaps the most telling difference is the depth and popularity of CP markets, as best measured by the amount outstanding. The United States dwarfs all other national markets. At the end of 1994, U.S. commercial paper outstanding was almost $600 billion. By contrast, CP outstanding in France (the largest domestic market in Europe) was only FF 150 billion (about $29 billion) at the end of 1994.

Available maturities are fairly standard across the spectrum, but average maturities—reflecting the terms that companies actually use—vary from 20 to 25 days in the United States to more than three months in the Netherlands. The minimum denomination of paper also varies widely: In Australia, Canada, Sweden, and the United States, firms can issue CP in much smaller amounts than in other markets. In most countries, the instrument is issued at a discount, with the full face value of the note redeemed upon maturity. In other markets, however, interest-bearing instruments are also offered.

By going directly to the market rather than relying on a financial intermediary such as a bank, large, well-known corporations can save substantial interest costs, often on the order of 1% or more. In addition, because commercial paper is sold directly to large institutions, U.S. CP is exempt from SEC registration requirements. This exemption reduces the time and expense of readying an issue of commercial paper for sale.

There are three major noninterest costs associated with using commercial paper as a source of short-term funds: (1) back-up lines of credit, (2) fees to commercial banks, and (3) rating service fees. In most cases, issuers back their paper 100% with lines of credit from commercial banks. Because its average maturity is very short, commercial paper poses the risk that an issuer might not be able to pay off or roll over maturing paper. Consequently, issuers use back-up lines as insurance against periods of financial stress or tight money, when lenders ration money directly rather than raise interest rates. For example, the market for Texaco paper, which provided the bulk of its short-term financing, disappeared after an $11.1 billion judgment against it. Texaco replaced these funds by drawing on its bank lines of credit.

Back-up lines usually are paid for through compensating balances, typically about 10% of the unused portion of the credit line plus 20% of the amount of credit actually used. As an alternative to compensating balances, issuers sometimes pay straight fees ranging from 0.375% to 0.75% of the line of credit; this explicit pricing procedure has been used increasingly in recent years.

Another cost associated with issuing commercial paper is fees paid to the large commercial and investment banks that act as issuing and paying agents for the paper issuers and handle all the associated paperwork. Finally, rating services charge fees ranging from $5,000 to $25,000 per year for ratings, depending on the rating service. Credit ratings are not legally required by any nation, but they are often essential for placing paper.

Euronotes and Euro-Commercial Paper

A recent innovation in nonbank short-term credits that bears a strong resemblance to commercial paper is the so-called Euronote. **Euronotes** are short-term notes, usually denominated in dollars and issued by corporations and governments. The prefix *Euro* indicates that the notes are issued outside the country in whose currency they are denominated. The interest rates are adjusted each time the notes are rolled over. Euronotes are often called **Euro-commercial paper** (Euro-CP, for short). Typically, though, the name Euro-CP is reserved for those Euronotes that are not underwritten.

There are some differences between the U.S. commercial paper and the Euro-CP markets. For one thing, the average maturity of Euro-CP is about twice as long as the average maturity of U.S. CP. Also, Euro-CP is actively traded in a secondary market, but most U.S. CP is held to maturity by the original investors. Central banks, commercial banks, and corporations are important parts of the investor base for particular segments of the Euro-CP market; the most important holders of U.S. CP are money market funds, which are not very important in the Euro-CP market. In addition, the distribution of U.S. issuers in the Euro-CP market is of significantly lower quality than the distribution of U.S. issuers in the U.S.-CP market. An explanation of this finding may lie in the importance of banks as buyers of less-than-prime paper in the Euro-CP market.

Another important difference between the two markets historically has been in the area of ratings. For example, at year end 1986, only about 45% of active Euro-CP issuers were rated. Credit ratings in the United States, on the other hand, are ubiquitous. This difference proved transitory, however, as investors became accustomed to the concept and the rating agencies facilitated the use of their services. For example, Standard and Poor's Corporation charges an entity with a U.S. rating only $5,000 on top of the $25,000 annual U.S. fee for a Euro-CP rating. Moody's has gone a step further by making its CP ratings global paper ratings that are applicable in any market or currency. By 1994, only 4% of Euro-CP issuers were unrated.

Although still dwarfed by the U.S. CP market, the Euro-CP market has grown rapidly since its founding in 1985 (see Exhibit 13.8). By the end of 1997, Euro-CP outstandings stood at $110 billion. As one indication of its growing maturity, in September 1994, British Telecom took less than 48 hours to raise a record $635 million via Euro-CP. Encouraged by such signs of the market's increasing depth and liquidity, many European companies have increased Euro-CP issuance. Moreover, despite their access to their huge domestic CP market, U.S. corporations have also started to take an interest. Two months after British Telecom broke Euro-CP records, AT&T extended the $200 million Euro-CP program it set up in 1987 to $1 billion. Despite these encouraging signs, most treasurers, whether from the United States or not, still regard the U.S. commercial paper market as their main source of low-cost funds to cover day-to-day working capital requirements because of its tremendous liquidity and depth.

However, the Euro-CP market has one advantage: flexibility. Unlike its U.S. counterpart, Euro-CP is multidenominational, allowing issuers to borrow in a range of currencies. For example, AT&T's $1 billion program allows it to issue CP in several currencies,

EXHIBIT 13.8 Euro-CP Outstandings, 1985–1997

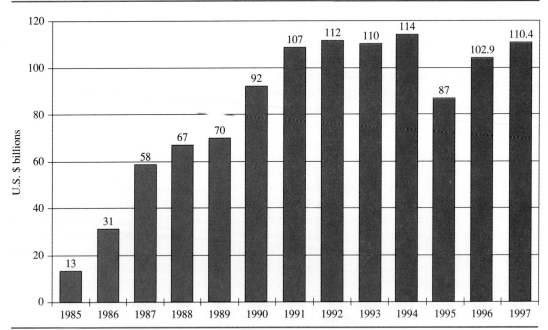

Source: Financial Market Trends, Organization for Economic Cooperation and Development, various issues.

including dollars, Swedish kronor, Dutch guilders, and Italian lire. Although 77% of all Euro-CP are still denominated in dollars, the nondollar portion has been growing over time.

Aside from the obvious benefit of allowing overseas subsidiaries to borrow in their local currencies, a multicurrency Euro-CP program also allows considerable scope for swap arbitrage. By combining a commercial paper issue with a cross-currency swap (Chapter 16 discusses swaps at length), borrowers are able to achieve significant interest savings. As of late 1994, corporate borrowers were able to achieve 10 to 15 basis point (100 basis points equal 1%) savings relative to their U.S. CP funding costs by arbitraging between different forward rates and interest rates. About 40% of all Euro-CP issues are now swapped. The development of computer technology that permits both dealer and issuer easier access to swap arbitrage opportunities and investor demand presages more growth ahead for swap-driven Euro-CP programs.

Calculating the Dollar Costs of Alternative Financing Options

This section presents explicit formulas to compute the effective dollar costs of a local currency loan and a dollar loan.[9] These cost formulas can be used to calculate the least expensive financing source for each future exchange rate. A computer can easily perform this analysis—called *breakeven analysis*—and determine the range of future exchange rates within which each particular financing option is cheapest.

With this breakeven analysis, the treasurer can readily see the amount of currency appreciation or depreciation necessary to make one type of borrowing less expensive than another. The treasurer will then compare the firm's actual forecast of currency change, determined objectively or subjectively, with this benchmark.

To illustrate breakeven analysis and show how to develop cost formulas, suppose that Du Pont's Mexican affiliate requires funds to finance its working capital needs for one year. It can borrow pesos at 45% or dollars at 11%. To determine an appropriate borrowing strategy, this section will develop explicit cost expressions for each of these loans using the numbers given above. These expressions then will be generalized to obtain analytical cost formulas that are usable under a variety of circumstances.

Case 1: No Taxes

Absent taxes and forward contracts, costing these loans is relatively straightforward.[10]

1. *Local currency loan*: Suppose the current exchange rate e_0 ($\$e_0$ = Mex\$1) is Mex\$1 = \$0.125, or Mex\$8 = \$1. Then, the peso cost of repaying the principal plus interest (45%) at the end of one year on one dollar's worth of pesos is $8(1.45)$ = Mex\$11.60. The dollar cost is $8(1.45)e_1$, where e_1 is the (unknown) ending exchange rate. Subtracting the dollar principal yields an effective dollar cost of $8(1.45)e_1 - 1$. For exam-

---

[9] This section draws on material in Alan C. Shapiro, "Evaluating Financing Costs for Multinational Subsidiaries," *Journal of International Business Studies*, Fall 1975, pp. 25–32.

[10] Loan cost formulas that factor in taxes are contained in the next section of this chapter titled "Case 2: Taxes."

ple, if the ending exchange rate is $0.10 (or Mex$10 = $1), then the cost per dollar borrowed equals

$$(8)(1.45)(0.10) - 1 = 0.16, \text{ or } 16\%$$

A simpler expression for the borrowing cost can be found by substituting $1/e_0$ for 8, yielding $1.45(e_1/e_0) - 1$. This expression equals $0.45(1 - d) - d$, where d is the (unknown) peso devaluation and is defined as $d = (e_0 - e_1)/e_0$. Thus, the effective dollar interest rate on borrowed pesos equals $0.45(1 - d) - d$. In general, the dollar (HC) cost of borrowing local currency at an interest rate of r_L and a currency change of d is the sum of the dollar interest cost less the exchange gain (loss) on repaying the principal:

$$\frac{\text{Dollar cost}}{\text{of LC loan}} = \text{Interest cost} - \text{Exchange gain (loss)} \quad\quad (13.1)$$

$$= r_L(1 - d) - d$$

The first term in Equation 13.1 is the dollar interest cost (paid at year end after an LC devaluation of d); the second term is the exchange gain or loss involved in repaying an LC loan valued at $1 at the beginning of the year with local currency worth $(1 - d)$ dollars at year end. As before, $d = (e_0 - e_1)/e_0$, where e_0 and e_1 are the beginning and ending exchange rates (LC 1 = e).

2. *Dollar loan*: The Mexican affiliate can borrow dollars at 11%. In general, the cost of a dollar (HC) loan to the affiliate is the interest rate on the dollar (HC) loan r_H.

ANALYSIS ■ The peso loan costs $0.45(1 - d) - d$, and the dollar loan costs 11%. To find the breakeven rate of currency depreciation at which the dollar cost of peso borrowing is just equal to the cost of dollar financing, equate the two costs and solve for d:

$$0.45(1 - d) - d = 0.11$$

or

$$d = \frac{(0.45 - 0.11)}{1.45} = 0.2345$$

In other words, the Mexican peso must devalue by 23.45%, to Mex$10.44 = $1, before it is less expensive to borrow pesos at 45% than dollars at 11%. Ignoring the factor of exchange risk, the borrowing decision rule is as follows:

If $d < 23.45\%$, borrow dollars.
If $d > 23.45\%$, borrow pesos.

Each of these cost formulas can be represented as a straight line with a positive intercept (equal to the effective interest rate when $d = 0$) and a negative (or zero) slope equal to the coefficient of the d term. The points at which these lines intersect (unless they are parallel) provide the breakeven values of d. These straight-line equations are

EXHIBIT 13.9 Low Cost Comparisons

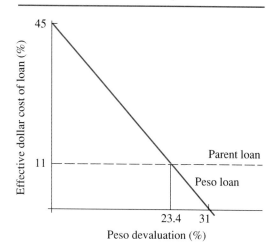

plotted in Exhibit 13.9. A peso devaluation of 31% will yield an effective interest rate of zero on the peso loan.

In the general case, the breakeven rate of currency change is found by equating the dollar costs of dollar and local currency financing:

$$r_H = r_L(1 - d) - d \qquad (13.2)$$

The solution to Equation 13.2 is

$$d^* = \frac{r_L - r_H}{1 + r_L} \qquad (13.3)$$

If the international Fisher effect holds, then we saw in Chapter 7 (Equation 7.14) that d^*, the breakeven amount of currency change, also equals the expected LC devaluation (revaluation); that is, the expected peso devaluation should equal 23.45% unless there is reason to believe that some form of market imperfection is not permitting interest rates to adjust to reflect anticipated currency changes.

Case 2: Taxes

Taxes complicate the calculating of various loan costs. Suppose the effective tax rate on the earnings of Du Pont's Mexican affiliate is 40%.

1. *Local currency loan*: The interest expense on one dollar's worth of pesos is $0.45 \times 8 =$ Mex\$3.6. After Mexican tax, this cost is $3.6 \times (1 - 0.40) =$ Mex\$2.16. The total after-tax peso cost of repaying the loan plus interest equals Mex\$10.16 (the principal amount of Mex\$8 plus the after-tax interest expense of Mex\$2.16). The dollar cost equals $10.16e_1$. Subtracting the dollar principal produces the after-tax dollar cost of

$10.16e_1 - 1$. This cost can be separated into two terms:

$$0.45 \times \frac{e_1}{e_0} \times 0.6 + \frac{e_1}{e_0} - 1$$

Substituting d for $(e_0 - e_1)/e_0$ yields

$$0.45(1 - d)0.6 - d = 0.27(1 - d) - d$$

The first term is the after-tax dollar interest cost of the peso interest expense, and the second term is the gain from the reduced dollar cost of repaying the peso principal.

In general, the after-tax dollar cost of borrowing in the local currency for a foreign affiliate equals the after-tax interest expense less the exchange gain (loss) on principal repayment, or

$$\frac{\text{After-tax dollar}}{\text{cost of LC loan}} = \text{Interest cost} - \text{Exchange gain (loss)}$$

$$= r_L(1 - d)(1 - t_a) - d \qquad (13.4)$$

where t_a is the affiliate's marginal tax rate. The first term in Equation 13.4 is the after-tax dollar interest cost paid at year end after an LC devaluation (revaluation) of d; the second is the exchange gain or loss in dollars of repaying a local currency loan valued at one dollar with local currency worth $(1 - d)$ dollars at the end of the year. The gain or loss has no tax effect for the affiliate because the same amount of local currency was borrowed and repaid.

2. *Dollar loan*: The after-tax cost of a dollar loan equals the Mexican affiliate's after-tax interest expense, $0.11(1 - 0.4)$, minus the dollar value to the Mexican affiliate of the tax write-off on the increased number of pesos necessary to repay the dollar principal after a peso devaluation, $0.4d$. This latter term is calculated in the following way: The dollar loan is converted to $1/e_0$ pesos, or Mex\$8. The number of pesos needed to repay this principal equals $1/e_1$ or an increase of

$$\frac{1}{e_1} - 8 = \frac{1}{e_1} - \frac{1}{e_0}$$

This extra expense is tax-deductible. The peso value of this tax deduction is $0.40(1/e_1 - 1/e_0)$, with a dollar value equal to $0.4(1/e_1 - 1/e_0)e_1$, or $0.4d$. Adding these two components yields a cost of dollar financing equal to

$$0.11(1 - 0.4) - 0.4d = 0.066 - 0.4d$$

We can generalize this analysis as follows: The total cost of the dollar loan is the after-tax interest expense less the tax write-off associated with the dollar principal repayment, or

$$\frac{\text{After-tax cost}}{\text{of dollar loan}} = \frac{\text{Interest cost}}{\text{to subsidiary}} - \text{Tax gain (loss)} \qquad (13.5)$$

$$= r_H(1 - t_a) - dt_a$$

The only change from the no-tax case, other than using after-tax instead of before-tax interest rates, is the addition of the second term, dt_a. This term equals the dollar value of the tax write-off (cost) associated with the increased (decreased) amount of local currency required to repay the dollar principal after an LC depreciation (appreciation).

Similarly, an LC revaluation will lead to a foreign exchange gain that is taxable, resulting in an increase in taxes equaling dt_a dollars. Hence, due to the tax effects, an LC devaluation will decrease the total cost of the dollar (HC) loan, and an LC revaluation will increase it.

ANALYSIS ■ As in case 1, we will set the cost of dollar financing, $0.066 - 0.4d$, equal to the cost of local currency financing, $0.27(1 - d) - d$, in order to find the breakeven rate of peso depreciation necessary to leave the firm indifferent between borrowing in dollars or pesos.

The breakeven value of d occurs when

$$0.066 - 0.4d = 0.27(1 - d) - d$$

or

$$d^* = 0.2345$$

Thus, the peso must devalue by 23.45% to Mex\$10.44 = \$1 before it is cheaper to borrow pesos at 45% than dollars at 11%. This is the same breakeven exchange rate as in the before-tax case. Although taxes affect the after-tax costs of the dollar and LC loans, they do not affect the relative desirability of the two loans. That is, if one loan has a lower cost before tax, it will also be less costly on an after-tax basis.

As in the no-tax case, effective dollar interest rates can be graphically illustrated as a function of alternative rates of peso devaluation. The intersection of the lines in Exhibit 13.10 provides the breakeven values of peso devaluation—that is, the devaluation percentages at which the firm would just be indifferent between one form of financing and another. Surprisingly, perhaps the most visible effect of taxes is the negative after-tax costs of the two loans at the breakeven rate of currency depreciation.[11] In contrast, the interest rate of 11% at the breakeven level of peso depreciation in the no-tax case is highly positive.

In general, the breakeven rate of currency appreciation or depreciation can be found by equating the dollar costs of local currency and dollar financing and solving for d:

$$r_H(1 - t_a) - dt_a = r_L(1 - d)(1 - t_a) - d$$

---

[11] The negative after-tax loan costs here are analogous to the negative real after-tax interest expense often encountered during periods of high inflation.

EXHIBIT 13.10 Low Cost Comparisons: Taxes

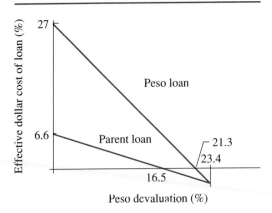

or

$$d* = \frac{r_L(1 - t_a) - r_H(1 - t_a)}{(1 + r_L)(1 - t_a)} = \frac{r_L - r_H}{1 + r_L} \qquad (13.6)$$

The tax rates cancel out and we are left with the same breakeven value for d as in the before-tax case (see Equation 13.3). This result is the same as in the case of Du Pont's Mexican affiliate and demonstrates that the earlier result was not a fluke.

Borrowing Strategy and Exchange Risk Management

This section provides a simple illustration of the interaction between financing choice and exchange risk management.[12] Assume that Du Pont's Mexican affiliate is going to invest the funds it borrows in a project that pays a relatively certain return of 50% annually in pesos.

Du Pont is concerned not only with its expected dollar profits from the investment but also with the effects of its loan choice on the project's exchange risk. It is assumed here that there are no taxes and that forward contracts are unavailable. If I is the local currency (peso) value of the investment, s is the percentage LC return on the investment, and σ refers to a standard deviation, then we can derive several important equations.

Operating Profit on Investment.   The dollar profit, π, equals the difference between dollar revenue and dollar cost, or

$$\pi = I(1 + s)e_1 - Ie_0$$

with a standard deviation for these profits of

$$\sigma(\pi) = I(1 + s)\sigma(e_1)$$

---

[12] This section is based on material in Lessard, "Currency and Tax Considerations in International Financing."

Cost of Dollar Financing. Using the dollar loan as an example, we have already seen that the cost (C_H) of dollar financing is

$$C_H = Ie_0 r_H$$

with a standard deviation of

$$\sigma(C_H) = 0$$

Cost of Local Currency Financing. The dollar cost of local currency financing, C_L, is

$$C_L = Ie_1(1 + r_L) - Ie_0$$

with a standard deviation of

$$\sigma(C_L) = I(1 + r_L)\sigma(e_1)$$

Profits Net of Financing Costs. By subtracting financing costs from the operating profit, we can compute net home-currency profits whether financing is done in home currency (π_H) or local currency (π_L):

$$\pi_H = [I(1 + s)e_1 - Ie_0] - Ie_0 r_H = I[(1 + s)e_1 - (1 + r_H)e_0]$$

with a standard deviation of

$$\sigma(\pi_H) = I(1 + s)\sigma(e_1)$$

and

$$\pi_L = [I(1 + s)e_1 - Ie_0] - [I(1 + r_L)e_1 - Ie_0] = I(s - r_L)e_1$$

with a standard deviation of

$$\sigma(\pi_L) = I(s - r_L)\sigma(e_1)$$

We saw that if Du Pont is concerned solely with minimizing expected cost, it will borrow pesos only if the peso is anticipated to devalue by more than 23.45%. If risk is an important consideration, however, then peso financing becomes relatively more attractive at any given exchange rate. The standard deviation of net profits with peso financing, $I(s - r_L)(e_1)$, is much smaller than the standard deviation with dollar (HC) financing, $I(1 + s)(e_1)$, because variation in dollar profits due to currency fluctuations is offset by equal variation in dollar financing costs. The difference in standard deviations of net profit between dollar and LC financing is

$$\sigma(\pi_H) - \sigma(\pi_L) = I(1 + r_L)\sigma(e_1) > 0$$

Assuming that I = Mex\$8 million = \$1 million at the current exchange rate of e_0 = \$0.125, with $\sigma(e_1)$ = 0.011375 and r_L = 0.45, then

$$\sigma(\pi_H) = \$136,500$$

and

$$\sigma(\pi_L) = \$4,550$$

If the expected value of e_1 is 10.45 (i.e., an expected devaluation of 23.45%, just large enough for the international Fisher effect to hold), then expected dollar profit in either case equals \$38,300. It seems obvious that, in this case at least, most firms would prefer to use peso financing even if they expected a peso devaluation of somewhat less than 23.45%. In fact, given the large downside risk with dollar financing and the relatively slender profit margin, it is unlikely that most firms would even consider this investment unless it could be financed with borrowed pesos.

13.5 ILLUSTRATION: AMERICAN EXPRESS

In early 1980, American Express (Amex) completed an eight-month study of the cash cycles of its travel, credit card, and traveler's check businesses operating in seven European countries.[13] On the basis of that project, Amex developed an international cash management system that was expected to yield cash gains—increased investments or reduced borrowing—of about \$35 million in Europe alone. About half of these savings were projected to come from accelerated receipts and better control of disbursements. The other half of projected gains represented improved bank-balance control, reduced bank charges, improved value-dating, and better control of foreign exchange.

The components of the system are collection and disbursement methods, bank-account architecture, balance targeting, and foreign exchange management. The worldwide system is controlled on a regional basis, with some direction from the corporate treasurer's office in New York. A regional treasurer's office in Brighton, England, controls cash, financing, and foreign exchange transactions for Europe, the Middle East, and Africa.

The most advantageous collection and disbursement method for every operating division in each country was found by analyzing the timing of mail and clearing floats. This analysis involved

- Establishing what payment methods were used by customers in each country because checks are not necessarily the primary method of payment in Europe
- Measuring the mail time between certain sending and receiving points
- Identifying clearing systems and practices, which vary considerably among countries
- Analyzing for each method of payment the value-dating practice, the times for processing check deposits, and the bank charges per item

---

[13] This section is adapted from Lars H. Thunell, "The American Express Formula," *Euromoney*, March 1980, pp. 121–127.

Using these data, Amex changed some of its collection and disbursement methods. For example, it installed interception points in Europe to minimize the collection float.

Next, Amex centralized the management of all its bank accounts in Europe on a regional basis. Allowing each subsidiary to set up its own independent bank account has the merit of simplicity, but it leads to a costly proliferation of different pools of funds. Amex restructured its bank accounts, eliminating some and ensuring that funds could move freely among the remaining accounts. By pooling its surplus funds, Amex can invest them for longer periods and also cut down on the chance that one subsidiary will be borrowing funds while another has surplus funds. Conversely, by combining the borrowing needs of various operations, Amex can use term financing and dispense with more expensive overdrafts. Reducing the number of accounts made cash management less complicated and also reduced banking charges.

The particular form of bank-account architecture used by Amex is a modular account structure that links separate accounts in each country with a master account. Management, on a daily basis, has only to focus on the one account through which all the country accounts have access to borrowing and investment facilities.

Balance targeting is used to control bank-account balances. The target is an average balance set for each account that reflects compensating balances, goodwill funds kept to foster the banking relationship, and the accuracy of cash forecasting. Aside from the target balance, the minimum information needed each morning to manage an account by balance targeting is the available opening balance and expected debits and credits.

Foreign exchange management in Amex's international cash management system focuses on its transaction exposure. This exposure, which is due to the multicurrency denomination of traveler's checks and credit card charges, fluctuates on a daily basis.

Procedures to control these exposures and to coordinate foreign exchange transactions center on how Amex finances its working capital from country to country, as well as the manner in which interaffiliate debts are settled. For example, if increased spending by cardholders creates the need for more working capital, Amex must decide whether to raise funds in local currency or in dollars. As a general rule, day-to-day cash is obtained at the local level through overdrafts or overnight funds.

To settle indebtedness among divisions, Amex uses interaffiliate settlements. For example, if a French cardholder uses her card in Germany, the French credit card office pays the German office, which in turn pays the German restaurant or hotel in Deutsche marks. Amex uses netting, coordinated by the regional treasurer's office in Brighton, to reduce settlement charges. For example, suppose that a German cardholder used his card in France at the same time the French cardholder charged with her card in Germany. Instead of two transactions, one foreign exchange transaction settles the differences between the two offices.

❧ 13.6 SUMMARY AND CONCLUSIONS

This chapter examined the diverse elements involved in international cash, accounts receivable, and inventory management, as well as the short-term financing of foreign affiliates. With regard to cash management, we saw that although the objectives are the same for the MNC as for the domestic firm — to accelerate the collection of funds and optimize their use — the key ingredients to successful management differ. The wider investment options available to the multinational firm were discussed, as were the concepts of multilateral netting, cash pooling, and multinational cash mobilization. As multinational firms develop more efficient and comprehensive information-gathering systems, the international cash management options available to them will increase. Accompanying these options will be even more sophisticated management techniques than currently exist.

Similarly, we saw that inventory and receivable management in the MNC involve the familiar cost-minimizing strategy of investing in these assets up to the point at which the marginal cost of extending another dollar of credit or purchasing one more unit of inventory is just equal to the additional expected benefits to be derived. These benefits accrue in the form of maintaining or increasing the value of other current assets—such as cash and marketable securities—increasing sales revenue, or reducing inventory stock-out costs.

We also have seen that most of the inventory and receivables management problems that arise internationally have close parallels in the purely domestic firm. Currency changes have effects that are similar to those of inflation, and supply disruptions are not unique to international business. The differences that do exist are more in degree than in kind.

The major reason why inflation, currency changes, and supply disruptions generally cause more concern in the multinational than in the domestic firm is that multinationals often are restricted in their ability to deal with these problems because of financial market constraints or import controls. Where financial markets are free to reflect anticipated economic events, there is no need to hedge against the loss of purchasing power by inventorying physical assets; financial securities or forward contracts are cheaper and more effective hedging media. Similarly, there is less likelihood that government policies will disrupt the flow of supplies among regions within a country than among countries.

We also examined the various short-term financing alternatives available to a firm, focusing on parent company loans, local currency bank loans, and commercial paper. We saw how factors such as relative interest rates, anticipated currency changes, the existence of forward contracts, and economic exposure combine to affect a firm's short-term borrowing choices. Various objectives that a firm might use to arrive at its borrowing strategy were evaluated. It was concluded that if forward contracts exist, the only valid objective is to minimize covered interest costs. In the absence of forward contracts, firms can either attempt to minimize expected costs or establish some trade-off between reducing expected costs and reducing the degree of cash-flow exposure. The latter goal involves offsetting operating cash inflows in a currency with financing cash outflows in that same currency.

This chapter also developed formulas to compute effective dollar costs of loans denominated in dollars (home currency) or local currency. These formulas were then used to calculate the breakeven rates of currency appreciation or depreciation that would equalize the costs of borrowing in the local currency or in the home currency.

➤Questions

1. High interest rates put a premium on careful management of cash and marketable securities.
   a. What techniques are available to an MNC with operating subsidiaries in many countries to economize on these short-term assets?
   b. What are the advantages and disadvantages of centralizing the cash management function?
   c. What can the firm do to enhance the advantages and reduce the disadvantages described in part b?

2. Standard advice given to firms exporting to soft-currency countries is to invoice in their own currency. Critically analyze this recommendation and suggest a framework that will help a financial manager decide whether to stipulate hard-currency invoicing in export contracts.
   a. Under what circumstances does this advice make sense?
   b. Are these circumstances consistent with market efficiency?

c. Are there any circumstances under which importer and exporter will mutually agree on an invoicing currency?

3. Suppose a subsidiary is all equity-financed and, hence, has no interest expenses. Does it still make sense to charge local managers for the working capital tied up in their operations? Explain.

4. Comment on the following statement: "One should borrow in those currencies expected to depreciate and invest in those expected to appreciate."

5. How can taxes affect the choice of currency denomination for loans?

6. How can the choice of currency denomination of loans enable a firm to reduce its exchange risk?

7. What are the three basic types of bank loans? Describe their differences.

8. How does each of the following affect the relationship of stated and effective interest rates?
   a. The lending bank requires the borrower to repay principal and interest at the end of the borrowing period only.
   b. Interest is deducted from the amount borrowed before the borrower receives the proceeds.
   c. What is the likely ranking of the above from least to most expensive?

9. Explain the characteristics of commercial paper that tend to limit its use to financially sound firms.

➤ Problems

1. A $1.5 billion Italian multinational manufacturing company has a total of $600 million in intercompany trade flows and settles accounts in 13 currencies. It also has about $400 million in third-party trade flows. Intercompany settlements are all made manually, there are no predefined remittance channels for either intercompany or third-party payments, and the methods and currencies of payment are determined by each unit independently of the other units. Payment terms for intercompany and third-party accounts are identical. What techniques might help this company better manage its affairs?

2. A major U.S. conglomerate operates eight large, independent subsidiaries in France that regularly trade with each other on an arm's-length basis (that is, setting price and terms as if they were independent entities). Some of the units are relatively mature and are net generators of cash; others are growing rapidly and need cash. In addition, these units trade with a number of other units located in other countries. They all have dealings with third parties in other countries as well. A recent audit revealed that these units maintained eight separate accounts at the same bank. What potential areas of improvement are there in this company's cash management?

3. SmithKline Beckman, the health-care products multinational, has 105 affiliates worldwide. There is a great deal of intercompany sales, dividend flows, and fee and royalty payments. Each unit makes its intercompany credit, payments, and hedging decisions independently. What advantages might SKB realize from centralizing international cash management and foreign exchange management?

4. Pfizer, the pharmaceutical company, generates approximately 52% of its sales overseas. A consulting study of treasury management revealed that the international division had its own treasury group that reported to the president of the international division. Both the domestic and the international treasury groups managed sizable cash portfolios. Moreover, Pfizer Inc. was significantly increasing its issues of U.S. commercial paper, and Pfizer International had cash surpluses. Intercompany sales were made on an arm's length basis, with no coordination of payments or credit terms. Each foreign unit would report monthly on what its bank balances were. All banking relations were managed locally. What profitable opportunities has Pfizer overlooked?

5. A major food and beverage manufacturer with three major divisions, 150 countries of operation, and international revenues accounting for 15% of total revenues of $6 billion conducted a treasury audit. It gathered data in the following areas: (a) local reports put out by the subsidiaries; (b) cross-border reports prepared by regional headquarters; (c) the system's organization; (d) transmission of data between subsidiaries, regional headquarters, and parent headquarters; (e) possible computerization of local reporting systems; (f) local-bank-balance reports; and (g) the accuracy of cash forecasts. What information should the company be looking for in each of these areas and why?

6. Twenty different divisions of Union Carbide sell to thousands of customers in more than 50 countries throughout the world. The proceeds are received in the form of drafts, checks, and letters of credit. Controlling the flow of funds from each transaction is an extremely complex task. Union Carbide wants to reduce the collection float to improve its cash

flow. What are some techniques that might help to achieve this objective?

7. RJR Nabisco, the tobacco and consumer products company, sells in more than 160 countries around the world. RJR collects, disburses, or invests more than $50 million each day in up to 80 different currencies. Much of the fund flows involve interaffiliate flows. In the mid-1980s, RJR's Corporate Treasury group discovered that combined borrowing of all RJR units totaled approximately $120 million to $130 million on a daily basis. Simultaneously, short-term investments entered into by RJR units ranged from $90 million to $100 million daily. Moreover, there was no central management of the fund flows. What are your recommendations to improve RJR's international cash management? Where and how might savings be achieved?

8. Newport Circuits is trying to decide whether to shift production overseas of its relatively expensive integrated circuits (they average around $11 each). Offshore assembly would save about 11.1¢ per chip in labor costs. However, by producing offshore, it would take about five weeks to get the parts to customers, in contrast to one week with domestic manufacturing. Thus, offshore production would force Newport to carry another four weeks of inventory. In addition, offshore production would entail combined shipping and customs duty costs of 3.2¢. Suppose Newport's cost of funds is 15%. Will it save money by shifting production offshore?

9. Tiger Car Corp., a leading Japanese automaker, is considering a proposal to locate a factory abroad in Tennessee. Although labor costs would rise by ¥33,000 per car, the time in transit for the cars (to be sold in the United States) would be reduced by 65 days. Tigers sell for ¥825,000, and TCC's cost of funds is 12.5%. Should TCC locate the plant in Tennessee?

10. Apex Supplies borrows FF 1 million at 12%, payable in one year. If Apex is required to maintain a compensating balance of 20%, what is the effective percentage cost of its loan (in FF)?

11. The Olivera Corp., a manufacturer of olive oil products, needs to acquire Lit 100 million in funds today to expand a pimiento-stuffing facility. Banca di Roma has offered them a choice of an 11% loan payable at maturity or a 10% loan on a discount basis. Which loan should Olivera choose?

12. To finance production of its new F-16 bubble gum, Hong Kong-based Top Gum Co. has been offered a one-year loan of HK$1.25 million at 9% payable at maturity with a 10% compensating balance.

a. What is the effective interest rate on this loan (in HK$)?
b. If the compensating balance requirement is 20%, what will be the effective interest rate?
c. If the compensating balance is 10%, but the loan is on a discount basis, what will be the effective interest rate?
d. If the company requires HK$1.25 million, how much must it borrow in part c to receive this amount?

13. If Consolidated Corp. issues a Eurobond denominated in yen, the 7% interest rate on the $1 million, one-year borrowing will be 2% less than rates in the United States. However, ConCorp would have to pay back the principal and interest in Japanese yen. Currently, the exchange rate is ¥183 = $1. By how much could the yen rise against the dollar before the Euroyen bond would lose its advantage to ConCorp?

14. Although the one-year interest rate is 10% in the United States, one-year, yen-denominated corporate bonds in Japan yield only 5%.

a. Does this present a riskless opportunity to raise capital at low yen interest rates?
b. Suppose the current exchange rate is ¥140 = $1. What is the lowest future exchange rate at which borrowing yen would be no more expensive than borrowing U.S. dollars?

15. Ford can borrow dollars at 12% or pesos at 80% for one year. The peso:dollar exchange rate is expected to move from $1 = Ps 3300 currently to $1 = Ps 4500 by year's end.

a. What is the expected after-tax dollar cost of borrowing dollars for one year if the Mexican corporate tax rate is 53%?
b. What is Ford's expected after-tax dollar cost of borrowing pesos for one year?
c. At what end-of-year exchange rate will the after-tax peso cost of borrowing dollars equal the after-tax peso cost of borrowing pesos?

16. The manager of an English subsidiary of a U.S. firm is trying to decide whether to borrow, for one year, dollars at 7.8% or pounds sterling at 12%. If the current value of the pound is $1.70, at what end-of-year exchange rate would the firm be indifferent now between borrowing dollars and pounds?

17. Suppose that a firm located in Belgium can borrow dollars at 8% or Belgian francs at 14%.

a. If the Belgian franc is expected to depreciate from BF 58 = $1 at the beginning of the year to BF 61 = $1 at the end of the year, what is the expected before-tax dollar cost of the Belgian franc loan?

b. If the Belgian corporate tax rate is 42%, what is the expected after-tax dollar cost of borrowing dollars, assuming the same currency change scenario?

c. Given the expected exchange rate change, which currency yields the lower expected after-tax dollar cost?

Bibliography

Business International Corporation. *Financing Foreign Operations.* New York: BIC, various issues.

GOELTZ, RICHARD K. "Managing Liquid Funds Internationally." *Columbia Journal of World Business,* July–August 1972, pp. 59–65.

LESSARD, DONALD R. "Currency and Tax Considerations in International Financing." Teaching Note No. 3, Massachusetts Institute of Technology, Spring 1979.

PRINDL, ANDREAS R. "International Money Management II: Systems and Techniques." *Euromoney,* October 1971.

SHAPIRO, ALAN C. "Evaluating Financing Costs for Multinational Subsidiaries." *Journal of International Business Studies,* Fall 1975, pp. 25–32.

_____. "International Cash Management: The Determination of Multicurrency Cash Balances." *Journal of Financial and Quantitative Analysis,* December 1976, pp. 893–900.

_____. "Payments Netting in International Cash Management." *Journal of International Business Studies,* Fall 1978, pp. 51–58.

MANAGING THE MULTINATIONAL FINANCIAL SYSTEM

An injudicious tax offers a great temptation to smuggling. But the penalties of smuggling must rise in proportion to the temptation. The law, contrary to all the ordinary principles of justice, first creates the temptation, and then punishes those who yield to it.

Adam Smith (1776)

The Eiffel Tower is the Empire State Building after taxes.

Anonymous

*T*he multinational corporation possesses a unique characteristic: the ability to shift funds and accounting profits among its various units through internal financial transfer mechanisms. Collectively, these mechanisms make up the **multinational financial system**. As we saw in Chapter 1, *internal financial transactions* are inherent in the MNC's global approach to international operations, specifically the highly coordinated international interchange of goods (material, parts, subassemblies, and finished products), services (technology, management skills, trademarks, and patents), and capital (equity and debt) that is the hallmark of the modern multinational firm. Indeed, almost 40% of U.S. imports and exports are transactions between U.S. firms and their foreign affiliates or parents. Intercompany trade is not confined to U.S. multinationals. Nearly 80% of the two-way trade between the United States and Japan goes from parent to foreign subsidiary or vice versa. So too does 40% of U.S.-European Community trade and 55% of EC-Japan trade.

The purpose of this chapter is to analyze the benefits, costs, and constraints associated with the multinational financial system. This analysis includes (1) identifying the conditions under which use of this system will increase the value of the firm relative to what it would be if all financial transactions were made at arm's length (that is, between unrelated entities) through external financial channels, (2) describing and evaluating the various channels for moving money and profits internationally, and (3) specifying the design principles for a global approach to managing international fund transfers. We will examine the objectives of such an approach and the behavioral, informational, legal, and economic factors that help determine its degree of success.

14.1 THE VALUE OF THE MULTINATIONAL FINANCIAL SYSTEM

The value of the MNC's network of financial linkages stems from the wide variations in national tax systems and significant costs and barriers associated with international financial transfers. Exhibit 14.1 summarizes the various factors that enhance the value of internal, relative to external, financial transactions. These restrictions are usually imposed to allow nations to maintain artificial values (usually inflated) for their currencies. In addition, capital controls are necessary when governments set the cost of local funds at a lower-than-market rate when currency risks are accounted for—that is, when government regulations do not allow the international Fisher effect or interest rate parity to hold.

Consequently, the ability to transfer funds and to reallocate profits internally presents multinationals with three different types of arbitrage opportunities.

1. *Tax arbitrage.* MNCs can reduce their tax burden by shifting profits from units located in high-tax nations to those in lower-tax nations. Or, they may shift profits from units in a taxpaying position to those with tax losses.

EXHIBIT 14.1 Market Imperfections that Enhance the Value of Internal Financial Transactions

Formal Barriers to International Transactions
Quantitative restrictions (exchange controls) and direct taxes on international movements of funds
Differential taxation of income streams according to nationality and global tax situation of the owners
Restrictions by nationality of investor and/or investment on access to domestic capital markets

Informal Barriers to International Transactions
Costs of obtaining information
Difficulty of enforcing contracts across national boundaries
Transaction costs
Traditional investment patterns

Imperfections in Domestic Capital Markets
Ceilings on interest rates
Mandatory credit allocations
Limited legal and institutional protection for minority shareholders
Limited liquidity due to thinness of markets
High transaction costs due to small market size and/or monopolistic practices of key financial institutions
Difficulty of obtaining information needed to evaluate securities

Source: Donald R. Lessard, "Transfer Prices, Taxes and Financial Markets: Implications of Internal Financial Transfers within the Multinational Firm," in *The Economic Effects of Multinational Corporations,* ed. Robert G. Hawkins (Greenwich, CT: JAI Press, 1979). Reprinted with permission by Donald R. Lessard and JAI Press.

2. *Financial market arbitrage.* By transferring funds among units, MNCs may be able to circumvent exchange controls, earn higher risk-adjusted yields on excess funds, reduce their risk-adjusted cost of borrowed funds, and tap previously unavailable capital sources.

3. *Regulatory system arbitrage.* When subsidiary profits are a function of government regulations (for example, when a government agency sets allowable prices on the firm's goods) or union pressure, rather than the marketplace, the ability to disguise true profitability by reallocating profits among units may give the multinational firm a negotiating advantage.[1]

A fourth possible arbitrage opportunity is the ability to permit an affiliate to negate the effect of credit restraint or controls in its country of operation. If a government limits access to additional borrowing locally, then the firm able to draw on external sources of funds not only can achieve greater short-term profits but also may be able to attain a more powerful market position over the long term.

14.2 INTERCOMPANY FUND-FLOW MECHANISMS: COSTS AND BENEFITS

The MNC can be visualized as *unbundling* the total flow of funds between each pair of affiliates into separate components that are associated with resources transferred in the form of products, capital, services, and technology. For example, dividends, interest, and loan repayments can be matched against capital invested as equity or debt; fees, royalties, or corporate overhead can be charged for various corporate services, trademarks, or licenses.

The different channels available to the multinational enterprise for moving money and profits internationally include transfer pricing, fee and royalty adjustments, leading and lagging, intercompany loans, dividend adjustments, and investing in the form of debt versus equity. This section examines the costs, benefits, and constraints associated with each of these methods of effecting *intercompany fund flows*. It begins by sketching out some of the tax consequences for U.S.-based MNCs of interaffiliate financial transfers.

Tax Factors

Total tax payments on intercompany fund transfers are dependent on the tax regulations of both the host and the recipient nations. The host country ordinarily has two types of taxes that directly affect tax costs: corporate income taxes and withholding taxes on dividend, interest, and fee remittances. In addition, several countries, such as Germany and Japan, tax retained earnings at a different (usually higher) rate than earnings paid out as dividends.

---

[1] See Donald R. Lessard, "Transfer Prices, Taxes, and Financial Markets: Implications of Internal Financial Transfers Within the Multinational Firm," in *The Economic Effects of Multinational Corporations*, ed. Robert G. Hawkins (Greenwich, Conn.: JAI Press, 1979); and David P. Rutenberg, "Maneuvering Liquid Assets in a Multinational Company," *Management Science*, June 1970, pp. B671–684.

Many recipient nations, including the United States, tax income remitted from abroad at the regular corporate tax rate. Where this rate is higher than the foreign tax rate, dividend and other payments will normally entail an incremental tax cost. There are a number of countries, however—such as Canada, the Netherlands, and France—that do not impose any additional taxes on foreign-source income.

As an offset to these additional taxes, most countries, including the United States, provide tax credits for affiliate taxes already paid on the same income. For example, if a subsidiary located overseas has $100 in pre-tax income, pays $30 in local tax, and then remits the remaining $70 to its U.S. parent in the form of a dividend, the U.S. Internal Revenue Service (IRS) will impose a $35 tax ($0.35 \times \$100$) but will then provide a dollar-for-dollar **foreign tax credit** (FTC) for the $30 already paid in foreign taxes, leaving the parent with a bill for the remaining $5. Foreign tax credits from other remittances can be used to offset these additional taxes. For example, if a foreign subsidiary earns $100 before tax, pays $45 in local tax, and then remits the remaining $55 in the form of a dividend, the parent will wind up with an FTC of $10, the difference between the $35 U.S. tax owed and the $45 foreign tax paid.

Transfer Pricing

The pricing of goods and services traded internally is one of the most sensitive of all management subjects, and executives typically are reluctant to discuss it. Each government normally presumes that multinationals use **transfer pricing** to its country's detriment. For this reason, a number of home and host governments have set up policing mechanisms to review the transfer pricing policies of MNCs.

The most important uses of transfer pricing include (1) reducing taxes, (2) reducing tariffs, and (3) avoiding exchange controls. Transfer prices also may be used to increase the MNC's share of profits from a joint venture and to disguise an affiliate's true profitability.

TAX EFFECTS ■ The following scenario illustrates the tax effects associated with a change in transfer price. Suppose that affiliate A produces 100,000 circuit boards for $10 apiece and sells them to affiliate B. Affiliate B, in turn, sells these boards for $22 apiece to an unrelated customer. As shown in Exhibit 14.2, pre-tax profit for the consolidated company is $1 million regardless of the price at which the goods are transferred from affiliate A to affiliate B.

Nevertheless, because affiliate A's tax rate is 30% whereas affiliate B's tax rate is 50%, consolidated after-tax income will differ depending on the transfer price used. Under the low-markup policy, in which affiliate A sets a unit transfer price of $15, affiliate A pays taxes of $120,000 and affiliate B pays $300,000, for a total tax bill of $420,000 and a consolidated net income of $580,000. Switching to a high-markup policy (a transfer price of $18), affiliate A's taxes rise to $210,000 and affiliate B's decline to $150,000, for combined tax payments of $360,000 and consolidated net income of $640,000. The result of this transfer price increase is to lower total taxes paid by $60,000 and raise consolidated income by the same amount.

In effect, profits are being shifted from a higher to a lower tax jurisdiction. In the extreme case, an affiliate may be in a loss position because of high start-up costs, heavy

EXHIBIT 14.2 Tax Effect of High Versus Low Transfer Price ($ Thousands)

	AFFILIATE A	AFFILIATE B	AFFILIATES A + B
Low-Markup Policy			
Revenue	1,500	2,200	2,200
Cost of goods sold	1,000	1,500	1,000
Gross profit	500	700	1,200
Other expenses	100	100	200
Income before taxes	400	600	1,000
Taxes (30%/50%)	120	300	420
Net income	280	300	580
High-Markup Policy			
Revenue	1,800	2,200	2,200
Cost of goods sold	1,000	1,800	1,000
Gross profit	800	400	1,200
Other expenses	100	100	200
Income before taxes	700	300	1,000
Taxes (30%/50%)	210	150	360
Net income	490	150	640

depreciation charges, or substantial investments that are expensed. Consequently, it has a zero effective tax rate, and profits channeled to that unit can be received tax-free. The basic rule of thumb to follow if the objective is to minimize taxes is as follows: If affiliate A is selling goods to affiliate B, and t_A and t_B are the marginal tax rates of affiliate A and affiliate B, respectively, then

If $t_A > t_B$, set the transfer price as low as possible.

If $t_A < t_B$, set the transfer price as high as possible.

TARIFFS ■ The introduction of tariffs complicates this decision rule. Suppose that affiliate B must pay *ad valorem* import duties—tariffs that are set as a percentage of the value of the imported goods—at the rate of 10%. Then, raising the transfer price will increase the duties that affiliate B must pay, assuming that the tariff is levied on the invoice (transfer) price. The combined tax-plus-tariff effects of the transfer price change are shown in Exhibit 14.3.

Under the low-markup policy, import tariffs of $150,000 are paid. Affiliate B's taxes will decline by $75,000 because tariffs are tax-deductible. Total taxes plus tariffs paid are $495,000. Switching to the high-markup policy raises import duties to $180,000 and simultaneously lowers affiliate B's income taxes by half that amount, or $90,000. Total taxes plus tariffs rise to $450,000. The high-markup policy is still desirable, but its benefit has been reduced by $15,000 to $45,000. In general, the higher the ad valorem tariff relative to the income tax differential, the more likely it is that a low transfer price is desirable.

There are some costs associated with using transfer prices for tax reduction. If the price is too high, tax authorities in the purchaser's (affiliate B's) country will see revenues forgone; if the price is too low, both governments might intervene. Affiliate A's government may view low transfer prices as tax evasion at the same time that the tariff commis-

EXHIBIT 14.3 Tax-Plus-Tariff Effect of High Versus Low Transfer Price ($ Thousands)

	AFFILIATE A	AFFILIATE B	AFFILIATES A + B
Low-Markup Policy			
Revenue	1,500	2,200	2,200
Cost of goods sold	1,000	1,500	1,000
Import duty (10%)	—	150	150
Gross profit	500	550	1,050
Other expenses	100	100	200
Income before taxes	400	450	850
Taxes (30%/50%)	120	225	345
Net income	280	225	505
High-Markup Policy			
Revenue	1,800	2,200	2,200
Cost of goods sold	1,000	1,800	1,000
Import duty	—	180	180
Gross profit	800	220	1,020
Other expenses	100	100	200
Income before taxes	700	120	820
Taxes (30%/50%)	210	60	270
Net income	490	60	550

sion in affiliate B's country sees dumping or revenue forgone or both. These costs must be paid for in the form of legal fees, executive time, and penalties.

Most countries have specific regulations governing transfer prices. For instance, **Section 482** of the U.S. Revenue Code calls for **arm's-length prices**—prices at which a willing buyer and a willing unrelated seller would freely agree to transact. The four alternative methods for establishing an arm's-length price, in order of their general acceptability to tax authorities, are as follows:

1. *Comparable uncontrolled price method.* Under this method, the transfer price is set by direct references to prices used in comparable bona fide transactions between enterprises that are independent of each other or between the multinational enterprise group and unrelated parties. In principle, this method is the most appropriate to use; in theory, it is the easiest. In practice, however, it may be impractical or difficult to apply. For example, differences in quantity, quality, terms, use of trademarks or brand names, time of sale, level of the market, and geography of the market may be grounds for claiming that the sale is not comparable. There is a gradation of comparability: Adjustments can be made easily for freight and insurance but cannot be made accurately for trademarks.

2. *Resale price method.* Under this method, the arm's-length price for a product sold to an associate enterprise for resale is determined by reducing the price at which it is resold to an independent purchaser by an appropriate markup (that is, an amount that covers the reseller's costs and profit). This method is probably most applicable to marketing operations. However, determining an appropriate markup can be difficult, especially where the reseller adds substantially to the value of the product. Thus, there is often quite a bit of leeway in determining a standard markup.

3. *Cost-plus method.* This method adds an appropriate profit markup to the seller's cost to arrive at an arm's-length price. This method is useful in specific situations, such as where semifinished products are sold between related parties or where one entity is essentially acting as a subcontractor for a related entity. However, ordinarily it is difficult to assess the cost of the product and to determine the appropriate profit markup. In fact, no definition of full cost is given, nor is there a unique formula for prorating shared costs over joint products. Thus, the markup over cost allows room for maneuver.

4. *Another appropriate method.* In some cases, it may be appropriate to use a combination of the above methods, or use still other methods (e.g., comparable profits and net yield methods) to arrive at the transfer price. In addition, the Treasury regulations are quite explicit that while a new market is being established, it is legitimate to charge a lower transfer price.

In light of Section 482, and the U.S. government's willingness to use it, and similar authority by most other nations, current practice by MNCs appears to be setting standard prices for standardized products. However, the innovative nature of the typical multinational ensures a continual stream of new products for which no market equivalent exists. Hence, some leeway is possible on transfer pricing. In addition, although finished products do get traded among affiliates, trade between related parties increasingly is in high-

---

ILLUSTRATION

PRESIDENT CLINTON SEEKS $45 BILLION FROM FOREIGN COMPANIES

One of the linchpins of President Bill Clinton's economic revival plan was a proposal to extract billions from foreign companies doing business in the United States. By closing loopholes and instituting more vigorous enforcement, Clinton said he believed that the federal government could raise $45 billion over four years from foreign companies, which take in nearly $1 trillion in U.S. revenue annually. Clinton's contention was based on a view prevalent in the IRS that foreign multinationals were manipulating their transfer prices on goods they sell to their U.S. subsidiaries to avoid paying U.S. tax. For example, IRS data for 1989 showed that the 44,480 domestic companies controlled by foreign entities generated $1 trillion of worldwide sales and had total assets of $1.4 trillion but reported net income of only $11.2 billion, a return on assets of under 1%.

However, many analysts are skeptical that there is all that much to be gained by tougher enforcement of Section 482. For example, one government study reportedly found that the IRS had only limited success in recovering additional taxes in transfer pricing cases involving foreign MNCS. From 1987 through 1989, the IRS got only 26.5% of the $757 million it sought.

In addition, the circumstantial evidence the IRS relied on to support its case was suspect. For example, in 1990, the IRS cited as evidence of transfer pricing manipulation the fact that, over a 10-year period, foreign-controlled companies' (1) gross income had more than doubled and (2) the taxes paid by these companies had hardly changed. However, this argument made no sense because on U.S. tax returns, "gross income" is "sales" less "cost of sales"; thus transfer price manipulation, if any, would reduce "gross income" (by raising the cost of goods sold). Therefore, a doubling of "gross income" could not indicate transfer price manipulation.

tech, custom-made components and subassemblies (for example, automobile transmissions and circuit boards) where there are no comparable sales to unrelated buyers. Firms also have a great deal of latitude in setting prices on rejects, scrap, and returned goods. Moreover, as trade in intangible services becomes more important, monitoring transfer prices within MNCs has become extraordinarily complex, creating plentiful opportunities for multinationals to use transfer pricing to shift their overall taxable income from one jurisdiction to another.

One means of dealing with the U.S. government's crackdown on alleged transfer pricing abuses is greater reliance on **advance pricing agreements** (APAs). The APA procedure allows the multinational firm, the IRS, and the foreign tax authority to work out, in advance, a method to calculate transfer prices. As of early 1994, approximately 40 APAs were in place and another 50 were in the pipeline. APAs are expensive, they can take quite a while to negotiate, and they involve a great deal of disclosure on the part of the MNC, but if a company wants assurance that its transfer pricing is in order, the APA is a useful tool.

EXCHANGE CONTROLS ■ Another important use of transfer pricing is to avoid currency controls. For example, in the absence of offsetting foreign tax credits, a U.S. parent will wind up with $\$0.65Q_0$ after tax for each dollar increase in the price at which it sells Q_0 units of a product to an affiliate with blocked funds (based on a U.S. corporate tax rate of 35%). Hence, a transfer price change from P_0 to P_1 will lead to a shift of $0.65(P_1 - P_0)Q_0$ dollars to the parent. The subsidiary, of course, will show a corresponding reduction in its cash balances and taxes, due to its higher expenses.

In fact, bypassing currency restrictions appears to explain the seeming anomaly whereby subsidiaries operating in less-developed countries (LDCs) with low tax rates are sold overpriced goods by other units. In effect, companies appear to be willing to pay a tax penalty to access otherwise unavailable funds.

JOINT VENTURES ■ Conflicts over transfer pricing often arise when one of the affiliates involved is owned jointly by one or more other partners. The outside partners are often suspicious that transfer pricing is being used to shift profits from the joint venture, where they must be shared, to a wholly owned subsidiary. Although there is no pat answer to this problem, the determination of fair transfer prices should be resolved before the establishment of a joint venture. Continuing disputes may still arise, however, over the pricing of new products introduced to an existing venture.

DISGUISING PROFITABILITY ■ Many LDCs erect high tariff barriers in order to develop import-substituting industries. However, because they are aware of the potential for abuse, many host governments simultaneously attempt to regulate the profits of firms operating in such a protected environment. When confronted by a situation where profits depend on government regulations, the MNC can use transfer pricing (buying goods from sister affiliates at a higher price) to disguise the true profitability of its local affiliate, enabling it to justify higher local prices. Lower reported profits also may improve a subsidiary's bargaining position in wage negotiations. It is probably for this reason that several international unions have called for fuller disclosure by multinationals of their worldwide accounting data.

EVALUATION AND CONTROL ■ Transfer price adjustments will distort the profits of reporting units and create potential difficulties in evaluating managerial performance. In addition, managers evaluated on the basis of these reported profits may have an incentive to behave in ways that are suboptimal for the corporation as a whole.

Reinvoicing Centers

One approach used by some multinationals to disguise profitability, avoid the scrutiny of governments, and coordinate transfer pricing policy is to set up reinvoicing centers in low-tax nations. The reinvoicing center takes title to all goods sold by one corporate unit to another affiliate or to a third-party customer, although the goods move directly from the factory or warehouse location to the purchaser. The center pays the seller and, in turn, is paid by the purchasing unit.

With price quotations coming from one location, it is easier and quicker to implement decisions to have prices reflect changes in currency values. The reinvoicing center also provides a firm with greater flexibility in choosing an invoicing currency. Affiliates can be ordered to pay in other than their local currency if required by the firm's external currency obligations. In this way, the MNC can avoid the costs of converting from one currency to another and then back again.

Having a reinvoicing center can be expensive, however. There are increased communications costs due to the geographical separation of marketing and sales from the production centers. In addition, tax authorities may be suspicious of transactions with an affiliated trading company located in a tax haven.

Before 1962, many U.S. multinationals had reinvoicing companies located in low- or zero-tax countries. By buying low and selling high, U.S. MNCs could siphon off most of the profit on interaffiliate sales with little or no tax liability because the U.S. government at that time did not tax unremitted foreign earnings. This situation changed with passage of the U.S. Revenue Act of 1962, which declared that reinvoicing-center income is **Subpart F** income and, hence, is subject to U.S. taxation immediately, whether remitted to the United States or not. For most U.S.-based multinationals, this situation negated the tax benefits associated with a reinvoicing center.

A 1977 ruling by the IRS, however, has increased the value of tax havens in general and reinvoicing centers in particular. That ruling, which allocates to a firm's foreign affiliates certain parent expenses that previously could be written off in the United States, has generated additional foreign tax credits that can be utilized only against U.S. taxes owed on foreign-source income, increasing the value of tax-haven subsidiaries.

A reinvoicing center, by channeling profits overseas, can create Subpart F income to offset these excess FTCs. In effect, foreign tax credits can be substituted for taxes that would otherwise be owed to the United States or to foreign governments. Suppose a firm shifts $100 in profit from a country with a 50% tax rate to a reinvoicing center where the tax rate is only 10%. If this $100 is deemed Subpart F income by the IRS, the U.S. parent will owe an additional $25 in U.S. tax (based on the U.S. corporate tax rate of 35% and the $10 foreign tax credit). However, if the company has excess foreign tax credits available, then each $100 shift in profits can reduce total tax payments by $25, until the excess FTCs are all expended.

Fees and Royalties

Management services such as headquarters advice, allocated overhead, patents, and trademarks are often unique and, therefore, are without a reference market price. The consequent difficulty in pricing these corporate resources makes them suitable for use as additional routes for international fund flows by varying the *fees* or *royalties* charged for using these intangible factors of production.

Transfer prices for services or intangible assets have the same tax and exchange control effects as those for transfer prices on goods. However, host governments often look with more favor on payments for industrial know-how than for profit remittances. Where restrictions do exist, they are more likely to be modified to permit a fee for technical knowledge than to allow for dividends.

For MNCs, these charges have assumed a somewhat more important role as a conduit for funneling remittances from foreign affiliates. To a certain extent, this trend reflects the fact that many of these payments are tied to overseas sales or assets that grew very rapidly during the 1960s and early 1970s, as well as the growing importance of tax considerations and exchange controls. For example, by setting low transfer prices on intangibles to manufacturing subsidiaries in low-tax locations such as Puerto Rico or Singapore, multinationals can receive profits essentially tax free.

In recognition of this possibility, the United States in 1986 amended Section 482 to provide that the transfer price of an intangible must be "commensurate with the income" the intangible generates. This means in practice that the IRS will not consider a related-party transfer price for an intangible arm's length unless it produces a split in profits between transferor and transferee that falls within the range of profits that unrelated parties realize (1) on similar intangibles, (2) in similar circumstances. Armed with this amendment, the IRS now carefully scrutinizes the pricing of intangibles.

The most common approach to setting fee and royalty charges is for the parent to decide on a desired amount of total fee remittances from the overseas operations, usually based on an allocation of corporate expenses, and then to apportion these charges according to subsidiary sales or assets. This method, which sometimes involves establishing identical licensing agreements with all units, gives these charges the appearance of a legitimate and necessary business expense, thereby aiding in overcoming currency restrictions.

Governments typically prefer prior agreements and steady and predictable payment flows; a sudden change in licensing and service charges is likely to be looked at with suspicion. For this reason, firms try to avoid abrupt changes in their remittance policies. However, where exchange controls exist or are likely, or if there are significant tax advantages, many firms will initially set a higher level of fee and royalty payments while still maintaining a stable remittance policy.

Special problems exist with joint ventures because the parent company will have to obtain permission from its partner(s) to be able to levy charges for its services and licensing contributions. These payments assure the parent of receiving at least some compensation for the resources it has invested in the joint venture, perhaps in lieu of dividends over which it may have little or no control.

Leading and Lagging

A highly favored means of shifting liquidity among affiliates is an acceleration (**leading**) or delay (**lagging**) in the payment of interaffiliate accounts by modifying the credit terms extended by one unit to another. For example, suppose affiliate A sells goods worth $1 million monthly to affiliate B on 90-day credit terms. Then, on average, affiliate A has $3 million of accounts receivable from affiliate B and is, in effect, financing $3 million of working capital for affiliate B. If the terms are changed to 180 days, there will be a one-time shift of an additional $3 million to affiliate B. Conversely, reducing credit terms to 30 days will create a flow of $2 million from affiliate B to affiliate A, as shown in Exhibit 14.4.

Shifting Liquidity.   The value of leading and lagging depends on the opportunity cost of funds to both the paying unit and the recipient. When an affiliate already in a surplus position receives payment, it can invest the additional funds at the prevailing local lending rate; if it requires working capital, the payment received can be used to reduce its borrowings at the borrowing rate. If the paying unit has excess funds, it loses cash that it would have invested at the lending rate; if it is in a deficit position, it has to borrow at the borrowing rate. Assessment of the benefits of shifting liquidity among affiliates requires that these borrowing and lending rates be calculated on an after-tax dollar (HC) basis.

Suppose, for example, that a multinational company faces the following effective, after-tax dollar borrowing and lending rates in Germany and the United States:

	BORROWING RATE (%)	LENDING RATE (%)
United States	3.8	2.9
Germany	3.6	2.7

Both the U.S. and German units can have either a surplus ($+$) or deficit ($-$) of funds. The four possibilities, along with the domestic interest rates (U.S./German) and

EXHIBIT 14.4 Fund-Transfer Effects of Leading and Lagging

	AFFILIATE A SELLS $1 MILLION IN GOODS MONTHLY TO AFFILIATE B		
		CREDIT TERMS	
BALANCE-SHEET ACCOUNTS	NORMAL (90 DAYS)	LEADING (30 DAYS)	LAGGING (180 DAYS)
Affiliate A			
Accounts receivable from B	$3,000,000	$1,000,000	$6,000,000
Affiliate B			
Accounts payable to A	$3,000,000	$1,000,000	$6,000,000
Net Cash Transfers			
From B to A		$2,000,000	—
From A to B		—	$3,000,000

interest differentials (U.S. rate − German rate) associated with each state, are as follows:

		GERMANY	
		+	−
UNITED STATES	+	2.9%/2.7% (0.2%)	2.9%/3.6% (−0.7%)
	−	3.8%/2.7% (1.1%)	3.8%/3.6% (0.2%)

For example, if both units have excess funds, then the relevant opportunity costs of funds are the U.S. and German lending rates of 2.9% and 2.7%, respectively, and the associated interest differential (in parentheses) is 0.2%. Similarly, if the U.S. unit requires funds while the German affiliate has a cash surplus, then the relevant rates are the respective U.S. borrowing and German lending rates of 3.8% and 2.7% and the interest differential is 1.1%.

If the interest rate differential is positive, the corporation as a whole—by moving funds to the United States—will either pay less on its borrowing or earn more interest on its investments. This move can be made by leading payments to the United States and lagging payments to Germany. Shifting money to Germany—by leading payments to Germany and lagging them to the United States—will be worthwhile if the interest differential is negative.

Based on the interest differentials in this example, all borrowings should be done in Germany, and surplus funds should be invested in the United States. Only if the U.S. unit has excess cash and the German unit requires funds should money flow into Germany.

For example, suppose the German unit owes $2 million to the U.S. unit. The timing of this payment can be changed by up to 90 days in either direction. Assume that the U.S. unit is borrowing funds, and the German unit has excess cash available. According to the prevailing interest differential of 1.1%, given the current liquidity status of each affiliate, the German unit should speed up, or lead, its payment to the U.S. unit. The net effect of these adjustments is that the U.S. firm can reduce its borrowing by $2 million, and the German unit has $2 million less in cash—all for 90 days. Borrowing costs for the U.S. unit are pared by $19,000 ($2,000,000 × 0.038 × 90/360), and the German unit's interest income is reduced by $13,500 ($2,000,000 × 0.027 × 90/360). There is a net savings of $5,500. The savings could be computed more directly by using the relevant interest differential of 1.1% as follows: $2,000,000 × 0.011 × 90/360 = $5,500.

ADVANTAGES ■ A leading and lagging strategy has several advantages over direct intercompany loans.

1. No formal note of indebtedness is needed, and the amount of credit can be adjusted up or down by shortening or lengthening the terms on the accounts. Governments do not always allow such freedom on loans.
2. Governments are less likely to interfere with payments on intercompany accounts than on direct loans.
3. Section 482 allows intercompany accounts up to six months to be interest free. In contrast, interest must be charged on all intercompany loans. The ability to set a zero interest rate is

valuable if the host government does not allow interest payments on parent company loans to be tax deductible or if there are withholding taxes on interest payments.

GOVERNMENT RESTRICTIONS ■ As with all other transfer mechanisms, government controls on intercompany credit terms are often tight and given to abrupt changes. Although appearing straightforward on the surface, these rules are subject to different degrees of government interpretation and sanction. For example, in theory Japan permits firms to employ leads and lags. In reality, however, leading and lagging is very difficult because regulations require that all settlements be made in accordance with the original trade documents unless a very good reason exists for an exception. On the other hand, Sweden, which prohibits import leads, will often lift this restriction for imports of capital goods.

Intercompany Loans

A principal means of financing foreign operations and moving funds internationally is to engage in intercompany lending activities. The making and repaying of **intercompany loans** is often the only legitimate transfer mechanism available to the MNC.

Intercompany loans are more valuable to the firm than arm's-length transactions only if at least one of the following market distortions exist: (1) credit rationing (due to a ceiling on local interest rates), (2) currency controls, or (3) differential tax rates among countries. This list is not particularly restrictive because it is the rare MNC that faces none of these situations in its international operations.

Although various types of intercompany loans exist, the most important methods at present are direct loans, back-to-back financing, and parallel loans. *Direct loans* are straight extensions of credit from the parent to an affiliate or from one affiliate to another. The other types of intercompany loans typically involve an intermediary.

BACK-TO-BACK LOANS ■ **Back-to-back loans**, also called *fronting loans* or *link financing*, are often employed to finance affiliates located in nations with high interest rates or restricted capital markets, especially when there is a danger of currency controls or when different rates of withholding tax are applied to loans from a financial institution. In the typical arrangement, the parent company deposits funds with a bank in country A that in

EXHIBIT 14.5 Structure of a Back-to-Back Loan

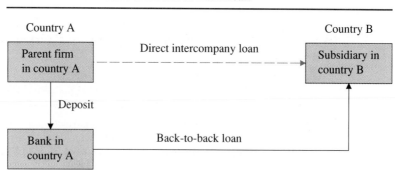

turn lends the money to a subsidiary in country B. These transactions are shown in Exhibit 14.5. By contrasting these transactions with a direct intercompany loan, the figure reveals that, in effect, a back-to-back loan is an intercompany loan channeled through a bank. From the bank's point of view, the loan is risk free because the parent's deposit fully collateralizes it. The bank simply acts as an intermediary or a front; compensation is provided by the margin between the interest received from the borrowing unit and the rate paid on the parent's deposit.

A back-to-back loan may offer several potential advantages when compared with a direct intercompany loan. Two of the most important advantages are as follows:

1. Certain countries apply different withholding-tax rates to interest paid to a foreign parent and interest paid to a financial institution. A cost saving in the form of lower taxes may be available with a back-to-back loan.

2. If currency controls are imposed, the government usually will permit the local subsidiary to honor the amortization schedule of a loan from a major multinational bank; to stop payment would hurt the nation's credit rating. Conversely, local monetary authorities would have far fewer reservations about not authorizing the repayment of an intercompany loan. In general, back-to-back financing provides better protection than does a parent loan against expropriation and/or exchange controls.

Some financial managers argue that a back-to-back loan conveys another benefit: The subsidiary seems to have obtained credit from a major bank on its own, possibly enhancing its reputation. However, this appearance is unlikely to be significant in the highly informed international financial community.

The costs of a back-to-back loan are evaluated in the same way as any other financing method (i.e., by considering relevant interest and tax rates and the likelihood of changes in currency value). To see how these calculations should be made, assume that the parent's opportunity cost of funds is 10%, and the parent's and affiliate's marginal tax rates are 34% and 40%, respectively. Then, if the parent earns 8% on its deposit, the bank charges 9% to lend dollars to the affiliate, and the local currency devalues by 11% during the course of the loan, the effective cost of this back-to-back loan equals

$$\begin{matrix} \text{Interest cost} \\ \text{to parent} \end{matrix} - \begin{matrix} \text{Interest income} \\ \text{to parent} \end{matrix} + \begin{matrix} \text{Interest cost} \\ \text{to subsidiary} \end{matrix} - \begin{matrix} \text{Tax gain on} \\ \text{exchange loss} \end{matrix} = \begin{matrix} \text{Effective} \\ \text{cost of} \\ \text{back-to-back loan} \end{matrix}$$

$$0.10(0.66) - 0.08(0.66) + 0.09(0.6) - 0.40(0.11) = 2.32\%$$

Variations on the back-to-back loan include the parent depositing dollars while the bank lends out local currency, or a foreign affiliate placing the deposit in any of several currencies with the bank loan being denominated in the same or a different currency. To calculate the costs of these variations would require some modification to the methodology shown previously, but the underlying rationale is the same: Include all interest, tax, and currency effects that accrue to both the borrowing and the lending units and convert these costs to the home currency.

Users of the fronting technique include U.S. companies that have accumulated sizable amounts of money in "captive" insurance firms and holding companies located in low-tax nations. Rather than reinvesting this money overseas (assuming that is the intent) by first paying dividends to the parent company and incurring a large tax liability, some of these companies attempt to recycle their funds indirectly via back-to-back loans.

For example, suppose affiliate A, wholly owned and located in a tax haven, deposits $2 million for one year in a bank at 7%; the bank, in turn, lends this money to affiliate B at 9%. If we assume that there are no currency changes, and if B has an effective tax rate of 50%, then its after-tax interest expense equals $90,000 ($2,000,000 × 0.09 × 0.5). The return to A equals $140,000 ($2,000,000 × 0.07), assuming that A pays no taxes. The net result of this transaction has been to shift $140,000 of income from B to A at a cost to B of only $90,000 after tax.

Back-to-back arrangements also can be used to access blocked currency funds without physically transferring them. Suppose Xerox wishes to use the excess reais (the plural of *real*, the Brazilian currency) being generated by its Brazilian operation to finance a needed plant expansion in the Philippines, where long-term money is virtually unobtainable. Xerox prefers not to invest additional dollars in the Philippines because of the high probability of a Philippine peso devaluation. Because of stringent Brazilian exchange controls, though, this movement of reais cannot take place directly. However, Xerox may be able to use the worldwide branching facilities of an international bank as a substitute for an internal transfer. For example, suppose the Brazilian branch of Chase Manhattan Bank needs *real* deposits to continue funding its loans in a restrictive credit environment. Chase may be willing to lend Xerox long-term pesos through its branch in the Philippines in return for a *real* deposit of equivalent maturity in Brazil.

In this way, Xerox gets the use of its funds in Brazil and at the same time receives locally denominated funds in the Philippines. Protection is provided against a peso devaluation, although the firm's *real* funds are, of course, still exposed. The value of this arrangement is based on the relative interest rates involved, anticipated currency changes, and the opportunity cost of the funds being utilized. Given the exchange and credit restrictions and other market imperfections that exist, it is quite possible that both the bank and its client can benefit from this type of arrangement. Negotiation between the two parties will determine how these benefits are to be shared.

PARALLEL LOANS ■ A **parallel loan** is a method of effectively repatriating blocked funds (at least for the term of the arrangement), circumventing exchange control restrictions, avoiding a premium exchange rate for investments abroad, financing foreign affiliates without incurring additional exchange risk, or obtaining foreign currency financing at attractive rates. As shown in Exhibit 14.6, it consists of two related but separate—that is, parallel—borrowings and usually involves four parties in at least two different countries. In Exhibit 14.6a, a U.S. parent firm wishing to invest in Spain lends dollars to the U.S. affiliate of a Spanish firm that wants to invest in the United States. In return, the Spanish parent lends pesetas in Spain to the U.S. firm's Spanish subsidiary. Drawdowns, payments of interest, and repayments of principal are made simultaneously. The differential between the rates of interest on the two loans is determined, in theory, by the cost of money in each country and anticipated changes in currency values.

Exhibit 14.6b shows how a parallel loan can be used to access blocked funds. In this instance, the Brazilian affiliate of ITT is generating reais that it is unable to repatriate. It lends this money to the local affiliate of Dow Chemical; in turn, Dow lends dollars to ITT in the United States. Hence, ITT would have the use of dollars in the United States

EXHIBIT 14.6 Structure of a Parallel Loan

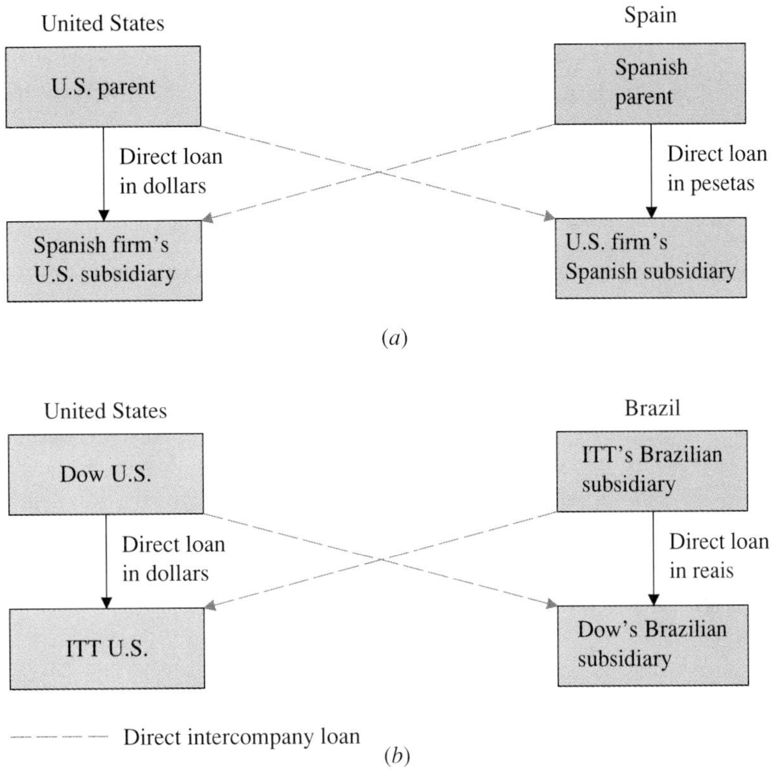

(a)

------- Direct intercompany loan

(b)

and Dow would obtain reais in Brazil. In both cases, the parallel transactions are the functional equivalent of direct intercompany loans.

Fees to banks brokering these arrangements usually run from 0.25% to 0.5% of the principal for each side. (Chapter 16 discusses currency swaps, which are an outgrowth of parallel loans.)

Dividends

Dividends are by far the most important means of transferring funds from foreign affiliates to the parent company, typically accounting for more than 50% of all remittances to U.S. firms. Among the various factors that MNCs consider when deciding on dividend payments by their affiliates are taxes, financial statement effects, exchange risk, currency controls, financing requirements, availability and cost of funds, and the parent's dividend payout ratio. Firms differ, though, in the relative importance they place on these variables, as well as in how systematically the variables are incorporated in an overall remittance policy.

The parent company's *dividend payout ratio* often plays an important role in determining the dividends to be received from abroad. Some firms require the same payout percentage as the parent's rate for each of their subsidiaries; others set a target payout rate as a percentage of overall foreign-source earnings without attempting to receive the same percentage from each subsidiary. The rationale for focusing on the parent's payout ratio is that the subsidiaries should contribute their share of the dividends paid to the stockholders. Thus, if the parent's payout rate is 60%, then foreign operations should contribute 60% of their earnings toward meeting this goal. Establishing a uniform percentage for each unit, rather than an overall target, is explained as an attempt to persuade foreign governments, particularly those of less-developed countries, that these payments are necessary rather than arbitrary.

TAX EFFECTS ■ A major consideration behind the dividend decision is the effective tax rate on payments from different affiliates. By varying payout ratios among its foreign subsidiaries, the corporation can reduce its total tax burden.

Once a firm has decided on the amount of dividends to remit from overseas, it can then reduce its tax bill by withdrawing funds from those locations with the lowest transfer costs. Here is a highly simplified example. Suppose a U.S. company, International Products, wishes to withdraw $1 million from abroad in the form of dividends. Each of its three foreign subsidiaries—located in Germany, the Republic of Ireland, and France— has earned $2 million before tax this year and, hence, all are capable of providing the funds. The problem for International Products is to decide on the dividend strategy that will minimize the firm's total tax bill.

The German subsidiary is subject to a split corporate tax rate of 50% on undistributed gross earnings and 36% on dividends, as well as a dividend withholding tax of 10%. As an export incentive, the Republic of Ireland grants a 15-year tax holiday on all export profits. Because the Irish unit receives all its profits from exports, it pays no taxes. There are no dividend withholding taxes. The French affiliate is taxed at a rate of 45% and must also pay a 10% withholding tax on its dividend remittances. It is assumed that there are no excess foreign tax credits available and that any credits generated cannot be used elsewhere. The U.S. corporate tax rate is 35%. Exhibit 14.7 summarizes the relevant tax consequences of remitting $1 million from each affiliate in turn.

These calculations indicate that it would be cheapest to remit dividends from Germany. In fact, by paying this $1 million dividend with its associated total worldwide tax cost of $1.86 million, International Products is actually reducing its worldwide tax costs by $40,000, compared with its total tax bill of $1.9 million in the absence of any dividend.[2] This result is due to the tax penalty that the German government imposes on retained earnings.

FINANCING REQUIREMENTS ■ In addition to their tax consequences, dividend payments lead to liquidity shifts. The value of moving these funds depends on the different opportunity costs of money among the various units of the corporation. For instance, an affiliate that must borrow funds will usually have a higher opportunity cost than a unit with excess cash available. Moreover, some subsidiaries will have access to low-cost financing sources, whereas others have no recourse but to borrow at a relatively high interest rate.

---

[2] In fact, paying a dividend from Germany could reduce International Products' tax bill even further. As shown in Exhibit14.7, International Products will receive a foreign tax credit of $240,351 on the German dividend. To the extent that International Products can use this credit, it can reduce its worldwide tax bill by an additional $240,351.

EXHIBIT 14.7 Tax Effects of Dividend Remittances from Abroad

Location of Foreign Affiliate	Dividend Amount	Host Country Income Tax If Dividend Paid	Host Country Withholding Tax	U.S. Income Tax	Total Taxes If Dividend Paid	Host Country Income Tax If No Dividend Paid	Worldwide Tax Liability If Dividend Paid*
Germany	$1,000,000	$360,000 500,000 $860,000	$100,000	0[1]	$ 960,000	$1,000,000	$1,860,000
Republic of Ireland	$1,000,000	0	0	$350,000	$ 350,000	0	$2,250,000
France	$1,000,000	$900,000	$100,000	0	$1,000,000	$ 900,000	$2,000,000

[1]Computation of U.S. tax owed

Profit before tax	$2,000,000
Tax ($1,000,000 × 0.50 + $1,000,000 × 0.36)	860,000
Profit after tax	$1,140,000
Dividend paid to U.S. parent company	1,000,000
Less withholding tax @ 10%	100,000
Net dividend received in United States	$ 900,000

Include in U.S. income

Gross dividend received	$1,000,000
Foreign indirect tax deemed paid[2]	754,386
U.S. gross dividend included	$1,754,386
U.S. tax @ 35%	614,035
Less foreign tax credit[2]	854,386
Net U.S. tax cost (credit)	$−240,351
U.S. tax payable	0

[2]Computation of indirect and total foreign tax credit

(a) Direct credit for withholding tax	$100,000
(b) Indirect foreign tax credit	

$$\frac{\text{Dividend paid}}{\text{Profit after tax}} \times \text{Foreign tax} = \frac{1,000,000}{1,140,000} \times 860,000 = 754,386$$

Total tax credit	$854,386

*Worldwide tax liability if dividend paid equals tax liability for foreign affiliate paying the dividend plus tax liabilities of non-dividend-paying affiliates plus any U.S. taxes owed.

All else being equal, a parent can increase its value by exploiting yield differences among its affiliates—that is, setting a high dividend payout rate for subsidiaries with relatively low opportunity costs of funds while requiring smaller dividend payments from those units facing high borrowing costs or having favorable investment opportunities.

EXCHANGE CONTROLS ■ Exchange controls are another major factor in the dividend decision. Nations with balance-of-payments problems are apt to restrict the payment of dividends to foreign companies. These controls vary by country, but in general they limit the size of dividend remittances, either in absolute terms or as a percentage of earnings, equity, or registered capital.

Many firms try to reduce the danger of such interference by maintaining a record of consistent dividends. The record is designed to show that these payments are part of an established financial program, rather than an act of speculation against the host country's currency. Dividends are paid every year, whether they are justified by financial and tax considerations or not, just to demonstrate a continuing policy to the local government and central bank. Even when they cannot be remitted, dividends are sometimes declared for the same reason, namely, to establish grounds for making future payments when these controls are lifted or modified.

Some companies even set a uniform dividend payout ratio throughout the corporate system to set a global pattern and maintain the principle that affiliates have an obligation to pay dividends to their stockholders. If challenged, the firm then can prove that its French or Brazilian or Italian subsidiaries must pay an equivalent percentage dividend. MNCs often are willing to accept higher tax costs to maintain the principle that dividends are a necessary and legitimate business expense. Many executives believe that a record of paying dividends consistently (or at least declaring them) helps in getting approval for further dividend disbursements.

JOINT VENTURES ■ The presence of local stockholders may constrain an MNC's ability to adjust its dividend policy in accordance with global factors. In addition, to the extent that multinationals have a longer-term perspective than their local partners, conflicts might arise, with local investors demanding a shorter payback period and the MNC insisting on a higher earnings-retention rate.

Equity versus Debt

Corporate funds invested overseas, whether they are called debt or equity, require the same rate of return, namely, the firm's marginal cost of capital. Nonetheless, MNCs generally prefer to invest in the form of loans rather than equity for several reasons.

First, a firm typically has wider latitude to repatriate funds in the form of interest and loan repayments than as dividends or reductions in equity because the latter fund flows are usually more closely controlled by governments. In addition, a reduction in equity may be frowned on by the host government and is likely to pose a real problem for a firm trying to repatriate funds in excess of earnings. Withdrawing these funds by way of dividend payments will reduce the affiliate's capital stock, whereas applying this money toward repayment of a loan will not affect the unit's equity account. Moreover, if the firm ever desired to increase its equity investment, it could relatively easily convert the loan into equity.

A second reason for the use of intercompany loans over equity investments is the possibility of reducing taxes. The likelihood of a tax benefit is due to two factors: (1) Interest paid on a loan is ordinarily tax-deductible in the host nation, whereas dividend payments are not; and (2) unlike dividends, loan principal repayments do not normally constitute taxable income to the parent company.

For example, suppose General Foods Corporation (GFC) is looking for a way to finance a $1 million expansion in working capital for its Danish affiliate, General Foods Denmark (GFD). The added sales generated by this increase in working capital promise to yield 20% after local tax (but before interest payments), or $200,000 annually, for the foreseeable future. GFC has a choice between investing the $1 million as debt, with an interest rate of 10%, or as equity. GFD pays corporate income tax at the rate of 50% as well as a 10% withholding tax on all dividend and interest payments. Other assumptions are that the parent expects all available funds to be repatriated and that any foreign tax credits generated are unusable.

If the $1 million is financed as an equity investment, the Danish subsidiary will pay the full return as an annual dividend to GFC of $200,000, of which GFC will receive $180,000 net of the withholding tax. Alternatively, if structured as a loan, the investment will be repaid in 10 annual installments of $100,000 each with interest on the remaining balance. Interest is tax-deductible, so the net outflow of cash from GFD is only half the interest payment. It is assumed that the parent does not have to pay additional tax to the United States on dividends received because of the high tax rate on GFD's income. In addition, the interest is received tax-free because of the availability of excess foreign tax credits. All funds remaining after interest and principal repayments are remitted as dividends. In year 5, for example, $100,000 of the $200,000 cash flow is sent to GFC as a loan repayment and $60,000 as interest (on a balance of $600,000). The $30,000 tax saving on the interest payment and the remaining $40,000 on the $200,000 profit are remitted as dividends. Hence, GFC winds up with $217,000 after the withholding tax of $13,000 (on total dividend plus interest payments of $130,000).

The evaluation of these financing alternatives is presented in Exhibit 14.8. If we assume a 15% discount rate, the present value of cash flows under the debt financing plan is $1,102,695. This amount is $199,275 more over the first 10 years of the investment's life than the $903,420 present value using equity financing. The reason for this disparity is the absence of withholding tax on the loan repayments and the tax deductibility of interest expenses. It is apparent that the higher the interest rate that can be charged on the loan, the larger the cash flows will be. After 10 years, of course, the cash flows are the same under debt and equity financing because all returns will be in the form of dividends.

Alternatively, suppose the same investment and financing plans are available in a country having no income or withholding taxes. This situation increases annual project returns to $400,000. Because no excess foreign tax credits are available, the U.S. government will impose corporate tax at the rate of 35% on all remitted dividends and interest payments. The respective cash flows are presented in Exhibit 14.9. In this situation, the present value under debt financing is $175,657 more (1,480,537 − 1,304,880) than with equity financing.

Firms do not have complete latitude in choosing their debt-to-equity ratios abroad, though. This subject is frequently open for negotiation with the host governments. In addition, dividends and local borrowings often are restricted to a fixed percentage of

EXHIBIT 14.8 Dollar Cash Flows under Debt and Equity Financing

		DEBT					EQUITY	
(1)	(2)	(3)	(4)	(5) Cash Flow to Parent	(6)	(7)	(8) Cash Flow to Parent
Principal Year Repayment	Interest	Dividend	Withholding Tax	(1 + 2 + 3 − 4)	Dividend	Withholding Tax	(6 − 7)
1 100,000	100,000	50,000	15,000	235,000	200,000	20,000	180,000
2 100,000	90,000	55,000	14,500	230,500	200,000	20,000	180,000
3 100,000	80,000	60,000	14,000	226,000	200,000	20,000	180,000
4 100,000	70,000	65,000	13,500	221,500	200,000	20,000	180,000
5 100,000	60,000	70,000	13,000	217,000	200,000	20,000	180,000
6 100,000	50,000	75,000	12,500	212,500	200,000	20,000	180,000
7 100,000	40,000	80,000	12,000	208,000	200,000	20,000	180,000
8 100,000	30,000	85,000	11,500	203,500	200,000	20,000	180,000
9 100,000	20,000	90,000	11,000	199,000	200,000	20,000	180,000
10 100,000	10,000	95,000	10,500	194,500	200,000	20,000	180,000
Present Value Discounted at 15%				$1,102,695			$903,420

equity. A small equity base also can lead to a high return on equity, exposing a company to charges of exploitation.

Another obstacle to taking complete advantage of parent company loans is the U.S. government. The IRS may treat loan repayments as constructive dividends and tax them if it believes the subsidiary is too thinly capitalized. Many executives and tax attorneys feel that the IRS is satisfied as long as the debt-to-equity ratio does not exceed 4 to 1.

Firms normally use guidelines such as 50% of total assets or fixed assets in determining the amount of equity to provide their subsidiaries. These guidelines usually lead to an equity position greater than that required by law, causing MNCs to sacrifice flexibility and pay higher taxes than necessary.

EXHIBIT 14.9 Dollar Cash Flows under Debt and Equity Financing

		DEBT					EQUITY	
(1)	(2)	(3)	(4)	(5) Cash Flow to Parent	(6)	(7)	(8) Cash Flow to Parent
Principal Year Repayment	Interest	Dividend	U.S. Tax (2 + 3) × 0.35	(1 + 2 + 3 − 4)	Dividend	U.S. Tax (6 × 0.35)	(6 − 7)
1 100,000	100,000	200,000	105,000	295,000	400,000	140,000	260,000
2 100,000	90,000	210,000	105,000	295,000	400,000	140,000	260,000
3 100,000	80,000	220,000	105,000	295,000	400,000	140,000	260,000
4 100,000	70,000	230,000	105,000	295,000	400,000	140,000	260,000
5 100,000	60,000	240,000	105,000	295,000	400,000	140,000	260,000
6 100,000	50,000	250,000	105,000	295,000	400,000	140,000	260,000
7 100,000	40,000	260,000	105,000	295,000	400,000	140,000	260,000
8 100,000	30,000	270,000	105,000	295,000	400,000	140,000	260,000
9 100,000	20,000	280,000	105,000	295,000	400,000	140,000	260,000
10 100,000	10,000	290,000	105,000	295,000	400,000	140,000	260,000
Present Value Discounted at 15%				$1,480,537			$1,304,880

The IRS also has gone after foreign companies operating in the United States that it accuses of "earnings stripping." Earnings stripping occurs when a foreign company uses parent loans instead of equity capital to fund its U.S. activities. The resulting interest payments, particularly on loans used to finance acquisitions in the United States, are deducted from pre-tax income, leaving minimal earnings to tax.

Congress tried to control earnings stripping with a law that took effect in 1989. Under this law, if a heavily indebted U.S. unit pays interest to its foreign parent, the interest deduction is limited to 50% of the unit's taxable income. But that still left a large loophole. The foreign-owned companies continued to take big interest deductions by arranging loans through banks rather than through their parent companies. To reassure the banks, the parent companies provided guarantees.

A law that took effect in 1994 attempts to close this loophole by treating guaranteed loans as if they were loans from the parent company. U.S. officials expected that foreign companies would convert their loans to their U.S. units into equity, thus generating taxable dividends. But most companies seem to have found new ways to avoid taxes. Some sold assets and leased them back, replacing loans with leases. Others factored accounts receivable or turned mortgages into securities. The net result so far has been few new taxes collected.

Invoicing Intercompany Transactions

Firms often have the option of selecting the currencies in which to invoice interaffiliate transactions. The choice of invoicing currency has both tax and currency-control implications.

TAX EFFECTS ■ The particular currency or currencies in which intercompany transactions are invoiced can affect after-tax profits if currency fluctuations are anticipated. For example, suppose a firm's Swedish subsidiary is selling subassemblies to a German affiliate. Assume that the firm's effective tax rate in Sweden is t_S, and in Germany it is t_G. Should the transaction be invoiced in Deutsche marks or kronor, if the Deutsche mark is expected to rise with respect to the krona and dollar?

If the invoice is in Deutsche marks, the German subsidiary will be unaffected (in terms of DM), but the Swedish subsidiary will show a foreign exchange gain. It will retain $1 - t_S$ of the gain after Swedish taxes, which is directly consolidated into the U.S. parent's account if the dollar:krona rate is assumed to remain unchanged.

If the invoice is in kronor, the Swedish subsidiary will be unaffected, but the German subsidiary will use fewer Deutsche marks to pay the invoice. After making more German profit, it will pay more tax, so it will retain $1 - t_G$ of its foreign exchange gain denominated in dollars. Therefore, to reduce taxes, if t_G is greater than t_S, invoice in Deutsche marks; otherwise invoice in kronor. A numerical example is provided in Exhibit 14.10.

Now suppose the subassemblies are flowing from Germany to Sweden (or from Sweden to Germany but with prepayment). If the invoice is in kronor, the Swedish subsidiary will be unaffected, but the German subsidiary will receive fewer Deutsche marks. On an after-tax basis, though, its dollar loss is only $1 - t_G$ of the original foreign exchange loss. Had the invoice been in Deutsche marks, the Swedish subsidiary would have incurred a loss after tax equal to just $1 - t_S$ of the before-tax loss. To minimize taxes, therefore, if t_G is

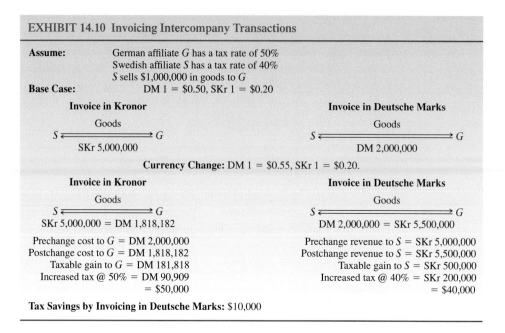

EXHIBIT 14.10 Invoicing Intercompany Transactions

Assume: German affiliate G has a tax rate of 50%
Swedish affiliate S has a tax rate of 40%
S sells $1,000,000 in goods to G
Base Case: DM 1 = $0.50, SKr 1 = $0.20

Invoice in Kronor	**Invoice in Deutsche Marks**
Goods	Goods
$S \rightleftharpoons G$	$S \rightleftharpoons G$
SKr 5,000,000	DM 2,000,000

Currency Change: DM 1 = $0.55, SKr 1 = $0.20.

Invoice in Kronor	**Invoice in Deutsche Marks**
Goods	Goods
$S \rightleftharpoons G$	$S \rightleftharpoons G$
SKr 5,000,000 = DM 1,818,182	DM 2,000,000 = SKr 5,500,000

Prechange cost to G = DM 2,000,000	Prechange revenue to S = SKr 5,000,000
Postchange cost to G = DM 1,818,182	Postchange revenue to S = SKr 5,500,000
Taxable gain to G = DM 181,818	Taxable gain to S = SKr 500,000
Increased tax @ 50% = DM 90,909	Increased tax @ 40% = SKr 200,000
= $50,000	= $40,000

Tax Savings by Invoicing in Deutsche Marks: $10,000

greater than t_S, invoice in kronor; otherwise, invoice in Deutsche marks. The general rule is to denominate intercompany transactions so that gains are taken in low-tax nations and losses in high-tax ones. Note finally that on a pre-tax consolidated basis, the parent's books would show a net gain of zero from these transactions (i.e., intercompany transactions net out on a before-tax basis but not after tax, unless effective tax rates are equal).

EXCHANGE CONTROLS ■ The choice of invoicing currency also can enable a firm to remove some blocked funds from a country that has currency controls. Suppose a subsidiary is located in a country that restricts profit repatriation. A forecasted local currency devaluation can provide this firm with an opportunity to shift excess funds to where they will earn a higher rate of return. This shift can be accomplished by invoicing exports from that subsidiary to the rest of the corporation in the local currency at a contracted price. As the local currency deteriorates, profit margins are squeezed in the subsidiary (compared with what they would have been with hard-currency billing) but improved elsewhere in the system. In effect, cost savings from the devaluation will be shifted elsewhere in the system. If that subsidiary were exporting $1 million worth of goods monthly to its parent, for example, then a 10% LC devaluation would involve a monthly shift of $100,000 to the parent.

Conversely, funds can be moved into a country, such as Germany, that imposes controls on capital inflows. They can be moved by invoicing exports to the German affiliate in a weak currency and invoicing exports from the German affiliate in a strong currency.

14.3 DESIGNING A GLOBAL REMITTANCE POLICY

The task facing international financial executives is to coordinate the use of the various financial linkages in a manner consistent with value maximization for the firm as a whole. This task requires the following four interrelated decisions: (1) how much money (if any) to remit; (2) when to do so; (3) where to transmit these funds; and (4) which transfer method(s) to use.

In order to take proper advantage of its internal financial system, the firm must conduct a systematic and comprehensive analysis of the available remittance options and their associated costs and benefits. It also must compare the value of deploying funds in affiliates other than just the remitting subsidiary and the parent. For example, rather than simply deciding whether to keep earnings in Germany or remit them to the U.S. parent, corporate headquarters must consider the possibility and desirability of moving those funds to, say, Italy or France via leading and lagging or transfer price adjustments. In other words, the key question to be answered is: Where and how in the world can available funds be deployed most profitably? Most multinationals, however, make their dividend remittance decision independently of, say, their royalty or leading and lagging decision, rather than considering what mix of transfer mechanisms would be best for the company overall.

In part, the decision to "satisfice" rather than optimize is due to the complex nature of the financial linkages in a typical multinational corporation. For instance, if there are 10 financial links connecting each pair of units in a multinational enterprise, then a firm consisting of a parent and two subsidiaries will have 30 intercompany links, three times as many as a parent with just one affiliate. A parent with three subsidiaries will have 60 links; a company with n units will have $10n(n + 1)/2$ financial linkages.

A real-life firm will have many more than three affiliates, so the exponential growth of potential intercompany relationships means that, unless the options are severely limited, system optimization will be impossible. It is not surprising, therefore, that surveys by David Zenoff and by Sidney Robbins and Robert Stobaugh found that few firms seemed to think in terms of a worldwide pool of funds to be allocated in accordance with global profit maximization.[3] Instead, most parents allowed their affiliates to keep just enough cash on hand to meet their fund requirements and required them to send the rest back home.

This limited approach to managing international financial transactions is understandable in view of the tangled web of interaffiliate connections that already has been depicted. Still, compromising with complexity ought not to mean ignoring the system's profit potential. A hands-off policy is not the only alternative to system optimization. Instead, the MNC should search for relatively high-yield uses of its internal financial system. This task often is made easier by choices that are generally more limited in practice than in theory.

First of all, many of the potential links will be impossible to use because of government regulations and the specifics of the firm's situation. For example, two affiliates might not trade with each other, eliminating the transfer pricing link. Other channels will be severely restricted by government controls.

---

[3] David B. Zenoff, "Remitting Funds from Foreign Affiliates," *Financial Executive*, March 1968, pp. 46–63; and Sidney M. Robbins and Robert B. Stobaugh, *Money in the Multinational Enterprise* (New York: Basic Books, 1973), p. 86.

Furthermore, in many situations, it is not necessary to develop an elaborate mathematical model to figure out the appropriate policy. For example, where a currency is blocked and investment opportunities are lacking or local tax rates are quite high, it normally will be in the company's best interest to shift its funds and profits elsewhere. Where credit rationing exists, a simple decision rule usually suffices: Maximize local borrowing. Moreover, most MNCs already have large staffs for data collection and planning, as well as some form of computerized accounting system. These elements can form the basis for a more complete overseas-planning effort.

The more limited, although still numerous, real-life options facing a firm and the existing nucleus of a planning system can significantly reduce the costs of centralizing the management of a firm's intercompany transactions. In addition, for most multinationals, fewer than 10 affiliates account for an overwhelming majority of intercompany fund flows. Recognizing this situation, several firms have developed systems to optimize flows among this limited number of units. The lack of global optimization (interactions with other affiliates are taken as given, rather than treated as decision variables) is not particularly costly because most of the major fund flows are already included. Realistically, the objective of such an effort should be profit improvement rather than system optimization.

Prerequisites

A number of factors strongly impact on an MNC's ability to benefit from its internal financial transfer system. These include the (1) number of financial links, (2) volume of interaffiliate transactions, (3) foreign-affiliate ownership pattern, (4) degree of product and service standardization, and (5) government regulations.

Because each channel has different costs and benefits associated with its use, the wider the range of choice, the greater a firm's ability to achieve specific goals. For example, some links are best suited to avoiding exchange controls, and others are most efficiently employed in reducing taxes. In this vein, a two-way flow of funds will give a firm greater flexibility in deploying its money than if all links are in only one direction. Of course, the larger the volume of flows through these financial arteries, the more funds that can be moved by a given adjustment in intercompany terms. A 1% change in the transfer price on goods flowing between two affiliates will have a 10 times greater absolute effect if annual sales are $10 million rather than $1 million. Similarly, altering credit terms by 30 days will become a more effective means of transferring funds as the volume of intercompany payables and receivables grows.

A large volume of intercompany transactions is usually associated with the worldwide dispersal and rationalization of production activities. As plants specialize in different components and stages of production, interaffiliate trade increases, as do the accompanying financial flows. Clearly, 100% ownership of all foreign affiliates removes a major impediment to the efficient allocation of funds worldwide. The existence of joint ventures is likely to confine a firm's transfer activities to a set of mutually agreed-upon rules, eliminating its ability to react swiftly to changed circumstances.

Also, the more standardized its products and services are, the less latitude an MNC has to adjust its transfer prices and fees and royalties. Conversely, a high-technology input, strong product differentiation, and short product life cycle enhance a company's ability to make use of its mechanisms for transfer pricing and fee adjustments. The latter situation is more typical of the MNC, so it is not surprising that the issue of transfer pricing is a bone of contention between multinationals and governments.

Last, and most important, government regulations exert a continuing influence on international financial transactions. It is interesting to consider that government tax, credit allocation, and exchange control policies provide the principal incentives for firms to engage in international fund maneuvers at the same time that government regulations most impede these flows.

Information Requirements

In order to take full advantage of its global financial system, a multinational firm needs detailed information on affiliate financing requirements, sources and costs of external credit, local investment yields, available financial channels, the volume of interaffiliate transactions, all relevant tax factors, and government restrictions and regulations on fund flows.

Without belaboring the points already made, it is clear that the costs and benefits of operating an integrated financial system depend on the funds and transfer options available, as well as on the opportunity costs of money for different affiliates and the tax effects associated with these transfer mechanisms. Hence, the implementation of centralized decision making requires information concerning all these factors.

Behavioral Consequences

Manipulating transfer prices on goods and services, adjusting dividend payments, and leading and lagging remittances lead to a reallocation of profits and liquidity among a firm's affiliates. Although the aim of this corporate intervention is to boost after-tax global profits, the actual result may be to destroy incentive systems based on profit centers and to cause confusion and computational chaos. Subsidiaries may rebel when asked to undertake actions that will benefit the corporation as a whole but will adversely affect their own performance evaluations. To counter this reaction, a firm must clearly spell out the rules, and adjust profit center results to reflect true affiliate earnings rather than the distorted remnants of a global profit-maximizing exercise.

14.4 ILLUSTRATION: TRANSFER PRICING AND TAX EVASION

On September 19, 1983, the Swiss-based commodities trading firm Marc Rich & Co. AG (the same one mentioned in Chapter 12 under countertrade), its U.S. unit, and its two principal officers, Marc Rich and Pincus Green, were indicted by the U.S. government for allegedly evading more than $100 million in U.S. taxes, making it the biggest tax-evasion case in history. The U.S. government charged that Marc Rich, his companies, and Pincus Green had the U.S. unit transfer profit to the Swiss parent by having the U.S. affiliate pay the Swiss company artificially high prices for oil.

In 1982, the United States subpoenaed from the Swiss parent documents that it thought would buttress its case—and that would make public a great deal of information about the company. Despite its size—annual revenue exceeding $10 billion—Marc Rich & Co. has a penchant for secrecy. Because of its refusal to give the documents to a grand jury, Marc Rich was cited in contempt of court and subject to a $50,000-a-day fine while it appealed a federal judge's refusal to vacate the contempt order. In September 1983, Marc Rich's internal documents were seized by the Swiss government on the ground that releasing them to U.S. authorities would violate Swiss secrecy laws.

Marc Rich settled with the U.S. government in October 1984. The back taxes plus interest, penalties, fines, and seized assets made the settlement worth almost $200 million—the most ever recovered in a criminal tax-evasion case.

Except for its magnitude, however, the Rich case is not unique. Since the late 1960s, the Department of Justice has been cracking down on the use of transfer pricing to evade U.S. taxes. One case was U.S. Gypsum Co. Strange things were happening to the price of gypsum rock that the company mined in Canada and shipped to the United States. The rock was sold by the company's Canadian unit at a low price, keeping Canadian profit and taxes down, and was resold to the U.S. unit at a high price, keeping U.S. profit and taxes down. The profit was siphoned into another U.S. Gypsum subsidiary that owned the rock only while it fell through the air from the Canadian conveyor belt down to the hold of a U.S. ship. This intermediate subsidiary was a paper company and was in a low tax bracket. The Department of Justice challenged this arrangement in court and won in a civil case.

In June 1983, the American sales subsidiary of Toyota Motor Co. was ordered by a federal judge to turn over to the Internal Revenue Service information about the prices its parent firm charged its car dealers in Japan. The IRS maintained that the Japanese data were necessary for it to determine whether the transfer prices it charged Toyota Motor Sales U.S.A. for Toyota products in the United States were being used to reduce its U.S. tax liability. The IRS claimed that the U.S. sales unit of Toyota trimmed its taxable U.S. income by paying its Japanese parent higher-than-reasonable prices for the vehicles sold in the United States. This case and a similar one brought against Nissan Motor Co. were settled in late 1987 when the U.S. sales units of both companies paid undisclosed amounts of additional income taxes to the IRS. Reportedly, the combined payments exceeded $600 million. These taxes were offset by tax refunds of an equivalent (though lesser) amount received from Japan's national tax agency for taxes paid on the income previously recorded in Japan but now reallocated to the United States. Currently, the IRS has cases pending against a number of Japanese and South Korean companies, including Hitachi, Mitsubishi Electric, Tokai Bank, and Daewoo.

Sometimes it is not the U.S. government that feels cheated. Amway of Canada and its U.S. parent, Amway Corp., were fined $25 million in November 1983 after pleading guilty in Ontario Supreme Court to using a complex transfer-pricing scheme to undervalue goods they were exporting to Canada, defrauding the Canadian government of more than $28 million in customs duties and sales tax.

❧ 14.5 SUMMARY AND CONCLUSIONS

This chapter examined a variety of fund-shifting mechanisms. Corporate objectives associated with the use of these techniques include financing foreign operations, reducing interest costs, reducing tax costs, and removing blocked funds.

It is apparent after examining these goals that there are trade-offs involved. For instance, removing blocked funds from a low-tax nation is likely to raise the firm's worldwide tax bill. Similarly, reducing exchange risk often results in higher interest expenses and adds to the financing needs of affiliates in soft-currency nations. The realistic weight that should be assigned to each of these goals depends on the individual impact of each goal on corporate profitability. Focusing on just one or two of these goals, such as avoiding exchange risk or minimizing taxes, to the exclusion of all others will probably lead to suboptimal decisions.

The recommended global approach to managing fund transfers is best illustrated by the creative use of financial linkages, whereby one unit becomes a conduit for the movement of funds elsewhere. For example, requiring affiliate A to remit dividends to its parent while financing this withdrawal by lowering transfer prices on goods sold to affiliate A by affiliate B will reduce income taxes and/or customs duties in the process. Or cash can be shifted from A to B through leading and lagging, with these same funds moved on to affiliate C by adjusting royalties or repaying a loan. Taking advantage of being multinational means remitting funds to the parent and other affiliates via royalties and licensing fees from some countries, dividend payments from other nations, and loan repayments from still others. All these maneuvers are to be coordinated with an eye toward maximizing corporate benefits.

It is apparent that the major benefit expected from engaging in these maneuvers comes from government actions that distort the risk-return trade-offs associated with borrowing or lending in different currencies or that alter after-tax returns because of tax asymmetries. The fact that a particular action is legal and profitable, however, does not necessarily mean it should be undertaken. When devising currency, credit, and tax regulations, governments obviously have other goals in mind besides creating profitable arbitrage opportunities for multinational firms. A company that consistently attempts to apply a "sharp pencil" and take maximum advantage of these arbitrage opportunities may optimize short-run profits, but this "penciling" is likely to be done at the expense of long-run profits.

The notion of being a good corporate citizen may be an amorphous concept, but firms that are perceived as being short-run profit-oriented may face questions regarding their legitimacy. More and more, multinationals are dependent on the goodwill of home and host governments, and actions that undermine this key factor may reduce the viability of their foreign operations.

Thus, it may well be worthwhile to pass up opportunities to make higher profits today if those profits are gained at the expense of the corporation's long-run international existence. As in all business decisions, of course, it is important to evaluate the costs and benefits associated with particular actions. Such evaluation has been the goal of this chapter in the area of intercompany fund flows.

➤Questions

1. In what aspect of an MNC's multinational financial system does its value reside?

2. California, like several other states, applies the unitary method of taxation to firms doing business within the state. Under the unitary method a state determines the tax on a company's worldwide profit

through a formula based on the share of the company's sales, assets, and payroll falling within the state. In California's case, the share of worldwide profit taxed is calculated as the average of these three factors.

a. What are the predictable corporate responses to the unitary tax?

b. What economic motives might help explain why Oregon, Florida, and several other states have eliminated their unitary tax schemes?

3. Under what circumstances is leading and lagging likely to be of most value?

4. What are the principal advantages of investing in foreign affiliates in the form of debt instead of equity?

5. In comparisons of a multinational firm's reported foreign profits with domestic profits, caution must be exercised. This same caution must also be applied when analyzing the reported profits of the firm's various subsidiaries. Only coincidentally will these reported profits correspond to actual profits.

a. Describe five different means that MNCs use to manipulate reported profitability among their various units.

b. What adjustments to its reported figures would be required to compute the true profitability of a

firm's foreign operations so as to account for these distortions?

c. Describe at least three reasons that might explain some of these manipulations.

6. In 1987, U.S.-controlled companies earned an average 2.09% return on assets, nearly four times their foreign-controlled counterparts. A number of American politicians have used these figures to argue that there is widespread tax cheating by foreign-owned multinationals.

a. What are some economically plausible reasons (other than tax evasion) that would explain the low rates of return earned by foreign-owned companies in the United States? Consider the consequences of the debt-financed U.S.-investment binge that foreign companies went on during the 1980s and the dramatic depreciation of the U.S. dollar beginning in 1985.

b. What are some of the mechanisms that foreign-owned companies can use to reduce their tax burden in the United States?

c. The corporate tax rate in Japan is 60%, whereas it is 35% in the United States. Are these figures consistent with the argument that big Japanese companies are overcharging their U.S. subsidiaries in order to avoid taxes? Explain.

➤Problems

1. Suppose Navistar's Canadian subsidiary sells 1,500 trucks monthly to the French affiliate at a transfer price of $27,000 per unit. Assume that the Canadian and French marginal tax rates on corporate income equal 45% and 50%, respectively.

a. Suppose the transfer price can be set at any level between $25,000 and $30,000. At what transfer price will corporate taxes paid be minimized? Explain.

b. Suppose the French government imposes an ad valorem tariff of 15% on imported tractors. How would this tariff affect the optimal transfer pricing strategy?

c. If the transfer price of $27,000 is set in French francs and the French franc revalues by 5%, what will happen to the firm's overall tax bill? Consider the tax consequences both with and without the 15% tariff.

d. Suppose the transfer price is increased from $27,000 to $30,000 and credit terms are extended from 90 days to 180 days. What are the fund-flow implications of these adjustments?

2. Suppose a U.S. parent owes $5 million to its English affiliate. The timing of this payment can be changed by up to 90 days in either direction. Assume the following effective annualized after-tax dollar borrowing and lending rates in England and the United States.

	LENDING (%)	BORROWING (%)
United States	3.2	4.0
England	3.0	3.6

a. If the U.S. parent is borrowing funds while the English affiliate has excess funds, should the parent speed up or slow down its payment to England?

b. What is the net effect of the optimal payment activities in terms of changing the units' borrowing costs and/or interest income?

3. Suppose that covered after-tax lending and borrowing rates for three units of Eastman Kodak—located in the United States, France, and Germany—are

	LENDING (%)	BORROWING (%)
United States	3.1	3.9
France	3.0	4.2
Germany	3.2	4.4

Currently, the French and German units owe $2 million and $3 million, respectively, to their U.S. parent. The German unit also has $1 million in payables outstanding to its French affiliate. The timing of these payments can be changed by up to 90 days in either direction. Assume that Kodak U.S. is borrowing funds while both the French and German subsidiaries have excess cash available.

a. What is Kodak's optimal leading and lagging strategy?

b. What is the net profit impact of these adjustments?

c. How will Kodak's optimal strategy and associated benefits change if the U.S. parent has excess cash available?

4. Suppose that in the section titled Dividends, International Products has $500,000 in excess foreign tax credits available. How will this situation affect its dividend remittance decision?

5. Suppose affiliate A sells 10,000 chips monthly to affiliate B at a unit price of $15. Affiliate A's tax rate is 45%, and affiliate B's tax rate is 55%. In addition, affiliate B must pay an ad valorem tariff of 12% on its imports. If the transfer price on chips can be set anywhere between $11 and $18, how much can the total monthly cash flow of A and B be increased by switching to the optimal transfer price?

6. Suppose GM France sells goods worth $2 million monthly to GM Denmark on 60-day credit terms. A switch in credit terms to 90 days will involve a one-time shift of how much money between the two affiliates?

7. Suppose that DMR SA, located in Switzerland, sells $1 million worth of goods monthly to its affiliate DMR Gmbh, located in Germany. These sales are based on a unit transfer price of $100. Suppose the transfer price is raised to $130 at the same time that credit terms are lengthened from the current 30 days to 60 days.

a. What is the net impact on cash flow for the first 90 days? Assume that the new credit terms apply only to new sales already booked but uncollected.

b. Assume that the tax rate is 25% in Switzerland and 50% in Germany and that revenues are taxed and costs deducted upon sale or purchase of goods, not upon collection. What is the impact on after-tax cash flows for the first 90 days?

8. Suppose a firm earns $1 million before tax in Spain. It pays Spanish tax of $0.52 million and remits the remaining $0.48 million as a dividend to its U.S. parent. Under current U.S. tax law, how much U.S. tax will the parent owe on this dividend?

9. Suppose a French affiliate repatriates as dividends all the after-tax profits it earns. If the French income tax rate is 50% and the dividend withholding tax is 10%, what is the effective tax rate on the French affiliate's before-tax profits, from the standpoint of its U.S. parent?

10.*Merck Mexicana SA, the wholly owned affiliate of the U.S. pharmaceutical firm, is considering alternative financing packages for its increased working-capital needs resulting from growing market penetration. Ps 250 million are needed over the next six months and can be financed as follows:

■ From the Mexican banking system at the semiannual rate of 50%

■ From the U.S. parent company at the semiannual rate of 6%

The parent company loan would be denominated in dollars and would have to be repaid through the floating-exchange-rate tier of the Mexican exchange market. The exchange loss would, thus, be fully incurred by the Mexican subsidiary. The exchange rate as of March 1, 1984, was Ps 250 = $1 and was widely expected to depreciate further.

a. If interest payments can be made through the stabilized tier of the Mexican exchange market where the dollar is worth Ps 125, what is the breakeven exchange rate on the floating tier that would make Merck Mexicana indifferent between dollar and peso financing?

b. Merck Mexicana imports from its U.S. parent $500,000 worth of chemical compounds monthly, payable on a 90-day basis. Suppose that the parent adjusts its transfer prices so that Merck Mexicana must now pay $700,000 monthly for its chemical supplies. All payments for imports of chemicals involved in the manufacture of pharmaceuticals are transacted through the stabilized tier of the exchange market. At the current exchange rate of Ps 250 = $1, what is the net before-tax annual benefit to Merck of this transfer price increase?

---

*Contributed by Laurent Jacque.

11. A well-known U.S. firm has a reinvoicing center (RC) located in Geneva. The reinvoicing center handles an annual sales volume of $1.2 billion–$700 million in interaffiliate sales and the rest in third-party sales. The RC buys goods manufactured by the parent company or other subsidiaries and reinvoices the product to other affiliates or third parties. Many of these trades are with "low-volume, highly complex countries." When buying the goods, the RC takes title to them, but it does not take actual possession of the goods. The RC pays the selling company in its own currency and receives payment from the purchasing company in its own currency. What benefits can such a center provide?

APPENDIX 14A Managing Blocked Currency Funds

A common policy of host governments facing balance-of-payments difficulties is to impose exchange controls that block the transfer of funds to nonresidents. As this chapter makes clear, the principal target of many of these controls is the multinational corporation with local operations. There are numerous types of currency controls, some more ingenious than others, but all with the goal of allocating foreign exchange via nonprice means. Often these exchange rate restrictions go hand in hand with substantial deviations from purchasing power parity and the international Fisher effect. Thus, currency controls are a major source of market imperfection, posing opportunities as well as risks for the multinational firm.

The purpose of this appendix is to identify and evaluate the major strategies, and associated tactics, that MNCs use to cope with actual, as well as potential, restrictions.

The management of blocked funds can be considered a three-stage process: (1) preinvestment planning, including analyzing the effects of currency controls on investment returns and structuring the operation so as to maximize the company's ability to access its funds; (2) developing a coordinated approach to repatriating blocked funds from an ongoing operation; and (3) maintaining the value of those funds that, despite all efforts, cannot be removed.

PREINVESTMENT PLANNING

To formulate an effective management plan, the MNC must (1) know what is at risk, and (2) devise a means to facilitate the use of blocked funds before they accumulate. These two elements form the nucleus of preinvestment planning.

In assessing the risks associated with currency controls, the firm must take into account the fact that the impact of these controls is unlikely to be uniform over the life of an investment. Rather, these effects are often favorable initially and then gradually turn unfavorable in the later stages of a project's life. For instance, a firm may be able to import capital goods at a very favorable exchange rate if this equipment is assigned a high priority. Many governments, in effect, subsidize the importation of certain "essential" products through the use of multiple exchange rates. Another advantage from exchange controls is that the company's affiliate may be able to arrange local currency financing at attractive (subsidized) rates by borrowing the blocked funds being held by subsidiaries of other multinationals. These effects of currency controls are particularly beneficial early on, when the project is a net user of funds. Later on, though, when the investment becomes a net generator of cash, or a cash cow, the imposition of remittance controls has an onerous impact. In other words, it is only when a project is throwing off excess cash that restrictions on profit repatriation are likely to be onerous. Until that time, controls may be advantageous.

Once a parent company has committed funds to another country, it has largely determined its ability to remit or utilize any resultant project cash flows. Thus, many firms have found it highly useful to structure their investments in advance, in a way that maximizes future remittance flexibility. The principal components of such a strategy include

- Establishing trading links with other units
- Charging separate fees for the use of trademarks, licenses, and other corporate services

- Employing local currency borrowing
- Utilizing special financing arrangements
- Investing parent company funds as debt rather than equity
- Negotiating special agreements with host governments

By establishing as many trading links as possible with other affiliates, multinationals can enhance their ability to repatriate funds via transfer price adjustments. Extensive intercompany transactions also allow for the leading and lagging of payments.

Charging affiliates for the use of corporate patents, licenses, trademarks, and other headquarters services has enabled many firms to continue receiving income from abroad in the form of fees and royalties, even when dividend remittances are controlled.

Companies normally borrow locally when investing in countries with currency controls, thereby reducing the amount of funds at risk. Cash flows that would otherwise remain blocked can be used to service local currency debt. Moreover, the greater the amount of local financing, the fewer profits that must be remitted to the parent company to ensure it a reasonable return on its investment. However, local credit restrictions often go hand in hand with exchange controls, requiring firms to explore alternative sources of funds. As we have previously seen, special financing arrangements such as currency swaps, back-to-back loans, and parallel loans provide an indirect means of borrowing locally. The ratio of parent company loans to parent equity can also affect its ability to withdraw funds from abroad and the cost of doing so.

Last, but not least, a company investing in a high-priority industry such as pharmaceuticals or computers may be able to bargain with the host government for authorization to repatriate a greater percentage of earnings if controls are currently in effect, or for an exemption from anticipated future controls. To be effective, these negotiations should take place before the investment.

REPATRIATING BLOCKED FUNDS

Firms operating in a country that has imposed exchange controls can transfer the funds being generated either directly as cash or else indirectly via special financial arrangements or in the form of goods and services purchased locally for use elsewhere. Direct transfer methods include

- Transfer price adjustments
- Fee and royalty charges
- Leads and lags in making payments abroad
- Dividends

Tactics for transferring funds indirectly include

- Parallel or back-to-back loans
- Purchase of commodities for transfer abroad
- Purchase of capital goods for corporatewide use
- Purchase of local services for worldwide use (e.g., engineering and architectural design services)

- Conducting corporate research and development
- Hosting corporate conventions, vacations, and so on

Most of these methods already have been discussed. Note, however, that, although paying dividends is the most-used means of repatriating earnings, it is also the most restricted. Two methods of increasing dividend payments in the face of these restrictions are becoming increasingly popular. Both involve increasing the value of the local investment base because the level of profit remittance often depends on the amount of a company's capital. One way to augment a unit's registered capital is to buy used equipment at artificially inflated values. A related technique is for an affiliate to acquire a bankrupt firm at a large discount from book value. The acquisition is then merged with the affiliate on the basis of the failed firm's original book value, thereby raising the affiliate's equity base.

One innovative use of blocked currencies is to create export equivalents by purchasing services locally, which can aid the firm in other countries. For example, an MNC with operations in Brazil might establish research and development facilities there and pay for them with blocked funds. Key research personnel would be transferred to Brazil to supplement local employees, with all salaries and expenses paid in reais. Similarly, Brazilian architectural and engineering firms can be engaged to design plants and buildings in California or Colombia, with their services paid for in reais.

A related, though more common, technique is to host conventions or business meetings in, say, Rio de Janeiro. In addition, employees can be sent on vacations in Brazil and provided with reais that would otherwise remain blocked. Similarly, employees of the firm may be asked, where possible, to fly Varig, the Brazilian national airline, with the tickets to be purchased with reais in Brazil by the local Brazilian affiliate. These activities benefit Brazil, as well as the MNC, because they help to create export-oriented jobs.

MAINTAINING THE VALUE OF BLOCKED FUNDS

Despite all efforts, a company may wind up with significant hoards of cash that cannot be repatriated. The company then has the choice of placing these funds in either long-term, illiquid investments such as new plant and equipment or else in fairly liquid, short-term assets such as local currency-denominated securities.

Notwithstanding the similarity of goals (namely, to maintain or increase the value of inconvertible funds), these short- and long-term investments are not necessarily substitutes for each other. Each form of investment is predicated on certain assumptions. Short-term placements implicitly assume at least one of the following:

1. The funds either are not necessary or yield too low a return in the current business.
2. There are no reasonable business opportunities in other fields.
3. Long-run prospects in the country are not favorable, so divestment is the best course.
4. Exchange controls are expected to be temporary.

Whatever the case may be, the firm is holding its funds in a liquid form, ready to repatriate them as soon as it is able to do so.

The situation is different with long-term investments. Here the premise is that the company intends to remain in the country and that, at some point, it will be able to repatriate the cash flows generated by its reinvested funds. Alternatively, it may be that because so much cash is available, there are not enough short-term investment possibilities, so the company is forced to seek out less-liquid repositories for its money.

BIBLIOGRAPHY

ARPAN, JEFFREY S. *International Intracorporate Pricing*. New York: Praeger, 1972.

LESSARD, DONALD R. "Transfer Prices, Taxes, and Financial Markets: Implications of International Financial Transfers within the Multinational Firm." In *The Economic Effects of Multinational Corporations*, edited by Robert G. Hawkins. Greenwich, Conn.: JAI Press, 1979.

OBERSTEINER, ERICH. "Should the Foreign Affiliate Remit Dividends or Reinvest?" *Financial Management*, Spring 1973, pp. 88–93.

ROBBINS, SIDNEY M., and ROBERT B. STOBAUGH. *Money in the Multinational Enterprise*. New York: Basic Books, 1973.

RUTENBERG, DAVID P. "Maneuvering Liquid Assets in a Multinational Company." *Management Science*, June 1970, pp. B671–684.

ZENOFF, DAVID B. "Remitting Funds from Foreign Affiliates." *Financial Executive*, March 1968, pp. 46–63.

Case Studies

CASE III.1

MOBEX INC.*

Mobex Inc., a U.S.-based firm engaged in the manufacture of instrument gauges for automobiles, was founded in 1949 by three former employees of Ford. The company began operations in Detroit, producing conventional mechanical instruments for major U.S. auto companies. The product line utilizes standard internal mechanisms with customized bezel (frontface) design to meet dashboard requirements of individual customers.

In 1971, the firm began design and testing of a new generation of instruments and gauges, utilizing custom, integrated circuits to perform functions heretofore performed by complex mechanical devices. The resulting product was cheaper to manufacture and had greater customer marketing appeal due to the digital readout feature. Mobex gained considerable competitive advantage by being the first to offer this new line in late 1975. The first of these products was installed in the 1977 luxury models.

The advent of the microprocessor marked another milestone in the evolution of Mobex. In 1983, Mobex began incorporating these so-called smart chips in a range of new products designed to continuously monitor certain performance attributes (e.g., fuel consumption and engine heat) and external conditions (e.g., temperature and altitude) and automatically adjust operating characteristics (e.g., the leanness of the fuel mixture and the timing of the spark-plug firing). These microprocessor-based products first appeared in the 1984-model cars and were an immediate success.

INTERNATIONAL OPERATIONS

Mobex set up a small manufacturing plant in West Germany in 1963 to supply Ford Europe with instrument gauges for its European models. Horst Stoffel, a native Bavarian and former employee of Beyrische Motor Werk (producers of BMW cars), had been hired to set up and manage Mobex AG (the new German subsidiary). Herr Stoffel's connections within the West German auto industry resulted in contracts with Opel and BMW by the end of 1967.

Results for 1967 showed foreign sales representing 6% of Mobex Inc.'s consolidated sales of $31 million. By 1970, as a result of Herr Stoffel's aggressive pursuit of new markets in Sweden and Switzerland (and a slump in the U.S. auto industry), Mobex AG accounted for 11% of total revenues of $43 million. In addition, productivity in the German subsidiary was higher than in the Detroit plant, causing foreign profits to represent 14% of the consolidated total.

These figures, plus informed opinions within the European auto industry, caused Mobex to commission a consulting study of the West European market. The company engaged the international project consulting group of Coopers & Lybrand, its certified public accountants, to perform the study. The group considered economic and psychological factors affecting auto sales, plus business environmental aspects, in England, France, Belgium, Spain, Switzerland, Italy, and Sweden.

The consultants identified two highly favored markets. Switzerland characterized a rapidly growing consumer market for automobiles, and it was believed that the new solid-state line of instruments Mobex offered would be in high demand. (Although Switzerland has no auto producers of its own, local dealers install many of the

*Source: This case was originally prepared by Carolyn Stevens, under the supervision of Alan C. Shapiro and Laurent Jacque. Revision and copyright © 1995 by Alan C. Shapiro.

PART THREE

gauges and instruments manufactured by Mobex as optional equipment or as part of the special-performance packages.) Mobex AG already exported the conventional line to that country. In addition, Switzerland offered an industrious workforce as well as a favorable tax environment.

Italy also presented a unique opportunity to Mobex. Labor problems in southern Italy had severely hampered deliveries by domestic suppliers to Fiat, the large Italian auto producer. Industry sources believed that the company could be induced to change suppliers to achieve a reliable delivery pattern. In addition, the promise of the new Mobex line of instruments was bound to hold high appeal for Fiat. It was further believed that Alfa Romeo would soon follow Fiat's lead, in order to obtain Mobex's high-quality product. The consultants recommended a manufacturing plant in northern Italy, where the labor force was considered to be more favorable.

Mobex had acted upon both recommendations, first locating and engaging foreign nationals with both technical and sales experience in the industry, and then providing the needed managerial assistance in start-up of operations. The Swiss subsidiary, Mobex Suisse, began operations in September 1971. A contract was signed by Fiat, leading to the start-up of operations in Italy (Mobex SpA) in February 1972. Shortly thereafter, Alfa Romeo also signed with Mobex SpA. Capital equipment for both of the subsidiaries was obtained in the United States, with local borrowing (using parent company guarantees) to finance start-up costs and to supply working capital.

Typical of practices within the local instrumentation industry, orders are placed annually for a 12-month supply, with delivery occurring every 90 days. Payment fell due 90 days after each delivery for the quantity supplied. This schedule enabled the company to engage in financial and production planning on a quarterly basis.

Growth in international operations continued to be rapid, because of Mobex's superior quality line, aggressive pricing, and strong marketing efforts by local managers. Foreign managers supplied a regular flow of information to the parent on product needs, thereby enabling the parent to maintain a centralized design and R&D effort in Detroit. Other than this centralized function, each subsidiary remained relatively autonomous. Local managers were free to hire local nationals, and parent management provided appropriate training. This policy of decentralization was maintained to encourage maximum adaptation to and penetration of local markets.

International production was rationalized in part during 1975, when it became clear that savings could be obtained by sourcing the flexible steel cable, fittings, plastic housings, and aluminum (deep-drawn) cups in Germany and assembling these for shipment to Switzerland and Italy. Savings were also possible by purchasing the silk-screen painted faceplates and clear glass covers in Switzerland for all three plants. Production of the line of instruments utilizing integrated circuits began at foreign locations in late 1978. The integrated circuits were sourced in the United States initially, but by 1982, about 80% of the circuits were being sourced in Europe. Manufacture of microprocessor-based products in Europe began in 1984. The microprocessors would have to be sourced in the United States for some time to come. Parts and subassemblies transferred among the foreign subsidiaries were priced at cost plus labor and handling, and payment was generally made upon delivery. No attempt has been made to determine if overall savings could be obtained by altering the transfer prices or payment schedules on these attractions.

INTERNATIONAL FINANCIAL PLANNING

By 1994, the coordination of international financial planning activities had become such a heavy burden on the parent's treasury staff that a special assistant treasurer in charge of international activities was hired. Kathryn Lee, formerly an assistant vice president in Northern Bank & Trust Company's international division, was selected for this position. Her role was to take charge of international financial planning activities.

Ms. Lee was introduced at the annual planning meeting in June 1994. Anticipating reluctance on the part of local managers and treasurers to relinquish control and flexibility, she scheduled an additional one-day session to highlight the need for coordination. In addition, she was sure that savings could be achieved by looking at the system as a whole rather than allowing each subsidiary to act independently. Now seemed to be an opportune time to conduct such a review because operations would be expanding owing to the growth supplied by the new products. In addition, owing to the decline in trade barriers associated with the advent of Europe 1992, Mobex intended to further rationalize its European production. The result was sure to be an increase in interaffiliate sales, in both absolute and relative terms, and, hence, greater advantages to rationalizing financial policy.

During this meeting, each of the foreign treasurers submitted projections for the relevant financing rates during the first two quarters of 1995 plus expected intersubsidiary purchases at present transfer price levels. This information is summarized in Exhibit III 1.1.

Exhibit III 1.1 also presents expected exchange rate changes against the U.S. dollar based upon weighted probability estimates. The lira was expected to depreciate 1% during each of the first two quarters in 1995, and the Deutsche mark was expected to appreciate 0.75% during the first quarter and 0.50% during the second quarter. The Swiss franc:dollar exchange rate was expected to remain constant throughout the first half of 1995.

Pro forma balance sheets, shown in Exhibit III 1.2, were prepared to highlight financing needs and exposure to exchange rate fluctuations. Interaffiliate payables and receivables were nonexistent because current payment terms were cash on delivery. Exhibit III 1.2 is expressed in U.S. dollar terms and separates out exposed and nonexposed transactions from a system standpoint. This means

EXHIBIT III 1.1 Financing and Hedging Costs (Annualized) and Amounts ($Millions)

	AFFILIATE QUARTER											
	MOBEX SPA – ITALY				MOBEX AG – GERMANY				MOBEX SUISSE – SWITZERLAND			
	1st		2nd		1st		2nd		1st		2nd	
	Limit	Cost	Limit	Cost	Limit	Cost	Limit	Cost	Limit	Cost	Limit	Cost
Overdraft	2.0	14%	2.0	15%	3.0	8%	3.0	10%	3.0	11%	3.0	12%
Two-quarter local loan	2.5	12	2.5	12	1.5	8	1.5	8	2.0	12	2.0	12
One-quarter Eurodollar loan*	2.0	10	2.0	12	3.0	10	3.0	12	2.0	10	2.0	12
Two-quarter Eurodollar loan*	2.0	10	2.0	10	3.0	10	3.0	10	2.0	10	2.0	10
Export financing	—	—	—	—	2.0	7	2.0	8	1.0	9	0.5	10
	Forward Discount				Forward Premium							
One-quarter forward premium (discount)		6%		6%		2.5%		2.5%		0%		0%
Two-quarter forward premium (discount)				10%				2%		0%		0%
Expected exchange rate changes		−1%		−1%		+0.75%		+0.50%		0%		0%

*The overall limit on Eurodollar borrowing is $5.0 million.

EXHIBIT III 1.2 Balance-Sheet Forecasts ($ Millions)

							AFFILIATE		
	MOBEX SPA – ITALY						MOBEX AG –		
	1st			2nd			1st		
	Total	Exposed	Not Exposed	Total	Exposed	Not Exposed	Total	Exposed	Not Exposed
Assets									
Cash	$1.0	$1.0	$—	$ 1.0	$1.0	$—	$ 1.5	$1.5	$—
Receivables	3.5	2.0	1.5	4.0	3.5	0.5	4.0	2.0	2.0
Intersubsidiary receivables*	—	—	—	—	—	—	—	—	—
Inventories	2.0	1.5	0.5	2.5	2.0	0.5	3.0	1.0	2.0
Plant and equipment	3.5	—	3.5	3.5	—	3.5	4.0	—	4.0
	$10.0	$4.5	$5.5	$11.0	$6.5	$4.5	$12.5	$4.5	$8.0
Liabilities									
Notes/loans	$—	$—	$—	$—	$—	$—	$—	$—	$—
Taxes payable	1.5	1.5	—	1.5	1.5	—	—	—	—
Accounts payable	0.5	0.5	—	1.0	0.5	0.5	2.0	1.5	0.5
Intersubsidiary payables*	—	—	—	—	—	—	—	—	—
Capital stock	2.5	—	2.5	2.5	—	2.5	3.0	—	3.0
Retained earnings	2.0	—	2.0	2.0	—	2.0	2.5	—	2.5
	$6.5	$1.5	$5.0	$ 7.0	$2.0	$5.0	$ 7.5	$1.5	$6.0
Financing needs	$3.5			$4.0			$5.0		
Exposure		$3.0			$4.5			$3.0	
Tax rate		$40%			40%			$45%	

* No interaffiliate credit is currently extended. However, purchase plans for the next two quarters include

A buys 0.5 from B in period 1   C buys 1.0 from B in period 1
A buys 0.5 from B in period 2   C buys 1.0 from B in period 2
A buys 0.5 from C in period 1   B buys 1.0 from C in period 1
A buys 0.5 from C in period 2   B buys 1.0 from C in period 2

Note: All intersubsidiary sales are denominated in seller's currency.

that if, contrary to current policy, Mobex SpA had an account payable denominated in Deutsche marks to Mobex AG, it would appear as a nonexposed item, even though it represented a foreign currency transaction, because it would be offset by a DM-denominated account receivable held by Mobex AG. From a system standpoint, therefore, it would not represent a foreign currency transaction exposure. Mobex's policy is to fully cover its estimated transaction exposure.

On the basis of these figures, each subsidiary treasurer had come up with a tentative plan to finance working-capital needs and cover its transaction exposure. These formulations were based on selection of the financing methods involving the lowest nominal cost and are shown in Exhibit III 1.3.

Ms. Lee suggested that savings might be achieved through a more rigorous examination of costs, considering the impact of currency changes

EXHIBIT III 1.2 Continued

QUARTER

| | GERMANY | | | MOBEX SUISSE – SWITZERLAND | | | | | |
| | 2nd | | | 1st | | | 2nd | | |
Total	Exposed	Not Exposed	Total	Exposed	Not Exposed	Total	Exposed	Not Exposed
$ 2.0	$2.0	$—	$ 1.5	$1.5	$—	$ 1.5	$1.5	$—
4.5	2.5	2.0	3.0	2.5	0.5	3.0	2.5	0.5
—	—	—	—	—	—	—	—	—
3.0	1.5	1.5	2.5	0.5	2.0	3.0	1.0	2.0
4.0	—	4.0	3.5	—	3.5	3.5	—	3.5
$13.5	$6.0	$7.5	$10.5	$4.5	$6.0	$11.0	$5.0	$6.0
$—	$—	$—	$—	$—	$—	$—	$—	$—
0.5	0.5	—	0.5	0.5	—	0.5	0.5	—
2.0	2.0	—	1.5	1.5	—	1.5	1.5	—
—	—	—	—	—	—	—	—	—
3.0	—	3.0	2.0	—	2.0	2.0	—	2.0
2.5	—	2.5	2.0	—	2.0	2.0	—	2.0
$ 8.0	$2.5	$5.5	$ 6.0	$2.0	$4.0	$ 6.0	$2.0	$4.0
$ 5.5			$ 4.5			$ 5.0		
	$3.5			$2.5			$3.0	
	45%			35%			35%	

and tax effects, in addition to the level of nominal interest rates. She proposed that an analysis of these additional factors be performed to determine the actual cost of the proposed financing methods. Two additional areas were also cited for investigation: (1) lagging of payments on intersubsidiary transactions and (2) adjusting the transfer prices on these transactions. With the exception of Italy, no country placed restrictions on leading or lagging of payments. Italy was primarily concerned with variations of 120 days or more. The treasurers themselves decided to limit lagging for 90 days, so Italy's restriction was not considered to be limiting (especially because the normal billing period within the industry was 90 days). The corporate tax counsel then expressed his opinion that transfer prices could be raised or lowered by 5% from their currently planned values, without provoking the attention of tax authorities in the various countries. Managers of the individual

EXHIBIT III 1.3 Affiliate Financial Plans for 1995 ($ Millions)

| | MOBEX SPA | | MOBEX AG | | MOBEX SUISSE | |
	1st	2nd	1st	2nd	1st	2nd
Working-capital requirements	$3.5	$4.0	$5.0	$5.5	$4.5	$5.0
Proposed financing:						
2-quarter Lit loan	2.0	2.0				
2-quarter DM loan			1.5	1.5		
Export financing			2.0	2.0	1.0	0.5
Overdraft		0.5	1.5	2.0		0.5
1-quarter Eurodollar loan					1.5	2.0
2-quarter Eurodollar loan	1.5	1.5			2.0	2.0
Total	$3.5	$4.0	$5.0	$5.5	$4.5	$5.0
Projected exposure	3.0	4.5	3.0	3.5	2.5	3.0
To be eliminated by						
Local currency loan(s)	2.0	2.0	5.0	5.5	1.0	1.0
Forward contracts	$1.0	$2.5	$(2.0)	$(2.0)	$1.5	$2.0

subsidiaries generally agreed that 5% was a conservative amount. To play it safe, it was decided to limit consideration of transfer pricing changes to the ±5% range.

The meeting concluded with an agreement to review possible savings through implementation of a system-financing approach at the final budgeting review session in November. In the interim, Ms. Lee's staff would be responsible for developing the cost formulations and taking an initial cut at determining the optional financing/transfer pricing solution, taking into account the costs and benefits and the constraints involved.

QUESTIONS

1. What are the expected after-tax dollar costs of the different financing alternatives facing each of Mobex's foreign affiliates?

2. What are the costs and benefits associated with lagging payments between the two members of each pair of affiliates?

3. What are the tax and financing consequences associated with adjusting transfer prices between the two members of each pair of affiliates?

4. What are the interactions between modifying the credit terms and changing the transfer prices on transactions?

5. In which currencies should the interaffiliate transactions be denominated, given the anticipated currency changes and tax considerations?

6. What is the optimal (i.e., the cost-minimizing) solution to the overall financing/transfer-pricing problem faced by Mobex's European subsidiaries, given the actual or self-imposed constraints they face? (*Hint:* This is a linear programming problem, and you can find the values of the dual variables, which will tell you how much it would be worth to Mobex to relax each of the constraints it faces.)

PART FOUR

FINANCING
FOREIGN
OPERATIONS

INTERNATIONAL FINANCING AND INTERNATIONAL FINANCIAL MARKETS

Money, like wine, must always be scarce with those who have neither wherewithal to buy it nor credit to borrow it.

Adam Smith (1776)

The growing internationalization of capital markets and the increased sophistication of companies means that the search for capital no longer stops at the water's edge. This reality is particularly true for multinational corporations. A distinctive feature of the financial strategy of MNCs is the wide range of external sources of funds that they use on an ongoing basis. General Motors packages car loans as securities and sells them in Europe and Japan. British Telecommunications offers stock in London, New York, and Tokyo, while Beneficial Corporation issues Euroyen notes that may not be sold in either the United States or Japan. Swiss Bank Corporation—aided by Italian, Belgian, Canadian, and German banks, as well as other Swiss banks—helps RJR Nabisco sell Swiss franc bonds in Europe and then swap the proceeds back into U.S. dollars.

This chapter explores the MNC's external medium- and long-term financing alternatives. Although many of the sources are internal to the countries in which the MNCs operate, more of their funds are coming from offshore markets, particularly the Eurocurrency and Eurobond markets. We will, therefore, study both national and international capital markets in this chapter and the links between the two. We begin by discussing trends in corporate financing patterns.

15.1 CORPORATE SOURCES AND USES OF FUNDS

Firms have three general sources of funds available: internally generated cash, short-term external funds, and long-term external funds. External finance can come from investors or lenders. Investors give a company money by buying the securities it issues in the financial markets. These securities, which are generally *negotiable* (tradable), usually take the form of publicly-issued debt or equity. Debt is the preferred alternative: Regardless of the country studied, debt accounts for the overwhelming share of external funds. By contrast, new stock issues play a relatively small and declining role in financing investment. Regardless of whether debt or equity is offered to the public, the issuer likely will turn to a financing specialist—the **investment banker**—to assist in designing and marketing the issue. The latter function usually requires purchasing, or **underwriting**, the securities and then distributing them. Investment bankers are compensated by the *spread* between the price at which they buy the security and the price at which they can resell it to the public.

The main alternative to issuing public debt securities directly in the open market is to obtain a loan from a specialized financial intermediary that issues securities (or deposits) of its own in the market. These alternative debt instruments usually are commercial bank loans—for short-term and medium-term credit—or privately placed bonds—for longer-term credit. Unlike publicly issued bonds, **privately placed bonds** are sold directly to only a limited number of sophisticated investors, usually life insurance companies and pension funds. Moreover, privately placed bonds are generally nonnegotiable and have complex, customized loan agreements—called *covenants*. The restrictions in the covenants range from limits on dividend payments to prohibitions on asset sales and new debt issues. They provide a series of checkpoints that permit the lender to review actions by the borrower that have the potential to impair the lender's position. These agreements have to be regularly renegotiated before maturity. As a result, privately placed bonds are much more like loans than publicly issued and traded securities.

National Financing Patterns

Companies in different countries have different financial appetites. British companies get an average of about 97% of their funds from internal sources, in comparison with about 91% for U.S. companies and 81% for German companies. In Japan, where their profitability has been low, companies have relied more heavily on external finance, getting more than 30% of their money from outside sources, primarily from banks.[1] The shortfall of funds reflects the Japanese strategy of making huge industrial investments and pursuing market share at the expense of profit margins. As Japanese firms emphasize profits over sales growth, they probably will rely more heavily on internal sources of funds. In the meantime, however, the combination of substantial bank borrowing and low profits means that Japanese firms have high debt/equity ratios.

In contrast to Japan, in Europe and the United States internal finance has consistently supplied the lion's share of financial requirements. The percentage of external finance fluctuates more or less in line with the business cycle: When profits are high,

---

[1] These figures are the net sources of finance during the period 1970–1989 and appear in Bert Scholtens, "Bank- and Market-Oriented Financial Systems: Fact or Fiction?" *BNL Quarterly Review*, September 1997, p. 304.

firms are even less reliant on external finance. Moreover, the predominance of internal financing is not accidental. After all, companies could pay out internal cash flow as dividends and issue additional securities to cover their investment needs.

Another empirical regularity about financing behavior is related to the composition of external finance. Regardless of the country studied, debt accounts for the overwhelming share of external funds. By contrast, new stock issues play a relatively small and declining role in financing investment.

Financial Markets versus Financial Intermediaries

Industry's sources of external finance differ widely from country to country. German, French, and Japanese companies rely heavily on bank borrowing, whereas U.S. and British industry raise much more money directly from financial markets by the sale of securities. In all these countries, however, bank borrowing is on the decline. There is a growing tendency for corporate borrowing to take the form of negotiable securities issued in the public capital markets rather than in the form of nonmarketable loans provided by financial intermediaries.

This process, termed **securitization**, is most pronounced among the Japanese companies. Securitization largely reflects a reduction in the cost of using financial markets at the same time that the cost of bank borrowing has risen. Until recently, various regulatory restrictions enabled banks to attract low-cost funds from depositors. With **financial deregulation**, which began in the United States in 1981 and in Japan in 1986, banks now must compete for funds with a wide range of institutions at market rates. In addition, regulatory demands for a stronger capital base have forced U.S. banks to use more equity financing, raising their cost of funds. Inevitably, these changes have pushed up the price of bank loans. Any top-flight company now can get money more cheaply by issuing commercial paper than it can from its banks. As a result, banks have a smaller share of the short-term business credit market. Japanese companies also are finding that issuing bonds and leasing equipment are cheaper sources of medium-term and long-term money as well.

At the same time, the cost of accessing the public markets is coming down, especially for smaller and less well-known companies. Historically, these companies found it more economical to obtain loans from banks or to place private bond issues with life insurance companies. These **private placements** proved cheaper because banks and life insurance companies specialize in credit analysis and assume a large amount of a borrower's debt. Consequently, they could realize important cost savings in several functions, such as gathering information about the condition of debtor firms, monitoring their actions, and renegotiating loan agreements.

Recent technological improvements in such areas as data manipulation and telecommunications have greatly reduced the costs of obtaining and processing information about the conditions that affect the creditworthiness of potential borrowers. Any analyst now has computerized access to a wealth of economic and financial information at a relatively low cost, along with programs to store and manipulate this information. Thus, investors are now more likely to find it cost-effective to lend directly to companies, rather than indirectly through *financial intermediaries*, such as commercial banks.

Financial Systems and Corporate Governance

Despite the apparent convergence of financial systems, there are still some notable differences among countries in terms of **corporate governance**, which refers to the means whereby companies are controlled. The United States and United Kingdom are often viewed as prototypes of a market-oriented financial system (frequently referred to as the Anglo-Saxon or AS model); whereas Germany, France, and Japan are generally regarded as typical representatives of bank-centered finance (the Continental European and Japanese or CEJ type of financial system). In AS countries, institutional investors (pension funds, mutual funds, university and other nonprofit endowments, and insurance companies) make up an important part of the financial system. In CEJ countries, banks dominate the picture. Equity finance is important in AS countries and institutional shareholders exert a great deal of corporate control. The accepted objective is to maximize shareholder value, and boosting the return on capital employed is stressed. The stress on shareholder value has recently been endorsed by an international advisory panel to the Organization for Economic Cooperation and Development (OECD). According to the head of the panel, the Asian economic crisis can be traced to weak corporate governance: "Nobody was watching [Asian] management; they were growing for the sake of growth with no concern for shareholder value."[2] The crisis could most likely have been avoided, he said, had American-style corporate governance been in place.

In CEJ countries, bank finance is prominent, share ownership and control are concentrated in banks and other firms, and corporate decisionmaking is heavily influenced by close personal relationships between corporate leaders who sit on each others' boards of directors. Individual shareholders have little voice, resulting in much less concern for shareholder value and relatively low returns on capital. However, in all countries, competitive pressures and the threat of hostile takeovers of underperforming companies are forcing greater managerial accountability and an increased focus on shareholder value.

The difference in financial systems has real consequences for financial structures. For example, as noted above, large Japanese companies employ a high degree of leverage, particularly compared with U.S. companies. The ability to take on such large amounts of debt stems in part from the vast mutual-aid networks that most large Japanese firms can tap. These are the fabled **keiretsu**, the large industrial groupings—often with a major bank at the center—that form the backbone of corporate Japan. Keiretsu ties constitute a complex web of tradition, cross-shareholdings, trading relationships, management, cooperative projects, and information swapping. The keiretsu provide financial backing, management advice, and favorable contracts to their members. A key mission of the keiretsu is to provide a safety net when corporate relatives get into trouble.

Partly because of the difference in industrial structures in the two countries, U.S. and Japanese firms relate to the banking system in a different way. Almost all big Japanese companies have one main bank—usually the bank around which the keiretsu is formed—that is their primary source of long-term loans. The main bank will have access

---

[2]Quoted in Robert L. Simison, "Firms Worldwide Should Adopt Ideas of U.S. Management, Panel Tells OECD," *Wall Street Journal*, April 2, 1998, p. A4.

to information about the company and have a say in its management that in most other countries would be unacceptable. Moreover, Japanese banks, unlike their U.S. counterparts, can hold industrial shares, so, the main bank often holds a sizable amount of the equity of its borrowers. For example, until recently, Japanese banks owned almost 40% of the outstanding stock of Japanese manufacturing companies. Thus, for Japanese compa-

---

ILLUSTRATION

TOYO KOGYO AND CHRYSLER EXPERIENCE FINANCIAL DISTRESS

The contrasting experiences of Toyo Kogyo (producer of Mazda cars) and Chrysler during recent periods of financial distress illustrate the unusual features of the Japanese financial system. In 1973, Toyo Kogyo (TK) was a successful producer of light and medium-sized cars. The energy crisis of 1974 precipitated a crisis at TK because of the high energy consumption of its rotary-engine, Wankel-powered Mazda models. Worldwide sales plunged 19%. To weather the storm, TK required a massive infusion of funds to develop new product offerings. Sumitomo Bank, its main bank, had the resources to rescue TK. On the basis of its thorough knowledge of TK's operations, Sumitomo decided that the company could be profitable with new product offerings and massive cost cutting. However, because it lacked confidence in TK's senior management, Sumitomo replaced them. With its own hand-picked executives in place, Sumitomo financed the simultaneous development of three new models and the overhaul of TK's production system, extended credit to suppliers, and had the vast group of related Sumitomo companies buy Mazda vehicles. The new models were highly successful and labor productivity grew by 118% over the next seven years.

In contrast to the situation at TK, Chrysler was able to persist with poor performance for more than two decades because its investors had no effective remedies. Chrysler could ignore the need to restructure its operations because of its continued ability to borrow money; it still had a substantial, though dwindling, amount of shareholders' equity to support these loans. Despite the activities of some dissident shareholders, Chrysler continued under a self-perpetuating management until the crisis of 1979. Although Chrysler faced bankruptcy, its banks refused to lend it more money. In contrast to TK's Japanese banks, Chrysler's banks had an incomplete understanding of its plight and no way to obtain all the essential information. Even if they had, they could not have sent in the same type of rescue team that Sumitomo sent into TK. In the end, Chrysler was rescued by the U.S. government, which offered loan guarantees sufficient for the company's survival at about half its former size.

In a man-bites-dog turn of events, however, Chrysler eventually restructured to become the most profitable auto company in the world, and Mazda Motor Company once again got into deep trouble. Hurt by the strong yen and by a high-risk expansion strategy that backfired when global recession struck, Mazda by 1994 was stuck with $8 billion in debt and a main bank (still Sumitomo) that refused to bail it out. Instead, Sumitomo invited Ford Motor Company to take effective management control of Mazda, thereby entrusting Ford with the role that the bank itself played in the 1970s.

nies, the strong relation with one main bank—along with close ties to the other members of their keiretsu—is their main method of minimizing the risk of financial distress.[3]

The same is true of Germany, where so-called **universal banking** is practiced; German commercial banks not only perform investment banking activities but also take major equity positions in companies. As both stockholder and creditor, German banks can reduce the conflicts between the two classes of investors, leading to lower costs and speedier action in "workouts" of financial problems. The resulting increase in organizational efficiency should mean less risk for German companies in taking on large amounts of debt. In the United States, where corporate bank relations are less intimate, companies rely primarily on equity as a shock absorber.

The price that Japanese and German companies pay for their heavy reliance on bank debt is less freedom of action. As the cost of accessing the capital markets directly has dropped, the main-bank relationship has gradually eroded in Japan and Germany, and Japanese and German companies have looked more to the equity market as their cushion against financial distress. The pace of change was accelerated in Japan because of a law that forced Japanese banks to reduce their shareholdings in individual companies to 5% or less by December 1987. In addition, in the wake of the Japanese stock market collapse and eroding corporate profits, some companies have begun to raise money by selling major chunks of their cross-shareholdings in the less important members of their keiretsu. Moreover, although relations between main banks and associated corporations remain intimate, companies are looking more carefully at second- and third-tier banks. Even main banks, however, are plagued with bad debts, declining bank capital, and a much higher cost of capital, forcing them to pay more attention to profitability and making them less reliable sources of capital for their corporate customers. In response, Japanese companies are turning increasingly to the corporate bond market to raise new capital.

Globalization of Financial Markets

The same advances in communications and technology that have lowered the cost of accessing financial markets directly, together with financial deregulation abroad—the lifting of regulatory structures that inhibit competition and protect domestic markets—have blurred the distinction between domestic and foreign financial markets. As the necessary electronic technology has been developed and transaction costs have plummeted, the world has become one vast, interconnected market. Markets for U.S. government securities and certain stocks, foreign exchange trading, interbank borrowing and lending—to cite a few examples—operate continuously around the clock and around the world and in enormous size. The *globalization of financial markets* has brought about an unprecedented degree of competition among key financial centers and financial institutions that has further reduced the costs of issuing new securities.

Growing competition also has led to increasing deregulation of financial markets worldwide. Deregulation is hastened by the process of **regulatory arbitrage**, whereby the users of capital markets issue and trade securities in financial centers with the lowest regu-

---

[3] For a detailed analysis of the governance structure of the keiretsu, see Steven N. Kaplan and Bernadette A. Minton, "Appointments of Outsiders to Japanese Boards: Determinants and Implications for Managers," *Journal of Financial Economics*, October 1994, pp. 225–258; and Erik Berglöf and Enrico Perotti, "The Governance Structure of the Japanese Financial Keiretsu," *Journal of Financial Economics*, October 1994, pp. 259–284.

latory standards and, hence, the lowest costs. In order to win back business, financial centers around the world are throwing off obsolete and costly regulations. For example, concerned that Tokyo had fallen behind London and New York as a global finance center, the Japanese government has developed a "Big Bang" financial reform program. This program would break down regulatory barriers between Japanese banks, insurance companies, and brokerage houses, and would also create opportunities in Japan for foreign financial companies by cutting red tape and barriers to the market. Deregulation—in Japan and elsewhere—is little more than an acknowledgment that the rules do not—and cannot—work.

ILLUSTRATION

CAPITAL MARKETS COMPETE FOR BUSINESS
In mid-1989, Germany abolished its 10% withholding tax on interest income, less than six months after it took effect. The tax led to record capital flight from Germany, putting downward pressure on the Deutsche mark. Upon announcement of the repeal, German bonds jumped in value.

Similarly, in early 1990, Sweden announced that it would scrap a turnover tax on bond trading and halve a similar tax on trading in equities that had driven most international trading of Swedish shares abroad. The taxes had been imposed one year earlier as political concessions to powerful Swedish unions, who resented the large amounts of money being made in Swedish financial markets. Almost everyone else vehemently opposed the levies. The dire consequences predicted by critics were borne out in decimated trading volumes. For example, the tax on bond trading slashed average daily trading volume of Swedish government securities by 78% from the year before. At the same time, volume on the Stockholm Stock Exchange plunged by 20% even though the market rose 119% in the period. The final straw was a study by the Finance Ministry that showed that the various levies actually resulted in a net reduction in tax collections for Sweden as trades shifted abroad. On the day the tax cuts were announced, stocks of the largest Swedish companies rose almost 2.5%. By 1995, the 1% equity turnover tax had been removed entirely and the fraction of trades in Swedish equities done in Stockholm had risen to 60% from 40% in 1990.

Financial deregulation also has been motivated by the growing recognition in nations with bank-centered financial systems that such systems are not providing adequately for the credit needs of the small and medium-sized firms that are the engines of growth and innovation. It has not escaped the notice of governments worldwide that corporate success stories of the past 20 years—companies like Microsoft, Dell, 3Com, Yahoo!, Oracle, and Amgen—have come predominantly from the United States. Germany and Japan can boast of few Netscapes, AOLs, Iomegas, Suns, Compaqs, or Genentechs. By deregulating their financial markets, policymakers hope that their countries will emulate the results of the U.S. system of corporate finance.

The combination of freer markets with widely available information has laid the foundation for global growth. Cross-border trading in financial assets was estimated in 1992 to be $35 trillion, up from $5 trillion in 1980.[4] Fund raising is global now as well.

---

[4] These figures appear in "The Global Capital Market: Supply, Demand, Pricing and Allocation," McKinsey Global Institute, November 1994.

EXHIBIT 15.1 Funds Raised on the International Capital Markets by Instrument

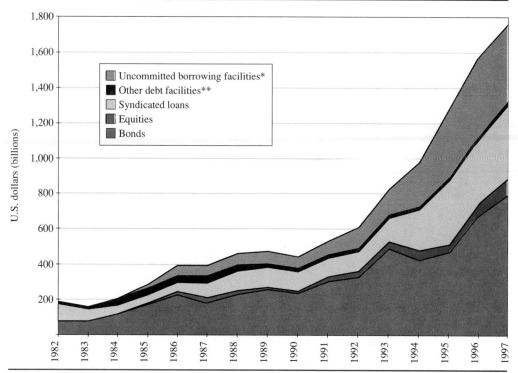

Source: Financial Market Trends, Organization for Economic Cooperation and Development, various issues.
*Euro-commercial paper and medium-term Euronote programs.
**Note issuance facilities and other back-up facilities.

According to OECD figures, the amount of money raised on the international capital markets has grown rapidly, rising to nearly $1.8 trillion in 1997 (see Exhibit 15.1). Treasurers are no longer confined to domestic markets as their source of funding and are now quick to exploit any attractive opportunity that occurs anywhere in the world.

Whereas competition drives the international financial system, innovation is its fuel. **Financial innovation** segments, transfers, and diversifies risk. It also enables companies to tap previously inaccessible markets and permits investors and issuers alike to take advantage of tax loopholes. More generally, financial innovation presents opportunities for value creation. To the extent that a firm can design a security that appeals to a special niche in the capital market, it can attract funds at a cost that is less than the market's required return on securities of comparable risk. However, such a rewarding situation is likely to be temporary, because the demand for a security that fits a particular niche in the market is not unlimited. On the other hand, the supply of securities designed to tap that niche is likely to increase dramatically once the niche is recognized. Even though financial innovation may not be a sustainable form of value creation, it nonetheless can enable the initial issuers to raise money at a below-market rate.

THE SWEDISH EXPORT
CREDIT CORPORATION
INNOVATES
The Swedish Export Credit
Corporation (SEK) borrows
about $2 billion annually. To
reduce its funding costs, SEK
relies heavily on financial innovation. For example, SEK issued a straight bond, stripped it
down to its two components—an annuity
consisting of the interest payments and a zero-coupon bond consisting of the principal repay-
ment at maturity—and sold the pieces to different investors. The annuity cash flow was tailored to meet the demands of a Japanese insurance company that was looking for an
interest-only security, whereas the zero-coupon
portion appealed to European investors who
desired earnings taxed as capital gains rather
than interest income. By unbundling the bond
issue into separate parts that appealed to distinct
groups of investors, SEK created a financial
transaction whose parts were worth more than
the whole.

Financial innovation has dramatically increased international capital mobility. As in
the domestic case, cross-border financial transfers can take place through international
securitization or international financial intermediation. The hypothetical case depicted in
Exhibit 15.2 illustrates the distinction between these two mechanisms for international
fund flows. A Belgian corporation with surplus funds seeks an investment outlet, and a
Japanese corporation requires additional funds. International securitization might involve
the Japanese firm issuing new bonds and selling them directly to the Belgian firm.

EXHIBIT 15.2 Securitization Versus Intermediation

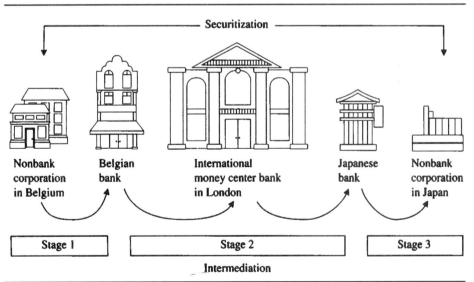

Nonbank corporation in Belgium — Belgian bank — International money center bank in London — Japanese bank — Nonbank corporation in Japan

Stage 1 | Stage 2 | Stage 3

Securitization

Intermediation

Source: Anthony Saunders, "The Eurocurrency Interbank Market: Potential for International Crises?" Reprinted from *Business Review*, January/February 1988, p. 19. Used with permission of Federal Reserve Bank of Philadelphia.

Alternatively, the Belgian firm's surplus funds could be transferred to the Japanese firm through international financial intermediation. This intermediation could involve three (or more) stages. First, the Belgian firm deposits its funds with a local Belgian bank. Second, the Belgian bank redeposits the money with an international money center bank in London that turns around and lends those funds to a Japanese bank. In stage 3, the Japanese bank lends those funds to the Japanese corporation.

Whether international fund flows take place through financial intermediation or securitization depends on the relative costs and risks of the two mechanisms. The key determinant here is the cost of gathering information on foreign firms. As these costs continue to come down, international securitization should become increasingly more cost-effective.

15.2 NATIONAL CAPITAL MARKETS AS INTERNATIONAL FINANCIAL CENTERS

The principal functions of a financial market and its intermediaries are to mobilize savings (which involves gathering current purchasing power in the form of money from savers and transferring it to borrowers in exchange for the promise of greater future purchasing power) and to allocate those funds among the potential users on the basis of expected risk-adjusted returns. Financial markets also facilitate both the transfer of risk (from companies to investors) and the reduction of risk (by investors holding a diversified portfolio of financial assets). Subsequent to the investment of savings, financial markets help to monitor managers (by gathering information on their performance) and exert corporate control (through the threat of hostile takeovers for underperforming firms and bankruptcy for insolvent ones). Financial markets also supply **liquidity** to investors by enabling them to sell their investments before maturity.

The consequences of well-functioning financial markets are: More and better projects get financed (owing to higher savings, a more realistic scrutiny of investment opportunities, and the lower cost of capital associated with risk diversification and increased liquidity); managers are compelled to run companies in accordance with the interests of investors (through active monitoring and the threat of bankruptcy or a hostile takeover for underperformers and by linking managerial compensation to stock prices); the rate of innovation is higher (by identifying and funding those entrepreneurs with the best chances of successfully initiating new goods and production processes); and individuals are able to select their preferred time pattern of consumption (saving consists of individuals deferring consumption in some periods so as to increase their consumption in later periods) and their preferred risk-return tradeoff. The result is stronger economic growth and greater consumer satisfaction. These factors are summarized in Exhibit 15.3.

Financial markets work best when property rights are secure, contracts are easily enforceable, meaningful accounting information is available, and borrowers and investors are accountable for their decisions and bear the economic consequences of their behavior. Absent these conditions, markets cannot allocate capital efficiently and economic growth suffers.

The financial disaster in Asia points out the dangers of allocating capital by cronyism and bureaucratic dictate rather than through a rational process governed by realistic

EXHIBIT 15.3 The Role and Consequences of Well-Functioning Financial Markets

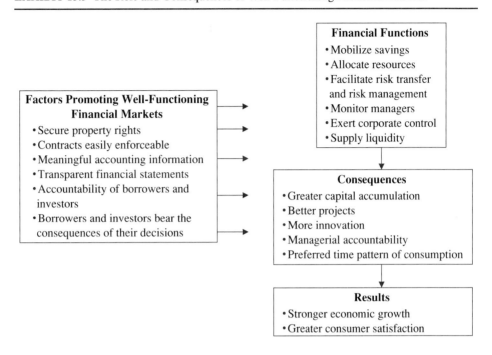

Financial Functions
- Mobilize savings
- Allocate resources
- Facilitate risk transfer and risk management
- Monitor managers
- Exert corporate control
- Supply liquidity

Factors Promoting Well-Functioning Financial Markets
- Secure property rights
- Contracts easily enforceable
- Meaningful accounting information
- Transparent financial statements
- Accountability of borrowers and investors
- Borrowers and investors bear the consequences of their decisions

Consequences
- Greater capital accumulation
- Better projects
- More innovation
- Managerial accountability
- Preferred time pattern of consumption

Results
- Stronger economic growth
- Greater consumer satisfaction

estimates of prospective risks and rewards. This command-and-control capitalism produced dysfunctional financial sectors that squandered hundreds of billions of dollars of hard-earned savings on unproductive investments and grandiose projects and begat corruption without end.

Conversely, a healthy dose of market discipline and the stringent credit standards it enforces can work wonders for an economy. A recent study by the McKinsey Global Institute examined **capital productivity**—the ratio of output (goods and services) to the input of physical capital (plant and equipment)—in Germany, Japan, and the United States.[5] Overall, U.S. capital productivity exceeded that of Japan and Germany by about 50%. As a result, the United States can simultaneously save less, consume more, and grow faster. This economic hat trick can be traced directly to: activist shareholders demanding management accountability and paying for performance; tough disclosure rules and financial transparency in corporate accounts that allow investors to make informed decisions; rigorous credit analysis that helps screen out bad risks; and a willingness to inflict pain on imprudent lenders, cut off capital to less competitive companies, and allow inefficient companies to fail. Although many politicians and others claim that U.S. financial markets have hindered U.S. productivity by forcing American companies to be short-term oriented, the McKinsey report suggests exactly the opposite—that the

[5] McKinsey Global Institute, *Capital Productivity*, Washington, D.C., June 1996. This report is summarized in Raj Agrawal, Stephen Findley, Sean Greene, Kathryn Huang, Aly Jeddy, William W. Lewis, and Markus Petry, "Capital Productivity: Why the U.S. Leads and Why It Matters," *The McKinsey Quarterly*, 1996, No. 3, pp. 38–55.

focus of U.S. financial markets on financial performance, reinforced by strong corporate governance, leads directly to improved business performance. In the process of rewarding success and penalizing failure, financial virtue creates its own reward.

International Financial Markets

Not surprisingly, most of the major financial markets attract both investors and fund raisers from abroad. That is, these markets are also *international financial markets*, where foreigners can both borrow and lend money. International financial markets can develop anywhere, provided that local regulations permit the market and that the potential users are attracted to it. The most important international financial centers are London, Tokyo, and New York. All the other major industrial countries have important domestic financial markets as well, but only some, such as Germany and—recently—France, are also important international financial centers. On the other hand, some countries that have relatively unimportant domestic financial markets are important world financial centers. The markets of those countries, which include Switzerland, Luxembourg, Singapore, Hong Kong, the Bahamas, and Bahrain, serve as financial *entrepôts*, or channels through which foreign funds pass. That is, these markets serve as financial intermediaries between nonresident suppliers of funds and nonresident users of funds.

Political stability and minimal government intervention are prerequisites for becoming and remaining an important international financial center, especially an entrepôt center. Historically, London's preeminence as an entrepôt for international finance comes from its being a lightly regulated offshore market in a world of financial rigidities. That is why it became home to the Euromarkets about 30 years ago. As financial markets deregulate, London's strength has shifted to its central location (including its central time zone), and financial infrastructure—its access to information by dint of its position astride huge international capital flows, its pool of financial talent, its well-developed legal system, and its telecommunications links. Even more important, financial firms need to be near big investors, and there is more money under management in London than anywhere else in Europe.

Foreign Access to Domestic Markets

Despite the increasing liberalization of financial markets, governments are usually unwilling to rely completely on the market to perform the functions of gathering and allocating funds. Foreigners particularly are often hampered in their ability to gain access to domestic capital markets because of government-imposed or government-suggested restrictions relating to the maturities and amounts of money that they can raise. They are hampered as well by the government-legislated extra costs, such as special taxes (for example, the U.S. interest equalization tax, or IET, that was in effect from 1963 to 1974), they must bear on those funds that they can raise. Nonetheless, the financial markets of many countries are open wide enough to permit foreigners to borrow or invest.

As a citizen of many nations, the multinational firm has greater leeway in tapping a variety of local money markets than does a purely domestic firm, but it, too, is often the target of restrictive legislation aimed at reserving local capital for indigenous companies or the local government. The capital that can be raised is frequently limited to local uses through the imposition of exchange controls. As we have seen previously, however, multi-nationals are potentially capable of transferring funds, even in the presence of currency controls, by using a variety of financial channels. To the extent, therefore, that local credits substitute for parent- or affiliate-supplied financing, the additional monies are available for removal.

THE FOREIGN BOND MARKET ■ The **foreign bond market** is an important part of the international financial markets. It is simply that portion of the domestic bond market that represents issues floated by foreign companies or governments. As such, foreign bonds are subject to local laws and must be denominated in the local currency. At times, these issues face additional restrictions as well. For example, foreign bonds floated in Switzerland, Germany, and the Netherlands are subject to a queuing system, where they must wait for their turn in line.

The United States and Switzerland contain the most important foreign bond markets (dollar-denominated foreign bonds sold in the United States are called **Yankee bonds**). Major foreign bond markets are also located in Japan and Luxembourg (yen bonds sold in Japan by a non-Japanese borrower are called **Samurai bonds**, in contrast to **Shogun bonds**, which are foreign currency bonds issued within Japan by Japanese corporations). Data on the amounts of foreign bond issues are presented in Exhibit 15.4.

Foreign bond issues, like their purely domestic counterpart, come in three primary flavors: fixed-rate issues, floating-rate notes (FRNS), and equity-related issues. **Fixed-rate issues** are similar to their domestic counterparts, with a fixed coupon, set maturity date, and full repayment of the principal amount at maturity. **Floating-rate notes** have variable coupons that are reset at fixed intervals, usually every three to six months. The new coupon is set at a fixed margin above a mutually agreed-upon reference rate such as the treasury bill rate or the commercial paper rate.

Equity-related bonds combine features of the underlying bond and common stock. The two principal types of equity-related bonds are convertible bonds and bonds with equity warrants. **Convertible bonds** are fixed-rate bonds that are convertible into a given number of shares before maturity. **Equity warrants** give their holder the right to buy a specified number of shares of common stock at a specified price during a designated time period. The relative amount of foreign bonds issued in the three different categories varies from year to year depending on market conditions.

After the economic crisis in Asia, Asian corporations scrambled for new equity to recapitalize and shore up balance sheets over-loaded with debt and unsupported by much remaining equity at the new exchange rates. The preferred method of providing new equity or its equivalent was through convertible bonds and preferred shares. Both give fresh money coming into a troubled company a higher claim on assets under liquidation than would straight equity while at the same time providing a significant upside opportunity for investors who thought the markets had overreacted. The benefit for struggling companies is that convertibles reduce debt service charges because they carry lower interest rates than the debt they replace. In addition, upon conversion (if it occurs), the debt becomes equity. Several such issues in early 1998 included Singapore Telephone's $1 billion five-year convertible bond; New World Infrastructure's $250 million five-year convertible bond; COSCO (Hong Kong) Group's $150 million exchangeable bond; and Guangdong Investment's $125 million in convertible preferred.

EXHIBIT 15.4 Foreign Bond Market (U.S. $ Billions)

Market	1984 $	1984 %	1985 $	1985 %	1986 $	1986 %	1987 $	1987 %	1988 $	1988 %	1989 $	1989 %	1990 $	1990 %
Austria	0.0	0.2	0.1	0.3	0.1	0.2	0.1	0.3	0.0		0.3	0.7	0.5	1.0
France	0.0	0.1	0.4	1.5	0.5	1.4	0.7	1.7	0.6	1.2	0.7	1.7	0.4	0.7
Japan	4.9	20.6	6.3	22.4	5.2	13.3	4.1	10.4	6.7	14.3	8.2	18.7	7.9	15.9
Luxembourg	0.2	0.9	0.4	1.4	0.8	2.1	1.4	3.5	1.8	3.8	1.7	3.8	4.4	8.9
Netherlands	1.2	4.9	1.0	3.5	1.7	4.4	1.0	2.6	0.8	1.6	0.3	0.6	0.6	1.3
Portugal	0.0		0.0				0.0		0.0	0.1	0.1	0.1	0.3	0.6
Spain	0.0		0.0				0.2	0.6	0.7	1.6	2.2	5.1	1.7	3.4
Switzerland	13.1	55.3	14.9	53.3	23.2	59.0	24.3	62.0	26.3	56.1	18.6	42.7	23.2	46.9
United States	4.3	18.1	4.9	17.6	6.8	17.2	7.4	18.9	10.1	21.4	9.4	21.6	9.9	20.0
Other	4.2	17.8	3.2	11.3	0.9	2.4	0.8	2.1	0.8	1.7	2.2	5.2	0.6	0.7
TOTAL	23.7	100.0	27.9	100.0	39.4	98.0	39.2	100.0	47.0	100.0	43.7	100.0	49.5	100.0

Note: A blank indicates no foreign bond market.

Source: Financial Market Trends, Organization for Economic Cooperation and Development, various issues.

THE FOREIGN BANK MARKET ■ The **foreign bank market** represents that portion of domestic bank loans supplied to foreigners for use abroad. As in the case of foreign bond issues, governments often restrict the amounts of bank funds destined for foreign purposes. Foreign banks, particularly Japanese banks, have become an important funding source for U.S. corporations.

One indication of the importance of foreign banks, and particularly Japanese banks, is the fact that six out of the world's 10 largest banks ranked by assets as of December, 1996, were Japanese, including four of the top five. The highest ranked U.S. bank, Chase, was 16th in size. (Of the world's 25 largest banks, only one was American; 10 were Japanese, four were French, four were German, three were British, with the Swiss, Dutch, and Chinese having one top-25 bank apiece.) The minimal representation by American banks among the world's largest owes primarily to prohibitions on interstate banking in the United States. Other factors include the high Japanese savings rate and relatively low U.S. savings rate.

THE FOREIGN EQUITY MARKET ■ The idea of placing stock in foreign markets has long attracted corporate finance managers. One attraction of the **foreign equity market** is the diversification of equity funding risk: A pool of funds from a diversified shareholder base insulates a company from the vagaries of a single national market. Some issues are too large to be taken up only by investors in the national stock market. For large companies located in small countries, foreign sales may be a necessity. When KLM, the Dutch airline, issued 50 million shares in 1986 to raise $304 million, it placed 7 million shares in Europe, 7 million in the United States, and 1 million in Japan. According to a spokesman for the company, "The domestic market is too small for such an operation."[6]

Selling stock overseas also can increase the potential demand for the company's shares, and hence its price, by attracting new shareholders. For example, a study by Gor-

---

[6]"International Equities: The New Game in Town," *Business International Money Report,* September 29, 1987, p. 306.

EXHIBIT 15.4 Continued

1991		1992		1993		1994		1995		1996		1997		
$	%	$	%	$	%	$	%	$	%	$	%	$	%	
0.4	0.8	0.6	1.1	0.4	0.5	0.4	0.7	0.3	0.3	0.3	0.2	0.4	0.4	Austria
0.7	1.4	0.4	0.6			0.3	0.5	0.0		0.0	0.0	0.0	0.0	France
5.2	10.2	7.4	12.9	15.2	17.7	11.2	18.6	18.0	18.8	35.5	29.8	17.4	18.0	Japan
5.5	10.9	5.5	9.6	3.5	4.0	11.0	18.3	13.8	14.4	8.4	7.1	2.9	3.0	Luxembourg
0.1	0.2	0.2	0.3	0.9	1.1	0.0		0.3	0.3	1.8	1.5	0.0	0.0	Netherlands
0.7	1.4	0.3	0.5	0.6	1.2	0.3	0.5	1.0	1.0	0.5	0.4	2.1	2.2	Portugal
2.7	5.3	1.6	2.9	3.0	3.5	1.6	2.7	2.3	2.4	6.3	5.3	5.7	5.9	Spain
20.2	39.9	18.1	31.4	27.0	31.4	20.0	33.2	27.1	28.2	25.0	21.0	21.0	21.8	Switzerland
14.4	28.5	23.2	40.2	35.4	41.1	15.0	24.9	32.4	33.8	40.5	34.0	45.5	47.2	United States
0.8	1.3	0.3	0.5	0.0	0.0	0.4	0.0	0.8	0.8	0.7	0.6	1.5	1.6	Other
50.6	100.0	57.6	100.0	86.1	100.0	60.3	100.0	96.0	100.0	119.0	100.0	96.5	100.0	TOTAL

don Alexander, Cheol Eun, and S. Janakiramanan found that foreign companies that listed their shares in the United States experienced a decline in their expected return.[7] Similarly, Dennis Logue and Anant Sundaram found that cross-listing foreign companies in the United States enhances the valuations for the listing companies by up to 10% relative to country and industry benchmarks.[8] This evidence is consistent with the theoretical work of Robert Merton, who has shown that a company can lower its cost of equity capital and, thereby, increase its market value by expanding its investor base.[9]

For a firm that wants to project an international presence, an international stock offering can spread the firm's name in local markets. In the words of a London investment banker, "If you are a company with a brand name, it's a way of making your product known and your presence known in the financial markets, which can have a knock-off effect on your overall business. A marketing exercise is done; it's just like selling soap."[10] According to Apple Computer's investor relations manager, Apple listed its shares on the Tokyo exchange and the Frankfurt exchange "to raise the profile of Apple in those countries to help us sell computers. In Japan, being listed there gets us more interest from the business press."[11]

To capture some of these potential benefits, more companies are selling stock issues overseas (see Exhibit 15.5 for data on this trend). For example, the largest common stock offering in the United States in 1991 was a $1.2 billion issue by Telephonos de Mexico

---

[7] Gordon J. Alexander, Cheol S. Eun, and S. Janakiramanan, "International Listings and Stock Returns: Some Empirical Evidence," working paper, University of Minnesota, May 1986.

[8] Dennis E. Logue and Anant K. Sundaram, "Valuation Effects of Foreign Company Listings on U.S. Exchanges," working paper, Amos Tuck, February 1994.

[9] Robert C. Merton, "A Simple Model of Capital Market Equilibrium with Incomplete Information," *Journal of Finance*, July 1987, pp. 483–510.

[10] "International Equities: The New Game in Town." *Business International Money Report*, September 29, 1987, p. 306.

[11] Quoted in Kathleen Doler, "More U.S. Firms See Foreign Shareholdings as a Plus," *Investors Business Daily*, February 17, 1994, p. A4.

ILLUSTRATION

WASTE MANAGEMENT LISTS ITS STOCK IN AUSTRALIA

Chicago-based Waste Management has been operating in Australia since 1984 and has gained a leading share of the garbage-collection market through expansion and acquisition. In 1986, the firm issued shares in Australia and then listed those shares on the Australian exchanges. A principal reason for the listing was to enhance its corporate profile. According to a Waste Management spokesman, "We view Australia as a growth market, and what we really wanted was to increase our visibility."[12]

Listing gets Waste Management better known in the financial community as well. This visibility in turn aids the expansion program, which hinges largely on mergers and acquisitions, by increasing contacts with potential joint-venture or acquisition candidates. Listing also facilitates stock-for-stock swaps.

An Australian listing also enhances the local profit-sharing package. Waste Management uses an employee stock program as an integral feature of its compensation. By listing locally, the firm increases the prominence of its shares and the program becomes more attractive to employees.

[12]"Waste Management Who? Why One U.S. Giant Is Now Listed Down Under," *Business International Money Report*, December 22, 1986, p. 403.

EXHIBIT 15.5 International Equity Issues

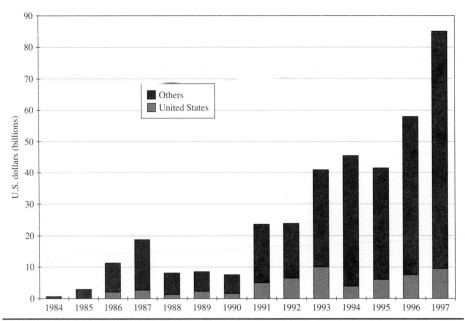

Source: *Financial Market Trends*, OECD, various issues.

EXHIBIT 15.6 The International Equity Offering by Deutsche Telekom

This advertisement appears as a matter of record only. All shares have been sold.

January 1997

Deutsche Telekom

Deutsche Telekom AG, Bonn, Germany

Global Initial Public Offering of 690,000,000 Ordinary Shares in the Form of Shares and American Depositary Shares at DM 28.50 per Share and at US$ 18.89 per ADS

Global Coordinators

Deutsche Bank
Aktiengesellschaft

Dresdner Bank
Aktiengesellschaft

Goldman, Sachs & Co.

The Shares were offered in the following regions

Germany – 462,300,000 Shares

Dresdner Bank
Aktiengesellschaft

Deutsche Bank
Aktiengesellschaft

DG BANK Deutsche Genossenschaftsbank	Westdeutsche Landesbank Girozentrale	Bayerische Landesbank Girozentrale	Commerzbank Aktiengesellschaft	Bayerische Hypotheken- und Wechsel-Bank Aktiengesellschaft	Bayerische Vereinsbank Aktiengesellschaft
	Helaba Landesbank Hessen-Thüringen Girozentrale	Bankgesellschaft Berlin Aktiengesellschaft	Norddeutsche Landesbank Girozentrale	SüdwestLB	
Baden-Württembergische Bank Aktiengesellschaft	BHF-BANK	CS First Boston Effectenbank Aktiengesellschaft	Goldman, Sachs & Co. oHG	Sal. Oppenheim jr. & Cie. Kommanditgesellschaft auf Aktien	M.M. Warburg & CO KGaA
Berenberg Bank Joh. Berenberg, Gossler & Co.	BfG Bank AG	Delbrück & Co., Privatbankiers	FLESSABANK Bankhaus Max Flessa & Co	Hamburgische Landesbank – Girozentrale –	Georg Hauck & Sohn Bankiers KGaA	IKB Deutsche Industriebank Aktiengesellschaft
Bankhaus Hermann Lampe Kommanditgesellschaft	Landesbank Rheinland-Pfalz –Girozentrale–	Landesbank Saar Girozentrale	Landesbank Sachsen Girozentrale	Landesbank Schleswig-Holstein Girozentrale	Landesgirokasse öffentliche Bank und Landessparkasse
	B. Metzler seel. Sohn & Co. Kommanditgesellschaft auf Aktien	SchmidtBank KGaA	Schröder Münchmeyer Hengst & Co	Trinkaus & Burkhardt Kommanditgesellschaft auf Aktien	

United Kingdom – 57,500,000 Shares

SBC Warburg
A DIVISION OF SWISS BANK CORPORATION

Dresdner Kleinwort Benson	NatWest Securities Limited	Deutsche Morgan Grenfell
ABN AMRO Rothschild	CS First Boston	Robert Fleming & Co. Limited
	Goldman Sachs International	
Bayerische Landesbank Girozentrale	Cazenove & Co.	Schroders

Americas – 97,750,000 Shares

Goldman, Sachs & Co.   **Deutsche Morgan Grenfell**   **Merrill Lynch & Co.**

Morgan Stanley & Co. Incorporated   Salomon Brothers Inc

ABN AMRO Rothschild   CS First Boston
A Division of ABN AMRO Securities (USA) Inc.

Dresdner Kleinwort Benson

Bear, Stearns & Co. Inc.	Alex. Brown & Sons Incorporated	Dean Witter Reynolds Inc.
Donaldson, Lufkin & Jenrette Securities Corporation	A.G. Edwards & Sons, Inc.	Everen Securities, Inc. J.P. Morgan & Co.
PaineWebber Incorporated	Prudential Securities Incorporated	RBC Dominion Securities Corporation
Scotia Capital Markets	Smith Barney Inc.	Toronto Dominion Securities
Advest, Inc.	Arnhold and S. Bleichroeder, Inc.	Robert W. Baird & Co. Incorporated
Sanford C. Bernstein & Co., Inc.	Dain Bosworth Incorporated	Edward D. Jones & Co., L.P.
Legg Mason Wood Walker Incorporated	McDonald & Company Securities, Inc.	Principal Financial Securities, Inc.
Stephens Inc.	Stifel, Nicolaus & Company Incorporated	Sutro & Co. Incorporated

Rest of Europe – 37,950,000 Shares

Paribas Capital Markets

	UBS Limited Dresdner Kleinwort Benson	
ABN AMRO Rothschild	Banque Nationale de Paris	CS First Boston
	Deutsche Morgan Grenfell Goldman Sachs International	
BBV Interactivos, S.V.B.	Bayerische Vereinsbank Aktiengesellschaft	Banque Générale du Luxembourg S.A.
Creditanstalt Investment Bank	Commerzbank Aktiengesellschaft Den Danske Bank	Enskilda Securities Skandinaviska Enskilda Banken
KB-Securities	Mediobanca - Banca di Credito Finanziario SpA	Société Générale

Asia-Pacific/Rest of World – 34,500,000 Shares

Daiwa Europe Limited

Nomura International	Lehman Brothers	Deutsche Morgan Grenfell
ABN AMRO Rothschild	CS First Boston	Dresdner Kleinwort Benson
	Goldman Sachs (Asia) L.L.C. HSBC Investment Banking	
Indosuez Capital	Nikko Europe Plc	WestLB Securities Pacific Ltd.
	Yamaichi International (Europe) Limited The Development Bank of Singapore Ltd	
	Rashid Hussain Securities Sdn Bhd Securities One	
	Ssangyong Investment & Securities Co., Ltd. J B Were & Son	
Kankaku (Europe) Limited	KOKUSAI Europe Limited	New Japan Securities Europe Limited
	Universal (U.K.) Limited Wako International (Europe) Limited	

Employee Share Ownership Programme of 23,700,000 Shares

UBS
Schweizerische Bankgesellschaft (Deutschland) AG

DG BANK
Deutsche Genossenschaftsbank

Verband der Post-, Spar- und Darlehnsvereine e.V.

Advisors to the Company
NM Rothschild & Sons
Coopers & Lybrand/Sietz & Partner Ltd.

Advisor to the Government of the Federal Republic of Germany
CS First Boston
KPMG Deutsche Treuhand-Gesellschaft

EXHIBIT 15.7 Equity Issuance in the United States by Foreign Companies

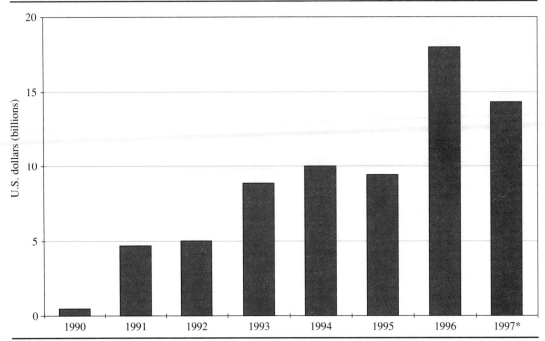

*Annualized, through August 12, 1997.

Source: Data from Securities Data Co.

(Telmex), part of a $2.2 billion global offering. In all, **Yankee stock issues**—stock sold by foreign companies to U.S. investors—totaled almost $5 billion in 1991. Exhibit 15.7 shows that that number is on the rise. For example, Deutsche Telekom's global equity offering (see Exhibit 15.6 for the annnouncement) included a $1.9 billion tranche to be sold in the United States because of its view that the size of its long-term capital needs required access to a larger and more liquid market than that provided by its German home market. Similarly, in October 1992, Roche Holding Ltd., the Swiss pharmaceutical giant, announced that it had completed a U.S. private placement of $275 million worth of stock. The move reflected Roche's desire to tap into the world's largest capital market because its rapid growth had "left Roche simply too big for the Swiss market."[13] Indeed, more foreign companies are selling their initial public offerings, or IPOs, in the United States because they "get a better price, a shareholder base that understands their business, and they can get publicity in a major market for their products."[14]

In 1993, Daimler-Benz, the industrial conglomerate that makes the Mercedes-Benz, became the first German company to list its shares on the New York Stock Exchange. To qualify for the listing, Daimler-Benz had to provide an onerous, by German standards,

---

[13] Quoted in "Roche Sells 100,000 Shares in U.S. Market," *Wall Street Journal*, October 2, 1992, p. A7.

[14] Quoted in Michael R. Sesit, "Foreign Firms Flock to U.S. for IPOs," *Wall Street Journal*, June 23, 1995, p. C1.

level of financial disclosure and undertake a costly revision of its accounting practices to conform to U.S. Generally Accepted Accounting Principles (GAAP). The difference between GAAP and German accounting rules is apparent in the publishing of Daimler-Benz's results for fiscal 1994. Although Daimler showed a profit of $636 million under German rules, it reported a loss of $748 million after conforming to GAAP and eliminating the impact of drawing down hidden reserves (which is allowed in Germany). Daimler-Benz undertook the arduous task of revising its financial statements because the company wanted access to the large and liquid pool of capital represented by the U.S. market. Daimler also felt that the positive image of its Mercedes cars would help it raise capital at a lower price from wealthy Americans. During the six weeks between the announcement of its plans for a New York Stock Exchange (NYSE) listing and actually receiving the listing, Daimler-Chrysler's shares rose more than 30%, in contrast to an 11% rise for the German stock market overall.

It's not just large foreign companies that are issuing stock in the United States these days. The existence of numerous U.S. investment analysts, entrepreneurs, and investors familiar with the nature and needs of emerging firms means that many medium-sized European firms are now finding it easier and quicker to do initial public offerings (IPOs) on the NASDAQ exchange in the United States than to raise capital in their underdeveloped domestic capital markets.

An important new avenue for foreign equity (and debt) issuers, ranging from France's Rhone-Poulenc to Korea's Pohang Iron & Steel, to gain access to the U.S. market was opened up in 1990 when the Securities and Exchange Commission (SEC) adopted **Rule 144A**, which allows qualified institutional investors to trade in unregistered private placements, making them a closer substitute for public issues. This rule greatly increases the liquidity of the private placement market and makes it more attractive to foreign companies, who are frequently deterred from entering the U.S. market by the SEC's stringent disclosure and reporting requirements.

The desire to build a global shareholder base also is inducing many American companies—which until recently issued stock almost exclusively in the United States—to sell part of their issues overseas. For example, in May 1992, General Motors raised $2.1 billion by selling 40 million shares in the United States, 6 million in Britain, 4.5 million in Europe, and 4.5 million in the Far East. Exhibit 15.8 shows the announcement of that issue. As usual, the benefits of expanded ownership must be traded off against the added costs of inducing more investors to become shareholders.

Most major stock exchanges permit sales of foreign issues provided they satisfy all the listing requirements of the local market. Some of the major stock markets list large numbers of foreign stocks. For example, Union Carbide, Black & Decker, Caterpillar, and General Motors are among the more than 200 foreign stocks listed on the German stock exchanges. Similarly, almost 500 foreign stocks—including ITT, Hoover and Woolworth—are listed on the British exchanges.

Despite the benefits of cross-listing, however, companies are now more selective about where they are listed, weighing the costs as well as the benefits. Exhibit 15.9 shows that although more foreign companies are listing on the NYSE, fewer are listing on the Tokyo Stock Exchange (foreign listings in Tokyo fell from 125 in 1991 to 97 in 1994). This exodus owes to a combination of Japan's turnover tax, its high brokerage commissions, and the high cost of maintaining a Tokyo exchange listing. More foreign compa-

EXHIBIT 15.8 General Motors' Global Equity Issue

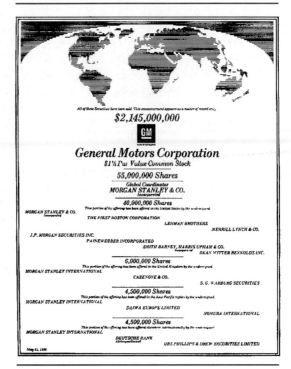

Source: Reprinted with permission from General Motors Corporation.

EXHIBIT 15.9 Number of Foreign
Companies Listed on Various Stock Markets

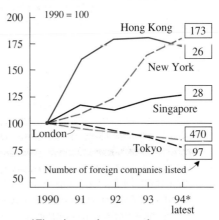

*First six months at annual rate;
London is first seven months

nies would list on the NYSE were it not for the SEC's demand that they first conform with tougher U.S. accounting and disclosure practices. The SEC, however, is under pressure to relax its stand by both foreign companies and the NYSE, the latter because it worries about falling behind in the race to list large non-U.S. companies. Nonetheless, in 1997, one out of every four new listings on the NYSE was a foreign company and such listings almost tripled between 1992 and 1997 to 320.

Globalization of Financial Markets Has Its Downside

The army of investors searching worldwide for the highest risk-adjusted returns wields a two-edged sword: It is likely to reward sound economic policies and swift to abandon countries whose economic fundamentals are questionable. As a result, countries such as Italy, Spain, Sweden, and Mexico with large public-sector or trade deficits or rapid money supply growth have earned harsh treatment from financial markets. By demanding bigger premiums for the risk of holding these nations' currencies, the markets force de facto devaluations of their currencies, thereby serving to punish the profligate and reward the virtuous. In the eyes of many international economists, markets have replaced the International Monetary Fund as the disciplinary force for the global economy.

A devaluation raises the cost of imports for a country, boosts its interest rates (to lure investors back), and forces the government to take steps to address the monetary, budget, or trade problems that led to the capital flight and devaluation in the first place. Of course, changing the policies that created these problems can impose significant costs on favored political constituents, which is why these problems were not addressed earlier.

Blaming financial markets for the political and economic disruptions caused by these policy changes misses the point. Financial markets are in the business of gathering and processing information from millions of savers and borrowers around the world in order to perform their real function, which is to price capital and allocate it to its most productive uses. In performing this function, markets reflect the perceptions of risk and reward of its participants. However, they do not create the underlying reality that caused those perceptions.

The long-run risk to the global economy caused by the abrupt shifts in capital flows and attendant waves of devaluations is that some politicians will seek to reimpose controls on capital and trade flows, particularly if the politicians manage to convince themselves that the markets are behaving in an irresponsible fashion. Such controls—whatever their motivation—would reverse the trend toward freer trade and capital markets and make the world worse off.

15.3 THE EUROMARKETS

This section discusses the Eurocurrency and Eurobond markets. It describes the functioning of these markets and then shows how each can be used to meet the multinational firm's financing requirements.

The Eurocurrency Market

A **Eurocurrency** is a dollar or other freely convertible currency deposited in a bank outside its country of origin. Thus, U.S. dollars on deposit in London become **Eurodollars**. These deposits can be placed in a foreign bank or in the foreign branch of a domestic U.S. bank.[15] The **Eurocurrency market** then consists of those banks—called **Eurobanks**—that accept deposits and make loans in foreign currencies.

The Eurobond and Eurocurrency markets are often confused with each other, but there is a fundamental distinction between the two. In the **Eurobond market, Eurobonds**, which are bonds sold outside the countries in whose currencies they are denominated, are issued directly by the final borrowers. The Eurocurrency market enables investors to hold short-term claims on commercial banks, which then act as intermediaries to transform these deposits into long-term claims on final borrowers. However, banks do play an important role in placing Eurobonds with the final investors.

The dominant Eurocurrency remains the U.S. dollar, but the importance of the Eurodollar waxes and wanes with the strength of the U.S. dollar. With dollar weakness in the latter parts of both the 1970s and 1980s, other currencies—particularly the Deutsche mark and the Swiss franc—increased in importance.

MODERN ORIGINS ▪ The origin of the post-World War II Eurodollar market is often traced to the fear of Soviet Bloc countries that their dollar deposits in U.S. banks might be attached by U.S. citizens with claims against Communist governments. Therefore, they left their dollar balances with banks in France and England.

Whatever its postwar beginnings, the Eurocurrency market has thrived for one reason: government regulation. By operating in Eurocurrencies, banks and suppliers of funds are able to avoid certain regulatory costs and restrictions that would otherwise be imposed. These costs and restrictions include:

1. Reserve requirements that lower a bank's earning asset base (that is, a smaller percentage of deposits can be lent out).
2. Special charges and taxes levied on domestic banking transactions, such as the requirement to pay Federal Deposit Insurance Corporation fees.
3. Requirements to lend money to certain borrowers at concessionary rates, thereby lowering the return on the bank's assets.
4. Interest rate ceilings on deposits or loans that inhibit the ability to compete for funds and lower the return on loans.
5. Rules or regulations that restrict competition among banks.

Although many of the most burdensome regulations and costs have been eased or abolished, the Eurocurrency market still exists. It will continue to exist as long as there are profitable opportunities to engage in offshore financial transactions. These opportunities persist because of continuing government regulations and taxes that raise costs and lower returns on domestic transactions. Owing to the ongoing erosion of domestic regulations, however, these cost and return differentials are much less significant today than they were in the past. As a consequence, the domestic money market and Eurocurrency

---

[15] The term *foreign* is relative to the operating unit's location, not to its nationality.

market are now tightly integrated for most of the major currencies, effectively creating a single worldwide money market for each participating currency.

EURODOLLAR CREATION ■ The creation of Eurodollars can be illustrated by using a series of T-accounts to trace the movement of dollars into and through the Eurodollar market.

First, suppose that Leksell AB, a Swedish firm, sells medical diagnostic equipment worth $1 million to a U.S. hospital. It receives a check payable in dollars drawn on Citibank in New York. Initially, Leksell AB deposits this check in its Citibank checking account for dollar-working-capital purposes. This transaction would be represented on the firm's and Citibank's accounts as follows:

CITIBANK		LEKSELL AB	
	Demand deposit due Leksell AB +$1M	Demand deposit in Citibank +$1M	

In order to earn a higher rate of interest on the $1 million account (U.S. banks cannot pay interest on corporate checking accounts), Leksell decides to place the funds in a time deposit with Barclays Bank in London. This transaction is recorded as follows:

CITIBANK		BARCLAYS	
	Demand deposit due Leksell AB −$1M Demand deposit due Barclays +$1M	Demand deposit in Citibank +$1M	Time deposit owed Leksell +$1M

LEKSELL AB	
Demand deposit in Citibank −$1M Demand deposit in Barclays +$1M	

One million Eurodollars have just been created by substituting a dollar account in a London bank for a dollar account held in New York. Notice that no dollars have left New York, although ownership of the U.S. deposit has shifted from a foreign corporation to a foreign bank.

Barclays could leave those funds idle in its account in New York, but the opportunity cost would be too great. If it cannot immediately loan those funds to a government or commercial borrower, Barclays will place the $1 million in the London interbank market. This involves loaning the funds to another bank active in the Eurodollar market. The interest rate at which such interbank loans are made is called the **London interbank offer rate (LIBOR)**. In this case, however, Barclays chooses to loan these funds to Ronningen SA, a Norwegian importer of fine wines. The loan is recorded as follows:

CITIBANK	BARCLAYS	
Demand deposit due Barclays −$1M	Demand deposit in Citibank −$1M	Time deposit owed Leksell +$1M
Demand deposit due Ronningen SA +$1M	Loan to Ronningen SA +$1M	

	RONNINGEN AB	
	Demand deposit in Citibank +$1M	Eurodollar loan from Barclays +$1M

We can see from this example that the Eurocurrency market involves a chain of deposits and a chain of borrowers and lenders, not buyers and sellers. One does not buy or sell Eurocurrencies. Ordinarily, an owner of dollars will place them in a time-deposit or demand-deposit account in a U.S. bank, and the owner of a French franc deposit will keep it in an account with a French bank. Until the dollar (or franc) deposit is withdrawn, control over its use resides with the U.S. (or French) bank. In fact, the majority of Eurocurrency transactions involve transferring control of deposits from one Eurobank to another Eurobank. Loans to non-Eurobanks account for fewer than half of all Eurocurrency loans. The net market size (subtracting off inter-Eurobank liabilities) is much smaller than the gross market size.

The example and data presented indicate that Eurocurrency operations differ from the structure of domestic banking operations in two ways:

1. There is a *chain of ownership* between the original dollar depositor and the U.S. bank.
2. There is a *changing control over the deposit* and the use to which the money is put.

It should be noted, however, that despite the chain of transactions, the total amount of foreign dollar deposits in the United States remains the same. Moreover, on the most fundamental level—taking in deposits and allocating funds—the Eurocurrency market operates much as does any other financial market, except for the absence of government regulations on loans that can be made and interest rates that can be charged. This section now examines some of the particular characteristics of Eurocurrency lending.

EUROCURRENCY LOANS ■ The most important characteristic of the Eurocurrency market is that loans are made on a floating-rate basis. Interest rates on loans to governments and their agencies, corporations, and nonprime banks are set at a fixed margin above LIBOR for the given period and currency chosen. At the end of each period, the interest for the next period is calculated at the same fixed margin over the new LIBOR. For example, if the margin is 75 **basis points** (100 basis points equal 1%) and the current LIBOR is 6%, then the borrower is charged 6.75% for the upcoming period. The reset period normally chosen is six months, but shorter periods such as one month or three months are possible. The LIBOR used corresponds to the maturity of the reset period (for example, six-month LIBOR, or LIBOR6, for a six-month reset period).

The *margin*, or spread between the lending bank's cost of funds and the interest charged the borrower, varies a good deal among borrowers and is based on the borrower's perceived riskiness. Typically, such spreads have ranged from as little as 15 basis points (0.15%) to more than 3%, with the median being somewhere between 1% and 2%.

The *maturity* of a loan can vary from approximately three to 10 years. Maturities have tended to lengthen over time, from a norm of about five years originally to a norm of eight to 10 years these days for prime borrowers. Lenders in this market are almost exclusively banks. In any single loan, there normally will be a number of participating banks that form a syndicate. The bank originating the loan usually will manage the syndicate. This bank, in turn, may invite one or two other banks to comanage the loan. The managers charge the borrower a once-and-for-all syndication fee of 0.25% to 2% of the loan value, depending on the size and type of the loan. Part of this fee is kept by the managers, and the rest is divided up among all the participating banks (including the managing banks) according to the amount of funds each bank supplies.

The **drawdown**—the period over which the borrower may take down the loan—and the repayment period vary in accordance with the borrower's needs. A commitment fee of about 0.5% per annum is paid on the unused balance, and prepayments in advance of the agreed-upon schedule are permitted but are sometimes subject to a penalty fee. Borrowers in the Eurocurrency market are concerned about the effective interest rate on the money they have raised. This rate compares the interest rate paid to the net loan proceeds received by the borrower. Its computation is shown in the following example.

ILLUSTRATION CALCULATING THE EFFECTIVE ANNUAL COST OF A EUROCURRENCY LOAN

San Miguel, the Filipino beer brewer, has arranged a DM 250 million, 5-year Euro-DM loan with a syndicate of banks led by Credit Suisse and Deutsche Bank. With an up-front syndication fee of 2.0%, net proceeds to San Miguel are:

DM 250,000,000 − (0.02 × DM 250,000,000)

$$= DM\ 245,000,000$$

The interest rate is set at LIBOR + 1.75%, with LIBOR reset every six months.

Assuming an initial LIBOR6 rate for DM of 5.5%, the first semi-annual debt service payment is

$$\frac{0.055 + 0.0175}{2} \times DM\ 250,000,000$$

$$= DM\ 9,062,500$$

San Miguel's effective annual interest rate for the first six months is thus

$$\frac{DM\ 9,062,500}{DM\ 245,000,000} \times 2 \times 100 = 7.40\%$$

Every six months, this annualized cost will change with LIBOR6.

MULTICURRENCY CLAUSES ■ Borrowing can be done in many different currencies, although the dollar is still the dominant currency. Increasingly, Eurodollars have a multicurrency clause. This clause gives the borrower the right (subject to availability) to switch from one currency to another on any rollover (or reset) date. The multicurrency option enables the borrower to match currencies on cash inflows and outflows (a potentially valuable exposure management technique as we saw in Chapter 9). Equally important, the option allows a firm to take advantage of its own expectations regarding currency changes (if they differ from the market's expectations) and shop around for those funds with the lowest effective cost.

An example of a typical multicurrency loan is a $100 million, 10-year revolving credit arranged by the Dutch firm Thyssen Bornemisza NV with nine Dutch, German, U.S., and Swiss banks, led by Amsterdam-Rotterdam Bank. Rates are fixed, at the company's discretion, at three-month, six-month, or 12-month intervals. At each rollover date, the firm can choose from any freely available Eurocurrency except Eurosterling, but only four different Eurocurrencies may be outstanding at any one time.

RELATIONSHIP BETWEEN DOMESTIC AND EUROCURRENCY MONEY MARKETS ■ The presence of arbitrage activities ensures a close relationship between interest rates in national and international (Eurocurrency) money markets. Interest rates in the U.S. and Eurodollar markets, for example, can differ only to the extent that there are additional costs, controls, or risks associated with moving dollars between, say, New York and London. Otherwise, arbitrageurs would borrow in the low-cost market and lend in the high-return market, quickly eliminating any interest differential between the two.

The cost of shifting funds is relatively insignificant, so we must look to currency controls or risk to explain any substantial differences between domestic and external

EXHIBIT 15.10 Interest Rate Relationships Between Domestic and Eurocurrency Credit Markets

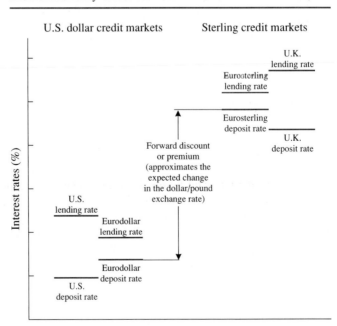

rates. To the extent that exchange controls are effective, the national money market can be isolated or segmented from its international counterpart. In fact, the difference between internal and external interest rates can be taken as a measure of the effectiveness of the monetary authorities' exchange controls.

Interest differentials also can exist if there is a danger of future controls. The possibility that at some future time either the lender or borrower will not be able to transfer funds across a border—also known as **sovereign risk**—can help sustain persistent differences between domestic and external money market rates.

In general, Eurocurrency spreads (a spread is the margin between lending and deposit rates) are narrower than in domestic money markets (see Exhibit 15.10). Lending rates can be lower because:

1. The absence of the previously described regulatory expenses that raise costs and lower returns on domestic transactions.
2. Most borrowers are well known, reducing the cost of information gathering and credit analysis.
3. Eurocurrency lending is characterized by high volumes, allowing for lower margins; transactions costs are reduced because most of the loan arrangements are standardized and conducted by telephone or telex.
4. Eurocurrency lending can and does take place out of tax-haven countries, providing for higher after-tax returns.

Eurocurrency deposit rates are higher than domestic rates for the following reasons:

1. They must be higher to attract domestic deposits.
2. Eurobanks can afford to pay higher rates based on their lower regulatory costs.
3. Eurobanks are able to pay depositors higher interest rates because they are not subject to the interest rate ceilings that prevail in many countries.
4. A larger percentage of deposits can be lent out.

EUROMARKET TRENDS ■ In recent years, the London interbank offer rate has started to fade as a benchmark for lending money in the Eurocurrency market, in much the same way that the prime rate is no longer the all-important benchmark in the U.S. bank loan market. Although Eurocurrency rates still are computed off LIBOR, a number of creditworthy borrowers—including Denmark, Sweden, several major corporations, and some banks—are obtaining financing in the Euromarkets at interest rates well below LIBOR. For example, high-quality borrowers can borrow at the **London interbank bid** rate (**LIBID**), the rate paid by one bank to another for a deposit, which is about 12.5 basis points (1/8 of 1%) below LIBOR. The highest-quality borrowers, such as the World Bank, can raise funds at below LIBID.

This trend largely reflects the fact that because many international bank loans soured in the early 1980s, banks lost much of their appeal to investors. As a result, banks' ability to impose themselves as the credit yardstick by which all other international borrowers are measured has faltered. What the Euromarket is saying in effect is that borrowers such as Denmark and the World Bank are considered better credit risks than are many banks.

Moreover, investor preferences for an alternative to bank Eurodollar certificates of deposit (whereby banks substitute their credit risk for their borrowers') have enabled investment banks to transform usual bank-syndicated lending into securities offerings, such as floating-rate notes. This preference for the ultimate borrower's credit risk, rather than the bank's credit risk, led to rapid growth in the Eurobond market, particularly the floating-rate segment of the market.

Eurobonds

Eurobonds are similar in many respects to the public debt sold in domestic capital markets, largely consisting of fixed-rate, floating-rate, and equity-related debt. Unlike domestic bond markets, however, the Eurobond market is almost entirely free of official regulation, but instead is self-regulated by the Association of International Bond Dealers. The prefix *Euro* indicates that the bonds are sold outside the countries in whose currencies they are denominated. For example, the General Motors issue shown in Exhibit 15.11 is a Eurobond. You can tell that because the tombstone says, "These securities have not been registered under the United States Securities Act of 1933 and may not be offered or sold in the United States or to United States persons as part of the distribution."

Borrowers in the Eurobond market are typically well known and have impeccable credit ratings (for example, developed countries, international institutions, and large

EXHIBIT 15.11 Announcement of a GMAC
Eurobond Issue

These securities have not been registered under the United States Securities Act of 1933 and may not be offered or sold in the United States or to United States persons as part of the distribution.

General Motors Acceptance Corporation

(Incorporated in the State of New York, United States of America)

U.S.$200,000,000

7⅝ per cent. Notes due September 3, 1991

Swiss Bank Corporation International Limited

Credit Suisse First Boston Limited	Deutsche Bank Capital Markets Limited
Merrill Lynch Capital Markets	Morgan Stanley International
Nomura International Limited	Salomon Brothers International Limited

Union Bank of Switzerland (Securities) Limited

Algemene Bank Nederland N.V.	BankAmerica Capital Markets Group
Bankers Trust International Limited	Banque Bruxelles Lambert S.A.
Banque Générale du Luxembourg S.A.	Banque Nationale de Paris
Banque Paribas Capital Markets Limited	Commerzbank Aktiengesellschaft
Crédit Lyonnais	Creditanstalt-Bankverein
Daiwa Europe Limited	IBJ International Limited
Leu Securities Limited	The Nikko Securities Co., (Europe) Ltd.
Shearson Lehman Brothers International	Société Générale
Sumitomo Trust International Limited	Swiss Volksbank
S.G. Warburg Securities	Wood Gundy Inc.

Yamaichi International (Europe) Limited

New Issue   This announcement appears as a matter of record only.   September 1986

Source: Reprinted with permission of General Motors Acceptance Corporation.

EXHIBIT 15.12 Eurobond New Issues: 1979–1997

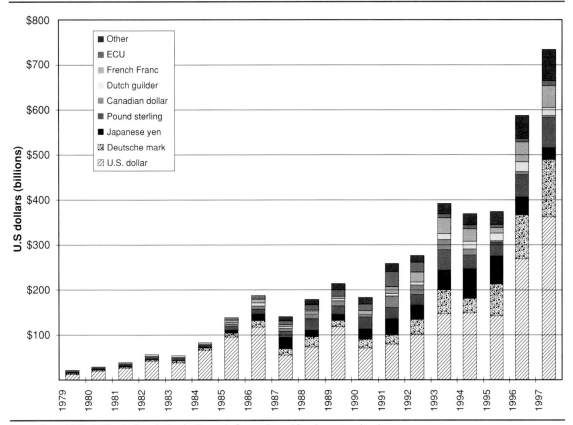

Source: Financial Market Trends, Organization for Economic Cooperation and Development, various issues.

multinational corporations like GM). Even then the amounts raised in the Eurobond market have historically been far less than those in the Eurocurrency market.

As can be seen in Exhibit 15.12, however, the Eurobond market has grown dramatically during the past 15 years, and its size now exceeds that of the Eurocurrency market. However, Exhibit 15.12 also shows that the Eurobond market has had its ups and downs.

As the market has grown, so have the numbers of large and small players, with about 60 financial institutions regularly competing for business, up from about 40 in 1990. The increased competition among underwriters—about 30 major banks are trying to be top-10 underwriters—has meant falling profit margins for them. But very few banks have dropped out or cut back yet, if only for overall marketing and public relations reasons. The Eurobond market is a key part of the global capital-raising business and most banks feel they have to be a part of it no matter what. At some point, of course, a shakeout will occur, but that has not happened just yet.

SWAPS ■ The major catalyst for growth in the Eurobond market in the last 15 years has been the emergence of a technique known as the **swap**, a financial transaction in which two counterparties agree to exchange streams of payments over time. It is now estimated that 70% of Eurobond issues are "swap-driven." Swaps allow borrowers to raise money in one market and to swap one interest rate structure for another (for example, from fixed to floating), or to go further and to swap principal and interest from one currency to another. These swaps allow the parties to the contract to arbitrage their relative access to different currency markets. A borrower whose paper is much in demand in one currency can obtain a cost saving in another currency sector by raising money in the former and swapping the funds into the latter currency. The next chapter discusses swaps in detail.

LINKS BETWEEN THE DOMESTIC AND EUROBOND MARKETS ■ The growing presence of sophisticated investors willing to arbitrage between the domestic dollar and Eurodollar bond markets — in part because the United States no longer imposes withholding taxes on foreign investors — has eliminated much of the interest disparity that once existed between Eurobonds and domestic bonds. Despite the closer alignment of the two markets, though, the Eurobond issuer may at any given time take advantage of Eurobond "windows" when a combination of domestic regulations, tax laws, and expectations of international investors enables the issuer to achieve a lower financing cost — often involving currency and interest swaps — than is available in domestic markets.[16] In addition to the possibility of reduced borrowing costs, the Eurobond issuer may diversify its investor base and funding sources by having access to the international Eurocapital markets of Western Europe, North America, and the Far East.

PLACEMENT ■ Issues are arranged through an underwriting group, often with a hundred or more underwriting banks involved for an issue as small as $25 million. A growing volume of Eurobonds is being placed privately because of the simplicity, speed, and privacy with which private placements can be arranged.

CURRENCY DENOMINATION ■ Historically, about 75% of Eurobonds have been dollar denominated. During the late 1970s, however, when the dollar was in a downward spiral, other currencies (particularly the Deutsche mark) became more important in the Eurobond market. The sharp increase in the share of dollar-denominated Eurobonds in the period up to mid-1985, shown in Exhibit 15.12, largely reflects the surging value of the dollar. The subsequent drop in the dollar's value again led to a rise in the share of nondollar issues, particularly yen and Deutsche mark issues. The absence of Swiss franc Eurobonds is due to the Swiss Central Bank's ban on using the Swiss franc for Eurobond issues.

As an alternative to issuing dollar, Deutsche mark, or other single-currency-denominated Eurobonds, several borrowers in recent years have offered bonds whose value is a weighted average, or "basket," of several currencies. The most successful of these currency "cocktails" is the European Currency Unit (ECU). From 1988 through 1997, corporations and government agencies sold about $135 billion of ECU bonds.

---

[16]The existence of such windows is documented by Yong-Cheol Kim and Rene M. Stulz, "The Eurobond Market and Corporate Financial Policy: A Test of the Clientele Hypothesis," *Journal of Financial Economics* 22 (1988): 189–205; and Yong-Cheol Kim and Rene M. Stulz, "Is There a Global Market for Convertible Bonds?" *Journal of Business* 65 (1992): 75–91.

INTEREST RATES ON FIXED-RATE EUROBONDS ■ Fixed-rate Eurobonds ordinarily pay their coupons once a year, in contrast to bonds issued in the U.S. market, where interest normally is paid on a semiannual basis. Issuers, of course, are interested in their **all-in cost**—that is, the effective interest rate on the money they have raised. This interest rate is calculated as the discount rate that equates the present value of the future interest and principal payments to the net proceeds received by the issuer. In other words, it is the *internal rate of return* on the bond. To compare a Eurobond issue with a U.S. domestic issue, therefore, the all-in cost of funds on an annual basis must be converted to a semiannual basis or vice versa.

The annual yield can be converted to a semiannual rate by use of the following formula:

$$\text{Semiannual yield} = (1 + \text{Annual yield})^{1/2} - 1 \qquad (15.1)$$

Alternatively, a semiannual yield can be annualized by rearranging terms in Equation 15.1 as follows:

$$\text{Annual yield} = (1 + \text{Semiannual yield})^2 - 1 \qquad (15.2)$$

For example, suppose that Procter & Gamble plans to issue a five-year bond with a face value of $100 million. Its investment banker estimates that a Eurobond issue would have to bear a 7.5% coupon and that fees and other expenses will total $738,000, providing net proceeds to P&G of $99,262,000. Exhibit 15.13a shows the cash flows associated with the Eurobond issue. The all-in cost of this issue, which is an annual rate, is shown as 7.68% (rounded to the nearest basis point). The computation was done using an Excel spreadsheet. To ensure that we have the correct all-in cost of funds, the third column shows that the present values of the cash flows, using a discount rate of 7.68%, sum to P&G's net proceeds of $99,262,000.

Alternatively, its investment banker tells Procter & Gamble that it can issue a $100 million five-year bond in the U.S. market with a coupon of 7.4%. With estimated issuance costs of $974,000, P&G will receive net proceeds of $99,026,000. Exhibit 15.13b shows the cash flows associated with this issue and its all-in cost of 3.82%. Notice that the cash flows are semiannual, as is the all-in cost. The third column performs the same check on the present value of these cash flows to ensure that we have the correct all-in cost.

According to Equation 15.1, the equivalent semiannual all-in cost for the Eurobond issue is $(1.0768)^{1/2} - 1 = 3.77\%$. These figures reveal that the all-in cost of the Eurobond is lower, making it the preferred issue assuming that other terms and conditions on the bonds are the same.

We could have converted the U.S. bond yield to its annual equivalent using Equation 15.2 and then compared that figure to the Eurobond yield of 7.68%. This computation would have yielded an annualized all-in cost of the U.S. bond issue equal to $(1.0382)^2 - 1 = 7.78\%$. As before, the decision is to go with the Eurobond issue because its all-in cost is 10 basis points lower.

INTEREST RATES ON FLOATING-RATE EUROBONDS ■ The interest rates on floating-rate Eurobonds (FRNs) are normally set in the same way as on Eurocurrency loans—as a

---

EXHIBIT 15.13 Comparison of All-In Costs of a Eurobond Issue and a U.S. Dollar Bond Issue

a. Calculation of Eurobond all-in cost

Year	Cash Flows	Present Value at 7.68%
0[1]	99,262,000	99,262,000
1[2]	(7,500,000)	(6,964,868)
2	(7,500,000)	(6,467,919)
3	(7,500,000)	(6,006,427)
4	(7,500,000)	(5,577,864)
5[3]	(107,500,000)	(74,244,921)
Internal rate of return	7.68%	Sum 0

[1]Proceeds of $100 million net of $738,000 in expenses.
[2]Coupon of 7.5% applied to $100 million principal.
[3]Repayment of $100 million principal plus last interest payment.

b. Calculation of U.S. dollar bond all-in cost

Six-Month Period		Present Value at 3.82%
0[1]	99,026,000	99,026,000
1[2]	−3,700,000	(3,563,895)
2	−3,700,000	(3,432,796)
3	−3,700,000	(3,306,520)
4	−3,700,000	(3,184,889)
5	−3,700,000	(3,067,732)
6	−3,700,000	(2,954,885)
7	−3,700,000	(2,846,189)
8	−3,700,000	(2,741,491)
9	−3,700,000	(2,640,644)
10[3]	−103,700,000	(71,286,961)
Internal rate of return	3.82%	Sum 0

[1]Proceeds of $100 million net of $974,000 in expenses.
[2]Semiannual coupon of 3.7% (7.4% annual) applied to $100 million principal.
[3]Repayment of $100 million principal plus last interest payment.

---

fixed spread over a reference rate, usually LIBOR. The reset period is usually three months or six months and the maturity of LIBOR corresponds to the reset period (for example, LIBOR6 for a six-month reset period). So, for example, if LIBOR3 on a reset date is 6.35%, an FRN that pays 50 basis points over LIBOR3 will bear an interest rate of 6.85% for the next 3 months.

Unlike the previous example, some FRNs have coupons that move opposite to the reference rate. Such notes are known as **inverse floaters**. An example would be a bond that pays 12% − LIBOR6. If LIBOR6 is, say, 7% on the reset date, the FRN's interest rate would be 5% (12% − 7%) for the coming six-month period. The risk to the issuer is that LIBOR will decline. Conversely, the investor's return suffers if LIBOR rises. There is a floor on the coupon at zero, meaning that the interest rate on an inverse floater is never negative.

EUROBOND SECONDARY MARKET ■ Historically, there has been a lack of depth in the Eurobond **secondary market** (the market where investors trade securities already bought). However, the growing number of institutions carrying large portfolios of Eurobonds for trading purposes has increased the depth and sophistication of this market, making it second only to the U.S. domestic bond market in liquidity, where liquidity refers to the ease of trading securities at close to their quoted price. Until recently, the liquidity of Eurobonds has not mattered because investors were usually willing to purchase and lock such issues away. However, because of heightened volatility in bond and currency markets, investors increasingly want assurance that they can sell Eurobonds before maturity at bid-ask spreads (the difference between the buy and the sell rate, which is a major determinant of liquidity) comparable to those in other capital markets.

One problem is that there is no central trading floor where dealers post prices. Hence, buyers sometimes have difficulty getting price quotes on Eurobonds. However, many commercial banks, investment banks, and securities trading firms act as *market-makers* in a wide range of issues by quoting two-way prices (buy and sell) and being prepared to deal at those prices. Another factor adding liquidity to the market is the presence of bond brokers. They act as middlemen, taking no positions themselves but transacting orders when a counterparty is found to match a buy or sell instruction. Brokers deal only with marketmakers and never with the ultimate investor.

EUROBOND RETIREMENT ■ Sinking funds or purchase funds are usually required if a Eurobond is of more than seven years' maturity. A *sinking fund* requires the borrower to retire a fixed amount of bonds yearly after a specific number of years. By contrast, a *purchase fund* often starts in the first year, and bonds are retired only if the market price is below the issue price. The purpose of these funds is to support the market price of the bonds, as well as reduce bondholder risk by assuring that not all the firm's debt will come due at once.

The desire for price support is reinforced by the past lack of depth in the secondary market. However, as the secondary market has become more liquid, this motivation for price support has become less important.

Most Eurobond issues carry *call provisions*, giving the borrower the option of retiring the bonds before maturity should interest rates decline sufficiently in the future. As with domestic bonds, Eurobonds with call provisions require both a call premium and higher interest rates relative to bonds without call provisions.

RATINGS ■ As noted earlier, Eurobond issuers are typically large multinational companies, government agencies, state-owned enterprises, or international organizations, all with familiar names and impeccable credit reputations. For this reason, in the past most Eurobonds carried no credit ratings, particularly because rating agencies charge a fee. However, as the market has expanded to include newer issuers who are less well-known and with lesser credit reputations, investors have demanded ratings to better assess an issuer's credit risk. The majority of Eurobond ratings are provided by Moody's and Standard and Poor's, the dominant U.S. rating agencies. Euroratings is another agency that specializes in the Euromarkets, particularly in the dollar-denominated debt of non-U.S. issuers.

Eurobond issues are rated according to their relative degree of default risk. Note that interest rate changes or currency movements can make even default-free Eurobonds very risky. Bond raters, however, stick to just credit risk. Their focus is on the issuer's ability to generate sufficient quantities of the currency in which their debt is denominated. This element is particularly important for borrowers issuing foreign-currency-denominated debt (for example, the government of Mexico issuing debt denominated in U.S. dollars).

RATIONALE FOR EXISTENCE OF EUROBOND MARKET ■ The Eurobond market survives and thrives because, unlike any other major capital market, it remains largely unregulated and untaxed. Thus, big borrowers, such as Texaco, IBM, and Sears Roebuck, can raise money more quickly and more flexibly than they can at home. And because the interest investors receive is tax free, these companies historically have been able to borrow at a rate that is below the rate at which the U.S. Treasury could borrow.

The tax-free aspect of Eurobonds is related to the notice in the tombstone for the GMAC Eurobond issue that it may not be offered to the U.S. public. U.S. tax law requires that for interest and principal to be payable in the United States, bonds must be in registered form. Eurobonds, however, are issued in bearer form, meaning they are unregistered, with no record to identify the owners. (Money can be considered to be a zero-coupon bearer bond.) This feature allows investors to collect interest in complete anonymity and, thereby, evade taxes. Although U.S. law discourages the sale of such bonds to U.S. citizens or residents, bonds issued in bearer form are common overseas.[17] As expected, investors are willing to accept lower yields on bearer bonds than on non-bearer bonds of similar risk.

Highly rated U.S. firms have long taken advantage of this opportunity to reduce their cost of funds by selling overseas Eurobonds in bearer form. Often corporations could borrow abroad below the cost at which the U.S. government could borrow at home. Exxon's issue of zero-coupon Eurobonds shows how companies were able to exploit the arbitrage possibilities inherent in such a situation. Zero-coupon bonds pay no interest until maturity. Instead, they are sold at a deep discount from their par value.

It is apparent that the Eurobond market, like the Eurocurrency market, exists because it enables borrowers and lenders alike to avoid a variety of monetary authority regulations and controls, as well as providing them with an opportunity to escape the payment of some taxes. As long as governments attempt to regulate domestic financial markets but allow a (relatively) free flow of capital among countries, the external financial markets will survive. If tax and regulatory costs rise, these markets will grow in importance.

Recent years have seen a reversal of some of these tax and regulatory costs. In 1984, the United States repealed its withholding tax on interest paid to foreign bondholders. That made domestic bonds, particularly U.S. Treasurys, more attractive to foreign investors. Whereas yields on top-rated Eurodollar bonds were about 40 basis points (100

---

[17] Americans can buy Eurobonds, but not until 40 days after they have been issued, which cuts down on availability because many Eurobond issues are bought and stashed away.

EXXON ENGAGES IN INTERNATIONAL TAX ARBITRAGE

In the fall of 1984, Exxon sold $1.8 billion principal amount of zero-coupon Eurobonds due November 2004 at an annual compounded yield of 11.65%, realizing net proceeds of about $199 million:

$$\text{Bond value} = \frac{\$1,800,000,000}{(1.1165)^{20}}$$

It then used part of the proceeds to buy $1.8 billion principal amount of U.S. Treasury bonds maturing in November 2004 from which the coupons had been removed and sold separately. The yield on these "stripped" Treasurys, which are effectively zero-coupon Treasury bonds, was around 12.20%.[18] At this yield, it would have cost Exxon $180 million to purchase the $1.8 billion in stripped Treasury bonds:

$$\text{Bond value} = \$1,800,000,000/(1.1220)^{20}$$
$$= \$180,000,000$$

At this price, Exxon earned the difference of about $19 million.

A peculiar quirk in Japanese law is largely responsible for the big difference in yield between zero-coupon Eurobonds and stripped Treasurys: Japanese investors—who were the principal buyers of the Eurobonds—did not have to pay tax on a zero-coupon bond's accrued interest if they sold the bond prior to maturity. Because of this tax advantage, they were willing to pay a premium price for zeros (relative to coupon-bearing bonds). Although in principle the Japanese would have preferred to purchase the higher-yielding (and safer) stripped U.S. Treasury bonds, they are prohibited by Japanese law from doing so. The threatened taxation of the accrued interest on zeros in Japan has eliminated this arbitrage opportunity.

---

[18] This case is discussed at greater length in John D. Finnerty, "Zero Coupon Bond Arbitrage: An Illustration of the Regulatory Dialectic at Work," *Financial Management*, Winter 1985, pp. 13–17.

basis points equals 1%) *below* similar-maturing Treasurys in 1984, the same Eurobonds in 1988 yielded about 70 basis points more than did Treasurys (see Exhibit 15.14).

At the same time, the United States began permitting well-known companies—precisely the ones that would otherwise have used the Eurobond market—to bypass complex securities laws when issuing new securities by using the *shelf registration* procedure. By lowering the cost of issuing bonds in the United States and dramatically speeding up the issuing process, shelf registration improved the competitive position of the U.S. capital market relative to the Eurobond market. More recently, the SEC's adoption of Rule 144A now enables companies to issue bonds simultaneously in Europe and the United States, further blurring the distinction between the U.S. bond market and its Eurobond equivalent. Other nations, such as Japan and England, are also deregulating their financial markets.

With repeal of the withholding tax and financial market deregulation in the United States and elsewhere, the Eurobond market lost some of the cost advantage that lured corporate borrowers in the past. Nonetheless, as long as Eurobond issuance entails low

EXHIBIT 15.14 Eurodollar Bond Yields Relative to Ten-Year U.S. Treasury Yields

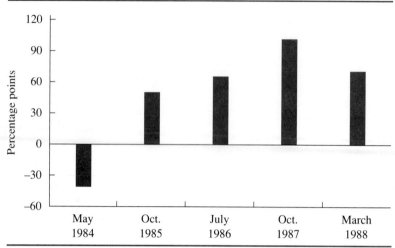

regulatory and registration costs relative to domestic bond issuance, the Eurobond market will continue to attract investors and borrowers from all over the world. Part of the lower registration costs stem from the less-stringent disclosure requirements in the Eurobond market, particularly as compared with the U.S. market. Even after the advent of shelf registration and Rule 144A, SEC disclosure requirements still are considered to be costly, time consuming, and burdensome by both U.S. and non-U.S. issuers.

Despite several forecasts of imminent death, the Eurobond market has survived, largely because its participants are so fleet of foot. As demand for one type of bond declines, quick-witted investment bankers seem to find other opportunities to create value for their customers. When the demand for fixed-rate Eurobonds fell, the Eurobond market led the boom in floating-rate note issues. When the FRN market collapsed in 1986, this business was replaced by issues of Japanese corporate bonds with equity warrants attached. In return for what is in effect a long-dated call option on the issuer's stock, the investor accepts a lower interest rate on the Eurobond to which the equity warrant is attached.

Demand for such issues soared as investors used the warrants to play Japan's rising stock market. In turn, Japanese companies found the Eurobond market easier and cheaper to use than the regulated domestic-yen market. However, the Tokyo stock market's plunge in 1990 triggered a 71.5% drop in the Japanese equity-warrant issue market and a 14.4% fall in the Eurobond market overall. The equity-related issue market continued to weaken through 1995.

The financial infrastructure in place in London should ensure the Eurobond market's survival. However, tax harmonization, financial deregulation, and the widespread loosening of capital controls mean that issuers have less incentive to borrow money offshore and are returning to their domestic markets to raise capital. If these trends persist, the Eurobond market may never regain its preeminence. However, it can still preserve its basic role—as the nimblest intermediary for international capital flows between domestic markets.

Eurobonds versus Eurocurrency Loans

Both Eurocurrency and Eurobond financing have their advantages and disadvantages. Although many of these factors are reflected in the relative borrowing costs, not all factors are so reflected. For a given firm, therefore, and for a specific set of circumstances, one method of financing may be preferred to the other. The differences are categorized in five ways.

1. *Cost of borrowing.* Eurobonds are issued in both fixed-rate and floating-rate forms. Fixed-rate bonds are an attractive exposure-management tool because known long-term currency inflows can be offset with known long-term outflows in the same currency. In contrast, the interest rate on a Eurocurrency loan is variable, making Eurocurrency loans better hedges for noncontractual currency exposures. The variable interest rate benefits borrowers when rates decline, but it hurts them when rates rise. Arbitrage between Eurobonds and Eurocurrencies, however, should not provide an automatic cost advantage to one or the other form of borrowing.

2. *Maturity.* Although the period of borrowing in the Eurocurrency market has tended to lengthen over time, Eurobonds still have longer maturities.

3. *Size of issue.* Historically, the amount of loanable funds available at any one time has been much greater in the interbank market than in the bond market. Now, however, the volume of Eurobond offerings exceeds global bank lending. In many instances, borrowers have discovered that the Eurobond market can easily accommodate financings of a size and at a price not previously thought possible. Moreover, although in the past the flotation costs of a Eurocurrency loan have been much lower than on a Eurobond (about 0.5% of the total loan amount versus about 2.25% of the face value of a Eurobond issue), competition has worked to lower Eurobond flotation costs.

4. *Flexibility.* In the case of a Eurobond issue, the funds must be drawn down in one sum on a fixed date and repaid according to a fixed schedule unless the borrower pays an often substantial prepayment penalty. By contrast, the drawdown in a floating-rate loan can be staggered to suit the borrower's needs with a fee of about 0.5% per annum paid on the unused portion (normally much cheaper than drawing down and redepositing) and can be prepaid in whole or in part at any time, often without penalty. Moreover, a Eurocurrency loan with a multicurrency clause enables the borrower to switch currencies on any rollover date, whereas switching the denomination of a Eurobond from currency A to currency B would require a costly combined refunding and reissuing operation. A much cheaper and comparable alternative, however, would be to sell forward for currency B an amount of currency A equal to the value of the Eurobond issue still outstanding. There is a rapidly growing market in such currency swaps that enable the proceeds from bonds issued in one currency to be converted into money in another currency.

5. *Speed.* Internationally known borrowers can raise funds in the Eurocurrency market very quickly, often within two to three weeks of first request. A Eurobond financing generally takes more time to put together, although here again the difference is becoming less significant.

Note Issuance Facilities and Euronotes

Eurobanks have responded to the competition from the Eurobond market by creating a new instrument: the **note issuance facility** (NIF). The NIF, which is a low-cost substitute for syndicated credits, allows borrowers to issue their own short-term **Euronotes**, which are then placed or distributed by the financial institutions providing the NIF. NIFs—sometimes also called *short-term note issuance facilities*, or SNIFs—have some features of the U.S. commercial paper market and some features of U.S. commercial lines of credit. Like commercial paper, notes under NIFs are unsecured short-term debt generally issued by large corporations with excellent credit ratings. Indeed, as noted in Chapter 13, notes issued under NIFs sometimes are referred to as Euro-commercial paper or Euro-CP. Like loan commitments in the United States, NIFs generally include multiple pricing components for various contract features, including a market-based interest rate and one or more fees known as participation, facility, and underwriting fees. Participation fees are paid when the contract is formalized and are generally about 10 basis points times the facility size. Other fees are paid annually and are sometimes based on the full size of the facility, sometimes on the unused portions.

Many NIFs include underwriting services as part of the arrangements. When they are included, the arrangement generally takes the form of a **revolving underwriting facility** (RUF). The RUF gives borrowers long-term continuous access to short-term money underwritten by banks at a fixed margin.

NIFs are more flexible than floating-rate notes and usually cheaper than syndicated loans. Banks eager to beef up their earnings without fattening their loan portfolios (which would then require them to add expensive equity capital) made NIFs an important new segment of the Euromarket. As in the case of floating-rate notes, the popularity of NIFs benefits from the market's current preference for lending to high-grade borrowers through securities rather than bank loans.

Here's how the basic facility works (although alternate methods exist in abundance in the marketplace). A syndicate of banks underwrites an amount—usually about $50–$200 million—for a specified period, typically for five to seven years. A LIBOR-based underwriting margin is set, determined by the credit rating of the borrower, the size of the issue, and market conditions. When the borrower decides to draw on the facility, the borrower can choose to issue promissory notes, called Euronotes, with one-month, three-month, six-month, or 12-month maturities. A tender panel of banks is then established, whose members submit competitive bids. Any bids above the agreed underwriting margin are not accepted but are automatically purchased by the underwriters at the agreed-upon margin over LIBOR.

In effect, NIFs are put options. They give borrowers the right to sell their paper to the bank syndicate at a price that yields the prearranged spread over LIBOR. Borrowers will exercise this right only if they cannot place their notes at a better rate elsewhere, a plight most likely to occur if their creditworthiness deteriorates. The primary risk to the banks, therefore, is that they might someday have to make good on their pledge to buy paper at a spread that is too low for the credit risks involved. Although Euronote issuers generally are firms with sound credit standing, a NIF may oblige the banks to keep rolling the notes over for five to 10 years—time enough for even the best credit risk to turn into a nightmare.

Most Euronotes are denominated in U.S. dollars and are issued with high face values (often $500,000 or more). They are intended for professional or institutional investors rather than private individuals.

The pricing of the Euro-CP issued under NIFs depends on two conventions. First, instead of carrying a coupon rate, Euro-CP are sold at a discount from face value. The return to the investor is the difference between the purchase price of the security and its face value. Second, the yield is usually quoted on a discount basis from its face value and is expressed in annual terms based on a 360-day year as follows:

$$\text{Discount rate} = \frac{\text{Discount}}{\text{Face value}} \times \frac{360}{n} \qquad (15.3)$$

where n is the number of days to maturity. For example, if a Euro-CP note with a face value of $100,000 and 87 days to maturity is sold by the Export-Import Bank of Japan (JEXIM) at a discount of $1,600 from face value, its discount rate would be computed as

$$\text{Discount rate} = \frac{1,600}{100,000} \times \frac{360}{87} = 6.62\%$$

Giving the discount pricing conventions for Euro-CP, the market price of an issue is calculated as

$$\text{Market price} = \text{Face value} \times \left[1 - \left(\text{Discount rate} \times \frac{n}{360} \right) \right] \qquad (15.4)$$

The application of Equation 15.4 would yield a market price of $98,400 for the JEXIM issue:

$$\text{Market price} = \$100,000 \times \left[1 - \left(0.0662 \times \frac{87}{360} \right) \right] = \$98,400$$

To compute the annual yield on a basis comparable to interest-bearing securities that also use a 360-day year, the discount must be divided by the market price rather than the face value:

$$\text{Annual yield} = \frac{\text{Discount}}{\text{Market price}} \times \frac{360}{n} \qquad (15.5)$$

In the case of the JEXIM issue, the annual yield would be 6.73%:

$$\text{Annual yield} = \frac{1,600}{98,400} \times \frac{360}{87} = 6.73\%$$

The relation between the annual yield and the discount rate, both based on a 360-day year, is as follows:

$$\text{Annual yield} = \text{Discount rate} \times \frac{\text{Face value}}{\text{Market price}} \qquad (15.6)$$

Euro-CP is sold at a discount to face value, so the annual yield will always exceed the discount rate.

The price of a Euro-CP issue can also be expressed in terms of its annual yield:

$$\text{Market price} = \frac{\text{Face value}}{1 + \left(\text{Annual yield} \times \dfrac{n}{360}\right)} \tag{15.7}$$

For example, the market price of a 114-day Euro-CP issue with a face value of $10,000 that is priced to yield 7.45% annually would be:

$$\text{Market price} = \frac{\$10,000}{1 + \left(0.0745 \times \dfrac{114}{360}\right)} = \$9,769.52$$

The yields on some Euro-CP, however, are based on a 365-day year. To reflect this difference, we would have to substitute 365 for 360 in the formula used to calculate the market price.

Note Issuance Facilities versus Eurobonds

In addition to their lower direct costs, NIFs offer several other benefits to the issuer relative to floating rate notes, their most direct competitor.

1. *Drawdown flexibility.*   Note issuers usually can opt to draw down all or part of their total credit whenever their need arises, and they can roll over portions at will. This option is especially valuable for borrowers with seasonal or cyclical needs.

2. *Timing flexibility.*   With FRNs, the borrower must live with the prevailing rate for the period's duration. By contrast, a Euronote borrower who thinks rates are going to fall can wait a month or so to issue. However, unless the financial director is better able to forecast interest rates, this option to wait is a dubious advantage.

3. *Choice of maturities.*   FRN issuers are generally locked into one maturity setting—three months or six months—over the life of the deal. NIFs, on the other hand, give borrowers the choice of issuing notes with different maturities whenever they choose to draw down new debt or roll over old.

Euro-Medium Term Notes

Securitization is rearing its head in the Euronote market too. A growing number of firms are now bypassing financial intermediaries and issuing **Euro-medium term notes (Euro-MTNs)** directly to the market. The Euro-MTN, which grew out of the medium-term notes issued in the U.S. market, is one of the most important developments in the Euromarkets in the past decade, as reflected in the fact that $242 billion in Euro-MTNs were issued in 1995, bringing total Euro-MTN outstandings to $594 billion. The three basic reasons for the success of the Euro-MTN market—speed, cost, and flexibility—ensure its continued growth and survival. Like their U.S. counterpart, Euro-MTNs are offered continuously rather than all at once like a bond issue. Euro-MTNs give issuers the flexibility to take advantage of changes in the shape and level of the yield curve and of the specific needs of investors with respect to amount, maturity, currency, and interest rate form (fixed or floating). They can be issued in maturities of as much as 30 years, although most are under five years, and in an ever-increasing range of currencies (32 by 1996, including the Polish zloty, Czech koruna, and South African rand, in addition to the standards, such as the dollar, yen, DM, sterling, and French and Swiss francs). Unlike conventional underwritten debt securities, a program of medium-term notes can be offered in small amounts; in different maturities, currencies, seniority, and security; and on a daily basis, depending on the issuer's needs and the investors' appetites. By contrast, it is not customary to issue underwritten securities in batches of less than $50 million. In this way, Euro-MTNs bridge the maturity gap between Euro-CP and the longer-term international bond. However, with more Euro-MTNs being issued with maturities of less than one year, this has cut into the Euro-CP market.

The costs of setting up a Euro-MTN program are estimated at $131,500.[19] In contrast, the estimated total cost of a Eurobond issue, including the printing and legal documentation associated with the prospectus (but excluding the underwriting fee), is about $100,000. Similar deals issued with Euro-MTN documentation incur only small one-time costs. Euro-MTN issuers thus save money after only two deals, and the savings increase with use.

Medium-term notes are not underwritten; securities firms place the paper as agents instead. Because issuers can change the price of their note offerings as often as several times a day, they can regularly fine-tune their liabilities to match the duration and amount of their assets and investors' demand. For example, a finance company that wants to match liabilities and assets could need, at a particular moment, $12 million of nine-month money, $15.3 million of 14-month money, and $19.1 million of 22-month money. The bond market—with its high issuance costs—could not economically supply such small or precise amounts of debt, but a company with a Euro-MTN program could post appealing rates at those maturities to create the demand for its notes.

That is precisely what General Motors Acceptance Corporation (GMAC) does. Through its global MTN program GMAC shifts its borrowings between the United States and Europe, depending on which market is cheapest. Like many MTN borrowers with a voracious appetite for debt, GMAC publishes rates daily; that is, it advertises the interest rates at which it will issue notes over the full range of the MTN maturity spectrum. Investors are free to approach GMAC, through its dealers, to buy MTNs. Depending on its own particular requirements, GMAC will shift the rates it advertises to encourage investors to lend at the maturities it most needs.

Moreover, unlike public bond issues, the amounts and timing of medium-term note sales are not disclosed. Such a lack of visibility is just fine with many companies, as it allows them to raise funds quickly and discreetly, without having to take the risk of a souring public offering. Euro-MTNs also can broaden an existing investor base to include bank trust departments, thrifts, and pension funds. Such a broadening of the investor base is not accidental. It is related to the most significant distinguishing feature of an MTN program, namely, that it is largely investor-driven, with notes tailored to meet the interests of particular investors.

---

[19] This estimate appears in "A Vision of the Future," *Euromoney*, March 1996, p. 114.

Medium-term notes are not new, but for many years the market consisted almost exclusively of the auto companies' finance arms, particularly GMAC, still the largest issuer with more than $10 billion outstanding. Until 1983, however, secondary trading in the notes, a necessary part of a liquid market, languished. What apparently brought the market to life was GMAC's decision that year to issue notes through dealers, rather than directly as it had been doing. The change convinced investors that there would be a strong secondary market, giving them the liquidity they sought.

The risk to the borrower of issuing Euro-MTNs rather than taking out a fixed-term Eurocurrency loan or Eurobond is that it might not be able to roll over its existing notes or place additional notes when necessary. This risk can be mitigated by extending the maturity of the MTNs issued or by issuing the notes under a revolving credit commitment, such as the note issuance facilities (NIFs) discussed earlier. In the event that market conditions at the time of any note issue are unfavorable and there is insufficient investor demand, the syndicate of banks underwriting the NIF provides the necessary funds by guaranteeing to purchase the notes or make advances in lieu of this, on pre-agreed terms. At maturity, the issuer repays the notes by either issuing new notes or drawing on other resources.

The Asiacurrency Market

Although dwarfed by its European counterpart, the **Asiacurrency (or Asiadollar) market** has been growing rapidly in terms of both size and range of services provided. Located in Singapore, because of the lack of restrictive financial controls and taxes there,

---

ILLUSTRATION

STEADY SAFE ISN'T

On January 12, 1998, the Asian financial crisis claimed another victim when Peregrine Investments Holdings Ltd., Hong Kong's premier investment bank and the largest Asian investment bank outside Japan, announced it would file for liquidation. Its collapse was triggered by the failure of a single large loan to an entrepreneur in Indonesia, where the dramatic decline of the rupiah brought a wave of loan defaults and bankruptcies and ultimately brought down President Suharto's authoritarian government. According to news reports, Peregrine lent $260 million—in the form of an unsecured bridge loan (a temporary loan to be repaid from the expected proceeds of a bond issue)—to a local taxicab operator named Yopie Widjara. Widjara, who reportedly enlisted President Suharto's eldest daughter as an equity investor, planned to create a system of car ferries linking the islands of Indonesia's sprawling archipelago. Peregrine's loan—which represented a third of its capital—was to be repaid through the sale of dollar-denominated bonds issued by Widjara's company, Steady Safe.

Peregrine's troubles started when the value of Asian currencies began falling during the summer of 1997, and it was unable to sell the Steady Safe bonds it had underwritten. Then the Indonesian government lifted trading curbs on the rupiah and the bottom fell out of the currency. From Rp 2,400:$1 in July 1997, the rupiah exchange rate fell to more than Rp 8,000:$1 in January 1998. With the rupiah cost of servicing its debt rising by more than 200%, Steady Safe was unable (or unwilling—it's not clear what happened to the $260 million it borrowed) to repay its dollar-denominated bridge loan and Peregrine collapsed.

the Asiadollar market was founded in 1968 as a satellite market to channel to and from the Eurodollar market the large pool of offshore funds, mainly U.S. dollars, circulating in Asia. Its primary economic functions these days are to channel investment dollars to a number of rapidly growing Southeast Asian countries and to provide deposit facilities for those investors with excess funds. As with all such investments, credit analysis is critical. The example of Steady Safe shows what could happen when lenders become too aggressive.

The Asiabond counterpart to the Asiadollar market is the dragon bond. A **dragon bond** is debt denominated in a foreign currency, usually dollars, but launched, priced, and traded in Asia. The first dragon bond was issued in November 1991 by the Asian Development Bank. After growing rapidly in the early 1990s, however, the bond portion of the non-Japanese Asiabond market has slumped despite the existence of plentiful Asian savers and borrowers. The market's fundamental problem is that Asian borrowers with a good international credit rating can raise money for longer, and for less, in Europe or the United States.

15.4 DEVELOPMENT BANKS

To help provide the huge financial resources required to promote the development of economically backward areas, the United States and other countries have established a variety of *development banks* whose lending is directed to investments that would not otherwise be funded by private capital. These investments include dams, roads, communication systems, and other infrastructure projects whose economic benefits cannot be completely captured by private investors, as well as projects such as steel mills or chemical plants whose value lies in perceived political and social advantages to the nation (or at least to its leaders). The loans generally are medium- to long-term and carry concessionary rates.

This type of financing has three implications for the private sector. First, the projects require goods and services, which corporations can provide. Second, establishing an infrastructure makes available new investment opportunities for multinational corporations. Third, even though most development-bank lending is done directly to a government, multinationals find that these banks are potential sources of low-cost, long-term, fixed-rate funds for certain types of ventures. The time-consuming nature of arranging financing from them, however—in part due to their insistence on conducting their own in-house feasibility studies—usually leaves them as a secondary source of funds. Their participation may be indispensable, however, for projects that require heavy infrastructure investments such as roads, power plants, schools, communications facilities, and housing for employees. These infrastructure investments are the most difficult part of a project to arrange financing for because they generate no cash flow of their own. Thus, loans or grants from an international or regional development bank are often essential to fill a gap in the project financing plan.

There are three types of development banks: the World Bank Group, regional development banks, and national development banks.

The World Bank Group

The *World Bank Group* is a multinational financial institution established at the end of World War II to help provide long-term capital for the reconstruction and development of member countries. It comprises three related financial institutions: the **International Bank for Reconstruction and Development** (IBRD), also known as the **World Bank**; the *International Development Association* (IDA); and the *International Finance Corporation* (IFC). The Group is important to multinational corporations because it provides much of the planning and financing for economic development projects involving billions of dollars for which private businesses can act as contractors and as suppliers of goods and engineering-related services.

IBRD ■ The IBRD, or World Bank, makes loans at nearly conventional terms for projects of high economic priority. To qualify for financing, a project must have costs and revenues that can be estimated with reasonable accuracy. A government guarantee is a necessity for World Bank funding. The bank's main emphasis historically has been on large infrastructure projects such as roads, dams, power plants, education, and agriculture. In recent years, however, the bank more and more has emphasized loans to help borrower countries alleviate their balance-of-payments problems. These loans are tied to the willingness of the debtor nations to adopt economic policies that will spur growth: freer trade, more open investment, lower budget deficits, and a more vigorous private sector. In addition to its members' subscriptions, the World Bank raises funds by issuing bonds to private sources.

IFC ■ The purpose of the IFC is to finance various projects in the private sector through loans and equity participations and to serve as a catalyst for flows of additional private capital investment to developing countries. In contrast to the World Bank, the IFC does not require government guarantees. It emphasizes providing risk capital for manufacturing firms that have a reasonable chance of earning the investors' required rate of return and that will provide economic benefits to the nation. Instead of focusing on the small and medium-size firms that may really need its help, however, the IFC tends to concentrate the bulk of its lending and equity investments in investment-grade conglomerates that regularly tap public markets.

IDA ■ The World Bank concentrates on projects that have a high probability of being profitable; consequently, many of the poorest of the less-developed countries (LDCs) are unable to access its funds. IDA was founded in 1960 to remedy this shortcoming. As distinguished from the World Bank, IDA is authorized to make soft (highly concessionary) loans (for example, 50-year maturity with no interest). It does require a government guarantee, however. The establishment of IDA illustrates a major unresolved issue for the World Bank Group: Should its emphasis be on making sound loans to developing countries, or should it concentrate on investing in those projects most likely to be of benefit to the host country? These goals are not necessarily in conflict, although many of a project's benefits may not be captured by the project itself but instead will appear elsewhere in the economy (for example, the benefits of an educational system).

Regional and National Development Banks

The past two decades have seen a proliferation of development banks. The functions of a development bank are to provide debt and equity financing to aid in the economic development of underdeveloped areas. This financing includes extending intermediate- to long-term capital directly, strengthening local capital markets, and supplying management consulting services to new companies. The professional guidance helps to safeguard, and thereby encourage, investments in a firm.

REGIONAL DEVELOPMENT BANKS ■ Regional development banks provide funds for the financing of manufacturing, mining, agricultural, and infrastructure projects considered important to development. They tend to support projects that promote regional cooperation and economic integration. Repayment terms for the loans, in most cases, are over a five- to 15-year period at favorable interest rates. The leading regional development banks include the following.

1. *European Investment Bank* (EIB).   The EIB offers funds for certain public and private projects in European and other nations associated with the Common Market. It emphasizes loans to the lesser-developed regions in Europe and to associated members in Africa.

2. *Inter-American Development Bank* (IADB).   The IADB is a key sources of long-term capital in Latin America. It lends to joint ventures, both minority and majority foreign-owned, and provides small amounts of equity capital. A recent initiative is to act as a catalyst for further private-sector funding for Latin American infrastructure projects. By partially guaranteeing commercial bank loans and directly lending to infrastructure projects, the IADB aims to bring funding to many projects for which commercial bank loans might not otherwise be available. For example, it recently lent $75 million to a private consortium led by General Electric and Bechtel to build and lease the second stage of the Samalayuca power project in Mexico.

3. *Atlantic Development Group for Latin America* (ADELA).   ADELA is an international private-investment company dedicated to the socioeconomic development of Latin America. Its objective is to strengthen private enterprise by providing capital and entrepreneurial and technical services.

4. *Asian Development Bank* (ADB).   The ADB guarantees or makes direct loans to member states and private ventures in Asian/Pacific nations and helps to develop local capital markets by underwriting securities issued by private enterprises.

5. *African Development Bank* (AFDB).   The AFDB makes or guarantees loans and provides technical assistance to member states for various development projects. Beneficiaries of ADB loans and activities are normally governments or government-related agencies.

6. *Arab Fund for Economic and Social Development* (AFESD).   The AFESD is a multilateral Arab fund that actively searches for projects (restricted to Arab League countries) and then assumes responsibility for project implementation by conducting feasibility studies, contracting, controlling quality, and supervising the work schedule.

7. *European Bank for Reconstruction and Development* (EBRD).   The EBRD, which was founded in 1990 with an initial capital of about $13 billion, is supposed to finance the privatization of Eastern Europe. Many critics are skeptical of its chances, however, because the person appointed to head it is a French socialist who masterminded the most sweeping program of nationalizations in French history. The EBRD's reputation was not helped when it was revealed in 1993 that in its first two years of operation, it had spent more than twice as much on its building, staff, and overhead (more than $300 million) as it had disbursed in loans (about $150 million) to its 25 client countries. Thanks to the nudging of the United States, one of the ground rules for the EBRD is that it must make 60% of its loans to the private sector, a target it is now approaching after an early tilt toward the public sector.

NATIONAL DEVELOPMENT BANKS ∎ Some national development banks concentrate on a particular industry or region; others are multipurpose. Although most are public institutions, there are several privately controlled development banks as well. The characteristics for success, however, are the same: They must attract capable, investment-oriented management; and they must have a large enough supply of economically viable projects to enable management to select a reasonable portfolio of investments.

The Proper Role of Development Banks

Critics of development banks claim that they tend to perpetuate economic stagnation by financing statist solutions, rather than free-market solutions. By giving money to governments, rather than providing capital to private firms, these banks support often-corrupt governments and bureaucratic planning over private enterprise. This support builds up the public sector at the expense of the private sector. In the absence of funds from development banks, recipient governments might have to reduce their spending and grant more freedom

---

ILLUSTRATION

THE WORLD BANK'S ROLE IN AFRICA'S AGRICULTURAL FAILURE The decline of African agriculture throughout the 1970s and 1980s is now widely understood to be due primarily to African governments' penchant for state marketing boards. These set the "official" and only price at which farmers may sell, and the government is frequently the sole legal buyer. Cheap food has been guaranteed to urban populations—the politically important constituency—by the rule of the African marketing boards: Buy low, sell low (or lower). Most African governments have run up large deficits for decades. Price-conscious farmers responded by selling their produce in black markets, smuggling it across the border, reverting to subsistence agriculture, or leaving the land entirely to join the urban swell. At the same time, low food prices led to increased demand and shortages.

The World Bank had a pivotal role in this situation. From 1974 to 1986 the bank channeled $5.5 billion into African agriculture—an eightfold increase in the lending rate over the previous period. Much of this money was used to set up and finance—that is, subsidize—state marketing boards.

to the private sector. Since the marketplace is a far more efficient mechanism for supplying goods and services, any policy that hampers the private sector will retard economic growth.

During their 50-year history, the World Bank and the IMF have lent an estimated $1.6 trillion, much of it to underwrite a development strategy that concentrated investment decisions in the hands of hugely unsuccessful parastatal companies. In short, these institutions have financed inefficient state monopolies at the expense of private, competitive development. For example, despite high levels of government-directed investment in Africa, per capita incomes in that region are actually lower today than they were in 1970.

Because they are government-owned and government-controlled institutions doing business with other governments, it is perhaps not surprising that development banks have been hostile in general to enforcing stringent free-market principles as conditions for receiving loans. Thus, they have been unable—and unwilling—to foster the prime conditions for development: sound economic policies and stable political environments. Economic success stories such as the United States, post-World War II Japan and West Germany, modern Taiwan, South Korea, Hong Kong, and Singapore all have a common factor: reliance on private enterprise to organize most economic activity. Even in other regions, such as Latin America, recent successes in achieving growth, economic stability, and rising standards of living have been due to free-market reforms: privatizing state-owned enterprises, letting currencies find their true market values, cutting government spending, lowering tax rates and tariffs, and welcoming foreign capital.

If one accepts the view that a vigorous private sector is key to economic growth, then development banks could play an important role in LDC development if they would condition their loans on the elimination in Third World economies of price controls, nationalized industries, high tax rates, government subsidies, state monopolies, trade restrictions, and impediments to private capital flows.

The Reagan administration recognized this role and tried to turn these banks into positive forces for LDC development. In 1985, Treasury Secretary James Baker outlined the so-called Baker Plan. The **Baker Plan** called for parceling out development aid only if borrowers pursue free-market reforms to encourage private enterprise. For example, the United States vetoed a concessional IADB loan to Guyana, which wanted a subsidized loan to aid its ailing rice farmers. The U.S. Treasury pointed out that rice farmers would be better served by eliminating Guyana's government price controls, a major cause of the poor rice production.

However, the Baker Plan and others like it face hostility from less market-oriented major members such as France, Canada, and Japan. Moreover, many of the banks' staffs are hostile to marketplace solutions and tend to emphasize capital flows from their banks as the key factor in development.

Private-Sector Alternatives

A recent report by the International Finance Corporation says that private-sector infrastructure financing has grown significantly in developing countries during the 1990s.[20] According to the IFC report, $26 billion in infrastructure projects that include at least

---

[20] Gary Bond and Laurence Carter, "Financing Private Infrastructure Projects," International Finance Corporation, 1994.

some private funds have been completed in recent years, and another $100 billion in such projects is highly probable. These amounts represent huge increases over privately financed infrastructure spending since 1988, when such spending was less than $500 million. Current examples of the trend include telecommunications-privatization projects in Hungary and Latvia; private power-generation efforts in India and Pakistan; toll roads in Argentina; the reconstruction and management of a Bulgarian airport; management of Malaysia's national sewage system; and an elevated mass-transit rail system in Thailand.

The report noted that the main reason for the shift toward private infrastructure was growing disenchantment with public monopoly ownership and provision of infrastructure services such as power, waste disposal, roads, and telecommunications. Underinvestment by many state utilities has resulted in shortages of these services, leading to constraints on growth. The surge in private infrastructure financing also is being fed by fiscal constraints on governments and external aid agencies, which are forcing governments to turn to private capital despite their preference for public ownership. In spite of the large risks, the private sector has responded with tens of billions of dollars through the use of innovative financing techniques—such as project financing, securitization, and asset-backed financing—that permit efficient risk bearing. Exhibit 15.15 shows the dramatic shift in

EXHIBIT 15.15 Private and Official Capital Flows to Developing Countries, 1989–1997

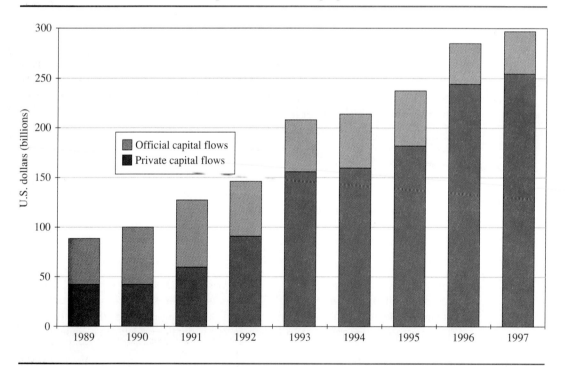

Note: Private capital flows include foreign direct investment, commercial loans, and equity and debt portfolio investments.
Source: The World Bank.

the importance of private capital flows to developing nations relative to official flows. As recently as 1991, official flows made up more than 50% of long-term financial flows to developing countries; by 1997, that figure was less than 15%. This evidence suggests that the proliferation of development banks and foreign aid may have actually stifled private infrastructure financing.

The World Bank and IFC are attempting to reposition themselves in this changed environment, where private capital flows dominate official flows. For example, in 1996, the IFC announced a program expanding its reach into 16 high-risk countries such as Cambodia, Guyana, Mongolia, Senegal, and Uzbekistan. Meanwhile, the World Bank is placing more emphasis on strengthening the legal and financial systems of developing countries so as to attract private investors. It is also seeking to stimulate private investment in these countries by providing more loan guarantees to businesses entering these markets.

❧ 15.5 SUMMARY AND CONCLUSIONS

Although there are significant differences among countries in their methods and sources of finance, corporate practice appears to be converging. Most significantly, more firms are bypassing financial intermediaries, mainly commercial banks, and going directly to the financial markets for funds. The convergence of corporate financing practice largely reflects the globalization of financial markets, the inextricable linkage—through arbitrage—of financial markets worldwide. In line with this trend, firms are finding that it pays to seek capital on a global basis, rather than restricting their search to any one nation or capital market.

We saw that the growth of the international capital markets, specifically the Eurocurrency and Eurobond markets, is largely a response to the restrictions, regulations, and costs that governments impose on domestic financial transactions. At the same time, capital flows between the international capital markets and domestic markets have linked domestic markets in a manner that increasingly makes such government intervention irrelevant. A principal means whereby markets are linked is the use of interest rate and currency swaps, whereby two counterparties agree to exchange streams of payments over time in order to convert from one interest rate structure and/or currency to another.

We also saw that, although some sources of funds are readily accessible (for example, the Eurocurrency market), others (such as government-subsidized export credits) are strictly limited in terms of availability. It is, of course, just this latter type of financing that is so appealing to corporate financial executives.

➤ Questions

1. What are some basic differences between the financing patterns of U.S. and Japanese firms? What might account for some of these differences?

2. How and why has the Japanese pattern of finance changed over time?

3. What is securitization? What forces underlie it, and how has it affected the financing policies of multinational corporations?

4. Why is bank lending on the decline worldwide? How have banks responded to their loss of market share?

5. What is meant by the globalization of financial markets? How has technology affected the process of globalization?

6. How has globalization affected government regulation of national capital markets?

7. Many financial commentators believe that bond owners and traders today have an enormous collective influence over a nation's economic policies. Explain why this might be correct.

8. Why are large multinational corporations located in small countries such as Sweden, Holland, and Switzerland interested in developing a global investor base?

9. Why are many U.S. multinationals seeking to improve their visibility with foreign investors, even going so far as to list their shares on foreign stock exchanges?

10. What is the difference between a Eurocurrency loan and a Eurobond?

11. What is the difference between a foreign bond and a Eurobond?

12. Why might investors and borrowers be attracted to an ECU bond?

13. What is the basic reason for the existence of the Eurodollar market? What factors have accounted for its growth over time?

14. Suppose the French government imposes an interest rate ceiling on French bank deposits. What is the likely effect on Eurofranc interest rates of this regulation?

15. List some reasons why a U.S.-based corporation might issue debt denominated in a foreign currency.

16. Why have Eurobonds traditionally yielded less than comparable domestic issues?

17. What factors account for the rise and recent decline of the Eurobond market as a source of financing for American companies?

18. It has been said that if other European interest rates converged toward German rates, ECU bonds would soar in value. Explain why this jump in value would occur.

19. a. What factors account for the growth of note issuance facilities?
    b. In what sense is the NIF part of the process of securitization?
    c. Why is the NIF described as a put option?

20. In an attempt to regain business lost to foreign markets, Swiss authorities abolished stamp duties on transactions between foreigners as well as on new bond issues by foreign borrowers. However, transactions involving Swiss citizens will still attract a 0.15% tax, and bond issues by Swiss borrowers were also made more expensive. What are the likely consequences of these changes for Swiss financial markets?

21. On October 14, 1993, Portugal's Ministry of Finance announced that it would scrap the 20% withholding tax imposed on the interest payments due foreigners holding government bonds. At present, foreigners whose governments have a double-taxation treaty with Portugal wait up to two years to claim back a portion of the tax. What might have been Portugal's motivations for scrapping the tax? What are the likely consequences of eliminating the withholding tax?

➤Problems

1. Suppose that the current 180-day interbank Eurodollar rate is 9% (all rates are stated on an annualized basis). If next period's rate is 9.5%, what will a Eurocurrency loan priced at LIBOR plus 1% cost?

2. Citibank offers to syndicate a Eurodollar credit for the government of Poland with the following terms:

Principal	US$1,000,000,000
Maturity	7 years
Interest rate	LIBOR + 1.5%, reset every six months
Syndication fee	1.75%

    a. What are the net proceeds to Poland from this syndicated loan?
    b. Assuming that six-month LIBOR is currently at 6.35%, what is the effective annual interest cost to Poland for the first six months of this loan?

3. Suppose that Zimbabwe has a choice of two possible $100 million, five-year Eurodollar loans. The first loan is offered at LIBOR + 1% with a 2.5% syndication fee, whereas the second loan is priced at LIBOR + 1.5% and a 0.75% syndication fee. Assuming that Zimbabwe has a 9% cost of capital, which loan is preferable? *Hint*: View this as a capital budgeting problem.

4. IBM wishes to raise $1 billion and is trying to decide between a domestic dollar bond issue and a Eurobond issue. The U.S. bond can be issued at a coupon of 6.75%, paid semianually, with underwriting and other expenses totaling 0.95% of the issue size. The Eurobond would cost only 0.55% to issue but would bear an annual coupon of 6.88%. Both issues would mature in 10 years.

a. Assuming all else is equal, which is the least expensive issue for IBM?

b. What other factors might IBM want to consider before deciding which bond to issue?

5. Refer to the example of Exxon's zero-coupon Eurobond issue in the section titled The Euromarkets.

a. How much would Exxon have earned if the yield on the stripped Treasurys had been 12.10%? 12.25%?

b. Suppose the Japanese government taxed the accretion in the value of zero-coupon bonds at a rate of 15%. Assuming the same 11.65% after-tax required yield, how would this tax have affected the price Japanese investors were willing to pay for Exxon's Eurobond issue? What would the pre-tax yield be at this new price? Would any arbitrage incentive still exist for Exxon?

c. Suppose Exxon had sold its zero-coupon Eurobonds to yield 11.5% and bought stripped Treasury bonds yielding 12.30% to meet the required payment of $1.8 billion. How much would Exxon have earned through its arbitrage transaction?

6. Daewoo Motors has been told it could issue $150 million face value in Euro-CP at a discount rate of 8.9% based on a 360-day year.

a. If the maturity of Daewoo's Euro-CP is 91 days, what will be its proceeds from the issue?

b. What will be its annual yield on this issue?

7. British Telecom (BT) has issued $1 billion in Euro-CP maturing in 75 days and priced to yield 5.8% annually based on a 360-day year.

a. What are BT's proceeds from the issue?

b. What is the discount rate on BT's issue?

8. Commerzbank is seeking to invest $100 million short term. It has the choice between buying Euro-CP yielding 6.34% annually and a U.S. bank deposit yielding 6.36% annually, both maturing in 150 days. The Euro-CP yield is calculated on a 360-day year, whereas the U.S. bank-deposit yield is calculated on a 365-day year.

a. How much Euro-CP in terms of face value can Commerzbank's $100 million buy?

b. Assuming that all else is equal, which is Commerzbank's preferred investment? Explain.

c. What would be the annual yield on the U.S. bank deposit if it were quoted on a 360-day year?

d. What would be the annual Euro-CP yield if it were quoted on a 365-day year?

9. A European company issues common shares that pay taxable dividends and bearer shares that pay an identical dividend but offer an opportunity to evade taxes. (Bearer shares come with a large supply of coupons that can be redeemed anonymously at banks for the current value of the dividend.)

a. Suppose taxable dividends are taxed at the rate of 10%. What is the ratio between market prices of taxable and bearer shares? If a new issue is planned, should taxable or bearer shares be sold?

b. Suppose, in addition, that it costs 10% of proceeds to issue a taxable dividend, whereas it costs 20% of the proceeds to issue bearer stocks because of the expense of distribution and coupon printing. What type of share will the corporation prefer to issue?

c. Suppose now that individuals pay 10% taxes on dividends and corporations pay no taxes but bear an administrative cost of 10% of the value of any bearer dividends. Determine the relative market prices for the two types of shares.

➤ Bibliography

Business International Corporation. *Financing Foreign Operations*, various issues.

DUFEY, GUNTER, and IAN H. GIDDY. "Innovation in the International Financial Markets." *Journal of International Business Studies*, Fall 1981, pp. 35–51.

_____. *The International Money Market*. Englewood Cliffs, N.J.: Prentice Hall, 1978.

GEORGE, ABRAHAM M., and IAN H. GIDDY, eds. *International Finance Handbook*. New York: John Wiley & Sons, 1983.

GRABBE, OREN J. *International Financial Markets*. New York: Elsevier, 1986.

KIM, YONG-CHEOL, and RENE M. STULZ. "The Eurobond Market and Corporate Financial Policy: A Test of the Clientele Hypothesis." *Journal of Financial Economics* 22 (1988): 189–205.

_____. "Is There a Global Market for Convertible Bonds?" *Journal of Business*. 65 (1992): 75–91.

SOLNIK, BRUNO H. *International Investments*. Reading, Mass.: Addison-Wesley, 1987.

SPECIAL FINANCING VEHICLES

Man is not the creature of circumstances, circumstances are the creatures of men.

Benjamin Disraeli (1826)

*T*he purpose of this chapter is to examine several special financing vehicles that multinational corporations can use to fund their foreign investments. These vehicles include interest rate and currency swaps, structured notes, international leasing, and bank loan swaps. Each of them presents opportunities to the multinational firm to reduce financing costs and/or risk.

16.1 INTEREST RATE AND CURRENCY SWAPS

Corporate financial managers can use swaps to arrange complex, innovative financings that reduce borrowing costs and increase control over interest rate risk and foreign currency exposure. As a result of the deregulation and integration of national capital markets and extreme interest rate and currency volatility, the relatively new swaps market has experienced explosive growth, with outstanding swaps by mid-1997 of $28.7 trillion. In fact, few Eurobonds are issued without at least one swap behind them to give the borrower cheaper or in some way more desirable funds.

This section discusses the structure and mechanics of the two basic types of swaps—interest rate swaps and currency swaps—and shows how swaps can be used to achieve diverse goals. Swaps have had a major impact on the treasury function, permitting firms to tap new capital markets and to take further advantage of innovative products without an increase in risk. Through the swap, they can trade a perceived risk in one market or currency for a liability in another. The swap has led to a refinement of risk management techniques, which in turn has facilitated corporate involvement in international capital markets.

Interest Rate Swaps

An **interest rate swap** is an agreement between two parties to exchange U.S. dollar interest payments for a specific maturity on an agreed upon notional amount. The term *notional* refers to the theoretical principal underlying the swap. Thus, the **notional principal** is simply a reference amount against which the interest is calculated. No principal ever changes hands. Maturities range from under a year to over 15 years; however, most transactions fall within a two-year to 10-year period. The two main types are coupon swaps and basis swaps. In a **coupon swap**, one party pays a *fixed rate* calculated at the time of trade as a spread to a particular Treasury bond, and the other side pays a *floating rate* that resets periodically throughout the life of the deal against a designated index. In a **basis swap**, two parties exchange floating interest payments based on different reference rates. Basically, using this relatively straightforward mechanism, interest rate swaps transform debt issues, assets, liabilities, or any cash flow from type to type and—with some variation in the transaction structure—from currency to currency.

THE CLASSIC SWAP TRANSACTION ■ Counterparties A and B both require $100 million for a five-year period. To reduce their financing risks, counterparty A would like to borrow at a fixed rate, whereas counterparty B would prefer to borrow at a floating rate. Suppose that A is a company with a BBB rating, while B is a AAA-rated bank. Although A has good access to banks or other sources of floating-rate funds for its operations, it has difficulty raising fixed-rate funds from bond issues in the capital markets at a price it finds attractive. By contrast, B can borrow at the finest rates in either market. The cost to each party of accessing either the fixed-rate or the floating-rate market for a new five-year debt issue is as follows:

BORROWER	FIXED-RATE AVAILABLE	FLOATING-RATE AVAILABLE
Counterparty A: BBB-rated	8.5%	6-month LIBOR + 0.5%
Counterparty B: AAA-rated	7.0%	6-month LIBOR
Difference	1.5%	0.5%

It is obvious that there is an anomaly between the two markets: One judges that the difference in credit quality between a AAA-rated firm and a BBB-rated firm is worth 150 basis points; the other determines that this difference is worth only 50 basis points (a basis point equals 0.01%). Through an interest-rate swap, both parties can take advantage of the 100 basis point spread differential.

To begin, A will take out a $100 million, five-year floating-rate Eurodollar loan from a syndicate of banks at an interest rate of LIBOR plus 50 basis points. At the same time, B will issue a $100 million, five-year Eurobond carrying a fixed rate of 7%. A and B will then enter into the following interest rate swaps with BigBank. Counterparty A agrees that it will pay BigBank 7.35% for five years, with payments calculated by multiplying that rate by the $100 million notional principal amount. In return for this payment, BigBank agrees to pay A six-month LIBOR over five years, with reset dates matching the reset dates on its floating rate loan. Through the swap, A has managed to turn a floating-rate loan into a fixed-rate loan costing 7.85%.

In a similar fashion, B enters into a swap with BigBank whereby it agrees to pay six-month LIBOR to BigBank on a notional principal amount of $100 million for five years in exchange for receiving payments of 7.25%. Thus, B has swapped a fixed-rate loan for a floating-rate loan carrying an effective cost of LIBOR minus 25 basis points.

Why would BigBank or any financial intermediary enter into such transactions? The reason BigBank is willing to enter into such contracts is more evident when looking at the transaction in its entirety. This classic swap structure is shown in Exhibit 16.1.

As a financial intermediary, BigBank puts together both transactions. The risks net out, and BigBank is left with a spread of 10 basis points:

Receive	7.35%
Pay	(7.25%)
Receive	LIBOR
Pay	(LIBOR)
Net	10 basis points

EXHIBIT 16.1 Classic Swap Structure

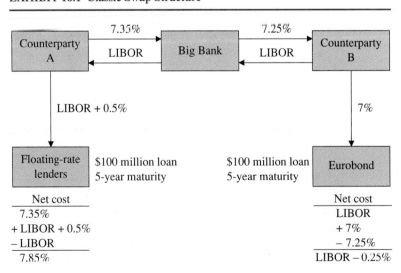

BigBank thus receives compensation equal to $100,000 annually for the next five years on the $100 million swap transaction:

$$\text{Swap Profit to BigBank} = 0.001 \times \$100,000,000$$

$$= \$100,000$$

COST SAVINGS ASSOCIATED WITH SWAPS ∎ The example just discussed shows the risk-reducing potential of interest rate swaps. Swaps also may be used to reduce costs. Their ability to do so depends on a difference in perceived credit quality across financial markets. In essence, interest rate swaps exploit the comparative advantages—if they exist—enjoyed by different borrowers in different markets, thereby increasing the options available to both borrower and investor.

Returning to the previous example, we can see that there is a spread differential of 100 basis points between the cost of fixed- and floating-rate borrowing for A and B that the interest rate swap has permitted the parties to share among themselves as follows:

PARTY	NORMAL FUNDING COST (%)	COST AFTER SWAP (%)	DIFFERENCE (%)
Counterparty A	8.50	7.85	0.65
Counterparty B	LIBOR	LIBOR − 0.25	0.25
BigBank	—	—	0.10
		TOTAL	1.00

In this example, A lowers its fixed-rate costs by 65 basis points, B lowers its floating-rate costs by 25 basis points, and BigBank receives 10 basis points for arranging the transaction and bearing the credit risk of the counterparties.

You might expect that the process of financial arbitrage would soon eliminate any such cost savings opportunities associated with a mispricing of credit quality. Despite this efficient markets view, many players in the swaps market believe that such anomalies in perceived credit risk continue to exist. The explosive growth in the swaps market supports this belief. It may also indicate the presence of other factors, such as differences in information and risk aversion of lenders across markets, that are more likely to persist.

Currency Swaps

Swap contracts can also be arranged across currencies. Such contracts are known as currency swaps and can help manage both interest rate and exchange rate risk. Many financial institutions count the arranging of swaps, both domestic and foreign currency, as an important line of business.

Technically, a **currency swap** is an exchange of debt-service obligations denominated in one currency for the service on an agreed upon principal amount of debt denominated in another currency. By swapping their future cash flow obligations, the counterparties are able to replace cash flows denominated in one currency with cash flows in a more desired currency. In this way, company A which has borrowed, say, Japanese

EXHIBIT 16.2 Diagram of a Fixed-for-Fixed Currency Swap

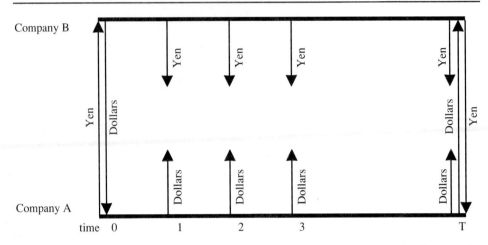

Company A issues yen debt and exchanges the yen principal amount for dollars with Company B, which has issued dollar debt. At each subsequent interest date, A pays dollar interest to B and receives yen interest from B. A uses the yen it receives from B to service its yen debt. At maturity, there is a final interest exchange and an exchange of principal amounts.

yen at a fixed interest rate can transform its yen debt into a fully hedged dollar liability by exchanging cash flows with counterparty B. As illustrated in Exhibit 16.2, the two loans that comprise the currency swap have parallel interest and principal repayment schedules. At each payment date, company A will pay a fixed interest rate in dollars and receive a fixed rate in yen. The counterparties also exchange principal amounts at the start and the end of the swap arrangement (denoted as time T in the diagram).

In effect, a U.S. firm engaged in a currency swap has borrowed foreign currency and converted its proceeds into dollars, while simultaneously arranging for a counterparty to make the requisite foreign currency payment in each period. In return for this foreign currency payment, the firm pays an agreed-upon amount of dollars to the counterparty. Given the fixed nature of the periodic exchanges of currencies, the currency swap is equivalent to a package of forward contracts. For example, in the dollar:yen swap discussed above, firm A has contracted to sell fixed amounts of dollars forward for fixed amounts of yen on a series of future dates.

The counterparties to a currency swap will be concerned about their **all-in cost**—that is, the effective interest rate on the money they have raised. This interest rate is calculated as the discount rate that equates the present value of the future interest and principal payments to the net proceeds received by the issuer.

Currency swaps achieve an economic purpose similar to the parallel loan arrangements discussed in Chapter 14. They have effectively displaced the use of parallel loans, however, because they solve two potential problems associated with parallel loans: (1) If there is no right of offset, default by one party does not release the other from making its contractually obligated payments; and (2) parallel loans remain on the balance sheet, even though they effectively cancel one another. With a currency swap, the **right of offset,** which gives each party the right to offset any nonpayment of principal or interest with a comparable nonpayment, is more firmly established. Moreover, because a currency swap is not a loan, it does not appear as a liability on the parties' balance sheets.

Although the structure of currency swaps differs from interest rate swaps in a variety of ways, the major difference is that with a currency swap, there is always an exchange of principal amounts at maturity at a predetermined exchange rate. Thus, the swap contract behaves like a long-dated forward foreign exchange contract, where the forward rate is the current spot rate.

The reason that there is always an exchange of principal amounts at maturity can be explained as follows. Assume that the prevailing coupon rate is 8% in one currency and 5% in the other currency. What would convince an investor to pay 8% and receive 300 basis points less? The answer lies in the spot and long-term forward exchange rates and how currency swaps adjust to compensate for the differentials. According to interest rate parity theory, forward rates are a direct function of the interest rate differential for the two currencies involved. As a result, a currency with a lower interest rate has a correspondingly higher forward exchange value. It follows that future exchange of currencies at the present spot exchange rate would offset the current difference in interest rates. This *exchange of principals* is what occurs in every currency swap at maturity based on the original amounts of each currency and, by implication, done at the original spot exchange rate.

In the classic currency swap, the counterparties exchange fixed-rate payments in one currency for fixed-rate payments in another currency. The following hypothetical example illustrates the structure of a fixed-for-fixed currency swap.

ILLUSTRATION

DOW CHEMICAL SWAPS FIXED-FOR-FIXED WITH MICHELIN

Suppose that Dow Chemical is looking to hedge some of its French franc exposure by borrowing in francs. At the same time, French tire manufacturer Michelin is seeking dollars to finance additional investment in the U.S. market. Both are seeking the equivalent of $200 million in fixed-rate financing for 10 years. Dow can issue dollar-denominated debt at a coupon rate of 7.5% or French franc-denominated debt at a coupon rate of 8.25%. Equivalent rates for Michelin are 7.7% in dollars and 8.1% in francs. Given that both companies have similar credit ratings, it is clear that the best way for them to borrow in the other's currency is to issue debt in their own currencies and then swap the proceeds and future debt-service payments.

Assuming a current spot rate of FF 4.91/$, Michelin would issue FF 982 million in 8.1% debt and Dow Chemical would float a bond issue of $200 million at 7.5%. The coupon payments on these bond issues are FF 79,542,000 (.081 × FF 982,000,000) and $15,000,000 (.075 × $200,000,000), respectively, giving rise to the following debt-service payments:

YEAR	MICHELIN	DOW CHEMICAL
1–10	FF 79,542,000	$15,000,000
10	FF 982,000,000	$200,000,000

After swapping the proceeds at time 0 (now), Dow Chemical winds up with FF 982 million in franc debt and Michelin has $200 million in debt to service. In subsequent years,

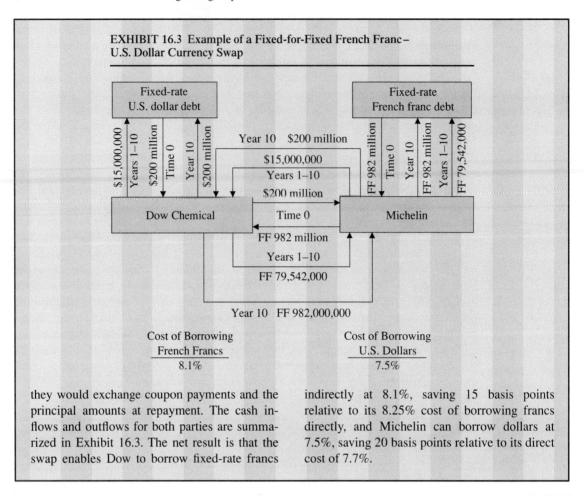

EXHIBIT 16.3 Example of a Fixed-for-Fixed French Franc–
U.S. Dollar Currency Swap

Cost of Borrowing
French Francs
8.1%

Cost of Borrowing
U.S. Dollars
7.5%

they would exchange coupon payments and the principal amounts at repayment. The cash inflows and outflows for both parties are summarized in Exhibit 16.3. The net result is that the swap enables Dow to borrow fixed-rate francs indirectly at 8.1%, saving 15 basis points relative to its 8.25% cost of borrowing francs directly, and Michelin can borrow dollars at 7.5%, saving 20 basis points relative to its direct cost of 7.7%.

INTEREST RATE/CURRENCY SWAPS ■ Although the currency swap market began with fixed-for-fixed swaps, most such swaps today are interest rate/currency swaps. As its name implies, an **interest rate/currency swap** combines the features of both a currency swap and an interest rate swap. This swap is designed to convert a liability in one currency with a stipulated type of interest payment into one denominated in another currency with a different type of interest payment. The most common form of interest rate/currency swap converts a fixed-rate liability in one currency into a floating-rate liability in a second currency. We can use the previous example of Dow Chemical and Michelin to illustrate the mechanics of a fixed-for-floating currency swap.

The two examples of Dow Chemical and Michelin have them dealing directly with one another. In practice, they would use a financial intermediary, such as a commercial bank or an investment bank, as either a broker or a dealer to arrange the swap. As a broker, the intermediary simply brings the counterparties together for a fee. In contrast, if the intermediary acts as a dealer, it not only arranges the swap, it also guarantees the swap payments that each party is supposed to receive. Because the dealer guarantees the

ILLUSTRATION

Dow Chemical Swaps Fixed-for-Floating with Michelin

Suppose that Dow Chemical decides it prefers to borrow floating-rate French francs instead of fixed-rate francs, whereas Michelin maintains its preference for fixed-rate dollars. Assume that Dow Chemical can borrow floating-rate francs directly at LIBOR + 0.35%, versus a cost to Michelin of borrowing floating-rate francs of LIBOR + 0.125%. As before, given Dow's cost of borrow-

ing dollars of 7.5%, versus Michelin's cost of 7.7%, the best way for them to achieve their currency and interest rate objectives is to issue debt in their own currencies and then swap the proceeds and future debt-service payments.

Exhibit 16.4 summarizes the cash inflows and outflows for both parties. The net result of the swap is that Dow Chemical can borrow francs indirectly at a floating rate of LIBOR + 0.125%, saving 22.5 basis points relative to its cost of borrowing floating-rate francs directly. Michelin's cost of borrowing fixed-rate dollars remains at 7.5%, a savings of 20 basis points.

EXHIBIT 16.4 Example of a Fixed-for-Floating Currency Swap

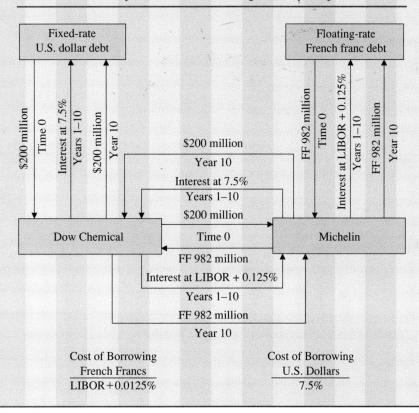

parties to the swap arrangement against default risk, both parties will be concerned with the dealer's credit rating. Financial intermediaries in the swap market must have high credit ratings since most intermediaries these days act as dealers.

Actual interest rate/currency swaps tend to be more complicated than the plain-vanilla Dow Chemical/Michelin swap shown on the previous page. The following example shows how intricate these swaps can be.

ILLUSTRATION

KODAK'S ZERO-COUPON AUSTRALIAN DOLLAR INTEREST RATE/CURRENCY SWAP

In late March 1987, Eastman Kodak Company, a AAA-rated firm, indicated to Merrill Lynch that it needed to raise U.S.$400 million.[1] Kodak's preference was to fund through nontraditional structures, obtaining U.S.$200 million for both five and 10 years. Kodak stated that it would spend up to two weeks evaluating nondollar financing opportunities for the five-year tranche, targeting a minimum size of U.S.$75 million and an all-in cost of U.S. Treasurys plus 35 basis points. In contrast, a domestic bond issue by Kodak would have to be priced to yield an all-in cost equal to about 50 basis points above the rate on U.S. Treasurys. At the end of the two-week period, the remaining balance was to be funded with a competitive bid.

After reviewing a number of potential transactions, the Capital Markets group at Merrill Lynch decided that investor interest in nondollar issues was much stronger in Europe than in the United States and that Merrill Lynch should focus on a nondollar Euroissue for Kodak. The London Syndicate Desk informed the Capital Markets Desk that it was a co-lead manager of an aggressively priced five-year, Australian dollar (A$) zero-coupon issue that was selling very well in Europe. The London Syndicate believed it could successfully underwrite a

similar five-year A$ zero-coupon issue for Kodak. It was determined that Merrill Lynch could meet Kodak's funding target if an attractively priced A$ zero-coupon swap could be found.

Meeting Kodak's minimum issue size of U.S.$75 million would necessitate an A$200 million zero-coupon issue, the largest A$ zero-coupon issue ever underwritten. Merrill Lynch then received a firm mandate on a five-year A$130 million zero-coupon swap with Australian Bank B at a semiannual interest rate of 13.39%. The remaining A$70 million was arranged through a long-dated forward foreign exchange contract with Australian Bank A at a forward rate of A$1 = U.S.$0.5286.

With the currency swap mandate and the long-dated forward contract, Merrill Lynch received final approval by Kodak for the transaction, and the five-year A$200 million zero-coupon issue was launched in Europe at a net price of 54 1/8%, with a gross spread of 1 1/8%. Net proceeds to Kodak were 53% of A$200 million, or A$106 million. Kodak converted this principal into U.S.$75 million at the spot rate of U.S.$0.7059. Simultaneously, Merrill Lynch entered into a currency swap with Kodak to convert the Australian dollar cash flows into U.S. dollar cash flows at 7.35% paid semiannually, or U.S. Treasurys plus 35 basis points (since five-year Treasury bonds were then yielding approximately 7%). That is, Kodak's all-in cost was 7.35%. As part of this swap, Merrill Lynch

---

[1] This example was supplied by Grant Kvalheim of Merrill Lynch, whose help is greatly appreciated. The actual interest rates and spot and forward rates have been disguised.

EXHIBIT 16.5 Kodak's A$200 Million Zero-Coupon Eurobond and Currency Swap

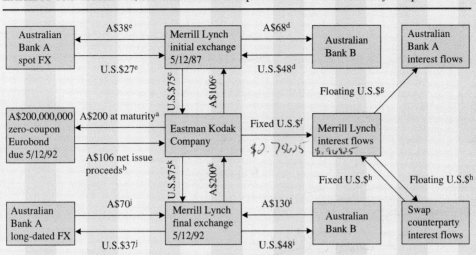

[a]Investors receive a single payment of A$200 million on 5/12/92, which represents both principal and interest.

[b]The bonds are priced at $54\frac{1}{8}\%$ less $1\frac{1}{8}\%$ gross spread. Net proceeds to Kodak at settlement on 5/12/87 are A$106 million.

[c]Kodak exchanges A$106 million with Merrill Lynch and receives U.S. $75 million at a fixed semiannual interest rate of 7.35%.

[d]Australian Bank B provides a 5-year A$130 million zero-coupon swap at a semiannual rate of 13.39%. In the currency swap's initial exchange on 5/12/87, Merrill Lynch pays Australian Bank B A$68 million (A$130,000,000 × $[1/(1 + (13.39\%/2))^{10}]$) and receives U.S. $48 million (A$68,000,000 × 0.7059) based on a spot exchange rate of U.S. $0.7059/A$1.

[e]Merrill Lynch sells the remaining A$38 million (A$106,000,000 − A$68,000,000) to Australian Bank A on 5/12/87 at a spot rate of U.S.$.7105/A$1, and receives U.S. $27 million.

[f]Kodak makes semiannual fixed-rate interest payments of U.S.$2,756,250 to Merrill Lynch [(7.35%/2) × U.S. $75,000,000)].

[g]Merrill Lynch makes semiannual floating-rate interest payments of LIBOR less 40 basis points on a notional principal amount of U.S.$48 million to Australian Bank B.

[h]Merrill Lynch makes semiannual interest payments of U.S. $1,884,000 based on a notional principal amount of U.S. $48 million and fixed interest rate of 7.85% and receives semiannual floating-rate interest payments of LIBOR flat in a fixed-floating rate swap with its book.

[i]Merrill Lynch receives A$130 million and pays U.S.$48 million in the Australian Bank B currency swap's final exchange on 5/12/92.

[j]In a long-dated forward foreign exchange transaction with Australian Bank A, Merrill Lynch purchases A$70 million on 5/12/92 for U.S.$37 million based on a forward exchange rate of U.S.$0.5286/A$1.

[k]On 5/12/92, Kodak pays U.S.$75 million to Merrill Lynch, receives A$200 million in return, and Kodak then pays the A$200 million to its zero-coupon bondholders.

agreed to make semiannual interest payments of LIBOR less 40 basis points to Australian Bank B. Merrill Lynch then arranged an interest rate swap to convert a portion of the fixed-rate payments from Kodak into floating-rate payments to Bank B. Exhibit 16.5 contains an annotated

EXHIBIT 16.6 MLCS Cash Flows—Eastman Kodak Transaction

| Date | Cash-Flow Type | KODAK CURRENCY SWAP | | AUSTRALIAN BANK B CURRENCY SWAP | |
		A$	U.S.$	A$	U.S.$
12 May 87	Initial exchange	106,000,000	(75,000,000)	(68,000,000)	48,000,000
12 Nov. 87	Interest	—	2,756,250[1]	—	(LIBOR − 40BPS)
12 May 88	Interest	—	2,756,250	—	(LIBOR − 40BPS)
12 Nov 88	Interest	—	2,756,250	—	(LIBOR − 40BPS)
12 May 89	Interest	—	2,756,250	—	(LIBOR − 40BPS)
12 Nov 89	Interest	—	2,756,250	—	(LIBOR − 40BPS)
12 May 90	Interest	—	2,756,250	—	(LIBOR − 40BPS)
12 Nov 90	Interest	—	2,756,250	—	(LIBOR − 40BPS)
12 May 91	Interest	—	2,756,250	—	(LIBOR − 40BPS)
12 Nov 91	Interest	—	2,756,250	—	(LIBOR − 40BPS)
12 May 92	Interest	—	2,756,250	—	(LIBOR − 40BPS)
12 May 92	Final exchange	(200,000,000)	75,000,000	130,000,000	(48,000,000)

[1](U.S.$75,000,000) × (0.0735) × (180 days/360 days).
[2](U.S.$48,000,000) × (0.0785) × (180 days/360 days).
[3](U.S.$2,756,250 − U.S.$1,884,000 + [U.S.$48,000 × 0.004 × (180 days/360 days)].

schematic diagram, based on a Merrill Lynch ad, of the currency and interest rate swaps and the long-dated foreign exchange purchase. Exhibit 16.6 summarizes the period-by-period cash flows associated with the transactions.

The final column of Exhibit 16.6 presents the net cash flows to Merrill Lynch from these transactions. The net present value (NPV) of these flows discounted at $r\%$ compounded semiannually is

$$NPV = \sum_{t=1}^{10} \frac{\$968,250}{(1 + r/2)^t} - \frac{\$10,000,000}{(1 + r/2)^{10}}$$

Discounted at the then risk-free, five-year Treasury bond rate of 7% compounded semiannually, the NPV of these flows is $963,365. Using a higher discount rate, say 7.5%, to reflect the various risks associated with these transactions results in a net present value to

DUAL CURRENCY BOND SWAPS ■ Another variant on the currency swap theme is a currency swap involving a **dual currency bond**—one that has the issue's proceeds and interest payments stated in foreign currency and the principal repayment stated in dollars. An example of a dual currency bond swap is the one involving the Federal National Mortgage Association (FNMA, or Fannie Mae). On October 1, 1985, FNMA agreed to issue 10-year, 8% coupon debentures in the amount of ¥50 billion (with net proceeds of ¥49,687,500,000) and to swap these yen for just over $209 million (an implied swap rate of ¥237.5479/$1). In return, Fannie Mae agreed to pay interest averaging about $21 million annually and to redeem these bonds at the end of 10 years at a cost of $240,400,000. Exhibit 16.7 shows the detailed yen and dollar cash flows associated with this currency swap. The net effect of this swap was to give Fannie Mae an all-in dollar cost of 10.67%

EXHIBIT 16.6 *(Continued)*

FOREIGN EXCHANGE MARKET		FIXED/FLOATING U.S. $ SWAP		
A$	U.S.$	Fixed	Floating	Net U.S.$ Flows
38,000,000	27,000,000	—	—	—
—	—	(1,884,000)[2]	LIBOR	968,250[3]
—	—	(1,884,000)	LIBOR	968,250
—	—	(1,884,000)	LIBOR	968,250
—	—	(1,884,000)	LIBOR	968,250
—	—	(1,884,000)	LIBOR	968,250
—	—	(1,884,000)	LIBOR	968,250
—	—	(1,884,000)	LIBOR	968,250
—	—	(1,884,000)	LIBOR	968,250
—	—	(1,884,000)	LIBOR	968,250
—	—	(1,884,000)	LIBOR	968,250
70,000,000	(37,000,000)			(10,000,000)

Merrill Lynch of $1,031,826. The actual NPV of these cash flows falls somewhere in between these two extremes.

By combining a nondollar issue with a currency swap and interest rate swap, Merrill Lynch was able to construct an innovative, lower-cost source of funds for Kodak. The entire package involved close teamwork and a complex set of transactions on three continents. In turn, through its willingness to consider non-traditional financing methods, Kodak was able to lower its cost of funds by about 15 basis points, yielding an annual savings of approximately $112,500 (0.0015 × $75,000,000). The present value of this savings discounted at 7.5% compounded semiannually (or 3.75% every six months) is

$$\sum_{t=1}^{10} \frac{\$56,250}{(1.0375)^t} = \$461,969$$

annually. In other words, regardless of what happened to the yen:dollar exchange rate in the future, Fannie Mae's dollar cost on its yen bond issue would remain at 10.67%. The 10.67% figure is the interest rate that just equates the dollar figures in column 2 to zero.

Let us illustrate the mechanics of this swap. Note that at the end of the first year, FNMA is obligated to pay its bondholders ¥4 billion in interest (an 8% coupon payment on a ¥50 billion face value debenture). To satisfy this obligation, FNMA pays $18,811,795 to Nomura, which in turn makes the ¥4 billion interest payment. As column 3 of Exhibit 16.7 shows, FNMA has effectively contracted with Nomura to buy ¥4 billion forward for delivery in one year at a forward rate of ¥212.6325.

Similarly, FNMA satisfies its remaining yen obligations (shown in column 1) by paying a series of dollar amounts (shown in column 2) to Nomura, which in turn makes the re-

EXHIBIT 16.7 Cash Flows Associated with Yen Debenture Currency Swap

PAYMENT DATE	PAYMENT ON DUAL CURRENCY DEBENTURE (1)	DOLLAR PAYMENT UNDER SWAP (2)	YEN: DOLLAR EXCHANGE RATE (3)
October 1, 1985	−¥49,687,500,000[1]	−$209,168,297[2]	¥237.3777/$1
October 1, 1986	¥4,000,000,000	$ 18,811,795	¥212.6325/$1
October 1, 1987	¥4,000,000,000	$ 19,124,383	¥209.1571/$1
October 1, 1988	¥4,000,000,000	$ 19,464,369	¥205.5037/$1
October 1, 1989	¥4,000,000,000	$ 19,854,510	¥201.4656/$1
October 1, 1990	¥4,000,000,000	$ 20,304,260	¥197.0030/$1
October 1, 1991	¥4,000,000,000	$ 20,942,380	¥191.0003/$1
October 1, 1992	¥4,000,000,000	$ 21,499,717	¥186.0490/$1
October 1, 1993	¥4,000,000,000	$ 22,116,875	¥180.8574/$1
October 1, 1994	¥4,000,000,000	$ 22,723,857	¥176.0265/$1
October 1, 1995	¥4,000,000,000	$ 23,665,098	¥169.0253/$1
October 1, 1995	$240,400,000	$240,400,000	¥207.9867/$1

[1]This number is the ¥50 billion face amount net of issue expenses.
[2]Net proceeds received after reimbursing underwriters for expenses of $150,000.

quired yen payments. The exchange of fixed dollar payments in the future for fixed yen payments in the future is equivalent to a sequence of forward contracts entered into at the forward exchange rates shown in column 3. Since the actual spot rate at the time the swap was entered into (August 29, 1985) was about ¥240/$1, the implicit forward rates on these forward contracts reveal that the yen was selling at a forward premium relative to the dollar; that is, it cost fewer yen to buy a dollar in the forward market than in the spot market. The reason the yen was selling at a forward premium was the same reason that Fannie Mae was borrowing yen: At this time, the interest rate on yen was below the interest rate on dollars.

Since this particular issue was a dual currency bond, with the issue's proceeds and interest payments stated in yen and the principal repayment stated in dollars, the final payment is stated in dollars only. However, it should be noted that by agreeing to a principal repayment of $240,400,000, instead of ¥50 billion, Fannie Mae was actually entering into the equivalent of a long-dated forward contract at an implicit forward rate of ¥207.9867/$1 (¥50,000,000,000/$240,400,000).

Economic Advantages of Swaps

Swaps provide a real economic benefit to both parties only if a barrier exists to prevent arbitrage from functioning fully. Such impediments may include legal restrictions on spot and forward foreign exchange transactions, different perceptions by investors of risk and creditworthiness of the two parties, appeal or acceptability of one borrower to a certain class of investor, tax differentials, and so forth.[2]

Swaps also allow firms to lower their cost of foreign exchange risk management. A U.S. corporation, for example, may want to secure fixed-rate funds in Deutsche marks in order to reduce its DM exposure, but it may be hampered in doing so because it is a rela-

---

[2]This explanation is provided in Clifford W. Smith, Jr., Charles W. Smithson, and Lee M. Wakeman, "The Evolving Market for Swaps," *Midland Corporate Finance Journal*, Winter 1986, pp. 20–32.

tively unknown credit in the German financial market. In contrast, a German company that is well established in its own country may desire floating-rate dollar financing, but is relatively unknown in the U.S. financial market.

In such a case, a bank intermediary familiar with the funding needs and "comparative advantages" in borrowing of both parties may arrange a currency swap. The U.S. company borrows floating-rate dollars, and the German company borrows fixed-rate DM. The two companies then swap both principal and interest payments. When the term of the swap matures, say, in five years, the principal amounts revert to the original holders. Both parties receive a cost savings because they borrow initially in the market where they have a comparative advantage and then swap for their preferred liability.

Currency swaps, thus, are often used to provide long-term financing in foreign currencies. This function is important because in many foreign countries, long-term capital and forward foreign exchange markets are notably absent or not well developed. Swaps are one type of vehicle providing liquidity to these markets.

In effect, swaps allow the transacting parties to engage in some form of tax, regulatory-system, or financial-market arbitrage. If the world capital market were fully integrated, the incentive to swap would be reduced because fewer arbitrage opportunities would exist. However, even in the United States, where financial markets function freely, interest rate swaps are very popular and are credited with cost savings.

16.2 STRUCTURED NOTES

In recent years, a new breed of financial instrument—the structured note—has become increasingly popular. **Structured notes** are interest-bearing securities whose interest payments are determined by reference to a formula set in advance and adjusted on specified *reset* dates. The formula can be tied to a variety of different factors, such as LIBOR, exchange rates, or commodity prices. Sometimes the formula includes multiple factors, such as the difference between three-month dollar LIBOR and three-month Swiss franc LIBOR. The common characteristic is one or more embedded derivative elements, such as swaps, forwards, or options. The purpose of this section is not to describe every type of structured note available since there are literally hundreds, with the design of new ones limited only by the creativity and imagination of the parties involved. Rather, it is to describe the general characteristics of these debt instruments and their uses.

We have already seen one of the earliest types of structured notes—a floating rate note whose interest payment is tied to LIBOR (the equivalent of swapping a fixed-rate for a floating-rate coupon). While the FRN formula is quite simple, the formulas on subsequent structured notes have become more complex to meet the needs of users who want to take more specific positions against interest rates or other prices. Structured notes allow companies and investors to speculate on the direction, range, and volatility of interest rates, the shape of the *yield curve* (which relates the yield to maturity on bonds to their time to maturity and is typically upward sloping), and the direction of equity, currency, and commodity prices. For example, a borrower who felt that the yield curve would flatten (meaning that the gap between short-term and long-term rates would narrow) might issue a note that pays an interest rate equal to 2% plus three times the difference between the six-month and 20-year interest rates.

Structured notes can also be used for hedging purposes. Consider, for example, a gold mine operator that would like to borrow money but whose cash flow is too volatile (owing to fluctuations in the price of gold) to be able to service ordinary fixed-rate debt. One solution for the operator is to issue a structured note whose interest payments are tied to the price of gold. If the price of gold rises, the operator's cash flows increase and it finds it easier to make the interest payments. When gold prices go down, its interest burden is lower. Not only does the note hedge the operator's gold price risk, but the greater ease of servicing this note lowers the operator's risk of default and hence the risk premium it must pay.

Inverse Floaters

One structured note that has received negative publicity in the past is the inverse floater. For example, the large quantity of inverse floaters held by Orange County in its investment portfolio exacerbated the disaster that overtook it when interest rates rose in 1994. An **inverse floater** is a floating-rate instrument whose interest rate moves inversely with market interest rates.[3] In a typical case, the rate paid on the note is set by doubling the fixed rate in effect at the time the contract is signed, and subtracting the floating reference index rate for each payment period. Suppose the coupon on a five-year, fixed-rate note is 6.5%. An inverse floater might have a coupon of 13% − LIBOR6, with the rate reset every six months. In general, an inverse floater is constructed by setting the payment equal to $nr - (n - 1)$LIBOR, where r is the market rate on a fixed-rate bond and n is the multiple applied to the fixed rate. If interest rates fall, this formula will yield a higher return on the inverse floater. If rates rise, the payment on the inverse floater will decline. In both cases, the larger n is, the greater the impact of a given interest rate change on the inverse floater's interest payment.

Issuers, such as banks, can use inverse floaters to hedge the risk of fixed-rate assets, such as a mortgage portfolio. If interest rates rise, the value of the bank's mortgage portfolio will fall, but this loss will be offset by a simultaneous decline in the cost of servicing the inverse floaters used to finance the portfolio.

The value of an inverse floater (for example, 13% − LIBOR6) is calculated by deducting the value of a floating-rate bond (for example, one priced at LIBOR6) from the value of two fixed-rate bonds, each with half of the fixed-coupon rate of the inverse floater (for example, two 6.5% fixed-rate bonds).[4] Mathematically, this valuation formula is represented as:

$$B(13\% - \text{LIBOR6}) = 2 \times B(6.5\%) - B(\text{LIBOR6})$$

where $B(x)$ represents the value of a bond paying a rate of x. That is, the value of the inverse floater is equal to the sum of two fixed-rate bonds paying a 6.5% coupon minus the value of a floating-rate bond paying LIBOR6.

---

[3] The interest payment has a floor of zero, meaning that the lender will never owe interest to the borrower.

[4] The object is to ensure that there are as many principal repayments as bonds (otherwise, if we priced a 13% coupon bond and subtracted off the value of a floating-rate bond, the net would be zero principal repayments— the principal amount on the 13% coupon bond − the principal on the floating-rate bond).

At the issue date, assuming that 6.5% is the issuer's market rate on a fixed-rate bond and LIBOR6 is the appropriate floating rate for the borrower's creditworthiness, the market value of each $100 par value inverse floater is $100 (2 × $100 − $100) since the fixed-rate and floating-rate bonds are worth $100 apiece.

To take another, somewhat more complicated example:

$$B(19.5\% - 2 \times LIBOR6) = 3 \times B(6.5\%) - 2 \times B(LIBOR6)$$

In effect, an inverse floater is equivalent to buying fixed-rate bonds partially financed by borrowing at LIBOR. For example, the cash flows on a $100 million inverse floater that pays 13% − LIBOR6 is equivalent to buying $200 million of fixed-rate notes bearing a coupon of 6.5% financed with $100 million of money borrowed at LIBOR6.

The effect of an inverse-floater structure is to magnify the bond's interest rate volatility. Specifically, the volatility of an inverse floater with a payment structure equal to $nr - (n - 1)$LIBOR is equal to n times the volatility of a straight fixed-rate bond. The reason is that the floating-rate portion of the inverse floater trades at or close to par while the fixed-rate portion—given its structure—changes in value with interest rate fluctuations at a rate that is n times the rate at which a single fixed-rate bond changes in value.

CALLABLE STEP-UP NOTE ■ **Step-ups** are callable debt issues that feature one or more increases in a fixed rate or a step-up in a spread over LIBOR during the life of the note. Most issuers of these notes have low credit ratings. Consequently, the purpose of the step up is usually to encourage the issuer to refinance. If the issuer does not refinance, the higher rate is designed to be compensation for the investor's acceptance of credit risk. Highly-rated issuers sometimes issue step-up bonds if they believe that interest rates will decline and they can issue a replacement bond at a lower rate.

STEP-DOWN COUPON NOTE ■ **Step-downs** are debt instruments with a high coupon in earlier payment periods and a lower coupon in later payment periods. This structure is usually motivated by a low short-term rate environment and regulatory or tax considerations. Investors seeking to front-load their interest income would be interested in such notes.

16.3 INTEREST RATE FORWARDS AND FUTURES

In addition to swaps and structured notes, companies also can use a variety of forward and futures contracts to manage their interest rate expense and risk. These contracts include forward forwards, forward rate agreements, and Eurodollar futures. All of them allow companies to lock in interest rates on future loans and deposits.

Forward Forwards

A **forward forward** is a contract that fixes an interest rate today on a future loan or deposit. The contract specifies the interest rate, the principal amount of the future deposit or loan, and the start and ending dates of the future interest rate period.

TELECOM ARGENTINA FIXES A FUTURE LOAN RATE

Suppose that Telecom Argentina needs to borrow $10 million in six months for a three-month period. It could wait six months and borrow the money at the then-current interest rate. Rather than risk a significant rise in interest rates over the next six months, however, Telecom Argentina decides to enter into a forward forward with Daiwa Bank that fixes this rate at 8.4% per annum. This contract guarantees that six months from today, Daiwa Bank will lend Telecom Argentina $10 million for a three-month period at a rate of 2.1% (8.4%/4). In return, nine months from today, Telecom Argentina will repay Daiwa the principal plus interest on the loan, or $10,210,000 ($10,000,000 × 1.021).

The forward forward rate on a loan can be found through arbitrage. For example, suppose that a company wishes to lock in a six-month rate on a $1 million Eurodollar deposit to be placed in three months. It can buy a forward forward or it can create its own. To illustrate this process, suppose that the company can borrow or lend at LIBOR. Then the company can derive a three-month forward rate on LIBOR6 by simultaneously borrowing the present value of $1 million for three months and lending that same amount of money for nine months. If LIBOR3 is 6.7%, the company will borrow $1,000,000/(1 + 0.067/4) = $983,526 today and lend that same amount for nine months. If LIBOR9 is 6.95%, at the end of nine months the company will receive $983,526 × (1 + 0.0695 × 3/4) = $1,034,792. The cash flows on these transactions are

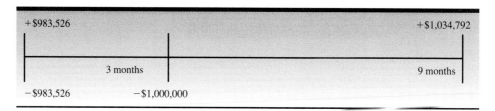

Notice that the borrowing and lending transactions are structured so that the only net cash flows are the cash outlay of $1,000,000 in three months and the receipt of $1,034,792 in nine months. These transactions are equivalent to investing $1,000,000 in three months and receiving back $1,034,792 in nine months. The interest receipt of $34,792, or 3.479% for six months, is equivalent to a rate of 6.958% per annum.

The process of arbitrage will ensure that the actual forward rate for LIBOR6 in three months will almost exactly equal the "homemade" forward forward rate.

Forward Rate Agreement

In recent years, forward forwards have been largely displaced by the forward rate agreement. A **forward rate agreement** (FRA) is a cash-settled, over-the-counter forward contract that allows a company to fix an interest rate to be applied to a specified future interest period on a notional principal amount. It is analogous to a forward foreign currency contract but instead of exchanging currencies, the parties to an FRA agree to exchange interest payments. The formula used to calculate the interest payment on a LIBOR-based FRA is

$$\text{Interest payment} = \text{notional principal} \times \frac{(\text{LIBOR} - \text{forward rate})\left(\dfrac{\text{days}}{360}\right)}{1 + \text{LIBOR} \times \left(\dfrac{\text{days}}{360}\right)} \qquad \textbf{(16.1)}$$

where *days* refers to the number of days in the future interest period. The discount reflects the fact that the FRA payment occurs at the start of the loan period whereas the interest expense on a loan is not paid until the loan's maturity. To equate the two, the differential interest expense must be discounted back to its present value using the actual interest rate. The following example of Unilever shows how a borrower can use an FRA to lock in the interest rate applicable for a future loan.

ILLUSTRATION

UNILEVER USES AN FRA TO FIX THE INTEREST RATE ON A FUTURE LOAN Suppose that Unilever needs to borrow $50 million in two months for a six-month period. To lock in the rate on this loan, Unilever buys a "2 × 6" FRA on LIBOR at 6.5% from Bankers Trust for a notional principal of $50 million. This means that Bankers Trust has entered into a two-month forward contract on six-month LIBOR. Two months from now, if LIBOR6 exceeds 6.5%, Bankers Trust will pay Unilever the difference in interest expense. If LIBOR6 is less than 6.5%, Unilever will pay Bankers Trust the difference.

Assume that in two months LIBOR6 is 7.2%. Because this rate exceeds 6.5%, Unilever will receive from Bankers Trust a payment determined by Equation 16.1 of

$$\text{Interest payment} = \$50,000,000$$

$$\times \frac{(0.072 - 0.065)\left(\dfrac{182}{360}\right)}{1 + 0.072 \times \left(\dfrac{182}{360}\right)} = \$170,730$$

In addition to fixing future borrowing rates, FRAs can also be used to fix future deposit rates. Specifically, by selling an FRA, a company can lock in the interest rate applicable for a future deposit.

Eurodollar Futures

A **Eurodollar future** is a cash-settled futures contract on a three-month, $1,000,000 Eurodollar deposit that pays LIBOR. These contracts are traded on the Chicago Mercantile Exchange (CME), the London International Financial Futures Exchange (LIFFE) and the Singapore International Monetary Exchange (SIMEX). Eurodollar futures contracts are traded for March, June, September, and December delivery. Contracts are traded out to three years, with a high degree of liquidity out to two years.

Eurodollar futures act like FRAs in that they help lock in a future interest rate and are settled in cash. However, unlike FRAs, they are marked to market daily (as in currency futures, this means that gains and losses are settled in cash each day). The price of a Eurodollar futures contract is quoted as an index number equal to 100 minus the annualized forward interest rate. For example, suppose the current futures price is 91.68. This price implies that the contracted-for LIBOR3 rate is 8.32%, that is, 100 minus 91.68. The value of this contract at inception is found by use of the following formula:

$$\text{Initial value of Eurodollar futures contract} - \$1,000,000 \left[1 - 0.0832 \left(\frac{90}{360} \right) \right] = \$979,200$$

The interest rate is divided by four to convert it into a quarterly rate. At maturity, the cash settlement price is determined by subtracting LIBOR3 on that date from 100. Whether the contract gained or lost money depends on whether cash LIBOR3 at settlement is greater or less than 8.32%. If LIBOR3 at settlement is 7.54%, the Eurodollar future on that date is valued at $981,150:

$$\text{Settlement value of Eurodollar futures contract} = \$1,000,000 \left[1 - 0.0754 \left(\frac{90}{360} \right) \right] = \$981,150$$

At this price, the buyer has earned $1,950 ($981,150 − $979,200) on the contract. As can be seen from the formula for valuing the futures contract, each basis point change in the forward rate translates into $25 for each contract ($1,000,000 × 0.0001/4), with increases in the forward rate reducing the contract's value and decreases raising its value. For example, if the forward rate rose three basis points, a long position in the contract would lose $75.

Prior to the settlement date, the forward interest rate imbedded in the futures contract is unlikely to equal the prevailing LIBOR3. For example, on January 6, 1995, the June 1995 Eurodollar futures contract closed at an index price of 92.22, implying a forward rate of 7.78% (100 − 92.22). Actual LIBOR3 on January 6 was 6.4375%. The discrepancy between the two rates reflects the fact that the 7.78% rate represented a 3-month implied forward rate as of June 19, 1995, which was 164 days in the future. The forward rate is based on the difference between 164-day LIBOR and LIBOR on a 254-day deposit (which matures on September 17, 1995, 90 days after the 164-day deposit).

The actual LIBOR3 used is determined by the respective exchanges. Both the CME and LIFFE conduct a survey of banks to establish the closing value for LIBOR3. Accordingly, contracts traded on the two exchanges can settle at slightly different values. SIMEX uses the CME's settlement price for its contracts.

Contracts traded on the CME and SIMEX have identical contractual provisions. Those two exchanges have an offset arrangement whereby contracts traded on one exchange can be converted into equivalent contracts on the other exchange. Accordingly, the two contracts are completely fungible. LIFFE does not participate in this arrangement.

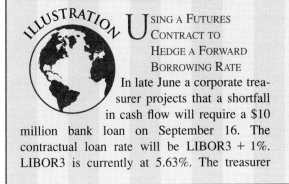

USING A FUTURES CONTRACT TO HEDGE A FORWARD BORROWING RATE
In late June a corporate treasurer projects that a shortfall in cash flow will require a $10 million bank loan on September 16. The contractual loan rate will be LIBOR3 + 1%. LIBOR3 is currently at 5.63%. The treasurer can use the September Eurodollar futures, which are currently trading at 94.18, to lock in the forward borrowing rate. This price implies a forward Eurodollar rate of 5.82% (100 − 94.18). By selling ten September Eurodollar futures contract, the corporate treasurer ensures a borrowing rate of 6.82% for the three-month period beginning September 16. This rate reflects the bank's 1% spread above the rate locked in through the futures contract.

16.4 INTERNATIONAL LEASING

Cross-border or international leasing can be used to both defer and avoid tax. It also can be used to safeguard the assets of a multinational firm's foreign affiliates and avoid currency controls.

Operating versus Financial Leases

Leases can be designated as either operating or financial leases. The tax advantages of international leasing typically turn on this distinction. An *operating lease* is a true lease in that ownership and use of the asset are separated. The operating lease agreement typically covers only part of the useful life of the asset; it may be renewed on a period-by-period basis. By contrast, a *financial lease* is one that extends over most of the economic life of the asset and is noncancelable or is cancelable only upon payment of a substantial penalty to the lessor. Normally, the payments under a financial lease amortize most of the economic value of the asset.

Noncancelability means that the firm has a contractual obligation to make all the lease payments specified in the agreement, regardless of whether—at a later date—it needs or wants the asset. In effect, economic ownership in a financial lease resides with the lessee. The contractual nature of a financial lease means that entering into one is equivalent to borrowing money and buying the asset outright. Thus, although it is in form a lease, in substance the lessor in a financial lease is lending money to the lessee, with the loan secured by the asset. The equivalence between a financial lease and debt financing extends further: The lessor's profit comes from interest, while default by the lessee can lead to bankruptcy. Leasing and borrowing can and should be considered as alternative financing techniques and can be compared as such.

Tax Factors

Vital issues in any leasing transaction are the tax status of lease payments and who gets to deduct depreciation and to claim any investment tax credit. In the United States, the answer depends on whether the transaction is considered an operating lease (also called a true lease) or a financial lease. The IRS makes this distinction to ensure that the lease transaction is not a disguised installment sale. A lease that qualifies as a true lease for tax purposes is called a tax-oriented lease.

In a *tax-oriented lease*, the lessor receives the tax benefits of ownership, and the lessee gets to deduct the full value of lease payments. But if the lease is considered a financial lease, lease payments are treated as installments of the purchase price plus interest and, therefore, are not fully deductible by the lessee. As the stipulated owner, the lessee is allowed tax depreciation for the purchase price and a tax deduction for the interest factor. The lessor is taxed on the interest imputed to the lease payment and realizes none of the tax benefits of ownership.

DOUBLE DIPPING ■ The principal tax advantage from international leasing arises when it is possible to structure a "double-dip" lease. In a **double-dip lease**, the disparate leasing rules of the lessor's and lessee's countries let both parties be treated as the owner of the leased equipment for tax purposes. Thus, both the lessee and the lessor are entitled to benefits such as fast depreciation and tax credits. This benefit to the lessor can be passed to the lessee in the form of lower rentals. In the absence of double dipping, the lessee's deductions would be limited to the rent paid.

An example of extraordinary benefits used to be U.K. lessors' financing U.S. projects (especially aircraft acquisitions) through financial leases. The U.K. lessor could claim a 100% first-year depreciation write-off and pass its tax savings via reduced rentals to the U.S. lessee, who would also claim the U.S. investment tax credit and depreciation for its own account. Another popular arrangement involved structuring double-dip leases with the United Kingdom and Ireland to get 100% first-year write-offs in both countries. Although this route is now less attractive (the United Kingdom has scaled down its first-year depreciation write-off for assets used outside the United Kingdom to 10%) the principles remain and can be applied to other countries.

Double dipping is most often achieved with lessees in countries that look to the economic reality of the arrangement (e.g., the United States, Japan, Germany, and the Netherlands) and lessors in countries that characterize leases solely on the basis of legal ownership (e.g., Switzerland, France, Sweden, and the United Kingdom). Readily recognizable financial leases will achieve depreciation allowances in both countries when the lease is considered an operating lease in the lessor's country but a financial lease in the lessee's country.

For example, a double dip from Sweden to Germany is relatively straightforward. As long as the lease does not require the lessee to purchase the asset, the Swedish lessor will get the normal depreciation allowance. The German lessee will be entitled to German allowances if, for example, the leased asset is limited-use property.

However, fine tuning and considerable skill are needed to double dip between two countries that make the economic ownership distinction. In such a case, it is necessary to structure a lease that fits into the perhaps very narrow crack between what the lessor's country considers an operating lease and the lessee's country considers a finance lease. For example, the Dutch airline KLM leased planes from the United States in an arrangement that gave both the U.S. lessor and KLM depreciation write-offs in their home countries. In that deal, the planes were registered in both the United States and the Netherlands. U.S. registry of the aircraft made them eligible for fast write-offs and an investment tax credit, even though they were used outside the United States.

It can also be beneficial to double dip from a captive leasing company in a low-tax country. For example, a multinational firm's leasing company located in Switzerland can lease an asset to an affiliate located in the United Kingdom, under an arrangement that permits both lessor and lessee to receive the tax benefits of ownership. Both benefits are at the 35% (previously 52%) corporate tax rate. Until 1986, the U.K. subsidiary would get its 100% first-year allowance (it's now 25%) and a deduction for the interest factor in the lease payment. The Swiss leasing affiliate's taxable income from lease receipts will be reduced by its depreciation allowances, with the balance taxed at approximately 10%. The after-tax earnings remain within the multinational group.

ADDITIONAL DIPS ■ Where additional parties are involved, it is possible that each will be entitled to capital allowances. For example, a triple dip can be achieved by arranging a lease with a Swiss lessor (always entitled to allowances), a U.K. lessee with a purchase option (qualifying it for the 25% depreciation allowance in the first year), and a German sublessee who satisfies German economic ownership rules (qualifying it for a depreciation deduction as well).

International Leasing Companies

By incorporating a captive international leasing company (for interaffiliate transactions) in an appropriate location, the MNC can shift income from high-tax to low-tax jurisdictions and reduce or eliminate withholding tax on lease payments. It also may be able to receive lease income tax free. For example, a multinational with excess foreign tax credits can utilize these FTCs by placing a lease that generates foreign-source income in a low-tax country. Similarly, by placing a lease that generates losses in a high-tax country, a multinational that has excess foreign tax credits can cut its taxes in that high-tax country and thereby reduce its excess FTCs.

Multinationals can also reduce political risk by investing in politically risky countries via a captive international leasing company incorporated in an appropriate location. Lease financing limits the ownership of assets by subsidiaries in politically unstable countries. Leasing also enables the firm to more easily extract cash from affiliates located in countries where there are exchange controls. Lease payments are often a more acceptable method of extracting funds than dividends, interest, or royalty payments. Similarly, there will be more chance of recovering assets (or at least obtaining compensation for them) in the case of nationalization if they are not owned by the local subsidiary.

The ideal characteristics of an international leasing company location include (1) no exchange control restrictions, (2) a stable currency, (3) political stability, and (4) a wide network of tax treaties to eliminate withholding tax on lease payments and payment of dividends by the leasing company to its parent. The end result, ideally, would be that profits arise in a low-tax country, tax-deductible expenses such as depreciation and lease payments arise in a high-tax country, and there are no withholding taxes on rent payments and on dividends to the parent of the leasing company.

Japanese Yen-Based Leasing

Because of Japan's huge trade surpluses, it has become a major source of international financing, including international lease financing. An important Japanese development in the international leasing business was the introduction in early 1981 of yen-based leases, known as Shoguns. **Shogun leases** allow leasing companies, usually with the help of U.S. banks, to bypass restrictions imposed by Japan's Ministry of Finance on long-term yen loans. Whereas (in principle at least) it is not possible to lend yen for more than ten years or for more than certain amounts, it is possible to provide both operating leases and conditional-sale leases for longer periods and for greater amounts and, thus, accommodate big-ticket items such as planes and ships.

During 1994, the Japanese financed over $5 billion in aircraft, representing almost 40% of the world's new aircraft deliveries, through Japan-leveraged leases. The Japanese demand for aircraft leases reflects a desire to diversify away from the bad-debt troubles on balance sheets, which stem largely from real-estate and securities losses. In contrast, aircraft are viewed as solid-performing assets that are less prone to trouble than movie studios or real estate. Moreover, Japanese tax authorities have encouraged leveraged lease transactions by allowing companies holding equity in an aircraft to defer taxes equal to their stakes in the aircraft's depreciation. This tax-sheltered structure enhances returns on the leveraged leases to above-market rates, allowing after-tax returns on some leases to reach 35% or more.

16.5 LDC DEBT-EQUITY SWAPS

Beginning in 1985, a market developed that enables investors to purchase the external debt of less-developed countries (LDCs) to acquire equity or domestic currency in those same countries. The market for **LDC debt-equity swaps**, as the transactions are called, grew rapidly in the late 1980s and early 1990s. Between 1985 and 1988, for example, about $15 billion worth of LDC loans were swapped. Although this amount was small in relation to the $437 billion that the 15 most troubled debtor nations owed the world's commercial banks, it provided a number of opportunities for multinational corporations and other investors. The most important swap markets were initiated by six major debtor nations—Chile, Brazil, Mexico, Venezuela, Argentina, and the Philippines. Many other nations initiated similar programs as well.

Types of Debt Swaps and Their Rationale

Swaps can be quite complex, but the basics are fairly simple. For several years, European and regional U.S. commercial banks have been selling troubled LDC loans in the so-called secondary market—an informal network of large banks, big multinational corporations, and some Wall Street investment banks that trade loans of troubled debtor nations over the telephone and by telex. The trading, centered in New York, has grown steadily since the international debt crisis broke in 1982.

The loans typically have traded at deep discounts to their face value, reflecting the market's opinion that they would not be repaid in full. These discounts vary by country and time period. In mid-1987, for example, average prices ranged from Chile at about

70% of par, or 70 cents on the dollar, to Bolivia at about 10% of par. There was also a tendency for prices to drift down over this time period, as the dimensions of the debt problem became clearer. Prices bottomed in 1989 and then began to move up as Latin American countries began to restructure their economies. Chile, which has generally followed sound economic policies, and bought back substantial amounts of its debt, has seen the greatest appreciation in the price of its debt. However, debt prices fell again in 1994 in line with the general decline in emerging markets, as U.S. interest rates rose and investors became less sanguine about the risks associated with investments in these markets. By 1998, however, prices of emerging market debt issued by Latin American nations had recovered quite substantially.

With good deals disappearing in the more creditworthy LDCs, a new market is developing for the bottom tier of LDC debt, such as Zaire, Gabon, Angola, Myanmar, North Korea, Cuba, Paraguay, Albania, and Russia. For example, in mid-1998, Zairean and Albanian debt was selling at about 10% of par, Cuban debt at 25%, North Korean debt at 35%, Paraguayan debt at 40%, and Russian debt at 75%.

Usually, discount market quotations are cast in terms of bids and offers, not single market-clearing transaction prices. The spread between bid and offered prices is frequently rather wide—one indication of a thin market—although it seems to have narrowed over time.

In a typical deal, a multinational that wants to invest in, say, Chile hires an intermediary (usually a bank) to buy Chilean loans in the secondary market. The company (again through a middleman) presents the loans, denominated in dollars, to the Chilean central bank, which redeems them for pesos. The central banks pay less than face value but more than the loans trade for in the secondary market. Chile pays about 92 cents on the dollar, and Mexico an average of 88 cents.

Thus, a company that wants to expand in Chile can pick up $100 million of loans in the secondary market for $70 million and swap them for $92 million in pesos. Chile gets $100 million of debt off its books and doesn't have to part with precious dollars. The company gets $92 million of investment for $70 million, which amounts to a 24% subsidy (22/92).

ILLUSTRATION: CITICORP STRUCTURES A DEBT SWAP FOR NISSAN

In 1986, Citicorp learned that Nissan Motors wanted to invest the equivalent of $54 million to expand its truck factory in Mexico. When Citicorp offered to get them the $54 million in pesos for much less than $54 million through a debt-equity swap, Nissan liked the idea.

Citicorp went to the Mexican government and got their approval for the swap. The Mexican government agreed to pay $54 million in pesos for approximately $60 million of their external debt, a 10% discount. Citicorp went out and bought the $60 million in Mexican bank debt for about $38 million. Nissan wound up with the peso equivalent of $54 million at a price of only $40 million. Mexico retired about $60 million in bank debt, and Citicorp was paid about $2 million for structuring the deal and assuming the risk.

EXHIBIT 16.8 How a Debt-Equity Swap Works

Manufacturers Hanover takes $115 million in Brazilian government loans to Multplic, a São Paulo broker. ⟶	The broker takes the loans to the Brazilian central bank's monthly debt auction, where they are valued at an average of 86¢ on the dollar.
With its cruzados, Manufacturers Hanover buys 10% of Companhia Suzano de Papel e Celulose's stock, and Suzano uses the bank's cash to expand production and exports. ⟵	Through the broker, Manufacturers Hanover exchanges the loans at the central bank for $100 million worth of Brazilian cruzados.* It pays the broker a $150,000 commission. The central bank retires the loans.

---

*At the time of this transaction, the cruzado was the Brazilian currency. It is now the *real*.

The variations on this theme are endless. Chrysler has used some of the pesos it got in swaps to pay off local debt owed by its Mexican subsidiary. Kodak and Unisys have used swaps to expand their operations in Chile. Club Med is building a new beach resort in Mexico. Big U.S. banks usually act only as intermediaries in these deals, but a few have swapped for themselves, trading their LDC loans for equity in Latin businesses. In 1988, Manufacturers Hanover used a debt-equity swap to exchange some of its $2 billion in Brazilian debt for a 10% equity stake in Companhia Suzano de Papel e Celulose, a Brazilian paper and pulp company (see Exhibit 16.8 for details).

In addition to debt-equity swaps, there are also "debt-peso" swaps. *Debt-peso swaps* enable residents of a debtor country to purchase their country's foreign debt at a discount and to convert this debt into domestic currency. To finance these purchases, residents use funds held abroad or hard currency acquired from international trade or in the exchange market.

By arrangement with the debt country, domestic-currency assets obtained via debt swaps are acquired at closer to the original face value of the debt. For example, in 1986, a purchaser who acquired Mexican debt for 57 cents on the dollar could obtain equity worth 82 cents. Even after accounting for fees and redemption discounts applied by debtor countries to convert the debt into domestic currency, debt-swaps allow investors to acquire the domestic currency of debtor countries much more cheaply than do official exchange markets. In effect, investors resorting to the debt-swap market enjoy a preferential exchange rate.

Because the debt swap market offers a more favorable exchange rate than do official exchange markets, it creates an incentive for arbitrage. For example, a resident of a debtor country may exchange 100 pesos in domestic currency in the official exchange market to acquire a dollar. The resident may then use this dollar to acquire, via the debt swap market, 125 pesos in domestic currency and thus gain a 25 peso profit. Exchange controls will not necessarily be effective in preventing this "round-trip" process because domestic residents may elude exchange controls in a number of ways (for example, by overstating imports or understating exports).

Typically, access to swap programs involves wading through a lot of government red tape. For example, rules for foreign investors to participate in Mexico's swap program are outlined, in sometimes indecipherable language, in a 44-page *Manual Operativo*. Mexico also tailors its swap program to promote industrial policy goals. It does that by redeeming loans at different prices, depending on how the proceeds will be used. If a foreign investor wants to buy shares in a nationalized company that the government is trying to privatize, the central bank redeems loans at full face value. It pays as much as 95 cents on the dollar if the pesos will be invested in tourism and other businesses that help the trade balance. However, it pays only 75 cents on the dollar on deals that create no jobs, no exports, and no new technology.

Costs and Benefits of Debt Swaps

Governments have found it difficult to assess the desirability of debt swaps. This section discusses some of the costs and benefits associated with these swaps.

DEBT SWAPS AND INFLATION ■ One problem with debt swaps is that there is no free lunch. All too frequently, governments finance their purchases of bank debt by simply printing more currency. Consider Brazil. Suppose it buys back $100 million of its debt during a particular month. If the government isn't running a surplus, which it rarely is, it must print money to pay for the debt. The net result is fewer dollars of foreign debt outstanding but higher inflation. In effect, Brazil has taxed its citizens (by means of the inflation tax) and turned over the proceeds to banks, which then use them to acquire local assets.

IMPACT ON CAPITAL FORMATION ■ One advantage of debt swaps is that they may lead to additional capital formation in the country. However, the incremental capital formation associated with the swap program depends on how much foreign investment would have occurred without the swap. In general, if swaps lower the cost of financing direct investment projects, more of them will be forthcoming. On the other hand, if swaps are permitted simply for the purchase of equity in existing local businesses without bringing in any additional capital investment, the impact on new capital formation may be minimal.

EFFECT ON PRIVATIZATION ■ One way in which to engage in a noninflationary swap program is to carry it out in conjunction with a privatization program. By trading existing debt for shares in a state-owned company, the government can swap its debts without printing more money. At the same time, the government will realize the benefits of privatization.

16.6 VALUING LOW-COST FINANCING OPPORTUNITIES

Sharp-eyed firms are always on the lookout for financing choices that are "bargains"— that is, financing options priced at below-market rates. The value of arranging *below-market financing* can be illustrated by examining a case involving Sonat, the energy and energy services company based in Birmingham, Alabama. In late 1984, Sonat ordered from Daewoo Shipbuilding, a South Korean shipyard, six drilling rigs that can be partly submerged. Daewoo agreed to finance the $425 million purchase price with an 8.5-year loan at an annual interest rate of 9% paid semiannually. The loan is repayable in 17 equal semiannual installments. How much is this loan worth to Sonat? That is, what is its net present value?

At 9% interest paid semiannually, Sonat must pay interest equal to 4.5% of the loan balance plus $25 million in principal repayment every six months for the next 8.5 years. In return, Sonat receives $425 million today. Given these cash inflows and outflows, we can calculate the loan's NPV just as we would for any project analysis. Note, however, that unlike the typical capital-budgeting problem we looked at, the cash inflow occurs immediately and the cash outflows later. But the principle is the same. All we need now is the required return on this deal and Sonat's marginal tax rate.

The required return is based on the opportunity cost of the funds provided, that is, the rate that Sonat would have to pay to borrow $425 million in the capital market. At the time the loan was arranged, in late 1984, the market interest rate on such a loan would have been about 16%. If the marginal tax rate at which the interest payments are written off is 50% (the federal plus state corporate tax rate at that time), then the after-tax semiannual required return is 4% (8% annually), and the after-tax semiannual interest payments are $0.0225 \times P_t$, where P_t is the loan balance in period t and 2.25% is the after-tax interest rate ($0.5 \times 4.5\%$). Now we can calculate the NPV of Sonat's financing bargain:

$$NPV = \$425,000,000 - \sum_{t=1}^{17} \frac{0.0225P_t}{(1.04)^t} - \sum_{t=1}^{17} \frac{\$25,000,000}{(1.04)^t}$$

$$= \$425,000,000 - \$372,210,000$$

$$= \$52,790,000$$

EXHIBIT 16.9 Calculating the Value of Sonat's Low-Cost Loan Arrangement (U.S. $ Millions)

PERIOD (1)	PRINCIPAL BALANCE (1) =	INTEREST (1) × 0.0225 (2) +	PRINCIPAL REPAYMENT (3) =	TOTAL PAYMENT (4) ×	PV FACTOR @ 4% (5) =	PRESENT VALUE (6)
1	$425	$9.56	$25	$34.56	0.962	$ 33.25
2	400	9.00	25	34.00	0.925	31.45
3	375	8.44	25	33.44	0.889	29.73
4	350	7.88	25	32.88	0.855	28.11
5	325	7.31	25	32.31	0.822	26.56
6	300	6.75	25	31.75	0.790	25.08
7	275	6.19	25	31.19	0.760	23.70
8	250	5.63	25	30.63	0.731	22.39
9	225	5.06	25	30.06	0.703	21.13
10	200	4.50	25	29.50	0.676	19.94
11	175	3.94	25	28.94	0.650	18.81
12	150	3.38	25	28.38	0.625	17.73
13	125	2.81	25	27.81	0.601	16.71
14	100	2.25	25	27.25	0.578	15.75
15	75	1.69	25	26.69	0.555	14.81
16	50	1.13	25	26.13	0.534	13.95
17	25	0.56	25	25.56	0.513	13.11
						Sum $372.21

These calculations are shown in Exhibit 16.9. You don't need a degree in financial economics to realize that borrowing money at 9% when the market rate is 16% is a good deal. But what the NPV calculations tell you is just how much a particular below-market financing option is worth.

Raising funds at a below-market rate is easier said than done, however. A company selling securities is competing for funds on a global basis, not only with other firms in its industry but with all firms, foreign and domestic, and with numerous government units and private individuals as well. The fierce competition for funds makes it highly unlikely that the firm can find bargain-priced funds. But, as we shall see, the task is not impossible. *Financial market distortions* arising from taxes, *government credit and capital controls*, and *government subsidies and incentives* sometimes enable firms to raise funds at below-market rates. Companies may also be able to raise low-cost money by devising securities for which specific investors are willing to pay a higher price.

Taxes

The asymmetrical tax treatment of various components of financial cost—such as dividend payments versus interest expenses and exchange losses versus exchange gains—often causes equality of before-tax costs to lead to inequality in after-tax costs. This asymmetry holds out the possibility of reducing after-tax costs by judicious selection of securities. Yet, everything is not always what it seems.

For example, many firms consider debt financing to be less expensive than equity financing because interest expense is tax-deductible, whereas dividends are paid out of after-tax income. But this comparison is too limited. In the absence of any restrictions, the supply of corporate debt can be expected to rise. Yields will also have to rise in order to attract investors in higher and higher tax brackets. Companies will continue to issue debt up to the point at which the marginal investor tax rate will equal the marginal corporate tax rate.[5] At this point, the necessary yield would be such that there would no longer be a tax incentive for issuing more debt.

The tax advantage of debt can be preserved only if the firm can take advantage of some tax distortion, issue tax-exempt debt, or sell debt to investors in marginal tax brackets below 34%. The example of zero-coupon bonds illustrates all of these categories.

ZERO-COUPON BONDS ■ In 1982, PepsiCo issued the first long-term *zero-coupon bond*. Although they have since become a staple of corporate finance, zero-coupon bonds initially were a startling innovation. They do not pay interest, but are sold at a deep discount to their face value. For example, the price on PepsiCo's 30-year bonds was around $60 for each $1,000 face amount of the bonds. Investors gain from the difference between the discounted price and the amount they receive at redemption.

Between 1982 and 1985, investors paid $4 billion for $18.9 billion worth of zero-coupon bonds, about half of which were purchased by Japanese investors. The offerings were attractive in Japan because the government does not tax the capital gain on bonds sold prior to maturity. Catering to this tax break, a number of companies—including Exxon (see the illustration in Chapter 15) and IBM—were able to obtain inexpensive

---

[5]This insight first appeared in Merton Miller, "Debt and Taxes," *Journal of Finance*, May 1977, pp. 261–276.

financing by targeting Japanese investors for zero-coupon bonds offered on international markets.

The ability to take advantage quickly of such tax windows is evident considering subsequent developments in Japan. Japan's Finance Ministry, embarrassed at this tax break, has effectively ended the tax exemption for zero-coupon bond gains; Japanese investors have accordingly demanded higher yields to compensate for their anticipated tax liability. The reaction by the Japanese government to the proliferation of zero-coupon debt illustrates a key point: If one devises a legal way to engage in unlimited tax arbitrage through the financial markets, the government will change the law.

This example also points out that even though the world's capital markets are highly integrated, companies can still profit from tax differentials and government restrictions on capital flows between countries. But the benefits go to those who are organized to quickly take advantage of such windows of opportunity.

DEBT VERSUS EQUITY FINANCING ■ Interest payments on debt extended by either the parent or a financial institution generally are tax-deductible by an affiliate, but dividends are not. In addition, principal repatriation is tax-free, whereas dividend payments may lead to further taxation. Thus, as we saw in Chapter 14, parent company financing of foreign affiliates in the form of debt rather than equity has certain tax advantages. The consequences of these and other considerations for an MNC's cost of capital are discussed in Chapter 18 in the section titled Establishing a Worldwide Capital Structure.

Government Credit and Capital Controls

Governments intervene in their financial markets for a number of reasons: to restrain the growth of lendable funds, to make certain types of borrowing more or less expensive, and to direct funds to certain favored economic activities. In addition, corporate borrowing is often restricted in order to hold down interest rates (thereby providing the finance ministry with lower-cost funds to meet a budget deficit). When access to local funds markets is limited, interest rates in them are usually below the risk-adjusted equilibrium level. There is often an incentive to borrow as much as possible where nonprice credit rationing is used.

Restraints on, or incentives to promote, overseas borrowing are often employed as well. There are numerous examples of restraints and incentives affecting overseas borrowing. Certain countries have limited the amount of local financing the subsidiary of a multinational firm can obtain to that required for working-capital purposes; any additional needs will have to be satisfied from abroad. A prerequisite condition for obtaining official approval for a new investment or acquisition often is a commitment to inject external funds. Capital-exporting nations may attempt to control balance-of-payments deficits by restricting overseas investment flows—as the United States did from 1968 to 1974 under the Office of Foreign Direct Investment (OFDI) regulations.

Conversely, when a nation is concerned about excess capital inflows, a portion of any new foreign borrowing might have to be placed on deposit with the government, thereby raising the effective cost of external debt. Ironically, the effect of many of these government credit allocation and control schemes has been to hasten the development of the external financial markets—the Eurocurrency and Eurobond markets—further reducing government ability to regulate domestic financial markets.

The multinational firm with access to a variety of sources and types of funds and the ability to shift capital with its internal transfer system has more opportunities than does a single-nation company to secure the lowest risk-adjusted cost money and to circumvent credit restraints. These attributes should give it a substantial advantage over a purely domestic company.

Government Subsidies and Incentives

Despite their often-hostile rhetoric directed against the multinational firm, many governments offer a growing list of incentives to MNCs to influence their production and export-sourcing decisions. Direct investment incentives include interest rate subsidies, loans with long maturities, official repatriation guarantees, grants related to project size, attractive prices for land, and favorable terms for the building of plants. For example, new investments located in Italy's Mezzogiorno region can qualify for cash grants that cover up to 40% of the cost of plant and equipment, in addition to low interest rate loans.

Governments sometimes will make the infrastructure investments as well by building the transportation, communication, and other links to support a new industrial project. Some indirect incentives include corporate income tax holidays, accelerated depreciation, and a reduction or elimination of the payment of other business taxes and import duties on capital equipment and raw materials.

In addition, as we saw in Chapter 12, all governments of developed nations have some form of export financing agency whose purpose is to boost local exports by providing loans with long repayment periods at interest rates below the market level and with low-cost political and economic risk insurance.

❧ 16.7 SUMMARY AND CONCLUSIONS

Multinational corporations can use creative financing to achieve various objectives. These include reducing their cost of funds, cutting taxes, and reducing political risk. This chapter focused on three such techniques—interest rate and currency swaps, international leasing, and LDC loan swaps.

Interest and currency swaps involve a financial transaction in which two counterparties agree to exchange streams of payments over time. In an interest rate swap, no actual principal is exchanged either initially or at maturity, but interest payment streams are exchanged according to predetermined rules and are based on an underlying notional amount. The two main types are coupon swaps (or fixed rate to floating rate) and basis swaps (from floating rate against one reference rate to floating rate with another reference rate).

Currency swap refers to a transaction in which two counterparties exchange specific amounts of two currencies at the outset and repay over time according to a predetermined rule that reflects both interest payments and amortization of principal. A cross-currency interest rate swap involves swapping fixed-rate flows in one currency to floating-rate flows in another.

Structured notes are complex debt instruments whose payments are tied to a reference index, such as LIBOR, and that have one or more embedded derivative elements, such as swaps, forwards, or options. However, they do perform a valuable function. They allow corporations and financial institutions to function more efficiently by enabling them to tailor financial products to meet their individual needs.

International lease transactions, or cross-border leasing, may provide certain tax and political risk management advantages. The principal tax advantage from international leasing arises when the disparate leasing rules of the lessor's and lessee's countries enable both parties to retain the tax advantages of ownership. This lease is known as a double-dip lease. International leasing also enables MNCs to reduce their assets at risk in politically unstable countries and may permit greater access to local profits. The latter benefit arises because host governments are more likely to permit lease payments than dividends or royalties.

Under a debt-equity program, a firm buys a country's dollar debt on the secondary loan market at a discount and swaps it into local equity. Although such programs are still in their infancy, debt-equity swaps can provide cheap financing for expanding plant and for retiring local debt in hard-pressed LDCs. However, debt swaps are not an unmixed blessing for the participating countries, since they tend to exacerbate already high rates of inflation.

➤Questions

1. What is the difference between a basis swap and a coupon swap?

2. What is a currency swap?

3. What factors underlie the economic benefits of swaps?

4. Comment on the following statement. "In order for one party to a swap to benefit, the other party must lose."

5. As noted in Chapter 15, the Swiss Central Bank bans the use of Swiss franc for Eurobond issues. Explain how currency swaps can be used to enable foreign borrowers who want to raise Swiss francs through a bond issue outside of Switzerland to get around this ban.

6. Comment on the following statement. "During the period 1987–1989, Japanese companies issued some $115 billion of bonds with warrants attached. Nearly all were issued in dollars. The dollar bonds usually carried coupons of 4% or less; by the time the Japanese companies swapped that exposure into yen (whose interest rate was as much as 5 percentage points lower than the dollar's), their cost of capital was zero or negative."

7. How can international lease transactions enable multinational firms to reduce taxes and political risk?

8. Global Industries (GI) is looking for a place in which to locate its international leasing company. What factors should GI take into account in selecting the location? Explain.

9. Explain the benefits of a debt-equity swap program to
   a. A multinational firm seeking to expand its investment in the country
   b. The host government
   c. The bank that sells its LDC debt in the secondary market

10. How can debt swaps cause inflation?

11. Jose Angel Gurria, Mexico's chief debt negotiator and the architect of its swap program, questions the gain to Mexico from its swap program: "The latest restructuring package on our debt gives us 20 years to repay principal, with no principal payments for seven years. So why are we giving money away? Why pay 88 cents for debt that is worth only 60 cents?" Comment on Señor Gurria's statement.

12. Why do governments provide subsidized financing for some investments?

➤Problems

1. In May 1988, Walt Disney Productions sold to Japanese investors a 20-year stream of projected yen royalties from Tokyo Disneyland. The present value of that stream of royalties, discounted at 6% (the return required by the Japanese investors), was ¥93 billion. Disney took the yen proceeds from the sale, converted them to dollars, and invested the dollars in bonds yielding 10%. According to Disney's chief financial officer, Gary Wilson, "In effect, we got money at a 6% discount rate, reinvested it at 10%, and hedged our royalty stream against yen fluctuations—all in one transaction."
   a. At the time of the sale, the exchange rate was ¥124 = $1. What dollar amount did Disney realize from the sale of its yen proceeds?
   b. Demonstrate the equivalence between Walt Disney's transaction and a currency swap. (*Hint*: A diagram would help.)

$$-4.9 + 4.5$$
$$(.008 - .0025)$$

c. Comment on Gary Wilson's statement. Did Disney achieve the equivalent of a free lunch through its transaction?

2. Suppose that IBM would like to borrow fixed-rate yen, whereas Korea Development Bank (KDB) would like to borrow floating-rate dollars. IBM can borrow fixed-rate yen at 4.5% or floating-rate dollars at LIBOR + 0.25%. KDB can borrow fixed-rate yen at 4.9% or floating-rate dollars at LIBOR + 0.8%.
   a. What is the range of possible cost savings that IBM can realize through an interest rate/currency swap with KDB?
   b. Assuming a notional principal equivalent to $125 million, and a current exchange rate of

5. At time t, 3M borrows ¥12.8 billion at an interest rate of 1.2%, paid semiannually, for a period of two years. It then enters into a two-year yen/dollar swap with Bankers Trust (BT) on a notional principal amount of $100 million (¥12.8 billion at the current spot rate). Every six months, 3M pays BT U.S. dollar LIBOR6, while BT makes payments to 3M of 1.3% annually in yen. At maturity, BT and 3M reverse the notional principals.
   a. Assume that LIBOR6 (annualized) and the ¥/$ exchange rate evolve as follows. Calculate the net dollar amount that 3M pays to BT ("−") or receives from BT ("+") each six-month period.

Time (months)	LIBOR6	¥/$ (spot)	Net $ Receipt (+)/Payment (−)
t	5.7%	128	
$t + 6$	5.4%	132	
$t + 12$	5.3%	137	
$t + 18$	5.9%	131	
$t + 24$	5.8%	123	

¥105/$, what do these possible cost savings translate into in yen terms?
   c. Redo parts a and b assuming that the parties use Bank of America, which charges a fee of 8 basis points to arrange the swap.

3. Company A, a low-rated firm, desires a fixed-rate, long-term loan. Company A presently has access to floating interest rate funds at a margin of 1.5% over LIBOR. Its direct borrowing cost is 13% in the fixed-rate bond market. In contrast, company B, which prefers a floating-rate loan, has access to fixed-rate funds in the Eurodollar bond market at 11% and floating-rate funds at LIBOR + 1/2%.
   a. How can A and B use a swap to advantage?
   b. Suppose they split the cost savings. How much would A pay for its fixed-rate funds? How much would B pay for its floating-rate funds?

4. Square Corp. has not tapped the Swiss franc public debt market because of concern about a likely appreciation of that currency and only wishes to be a floating-rate dollar borrower, which it can be at LIBOR + 3/8%. Circle Corp. has a strong preference for fixed-rate Swiss franc debt, but it must pay 1/2 of 1% more than the 5 1/4% coupon that Square Corp.'s notes would carry. Circle Corp., however, can obtain Eurodollars at LIBOR flat (a zero margin). What is the range of possible cost savings to Square from engaging in a currency swap with Circle?

   b. What is the all-in dollar cost of 3M's loan?
   c. Suppose 3M decides at $t + 18$ to use a six-month forward contract to hedge the $t + 24$ receipt of yen from BT. Six-month interest rates (annualized) at $t + 18$ are 5.9% in dollars and 2.1% in yen. With this hedge in place, what fixed dollar amount would 3M have paid (received) at time $t + 24$? How does this amount compare to the $t + 24$ net payment computed in part (a)?
   d. Does it make sense for 3M to hedge its receipt of yen from BT? Explain.

6. In the Kodak Australian dollar swap example, suppose Merrill Lynch had been able to arrange a forward contract for the A$70 million at a rate of A$1 = U.S.$0.49.
   a. If Merrill Lynch had retained full benefits from the better forward rate, what would have been the present value of its profit on the deal?
   b. If Merrill Lynch had passed these savings on to Kodak, what would have been Kodak's annualized all-in rate on the swap? (Hint: Take the internal rate of return on all of Kodak's cash flows.)
   c. Suppose the interest rate swap with Australian bank B had been at LIBOR − 20 basis points. If Merrill Lynch passed this higher cost along to Kodak, what would Kodak's all-in cost have been?

7. At present, LIBOR3 is 7.93% and LIBOR6 is 8.11%. What is the forward forward rate for a LIBOR3 deposit to be placed in three months?

8. Nestlé rolls over a $25 million loan priced at LIBOR3 on a three-month basis. The company feels that interest rates are rising and that rates will be higher at the next roll-over date in three months time. Suppose the current LIBOR3 is 5.4375%.
   a. Explain how Nestlé can use an FRA at 6% from Credit Suisse to reduce its interest rate risk on this loan.
   b. In three months time interest rates have risen to 6.25%. How much will Nestlé receive/pay on its FRA? What will be Nestlé's hedged interest expense for the upcoming three-month period?
   c. After three months, interest rates have fallen to 5.25%. How much will Nestlé receive/pay on its FRA? What will be Nestlé's hedged interest expense for the next three-month period?

9. Suppose that Skandinaviska Ensilden Banken (SEB), the Swedish bank, funds itself with three-month Eurodollar time deposits at LIBOR. Assume that Alfa Laval comes to SEB seeking a one-year, fixed-rate loan of $10 million, with interest to be paid quarterly. At the time of the loan disbursement, SEB raises three-month funds at 5.75%, but has to roll over this funding in three successive quarters. If he does not lock in a funding rate and interest rates rise, the loan could prove to be unprofitable. The three quarterly re-funding dates fall shortly before the next three Eurodollar futures-contract expirations in March, June, and September.
   a. At the time the loan is made, the price of each contract is 94.12, 93.95, and 93.80. Show how SEB can use Eurodollar futures contracts to lock in its cost of funds for the year. What is SEB's hedged cost of funds for the year?
   b. Suppose that the settlement prices of the March, June, and September contracts are, respectively, 92.98, 92.80, and 92.66. What would have been SEB's unhedged cost of funding the loan to Alfa Laval?

10. Ford has a $20 million Eurodollar deposit maturing in two months time that it plans to roll over for a further six months. The company's treasurer feels that interest rates will be lower in two months when rolling over the deposit. Suppose the current LIBOR6 is 7.875%.
   a. Explain how Ford can use an FRA at 7.65% from Banque Paribas to lock in a guaranteed six-month deposit rate when it rolls over its deposit in two months.
   b. In two months, LIBOR6 has fallen to 7.5%. How much will Ford receive/pay on its FRA? What will be Ford's hedged deposit rate for the upcoming six-month period?
   c. In two months, LIBOR6 has risen to 8%. How much will Ford receive/pay on its FRA? What will be Ford's hedged deposit rate for the next six months?

11. Chrysler has decided to make a $100 million investment in Mexico via a debt-equity swap. Of that $100 million, $20 million will go to pay off high-interest peso loans in Mexico. The remaining $80 million will go for new capital investment. The government will pay 86 cents on the dollar for debt used to pay off peso loans and 92 cents on the dollar for debt used to finance new investment. If Chrysler can buy Mexican debt in the secondary market for 60 cents on the dollar, how much will it cost Chrysler to make its $100 million investment?

12. Suppose that the cost of borrowing restricted French francs is 7% annually, whereas the market rate for these funds is 12%. If a firm can borrow FF 10 million of restricted funds, how much will it save annually in before-tax franc interest expense?

13. Suppose that one of the inducements provided by Taiwan to woo Xidex into setting up a local production facility is a 10-year, $12.5 million loan at 8% interest. The principal is to be repaid at the end of the tenth year. The market interest rate on such a loan is about 15%. With a marginal tax rate of 40%, how much is this loan worth to Xidex?

➤ Bibliography

FIERMAN, JACLYN. "Fast Bucks in Latin Loan Swaps." *Fortune*, August 3, 1987, pp. 91–99.

MORENO, RAMON. "LDC Debt Swaps." *FRBSF Weekly Letter*, Federal Reserve Bank of San Francisco, September 4, 1987.

SMITH, CLIFFORD W., JR., CHARLES W. SMITHSON, and LEE M. WAKEMAN. "The Evolving Market for Swaps." *Midland Corporate Finance Journal*, Winter 1986, pp. 20–32.

INTERNATIONAL BANKING TRENDS AND STRATEGIES

Whether we like it or not, the globalization of financial markets and institutions is a reality.

E. Gerald Corrigan
President, Federal Reserve
Bank of New York

$\mathcal{T}$he growth and increasing integration of the world economy since the end of World War II have been paralleled by expansion of global banking activities. Banks followed their customers overseas and lent to governments presiding over promising national economies. One indication of the worldwide scope of banking today is suggested by the fact that international bank loans extended by commercial banks located in major financial centers around the world have increased year after year during the past decade. The amount of these loans extended reached a total of $390 billion in 1997. Underneath the facade of unbroken growth, however, lie many divergent trends that have been profoundly influencing the direction of international banking activities during the past 30 years.

This chapter provides an overview of international banking. It focuses on recent trends in the expansion of international banking activities and the organizational forms and strategies associated with overseas bank expansion. The chapter also discusses the regulatory changes that have influenced and been influenced by the growth of international banking. Some of these regulatory changes have been prompted by the international debt crisis, a topic that this chapter discusses as well. More generally, the chapter

589

explores some of the factors that determine a country's ability to repay its foreign debts. This is the subject called *country risk analysis*.

17.1 RECENT PATTERNS OF INTERNATIONAL BANKING ACTIVITIES

International banking has grown notably in both complexity and risk during the past two decades. Until recently, international banking was confined largely to providing foreign exchange and to financing specific export and import transactions through letters of credit and acceptances. This limitation is no longer the case.

International banking has grown steadily throughout the post-World War II period. Expansion of international trade in the 1950s and the effective emergence of the MNC in the 1960s sharply increased the demand for international financial services. Banks located in the traditional financial centers responded by extending loans and developing new, highly innovative financial techniques (such as the Eurocurrency markets) that laid the foundation for totally new approaches to the provision of international banking services. The initial ventures overseas of many of these banks were defensive in nature, designed to retain the domestic business of customers who invested abroad by expanding and improving the scope of their activities abroad. In the words of an economist at Deutsche Bank, "In a global market, if you can't serve your multinational customers in all the major cities of the world, they won't need you even in their home country."[1] However, it was the onset of the "energy crisis" that launched international banks into a period of phenomenal growth.

The Era of Growth

The energy crisis, brought about by the quadrupling of oil prices in late 1973, created a great need for global financial intermediation—for recycling *OPEC's* (Organization of Petroleum Exporting Countries) surplus revenues back to deficit-plagued oil-importing countries. Without such *"petrodollar" recycling*, the balance-of-payments deficits of the oil-importing countries would not have been financed, threatening dire consequences for the entire world economy. The alternative would have entailed massive economic dislocation to speedily adapt to the changing relative price of oil.

International banks were able to recycle funds from oil-exporting to oil-importing nations because (1) they had broad experience in international lending, backed by capable staffs and worldwide facilities; and (2) they were the recipients of large shares of OPEC's surplus revenues in the form of deposits placed with them by OPEC's central bankers.

Flush with OPEC money, the banks embarked upon a rapid lending expansion, often with the active encouragement of their governments. The net claims outstanding of banks in the reporting area of the **Bank for International Settlements** (BIS; see Chapter 3) increased from $155 billion at year-end 1973 to $665 billion at year end 1979—a more than fourfold increase in loans to final borrowers in just six years.

---

[1] Quoted in *Fortune*, February 26, 1990, p. 95.

Of this amount, $157 billion (24%) was loans to less-developed countries (LDCs) and $60 billion (9%) was loans to Communist countries. Thus, about one-third of these loans, aggregating $217 billion, were to nations that turned out to be very risky credits.

Banks located in major financial centers throughout the world participated in this expansion of international lending. The largest share of the total was booked by banks located in major European centers, particularly in London, where foreign branches of major banks throughout the world (including U.S. banks) were operating.

Loans to the less-developed countries were the fastest growing category of international bank loans during the 1970s. A combination of sharply increased oil-import bills and a recession in the industrial countries that cut into the LDCs' export earnings—compounded by unrealistic exchange rate policies leading to overvalued exchange rates—sharply raised these countries' aggregate balance-of-payments deficits from an annual average of about $7 billion in the 1970–1973 period to $21 billion in 1974 and $31 billion in 1975. The banks, replete with funds and faced with declining domestic loan demand, were willing and able to provide financing in the forms of direct government loans and development financing.

The International Banking Crisis of 1982

Bank lending to LDCs continued to grow rapidly during the early 1980s. In the summer of 1982, however, the international financial markets were shaken when a number of developing countries found themselves unable to meet payments to major banks around the world on debt amounting to several hundred billion dollars. With the onset of the debt crisis, lending to LDCs quickly dried up.

Several developments set the stage for the *international banking crisis of 1982*. One of these was the growing trend in overseas lending to set interest rates on a floating basis—that is, at a rate that would be periodically adjusted on the basis of the rates prevailing in the market. Floating-rate loans made borrowers vulnerable to increases in real interest rates. Borrowers were also vulnerable to increases in the real value of the dollar because more than 80% of these loans were denominated in dollars.

The catalyst of the crisis was provided by the economic policies pursued by the industrial countries in general, and by the United States in particular, in their efforts to deal with rising domestic inflation. The combination of an expansionary fiscal policy and tight monetary policy led to sharply rising real interest rates in the United States—and in the Euromarkets where the banks funded most of their international loans. The variable-rate feature of the loans combined with rising indebtedness boosted the LDCs' net interest payments to banks from $11 billion in 1978 to $44 billion in 1982. Furthermore, the dollar's sharp rise in the early 1980s increased the real cost to the borrowers of meeting their debt payments.

The final element setting the stage for the crisis was the onset of a recession in industrial countries. The recession reduced the demand for the LDCs' products and, thus, the export earnings needed to service their bank debt. The interest payments/export ratio reached 50% for some of these countries in 1982. This ratio meant that more than half of these countries' exports were needed to maintain up-to-date interest payments, leaving less than half of the export earnings to finance essential imports and to repay principal on their bank loans. These trends made the LDCs highly vulnerable.

ONSET OF THE CRISIS ■ The first major blow to the international banking system came in August 1982, when Mexico announced that it was unable to meet its regularly scheduled payments to international creditors. Shortly thereafter, Brazil and Argentina (the second- and third-largest debtor nations) found themselves in a similar situation. By the spring of 1983, about 25 LDCs—accounting for two-thirds of the international banks' claims on this group of countries—were unable to meet their debt payments as scheduled and had entered into loan-rescheduling negotiations with the creditor banks.

Compounding the problems for the international banks was a sudden drying up of funds from OPEC. A worldwide recession that reduced the demand for oil put downward pressure on oil prices and on OPEC's revenues. In 1980, OPEC contributed about $42 billion to the loanable funds of the BIS-reporting banks. By 1982, the flow had reversed, as OPEC nations became a net drain of $26 billion in funds. At the same time, banks cut back sharply on their lending to LDCs.

THE BAKER PLAN ■ In October 1985, U.S. Treasury Secretary James Baker called on 15 principal middle-income debtor LDCs—the *"Baker 15 countries"*—to undertake growth-oriented structural reforms that would be supported by increased financing from the World Bank, continued modest lending from commercial banks, and a pledge by industrial nations to open their markets to LDC exports. The goal of the **Baker Plan** was to buttress LDC economic growth, making these countries more desirable borrowers and restoring their access to international capital markets.

Achievement fell far short of these objectives. Most of the Baker 15 lagged in delivering on promised policy changes and economic performance. Instead of getting their economic houses in order, many of the LDCs—feeling they had the big banks on the hook—sought to force the banks to make more and more lending concessions. The implied threat was that the banks would otherwise be forced to take large write-offs on their existing loans.

In May 1987, Citicorp's new chairman, John Reed, threw down the gauntlet to big debtor countries by adding $3 billion to the bank's loan-loss reserves. Citicorp's action was quickly followed by large additions to loss reserves by most other big U.S. and British banks. The banks who boosted their loan-loss reserves became tougher negotiators, arguing against easing further the terms for developing countries' debt settlements. The decisions of these banks to boost their reserves—plus the announcement by some of their intention, one way or another, to dispose of a portion of their existing LDC loans— precipitated fresh questioning of the LDC debt strategy, in particular of the Baker initiative. The problem, however, was not with the Baker initiative, but rather with its implementation. By adding to their loan-loss reserves, Citicorp and the other banks that followed suit put themselves in a stronger position to demand reforms in countries to which they lend—a key feature of the Baker Plan.

THE BRADY PLAN ■ Faced with the Baker Plan's failure to resolve the ongoing debt crisis, Nicholas Brady, James Baker's successor as U.S. Treasury Secretary, put forth a new plan in 1989 that emphasized debt relief through forgiveness instead of new lending. Under the **Brady Plan**, banks had a choice: They could either make new loans or write off portions of their existing loans in exchange for new government securities—so-called **Brady bonds**—whose interest payments were backed with money from the International

EXHIBIT 17.1 Amount of Bank Debt Converted into Brady Bonds by the End of 1993

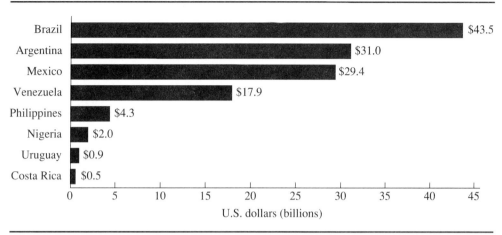

U.S. dollars (billions)

Source: Copyright Salomon Brothers, Inc.

Monetary Fund. The problem with the Brady Plan was that, for it to work, commercial banks would have to do both: make new loans at the same time that they were writing off existing loans. Instead, many banks used the Brady Plan as an opportunity to exit the LDC debt market. They added billions to their loan-loss reserves to absorb the necessary loan write-downs, virtually guaranteeing that they would slash new LDC lending to the bone. In the aggregate, banks said goodbye to $26 billion of their LDC loans. The flip side of this write-off, as shown in Exhibit 17.1, is the large amount of debt converted to Brady bonds.

ILLUSTRATION

MEXICO IMPLEMENTS THE BRADY PLAN

Mexico's Brady Plan agreement in February 1990 offered three options to the banks: (1) convert their loans into salable bonds with guarantees attached, but worth only 65% of the face value of the old debt; (2) convert their debt into new guaranteed bonds whose yield was just 6.5%; or (3) keep their old loans but provide new money worth 25% of their exposure's value. Only 10% of the banks risked the new money option, whereas 41% agreed to debt reduction and 49% chose a lower interest rate. This agreement saved Mexico $3.8 billion a year in debt-servicing costs, but hurt its access to future funds.

As part of this restructuring, the U.S. government agreed to sell Mexico 30-year, zero-coupon Treasury bonds in a face amount of $33 billion. Mexico bought the zeros to serve as collateral for the bonds it planned to issue as a substitute for the bank loans. When the Mexican deal was struck, 30-year Treasury zeros were selling in the open market to yield 7.75%. But the U.S. Treasury priced the zeros it sold to Mexico to yield 8.05%, reflecting the price not of zeros but of conventional 30-year Treasury bonds. The Treasury also charged Mexico a service fee of 0.125% per annum, as is customary in such special Treasury sales. As a result,

Mexico received an effective annual interest yield of 7.925% (8.05% − 0.125%).

The sale of zero-coupon bonds to Mexico at a cut-rate price brought criticism from Congress and Wall Street. Many critics argued that any U.S. aid to Mexico ought to flow through normal channels and not be done in a back-door financing scheme. Without taking sides in this dispute, what is the value of the subsidy the Treasury provided to Mexico?

Solution. If the Treasury had priced the zeros to yield 7.75%, less the 0.125% service fee, or 7.625%, the $33 billion face value of

bonds would have cost Mexico

$$\frac{\$33 \text{ billion}}{(1.07625)^{30}} = \$3.64 \text{ billion}$$

By pricing the bonds to yield 7.925%, the Treasury sold the bonds to Mexico at a price of

$$\frac{\$33 \text{ billion}}{(1.07625)^{30}} = \$3.35 \text{ billion}$$

The result was a savings to Mexico of about $290 million.

Confronted with interruptions in inflows of funds due to repayment problems on their past loans, with the drying up of new sources, and with the growing uncertainties as to the capacity of their borrowers to service their debt, the international banks pulled back sharply on their lending to LDCs. Simply put, the banks washed their hands of the LDC debt mess. They were saying, in effect, that there were few good loans to be made in Latin America or Africa, because the politics and systems are so bad that no loan would be good even if the old debts were all canceled.

THE INTERNATIONAL DEBT CRISIS EASES ■By late 1983, the intensity of the international debt crisis began to ease as the world's economic activities picked up—boosting the LDCs' export earnings—and as the orderly rescheduling of many overdue international loans was completed. However, although lending by international banks picked up,

EXHIBIT 17.2 Net Change in Commercial Bank Lending to Developing Countries

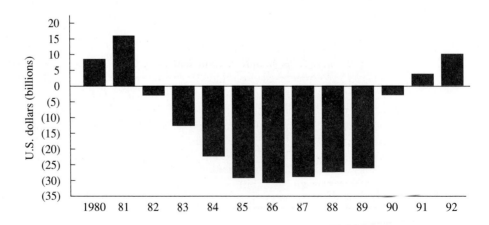

it continued to be depressed compared with the high-growth period of the late 1970s. Indeed, as Exhibit 17.2 shows, it was not until 1991 that net international bank lending (new loans net of loans repaid) turned positive again, following on the heels of the dramatic economic reforms under way in many LDCs. The major reason for the drop in international lending activity through the 1980s was the difficulty the LDC debtor nations had in achieving sustained economic growth. The LDCs required capital formation to grow but the banks first wanted to see economic reforms that would improve the odds that the LDCs would be able to service their debts.

Many of the LDCs pushed for **debt relief**—that is, reducing the principal or interest payments or both on loans. One argument for debt relief is that it may reduce the probability of default.[2] This argument rests on the view that the honoring of debt obligations is a matter of cost/benefit analysis. At high levels of debt, the benefits of default may outweigh its costs, increasing the chance that governments will favor this option. Conversely, reducing a country's debt obligations lowers the country's expected gains from default, thereby reducing its incentive to default. This proposition has found no empirical support so far.

Moreover, the middle-income debtor nations addressed by the Baker and Brady Plans were neither very poor nor insolvent. They possess considerable human and natural resources, reasonably well-developed infrastructures and productive capacities, and the potential for substantial growth in output and exports given sound economic policies. In addition, many of these countries possess considerable wealth—much of it invested abroad.

Although debt burdens exacerbated the economic problems faced by these countries, all too often the underlying causes were to be found in patronage-bloated bureaucracies, overvalued currencies, massive corruption, and politically motivated government investments in money-losing ventures. The Baker and Brady countries also suffered from markets that were distorted by import protection for inefficient domestic producers and government favors for politically influential groups. For these countries, debt relief was at best an ineffectual substitute for sound macroeconomic policy and major structural reform; relief would only have weakened discipline over economic policy and undermined support for structural reform.

THE INTERNATIONAL DEBT CRISIS ENDS ■ Ten years after it began, the decade-long Latin American debt crisis ended in July 1992, with the signing of an agreement with Brazil to restructure the $44 billion it owed foreign banks. In the end, however, it was not negotiation that cut the Gordian knot of Latin American debt—no new money until Latin America showed economic growth and no growth without new money—but genuine economic reforms, forced on unwilling governments by the unrelenting pressures of the debt crisis. Mexico and Chile, hopelessly mired in debt, so thoroughly reformed their economies, spurring economic growth in the process, that they were able to raise new money from the international capital markets. These economic reforms included opening their markets to imports, tearing down barriers to foreign investment, selling off state-owned companies, instituting tight money policies, and cutting government deficits. The

---

[2] See Paul Krugman, "Market-Based Debt-Reduction Schemes," NBER Working Paper, No. 2587, National Bureau of Economic Research, Cambridge, 1988.

examples of Mexico and Chile led others, such as Argentina and Venezuela, to change their policies as well.

These bold reforms in Latin America reversed the flow of capital there. And this new capital will not create another debt crisis because most of it is equity, not debt. In contrast to the floating-rate loans of the 1970s and early 1980s, much of the new capital flowing into Latin America comes from investors buying shares in local companies and direct investment by multinational firms. Where a continuing inflow of credit is still important, as in the case of Mexico, however, the potential for another debt crisis remains.

The experiences of Mexico, Chile, and Argentina—countries that appear to have successfully implemented economic reform programs—show that they all met the following criteria:[3]

- A head of state who demonstrates strong will and political leadership
- A viable and comprehensive economic plan that is implemented in a proper sequence
- A motivated and competent economic team working in harmony
- Belief in the plan from the head of state, the cabinet, and other senior officials
- An integrated program to sell the plan to all levels of society through the media

These criteria are applicable not only to Latin America, but to all LDCs anywhere in the world. Even with the best of intentions, however, economic reform is painful and difficult to make stick unless all levels of society are convinced that instituting free-market policies will bring the long-term benefits of sustained growth.

International Bank Regulation

As the world's financial markets globalize, bank regulators are increasingly concerned about establishing a well-defined official framework that sets ground rules for safe operation of all banks. The fear is that bank failures in one nation could spill across borders and affect the world banking system. Up to now, however, the standards applied by the various national authorities have differed considerably, artificially creating competitive advantages—and competitive disadvantages—between international banks. One concern is that banks will take advantage of the different degrees of regulatory scrutiny by engaging in regulatory arbitrage—transferring activities across borders to those nations that impose the laxest standards. Two additional and related issues that arise in regulating internationally active banks are determining (1) who is responsible for guaranteeing the safety of these banks—that is, who is the *lender of last resort* if they get in trouble; and (2) who is responsible for monitoring these banks—regulators in their home market or regulators in the foreign markets in which they operate? The danger is that without an overarching supervisory structure some banks could fall between the cracks of the regulatory system.

To some extent, this is the problem that arose in the case of the Bank of Credit and Commerce International (BCCI), a fraud-ridden Pakistani bank (also known as the Bank

---

[3] These criteria appear in William R. Rhodes, "Third World Debt: The Disaster That Didn't Happen," *The Economist*, September 12, 1992, pp. 21–23.

of Crooks and Criminals International) that operated on a global scale. A small group within BCCI stole billions of dollars from the institution, laundered drug money, faked loans, and hid losses under the noses of regulators for years. To some extent, each of the different national regulators thought that someone else was supervising BCCI.

However, the problems of BCCI are simply the tip of the iceberg. The major national banking systems in the 1980s experienced a decade of hasty deregulation, globalization, and innovation, which ended with a long list of sins, scandals, and financial excesses. Bankers around the world bet the bank on ever-increasing asset values—real estate, oil, stocks—but instead they have been hit with asset deflation—a decline in the value of the assets that secure their loans or constitute bank capital. In response to possibilities like BCCI, as well as reckless lending practices of banks worldwide, bank supervisors now are trying to work together more systematically to reduce the risks to the international banking system. One outcome is new *risk-based capital standards.*

INTERNATIONAL RISK-BASED CAPITAL STANDARDS ■ One of the most important factors currently affecting international bank expansion is the new capital requirements set down by the Basle Committee under the auspices of the Bank for International Settlements in Basle, Switzerland, which took effect in 1993. In late 1987, the Basle Committee developed a risk-based framework for measuring the adequacy of **bank capital**—the equity capital and other reserves available to protect depositors against credit losses. Its objective was to strengthen the international banking system and to reduce competitive inequalities arising from differences in capital requirements across nations. In part, the Basle Committee's requirements were aimed at Japanese banks, because their reserves were so much smaller than those required by most Western central banks. Under the *Basle agreement*, by the end of 1992 banks had to achieve a minimum 8% risk-based ratio of capital to assets.

There was disagreement in Basle on how to define bank capital, with a political compromise eventually producing a two-tier system. Core, or Tier 1, capital consists of shareholders' funds and retained earnings. Supplemental, or Tier 2, capital consists of internationally accepted noncommon-equity items—such as hidden reserves, preferred shares, and subordinated debt—to add to core capital. However, Tier 2 capital can make up only 50% of a bank's capital. Thus, core capital must amount to at least 4% of risk-weighted assets.

The Basle Committee recently revised its capital adequacy standards to incorporate market risk, the risk that asset prices will change in response to movements in broad market factors such as interest rates and exchange rates. The committee's capital standards permit banks to use their own models to determine the amount of capital necessary to protect themselves from market risk. At the heart of these internal models is the **value-at-risk** (VAR) calculation, which allows a bank to estimate the maximum amount it might expect to lose in a given time period with a certain probability. For example, a 2% VAR of $100 million means that a loss exceeding $100 million is expected to occur no more than one period out of 50. These models are based on asset pricing theory, involve very sophisticated mathematics (especially for valuing derivative products), and their implementation involves complex statistical issues.

To meet the capital standards, banks are focusing more on profits and less on growth. This is especially true for Japanese banks, which are finally developing credit-rating systems and lending more prudently, although none has tried to raise profitability by shrinking its loan book.

Japanese International Bank Expansion

Perhaps the most significant phenomenon in recent years has been the aggressive international expansion of Japanese banks. In the mid-1980s, Japanese banks—flush with the proceeds of Japan's trading surpluses—set out to gain market share worldwide through low-cost loans.[4] With more cash than caution, Japanese banks tripled their overseas loan portfolio by 1990, accounting for 38% of international banking assets and more than 12% of total lending in the United States.

However, financial adversity at home now is changing the way Japanese banks do business abroad. No longer can they push for market share at the expense of profits. Instead, Japanese banks are making fewer overseas loans, and on those fewer loans they are demanding higher returns, leading to higher interest rates for borrowers. In part, this change is a response to financial market deregulation in Japan, which has raised the cost of funds to the banks. The new focus on profit is also affected by the Basle capital requirements.

Like all banks, Japanese banks are trying to meet the new Basle standards, but the Tokyo stock market's collapse during 1990 has made that harder in two ways. For Japanese banks, Tier 2 capital consisted mainly of 45% of the unrealized gains on their huge equity stakes in other companies. The sharp drop in Japanese equity values means that the value of these "hidden reserves" has plunged. By 1998, the continuing fall in the Japanese stock market had, for many banks, wiped out the last of these hidden gains.

The stock market fall also hurt the ability of Japanese banks to issue equity-linked securities. These issues have been crucial sources of capital in the past, providing Japanese banks with ¥6 trillion ($40 billion) of fresh money between 1987 and 1989.

Moreover, a sharp rise in Japanese interest rates inflicted huge losses on the banks' bond portfolios and lowered their capital ratios still further. By the end of 1990, all the large Japanese banks had slipped below the required 8% capital adequacy ratio. To meet this standard, banks must either reduce lending growth, which reduces the amount of equity capital they must hold, or boost profits, which adds to equity. Although the Bank of Japan engineered a steep drop in interest rates to boost bank profits, that has been offset by a proliferation of bad loans stemming from plunging Japanese real estate prices. According to some estimates, Japanese banks in 1998 had to contend with more than *$1 trillion* in problem loans.

Not all of these problem loans are in Japan. Between 1986 and 1990, Japanese banks boosted their share of property lending in California from 10% to 20%, lending when property values were at their peak. Prices since have fallen sharply. Their problems are most evident in downtown Los Angeles, where virtually every new office project in the late 1980s was financed by the Japanese. The recovery in California real estate has eased this problem somewhat but the Japanese banks still suffer from a severe diminution of their capital.

---

[4]One indication of the Japanese drive for market share is reflected in the fact that overseas assets account for nearly half of the major banks' total assets, but the margins on these assets are less than one-tenth of those achieved in Japan.

In response to these pressures, Japanese banks are pulling back from "commodity" lending, the business of lending large volumes of money at low margins that fueled their expansion during the 1980s. In so doing, Japanese banks are no longer as dominant a force in international finance. For example, the number of Japanese banks leading syndicated loans in the low-margin Eurocurrency market dropped to just four out of the top 20 in 1990. In 1989, they led seven of the top 20 syndicates. After accounting for a peak of 39.4% of all cross-border bank transactions by March 1989, Japanese banks' market share shrank in 1990 for the first time since 1984 to 36%. That compares with 12% for U.S. banks and 9.6% for German banks. This international downsizing has continued.

High capital requirements combined with the huge quantity of bad loans should mean that the days of lending money at cheap rates simply to grab market share are gone for Japanese banks. However, just when it appeared that some of Japan's largest banks were cleaning up their balance sheets, the Asian crisis hit. It revealed that Japanese banks had been up to some of their old tricks, making enormous volumes of relatively low-margin loans to Thailand and other East Asian countries ($261 billion at year end 1996). Partly, this lending was designed to gain market share, but for most of the Japanese banks it was an attempt to boost profits because these loans carried higher margins than did loans in Japan, even if they were too low to compensate for risk.

The basic problem is that there is too much capital tied up in Japan's banking system, provoking fierce competition to lend and minuscule margins. The obvious solution is for Japanese banks to shrink their lending and boost their profits. This solution could be helped along if Japan's Ministry of Finance would let sick banks go bust, something it has been unwilling to tolerate up to now.

17.2 ORGANIZATIONAL FORMS AND STRATEGIES IN BANK EXPANSION OVERSEAS

Decisions by banks as to how to approach foreign markets are influenced by a number of variables, such as overall financial resources, level of experience with the markets, knowledge of the markets, volume of international business, and the strategic plans of the bank, as well as the banking structure of the foreign countries in which business is done. Possible entry strategies include branching, local bank acquisitions, and representative offices. However, until the volume of business in another country is substantial, most banks will choose to rely on correspondent banking relationships to handle their needs in that country. U.S. banks also make use of domestic organizational forms for carrying on international banking activities, including Edge Act and Agreement corporations and international banking facilities. Each of these forms and strategies is described in more detail in this section.

Correspondent Banking

A **correspondent bank** is a bank located elsewhere that provides a service for another bank. U.S. banks without branches abroad have relied on their foreign correspondents to help finance their multinational corporate clients' local foreign subsidiaries that need local currency funding. Foreign correspondents also can provide other services, such as foreign exchange conversions and letters of credit.

ADVANTAGES ■ The major advantage of taking the correspondent route is that the cost of market entry is minimal and can be adjusted to the scale of service required in a given locale; no investment in staff or facilities is required. Yet the bank can still enjoy the benefits derived from having multiple sources of business given and received, as well as referrals of local banking opportunities. Moreover, correspondents' local knowledge and contacts may be extensive and highly useful in rendering services to the bank's clients doing business in that country.

DISADVANTAGES ■ One problem with relying on correspondent banks to provide all necessary services is that correspondents may assign low priority to the needs of the U.S. bank's customers. In addition, due to legal restrictions on traditional banking policies, certain types of credits may be difficult to arrange. Correspondents also may be reluctant to provide credits on a more regular and extensive basis.

Representative Offices

Representative offices are small offices opened up to provide advisory services to banks and customers and to expedite the services of correspondent banks. They also serve as foreign loan production offices able to negotiate various business transactions. Representative offices are not authorized to obtain and transfer deposits and do not provide on-site operating services. The assets and liabilities attributable to a representative office are booked elsewhere in the parent bank's system.

Such offices are regarded as excellent sources of economic and political intelligence on the host country and the local market. They also provide financial contacts with local institutions, commercial contacts for the bank's domestic customers, and assistance to customers in obtaining government approvals or understanding government regulations. Representative offices are especially appropriate when the expected business volume in a market is too small to justify the investment required to establish a branch, or local opportunities are uncertain, and the bank wants to learn more about the market at minimal cost before deciding whether further expansion is warranted.

ADVANTAGES ■ As with exporting, representative offices provide a low-cost means of scouting out the local market. They can deliver certain services more efficiently than can a branch, especially if the required volume is small. They can help the bank attract additional business or prevent the loss of current business.

DISADVANTAGES ■ Taking the analogy to exporting further, the benefits may at times be outweighed by the inability to effect more substantial market penetration. Also, despite the fact that they are not capital-intensive when compared with branches or local acquisitions, representative offices can be expensive. Moreover, it is more difficult to attract qualified personnel to work in a representative office overseas than in a foreign branch.

Foreign Branches

The principal service offered by *foreign bank branches*, as with commercial banking anywhere, is the extension of credit, primarily in the form of lending money. The major portion of the lending done by branches in important international money centers such as London and Singapore involves cross-border loans, primarily because these locations are the major trading and booking centers for the Eurocurrency and Asiacurrency markets. These branches also serve as deposit-taking institutions.

Despite government regulations that have held down bank branching in foreign markets, the phenomenal growth of international banking over the past 25 years has been paralleled by an explosive expansion in overseas branching. Before 1960, only seven U.S. banks maintained a total of 132 branches abroad. By the end of 1979, those numbers had grown to 130 banks with just under 800 foreign branches. Major banks from Canada, Japan, and the Western European countries also have jumped on the branching bandwagon.

There are several reasons for this massive proliferation of overseas branching. First, there is the "follow the customer" rationale. Unless domestic customers that expand abroad are serviced overseas, the bank is likely to lose its clients' domestic, as well as foreign, business. Yet, it turns out that only a minor share of foreign branch business is with head-office customers. Increasingly, the business of overseas branches is with purely local, indigenous enterprises.

This trend is related to the second reason for having foreign branches: the direct contribution to bank earnings that the branches provide, quite aside from any indirect contribution associated with protecting the domestic customer base. Specifically, business with local companies has turned out to be profitable on a stand-alone basis. Moreover, foreign earnings help to diversify the bank's earnings base, thereby moderating swings in domestic earnings.

The third reason for establishing foreign branches is that these branches provide access to overseas money markets. Large international banks have a need for branches located in international money markets abroad, such as London and Singapore, to fund their international assets. At the same time, these international money markets often offer opportunities to invest funds at more attractive rates than can be done in a bank's own domestic money markets.

ADVANTAGES ■ In addition to the above-mentioned advantages, a bank can exert maximum control over its foreign operations through a branch. For one thing, operating and credit policies can be closely integrated. More important, a foreign branch network allows the parent to offer its customers—both domestic and foreign—direct and integrated service, such as the rapid collection and transfer of funds internationally in a number of countries on a consistent policy basis. Thus, Citibank may require a branch in France as much to accommodate Siemens and Sony as Coca-Cola. Foreign branches also allow a bank to better manage its customer relationships, providing services to a customer on the basis of the value of its worldwide relationship rather than its relationship in a specific country.

DISADVANTAGES ■ The cost of establishing a branch can be quite high, running to several hundred thousand dollars annually for a typical European branch, plus the fixed cost of remodeling the new facilities. An indirect cost is the possibility of alienating correspondent banks when a new branch is opened. Developing and training management to staff these branches are also difficult and expensive. On the plus side, having foreign branches offers the chance for junior officers to gain valuable overseas experience and makes it easier to attract good personnel eager for that type of experience.

Acquisitions

The alternative to expanding by opening new branches is to grow through acquisitions. This approach is followed by most foreign banks trying to penetrate the U.S. market.

ADVANTAGES ■ Acquiring a local bank has two main advantages. First, buying an existing retail bank will afford immediate access to the local deposit market, eliminating the problem of funding local loans. Second, the existing management will have an established network of local contacts and clients that would be difficult (if not impossible) to duplicate.

DISADVANTAGES ■ Despite these advantages, the history of bank acquisitions abroad is littered with examples of ill-fated investments—particularly in the United States. The acquisition by Britain's Midland Bank of a 57% interest in Crocker National, a California bank, must rank as one of the most expensive entrance tickets to a market; Midland's stake, which cost it $820 million in 1981, was worth about $300 million in early 1985. At that time, Midland agreed to buy the remaining 43% of Crocker for about $250 million. In the interim, however, Crocker's continued troubles (it lost $325 million in 1984 alone) forced Midland to invest an additional $250 million in Crocker at the end of 1983 and to lend it $125 million more. In 1986, Midland sold Crocker to Wells Fargo. The other big British banks (National Westminster, Lloyds, and Barclays) also have had their share of troubles in the United States.

These troubles are not confined to British banks entering the U.S. market. In August 1983, Bank of America paid $147 million for Banco Internacional in Argentina—just before the Argentine peso dropped through the floor and the country's seemingly intractable debt problems blew up France's Credit Lyonnais has been obliged to pump large amounts of fresh capital into its Dutch subsidiary. Sweden's Skandinaviska Enskilda Banken's 24% stake in Banque Scandinaive of Switzerland was followed by two years of large losses.

The basic problem with these troubled investments seems to be that the acquiring banks spent too much energy on making the deal, often to the exclusion of developing future strategy. The usual rationale for the acquisition—synergy—is overworked. In the case of its acquisition of Crocker, for example, Midland brought little more to the deal than new money. The absence of any other contribution by the acquirer is usually a warning sign, for if the business being acquired is healthy apart from needing more capital, why is it choosing to sacrifice its independence rather than to go to the capital markets for additional funds?

The message seems to be that banking acquisitions are expensive, highly risky, and difficult to make work effectively, especially if all the acquirer brings to the deal is money. Unfortunately, most acquirers don't bring more than money to the deal. Not surprisingly, acquirers typically lose in such deals.

Edge Act and Agreement Corporations

Edge Act and Agreement corporations are subsidiaries of U.S. banks that are permitted to carry on international banking and investment activities. The practical effect of the various restrictions they operate under is to limit Edge Act institutions to handling foreign customers and to handling the international business of domestic customers. The list of permissible international activities for an Edge Act corporation includes deposit taking from outside the United States, lending money to international businesses, and making equity investments in foreign corporations, a power denied to its U.S. parent. An Agreement corporation is functionally similar to an Edge Act corporation. Usually it is a state-chartered corporation that enters into an agreement with the Federal Reserve to limit its activities to those of an Edge Act corporation.

Edge Act corporations are physically located in the United States, usually in a state other than where the head office is located to get around the prohibition on interstate branch banking. For example, a California bank can set up an Edge Act subsidiary in New York City to compete with New York banks for corporate business related to international activities—foreign exchange trading, export and import financing, accepting deposits associated with such operations, international fund remittances, and buying and selling domestic securities for foreign customers or foreign securities for domestic customers. Since June 1979, when the Federal Reserve permitted interstate branching by Edge Act corporations, these corporations have rivaled loan production offices in giving money-center banks and major regional banks on-site access to otherwise restricted markets.

A growing share of Edge Act business involves maintaining individual accounts for U.S. citizens living abroad and foreign citizens wanting their deposits in New York, Miami, Los Angeles, or some other U.S. city. Although they don't openly acknowledge the fact, an important source of relatively low-cost deposits for Edge Act corporations is in the form of "flight capital" from wealthy Central and South American depositors concerned with the lack of political stability in their own countries.

International Banking Facilities

Because of the unfavorable regulatory and tax environment in the United States, many domestic banks conduct their international operations through offshore branches. Frequently, these offices are simply "shell" operations used solely for booking purposes. By using foreign offices, banks are able to avoid state and local taxes on their foreign business profits, as well as costly reserve requirements and interest rate ceilings that would apply if the deposits were placed in the United States.

As a result of these advantages, London became the center of international banking activities. In addition, the number of branches in such places as the Cayman Islands and the Bahamas has grown significantly. Late in 1981, in an attempt to attract Eurodollar business back to the United States from these offshore locations, the Federal Reserve authorized U.S. financial institutions, including U.S. branches and agencies of foreign banks, to establish **international banking facilities** (IBFs). IBFs are permitted to conduct international banking business (such as receiving foreign deposits and making foreign loans) largely exempt from domestic regulatory constraints.

IBFs are merely bookkeeping entities that represent a separate set of asset and liability accounts of their establishing offices. The only requirement to establish an IBF is that the establishing institution must give the Federal Reserve two weeks' notice. The major activities of an IBF are deposit taking and lending to statutorily defined foreign persons, subject to certain restrictions. The major restrictions are (1) engaging only in foreign deposit taking and lending, (2) a minimum transaction size of $100,000, and (3) prohibition from issuing negotiable instruments, such as certificates of deposit (CDs).

Despite these restrictions, IBFs are popular because they are accorded most of the advantages of offshore banking without the need to be physically offshore. Most important is waiver of the regulation requiring that banks keep a percentage of their deposits in noninterest-earning accounts at the Federal Reserve. In addition, deposits in IBFs are not subject to interest rate ceilings or deposit insurance assessment.

IBFs are located mainly in the major financial centers; almost half of the nearly 500 IBFs are in New York, with the remainder being primarily in California, Florida, and Illinois. Although the geographical distribution of IBFs largely reflects the preexisting distribution of international banking business, differences in tax treatment have had some effect on the location decision. Florida, for example, exempts IBFs from state taxes and also ranks first in the number of Edge Act IBFs and second (after New York) in the number of IBFs set up by U.S.-chartered banks.

IBFs appear to have a very high proportion of both assets and liabilities due to other banking institutions. This interbank activity reinforces the belief that IBFs are now an integral part of the Eurodollar market. As we saw in Chapter 15, a high proportion of interbank business is one of the characteristics of the Eurocurrency markets because there may be several interbank transactions between ultimate borrowers and ultimate lenders. Thus far, IBFs seem to have had the intended effect of shifting international banking business from offshore locations back to the United States. They have not expanded the total volume of international banking business.

Foreign Banks in the United States

Foreign bank activity in the United States grew dramatically during the 1980s. This expansion reflects several key factors, including (1) the rapid growth of U.S. international trade, (2) the size and importance of U.S. financial markets, (3) the growth of foreign direct investment in the United States, (4) the role of the U.S. dollar as an international medium of exchange, and (5) a willingness to trade lower profits for increased market share. However, foreign bank growth in the United States has slowed down during the 1990s.

Foreign banks operating in the U.S. market typically concentrate their activities heavily on the wholesale market; they are generally not major factors in retail banking markets. In addition, most of the foreign banks that have a sizable presence in the United States are affiliated with well-known major banks abroad (for example, Barclays, Mitsubishi Bank, Swiss Bank Corporation, and Bank of Montreal).

As of mid-1994, the total number of foreign banking offices in the United States stood at 652, a mere 6% of the U.S. total. But these foreign offices held $952 billion in assets, or 21% of total U.S. banking assets (see Exhibit 17.3). Foreign banks also account for more than 30% of all commercial and industrial loans outstanding to U.S. companies. Most of the foreign-bank expansion in the United States during the 1980s was by Japan-

EXHIBIT 17.3 Foreign-Owned Bank Market Share in the United States

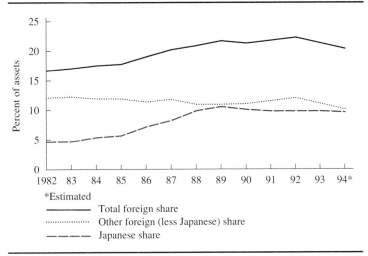

*Estimated

——————— Total foreign share
.... Other foreign (less Japanese) share
— — — — Japanese share

Source: Gary C. Zimmerman, "Slower Growth for Foreign Banks?" *FRBSF Weekly Newsletter,* Federal Reserve Bank of San Francisco, January 25, 1991, p. 1; and Linda M. Aguilar, "A Current Look at Foreign Banking in the U.S. and Seventh District," *Economic Perspectives,* Federal Reserve Bank of Chicago, Fall 1994, pp. 23, 26.

ese banks (as Exhibit 17.3 shows, other foreign banks actually lost U.S. market share). As a result of their rapid expansion, Japanese-owned banks controlled about 25% of the banking market in California, largely owing to pricing so razor-thin that American banks formally ceded certain types of lending. In certain markets, such as standby letters of credit associated with municipal bond offerings, Japanese banks accounted for up to 70% of the total U.S. market.

All this is changing, however, as the push for growth over profits appears to be at an end. More stringent international capital standards and risk-based capital requirements are constraining the growth prospects of foreign and domestic banks alike. These tightened standards, added to the reduction in the value of Japanese banks' "hidden reserves," mean that Japanese banks no longer can afford to make cheap loans to corporate borrowers to build market share. And they no longer seek to issue every letter of credit written in America. These changes are reflected in the U.S. market share of Japanese banks. After peaking in 1989 at about 11%, the Japanese market share then began falling, reaching about 9.5% in 1994. Faced with their own problems, foreign banks have not rushed in to fill the gap left by the Japanese, leading to a decline in the foreign-bank share of U.S. assets. This reluctance probably is explained by the low profitability of foreign banks in the U.S. market. Specifically, over the decade 1983–1992, the return on assets for foreign banks averaged less than one-fifth that of U.S. banks.[5] These developments signal slower growth ahead for foreign banks in the United States.

---

[5] These conclusions appear in a study by the Office of the Comptroller of the Currency, cited by Kenneth H. Bacon, "Study Shows While Foreign Banks Lend Widely in U.S., They're Behind in Profit," *Wall Street Journal,* June 13, 1994, p, A5A.

Implications of Europe 1992 for Banks

With the creation of a single European market on January 1, 1993, banks can now branch throughout the European Community while providing a greater range of financial services.[6] The single license should enable all banks in Europe to realize a number of cost benefits, primarily by being able to operate throughout the EC using common distribution networks, managers, and support systems. A single banking license should also lower bank costs in Europe by eliminating overlapping or conflicting standards and regulatory procedures.

Despite the potential cost advantages of establishing an EC-wide bank network, however, the empirical evidence suggests that the scope for economies of scale in banking is rather limited. Indeed, the correlation between size and profitability for the largest EC banks is either negative or minimal. Economies of scale in banking appear to be limited to instances where the bank's strategy involves common customers, information systems, skills, or processing facilities. One such business is credit cards, which benefits from economies of scale in issuance, centralized data processing, and design standardization.

Absent these commonalities, it appears unlikely that cross-border mergers between large institutions will create value: The cost of merging operations will probably exceed any economies of scale or scope they might achieve. Moreover, the ever-present danger is that as entry barriers start falling, new entrants might overpay in trying to get established and gain market share. Value-creating mergers are most likely to be between relatively small local banks in countries characterized by relatively low concentration ratios.

Banks seeking to benefit from Europe 1992 should focus on pursuing businesses in which they have a sustainable competitive advantage rather than competing directly against dominant local banks that have strongly entrenched positions in delivering basic banking services.

ILLUSTRATION

CITICORP'S STRATEGY FOR EUROPE 1992

Citicorp's EC strategy is to utilize a cadre of experienced professionals, with expertise in a number of product categories and management techniques, to develop and deliver products in multiple markets. Although each market presents a unique set of challenges, Citicorp employees have already experienced most, if not all, of these situations elsewhere in the world and, thus, have ideas and procedures for dealing with them. Citicorp's large EC credit-card operation, for example, relies heavily on its extensive U.S. experience in the business. It also has employed its experience in circumventing U.S. restrictions on interstate banking to build a 700-branch network throughout Europe.

---

[6]This section is based on Alan C. Shapiro, "Economic Import of Europe 1992," *Journal of Applied Corporate Finance*, Winter 1991, pp. 25–36.

The single banking market is leading to growing competition, especially for retail banking services, such as checking accounts, consumer credit, credit cards, mortgage loans, foreign exchange, and travelers checks.[7] The winners from vigorous competition will be consumers, who should pay lower prices for higher quality services, and banks with the capital, technical expertise, and efficient operations and marketing to take advantage of market opportunities to expand their activities. The losers will be inefficient banks operating in overbanked and protected national markets. Not surprisingly, the nations whose banks are expected to face the sharpest price declines from deregulation have been the slowest to implement the EC's banking directives.

17.3 VALUE CREATION IN INTERNATIONAL BANKING

This section examines some of the strategies that international banks can use to create value in their marketplace. It also examines the likely shape of future international banking competition.

Capital Adequacy

Banks with a large capital base can gain competitive advantage in underwriting securities and reducing the cost of funding loans. They also can develop more profitable business relations because companies today are paying added attention to the financial strength of the banks they deal with. Given the protracted investment of time and money in building a relationship with a bank, many chief financial officers will deal only with well-capitalized banks, whom they naturally regard as being more reliable financial partners. Such banks also can serve as a credible counterparty where customers assume bank counterparty risk, including L/Cs, swaps, and forwards.

Human Resources and the Banking Organization

International banking is a fiercely competitive business. A corollary is that it contains no obvious, unexploited profitable niches. Any new insight in developing competitive advantage, such as a novel product or service, is quickly imitated and the extra profits competed away. The real source of competitive advantage is having an organization that can exploit information to learn and to innovate more quickly than competitors. Unfortunately, the advice to be continually innovative and creative is easier said than done.

For a bank to be a successful innovator, its front line people must have—in addition to the necessary information and skills—the authority, accountability, and motivation to use their minds to exploit that information. That means that career systems—promotions, reward, and development—have to be geared to creating and sustaining a learning environment. In the competitive world of international banking, the returns will come not by lending money but by selling companies on the belief that the bank knows how to solve their corporate finance problems and can do so quickly. That is, CFOs expect their bankers

---

[7] Wholesale banking services typically are provided to large multinational corporations and financial institutions in competitive financial markets that are already regionally integrated.

to do more than evaluate loan requests. CFOs expect bankers to create tailor-made solutions to corporate financial problems. The demand by large companies for their banks to serve as consultants and problem-solvers offers banks the opportunity to perform higher margin, fee-generating services. A bank will succeed in this world only if the people meeting the customers can identify their problems and willingly share that information with others who can develop profitable solutions to those problems.

Bank officers require a lot of training in order to offer such high-quality service. This training must emphasize recognizing market opportunities, understanding the corporate environment, and clearly communicating to customers the often-complex options available to solve problems.

This view of value creation in international banking suggests that bank strategies should not rely on hardware or expensive acquisitions. Rather, banks should invest in preparing and enabling people to recognize and respond to customer problems. This process is ongoing because today's creative solution to a problem—such as currency swaps—is tomorrow's commodity product. Banks also must invest in the information technology necessary to identify customer problems and create new services and products to solve them.

Information Systems

Perhaps the most pervasive challenge facing the banker in the international market is the requirement for investment in technology. The value of good information is as important today as it was in Baron Rothschild's day when he used carrier pigeons to bring him the news that Wellington had triumphed at Waterloo. Rothschild made his fortune on state-of-the-art communications. Today, bankers rely on computers and telecommunications, not carrier pigeons, but the quest is much the same—seeking tools to carry out banking faster and better than the competition.

New technology also is required to supply customers with the increasingly complex financial instruments that they are now demanding to implement sophisticated financial strategies. At the same time, a bank's risk management system must keep pace with these new developments. In response, many banks have invested heavily in the value-at-risk models mentioned earlier.

Thus, merely to compete in complex global markets requires substantial investment in information systems. A bank can gain by being a purveyor of information transfer systems, either because it has better information to influence business decisions or because it is able to sell information to others. However, the scale of investment in technology to fully exploit the benefits of information acquisition and transfer is bound to be so great that only those banks with a firm commitment to the concept of providing a global service are likely to undertake it.

Another use of information systems is to keep track of customer profitability, an increasingly important task as fierce competition makes lending to companies, especially large ones, a business with margins too low to compensate for the risks being borne. In response, more banks are trying to determine which of their low-margin customers to let go. Such culling, however, depends on the ability to measure which customer relationships are profitable overall, taking into account the profits the bank is earning on the non-credit services being bought by their customers. The more diversified a bank's operations, and the more interactions it has with its multinational customers in multiple locations, the more difficult it is for the bank to value every aspect of its dealings with them. A good information system allows the bank to separate its good customers from bad—and to show the bad ones why they are being let go. At the same time, the bank can focus on its good customers in order to deepen its relationships with them.

The information system also can serve as the basis of a control system to closely track what is going on in the various nooks and crannies of a sprawling organization and supply the necessary inputs to the risk management models mentioned earlier. This is essential for banks with overseas operations, as Daiwa Bank found to its cost in 1995 when it discovered that one of the traders in its New York branch had managed to conceal a $1.1 billion trading loss from his bosses in Tokyo for more than a decade.

Transaction-Processing Services

Companies that are globalizing their operations are demanding seamless service around the world, including a transfer agent who can pull together all transactions on a global, integrated basis. Thus, transaction-processing services, such as funds movement, foreign exchange, international cash management, lock boxes, disbursements, and global custody of securities can be very profitable bank services. However, banks must invest huge amounts of money in a state-of-the-art information system to be a credible player in these businesses. Given the significant economies of scale, market share is critical.

Instead of trying to provide global cash management services, a bank could focus on particular niches. An example would be PC-based treasury workstations that integrate foreign exchange, funds movements, and cash management, instead of mainframe-based bank systems with dumb terminals. Banks are starting to sell the same cross-border payments services to smaller companies that larger companies are now using. With an increasing number of smaller companies doing international business, this is a growing market.

Global Competition in International Banking

The 1990s is an era of intensified global competition in international banking. Already, big banks are battling each other on many fronts: lending, underwriting, leasing, financial advice and risk management services, currency and securities trading, insurance, money management, and consumer banking. Some experts predict that of about 40 to 50 banks now aspiring to be global banking powerhouses, only six to 10 will actually make the grade by the end of the 1990s. Survival will demand an ability to serve the financing and risk management needs of corporations, big and small, anywhere in the world. They also must be able to intermediate a large share of the growing cross-border flow of capital and be flexible enough to shift resources quickly to fast-growing areas and high-return businesses. At the same time, they must cope with deregulation, competition from nonbanking firms (such as General Electric Capital) and securities markets, and the industry's overcapacity—too many banks chasing too few customers.

The key to success is likely to be banks' ability to shift from their traditional business of deposit taking, lending, and money transmission to the new sources of financial profits: the trading of currencies, securities, and derivative products. To be successful here, banks must be able to manage the complex risks inherent in these markets. They also must be able to position themselves properly in the marketplace, including segmenting their customer base to identify needs and provide new products and financial and risk-management solutions tailored for specific groups.

The major competitors include European, Japanese, and American banks. Each has important competitive strengths and weaknesses.

EUROPEAN BANKS ■ Some of the strongest contenders for global dominance are the large European banks. Their strengths include solid capital bases, strong balance sheets, and dominant shares of home markets. However, they have little experience with head-to-head competition. Moreover, European banking's profit sanctuaries—local markets protected from price-cutting competition by collusive arrangements—are being invaded in the post-1992 world. In addition, European bankers are facing a number of other challenges: slow economic growth and weak loan demand, which is making it hard for banks to grow their earnings; the requirements for massive investments in technology; strong competition from nonbank financial-service providers, as well as from other banks; and the massive overcapacity in many countries with branches on nearly every street corner, which has led to reduced credit quality, low margins, and poor stock market returns. These conditions all suggest that a considerable consolidation—facilitated, as in the United States, by a wave of mergers and acquisitions—should take place over the next decade. The process is already under way among Europe's smaller, weaker banks and beginning among the larger banks as well.

The resulting shakeout could leave the strongest European banks—Germany's Deutsche Bank, Union Bank of Switzerland, Credit Suisse, Swiss Bank Corporation, and England's Barclays—stronger than ever. Domestic market dominance will enable them to subsidize forays abroad, but as we saw earlier they will be able to capitalize on their financial strength only if they can bring more to the table than money. So far they have not demonstrated transferable skills that will confer on them a competitive advantage abroad. So, many of them are looking to buy new competencies by acquiring investment banking houses, mortgage processors, stock brokers, insurance companies, and money management firms. The jury is out on how successful these financial diversification strategies will be, but history is not on their side.

JAPANESE BANKS ■ The giant Japanese banks, historically backed by a strong economy, diligent savers, and powerful corporate customers, are rich in assets. But they are poor in capital and innovativeness and are learning that size does not necessarily translate into competitive advantage. Also, as noted earlier, Japanese banks are now under unprecedented financial stress as financial deregulation narrows their spreads between borrowing and lending money, the weak economy reduces demand for their services, and huge losses on real estate and other loans deplete their capital. Concerns about the health of Japan's banking system have forced even top Japanese banks to pay a premium (relative to foreign banks) to borrow in the interbank market, making Japanese banks less competitive both at home and abroad. Even the main-bank relationship that has historically locked Japanese customers to their primary bank is threatened under the strain of competition from securities markets and foreign banks with lower-cost funds.

The problems facing Japanese banks are not accidental. They are the natural outcome of a corporate mentality that invariably subordinated profitability to growth and market share and a government that tried to soften the rough edges of capitalism by propping up failing banks and merging failing banks with healthy ones. But the Japanese government's actions simply bottled up problem loans and tempted bankers into risky credit practices.

The Japanese government has attempted once again to help solve the banks' bad-debt problems, this time by boosting their profits. By forcing short-term interest rates down, the Bank of Japan has cut the banks' cost of funds to almost zero, improving their lending margins and giving them large capital gains on their government bond portfolios. This is only a temporary answer to the problems facing the Japanese banking system, however, because its structural weaknesses endure: It is rigid, impervious to change, unwilling to innovate, and pampered by a government bureaucracy that refuses to admit mistakes. Instead of sheltering banks from failure, the government must allow the cleansing action of the marketplace to operate. Unless and until Japanese banks are subject to the same market forces facing their manufacturing counterparts, they will be unable to compete on a level playing field. Thus, the recent rash of Japanese bank failures is actually a positive development, giving the remaining banks a strong incentive to strengthen their balance sheets and focus on profits rather than market share.

U.S. BANKS ■ Most large American banks have, until recently, been plagued with weak capital bases and the threat of huge write-downs on their loans to LDCs, real estate developers, and leveraged buyouts. But after two decades of competing domestically and internationally against each other, as well as against insurers, investment banks, and money managers, U.S. banks have developed superior creative skills. By investing heavily in computers, information systems, software, and personnel training, commercial banks such as Citicorp, Bankers Trust, and J.P. Morgan, along with their investment banking counterparts such as Goldman Sachs, Merrill Lynch, and Morgan Stanley, have effectively managed to institutionalize the process of innovation. Their innovativeness and advanced technology give them the edge in currency and securities trading and in designing and distributing myriad new products—capabilities widely expected to enable them to seize profitable opportunities in the deregulating European and Japanese markets. They also have added a whole new dimension to banking—the aggressive restructuring of industry in addition to the old-fashioned raising of funds. Naturally, the former is far more profitable than the latter. Moreover, U.S. banks have by now rebuilt their capital bases.

With intensifying competition, overcapacity, and declining bank margins, the expectation is for further big bank mergers and acquisitions, first within country borders and then across them. For example, in 1996, Wells Fargo acquired First Interstate, Chase Manhattan merged with Chemical Bank, and Mitsubishi Bank and Bank of Tokyo combined to form the world's largest bank. Mega-mergers in Europe have been rarer. The aim of these mergers varies, with some (such as the Chase-Chemical merger) designed to produce huge cost savings from combining overlapping businesses and others (such as the Mitsubishi-Bank of Tokyo merger) designed to bring together banks with complementary skills and businesses.

17.4 COUNTRY RISK ANALYSIS IN INTERNATIONAL BANKING

As we saw earlier, the big money-center banks, as well as many regional banks, rechanneled hundreds of billions of petrodollars during the 1970s and early 1980s to less-developed countries and Communist countries. Major banks earned fat fees for arranging loans to Poland, Mexico, Brazil, and other such borrowers. The regional banks earned the spreads between the rates at which they borrowed Eurodollars and the rates at which those loans were syndicated. These spreads were minimal, usually on the order of 0.5% to 0.75%.

Most of the money was squandered on all sorts of extravagant and profitless projects, from Amazon highways to highly automated steel mills in Africa. All this made sense, however, only as long as banks and their depositors were willing to suspend their disbelief about the risks of international lending. By now, the risks are big and obvious. The purpose of this section is to explore **country risk** from a bank's standpoint, the possibility that borrowers in a country will be unable to service or repay their debts to foreign lenders in a timely manner. The essence of country risk analysis at commercial banks, therefore, is an assessment of factors that affect the likelihood that a country, such as Mexico, will be able to generate sufficient dollars to repay foreign debts as these debts come due.

These factors are both economic and political. Among economic factors often pointed to are the country's resource base and its external financial position. Most important, however, are the quality and effectiveness of a country's economic and financial management policies. As we saw earlier, the evidence from those countries that suffered through the international debt crisis is that all too often the underlying causes of country risk are homegrown, with massive corruption, bloated bureaucracies, and government intervention in the economy. This leads to inefficient and uncompetitive industries and huge amounts of capital squandered on money-losing ventures. In addition, poor macroeconomic policies have made their own contribution to economic instability. Many of these countries have large budget deficits that they monetized, leading to high rates of inflation, overvalued currencies, and periodic devaluations. The deficits stem from too many promises that cannot be met from available resources and high tax rates that result in tax evasion, low revenue collections, and further corruption.

Political factors that underlie country risk include the degree of political stability of a country and the extent to which a foreign entity, such as the United States, is willing to implicitly stand behind the country's external obligations. Lending to a private-sector borrower also exposes a bank to commercial risks, in addition to country risk. Because these commercial risks are generally similar to those encountered in domestic lending, they are not treated separately.

It is worthwhile to note some differences between international loans and domestic loans because these differences have a lot to do with the nature of country risk. Ordinarily banks can control borrowers by means of loan covenants on their dividend and financing decisions. Quite often, however, the borrowers in international loan agreements are sovereign states, or their ability to repay depends on the actions of a sovereign state. The unique characteristics of a sovereign borrower render irrelevant many of the loan covenants—such as dividend or merger restrictions—normally imposed on borrowers. Moreover, sovereign states ordinarily refuse to accept economic or financial policy restrictions imposed by foreign banks.

The key issue posed by international loans, therefore, is: How do banks ensure the enforceability of these debt contracts? What keeps borrowers from incurring debts and then defaulting voluntarily? In general, seizure of assets is not useful unless the debtor has substantial external assets, as in the case of Iran. Ordinarily, though, the borrower has few external assets.

One answer to the question of enforceability is that it is difficult for borrowers that repudiate their debts to reenter private capital markets.[8] That is, banks find it in their best interest, ex post, to deny further credit to a borrower that defaults on its bank loans. If the bank fails to adhere to its announced policy of denying credit to its defaulters, some of its other borrowers may decide to default on their debts because they realize that default carries no penalty. Having a reputation for being a tough bank is a valuable commodity in a hard world that has no love for moneylenders.

The presence of **loan syndications**—in which several banks share a loan—and **cross-default clauses**—which ensure that a default to one bank is a default to all banks—means that if the borrower repudiates its debt to one bank, it must repudiate its debt to *all* the banks. This constraint makes the penalty for repudiation much stiffer because a large number of banks will now deny credit to the borrower in the future. After a few defaults, the borrower will exhaust most of the potential sources of credit in the international financial markets. For many borrowers, this penalty is severe enough that they do not voluntarily default on their bank loans. Thus, countries will sometimes go to extraordinary lengths to continue servicing their debts. Bank country risk analysis can, therefore, focus largely on ability to repay rather than willingness to repay.

Notice, however, that the threat to cut off credit to borrowers who default is meaningful only so long as the banks have sufficient resources to reward with further credit those who do not default. If several countries default simultaneously, then the banks' promise to provide further credit to borrowers who repay their debts is no longer credible; the erosion in their capital bases caused by the defaults will force the banks to curtail their loans. Under these circumstances, even those borrowers who did not default on their loans will suffer a reduction of credit. The lesser penalty for defaulting may induce borrowers to default en masse. The possibility of mass defaults is the real *international debt crisis.*

Country Risk and the Terms of Trade

What ultimately determines a nation's ability to repay foreign loans is that nation's ability to generate U.S. dollars and other hard currencies. This ability, in turn, is based on the nation's **terms of trade**, the weighted average of the nation's export prices relative to its import prices—that is, the exchange rate between exports and imports. Most economists would agree that these terms of trade are largely independent of the nominal exchange

---

[8] The discussion of debt repudiation and the nature of international loans comes from Bhagwan Chowdry, "What Is Different about International Lending?" working paper, University of Chicago, November 1987.

rate, unless the observed exchange rate has been affected by government intervention in the foreign exchange market.

In general, if its terms of trade increase, a nation will be a better credit risk. Alternatively, if its terms of trade decrease, a nation will be a poorer credit risk. This *terms-of-trade risk*, however, can be exacerbated by political decisions. When a nation's terms of trade improve, foreign goods become relatively less expensive, the nation's standard of living rises, and consumers and businesses become more dependent on imports. However, because there is a large element of unpredictability to relative price changes, shifts in the terms of trade also will be unpredictable. When the nation's terms of trade decline, as must inevitably happen when prices fluctuate randomly, the government will face political pressure to maintain the nation's standard of living.

A typical response is for the government to fix the exchange rate at its former (and now overvalued) level—that is, to subsidize the price of dollars. Loans made when the terms of trade improved are now doubly risky: first, because the terms of trade have declined and, second, because the government is maintaining an *overvalued currency*, further reducing the nation's net inflow of dollars. The deterioration in the trade balance usually results in added government borrowing. This was the response of the Baker 15 countries to the sharp decline in their terms of trade shown in Exhibit 17.4. Capital flight exacerbates this problem, as residents recognize the country's deteriorating economic situation.

To summarize, a terms-of-trade risk can be exacerbated if the government tries to avoid the necessary drop in the standard of living when the terms of trade decline by maintaining the old and now-overvalued exchange rate. In reality, of course, this element of country risk is a political risk. The government is attempting by political means to hold off the necessary economic adjustments to the country's changed wealth position.

A key issue, therefore, in assessing country risk is the speed with which a country adjusts to its new wealth position. In other words, how fast will the necessary austerity

EXHIBIT 17.4 Decline in Terms of Trade in Baker 15 Countries

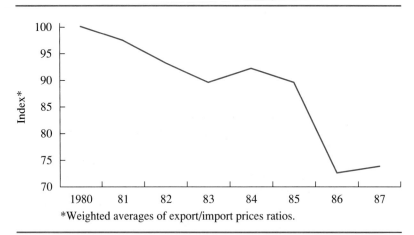

*Weighted averages of export/import prices ratios.

Source: Data from Organization for Economic Cooperation and Development, International Monetary Fund, and Federal Reserve Bank of New York.

policy be implemented? The speed of adjustment will be determined in part by the government's perception of the costs and benefits associated with austerity versus default.

The Government's Cost/Benefit Calculus

The cost of austerity is determined primarily by the nation's external debts relative to its wealth, as measured by its gross domestic product. The lower this ratio, the lower the relative amount of consumption that must be sacrificed to meet a nation's foreign debts.

The cost of default is the likelihood of being cut off from international credit. This possibility brings with it its own form of austerity. Most nations will follow this path only as a last resort, preferring to stall for time in the hope that something will happen in the interim. That something could be a bailout by the IMF, the Bank for International Settlements, the Federal Reserve, or some other major central bank. The bailout decision is largely a political decision. It depends on the willingness of citizens of another nation, usually the United States, to tax themselves on behalf of the country involved.[9] This willingness is a function of two factors: (1) the nation's geopolitical importance to the United States, and (2) the probability that the necessary economic adjustments will result in unacceptable political turmoil.

The more a nation's terms of trade fluctuate and the less stable its political system, the greater the odds the government will face a situation that will tempt it to hold off on the necessary adjustments. Terms-of-trade variability probably will be inversely correlated with the degree of product diversification in the nation's trade flows. With limited diversification—for example, dependence on the export of one or two primary products or on imports heavily weighted toward a few commodities—the nation's terms of trade are likely to be highly variable. This characterizes the situation facing many Third World countries. It also describes, in part, the situation of those OECD (Organization for Economic Cooperation and Development) nations heavily dependent on oil imports.

❧ 17.5 SUMMARY AND CONCLUSIONS

This chapter examined the various means and reasons whereby banks have expanded their international operations and loan portfolios in the post-World War II period. Banks have several options in their overseas expansion, including foreign branches, correspondent banking, representative offices, acquisitions of local banks, Edge Act and Agreement corporations, and international banking facilities. We explored the advantages and disadvantages of each of these vehicles for international expansion. In general, those banks that brought only money to the international marketplace have not earned sufficient profits to justify their presence there.

The pattern of bank lending overseas has been one of rapid expansion, beginning in the early 1960s and sharply accelerating after the first OPEC oil price shock in late 1973. This expansion has been followed by a sharp contraction in international bank lending on the heels of the great debt crisis of 1982. More recently, Japanese banks greatly expanded their overseas loan portfolios but are now in the process of retrenching. One reason for this pull-back is the new international agreement setting minimum risk-based capital standards for banks that became effective in 1993. Marginal banks will find international lending to be uneconomical under these standards.

---

[9] See Tamir Agmon and J. K. Dietrich, "International Lending and Income Redistribution: An Alternative View of Country Risk," *Journal of Banking and Finance*, December 1983, pp. 483–495, for a discussion of this point.

The end of "let's pretend" in international banking has led to a new emphasis on country risk analysis. From the bank's standpoint, country risk—the credit risk on loans to a nation—is largely determined by the real cost of repaying the loan versus the real wealth that the country has to draw on. These parameters, in turn, depend on the variability of the nation's terms of trade and the government's willingness to allow the nation's standard of living to adjust rapidly to changing economic fortunes.

The experience of those countries that have made it through the international debt crisis suggests that others in a similar situation can get out only if they institute broad systemic reforms. Their problems are caused by governments spending too much money they do not have to meet promises they should not make. They create public-sector jobs for people to do things they should not do and subsidize companies to produce high-priced goods and services. These countries need less government and fewer bureaucratic rules. Debt forgiveness or further capital inflows would only tempt these nations to postpone economic adjustment further.

➤ Questions

1. What are the relative advantages and disadvantages of expanding overseas via the following:
   a. Foreign branches
   b. Representative offices
   c. Acquisitions
   d. Correspondent banking

2. What impact will the new capital requirements have on international bank expansion? On the Eurocurrency market?

3. Japanese banks' share of total bank assets in Los Angeles jumped from 11% to 24% between 1983 and 1988.
   a. What factors might account for the large increase in Japanese bank activity in southern California?
   b. How will the new capital requirements likely affect Japanese bank expansion in the United States?

4. According to John Reed, Citicorp's chairman, money is "information on the move." Explain what Mr. Reed might mean by this statement.

5. Why have international bank mergers generally been failures?

6. Assess the competitive strengths and weaknesses of European, U.S., and Japanese banks. How is financial deregulation in Japan and Europe likely to affect the competitive balance in international banking?

7. What were the key contributing factors to the international debt crisis that began in 1982?

8. What economic and political factors account for the fact that the foreign loans of major banks were in such bad shape during the 1980s?

9. How are U.S. international banking practices likely to be influenced by the belief that the Federal Reserve would not permit a major U.S. bank to fail?

10. Before 1983, U.S. banks did not have to, and generally did not, provide data to the public on the geographical distribution of their loan portfolios. Assuming the stock market is strongly efficient, what response would you expect to the public release of information in early 1983 that concerned the extent of bank loans to various Latin American countries? Recall that the international debt crisis struck in mid-1982.

11. Why is it crucial for banks to prevent several defaults at once?

12. What incentive do borrowers have to form a debtors' cartel and simultaneously default? Who would choose not to belong to such a cartel?

13. In the week after Citicorp's $3 billion write-down of its Latin American debt, its stock price rose by 10%. What might explain this rise in Citicorp's stock price?

14. How did capital flight contribute to the international debt crisis?

15. According to one Japanese banker, the $37.5 billion in Japanese bank loans outstanding to Thailand in 1997 were hard currency loans, so the banks did not suffer an exchange loss when the baht devalued. Comment.

Problems

1. In the Mexican debt restructuring discussed in the chapter, suppose that when the deal was struck 30-year Treasury zeros were selling in the open market at a yield of about 7.62%, but the U.S. Treasury priced the zeros that it sold to Mexico to yield 8.15%. The Treasury also charged Mexico its customary service fee of 0.125% per annum, giving Mexico an effective annual interest yield of 8.025%.
   a. Under these circumstances, what would have been the amount of the subsidy provided by the Treasury (relative to the cost to Mexico of buying the zeros to yield 7.62% less 0.125%)?
   b. What does this complicated deal accomplish? Who benefits? Who loses?

2. In early 1988, the Mexican government sought to swap about $15 billion of its outstanding bank debt for $10 billion in bonds. The principal amount of the Mexican bonds would be secured by 20-year, zero-coupon U.S. Treasury bonds having a face value of $10 billion. Because of doubts about Mexico's creditworthiness, Mexican bank debt was then valued by the market at about 50 cents on the dollar.
   a. How should Mexico's offer be evaluated?
   b. Suppose the bond interest payments are more certain than bank debt payments. Before the bond swap is worthwhile, how much would a Mexican promise to pay $1 in bond interest have to be worth relative to a Mexican promise to pay $1 in bank debt?

Bibliography

CHRYSTAL, K. ALEC. "International Banking Facilities." *Federal Reserve Bank of St. Louis Review*, April 1984, pp. 5–11.

Federal Reserve Bank of Chicago. *International Letter*, various issues.

LESSARD, DONALD R. "North-South: The Implications for Multinational Banking." *Journal of Banking and Finance* 7 (1983): pp. 521–536.

LESSARD, DONALD R., and JOHN WILLIAMSON, eds. *Capital Flight and Third World Debt*. Washington, D.C.: Institute for International Economics, 1987.

SHAPIRO, ALAN C. "Currency Risk and Country Risk in International Banking." *Journal of Finance*, July 1985, pp. 881–891.

_____. "Risk in International Banking." *Journal of Financial and Quantitative Analysis*, December 1982, pp. 727–739.

UNGER, BROOKE. "A Survey of World Banking." *The Economist*, May 2, 1992, special report.

WIHLBORG, CLAS. "Currency Risks in International Financial Markets." *Princeton Studies in International Finance* No. 44, December 1978.

THE COST OF CAPITAL FOR FOREIGN INVESTMENTS

Traders and other undertakers may, no doubt, with great propriety, carry on a very considerable part of their projects with borrowed money. In justice to their creditors, however, their own capital ought to be, in this case, sufficient to ensure, if I may say so, the capital of those creditors; or to render it extremely improbable that those creditors should incur any loss, even though the success of the project should fall very short of the expectations of the projectors.

Adam Smith (1776)

$\mathcal{A}$ central question for the multinational corporation is whether the required rate of return on foreign projects should be higher, lower, or the same as that for domestic projects. To answer this question, we must examine the issue of cost of capital for multinational firms, one of the most complex issues in international financial management. Yet, it is an issue that must be addressed, because the foreign investment decision cannot be made properly without knowledge of the appropriate cost of capital.

In this chapter we will seek to determine the cost-of-capital figure(s) that should be used in appraising the profitability of foreign investments. By definition, the *cost of capital* for a given investment is the minimum risk-adjusted return required by shareholders of the firm for undertaking that investment. As such, it is the basic measure of financial performance. Unless the investment generates sufficient funds to repay suppliers of capital, the firm's value will suffer. This return requirement is met only if the net present value of future project cash flows, using the project's cost of capital as the discount rate, is positive.

The development of appropriate cost-of-capital measures for multinational firms is closely bound to how those measures will be used. Because they are to be used as discount rates to aid in the global resource-allocation process, the rates must reflect the value to firms of engaging in specific activities. Thus, the emphasis here is on the cost of capital or required rate of return for a specific foreign project rather than for the firm as a whole. Unless the financial structures and commercial risks are similar for all projects engaged in, the use of a single overall cost of capital for project evaluation is incorrect. Different discount rates should be used to value projects that are expected to change the risk complexion of the firm.

The chapter also examines the factors that are relevant in determining the appropriate mix of debt and equity financing for the parent and its affiliates. In selecting financial structures for its various units, the multinational corporation must consider the availability of different sources of funds and the relative cost and effects of these sources on the firm's operating risks.

18.1 THE COST OF EQUITY CAPITAL

The **cost of equity capital** for a firm is the minimum rate of return necessary to induce investors to buy or hold the firm's stock. This required return equals a basic yield covering the time value of money plus a premium for risk. Because owners of common stock have only a residual claim on corporate income, their risk is the greatest, and so also are the returns they demand.

Alternatively, the cost of equity capital is the rate used to capitalize total corporate cash flows. As such, it is just the weighted average of the required rates of return on the firm's individual activities. From this perspective, the corporation is a mutual fund of specified projects, selling a compound security to capital markets. According to the principle of value additivity, the value of this compound security equals the sum of the individual values of the projects.

Although the two definitions are equivalent, the latter view is preferred from a conceptual standpoint because it focuses on the most important feature of the cost of equity capital—namely, that this cost is not an attribute of the firm per se but is a function of the riskiness of the activities in which it engages. Thus, the cost of equity capital for the firm as a whole can be used to value the stream of future equity cash flows—that is, to set a price on equity shares in the firm. It cannot be used as a measure of the required return on equity investments in future projects unless these projects are of a similar nature to the average of those already being undertaken by the firm.

One approach to determining the project-specific required return on equity is based on modern capital market theory. According to this theory, an equilibrium relationship exists between an asset's required return and its associated risk, which can be represented by the **capital asset pricing model** or CAPM:

$$r_i = r_f + \beta_i(r_m - r_f) \tag{18.1}$$

WHERE

r_i = equilibrium expected return for asset i

r_f = rate of return on a risk-free asset, usually measured as the yield on a 30-day U.S. government Treasury bill

r_m = expected return on the market portfolio consisting of all risky assets

β_i = $\text{cov}(r_i, r_m)/\sigma^2(r_m)$, where $\text{cov}(r_i, r_m)$ refers to the covariance between returns on security i and the market portfolio and $\sigma^2(r_m)$ is the variance of returns on the market portfolio

The CAPM is based on the notion that intelligent, risk-averse shareholders will seek to diversify their risks, and, as a consequence, the only risk that will be rewarded with a risk premium will be systematic risk. As can be seen from Equation 18.1, the risk premium associated with a particular asset i is assumed to equal $\beta_i(r_m - r_f)$, where β_i is the **systematic**, or **nondiversifiable risk** of the asset. In effect, β (beta) measures the correlation between returns on a particular asset and returns on the market portfolio. The term $r_m - r_f$ is known as the **market risk premium**.

Where the returns and financial structure of an investment are expected to be similar to those of the firm's typical investment, the corporatewide cost of equity capital may serve as a reasonable proxy for the required return on equity of the project. In this case, estimates of the value of the project's beta can be found either by direct computation using the CAPM or through professional investment companies that keep track of company betas.

It should be emphasized again that using a company beta to estimate the required return on a project's equity capital is valid only for investments with financial characteristics typical of the "pool" of projects represented by the corporation. This cost-of-equity-capital estimate is useless in calculating project-specific required returns on equity when the characteristics of the project diverge from the corporate norm.

18.2 THE WEIGHTED AVERAGE COST OF CAPITAL FOR FOREIGN PROJECTS

As commonly used, the required return on equity for a particular investment assumes that the financial structure and risk of the project is similar to that for the firm as a whole. This cost of equity capital, k_e, is then combined with the after tax cost of debt, $i_d(1 - t)$, to yield a **weighted average cost of capital** (WACC) for the parent and the project, k_0, computed as

$$k_0 = (1 - L)k_e + Li_d(1 - t) \qquad (18.2)$$

where L is the parent's debt ratio (debt to total assets). This cost of capital is then used as the discount rate in evaluating the specific foreign investment. It should be stressed that k_e is the required return on the firm's stock given the particular debt ratio selected.

Two caveats in employing the weighted average cost of capital are appropriate here. First, the weights must be based on the proportion of the firm's capital structure accounted for by each source of capital using *market*, not *book*, values. Second, in calculating the WACC, the firm's historical debt-equity mix is not relevant. Rather, the weights must be marginal weights that reflect the firm's *target capital structure*, that is, the proportions of debt and equity the firm plans to use in the future.

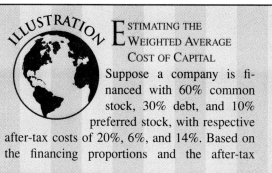

ESTIMATING THE WEIGHTED AVERAGE COST OF CAPITAL Suppose a company is financed with 60% common stock, 30% debt, and 10% preferred stock, with respective after-tax costs of 20%, 6%, and 14%. Based on the financing proportions and the after-tax costs of the various capital components and Equation 18.2, the WACC for this firm is calculated as 15.2% (0.6 × 0.20 + 0.3 × 0.06 + 0.1 × 0.14). If the net present value of those cash flows—discounted at the weighted average cost of capital—is positive, the investment should be undertaken; if it is negative, the investment should be rejected.

However, both project risk and project financial structure can vary from the corporate norm. It is necessary, therefore, to adjust the costs and weights of the different cost components to reflect their actual values.

Costing Various Sources of Funds

Suppose a foreign subsidiary requires I dollars to finance a new investment to be funded as follows: P dollars by the parent; E_f dollars by the subsidiary's retained earnings; and D_f dollars by foreign debt, with $P + E_f + D_f = I$. To compute the project's weighted cost of capital, we first must compute the individual cost of each component.

PARENT COMPANY FUNDS ■ The required rate of return on parent company funds is the firm's marginal cost of capital, k_0. Hence, parent funds invested overseas should yield the parent's marginal cost of capital provided that the foreign investments undertaken do not change the overall riskiness of the MNC's operations. (The effect of risk will be addressed later.)

RETAINED EARNINGS ■ The cost of retained earnings overseas, k_s, is a function of dividend withholding taxes, tax deferral, and transfer costs. In general, if T equals the incremental taxes owed on earnings repatriated to the parent, then $k_s = k_e(1 - T)$.

LOCAL CURRENCY DEBT ■ The after-tax dollar cost of borrowing locally, i_f, equals the sum of the after-tax interest expenses plus the expected exchange gain or loss.

Computing the Weighted Average Cost of Capital

With no change in risk characteristics, the parent's after-tax cost of debt and equity remain at $i_d(1 - t)$ and k_e, respectively. As introduced above, the subsidiary's cost of retained earnings equals k_s and its expected after-tax dollar cost of foreign debt equals i_f.

Under these circumstances the weighted cost of capital for the project equals

$$k_I = k_0 - a(k_e - k_s) - b[i_d(1 - t) - i_f] \qquad (18.3)$$

where $a = E_f/I$ and $b = D_f/I$.[1] If this investment changes the parent's risk characteristics in such a way that its cost of equity capital is k'_e, rather than k_e, and its cost of debt i'_d, rather than i_d, Equation 18.3 becomes instead

$$k_I = k_0 + (1 - L)(k'_e - k_e) + L(i'_d - i_d)(1 - t) - a(k'_e - k_s) - b[i'_d(1 - t) - i_f] \quad \text{(18.4)}$$

ILLUSTRATION

ESTIMATING A FOREIGN PROJECT'S WEIGHTED AVERAGE COST OF CAPITAL

Suppose that a new foreign investment requires $100 million in funds. Of this total, $20 million will be provided by parent company funds, $25 million by retained earnings in the subsidiary, and $55 million through the issue of new debt by the subsidiary. The parent's cost of equity equals 14%, and its after-tax cost of debt is 5%. If the firm's current debt ratio, which is considered to be optimal, is 0.3, then k_0 equals 11.3% ($0.14 \times 0.7 + 0.05 \times 0.3$). However, this project has higher systematic risk than the typical investment undertaken by the firm, thereby requiring a rate of return of 16% on new parent equity and 6% on new parent debt. Based

on an incremental tax of 8% on repatriated earnings, the cost of retained earnings is estimated to be 14.7% [$0.16 \times (1 - 0.08)$]. Let the nominal LC rate of interest be 20%, with an anticipated average annual devaluation of 7%. Then with a foreign tax rate of 40%, the expected after-tax dollar cost of the LC debt is 4.2% [$0.20 \times (1 - 0.4)(1 - 0.07) - 0.07$].

Applying Equation 18.4, the project's weighted average cost of capital is

$$\begin{aligned} k_I = {}& 0.113 + 0.7(0.16 - 0.14) + 0.3(0.06 - 0.05) \\ & - [25/100(0.16 - 0.147) \\ & \quad - 55/100(0.06 - 0.042)] = 0.117 \end{aligned}$$

The parent's weighted average cost of capital for this project would have been 13% ($0.16 \times 0.7 + 0.06 \times 0.3$) in the absence of the retained earnings and foreign debt financing.

18.3 THE ALL-EQUITY COST OF CAPITAL FOR FOREIGN PROJECTS

The various adjustments needed to go from the weighted average cost of capital for the firm to the weighted average cost of capital for the project makes it a somewhat awkward technique to use at times. An alternative is the use of an **all-equity discount rate**, $k*$, that abstracts from the project's financial structure and that is based solely on the riskiness of the project's anticipated cash flows. In other words, the all-equity cost of capital equals the company's cost of capital if it were all-equity financed, that is, with no debt.

To calculate the all-equity rate, we rely on the CAPM introduced earlier in Equation 18.1:

$$k* = r_f + \beta* (r_m - r_f) \quad \text{(18.5)}$$

---

[1] See Alan C. Shapiro, "Financial Structure and the Cost of Capital in the Multinational Corporation," *Journal of Financial and Quantitative Analysis*, June 1978, pp. 211–226.

where β^* is the **all-equity beta**—that is, the beta associated with the unleveraged cash flows.

ILLUSTRATION

ESTIMATING A FOREIGN PROJECT'S COST OF CAPITAL

Suppose that a foreign project has an all-equity beta of 1.15, the risk-free return is 7%, and the required return on the market is estimated at 15%. Then based on Equation 18.5, the project's cost of capital is

$$k^* = 0.07 + 1.15(0.15 - 0.07)$$
$$= 16.2\%$$

In reality, of course, the firm will not be able to estimate β^* with the degree of precision implied here. Instead, it will have to use guesswork based on theory. The considerations involved in the estimation process are discussed in the following section.

If the project is of similar risk to the average project selected by the firm, it is possible to estimate β^* by reference to the firm's stock price beta, β_e. In other words, β_e is the beta that appears in the estimate of the firm's cost of equity capital, k_e, given its current capital structure.

To transform β_e into β^*, we must separate out the effects of debt financing. This operation is known as *unlevering*, or converting a levered equity beta to its unlevered or all-equity value. Unlevering can be accomplished by using the following approximation:

$$\beta^* = \frac{\beta_e}{1 + (1 - t)D/E} \qquad (18.6)$$

where t is the firm's marginal tax rate, and D/E is its current debt-to-equity ratio. Thus, for example, if a firm has a stock price beta of 1.1, a debt/equity ratio of 0.6, and a marginal tax rate of 35%, Equation 18.6 estimates its all-equity beta as 0.79 [1.1/(1 + 0.65 × 0.6)].

18.4 DISCOUNT RATES FOR FOREIGN INVESTMENTS

The importance of the CAPM for the international firm is that the relevant component of risk in pricing a firm's stock is its systematic risk—that is, that portion of return variability that cannot be eliminated through diversification. Evidence suggests that most of the economic and political risk faced by MNCs is unsystematic risk, which therefore can be eliminated through diversification on the level of the individual investor. Although these risks may be quite large, they should not affect the discount rate to be used in valuing foreign projects.

On the other hand, much of the systematic or general market risk affecting a company, at least as measured using a domestic stock index such as the Standard and Poor's 500, is related to the cyclical nature of the national economy in which the company is domiciled. Consequently, the returns on a project located in a foreign country whose economy is not perfectly synchronous with the home country's economy should be less highly correlated with domestic market returns than the returns on a comparable domestic project. If this is the case, then the systematic risk of a foreign project actually could be lower than the systematic risk of its domestic counterpart.

Paradoxically, it is the less developed countries (LDCs), where political risks are greatest, that are likely to provide the largest diversification benefits. This is because the economies of LDCs are less closely tied to that of the United States, or of any other Western economy. By contrast, the correlation among the economic cycles of developed countries is considerably stronger, so the diversification benefits from investing in industrialized countries, from the standpoint of a Western investor are proportionately less.

It should be noted, however, that the systematic risk of projects even in relatively isolated LDCs is unlikely to be far below the average for all projects because these countries are still tied into the world economy. The important point about projects in LDCs, then, is that their ratio of systematic to total risk generally is quite low; their systematic risk, although perhaps slightly lower, is probably not significantly less than that of similar projects located in industrialized countries.

Even if a nation's economy is not closely linked to the world economy, the systematic risk of a project located in that country might still be rather large. For example, a foreign copper-mining venture probably will face systematic risk very similar to that faced by an identical extractive project in the United States, whether the foreign project is located in Canada, Chile, or Zaire. The reason is that the major element of systematic risk in any extractive project is related to variations in the price of the mineral being extracted, which is set in a world market. The world market price, in turn, depends on worldwide demand, which itself is systematically related to the state of the world economy. By contrast, a market-oriented project in an LDC, whose risk depends largely on the evolution of the domestic market in that country, is likely to have a systematic risk that is small in both relative and absolute terms.

An example of the latter would be a Ford plant in Brazil whose profitability is closely linked to the state of the Brazilian economy. The systematic risk of the project, therefore, largely depends on the correlation between the Brazilian economy and the U.S. economy. Although positive, this correlation is much less than 1.

Thus, *corporate international diversification* should prove beneficial to shareholders, particularly where there are barriers to *international portfolio diversification*. To the extent that multinational firms are uniquely able to supply low-cost international diversification, investors may be willing to accept a lower rate of return on shares of MNCs than on shares of single-country firms. By extension, the risk premium applied to foreign projects may be lower than the risk premium for domestic ones; that is, the required return on foreign projects may be less than the required return on comparable domestic projects. The net effect may be to enable MNCs to undertake overseas projects that would otherwise be unattractive.

However, if international portfolio diversification can be accomplished as easily and as cheaply by individual investors, then, although required rates of return on MNC securities would be lower to reflect the reduced covariability of MNC returns caused by international diversification, the discount rate would not be reduced further to reflect investors' willingness to pay a premium for the indirect diversification provided by the shares of MNCs. In fact, though, American investors actually undertake very little foreign portfolio investment. The lack of widespread international portfolio diversification has an important implication for estimating the beta coefficient.

Key Issues in Estimating Foreign Project Betas

Although the CAPM is the model of choice for estimating the cost of capital for foreign projects, the type of information that is needed to estimate foreign subsidiary betas directly—a history of past subsidiary returns or future subsidiary returns relative to predicted market returns—does not exist. About the only practical way to get around this problem is to find publicly traded firms that share similar risk characteristics and use the average beta for the portfolio of corporate surrogates to proxy for the subsidiary's beta. This approach, however, introduces three additional questions for a U.S. multinational:

1. *Should the corporate proxies be U.S. or local (i.e., foreign) companies?* Although local companies should provide a better indication of risk, such companies may not exist. By contrast, selecting U.S. proxies ensures that such proxies and their data exist, but their circumstances—and hence their betas—may be quite different than those facing the foreign subsidiaries. In addition, it is important to differentiate between the unsystematic risks faced by a foreign project—which individual investors can eliminate through diversification—and the systematic risks affecting that project, which may be small relative to the project's total risk.

2. *Is the relevant base portfolio against which the proxy betas are estimated the U.S. market portfolio, the local portfolio, or the world market portfolio?* Selecting the appropriate portfolio matters because a risk that is systematic in the context of the local market portfolio may well be diversifiable in the context of the U.S. or world portfolio. If this is the case, using the local market portfolio to calculate beta will result in a higher required return—and a less desirable project—than if beta were calculated using the U.S. or world market portfolio.

3. *Should the market risk premium be based on the U.S. market or the local market?* One argument in favor of using the local-market risk premium is that this is the risk premium demanded by investors on investments in that market. On the other hand, estimates of the local-market risk premium may be subject to a good deal of statistical error. Moreover, such estimates may be irrelevant to the extent that an MNC's investors are not the same as the investors in the local market and the two sets of investors measure risk differently.

Let us now address those three questions and their related issues. As in any application of a theoretical model, the suggested answers are not precisely right but rather are based on a mix of theory, empirical evidence, and judgment.

PROXY COMPANIES ■ Three alternatives for estimating proxy betas are proposed here. These alternatives are presented in order of their desirability. Other approaches are also mentioned.[2]

---

[2] This section is based on a discussion with Rene Stulz.

Local Companies.   As much as possible, the corporate proxies should be local companies. The returns on an MNC's local operations are likely to depend in large measure on the evolution of the local economy. Inevitably, therefore, the timing and magnitude of these returns will differ from those of the returns generated by comparable U.S. companies. This means that the degree of systematic risk for a foreign project, at least as measured from the perspective of an American investor, may well be lower than the systematic risk of comparable U.S. companies. Put differently, using U.S. companies and their returns to proxy for the returns of a foreign project will likely lead to an upward-biased estimate of the risk premium demanded by the MNC's investors.

Some indication of the upward bias in the estimate of beta imparted by using U.S. proxy companies to estimate the betas for foreign projects is provided by presenting the foreign market betas relative to the U.S. index for some foreign countries. The betas for the foreign markets from a U.S. perspective are calculated in the same way that individual asset betas are calculated:

$$\text{Foreign market beta} = \frac{\genfrac{}{}{0pt}{}{\text{Correlation with}}{\text{U.S. market}} \times \genfrac{}{}{0pt}{}{\text{Standard deviation}}{\text{of foreign market}}}{\text{Standard deviation of U.S. market}}$$

According to Equation 18.7, in conjunction with data from the 27-year period 1970–1996, the beta for the Australian market relative to the U.S. market was 0.80 (0.46 × 0.8646/0.4952). The corresponding betas for Hong Kong and Singapore were 0.67 and 0.78, respectively:

COUNTRY	CORRELATION WITH U.S. MARKET	STANDARD DEVIATION OF RETURNS (%)	BETA FROM U.S. PERSPECTIVE
Australia	0.46	86.46	0.80
Hong Kong	0.30	110.87	0.67
Singapore	0.45	85.29	0.78
United States	1.00	49.52	1.00

Of course, it may be that some U.S. companies operating overseas would have betas in the foreign markets in excess of 1.0, thereby raising their betas relative to the estimated foreign market betas. Nonetheless, this evidence does suggest the possibility that the average beta of U.S. proxy companies overstates the betas for foreign subsidiaries from a U.S. perspective.

Notice also that despite investment risks associated with the Hong Kong and Singapore markets (standard deviations of 110.87% and 85.29%, respectively) of about twice that of the U.S. market (a standard deviation of 49.52%), both markets had betas that were substantially lower than the U.S. market beta of 1.0. The reason, of course, is that much of the risk associated with markets in individual countries is unsystematic and so can be eliminated by diversification, as indicated by the relatively low betas of these markets.

Proxy Industry.   If foreign proxies are not directly available, a second alternative is to find a proxy industry in the local market, that is, one whose U.S. industry beta is similar to that of the project's U.S. industry beta. One way to analyze the empirical validity of this approach is to check whether the betas of the two industries (the project's and the proxy's) are also similar in other national markets that contain both industries (e.g., Britain, Germany, and Japan).

Adjusted U.S. Industry Beta.   The third—and least preferred—alternative is to estimate the foreign project's beta by computing the U.S. industry beta for the project, $\beta_{USPROXY}$, and multiplying it by the *unlevered* foreign market beta relative to the U.S. index. Specifically, suppose that β_{AUS} is the unlevered beta for the Australian market relative to the U.S. market. Then, under this proposed methodology, the unlevered beta for the Australian project, β_{AUSSUB}, would be estimated as

$$\beta_{AUSSUB} = \beta_{USPROXY} \times \beta_{AUS} \qquad (18.8)$$

The reason this approach is the least preferred of the three alternatives is that implicit in it are two questionable assumptions:

1. *The beta for an industry in the United States will have the same beta in each foreign market.* That is a heroic assumption considering that national markets have different industries and different weightings of industries in their indices.

2. *The only correlation with the U.S. market of a foreign company in the project's industry comes through its correlation with the local market and the local market's correlation with the U.S. market.* However, it is conceivable that, say, an oil firm could have a low correlation with the local market but a high correlation with the U.S. market.

That being said, to the extent that returns for a foreign project depend largely on the evolution of the local economy it operates in, then these two assumptions are likely to be satisfied. In that case, this approach would be an appropriate compromise.

Other Approaches.   In a country such as China, which has no historical stock market data, there are no proxy companies or industries that can be used to estimate a proxy beta. At the same time, the use of a U.S. proxy beta is likely to be inappropriate because the degree of systematic risk should be quite small. A proxy beta for the local subsidiary can be constructed by first estimating a proxy beta for the local market relative to the U.S. market and then assuming that the local industry beta will be similar to that in other countries at a similar stage of development. One way to generate market betas for foreign countries is to calculate the betas of closed-end country funds relative to the U.S. market. For example, there are several closed-end China funds traded on the New York Stock Exchange. One advantage of this approach is that it provides a direct measure of how U.S. investors perceive the risk of investments in that country. This approach to calculating a foreign market beta relative to the U.S. market would be appropriate even if a local stock market exists. Of course, it would be necessary to delever the fund's beta to estimate the unlevered equity beta for that country's market.

Other means of calculating proxy market betas are to study GNP correlations or to use as a proxy for the China market beta the market beta of a nation such as India that is at a similar stage of development and has a functioning stock market. The advantage of this method is that data are more current and are based on actual investor perceptions. However, India may be a poor proxy for China.

Once a proxy country beta is calculated and unlevered, the MNC then can apply the same approach previously recommended. Specifically, the MNC would multiply this figure, call it $\beta_{CHINAPROXY}$ by its U.S. industry beta, $\beta_{USPROXY}$, and the resulting number would serve as the proxy beta for the MNC's China project.

Although these approaches to estimating foreign subsidiary betas all involve a variety of assumptions, these assumptions appear to be no less plausible than the assumption that foreign operations are inherently riskier than comparable domestic operations and should be assessed an added risk premium. This is the consequence, unintended though it may be, of adding a sovereign risk premium to an estimated U.S. cost-of-capital figure translated into local currency.

THE RELEVANT BASE PORTFOLIO ■ The appropriate market portfolio to use in measuring a foreign subsidiary's beta depends on one's view of world capital markets. More precisely, it depends on whether or not capital markets are globally integrated. If they are, then the world portfolio is the correct choice; if they are not, the correct choice is the domestic portfolio. The test of capital market integration depends on whether these assets are priced in a common context; that is, capital markets are integrated to the extent that security prices offer all investors worldwide the same trade-off between systematic risk and real expected return.

The truth probably lies somewhere in between. Capital markets now are integrated to a great extent, and they can be expected to become ever more so with time. However, because of various government regulations and other market imperfections, that integration is not complete. Unfortunately, it is not currently within our power, if indeed it ever will be, to empirically determine the relevant market portfolio and, hence, the correct beta to use in project evaluation. (The problem of determining the appropriate market portfolio to use in estimating beta arises domestically as well as internationally.)

A pragmatic recommendation is to measure the betas of international operations against the U.S. market portfolio. This recommendation is based on the following two reasons:

1. It ensures comparability of foreign with domestic investments, which are evaluated using betas that are calculated relative to a U.S. market index.
2. The relatively minor amount of international diversification attempted (as yet) by American investors suggests that the relevant portfolio from their standpoint is the U.S. market portfolio.

This reasoning suggests that the required return on a foreign project may well be lower, and is unlikely to be higher, than the required return on a comparable domestic project. Thus, applying the same discount rate to an overseas project as to a similar domestic project probably will yield a conservative estimate of the relative systematic riskiness of the project.

Using the domestic cost of capital to evaluate overseas investments also is likely to understate the benefits that stem from the ability of foreign activities to reduce the firm's total risk. As we saw in Chapter 1, reducing total risk can increase a firm's cash flows. By confining itself to its domestic market, a firm will be sensitive to periodic downturns associated with the domestic business cycle and other industry-specific factors. By operating in a number of countries, the MNC can trade off negative swings in some countries against positive ones in others. This option is especially valuable for non-U.S. firms whose local markets are small relative to the efficient scale of operation.

Evidence from the Stock Market.   The most careful study to date of the effects of foreign operations on the cost of equity capital is by Ali Fatemi.[3] That study compared the performance of two carefully constructed stock portfolios: a portfolio of 84 MNCs, each with at least 25% of its annual sales generated from international operations; and a portfolio of 52 purely domestic firms. Monthly performance comparisons were made over the five-year period January 1976–December 1980.

Although the validity of the study is limited by the relatively short time period involved, the difficulty in properly matching MNCs with their purely domestic counterparts (most firms do business in more than one industry), and the difficulty in calculating the degree of sales from abroad (consider the transfer pricing problem, for example), its conclusions are nonetheless of interest.

1. The rates of return on the two portfolios are statistically identical. Ignoring risk, MNCs and uninational (purely domestic) corporations (UNCs) provide shareholders the same returns.
2. Consistent with our expectations, the rates of return on the MNC portfolio fluctuate less than those on the UNC portfolio. Thus, corporate international diversification seems to reduce shareholder total risk and may do the same for the firm's total risk.
3. The betas of the multinational portfolio are significantly lower and more stable than are those of the purely domestic portfolio, indicating that corporate international diversification reduces the degree of systematic risk, at least if systematic risk is calculated relative to the domestic portfolio. It was also found that the higher the degree of international involvement, the lower the beta.

Despite the apparent benefits of corporate international diversification for shareholders, however, research by Bertrand Jacquillat and Bruno Solnik concluded that, although multinational firms do provide some diversification for investors, they are poor substitutes for international portfolio diversification.[4] Their results indicate that an internationally diversified portfolio leads to a much greater reduction in variance than does one comprising firms with internationally diversified activities. Thus, the advantages of international portfolio diversification remain.

THE RELEVANT MARKET RISK PREMIUM ■ In line with the basic premise that multinationals should use a methodology that is as consistent as possible with the methodology used to calculate the cost of capital for U.S. investments, the recommended market risk premium to be used is the U.S. market risk premium. There are several reasons for believing that this is the appropriate market risk premium to use. First, the U.S. market risk premium is the one likely to be demanded by a U.S. company's mostly American

---

[3] Ali M. Fatemi, "Shareholder Benefits from Corporate International Diversification," *Journal of Finance*, December 1984, pp. 1325–1344.

[4] Bertrand Jacquillat and Bruno H. Solnik, "Multinationals Are Poor Tools for Diversification," *Journal of Portfolio Management*, Winter 1978, pp. 8–12.

investors. A second reason for preferring the U.S. market risk premium is the earlier recommendation that the betas of foreign subsidiaries be estimated relative to the U.S. market. Using the U.S. market risk premium will ensure consistency between the measure of systematic risk and price per unit of this systematic risk. Finally, the quality, quantity, and length of U.S. capital market data are by far the best in the world, increasing the statistical validity of the estimated market risk premium.

Conversely, no other country has a stock market data series of the same length and quality as that of the United States. In addition, virtually all foreign countries have undergone dramatic changes in their economic and political regimes since the end of World War II—changes that inevitably will affect the required risk premium for those markets. To the extent that such regime changes have altered the market risk premium in foreign countries, estimates of these risk premiums based on historical data are less useful as forecasts of required risk premiums going forward.

The bottom line is that U.S. capital markets have the best data available on the required return that investors demand per unit of risk. Moreover, as national capital markets become increasingly integrated globally, the market price of risk becomes the same worldwide. Add to these points the fact that shareholders of U.S. firms are mostly American and a strong case can be made that the U.S. market risk premium is the appropriate price of risk for a foreign project.

Recommendations

In summary, the recommended approach to estimating the cost of equity capital for a foreign subsidiary is to find a proxy portfolio in the country in which that subsidiary operates and calculate its beta relative to the U.S. market. That beta should then be multiplied by the risk premium for the U.S. market. This estimated equity risk premium for the foreign subsidiary would then be added to the U.S. (home country) nominal risk-free rate to compute a dollar (home currency) cost of equity capital.

An alternative, but problematic approach used by many investment bankers these days is to estimate a sovereign risk premium for the foreign country (by taking the difference between the interest rate on U.S. dollar-denominated debt issued by the foreign government and the rate on U.S. government debt of the same maturity) and adding that figure to the estimated U.S. cost of capital. In particular, to the extent that the estimated sovereign risk premium measures risk (it may measure a liquidity premium), it is not systematic risk but rather default risk that is being measured. And default risk does not enter into the cost of equity capital. Of course, default risk is likely to be closely linked to political risk, but adjusting the cost of capital is not necessarily the best way to factor political risk into a foreign investment analysis. As recommended in Chapter 21, a better approach for dealing with political risk is to first identify its likely cash flow consequences and then adjust projected cash flows to incorporate those consequences.

18.5 ESTABLISHING A WORLDWIDE CAPITAL STRUCTURE

In estimating the weighted average cost of capital for an MNC or its affiliates, we took the capital structure as given. However, the capital structure itself should be the outcome of an optimal global financial plan. This plan requires consideration not only of the component costs of capital, but also of how the use of one source affects the cost and availability of other sources. A firm that uses too much debt might find the cost of equity (and new debt) financing prohibitive. The capital structure problem for the multinational enterprise, therefore, is to determine the mix of debt and equity for the parent entity and for all consolidated and unconsolidated subsidiaries that maximizes shareholder wealth.

The focus is on the consolidated, *worldwide capital structure* because suppliers of capital to a multinational firm are assumed to associate the risk of default with the MNC's worldwide debt ratio. This association stems from the view that bankruptcy or other forms of financial distress in an overseas subsidiary can seriously impair the parent company's ability to operate domestically. Any deviations from the MNC's target capital structure will cause adjustments in the mix of debt and equity used to finance future investments.

Another factor that may be relevant in establishing a worldwide debt ratio is the empirical evidence that earnings variability appears to be a decreasing function of foreign-source earnings. Because the risk of bankruptcy for a firm is dependent on its total earnings variability, the earnings diversification provided by its foreign operations may enable the multinational firm to leverage itself more highly than can a purely domestic corporation, without increasing its default risk.

Foreign Subsidiary Capital Structure

After a decision has been made regarding the appropriate mix of debt and equity for the entire corporation, questions about individual operations can be raised. How should MNCs arrange the capital structures of their foreign affiliates? And what factors are relevant in making this decision? Specifically, the problem is whether foreign subsidiary capital structures *should*

- Conform to the capital structure of the parent company
- Reflect the capitalization norms in each foreign country
- Vary to take advantage of opportunities to minimize the MNC's cost of capital

Disregarding public and government relations and legal requirements for the moment, the parent company could finance its foreign affiliates by raising funds in its own country and investing these funds as equity. The overseas operations would then have a zero debt ratio (debt/total assets). Alternatively, the parent could hold only one dollar of share capital in each affiliate and require all to borrow on their own, with or without guarantees; in this case, affiliate debt ratios would approach 100%. Or the parent could itself borrow and relend the monies as intracorporate advances. Here again, the affiliates' debt ratios would be close to 100%. In all these cases, the total amount of borrowing and the debt/equity mix of the consolidated corporation are identical. Thus, the question of an optimal capital structure for a foreign affiliate is completely distinct from the corporation's overall debt/equity ratio.

Moreover, any accounting rendition of a separate capital structure for the subsidiary is wholly illusory *unless* the parent is willing to allow its affiliate to default on its debt.[5]

---

[5] See, for example, Michael Adler, "The Cost of Capital and Valuation of a Two-Country Firm," *Journal of Finance*, March 1974, pp. 119–132; and Alan C. Shapiro, "Financial Structure and Cost of Capital in the Multinational Corporation," *Journal of Financial and Quantitative Analysis*, June 1978, pp. 211–226.

EXHIBIT 18.1 Subsidiary Capital Structure: Debt-to-Equity Ratios

I. 100% PARENT FINANCED		II. 100% PARENT FINANCED	
$100	D = $50 E = 50	$100	D = $100 E = 0
	D/E = 1 : 1		D/E = Infinity

III. 100% PARENT FINANCED		IV. 100% BANK FINANCED	
$100	D = $ 0 E = 100	$100	D = $100 E = 0
	D/E = 0		D/E = Infinity

As long as the rest of the MNC group has a legal or moral obligation or sound business reasons for preventing the affiliate from defaulting, the individual unit has no independent capital structure. Rather, its true debt/equity ratio is equal to that of the consolidated group. Exhibits 18.1 and 18.2 show the stated and the true debt-to-equity ratios for a subsidiary and its parent for four separate cases. In cases I, II, and III, the parent borrows $100 to invest in a foreign subsidiary, in varying portions of debt and equity. In case IV, the subsidiary borrows the $100 directly from the bank. Depending on what the parent calls its investment, the subsidiary's debt-to-equity ratio can vary from zero to infinity. Despite this variation, the consolidated balance sheet shows a debt-to-equity ratio after the foreign investment of 4 : 7, regardless of how the investment is financed and what it is called.

Exhibit 18.3 shows that the financing mechanism does affect the pattern of returns, whether they are called dividends or interest and principal payments. It also determines the initial recipient of the cash flows. Are the cash flows from the foreign unit paid directly to the outside investor (the bank) or are they first paid to the parent, which then turns around and repays the bank?

The point of this exercise is to show that unlike the case for the corporation as a whole, an affiliate's degree of leverage does not determine its financial risk. Therefore,

EXHIBIT 18.2 Consolidated Parent Balance Sheet: Debt-to-Equity Ratios

BEFORE FOREIGN INVESTMENT

$1000	D = $300 E = 700
	D/E = 3 : 7

AFTER FOREIGN INVESTMENT

Case I, II, and III Parent Financed with 100% Bank Debt			Case IV Subsidiary Financed with 100% Bank Debt		
Domestic Foreign	$1,000 100	D = $400 E = 700	Domestic Foreign	$1,000 100	D = $400 E = 700
	D/E = 4 : 7			D/E = 4 : 7	

EXHIBIT 18.3 Subsidiary Capital Structure Depends on What Its Funds Are Called

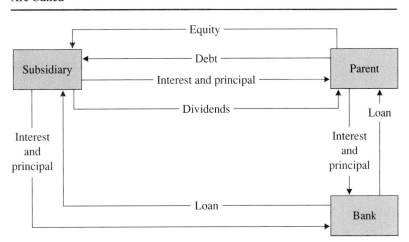

the first two options—having affiliate financial structures conform to parent or local norms—are unrelated to shareholder wealth maximization.

The irrelevance of subsidiary financial structures seems to be recognized by multinationals. In a 1979 survey by Business International of eight U.S.-based MNCs, most of the firms expressed little concern with the debt/equity mixes of their foreign affiliates.[6] (Admittedly, for most of the firms interviewed, the debt ratios of affiliates had not significantly raised the MNCs' consolidated indebtedness.) Their primary focus was on the worldwide, rather than individual, capital structure. The third option of varying affiliate financial structures to take advantage of local financing opportunities appears to be the appropriate choice. Thus, within the constraints set by foreign statutory or minimum equity requirements, the need to appear to be a responsible and good guest, and the requirements of a worldwide financial structure, a multinational corporation should finance its affiliates to minimize its incremental average cost of capital.

A subsidiary with a capital structure similar to its parent may forgo profitable opportunities to lower its cost of funds or its risk. For example, rigid adherence to a fixed debt/equity ratio may not allow a subsidiary to take advantage of government-subsidized debt or low-cost loans from international agencies. Furthermore, it may be worthwhile to raise funds locally if the country is politically risky. In the event the affiliate is expropriated, for instance, it would default on all loans from local financial institutions. Similarly, borrowing funds locally will decrease the company's vulnerability to exchange controls. Local currency (LC) profits can be used to service its LC debt.

Borrowing in the local currency also can help a company reduce its foreign exchange exposure. As we saw in Chapter 13, if financing opportunities in various currencies are fairly priced, firms can structure their liabilities to reduce their exposure to

[6]"Policies of MNCs on Debt/Equity Mix," *Business International Money Report*, September 21, 1979, pp. 319–320.

foreign exchange risk at no added cost to shareholders. The basic rule is to finance assets that generate foreign-currency cash flows with liabilities denominated in those same foreign currencies.

On the other hand, forcing a subsidiary to borrow funds locally to meet parent norms may be quite expensive in a country with a high-cost capital market or if the subsidiary is in a tax-loss-carryforward position. In the latter case, because the subsidiary cannot realize the tax benefits of the interest write-off, the parent should make an equity injection financed by borrowed funds. In this way, the interest deduction need not be sacrificed.

LEVERAGE AND THE TAX REFORM ACT OF 1986 ■ The choice of where to borrow to finance foreign operations has become more complicated with passage of the Tax Reform Act of 1986 because the distribution of debt between U.S. parents and their foreign subsidiaries affects the use of foreign tax credits. Moreover, the Tax Reform Act has put many U.S.-based MNCs in a position of excess foreign tax credits. One way to use up these FTCs is to push expenses overseas—and thus lower overseas profits—by increasing the leverage of foreign subsidiaries. In the aforementioned example, the U.S. parent may have one of its taxpaying foreign units borrow funds and use them to pay a dividend to the parent. The parent can then turn around and invest these funds as equity in the non-taxpaying subsidiary. In this way, the worldwide corporation can reduce its taxes without being subject to the constraints imposed by the Tax Reform Act.

LEASING AND THE TAX REFORM ACT OF 1986 ■ As an alternative to increasing the debt of foreign subsidiaries, U.S. multinationals could expand their use of leasing in the United States. Although leasing an asset is economically equivalent to using borrowed funds to purchase the asset, the international tax consequences differ. Before 1986, U.S. multinationals counted virtually all their interest expense as a fully deductible U.S. expense. Under the new law, firms must allocate interest expense on general borrowings to match the location of their assets, even if all the interest is paid in the United States. This allocation has the effect of reducing the amount of interest expense that can be written off against U.S. income. Rental expense, on the other hand, can be allocated to the location of the leased property. Lease payments on equipment located in the United States, therefore, can be fully deducted.

At the same time, leasing equipment to be used in the United States, instead of borrowing to finance it, increases reported foreign income (because there is less interest expense to allocate against foreign income). The effect of leasing, therefore, is to increase the allowable foreign tax credit to offset U.S. taxes owed on foreign source income, thereby providing another tax advantage of leasing for firms that owe U.S. tax on their foreign source income.

COST MINIMIZING APPROACH TO GLOBAL CAPITAL STRUCTURE ■ The cost-minimizing approach to determining foreign-affiliate capital structures would be to allow subsidiaries with access to low-cost capital markets to exceed the parent-company capitalization norm, while subsidiaries in higher-capital-cost nations would have lower target debt ratios. These costs must be figured on an after-tax basis, taking into account the company's worldwide tax position.

The basic hypothesis proposed in this section is that a subsidiary's capital structure is relevant only insofar as it affects the parent's consolidated worldwide debt ratio. Nonetheless, some companies have a general policy of "every tub on its own bottom." Foreign units are expected to be financially independent after the parent's initial investment. The rationale for this policy is to "avoid giving management a crutch." By forcing foreign affiliates to stand on their own feet, affiliate managers presumably will be working harder to improve local operations, thereby generating the internal cash flow that will help replace parent financing. Moreover, the local financial institutions will have a greater incentive to monitor the local subsidiary's performance because they can no longer look to the parent company to bail them out if their loans go sour.

However, companies that expect their subsidiaries to borrow locally had better be prepared to provide enough initial equity capital or subordinated loans. In addition, local suppliers and customers are likely to shy away from a new subsidiary operating on a shoestring if that subsidiary is not receiving financial backing from its parent. The foreign subsidiary may have to show its balance sheet to local trade creditors, distributors, and other stakeholders. Having a balance sheet that shows more equity demonstrates that the unit has greater staying power.

It also takes more staff time to manage a highly leveraged subsidiary in countries like Brazil and Mexico, where government controls and high inflation make local funds scarce. One treasury manager complained, "We spend 75–80% of management's time trying to figure out how to finance the company. Running around chasing our tails instead of attending to our basic business—getting production costs lower, sales up, and making the product better."[7]

Joint Ventures

Because many MNCs participate in joint ventures, either by choice or necessity, establishing an appropriate financing mix for this form of investment is an important consideration. The previous assumption that affiliate debt is equivalent to parent debt in terms of its impact on perceived default risk may no longer be valid. In countries such as Japan and Germany, increased leverage will not necessarily lead to increased financial risks, due to the close relationship between the local banks and corporations. Thus, debt raised by a joint venture in Japan, for example, may not be equivalent to parent-raised debt in terms of its impact on default risk. The assessment of the effects of leverage in a joint venture requires a qualitative analysis of the partner's ties with the local financial community, particularly with the local banks.

Unless the joint venture can be isolated from its partners' operations, there are likely to be some significant conflicts associated with this form of ownership. Transfer pricing, setting royalty and licensing fees, and allocating production and markets among plants are just some of the areas in which each owner has an incentive to engage in activities that will harm its partners. These conflicts explain why bringing in outside equity investors is generally such an unstable form of external financing.

Because of their lack of complete control over a joint venture's decisions and its profits, most MNCs will, at most, guarantee joint-venture loans in proportion to their

---

[7] "Determining Overseas Debt/Equity Ratios," *Business International Money Report*, January 27, 1986, p. 26.

share of ownership. But where the MNC is substantially stronger financially than its partner, the MNC may wind up implicitly guaranteeing its weaker partner's share of any joint-venture borrowings, as well as its own. In this case, it makes sense to push for as large an equity base as possible; the weaker partner's share of the borrowings is then supported by its larger equity investment.

18.6 ILLUSTRATION: NESTLÉ

Nestlé, the $17 billion Swiss foods conglomerate, is about as multinational as a company can be. About 98% of its sales take place overseas, and the group's diversified operations span 150 countries. Nestlé's numerous (and generally wholly owned) subsidiaries are operationally decentralized. However, finances are centralized in Vevey, Switzerland. Staffed by only 12 people, the finance department makes all subsidiary funding decisions, manages the resulting currency exposures, determines subsidiary dividend amounts, sets the worldwide debt/equity structure, and evaluates subsidiary performance.

Nestlé's centralized finance function plays the pivotal role in the firm's intricate web of subsidiary-to-headquarters profit remittances and headquarters-to-subsidiary investment flows. Profits and excess cash are collected by the treasury department in Vevey and then channeled back to overseas subsidiaries in the form of equity and debt investments. Nestlé considers this approach to be the best possible investment for the group's wealth.

When a subsidiary is first established, its fixed assets—which form about half of the total investment—are financed by the Nestlé group, generally with equity. Later on, the group may supply long-term debt as needed to support operations. The local subsidiary manager handles all the marketing and production decisions, but decisions regarding long-term debt and equity funding are managed solely by Vevey headquarters.

The other half of the investment—working capital—is then acquired locally, usually via bank credit or commercial paper. However, Nestlé varies this general approach to suit each country. In certain countries—those that permit free transfers of funds—Nestlé finances part of the working capital from Vevey instead of using local bank credits.

Central control over affiliate capital structures is facilitated by the policy of forcing local managers to dividend out almost 100% of their profits to Switzerland. The particular capital structure chosen for an affiliate depends on various considerations, including taxes, political risk, and currency risk.

To ensure that it borrows at the lowest possible cost, Nestlé takes considerable care to structure its capital base to keep a top credit rating. The desire for a low-risk capital structure is also consistent with Nestlé's business strategy. According to Senior Vice President, Finance, Daniel Regolatti, "Our basic strategy is that we are an industrial company. We have a lot of risks in a lot of countries, so we should not add high financial risks."[8]

---

[8] "The Nestlé Approach to Capital Markets and Innovation," *Business International Money Report*, October 27, 1986, p. 337.

18.7 ILLUSTRATION: INTERNATIONALIZING THE COST OF CAPITAL OF NOVO INDUSTRI

Capital market segmentation implies that the same firm raising debt or equity funds in different national capital markets may face a different cost of capital as a result of diverging investor perceptions between domestic and foreign shareholders or of asymmetry in tax policies, exchange controls, and political risks. Indeed, a firm based in a fully segmented capital market is likely to have a higher cost of capital due to a relatively depressed price for its stock than if it had access to fully integrated capital markets. A good illustration of how a company can overcome such segmentation barriers in order to effectively reduce its cost of capital is provided by Novo Industri, the Danish multinational firm that is a recognized industry leader in the manufacturing of industrial enzymes and pharmaceuticals (mostly insulin) in Western Europe.[9]

Novo perceived it had a high cost of capital relative to its foreign competitors for several reasons. First, Danish investors were prohibited from investing in foreign stocks. Because Danish stock price movements are closely correlated with each other, Danish investors bore a great deal of systematic risk, raising their required returns. Second, Denmark taxed capital gains on stocks at prohibitive rates, reducing stock turnover and hence liquidity. These effects combined to greatly increase the pre-tax return required by Danish investors. Finally, foreign investors were quite unfamiliar with the Danish market, thereby reducing their incentive to arbitrage away the high returns available on Danish stocks.

In 1977, Novo embarked on an ambitious strategy aimed at internationalizing its cost of capital in order to be in a position to better compete with its major multinational rivals such as Eli Lilly (United States), Miles Laboratory (United States-based but a subsidiary of the giant chemical conglomerate Bayer, headquartered in Germany), and Gist Brocades (the Netherlands). The first step was for Novo to float a $20 million convertible Eurobond issue (1978). In connection with this offering, it listed its shares on the London Stock Exchange (1979) to facilitate conversion and to gain visibility among foreign investors. Next, Novo decided to capitalize on the emerging interest among U.S. investors for biotechnology companies. It ran a seminar in New York (1980) and then sponsored an American Depository Receipts system and listed its shares on the U.S. over-the-counter market (NASDAQ, in 1981).

Having gained significant visibility among both the London and New York investment communities, Novo was ready to take the final and most difficult step—floating an equity issue on the New York Stock Exchange. Under the guidance of Goldman Sachs, a prospectus was prepared for SEC registration of a U.S. stock offering and eventual listing on the NYSE. On July 8, 1981, Novo became the first Scandinavian firm to successfully sell stock through a public issue in the United States.

Exhibit 18.4 illustrates how the price of Novo's B shares increased dramatically between the issue of the convertible Eurobond (1978) and the equity issue on the New York Stock Exchange (1981). This gain in share price is highly correlated with, and is

---

[9] Adapted from Arthur I. Stonehill and Kare B. Dullum, *Internationalizing the Cost of Capital* (New York: Wiley, 1982) by Laurent L. Jacque and Alan C. Shapiro.

EXHIBIT 18.4 Novo's B Share Prices Compared with Stock Market Indices, 1977–1982

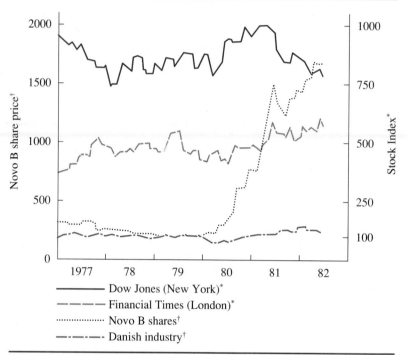

Source: A. Stonehill and K. Dullum, *Internationalizing the Cost of Capital* (Copenhagen Business School Press, 1982). Reproduced by permission of the Copenhagen Business School Press.

probably a result of, steady foreign buying. Indeed, by July 1981, foreign ownership of Novo's B shares exceeded 50% as Danish investors were more than willing to sell a stock that many considered to be grossly and increasingly overvalued. At the same time, foreign investors—mostly from the United States—were ready to step up their investment in a stock that they considered to be either grossly undervalued or a suitable vehicle for international diversification. As its P/E ratio had more than tripled from 9 to 31, Novo was successful in sourcing much-needed capital to better compete with its foreign rivals.[10]

≈ 18.8 SUMMARY AND CONCLUSIONS

Analysis of the available evidence on the impact of foreign operations on firm riskiness suggests that if there is an effect, that effect is generally to reduce both actual and perceived riskiness. These results indicate that corporations should continue investing abroad as long as there are profitable opportunities there. Retrenching because it is believed that investors desire smaller international

---

[10]This is, of course, an illustration, and not a proof, of how selling securities to foreign investors can affect the cost of capital for a firm.

operations is likely to lead to the forgoing of profitable foreign investments that would be rewarded, instead of penalized, by the firm's shareholders. At the very least, executives of multinational firms should seriously question the use of a risk premium to account for the added political and economic risks of overseas operations when evaluating prospective foreign investments.

The use of any risk premium ignores the fact that the risk of an overseas investment in the context of the firm's other investments, domestic as well as foreign, will be less than the project's total risk. How much less depends on how highly correlated are the outcomes of the firm's different investments. Thus, the automatic inclusion of a premium for risk when evaluating a foreign project is not a necessary element of conservatism; rather, it is a management shortcut that is unlikely to benefit the firm's shareholders. Some investments, however, are more risk-prone than are others, and these risks must be accounted for. Chapter 21, on capital budgeting, presents a method for conducting the necessary risk analysis for foreign investments when the foreign risks are unsystematic. This chapter showed how the necessary adjustments in project discount rates can be made, using the capital asset pricing model, when those additional foreign risks are systematic in nature. Specifically, the recommended approach to estimating the cost of equity capital for a foreign subsidiary is to find a proxy portfolio in the country in which that subsidiary operates and calculate its beta relative to the U.S. market. That beta should then be multiplied by the risk premium for the U.S. market. This estimated equity risk premium for the foreign subsidiary would then be added to the U.S. (home country) risk-free rate to compute a dollar (home currency) cost of equity capital.

Lastly, we assessed the factors that are relevant in determining appropriate parent, affiliate, and worldwide capital structures, taking into account the unique attributes of being a multinational corporation. We saw that the optimal global capital structure entails that mix of debt and equity for the parent entity and for all consolidated and unconsolidated subsidiaries that maximizes shareholder wealth. At the same time, affiliate capital structures should vary to take advantage of opportunities to minimize the MNC's cost of capital.

➤ Questions

1. What factors should be considered in deciding whether the cost of capital for a foreign affiliate should be higher, lower, or the same as the cost of capital for a comparable domestic operation?

2. Comment on the following statement: "There is a curious contradiction in corporate finance theory: Because equity is more expensive than debt, highly leveraged subsidiaries should be assigned a low hurdle rate. But when the highly leveraged subsidiaries are in risky nations, country risk dictates just the opposite: a high hurdle rate."

3. Comment on the following statement: "Our conglomerate recognizes that foreign investments have a very low covariance with our domestic operations and, thus, are a good source of diversification. We do not 'penalize' potential foreign investments with a high discount rate but, rather, use a discount rate just 3% above the prevailing riskless rate."

4. According to an article in *Forbes*, "American companies can and are raising capital in Japan at relatively low rates of interest. Dow Chemical, for instance, has raised $500 million in yen. That cost the company over 50% less than it would have at home." Comment on this statement.

5. In early 1990, major Tokyo Stock Exchange issues sold for an average 60 times earnings, more than four times the 13.8 price/earnings ratio for the S&P 500 at that time. According to *Business Week* (February 12, 1990, p. 76), "Since p-e ratios are a guide to a company's cost of equity capital, this valuation gap implies that raising new equity costs Japanese companies less than 2% a year, vs. an average 7% for the U.S." Comment on this statement.

6. What are some of the advantages and disadvantages of having highly leveraged foreign subsidiaries?

7. How has the Tax Reform Act of 1986 affected the capital-structure choice for foreign subsidiaries?

8. What financing problems might be associated with joint ventures?

9. Under what circumstances does it make sense for a company to not guarantee the debt of its foreign affiliates?

10. How can financing strategy be used to reduce foreign exchange risk?

11. How can financial strategy be used to reduce political risk?

➤Problems

1. A firm with a corporatewide debt-to-equity ratio of $1:2$, an after-tax cost of debt of 7%, and a cost of equity capital of 15% is interested in pursuing a foreign project. The debt capacity of the project is the same as for the company as a whole, but its systematic risk is such that the required return on equity is estimated to be about 12%. The after-tax cost of debt is expected to remain at 7%.
   a. What is the project's weighted average cost of capital? How does it compare with the parent's WACC?
   b. If the project's equity beta is 1.21, what is its unlevered beta?

2. Suppose that a foreign project has a beta of 0.85, the risk-free return is 12%, and the required return on the market is estimated at 19%. What is the cost of capital for the project?

3. Compania Troquelados ARDA is a medium-sized Mexico City auto parts maker. It is trying to decide whether to borrow dollars at 9% or Mexican pesos at 75%. What advice would you give it? What information would you need before you gave the advice?

4. Boeing Commercial Airplane Co. manufactures all its planes in the United States and prices them in dollars, even the 50% of its sales destined for overseas markets. What financing strategy would you recommend for Boeing? What data do you need?

5. All-Nippon Airways, a Japanese airline, flies exclusively within Japan. It is looking to finance a recent purchase of Boeing 737s. The director of finance for All-Nippon is attracted to dollar financing because he expects the yen to keep appreciating against the dollar. What is your advice to him?

6. United Airlines recently inaugurated service to Japan and now wants to finance the purchase of Boeing 747s to service that route. The CFO for United is attracted to yen financing because the interest rate on yen is 300 basis points lower than the dollar interest rate. Although he does not expect this interest differential to be offset by yen appreciation over the 10-year life of the loan, he would like an independent opinion before issuing yen debt.

   a. What are the key questions you would ask in responding to UAL's CFO?
   b. Can you think of any other reason for using yen debt?
   c. What would you advise him to do, given his likely responses to your questions and your answer to part b?

7. The CFO of Eastman Kodak is thinking of borrowing Japanese yen because of their low interest rate, currently at 4.5%. The current interest rate on U.S. dollars is 9%. What is your advice to the CFO?

8. Rohm & Haas, a Philadelphia-based specialty chemicals company, traditionally finances its Brazilian operations from outside that country because it's "too expensive" to borrow local currency in Brazil. Brazilian interest rates vary from 50% to more than 100%. Rohm & Haas is now thinking of switching to cruzeiro financing because of a pending cruzeiro devaluation. Assess Rohm & Haas's financing strategy.

9. In order to develop large agricultural estates, the Republic of Coconutland offers the following financing deal: If an investor agrees to purchase a plantation and put up half the cost in U.S. dollars, the government will make a 20-year, zero-interest loan of U.S. dollars to cover the other half.
   a. What risks does the scheme entail?
   b. How can an investor use financing to reduce these risks?

10. Nord Resources's Ramu River property in Papua New Guinea contains one of the world's largest deposits of cobalt and chrome outside of the Soviet Union and South Africa. The cost of developing a mine on this property is estimated to be around $150 million.
   a. Describe three major risks in undertaking this project.
   b. How can Nord structure its financing to reduce these risks?
   c. How can Nord use financing to add value to this project?

➤ Bibliography

ADLER, MICHAEL, and BERNARD DUMAS. "International Portfolio Choice and Corporation Finance: A Synthesis." *Journal of Finance*, June 1983, pp. 925–984.

AGMON, TAMIR, and DONALD R. LESSARD. "Investor Recognition of Corporate International Diversification." *Journal of Finance*, September 1977, pp. 1049–1056.

BLACK, FISCHER. "International Capital Market Equilibrium with Investment Barriers." *Journal of Financial Economics*, December 1974, pp. 337–352.

CHAN, K. C., G. ANDREW KAROLYI, and RENE M. STULZ. "Global Financial Markets and the Risk Premium on U.S. Equity." *Journal of Financial Economics*, October 1992, pp. 137–167.

CHO, CHINHYUNG D., CHEOL S. EUN, and LEMMA W. SENBET, "International Arbitrage Pricing Theory: An Empirical Investigation," *Journal of Finance* 41(1986): 313–330.

GRAUER, FREDERICK L. A., ROBERT H. LITZENBERGER, and RICHARD E. STEHLE. "Sharing Rules and Equilibrium in an International Capital Market under Uncertainty." *Journal of Financial Economics*, June 1976, pp. 233–257.

HARVEY, CAMPBELL R. "The World Price of Covariance Risk." *Journal of Finance* 46 (1991): 111–157.

HUGHES, JOHN S., DENNIS E. LOGUE, and RICHARD J. SWEENEY. "Corporate International Diversification and Market Assigned Measures of Risk and Diversification." *Journal of Financial and Quantitative Analysis*, November 1975, pp. 627–637.

JACQUILLAT, BERTRAND, and BRUNO H. SOLNIK. "Multinationals Are Poor Tools for Diversification." *Journal of Portfolio Management*, Winter 1978, pp. 8–12.

KESTER, W. CARL, and TIMOTHY A. LUEHRMAN. "The Myth of Japan's Low-Cost Capital." *Harvard Business Review*, May-June 1992, pp. 130–138.

SHAPIRO, ALAN C. "Financial Structure and the Cost of Capital in the Multinational Corporation." *Journal of Financial and Quantitative Analysis*, June 1978, pp. 211–226.

SOLNIK, BRUNO H. "Testing International Asset Pricing: Some Pessimistic Views." *Journal of Finance*, May 1977, pp. 503–512.

STONEHILL, ARTHUR I., and KARE B. DULLUM. *Internationalizing the Cost of Capital*. New York: Wiley, 1982.

STULZ, RENE M. "A Model of International Asset Pricing." *Journal of Financial Economics*, December 1981, pp. 383–406.

WHEATLEY, SIMON. "Some Tests of International Equity Integration." *Journal of Financial Economics* 21 (1988): 177–212.

◆APPENDIX 18A Calculating Long-Term Debt Costs

Chapter 13 showed that the after-tax expected dollar cost to a foreign affiliate of a one-year foreign currency (FC) loan equals $r_f(1 - d)(1 - t) - d$, where r_f is the FC interest rate, d is the expected FC devaluation relative to the dollar, and t is the local tax rate. The after-tax cost of borrowing dollars at an interest rate of r_{us} was similarly shown to equal $r_{us}(1 - t) - dt$, assuming that foreign exchange losses are tax-deductible locally. When $r_{us} = r_f(1 - d) - d$, the company is indifferent between borrowing dollars or the foreign currency.

This appendix shows how to calculate the dollar costs of long-term debt, both before and after tax. While the tax factor is often crucial, governments and other nontaxpaying borrowers are important users of Eurobonds and Eurocredits, and taxation is not relevant in their case.[1]

NO TAXES

Assume that a firm can borrow dollars or the local (foreign) currency for n years at fixed interest rates of r_{us} and r_f, respectively. Interest is to be paid at the end of each year, and the principal will be repaid in a lump sum at the end of year n. If P is the principal amount in local currency of the foreign currency loan and e_i is the dollar value of the foreign currency at the end of year i, then the effective dollar cost of the foreign currency debt, in the absence of taxes, is the solution, r, to Equation 18A.1:

$$-Pe_0 + \sum_{i=1}^{n} \frac{r_f Pe_i}{(1 + r)^i} + \frac{Pe_n}{(1 + r)^n} = 0 \tag{18A.1}$$

In other words, r is the internal rate of return, or yield, on the foreign currency-denominated bond. The yield on the dollar debt remains at r_{us}. With flotation costs of s per dollar's worth of foreign currency, the first term in Equation 18A.1 would become $-Pe_0(1 - s)$. In general, the equation can be solved for r only by using techniques of numerical analysis (unless you happen to have a calculator with an internal rate of return function). Its application is illustrated in the following example.

ILLUSTRATION

EVALUATING DMR'S **L**ONG-TERM FINANCING CHOICES
Suppose DMR Inc. is planning to float a seven-year, $30 million bond issue. It has the choice of having its Swiss subsidiary borrow dollars at a coupon rate of 9.625% or Swiss francs at 3.5%. Both bond issues are sold at par. The flotation costs are 3% for the Swiss franc issue and 1.2% for the dollar issue, leading to an effective rate of 4% for the Swiss franc debt and 9.87% for the dollar debt. Repayment is in a lump sum at the end of year 7.

---

[1] The equations in this appendix are elaborated upon in Alan C. Shapiro, "The Impact of Taxation on the Currency-of-Denomination Decision for Long-Term Borrowing and Lending," *Journal of International Business Studies,* Spring–Summer 1984, pp. 15–25.

The current exchange rate is 1.75 Swiss francs to the dollar. Thus, DMR can borrow either $30 million or SFr 52,500,000. If the exchange rates and dollar servicing requirements listed in Exhibit 18A.1 are forecast for the coming seven years, which issue is preferable?

Using Equation 18A.1 (and adjusting for flotation costs), the effective cost of the Swiss franc issue, given the expected dollar deprecia-

tion of approximately 6.1% compounded annually, turns out to equal 10.31%. The effective cost of the dollar debt remains at 9.87%. To minimize expected dollar costs, therefore, DMR should issue dollar debt. The breakeven rate of annual dollar decline at which DMR should just be indifferent between borrowing dollars or Swiss francs equals 5.64%.

Annual Revaluation

Making such detailed currency projections generally is not done, given the uncertainties involved. Instead, it is simpler to project an average rate of currency change over the life of the debt and to calculate effective dollar costs on that basis. For example, suppose the foreign currency is expected to revalue (devalue) relative to the dollar at a steady rate of g per annum (i.e., one dollar's worth of foreign currency today will be worth $(1 + g)^i$ dollars at the end of i years). Then, the interest expense in year i per dollar's worth of foreign currency borrowed today equals $r_f(1 + g)^i$, while the principal repayment is $(1 + g)^n$.

The present value of the cash flow per dollar of foreign currency financing discounted at r equals

$$-1 + \sum_{i=1}^{n} \frac{r_f(1 + g)^i}{(1 + r)^i} + \frac{(1 + g)^n}{(1 + r)^n} = 0 \qquad (18A.2)$$

The effective yield, r, equals $r_f(1 + g) + g$. This is the same as the cost of a one-period foreign currency loan that appreciates at an annual rate of g during the

EXHIBIT 18A.1 Cash Flows Associated with Swiss Franc Debt

YEAR	CASH-FLOW CATEGORY	SWISS FRANC CASH FLOW (1)	÷	RATE OF EXCHANGE (2)	=	DOLLAR CASH FLOW (3)
0	Bond Sale	−52,500,000		1.75		−30,000,000.00
	Flotation charge	1,575,000		1.75		900,000.00
1	Interest	1,837,500		1.665		1,103,603.60
2	Interest	1,837,500		1.580		1,162,974.68
3	Interest	1,837,500		1.495		1,229,097.00
4	Interest	1,837,500		1.410		1,303,191.49
5	Interest	1,837,500		1.325		1,386,792.45
6	Interest	1,837,500		1.240		1,481,854.84
7	Interest	1,837,500		1.155		1,590,909.09
	Principal repayment	SFr 52,500,000		1.155		$45,454,545.45

period. Thus, in order for the yield on the foreign currency-denominated bond to equal r_{us}, it is necessary that

$$r_{us} = r_f(1 + g) + g \tag{18A.3}$$

For instance, if $r_{us} = 9\%$ and the currency is expected to appreciate at a rate of 3% annually (i.e., $g = 3\%$), then the breakeven value of r_f is 5.83%. In other words, if r_f is greater than 5.83%, it would be cheaper to borrow dollars, and vice versa if r_f is less than 5.83%. If $r_f = 5.83\%$, the firm should be indifferent between the two currencies.

TAXES

Chapter 5 demonstrated that international covered interest arbitrage normally ensures that the annualized forward exchange premium or discount equals the nominal yield differential between debt denominated in different currencies. Moreover, in an efficient market, the forward premium or discount should equal the expected rate of change of the exchange rate (adjusted for risk). Therefore, in the absence of taxes, corporations willing to base decisions solely on expected costs should be indifferent between issuing debt in one currency or another.

The presence of taxes, however, distorts the interest arbitrage relationships that have already been developed, since interest rates that were at parity before tax may no longer be so after tax. This presents a new decision problem for international financial executives. The discussion now turns to some of the alternative tax treatments of exchange gains and losses arising from foreign currency loans and how these tax effects can be integrated into the computation of effective after-tax differences in the costs of borrowing in different currencies.

In general, using the same notation as before and letting t be the foreign tax rate, the after-tax yield on a foreign currency-denominated bond issued by a local affiliate can be found as the solution, r, to Equation 18A.4:

$$-Pe_0 + \sum_{i=1}^{n} \frac{r_f Pe_i(1 - t)}{(1 + r)^i} + \frac{Pe_n}{(1 + r)^n} = 0 \tag{18A.4}$$

Similarly, the effective after-tax cost of dollar debt is the solution, k, to Equation 18A.5:

$$-L + \sum_{i=1}^{n} \frac{r_{us}L(1 - t)}{(1 + k)^i} + \frac{L + t(Le_n/e_0 - L)}{(1 + k)^n} = 0 \tag{18A.5}$$

where $L = Pe_0$ is the dollar equivalent of the foreign currency loan. The final term in Equation 18A.5, $t(Le_n/e_0 - L)$, equals the tax on the LC gain on repaying the dollar loan if the currency appreciates (because it now costs fewer units of LC to repay a given dollar principal) or tax deduction on the LC loss if the LC depreciates. As before, with flotation costs of s per dollar, the first term in Equations 18A.4 and 18A.5 would be multiplied by $(1 - s)$.

EXHIBIT 18A.2 After-Tax Cash Flows Associated with Swiss Franc Debt

YEAR	CASH-FLOW CATEGORY	SWISS FRANC CASH FLOWS (1)	÷	RATE OF EXCHANGE (2)	×	AFTER-TAX FACTOR (3)	=	AFTER-TAX DOLLAR CASH FLOWS (4)
0	Bond Sale	−52,500,000		1.75		1		−30,000,000.00
	Flotation charge	1,575,000		1.75		0.55		495,000.00
1	Interest	1,837,500		1.665		0.55		606,981.98
2	Interest	1,837,500		1.580		0.55		639,636.09
3	Interest	1,837,500		1.495		0.55		676,003.35
4	Interest	1,837,500		1.410		0.55		716,755.33
5	Interest	1,837,500		1.325		0.55		762,735.88
6	Interest	1,837,500		1.240		0.55		815,020.14
7	Interest	1,837,500		1.155		0.55		875,000.01
	Principal repayment	SFr 52,500,000		1.155		1		45,454,545.45

Equations 18A.4 and 18A.5 (adjusted for flotation costs) can be applied to the previous example of dollar versus Swiss franc debt. Assume that the tax rate is 45%, all flotation costs are tax-deductible as soon as they are incurred, and the debt is issued by DMR's Swiss affiliate. Exhibits 18A.2 and 18A.3 contain the year-by-year Swiss franc cash flows and dollar cash flows associated with both issues.

The effective after-tax yield on the Swiss franc issue is now 8.40%, and on the dollar debt issue, 8.02%. These contrast with the respective no-tax yields of 10.31% and 9.87%.

Annual Revaluation

If a steady appreciation of the foreign currency at a rate of *g* per annum is anticipated, then the effective after-tax dollar yield on the foreign currency bond issued by a local affiliate can be found by solving Equation 18A.6:

EXHIBIT 18A.3 After-Tax Cash Flows Associated with Dollar Debt

YEAR	CASH-FLOW CATEGORY	DOLLAR CASH FLOWS (1)	×	AFTER-TAX FACTOR (2)	=	AFTER-TAX DOLLAR CASH FLOWS (3)
0	Bond Sale	−30,000,000		1		−30,000,000.00
	Flotation charge	360,000		0.55		198,000.00
1	Interest	2,887,500		0.55		1,588,125.00
2	Interest	2,887,500		0.55		1,588,125.00
3	Interest	2,887,500		0.55		1,588,125.00
4	Interest	2,887,500		0.55		1,588,125.00
5	Interest	2,887,500		0.55		1,588,125.00
6	Interest	2,887,500		0.55		1,588,125.00
7	Interest	2,887,500		0.55		1,588,125.00
	Principal repayment	$30,000,000		1		$30,000,000.00
	Capital gain recognized by Swiss tax authorities (SFr 17,850,000 at $0.87)	$15,454,545		0.45		$ 6,954,545.30

$$-1 + \sum_{i=1}^{n} \frac{r_f(1 + g)^i(1 - t)}{(1 + r)^i} + \frac{(1 + g)^n}{(1 + r)^n} = 0 \qquad (18A.6)$$

The solution to Equation 18A.6 is $r = r_f(1 + g)(1 - t) + g$, the same as in the single-period case.

Assuming that $r_f = 6\%$, $t = 45\%$, and $g = 3\%$, the effective cost of foreign currency borrowing equals $0.06 \times 1.03 \times 0.55 + 0.03$, or 6.4%. In the absence of taxes, this cost would equal 9.18% ($0.06 \times 1.03 + 0.03$).

England does not allow the tax deductibility of exchange losses on foreign currency debt. In other words, the term $t(Le_n/e_0 - L)$ in Equation 18A.5 goes to zero. It can be shown that if debt is issued in the United Kingdom, and assuming the pound sterling will devalue relative to the dollar, there is a simple after-tax equilibrium relationship between the dollar interest rate, r_{us}, and the pound sterling rate, r_{uk}:

$$r_{us}(1 - t) = r_{uk}(1 - u)(1 - t) - u \qquad (18A.7)$$

where u is the anticipated annual devaluation of the pound relative to the dollar. When Equation 18A.7 holds, the after-tax dollar costs of borrowing pounds and dollars are the same.

To illustrate Equation 18A.7, suppose $r_{us} = 9\%$, $t = 40\%$, and the pound is expected to devalue by 2% annually. According to Equation 18A.7 then, the equilibrium pound interest rate is $r_{uk} = 12.59\%$.

➤Problems

1. Multicountry, Inc. has a $200 million principal value Eurobond with two more 10% coupon interest payments due at the end of the next two years. Multicountry would like to switch currencies on the bond. The issue is currently denominated in yen, but Multicountry believes that the Deutsche mark would be more advantageous. Given the following current and expected currency rates, what should Multicountry do, assuming it wishes to minimize its expected financing cost? The interest rate will remain at 10%.

	$/DM	$/¥
Current exchange rate	0.654	0.00761
Expected, one year	0.665	0.00772
Expected, two year	0.685	0.00799

2. Suppose the current rate of exchange between the U.S. dollar and the pound sterling is £1 = $2. The English affiliate of Global Industries, GI Ltd, is contemplating raising $12 million by issuing bonds denominated in either dollars or pounds sterling. The dollar bonds would carry a coupon rate of 10%, and the pound sterling bonds would carry a coupon rate of 13%. In either case, the bonds would have annual interest payments and mature in five years.

a. Suppose GI Ltd is interested only in minimizing its expected financing costs. In the absence of taxes, what annual rate of pound devaluation or revaluation would leave GI Ltd indifferent between borrowing either pounds or dollars? What would be the expected exchange rate at the end of year 5, given these currency changes?

b. Suppose the British tax rate is 45% and exchange losses on foreign currency principal repayments are not tax-deductible, but all interest expenses, including exchange losses, are tax-deductible. Rework part a on an after-tax basis.

c. Suppose the international Fisher effect holds on an after-tax basis. Which currency should GI Ltd borrow, given the tax scenario in part b? Explain your answer.

d. What other factors besides expected borrowing costs might the parent corporation be concerned about in deciding whether to approve GI Ltd's currency selection for this bond issue?

Case Studies

PART FOUR

CASE IV.1
PLANO CRUZADO

On February 28, 1986, President Jose Sarnay of Brazil announced the Plano Cruzado. At the time, Brazilian inflation was running at an annualized rate of more than 400%. The plan slashed inflation by freezing prices and wages. The purpose of the plan was to impose "shock treatment" on the economy and break the cycle of "inertial inflation" caused by high inflationary expectations. However, in a move that foreshadowed the splits that bedeviled the Plan, workers were granted pay hikes of 8% to 15%, just before the freeze. At the same time, government spending went largely unchecked, and the public-sector deficit—financed largely by printing more cruzados—grew to 4.5% of gross domestic product.

In November, the Plan achieved its first incontrovertible success: Government parties swept the congressional and gubernatorial races. Price controls were eased just after the election. However, the government found it politically impossible to remove subsidies to state industries because these industries formed the base for political power. Instead, large price hikes for state companies were granted by imposing huge increases in indirect taxes and tariffs on their products, and an attempt was made to disguise the effect of these increases on inflation by altering the basket of goods on which inflation was calculated. "They wanted me to tamper with inflation—simple as that," commented the head of the National Statistics Office, who immediately resigned.

QUESTIONS

1. What were the likely consequences for Brazil of controlling prices while gunning the money supply? Consider the effect on production and the availability of products in the stores.

2. How did the Plano Cruzado affect Brazil's huge trade surplus?

3. What would be your forecast of the Plan's effect on Brazil's ability to service its foreign debts?

4. President Sarnay terminated Plano Cruzado in February 1987, one year after it began. What impact do you think the Plan had in reducing inflation expectations? How would you go about measuring the effect of the Plan on inflation expectations?

5. What was the likely price response to the removal of price controls?

6. If you were a banker, how would seeing such a Plan put into effect affect your willingness to lend money to Brazil? Explain.

CASE IV.2
MULTINATIONAL MANUFACTURING, INC.

PART I

Multinational Manufacturing, Inc. (MMI) is a large manufacturing firm engaged in the production and sale of a widely diversified group of products in a number of countries throughout the world. Some product lines enjoy outstanding success in new fields developed on the basis of an active research and development program; other product lines, whose innovative leads have disappeared, face very severe competition.

Each domestic product line and foreign affiliate is a separate profit center. Headquarters influences these centers primarily by evaluating their

Source: Based on a report in the *Wall Street Journal*, February 13, 1987, p. 27.

managers on the basis of certain financial criteria, including return on investment, return on sales, and growth in earnings.

Division and affiliate executives are held responsible for planning and evaluating possible new projects. Each project is expected to yield at least 15%. Projects requiring an investment below $250,000 (about one-third of the projects) are approved at the division or affiliate level without formal review by headquarters management. The present cutoff rate was established three years ago as part of a formal review of capital budgeting procedures. The conclusion at that time was that the company's weighted average cost of capital was 15%, and it should be applied when calculating net present values of proposed projects. In announcing the policy, Mr. Thomas Black, Vice President-Finance, said, "It's about time that we introduced some modern management techniques in allocating our capital resources."

Now Mr. Black is concerned that the policy introduced three years ago is having some unintended consequences. Specifically, top management gets to review only obvious investment candidates. Low-risk, low-return projects and high-risk, high-return projects seem to be systematically screened out along the way. The basis for this screening is not entirely clear, but it appears to be related to the way in which managerial performance is evaluated. Local executives seem to be concerned that low-potential projects will hurt their performance appraisal, while high-potential projects can turn out poorly. The president of one foreign affiliate said privately when asked why he never submitted projects at the extremes of risk and return, "Why should I take any chances? When headquarters say it wants 15%, it means 15% and nothing less. My crystal ball isn't good enough to allow me to accurately estimate sales and costs in this country, especially when I never know what the government is going to do."

QUESTIONS: PART I

Make recommendations to Mr. Black concerning the following points:

1. Should MMI lower the hurdle rate in order to encourage the submission of more proposals,

or should it drop the hurdle rate concept completely?

2. Should MMI invest in lower-return projects that are less risky and/or in high-risk projects that appear promising? What is the relevant measure of risk?

3. How should MMI factor in the additional political and economic risks it faces overseas in conducting these project analyses?

4. Why are projects at the extremes of risk and return not reaching top management for review?

5. What actions, if any, should Mr. Black take to correct the situation?

PART II

In line with this current review of capital budgeting procedures, Mr. Black is also reconsidering certain financial policies that he recently recommended to MMI's board of directors. These policies include the maintenance of a debt/total assets ratio of 35% and a dividend payout rate equal to 60% of consolidated earnings. In order to achieve these ratios for the firm overall, each affiliate has been directed to use these ratios as guidelines in planning its own capital structure and payout rate.

This directive has been controversial. The executives of several foreign affiliates have raised questions about the appropriateness of applying these guidelines at the local level. The general managers of some of the largest affiliates have been particularly vocal in their objections, stating that it simply was not possible for overall policies relative to capital structure proportions to be given much consideration in financial planning at the local level. They pointed out that differences in the economic and political environment in which the various affiliates operate are far too great to force them into a financial straitjacket designed by headquarters. In their view, they must be left free to respond to their own unique set of circumstances.

The executives of the Brazilian affiliate, for example, felt that their financing should not follow the same pattern as that of the overall firm because inflationary conditions made local borrowing especially advantageous in Brazil. Executives of other foreign affiliates stressed the need for varying capital structures in order to cope with the exchange risks posed by currency fluctuations. The general manager of the Mexican affiliate, which is owned on a 50-50 basis with local investors, has argued forcefully that, despite effective headquarters control over the policies of this operation, joint ventures such as his cannot and should not be financed in the same manner as firms wholly owned by MMI. In addition, the tax manager of MMI has expressed his concern that implementing a rigid policy of repatriating 60% of each affiliate's earnings in the form of dividends will impose substantial tax costs on MMI. Morever, Mr. Black recently attended a seminar at which it was pointed out that overseas affiliates can sometimes be financed in such a way that their susceptibility to political and economic risks is diminished.

QUESTIONS: PART II

Make recommendations to Mr. Black concerning the policies that should be adopted as guides in planning the capital structure and dividend payout policies of foreign affiliates, taking into account the following key questions:

1. What are the pros and cons of using the following sources of funds to finance the operations of the foreign affiliates: equity funds versus loans from MMI, retained earnings of the affiliates, and outside borrowings? Consider cost, political and economic risks, and tax consequences in your answer.

2. Given these considerations, under what circumstances, if any, should the capital structure of foreign affiliates include more or less debt than the 35% considered desirable for the firm as a whole?

3. How will the resultant capital structures affect the required rates of return on affiliate projects? The actual rates of return?

4. How should MMI's dividend policy be implemented at the affiliate level?

PART FIVE

FOREIGN
INVESTMENT
ANALYSIS

INTERNATIONAL PORTFOLIO INVESTMENT

Capital now flows at the speed of light across national borders and into markets once deemed impregnable.

Citicorp Annual Report (1991)

$\mathcal{T}$here was a time when investors treated national boundaries as impregnable barriers, limiting their reach and financial options to predominantly domestic and regional markets. Times have changed. Just as companies and consumers are going global, so are increasing numbers of investors. American investors are buying foreign stocks and bonds and foreign investors are purchasing U.S. securities. The purpose of this chapter is to examine the nature and consequences of international portfolio investing. Although the chapter focuses on international investing from an American perspective, its lessons are applicable to investors from around the world.

19.1 THE BENEFITS OF INTERNATIONAL EQUITY INVESTING

The advantages of *international investing* are several. For one thing, an international focus offers far more opportunity than a domestic focus. More than half of the world's stock market capitalization is in non-U.S. companies, and this fraction generally has

653

EXHIBIT 19.1 Stock Market Capitalization as Percent of World Total (20 Main Stock Markets)

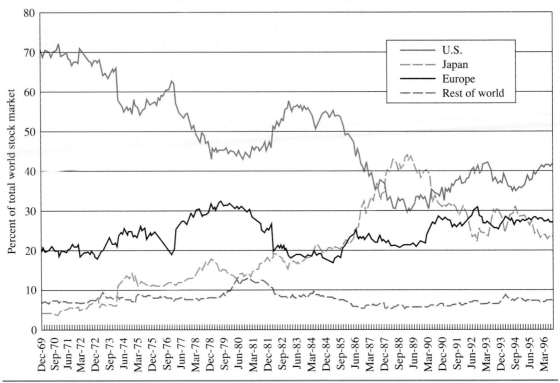

Source: Data from *Morgan Stanley Capital International.*

increased over time (see Exhibit 19.1). In fact, if you want to invest in certain products with huge global markets, you will find that most of the big, highly profitable manufacturers are overseas. For example, videotape recorders are the world's best-selling consumer electronics product, and 95% of them are made in Japan; more than 80% of all cars are made abroad, 85% of all stereo systems, and 99% of all 35mm cameras. The Japanese dominance of these and other consumer-product markets helps explain why by 1989 Japan's market capitalization actually exceeded that of the United States. The 60% plunge in the Tokyo Stock Exchange between 1990 and 1998 and the dramatic rise in the U.S. stock market during that same period has once again given the U.S. stock market— at about four times Japan's capitalization—the dominant share of the world's stock market capitalization. In fact, by 1998, Japan's market capitalization had fallen to third place, behind that of Great Britain.

International Diversification

The expanded universe of securities available internationally suggests the possibility of achieving a better *risk-return trade-off* than by investing solely in U.S. securities: That is, expanding the universe of assets available for investment should lead to higher returns for the same level of risk or less risk for the same level of expected return. This relation follows from the basic rule of portfolio diversification: *The broader the diversification, the more stable the returns and the more diffuse the risks.*

Prudent investors know that diversifying across industries leads to a lower level of risk for a given level of expected return. For example, a fully diversified U.S. portfolio is only about 27% as risky as a typical individual stock. Put another way, about 73% of the risk associated with investing in the average stock can be eliminated in a fully diversified U.S. portfolio. Ultimately, though, the advantages of such diversification are limited because all companies in a country are more or less subject to the same cyclical economic fluctuations. Through **international diversification**—that is, by diversifying across nations whose economic cycles are not perfectly in phase—investors should be able to reduce still further the variability of their returns. In other words, risk that is systematic in the context of the U.S. economy may be unsystematic in the context of the global economy. For example, an oil price shock that hurts the U.S. economy helps the economies of oil-exporting nations, and vice versa. Thus, just as movements in different stocks partially offset one another in an all-U.S. portfolio, so also do movements in U.S. and non-U.S. stock portfolios cancel each other out somewhat.

The possibility of achieving a better risk-return trade-off by investing internationally is supported by Exhibit 19.2, which shows the annualized returns and standard deviations of returns for a variety of developed and emerging stock markets over the

EXHIBIT 19.2 Annualized Monthly Returns and Standard Deviations of Returns: 1981–1996

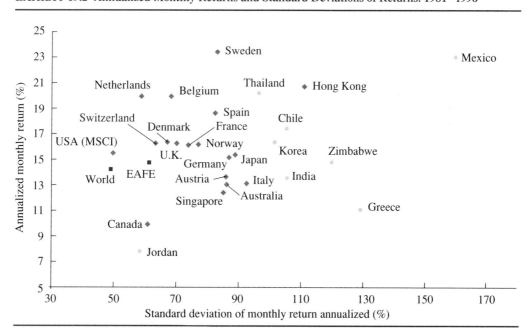

Note: Circles are emerging markets, diamonds are developed markets, and squares show EAFE and MSCI world indices.

16-year period 1981–1996.[1] Exhibit 19.2 illustrates three points:

1. Historically, national stock markets have wide differences in returns and risk (as measured by the standard deviation of annual returns).
2. **Emerging markets** (shown as circles)—a term that encompasses all of South and Central America; all of the Far East with the exception of Japan, Hong Kong, Singapore, Australia, and New Zealand; all of Africa; and parts of Southern Europe, as well as Eastern Europe and countries of the former Soviet Union—have had higher risk and return than the developed markets (shown as squares).
3. The Morgan Stanley Capital International **Europe, Australia, Far East (EAFE) Index** (which reflects all major stock markets outside of North America, 20 altogether) has had lower risk than most of its individual country components.

Empirical research bears out the significant benefits from international equity diversification suggested by Exhibit 19.2. Bruno Solnik and Donald Lessard, among others, have both presented evidence that national factors have a strong impact on security returns relative to that of any common world factor.[2] They also found that returns from the different national equity markets have relatively low correlations with one another. More recent research shows that differences in industrial structure and currency movements account for very little of the low correlation between national stock market returns.[3] The more likely explanation for the low degree of international return correlation is that local monetary and fiscal policies, differences in institutional and legal regimes, and regional economic shocks induce large country-specific variation in returns.

CORRELATIONS AND THE GAINS FROM DIVERSIFICATION ■ Exhibit 19.3 contains some data on correlations between the U.S. and non-U.S. stock markets. **Foreign market betas**, which are a measure of market risk derived from the capital asset pricing model (see Chapter 18), are calculated relative to the U.S. market in the same way that individual asset betas are calculated:

$$\frac{\text{Foreign market}}{\text{beta}} = \frac{\text{Correlation with}}{\text{U.S. market}} \times \frac{\text{Standard deviation of foreign market}}{\text{Standard deviation of U.S. market}}$$

For example, the Canadian market beta is $0.70 \times 60.7/49.5 = 0.86$. Market risk also is calculated from a world perspective, where the correlations are calculated relative to the world index. Notice that the betas calculated relative to the world index are higher than the betas calculated relative to the U.S. market for all but the U.S. market.

---

[1] The standard deviations used in this chapter are all monthly standard deviations annualized (multiplied by 12). They are significantly larger than annual standard deviations calculated using yearly data because monthly returns are more volatile than yearly returns (monthly ups and downs often cancel each other out).

[2] Bruno H. Solnik, "Why Not Diversify Internationally Rather Than Domestically?" *Financial Analysts Journal*, July–August 1974, pp. 48–54; and Donald R. Lessard, "World, Country, and Industry Relationships in Equity Returns: Implications for Risk Reduction Through International Diversification," *Financial Analysts Journal*, January–February 1976, pp. 32–38.

[3] Steven L. Heston and K. Geert Rouwenhorst, "Does Industrial Structure Explain the Benefits of International Diversification?" *Journal of Financial Economics*, August 1994, pp. 3–27.

EXHIBIT 19.3 How Foreign Markets Were Correlated
with U.S. Market and World Index, 1970–1996

COUNTRY	CORRELATION WITH U.S. MARKET	STANDARD DEVIATION OF RETURNS[1] (%)	MARKET RISK (BETA) FROM U.S. PERSPECTIVE	CORRELATION WITH WORLD INDEX	MARKET RISK (BETA) FROM WORLD PERSPECTIVE
United States	1.00	49.52	1.00	0.82	0.83
Canada	0.70	60.65	0.86	0.70	0.87
Australia	0.46	86.46	0.81	0.55	0.97
Hong Kong	0.30	110.87	0.68	0.41	0.92
Japan	0.25	86.37	0.44	0.68	1.21
Singapore	0.45	85.29	0.77	0.53	0.92
Austria	0.12	86.16	0.22	0.29	0.52
Belgium	0.42	68.25	0.57	0.62	0.86
Denmark	0.31	66.99	0.42	0.48	0.65
France	0.43	77.00	0.67	0.62	0.98
Germany	0.34	73.76	0.51	0.55	0.83
Italy	0.21	92.58	0.40	0.42	0.81
Netherlands	0.57	58.56	0.67	0.73	0.88
Norway	0.43	88.51	0.78	0.50	0.91
Spain	0.28	82.44	0.47	0.48	0.81
Sweden	0.41	83.02	0.68	0.55	0.94
Switzerland	0.49	63.22	0.62	0.67	0.87
United Kingdom	0.50	69.88	0.71	0.68	0.97
EAFE index[2]	0.47	61.27	0.58	0.87	1.09
World index[3]	0.82	48.75	0.81	1.00	1.00

[1] Monthly standard deviation annualized.
[2] The Morgan Stanley Capital International Europe, Australia, Far East (EAFE) Index is the non-North American part of the world index and consists of 20 major stock markets from these parts of the world.
[3] The Morgan Stanley Capital International World Index has a combined market value of $8.6 trillion, covers 22 countries including the United States, and includes about 1,600 of the largest companies worldwide.
Source: Data from *Morgan Stanley Capital International.*

Measured for the 27-year period 1970–1996, foreign markets in developed (EAFE) countries were correlated with the U.S. market from a high of 0.70 for Canada to a low of 0.12 for Austria. The relatively high correlation for Canada reveals that this market tracked the U.S. market's ups and downs. Austria's low correlation, on the other hand, indicates that the Austrian and U.S. markets have tended to move largely independently of each other.

Notice also that the investment risks associated with these different markets can be quite different—with the Hong Kong market showing the highest level and the Dutch market the lowest. Indeed, all the markets had a higher level of risk, as measured by the standard deviation of returns, than the U.S. market. Yet the internationally diversified **Morgan Stanley Capital International World Index** had the lowest level of risk—lower even than the U.S. market. The reason, of course, is that much of the risk associated with markets in individual countries is unsystematic and so can be eliminated by diversification, as indicated by the relatively low betas of these markets.

These results imply that international diversification may significantly reduce the risk of portfolio returns. In fact, the standard deviation of an internationally diversified portfolio appears to be as little as 11.7% of that of individual securities. In addition, as shown in Exhibit 19.4, the benefits from international diversification are significantly

EXHIBIT 19.4 The Potential Gains from International Diversification

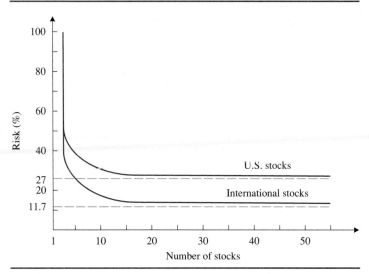

Source: Bruno H. Solnik, "Why Not Diversify Internationally Rather Than Domestically?" Reprinted with permission from Financial Analysts Federation, Charlottesville, VA. All rights reserved. *Financial Analysts Journal,* July–August 1974.

EXHIBIT 19.5 The U.S. Market Lags Behind the EAFE Index: 1970–1997

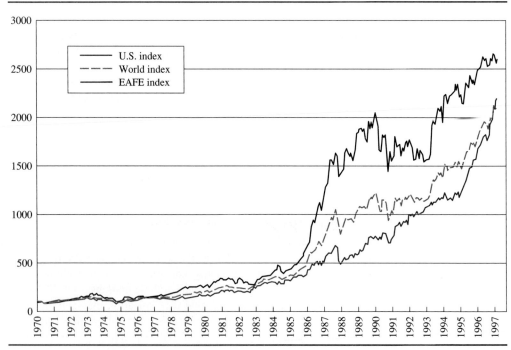

Source: Data from *Morgan Stanley Capital International.*

greater than those that can be achieved solely by adding more domestic stocks to a portfolio. Specifically, an internationally diversified portfolio appears to be less than half as risky as a fully diversified U.S. portfolio.

Moreover, from 1970 through 1996, the compound annual return for the EAFE Index was 12.9%, compared with 11.8% for the U.S. market. Further, the EAFE Index outpaced the U.S. market 19 times during the 30-year period 1961–1990. In the 27 years ended February 1997, the U.S. portion of Morgan Stanley Capital International's world stock index was up 2,082%, and the cumulative return for the EAFE Index was 2,542% (see Exhibit 19.5). Even more astounding, from 1949 to 1990, the Japanese market soared an incredible 25,000%. Even with the recent plunge in the Tokyo Stock Exchange, an investor in the Japanese stock market would still be way ahead.

The obvious conclusion is that international diversification pushes out the **efficient frontier** — the set of portfolios that has the smallest possible standard deviation for its level of expected return and has the maximum expected return for a given level of risk — allowing investors simultaneously to reduce their risk and increase their expected return. Exhibit 19.6 illustrates the effect of international diversification on the efficient frontier.

One way to estimate the benefits of international diversification is to consider the expected return and standard deviation of return for a portfolio consisting of a fraction a invested in U.S. stocks and the remaining fraction, $1 - a$, invested in foreign stocks. Define r_{us} and r_{rw} to be the expected returns on the U.S. and rest-of-world stock portfolios, respectively. Similarly, let σ_{us} and σ_{rw} be the standard deviations of the U.S. and rest-of-world portfolios. The expected return r_p can be calculated as

$$r_p = ar_{us} + (1 - a)r_{rw} \tag{19.1}$$

To calculate the standard deviation of this portfolio, it helps to know that the general formula for the standard deviation of a two-asset portfolio with weights w_1 and

EXHIBIT 19.6 International Diversification Pushes Out the Efficient Frontier

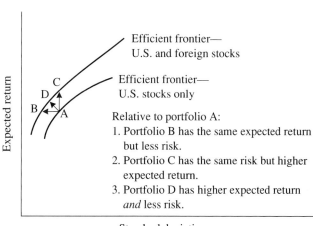

w_2 $(w_1 + w_2 = 1)$ is

$$\text{Portfolio standard deviation} = [w_1^2\sigma_1^2 + w_2^2\sigma_2^2 + 2\,w_1 w_2 \sigma_{12}\sigma_1\sigma_2]^{1/2} \qquad (19.2)$$

where σ_1^2 and σ_2^2 are the respective variances of the two assets; σ_1 and σ_2 are their standard deviations; and σ_{12} is their correlation. We can apply Equation 19.2 to our internationally diversified portfolio by treating the domestic and foreign portfolios as separate assets. This operation yields a portfolio standard deviation σ_p equal to

$$\sigma_p = [a^2\sigma_{us}^2 + (1 - a)^2\sigma_{rw}^2 + 2a(1 - a)\sigma_{us}\sigma_{rw}\sigma_{us,rw}]^{1/2} \qquad (19.3)$$

where $\sigma_{us,rw}$ is the correlation between the returns on the U.S. and foreign stock portfolios.

To see the benefits of international diversification, assume that the portfolio is equally invested in U.S. and foreign stocks, where the EAFE Index represents the foreign stock portfolio. Using data from Exhibit 19.3, we see that $\sigma_{us} = 49.5\%$, $\sigma_{rw} = 61.3\%$, and $\sigma_{us,rw} = 0.47$. According to Equation 19.3, these figures imply that the standard deviation of the internationally diversified portfolio is

$$\sigma_p = [0.5^2\,(49.5)^2 + 0.5^2\,(61.3)^2 + 0.5^2 \times 2 \times 49.5 \times 61.3 \times 0.47]^{1/2}$$

$$= 0.5(9{,}060.2)^{1/2}$$

$$= 47.6\%$$

EXHIBIT 19.7 Correlation of Monthly Returns with the U.S. Market: Developed Country Markets

	1981–1996	1981–1984	1985–1988	1989–1992	1993–1996
Canada	0.71	0.71	0.79	0.70	0.66
Australia	0.42	0.43	0.37	0.47	0.49
Hong Kong	0.34	0.08	0.55	0.41	0.49
Japan	0.25	0.38	0.19	0.31	0.16
Singapore	0.48	0.41	0.54	0.63	0.29
Austria	0.13	0.10	0.09	0.14	0.28
Belgium	0.42	0.30	0.27	0.28	0.26
Denmark	0.32	0.38	0.41	0.40	0.39
France	0.45	0.27	0.29	0.28	0.25
Germany	0.36	0.28	0.29	0.28	0.26
Italy	0.22	0.14	0.20	0.19	0.20
Netherlands	0.59	0.56	0.58	0.60	0.59
Norway	0.49	0.38	0.42	0.43	0.42
Spain	0.35	−0.10	0.01	0.01	0.00
Sweden	0.42	0.29	0.31	0.30	0.29
Switzerland	0.49	0.50	0.50	0.49	0.46
United Kingdom	0.56	0.47	0.47	0.46	0.43
World	0.76	0.88	0.89	0.89	0.88

Source: Data from *Morgan Stanley Capital International.*

Here the risk of the internationally diversified portfolio is below the risk of the U.S. portfolio. Moreover, as indicated earlier, the expected return is higher as well.

RECENT CORRELATIONS ∎ The benefits of diversification depend on relatively low correlations among assets. It is often assumed that, as their underlying economies become more closely integrated and cross-border financial flows accelerate, national capital markets will become more highly correlated, significantly reducing the benefits of international diversification. Indeed, the correlations between the U.S. and non-U.S. stock markets are generally higher today than they were during the 1970s. However, Exhibit 19.7 shows that, contrary to intuition, these correlations have, if anything, fallen in recent years. Using four-year periods—1981–1984, say, versus 1993–1996—correlations are almost uniformly lower in more recent periods, particularly during the 1990s, when economic integration has presumably increased.

In addition to the somewhat higher correlations today relative to the 1970s, the U.S. market has outperformed the Japanese and European markets in recent years (see Exhibit 19.5 over the period 1990–1996), reducing the expected return on the EAFE Index relative to the U.S. market.

Recent research points to another problematic aspect of international investing as well: When markets are the most volatile, and investors most seek safety, global diversification is of limited value.[4] In particular, Exhibit 19.8 shows that the correlations among markets appear to increase when market volatility is at its highest. Even worse, the markets appear to move in synchrony only when they are falling, not when they are rising. In other words, only bear markets seem to be contagious, not bull markets.

EXHIBIT 19.8 No Place to Hide?

World markets seem to be most in step with each other when volatility is greatest. Correlation among markets is measured here on a scale of 0 to 1, where 1 indicates that markets track each other perfectly and 0 means they are completely independent. Volatility is measured on the basis of "standard deviation"—how much prices vary from the mean.

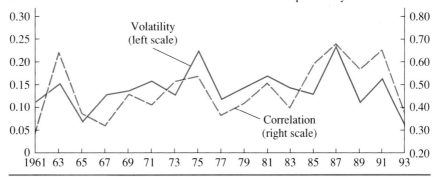

Source: Bruno Solnik, Hautes Etudes Commerciales in the *Wall Street Journal*, April 14, 1994, p. C1. Reprinted by permission of the *Wall Street Journal*, © 1994 Dow Jones & Company, Inc. All rights reserved worldwide.

[4]See Patrick Odier and Bruno Solnik, "Lessons for International Asset Allocation," *Financial Analysts Journal*, March–April 1993, pp. 63–77.

EXHIBIT 19.9 Average Returns and Standard Deviation of Returns, 1970–1996

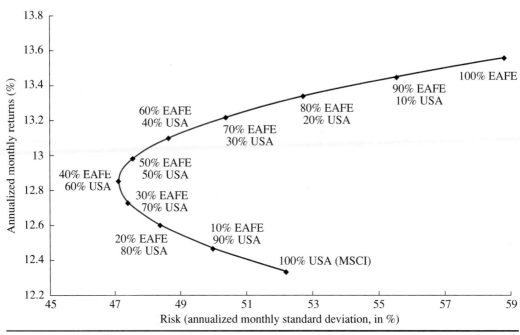

Source: Data from *Morgan Stanley Capital International* (MSCI).

Taken together, these changes have diminished the risk-return benefits of international investing. Nonetheless, these benefits still exist, particularly for those who have patience to invest for the long term. Exhibit 19.9 calculates the standard deviations and annualized returns of different mixes of the U.S. (MSCI) Index and the EAFE Index using quarterly data for the 27-year period 1970–1996. Shifting from a portfolio invested 100% in the U.S. Index to one that contains up to 40% invested in the EAFE Index reduces risk and at the same time return increases. As the percentage invested in the EAFE Index is increased past 40%, both portfolio risk and return rise.

Investing in Emerging Markets

We have already seen in Exhibit 19.2 some of the high risks and rewards historically associated with investing in emerging markets. So it should perhaps come as no surprise that it is often these countries, with volatile economic and political prospects, that offer the greatest degree of diversification and the highest expected returns.[5] Exhibit 19.10 shows how some of these "emerging markets" performed over the ten-year period ending

[5] See, for example, Vihang R. Errunza, "Gains from Portfolio Diversification into Less Developed Countries," *Journal of International Business Studies*, Fall–Winter 1977, pp. 83–99, and Warren Bailey and Rene M. Stulz, "Benefits of International Diversification: The Case of Pacific Basin Stock Markets," *Journal of Portfolio Management*, Summer 1990, pp. 57–62.

EXHIBIT 19.10 Risk and Return for Emerging Markets, 1987–1996

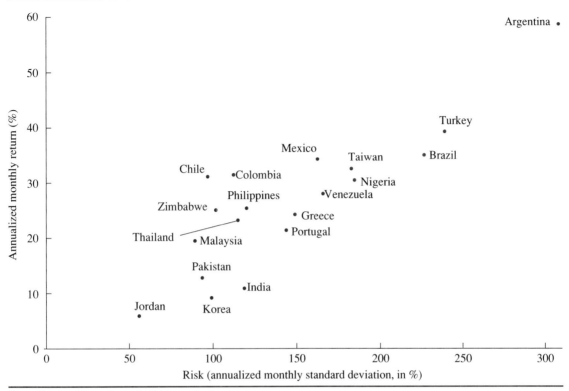

Source: Data from *Morgan Stanley Capital International.*

December 31, 1996. As can be seen, historically, no other stock markets so lavishly rewarded investors. Gains of 30% to 40% a year were not unusual. In response to figures such as these, since 1989, private capital flows to emerging stock markets have expanded from less than $10 billion a year to between $60 billion and $80 billion a year.

But the high returns possible in emerging markets are matched by some breathtaking risks. To begin, these are small markets, representing in the aggregate less than 10% of the world's stock market capitalization. Hence, they are subject to the usual volatility and lack of liquidity associated with small markets. But they face some unique risks as well: relatively unstable governments, the risk of nationalization of businesses, less protection of property rights, and the threat of abrupt price movements. For example, in February 1990, when the newly elected Brazilian president froze most personal bank accounts, the São Paulo exchange plummeted 70% in a few days. Similarly, Taiwan's market rose more than 1,000% from January 1987 to its peak in February 1990. It then gave back most of these gains, falling nearly 80% by October 1990. The Indian stock market also has taken a roller coaster ride. In the euphoria that accompanied the liberalization of the Indian economy, the Bombay stock index rose 458% between June 1991 and April 1992. It then fell 30% in April and May after disclosure of a scandal in which a

EXHIBIT 19.11 Correlations of Monthly Returns with the U.S. Market: Emerging Markets

	1985–1996	1985–1988	1989–1992	1993–1996
Greece	0.12	0.23	0.02	0.14
Jordan	0.03	−0.20	0.26	0.02
Nigeria	0.04	0.14	−0.04	0.00
Portugal	N/A	N/A	0.34	0.21
Turkey	N/A	N/A	−0.20	−0.05
Zimbabwe	−0.02	−0.12	0.08	0.05
India	−0.08	0.00	−0.19	−0.06
Indonesia	N/A	N/A	N/A	0.52
Korea	0.20	0.26	0.22	0.02
Malaysia	0.44	0.51	0.55	0.20
Pakistan	0.01	−0.07	0.05	0.00
Philippines	0.25	0.16	0.45	0.18
Taiwan	0.13	0.12	0.17	0.11
Thailand	0.30	0.27	0.39	0.31
Argentina	0.05	0.00	0.04	0.43
Brazil	0.11	0.04	0.20	0.11
Chile	0.26	0.37	0.14	0.22
Colombia	0.07	0.15	0.10	−0.10
Mexico	0.35	0.42	0.36	0.20
Venezuela	−0.06	−0.08	−0.08	0.00
Composite	0.30	0.33	0.28	0.27
Asia	0.25	0.31	0.22	0.22
Latin America	0.27	0.30	0.24	0.24

Source: Data from *Morgan Stanley Capital International.*

broker cheated financial institutions out of at least half a billion dollars to play the market. The more recent debacles in the Mexican Bolsa and Asian stock markets will remain fresh in the minds of investors for some time to come.

Despite their high investment risks, however, emerging markets can reduce portfolio risk because of their low correlations with returns elsewhere. That is, most of their high total risk is unsystematic in nature.

For example, Exhibit 19.11 shows how some emerging markets were correlated with the U.S. market for the same four-year periods shown in Exhibit 19.7. For the four-year period 1993–1996, these correlations range from a high of 0.52 for Indonesia to a low of −0.10 for Colombia. Moreover, as the composite indices show, these correlations have generally fallen in more recent years. Although not shown here, these emerging markets also have a low correlation with the MSCI World Index.

The fact that most of the emerging markets as a group as well as individually have low correlations with the U.S. market and the MSCI World Index indicates the potential for significant diversification benefits. Exhibit 19.12 shows for the ten-year period 1987–1996 the risk and return of a global portfolio that combines in varying proportions the MSCI World Index with the **IFC Emerging Markets Index**, published by the International Finance Corporation (an international lending organization discussed in Chapter 15).

We see that shifting from a portfolio invested 100% in the MSCI World Index to one that contains up to 20% invested in the IFC Emerging Markets Index reduces risk and

EXHIBIT 19.12 Risk and Return for Various Mixes of MSCI World and IFC Emerging Markets Indices, 1987–1996

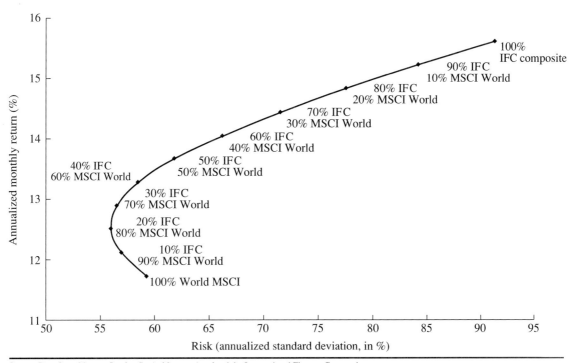

Source: Data from *Morgan Stanley Capital International* and the International Finance Corporation.

at the same time increases expected return. Beyond that point, portfolio risk starts increasing as the higher volatility of the Emerging Markets Index more than offsets the benefits of diversification. Thus, even if one did not expect the emerging markets to out-perform the developed country markets, risk reduction alone would dictate an investment of up to 20% in the emerging markets. Because the emerging markets outperformed the developed country markets during this period, a 20% investment in the IFC Index would have reduced the annualized standard deviation by 3.34% and increased the annual return by 0.78% as well.

One caveat to the data presented on the low correlations between U.S. and emerging-country stock market returns, which are based on monthly data, is that monthly return correlations have tended to understate the long-run interrelatedness of emerging markets and their developed-country counterparts. A recent study shows that correlations between developed and developing country markets are much higher when these correlations are computed using yearly data instead of monthly data.[6] In contrast, correlations between developed nation markets do not vary significantly when computed using yearly instead

[6] John Mullin, "Emerging Equity Markets in the Global Economy," *Federal Reserve Bank of New York Quarterly Review*, Summer 1993, pp. 54–83.

ILLUSTRATION

LATIN AMERICAN
STOCKS WERE HOTTER
THAN SALSA

A conversion to free-market
economics in much of Latin
America did wonders for
their stock markets in the early
1990s. Investors expected that tighter monetary
policies, lower tax rates, significantly lower
budget deficits, and the sale of money-losing
state enterprises would go a long way to cure
the sickly Latin economies. These expectations,
in turn, helped stock markets to soar during the
early 1990s in all the countries shown in Exhibit
19.13 except Brazil—the one country that has

backslid on instituting serious market-oriented
reforms. Overall, from 1984 through 1996, the
six Latin American stock markets depicted in
Exhibit 19.13 rose an average 2,300% in dollar
terms, far exceeding the returns available in
any other region of the world. It should be
noted, however, that these gains reflect the markets' low starting points as much as the success
of the economic reforms. Most Latin American
markets took a tumble in 1994, led by the Mexican market, as interest rates rose in the United
States and it became obvious that the payoffs
from these economic changes would take more
time to be realized and would be riskier than
expected.

EXHIBIT 19.13 Latin American Stock Markets Take Off, 1985–1996

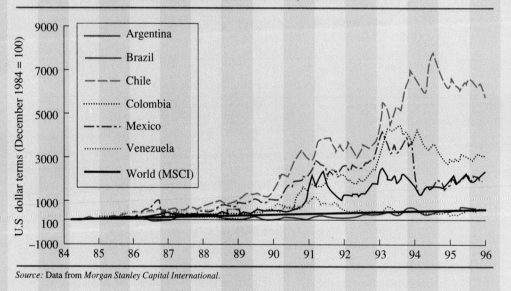

Source: Data from *Morgan Stanley Capital International.*

of monthly data. These results suggest that, because of various impediments to capital
mobility in emerging markets, events that affected developed country returns immediately
tended to affect emerging market returns with a lag. As these impediments to capital
mobility—which include government restrictions on capital flows and a lack of liquidity—are reduced, the low monthly correlations between developed and developing
markets are likely to rise in the future.

Questions also exist as to the returns that can reliably be expected from emerging market investments. A number of researchers have suggested that these high measured returns may owe more to selection and survivorship bias than to economic reality.[7] In particular, given the way in which the IFC data base is constructed, emerging markets that fail to reach a certain threshold capitalization are not included in the emerging market series. We don't observe emerging markets that fail to survive or to achieve the minimum size. Because high-return emerging markets are far more likely to survive and to reach the threshold, the IFC series is clearly biased against low-return markets. Similarly, emerging markets with historically low returns will enter the data base if their recent returns are high, again biasing measured emerging-market returns. For example, a number of Latin American stock markets were in existence with low returns for many years before they entered the IFC data base. These biases result in an overly optimistic picture of the future returns that can be expected from emerging-market investments.

Barriers to International Diversification

Despite the demonstrated benefits to international diversification, these benefits will be limited to the extent that there are barriers to investing overseas. Such barriers do exist. They include legal, informational, and economic impediments that serve to segment national capital markets, deterring investors seeking to invest abroad. The lack of *liquidity*—the ability to buy and sell securities efficiently—is a major obstacle on some overseas exchanges. Other barriers include currency controls, specific tax regulations, relatively less-developed capital markets abroad, exchange risk, and the lack of readily accessible and comparable information on potential foreign security acquisitions. The lack of adequate information can significantly increase the perceived riskiness of foreign securities, giving investors an added incentive to keep their money at home.

Some of these barriers are apparently being eroded. Money invested abroad by both large institutions and individuals is growing dramatically. Despite the erosion in barriers to foreign investing and the consequent growth in the level of foreign investing, however, these investments still represent a relatively minor degree of international diversification. For example, in 1993, Americans held 94% of their equity investments in domestic stocks. Nonetheless, discussions with U.S. institutional investors indicate that many intend to have 20%–25% of their funds invested overseas by the year 2000 (in contrast to only 7% in 1993).[8] This so-called **home bias**—the tendency to hold domestic assets in one's investment portfolio—is also apparent in other countries as well, with domestic residents holding a disproportionate share of the nation's stock market wealth.[9]

---

[7] See, for example, Stephen J. Brown, William N. Goetzmann, and Stephen A. Ross, "Survival," *Journal of Finance*, Vol. 50, pp. 853–873; Campbell Harvey, "Predictable Risk and Returns in Emerging Markets," *Review of Financial Studies*, Vol. 8, pp. 773–816; and William N. Goetzmann and Philippe Jorion, "Re-emerging Markets," Yale School of Management Working Paper, July 1996.

[8] These discussions were reported by Michael R. Sesit, "Foreign Investing Makes a Comeback," *Wall Street Journal*, September 1, 1989, pp. C1 and C14.

[9] The home bias has been documented by Kenneth R. French and James M. Poterba, "Investor Diversification and International Equity Markets," *American Economic Review, Papers and Proceedings*, 1991, pp. 222–226; Ian A Cooper and Evi Kaplanis, "What Explains the Home Bias in Portfolio Investment," *Review of Financial Studies*, Vol 7, pp. 45–60; and Linda Tesar and Ingrid M. Werner, "Home Bias and High Turnover," *Journal of International Money and Finance*, Vol. 14, pp. 467–493.

Several explanations for the home bias in portfolio investments have been put forth.[10] These include the existence of political and currency risks and the natural tendency to invest in the familiar and avoid the unknown. Whether these preferences are rational is another issue.

There are several ways in which U.S. investors can diversify into foreign securities. A small number of foreign firms—fewer than 100—have listed their securities on the New York Stock Exchange (NYSE) or the American Stock Exchange. Historically, a major barrier to foreign listing has been the NYSE requirements for substantial disclosure and audited financial statements. For firms that wished to sell securities in the United States, the U.S. Securities and Exchange Commission's (SEC) disclosure regulations also have been a major obstruction. However, the gap between acceptable NYSE and SEC accounting and disclosure standards and those acceptable to European multinationals has narrowed substantially. Moreover, Japanese and European multinationals that raise funds in international capital markets have been forced to conform to stricter standards. This change may encourage other foreign firms to list their securities and gain access to the U.S. capital market.

Investors can always buy foreign securities in their home markets. One problem with buying stocks listed on foreign exchanges is that it can be expensive, primarily because of steep brokerage commissions. Owners of foreign stocks also face the complications of foreign tax laws and the nuisance of converting dividend payments into dollars.

Instead of buying foreign stocks overseas, investors can buy foreign equities traded in the United States in the form of

1. **American Depository Receipts** (ADRs): These receipts are certificates of ownership issued by a U.S. bank as a convenience to investors in lieu of the underlying shares it holds in custody. The investors in ADRs absorb the handling costs through transfer and handling charges. ADRs for about 1,000 companies from 33 foreign countries are currently traded on U.S. exchanges.
2. **American shares**: These shares are securities certificates issued in the United States by a transfer agent acting on behalf of the foreign issuer. The foreign issuer absorbs part or all of the handling expenses involved.

The easiest approach to investing abroad is to buy shares in an internationally diversified mutual fund, of which a growing number are available. There are four basic categories of mutual fund that invest abroad:

1. **Global funds** can invest anywhere in the world, including the United States.
2. **International funds** invest only outside the United States.
3. **Regional funds** focus on specific geographical areas overseas, such as Asia or Europe.
4. **Single-country funds** invest in individual countries, such as Germany or Taiwan.

The greater diversification of the global and international funds reduces the risk for investors, but it also lessens the chances of a high return if one region (for example, Asia) or country (for example, Germany) suddenly gets hot. The problem with this approach is

---

[10] See, for example, Jun-Koo Kang and René M. Stulz, "Why Is There a Home Bias? An Analysis of Foreign Portfolio Equity Ownership in Japan," *Journal of Financial Economics*, October 1997, pp. 3–28.

that forecasting returns is essentially impossible in an efficient market. This suggests that most investors would be better off buying an internationally diversified mutual fund. Of course, it is possible to construct one's own internationally diversified portfolio by buying shares in several different regional or country funds.

19.2 INTERNATIONAL BOND INVESTING

The benefits of international diversification extend to bond portfolios as well. Barnett and Rosenberg started with a portfolio fully invested in U.S. bonds and then replaced them, in increments of 10%, with a mixture of foreign bonds from seven markets.[11] They then calculated for the period 1973–1983 the risk and return of the 10 portfolios they created. Their conclusions were as follows:

1. As the proportion of U.S. bonds fell, the portfolio return rose. This result reflects the fact that foreign bonds outperformed U.S. bonds over this 10-year period.
2. As the proportion of U.S. bonds fell from 100% to 70%, the volatility of the portfolio fell. This fact reflects the low correlation between U.S. and foreign bond returns.
3. By investing up to 60% of their funds in foreign bonds, U.S. investors could have raised their return substantially while not increasing risk above the level associated with holding only U.S. bonds.

Other studies examining different time periods and markets have similarly found that an internationally diversified bond portfolio delivers superior performance.

19.3 OPTIMAL INTERNATIONAL ASSET ALLOCATION

The evidence clearly indicates that both international stock diversification and international bond diversification pay off. Not surprisingly, expanding the investment set to include stocks and bonds, both domestic and foreign, similarly pays off in terms of an improved risk-return trade-off.

The most detailed study to date of the advantages of international stock and bond diversification is by Bruno Solnik and Bernard Noetzlin.[12] They compared the performances of various investment strategies over the period 1970–1980. Exhibit 19.14 shows the outcome of their analysis. The right-hand curve is the efficient frontier when investments are restricted to stocks only. The left-hand curve is the efficient frontier when investors can buy both stocks and bonds. All returns are calculated in U.S. dollars.

The conclusions of their study were as follows:

1. International stock diversification yields a substantially better risk-return trade-off than does holding purely domestic stock.

[11] G. Barnett and M. Rosenberg, "International Diversification in Bonds," *Prudential International Fixed Income Investment Strategy*, Second Quarter 1983.
[12] Bruno H. Solnik and Bernard Noetzlin, "Optimal International Asset Allocation," *Journal of Portfolio Management*, Fall 1982, pp. 11–21.

EXHIBIT 19.14 Efficient Frontiers, December 1970 to December 1980

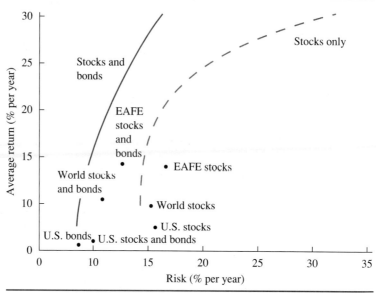

Source: Bruno H. Solnik and Bernard Noetzlin, "Optimal International Asset Allocation," *Journal of Portfolio Management*, Fall 1982. This copyrighted material is reprinted with permission from *Journal of Portfolio Management*, Institutional Investor, Inc., 488 Madison Avenue, New York, NY 10022.

2. International diversification combining stock and bond investments results in substantially less risk than international stock diversification alone.

3. A substantial improvement in the risk-return trade-off can be realized by investing in internationally diversified stock and bond portfolios whose weights do not conform to relative market capitalizations. In other words, the various market indices used to measure world stock and bond portfolios (e.g., Capital International's EAFE Index and World Index) do not lie on the efficient frontier.

As indicated by Exhibit 19.14, optimal *international asset allocation* makes it possible to double or even triple the return from investing in an index fund without taking on more risk. Although Solnik and Noetzlin had the advantage of hindsight in constructing their efficient frontier, they concluded that the opportunities for increased risk-adjusted returns are sizable and that the performance gap between optimal international asset allocations and passive investing in simple index funds is potentially quite large.

19.4 MEASURING THE TOTAL RETURN FROM FOREIGN PORTFOLIO INVESTING

This section shows how to measure the return associated with investing in securities issued in different markets and denominated in a variety of currencies. In general, the **total dollar return** on an investment can be decomposed into three separate elements: dividend/interest income, capital gains (losses), and currency gains (losses).

Bonds

The one-period total dollar return on a foreign bond investment $R_\$$ can be calculated as follows:

$$\begin{array}{c} \text{Dollar} \\ \text{return} \end{array} = \begin{array}{c} \text{Foreign currency} \\ \text{return} \end{array} \times \begin{array}{c} \text{Currency} \\ \text{gain (loss)} \end{array} \qquad (19.4)$$

$$1 + R_s = \left[1 + \frac{B(1) - B(0) + C}{B(0)} \right] (1 + g)$$

WHERE $B(t)$ = foreign currency (FC) bond price at time t
C = foreign currency coupon income
g = percent change in dollar value of the foreign currency

For example, suppose the initial bond price is FC 95, the coupon income is FC 8, the end-of-period bond price is FC 97, and the local currency appreciates by 3% against the dollar during the period. Then, according to Equation 19.4, the total dollar return is 13.8%:

$$\begin{aligned} R_s &= \left[1 + \frac{97 - 95 + 8}{95} \right] (1 + 0.03) - 1 \\ &= (1.105)(1.03) - 1 \\ &= 13.8\% \end{aligned}$$

Note that the currency gain applies to both the local-currency principal and to the local-currency return.

Stocks

Using the same terminology, the one-period total dollar return on a foreign stock investment $R_\$$ can be calculated as follows:

$$\begin{array}{c} \text{Dollar} \\ \text{return} \end{array} = \begin{array}{c} \text{Foreign currency} \\ \text{return} \end{array} \times \begin{array}{c} \text{Currency} \\ \text{gain (loss)} \end{array} \qquad (19.5)$$

$$1 + R_s = \left[1 + \frac{P(1) - P(0) + DIV}{P(0)} \right] (1 + g)$$

WHERE $P(t)$ = foreign currency stock price at time t
DIV = foreign currency dividend income

For example, suppose the beginning stock price is FC 50, the dividend income is FC 1, the end-of-period stock price is FC 48, and the foreign currency depreciates by 5% against the dollar during the period. Then according to Equation 19.5, the total dollar return is −6.9%:

$$\begin{aligned} R_s &= \left[1 + \frac{(48 - 50 + 1)}{50} \right] (1 - 0.05) - 1 \\ &= (0.98)(0.95) - 1 \\ &= -6.9\% \end{aligned}$$

In this case, the investor suffered both a capital loss on the FC principal and a currency loss on the investment's dollar value.

19.5 MEASURING EXCHANGE RISK ON FOREIGN SECURITIES

We have just seen that the dollar return on a foreign security can be expressed as

$$\frac{\text{Dollar}}{\text{return}} = \frac{\text{Foreign currency}}{\text{return}} \times \frac{\text{Currency}}{\text{gain (loss)}} \tag{19.6}$$

$$1 + R_s = (1 + R_f)(1 + g)$$

where R_f is the foreign currency rate of return. Ignoring the cross-product term, $R_f g$, which should be quite small relative to the other terms (because R_f and g are usually much less than 1) we can approximate Equation 19.6 by Equation 19.7:

$$R_s = R_f + g \tag{19.7}$$

Equation 19.7 says that the dollar rate of return is approximately equal to the sum of the foreign currency return plus the change in the dollar value of the foreign currency. Foreign currency fluctuations introduce exchange risk. As we have already seen (in Chapter 7), the prospect of exchange risk is one of the reasons that investors have a preference for home country securities.

Using Equation 19.7, we can see how exchange rate changes affect the risk of investing in a foreign security (or a foreign market index). Specifically, we can write the standard deviation of the dollar return, $\sigma_\$$, as

$$\sigma_s = [\sigma_f^2 + \sigma_g^2 + 2\sigma_f\sigma_g\sigma_{f,g}]^{1/2} \tag{19.8}$$

WHERE σ_f^2 = the variance (the standard deviation squared) of the foreign currency return
σ_g^2 = the variance of the change in the exchange rate
$\sigma_{f,g}$ = the correlation between the foreign currency return and the exchange rate change

Equation 19.8 shows that the foreign exchange risk associated with a foreign security depends on both the standard deviation of the foreign exchange rate change and the covariance between the exchange rate change and the foreign currency return on the security.

For example, suppose that the standard deviation of the return on Matsushita, a Japanese firm, in terms of yen is 23% and the standard deviation of the rate of change in the dollar:yen exchange rate is 17%. In addition, the estimated correlation between the yen return on Matsushita and the rate of change in the exchange rate is 0.31. Then, according to Equation 19.8, the standard deviation of the dollar rate of return on investing in Matsushita stock is 32.56%:

$$\sigma_s(\text{Matsushita}) = (0.23^2 + 0.17^2 + 2 \times 0.23 \times 0.17 \times 0.31)^{1/2} = 0.3256$$

Clearly, foreign exchange risk increases risk in this case. However, the foreign exchange risk is not additive; that is, the standard deviation of the dollar return—32.56%—is less than the sum of the individual standard deviations—23% + 17%, or 40%. It is conceivable that exchange risk could lower the risk of investing overseas. Lowering risk would require a sufficiently large negative correlation between the rate of exchange rate change and the foreign currency return.

HEDGING CURRENCY RISK ■ The existence of exchange risk leaves open the possibility of hedging to reduce it. Indeed, several studies have suggested that hedging currency risk can reduce the variability of returns on internationally diversified stock and bond portfolios while having little impact on or even enhancing expected returns.[13] These conclusions, however, rested mostly on data from the early 1980s. More recent data call these conclusions into question. These data show that, although during the period 1980–1985—when the dollar was rising—the risk-adjusted returns on hedged stock portfolios dominated those on unhedged portfolios, this result is reversed for the period 1986–1996—when the dollar was generally falling.[14] These reversals in the dominance of unhedged versus hedged efficient frontiers for international stock portfolios occurred because of changes over time in the standard deviations and correlation coefficients of national stock market returns expressed in dollars. That is, the covariance structure of national stock market returns and currency movements is unstable.

In contrast, the returns of hedged, internationally diversified, bond portfolios exhibited dramatically lower volatility than the returns of unhedged bond portfolios over both subperiods. Nonetheless, the case for hedging international bond portfolios is not decisive because the lower standard deviation of hedged bond returns is generally matched by lower returns. These results present investors with the familiar trade-off of risk versus return.

❧ 19.6 SUMMARY AND CONCLUSIONS

As the barriers to international capital flows come down and improved communications and data-processing technology provide low-cost information about foreign securities, investors are starting to realize the enormous potential in international investing. We saw in this chapter that international

[13] See, for example, Amdré F. Perold and Evan C. Schulman, "The Free Lunch in Currency Hedging: Implications for Investment Policy and Performance Standards," *Financial Analysts Journal*, May/June 1988, pp. 45–50; Mark R. Eaker and Dwight M. Grant, "Currency Risk Management in International Fixed-Income Portfolios," *Journal of Fixed Income*, December 1991, pp. 31–37; Jack Glen and Philippe Jorion, "Currency Hedging for International Portfolios," *Journal of Finance*, December 1993, pp. 1865–1886; and Richard M. Levich and Lee R. Thomas III, "Internationally Diversified Bond Portfolios: The Merits of Active Currency Management," NBER Working Paper No. 4340, April 1993.

[14] These results are contained in Peter A. Abken and Milind M. Shrikhande, "The Role of Currency Derivatives in Internationally Diversified Portfolios," *Federal Reserve Bank of Atlanta Economic Review*, Third Quarter 1997, pp. 34–59.

stock and bond diversification can provide substantially higher returns with less risk than investment in a single market. A major reason is that international investment offers a much broader range of opportunities than domestic investment alone, even in a market as large as the United States. An investor restricted to the U.S. stock market, for example, is cut off, in effect, from more than half of the available investment opportunities.

Even though a passive international portfolio—one invested in an index fund based on market capitalization weights—improves risk-adjusted performance, an active strategy can do substantially better. The latter strategy bases the portfolio proportions of domestic and foreign investments on their expected returns and their correlations with the overall portfolio.

➤ Questions

1. As seen in Exhibit 19.3, Hong Kong stocks are about twice as volatile as U.S. stocks. Does that mean that risk-averse American investors should avoid Hong Kong equities? Explain.

2. What characteristics of foreign securities lead to diversification benefits for American investors?

3. Will increasing integration of national capital markets reduce the benefits of international diversifications?

4. Studies show that the correlations between domestic stocks are greater than the correlations between domestic and foreign stocks. Explain why this is likely to be the case. What implications does this fact have for international investing?

5. An alternative to investing in foreign stocks is to invest in the shares of domestic multinationals. Are multinationals likely to provide a reasonable substitute for international portfolio investment?

6. Who is likely to gain more from investing overseas, a resident of the United States or of Mexico? Explain.

7. Why have Latin American stocks performed so well since the mid-1980s? Is this performance likely to continue? Explain.

8. Would you expect emerging markets on average to outperform developed country markets in the future? Explain.

9. Mexican bonds are currently yielding more than 100% annually. Does this high yield make them suitable for American investors looking to raise the return on their portfolios? Explain.

10. The Brazilian stock market rose by 165% during 1988. Are American investors likely to be pleased with that performance? Explain.

11. Comment on the following statement: "On October 19, 1987, the U.S. stock market crashed. As the globe turned the following day, the devastation spread from New York to Tokyo, Hong Kong, Sydney, and Singapore, and on to Frankfurt, Paris, and London, then back to New York. The domino-style spread of the crash from one market to the next accelerated as international investors attempted to outrun the wave of panic selling from Tokyo to London and back to New York. It is difficult to imagine that some investors thought they had been able to diversify their investment risks by spreading their money across different stock markets around the world, when in fact their downside risks were actually multiplying as one market followed another into decline."

12. Persian Gulf countries receive virtually all their income from oil revenues denominated in dollars. At the same time, they buy substantial amounts of goods and services from Japan and Western Europe. Their investment portfolios are heavily weighted toward short-term U.S. Treasury bills and other dollar-denominated money market instruments. Comment on their asset allocation.

13. According to one investment adviser, "I feel more comfortable investing in Western Europe or Canada. I would not invest in South America or other regions with a record of debt defaults and restructurings. The underwriters of large new issues of ADRs of companies from these areas assure us that things are different now. Maybe, but who can say that a government that has defaulted on debt won't change the rules again?" Comment on this statement.

14. Investors should avoid Hong Kong, given its problematic outlook now that Britain has surrendered the colony to China. Comment.

15. In deciding where to invest your money, you read that Germany looks like it's well positioned to capitalize on the opening of Eastern Europe. But Britain is troubled by weak growth and high inflation and interest rates. Which of these countries would it make sense to invest in? Explain.

16. As noted in the chapter, from 1949 to 1990, the Japanese market rose 25,000%.
   a. Given these returns, does it make sense for Japanese investors to diversify internationally?
   b. What arguments would you use to persuade a Japanese investor to invest overseas?

c. Why might Japanese (and other) investors still prefer to invest in domestic securities despite the potential gains from international diversification?

17. Does the high volatility of emerging markets lead to high expected returns for investors?

18. As more U.S. investors shift funds into emerging markets, what factors will drive expected returns?

19. Because ADRs are denominated in dollars and are traded in the United States, they present less foreign exchange risk to U.S. investors than do the underlying foreign shares of stock. Comment.

20. During 1995, the Morgan Stanley Capital International World Index of developed country stock markets rose by 18.7% in dollar terms. In contrast, the IFC emerging markets index fell by just over 17% in dollar terms. Many investment advisers point to this sorry performance of emerging markets as an expensive lesson to investors not to venture too far from home. Do the diverging performances of mature and emerging markets argue against investing in emerging markets?

➤ Problems

1. During the year the price of British gilts (government bonds) went from £102 to £106, while paying a coupon of £9. At the same time, the exchange rate went from £1:$1.76 to £1:$1.62. What was the total dollar return, in percent, on gilts for the year?

2. During the first half of 1990, Swiss government bonds yielded a local currency return of −1.6%. However, the Swiss franc rose by 8% against the dollar over this six-month period. Corresponding figures for France were 1.8% and 2.6%. Which bond earned the higher U.S. dollar return? What was the return on the higher bond?

3. During the year, Toyota Motor Company shares went from ¥9,000 to ¥11,200, while paying a dividend of ¥60. At the same time, the exchange rate went from $1 = ¥145 to $1 = ¥120. What was the total dollar return, in percent, on Toyota stock for the year?

4. During 1989, the Mexican stock market climbed 112% in peso terms while the peso depreciated by 28.6% against the U.S. dollar. What was the dollar return on the Mexican stock market during the year?

5. On February 14, 1994, the dollar fell from ¥106.85 to ¥102.65. Meanwhile, the Tokyo stock market fell 1.63% as measured in yen. What was the one-day dollar return on the Tokyo stock market?

6. During 1997, the Korean Stock Exchange's composite index fell by 42%, while the won lost half its value against the dollar. What was the combined effect of these two declines on the dollar return associated with Korean stocks during 1997?

7. a. In 1992, the Brazilian market rose by 1,117% in cruzeiro terms, while the cruzeiro fell by 91.4% in dollar terms. Meanwhile, the U.S. market rose by 8.5%. Which market did better?

   b. In 1993, the Brazilian market rose by 4,190% in cruzeiro terms, while the cruzeiro fell by 95.9% in dollar terms. Did the Brazilian market do better in dollar terms in 1992 or in 1993?

8. Here are data on stock market returns and exchange rate changes during 1988 for 12 stock markets. Determine the dollar return on each of these markets.

COUNTRY	RETURN IN LOCAL CURRENCY (%)	CURRENCY UNITS PER DOLLAR	
		12/31/87	12/31/88
Australia	14.5	1.41	1.17
Belgium	56.3	35.10	38.80
Canada	10.9	1.29	1.20
France	56.8	5.65	6.31
Germany	27.9	1.68	1.85
Holland	42.8	1.88	2.09
Italy	26.2	1230.00	1357.00
Japan	44.8	129.00	128.00
Spain	25.0	114.00	116.00
Sweden	60.5	6.03	6.30
Switzerland	31.9	1.37	1.58
United Kingdom	9.1	0.56	0.57

9. Suppose that the dollar is now worth DM 1.6372. If one-year German bonds are yielding 9.8% and one-year U.S. Treasury bonds are yielding 6.5%, at what end-of-year exchange rate will the dollar returns on the two bonds be equal? What amount of DM appreciation or depreciation does this equilibrating exchange rate represent?

10. Suppose that over a 10-year period the annualized peseta return of a Spanish bond has been 12.1%. If a comparable dollar bond has yielded an annualized return of 8.3%, what cumulative devaluation of the peseta over this period would be necessary for the return on the dollar bond to exceed the dollar return on the Spanish bond?

11. In 1990, Matsushita bought MCA Inc. for $6.1 billion. At the time of the purchase, the exchange rate was about ¥145/$. By the time that Matsushita sold an 80% stake in MCA to Seagram for $5.7 billion in 1995, the yen had appreciated to a rate of about ¥97/$.
    a. Ignoring the time value of money, what was Matsushita's dollar gain or loss on its investment in MCA?
    b. What was Matsushita's yen gain or loss on the sale?
    c. What did Matsushita's yen gain or loss translate into in terms of dollars? What accounts for the difference between this figure and your answer to part a?

12. The standard deviations of U.S. and Mexican returns over the period 1989–1993 were 12.7% and 29.7%, respectively. In addition, the correlation between the U.S. and Mexican markets over this period was 0.34. Assuming that these data reflect the future as well, what is the Mexican market beta relative to the U.S. market?

13. A portfolio manager is considering the benefits of increasing his diversification by investing overseas. He can purchase shares in individual country funds with the following characteristics:

    a. What are the expected return and standard deviation of return of a portfolio with 25% invested in the United Kingdom and 75% in the United States?
    b. What are the expected return and standard deviation of return of a portfolio with 25% invested in Spain and 75% in the United States?
    c. Calculate the expected return and standard deviation of return of a portfolio with 50% invested in the United States and 50% in the United Kingdom. With 50% invested in the United States and 50% invested in Spain.
    d. Calculate the expected return and standard deviation of return of a portfolio with 25% invested in the United States and 75% in the United Kingdom. With 25% invested in the United States and 75% invested in Spain.
    e. Plot these two sets of risk-return combinations (parts a through d), as in Exhibit 19.6. Which leads to a better set of risk-return choices, Spain or the United Kingdom?
    f. How can you achieve an even better risk-return combination?

14. Suppose that the standard deviations of the British and U.S. stock markets have risen to 38% and 22%, respectively, and the correlation between the U.S. and British markets has risen to 0.67. What is the new beta of the British market from a U.S. perspective?

15. Suppose that the standard deviation of the return on Nestlé, a Swiss firm, in terms of Swiss francs is 19% and the standard deviation of the rate of change in the dollar-franc exchange rate is 15%. In addition, the estimated correlation between the Swiss franc return on Nestlé and the rate of change in the exchange rate is 0.17. Given these figures, what is the standard deviation of the dollar rate of return on investing in Nestlé stock?

	UNITED STATES (%)	UNITED KINGDOM (%)	SPAIN (%)
Expected return	15	12	5
Standard deviation of return	10	9	4
Correlation with the United States	1.0	0.33	0.06

➤Bibliography

BAILEY, WARREN, and RENE M. STULZ, "Benefits of International Diversification: The Case of Pacific Basin Stock Markets." *Journal of Portfolio Management*, Summer 1990, pp. 57–62.

ERRUNZA, VIHANG R. "Gains from Portfolio Diversification into Less Developed Countries." *Journal of International Business Studies*, Fall–Winter 1977, pp. 83–99.

IBBOTSON, ROGER C., RICHARD C. CAR, and ANTHONY W. ROBINSON. "International Equity and Bond Returns." *Financial Analysts Journal*, July–August 1982, pp. 61–83.

LESSARD, DONALD R. "World, Country, and Industry Relationships in Equity Returns: Implications for Risk Reduction Through International Diversification." *Financial Analysts Journal*, January–February 1976, pp. 32–38.

SOLNIK, BRUNO H. *International Investments*. Reading, Mass.: Addison-Wesley, 1988.

_____. "Why Not Diversify Internationally Rather Than Domestically?" *Financial Analysts Journal*, July–August 1974, pp. 48–54.

SOLNIK, BRUNO H., and BERNARD NOETZLIN. "Optimal International Asset Allocation." *Journal of Portfolio Management*, Fall 1982, pp. 11–21.

CORPORATE STRATEGY AND FOREIGN DIRECT INVESTMENT

Luck. There isn't any. Just winners and losers.

The Silver Fox

*A*lthough investors are buying an increasing amount of foreign stocks and bonds, most still invest overseas indirectly, by holding shares of multinational corporations. MNCs create value for their shareholders by investing overseas in projects that have positive *net present values* (NPVs)—returns in excess of those required by shareholders. To continue to earn excess returns on foreign projects, multinationals must be able to transfer abroad their sources of domestic competitive advantage. This chapter discusses how firms create, preserve, and transfer overseas their competitive strengths.

The focus here on competitive analysis and value creation stems from the view that generating projects that are likely to yield *economic rent*—excess returns that lead to positive net present values—is a critical part of the capital budgeting process. This is the essence of corporate strategy—creating and then taking best advantage of imperfections in product and factor markets that are the precondition for the existence of economic rent.

The purpose of this chapter is to examine the phenomenon of **foreign direct investment** (FDI)—the acquisition abroad of plant and equipment—and identify those market imperfections that lead firms to become multinational. Only if these imperfections are well understood can a firm determine which foreign investments are likely ex ante to have positive NPVs. The chapter also analyzes corporate strategies for international expansion and presents a normative approach to global strategic planning and foreign investment analysis.

20.1 THEORY OF THE MULTINATIONAL CORPORATION

It has long been recognized that all MNCs are oligopolists (although the converse is not true), but it is only recently that oligopoly and multinationality have been explicitly linked via the notion of *market imperfections*. These imperfections can be related to product and factor markets or to financial markets.

Product and Factor Market Imperfections

The most promising explanation for the existence of multinationals relies on the theory of *industrial organization* (IO), which focuses on imperfect product and factor markets. *IO theory* points to certain general circumstances under which each approach—exporting, licensing, or local production—will be the preferred alternative for exploiting foreign markets.

According to this theory, multinationals have *intangible capital* in the form of trademarks, patents, general marketing skills, and other organizational abilities.[1] If this intangible capital can be embodied in the form of products without adaptation, then exporting generally will be the preferred mode of market penetration. Where the firm's knowledge takes the form of specific product or process technologies that can be written down and transmitted objectively, then foreign expansion usually will take the licensing route.

Often, however, this intangible capital takes the form of organizational skills that are inseparable from the firm itself. A basic skill involves knowing how best to service a market through new-product development and adaptation, quality control, advertising, distribution, after-sales service, and the general ability to read changing market desires and translate them into salable products. Because it would be difficult, if not impossible, to unbundle these services and sell them apart from the firm, this form of market imperfection often leads to corporate attempts to exert control directly via the establishment of foreign affiliates. However, internalizing the market for an intangible asset by setting up foreign affiliates makes economic sense if—and only if—the benefits from circumventing market imperfections outweigh the administrative and other costs of central control.

A useful means to judge whether a foreign investment is desirable is to consider the type of imperfection that the investment is designed to overcome.[2] *Internalization*, and hence FDI, is most likely to be economically viable in those settings where the possibility of contractual difficulties make it especially costly to coordinate economic activities via arm's-length transactions in the marketplace.

Such "market failure" imperfections lead to both vertical and horizontal direct investment. *Vertical integration*—direct investment across industries that are related to

---

[1] Richard E. Caves, "International Corporations: The Industrial Economics of Foreign Investment," *Economica*, February 1971, pp. 1–27.

[2] These considerations are discussed by William Kahley, "Direct Investment Activity of Foreign Firms," *Economic Review*, Federal Reserve Bank of Atlanta, Summer 1987, pp. 36–51.

different stages of production of a particular good—enables the MNC to substitute internal production and distribution systems for inefficient markets. For instance, vertical integration might allow a firm to install specialized cost-saving equipment in two locations without the worry and risk that facilities may be idled by disagreements with unrelated enterprises. *Horizontal direct investment*—investment that is cross-border but within an industry—enables the MNC to utilize an advantage such as know-how or technology and avoid the contractual difficulties of dealing with unrelated parties. Examples of contractual difficulties are the MNC's inability to price know-how or to write, monitor, and enforce use restrictions governing technology-transfer arrangements. Thus, foreign direct investment makes most sense when a firm possesses a valuable asset and is better off directly controlling use of the asset rather than selling or licensing it.

Yet the existence of market failure is not sufficient to justify FDI. Because local firms have an inherent cost advantage over foreign investors (who must bear, for example, the costs of operating in an unfamiliar, and possibly hostile, environment), multinationals can succeed abroad only if the production or marketing edge that they possess cannot be purchased or duplicated by local competitors. Eventually, though, all barriers to entry erode, and the firm must find new sources of *competitive advantage* or be driven back to its home country. Thus, to survive as multinational enterprises, firms must create and preserve effective barriers to direct competition in product and factor markets worldwide.

Financial Market Imperfections

An alternative, though not necessarily competing, hypothesis for explaining foreign direct investment relies on the existence of financial market imperfections. We have already seen in Chapter 14 that the ability to reduce taxes and circumvent currency controls may lead to greater project cash flows and a lower cost of funds for the MNC than for a purely domestic firm.

An even more important financial motivation for foreign direct investment is likely to be the desire to reduce risks through international diversification. This motivation may be somewhat surprising because the inherent riskiness of the multinational corporation is usually taken for granted. Exchange rate changes, currency controls, expropriation, and other forms of government intervention are some of the risks that are rarely, if ever, encountered by purely domestic firms. Thus, the greater a firm's international investment, the riskier its operations should be.

Yet, there is good reason to believe that being multinational may actually reduce the riskiness of a firm. Much of the systematic or general market risk affecting a company is related to the cyclical nature of the national economy in which the company is domiciled. Hence, the diversification effect due to operating in a number of countries whose economic cycles are not perfectly in phase should reduce the variability of MNC earnings. Several studies indicate that this result, in fact, is the case.[3] Thus, because foreign cash flows generally are not perfectly correlated with those of domestic invest-

---

[3] See, for example, Benjamin I. Cohen, *Multinational Firms and Asian Exports* (New Haven, Conn.: Yale University Press, 1975); and Alan Rugman, "Risk Reduction by International Diversification," *Journal of International Business Studies*, Fall 1976, pp. 75–80.

ments, the greater riskiness of individual projects overseas can well be offset by beneficial portfolio effects. Furthermore, because most of the economic and political risks specific to the multinational corporation are unsystematic, they can be eliminated through diversification.

The value of international diversification was made clear in Chapter 19. Thus, the ability of multinationals to supply an indirect means of international diversification should be advantageous to investors. However, this *corporate international diversification* will prove beneficial to shareholders only if there are barriers to direct *international portfolio investment* by individual investors. These barriers do exist and were described in Chapter 19. However, we also saw that many of these barriers are eroding.

Our present state of knowledge does not allow us to make definite statements about the relative importance of financial and nonfinancial market imperfections in stimulating foreign direct investment. Most researchers who have studied this issue, however, would probably agree that the nonfinancial market imperfections are much more important than the financial ones. In the remainder of this chapter, therefore, we will concentrate on the effects of nonfinancial market imperfections on overseas investment.

20.2 THE STRATEGY OF MULTINATIONAL ENTERPRISE

An understanding of the strategies followed by MNCs in defending and exploiting those barriers to entry created by product and factor market imperfections is crucial to any systematic evaluation of investment opportunities. For one thing, such an understanding would suggest those projects that are most compatible with a firm's international expansion. This ranking is useful because time and money constraints limit the investment alternatives that a firm is likely to consider. More important, a good understanding of multinational strategies should help to uncover new and potentially profitable projects; only in theory is a firm fortunate enough to be presented, with no effort or expense on its part, with every available investment opportunity. This creative use of knowledge about global corporate strategies is as important an element of rational investment decision making as is the quantitative analysis of existing project possibilities.

Linking strategic planning and capital allocation yields two other key advantages as well. First, the true economics of investments can be assessed more accurately for strategies than for projects. Second, the quality of the capital budgeting process typically improves greatly when capital expenditures are tied directly to the development and approval of business strategies designed to build or exploit competitive advantages.

Some MNCs rely on product innovation, others on product differentiation, and still others on cartels and collusion to protect themselves from competitive threats. We will now examine three broad categories of multinationals and their associated strategies.[4]

---

[4] These categories are described by Raymond Vernon, *Storm Over the Multinationals* (Cambridge, Mass.: Harvard University Press, 1977); and Ian H. Giddy, "The Demise of the Product Cycle Model in International Business Theory," *Columbia Journal of World Business*, Spring 1978, p. 93.

Innovation-Based Multinationals

Firms such as 3M (United States), N.V. Philips (Netherlands), and Sony (Japan) create barriers to entry by continually introducing new products and differentiating existing ones, both domestically and internationally. Firms in this category spend large amounts of money on research and development (R&D) and have a high ratio of technical to factory personnel. Their products typically are designed to fill a need perceived locally that often exists abroad as well. Similarly, firms such as Wal-Mart, Toys 'R' Us, and Price/Costco take advantage of unique process technologies—largely in the form of superior information gathering, organizational, and distribution skills—to sell overseas.

But technological leads have a habit of eroding. In addition, even the innovative multinationals retain a substantial proportion of standardized product lines. As the industry matures, other factors must replace technology as a barrier to entry; otherwise, local competitors may succeed in replacing foreign multinationals in their home markets.

The Mature Multinationals

What strategies have enabled the automobile, petroleum, paper and pulp, and packaged-foods industries, among others, to maintain viable international operations long after their innovative leads have disappeared and their products have become standardized? Simply put, these industries have maintained international viability by erecting the same barriers to entry internationally as those that allowed them to remain domestic

EXHIBIT 20.1 Inflows of Foreign Direct Investment to Developing Countries: 1980–1995

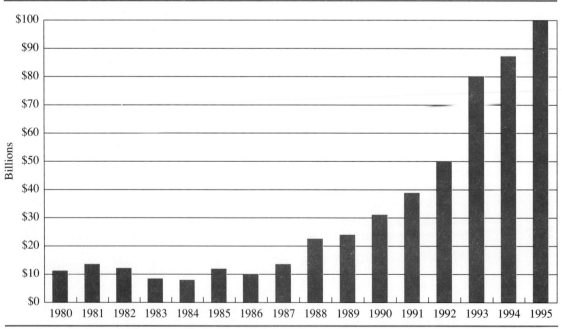

Source: Reprinted with permission from the World Bank.

oligopolists. A principal barrier is the presence of **economies of scale**, which exist whenever a given increase in the scale of production, marketing, or distribution results in a less-than-proportional increase in cost. The existence of scale economies means that there are inherent cost advantages to being large. The more significant these scale economies are, therefore, the greater will be the cost disadvantage faced by a new entrant to the market.

Some companies, such as Coca-Cola, MacDonald's, Nestlé, and Procter & Gamble, take advantage of enormous advertising expenditures and highly developed marketing skills to differentiate their products and keep out potential competitors that are wary of the high marketing costs of new product introduction. By selling in foreign markets, these firms can exploit the premium associated with their strong brand names. Increasingly, consumer goods firms that have traditionally stayed home also are going abroad in an attempt to offset slowing or declining domestic sales in a maturing U.S. market. Such firms, which include Anheuser-Busch (maker of Budweiser beer), Campbell Soup, and Philip Morris, find that selling overseas enables them to utilize their marketing skills and to take advantage of the popularity of American culture abroad.

Both the established multinationals and the newcomers now are moving into the emerging markets of Eastern Europe, Latin America, and Asia in a big way. Companies such as Nestlé and Procter & Gamble expect sales of brand-name consumer goods to soar as disposable incomes rise in the developing countries in contrast to the mature markets of Europe and the United States. The costs and risks of taking advantage of these profitable growth opportunities are also lower today now that their more free-market-oriented governments have reduced trade barriers and cut regulations. In response, foreign direct investment in emerging markets by multinationals has soared in recent years (see Exhibit 20.1). For example, Exhibit 20.2 shows the capital investments for soda production and bottling facilities in emerging markets announced in 1992 and early 1993 alone by PepsiCo and Coca-Cola as they raced to ensure that there would be a bottle of cola on every street corner around the world.

Other firms, such as Alcan and Exxon, fend off new market entrants by exploiting economies of scale in production and transportation. Economies of scale also explain why

EXHIBIT 20.2 Planned Capital Investments for Soda Production and Bottling in Emerging Markets Announced by PepsiCo and Coca-Cola in 1992 and 1993 (U.S. $ Millions)

COUNTRY	PEPSICO	COCA-COLA
Chile	110	100
China	7.5	250
Czech Republic	N/A	28
Hungary	115	59
India	N/A	20
Mexico	750	195
Poland	560	300
Russia	N/A	65
Vietnam*	2	25
Total	$1,544.5	$1,042.0

*Pending removal of U.S. sanctions.

Source: Company reports. Data reported in *Business Week,* August 30, 1993, p. 46.

so many firms invested in Western Europe in preparation for *Europe 1992*, when cross-border barriers to the movement of goods, services, labor, and capital were removed (see Appendix 20B). Their basic rationale was that once Europe becomes a single market, the opportunities to exploit economies of scale will be greatly expanded. Companies that were not well positioned in the key European markets feared that they would be at a cost disadvantage relative to multinational rivals that were better able to exploit these scale economies.

Still other firms take advantage of economies of scope. **Economies of scope** exist whenever the same investment can support multiple profitable activities less expensively in combination than separately. Examples abound of the cost advantages to producing and selling multiple products related by a common technology, set of production facilities, or distribution network. For example, Honda has leveraged its investment in small-engine technology in the automobile, motorcycle, lawn mower, marine engine, chain saw, and generator businesses. Similarly, Matsushita has leveraged its investment in advertising and distribution of Panasonic products in a number of consumer and industrial markets, ranging from personal computers to VCRs. Each dollar invested in the Panasonic name or distribution system aids sales of dozens of different products.

Production economies of scope are becoming more prevalent as flexible manufacturing systems allow the same equipment to produce a variety of products more cheaply in combination than separately. The ability to manufacture a wide variety of products—with little cost penalty relative to large-scale manufacture of a single product—opens up new markets, customers, and channels of distribution, and with them, new routes to competitive advantage.

A strategy that is followed by Texas Instruments, Hewlett-Packard, Sony, and others, is to take advantage of the *learning curve* in order to reduce costs and drive out actual and potential competitors. This latter concept is based on the old dictum that you improve with practice. As production experience accumulates, costs can be expected to decrease because of improved production methods, product redesign and standardization, and the substitution of cheaper materials or practices. Thus, there is a competitive payoff from rapid growth. By increasing its share of the world market, a firm can lower its production costs and gain a competitive advantage over its rivals.

The consequences of disregarding these economic realities are illustrated by U.S. television manufacturers, who (to their sorrow) ignored the growing market for color televisions in Japan in the early 1960s. The failure of the U.S. manufacturers to preempt Japanese color-TV development spawned a host of Japanese competitors—such as Sony, Matsushita, and Hitachi—who not only came to dominate their own market but eventually took most of the U.S. market. The moral seems to be that to remain competitive at home, it is often necessary to challenge potential rivals in their local markets.

To counter the danger that a foreign multinational will use high home-country prices to subsidize a battle for market share overseas, firms often will invest in one another's domestic markets. This strategy is known as *cross-investment*. The implied threat is that "if you undercut me in my home market, I'll do the same in your home market." Firms with high domestic market share, and minimal sales overseas, are especially vulnerable to the strategic dilemma illustrated by the example of Fiat.

ILLUSTRATION

FIAT'S STRATEGIC DILEMMA

Suppose Toyota, the Japanese auto company, cuts prices in order to gain market share in Italy. If Fiat, the dominant Italian producer with minimal foreign sales, responds with its own price cuts, it will lose profit on most of its sales. In contrast, only a small fraction of Toyota's sales and profits are exposed. Fiat is effectively boxed in: If it responds to the competitive intrusion with a price cut of its own, the response will damage it more than Toyota.

The correct competitive response is for the local firm (Fiat) to cut price in the intruder's (Toyota's) domestic market (Japan). Having such a capability will deter foreign competitors from using high home-country prices to subsidize marginal cost pricing overseas. However, this strategy necessitates investing in the domestic markets of potential competitors. The level of market share needed to pose a credible retaliatory threat depends on access to distribution networks and the importance of the market to the competitor's profitability. The easier distribution access is and the more important the market is to competitor profitability, the smaller the necessary market share.[5]

---

[5] The notion of undercutting competitors in their home market is explored in Gary Hamel and C. K. Prahalad, "Do You Really Have a Global Strategy?" *Harvard Business Review*, July–August 1985, pp. 139–148.

The Senescent Multinationals

Eventually, product standardization is far enough advanced or organizational and technological skills are sufficiently dispersed that all barriers to entry erode. What strategies do large multinationals follow when the competitive advantages in their product lines or markets become dissipated?

One possibility is to enter new markets where little competition currently exists. For example, Crown Cork & Seal, the Philadelphia-based maker of bottle tops and cans, reacted to slowing growth and heightened competition in its U.S. business by expanding overseas. It set up subsidiaries in such countries as Thailand, Malaysia, Zambia, Peru, and Ecuador, guessing—correctly, as it turned out—that in those developing and urbanizing societies, people would eventually switch from home-grown produce to food in cans and drinks in bottles. However, local firms soon are capable of providing stiff competition for those foreign multinationals that are not actively developing new sources of differential advantage.

One strategy often followed when senescence sets in is to use the firm's *global-scanning capability* to seek out lower-cost production sites. Costs can then be minimized by combining production shifts with *rationalization* and *integration* of the firm's manufacturing facilities worldwide. This strategy usually involves plants specializing in different stages of production—for example, in assembly or fabrication—as well as in particular components or products. Yet the relative absence of market imperfections confers a multinational production network with little, if any, advantage over production by purely local enterprises. For example, many U.S. electronics and textile firms shifted production facilities to Asian locations, such as Taiwan and Hong Kong, to take advantage of lower labor costs there. However, as more firms took advantage of this cost-reduction opportunity, competition in U.S. consumer electronics and textile markets—increasingly from Asian firms—intensified, causing domestic prices to drop and excess profits to be dissipated.

In general, the excess profits due to processing new information are temporary. Once new market or cost-reduction opportunities are recognized by other companies, the profit rate declines to its normal level. Hence, few firms rely solely on cost minimization or entering new markets to maintain competitiveness.

The more common choice is to drop old products and turn corporate skills to new products. Companies that follow this strategy of continuous product rollover are likely to survive as multinationals. Those who are unable to transfer their original competitive advantages to new products or industries must plan on divesting their foreign operations and returning home. But firms that withdraw from overseas operations because of a loss of competitive advantage should not count on a very profitable homecoming.

ILLUSTRATION

THE U.S. TIRE INDUSTRY GETS RUN OVER

The U.S. tire industry illustrates the troubles faced by multinational firms that have lost their source of differential advantage. Although Europe once was a profitable market for the Big Four U.S. tiremakers—Goodyear, Firestone, Goodrich, and Uniroyal—each of these firms has, by now, partially or completely eliminated its European manufacturing operations. The reason is the extraordinary price competition resulting from a lack of unique products or production processes and the consequent ease of entry into the market by new firms. Moreover, these firms then faced well-financed challenges in the U.S. market by, among others, the French tiremaker Michelin, the developer of the radial tire and its related production technology. Uniroyal responded by selling off its European tire-manufacturing operation and reinvesting its money in businesses that were less competitive there (and, hence, more profitable) than the tire industry. This reinvestment includes its chemical, plastics, and industrial-products businesses in Europe. Similarly, Goodrich stopped producing tires for new cars and expanded its operations in polyvinyl chloride resin and specialty chemicals. In 1986, Uniroyal and Goodrich merged their tire units to become Uniroyal Goodrich Tire, selling only in North America. Late in 1989, its future in doubt, Uniroyal Goodrich sold out to Michelin. The previous year, in early 1988, Firestone sold out to the Japanese tiremaker Bridgestone. Goodyear is now the only one of the Big Four tiremakers that is still a U.S. company.

Foreign Direct Investment and Survival

Thus far we have seen how firms are capable of becoming and remaining multinationals. However, for many of these firms, becoming multinational is not a matter of choice but, rather, one of survival.

COST REDUCTION ■ It is apparent, of course, that if competitors gain access to lower-cost sources of production abroad, following them overseas may be a prerequisite for domestic survival. One strategy that is often followed by firms for which cost is the key consideration is to develop a global-scanning capability to seek out lower-cost production sites or production technologies worldwide. In fact, firms in competitive industries have to continually seize new, nonproprietary, cost-reduction opportunities, not to earn excess returns but to make normal profits and survive.

ECONOMIES OF SCALE ■ A somewhat less-obvious factor motivating foreign investment is the effect of economies of scale. In a competitive market, prices will be forced close to marginal costs of production. Hence, firms in industries characterized by high fixed costs relative to variable costs must engage in volume selling just to break even. A new term describes the size that is required in certain industries to compete effectively in the global marketplace: *world-scale*. These large volumes may be forthcoming only if the firms expand overseas. For example, companies manufacturing products such as computers that require huge R&D expenditures often need a larger customer base than that provided by even a market as large as the United States in order to recapture their investment in knowledge. Similarly, firms in capital-intensive industries with enormous production economies of scale may also be forced to sell overseas in order to spread their overhead over a larger quantity of sales.

L.M. Ericsson, the highly successful Swedish manufacturer of telecommunications equipment, is an extreme case. The manufacturer is forced to think internationally when designing new products because its domestic market is too small to absorb the enormous R&D expenditures involved and to reap the full benefit of production scale economies. Thus, when Ericsson developed its revolutionary AXE digital switching system, it geared its design to achieve global market penetration.

These firms may find a foreign-market presence necessary in order to continue selling overseas. Local production can expand sales by providing customers with tangible evidence of the company's commitment to service the market. It also increases sales by improving a company's ability to service its local customers. For example, an executive from Whirlpool, explaining why the company decided to set up operations in Japan after exporting to it for 25 years, said, "You can only do so much with an imported product. We decided we needed a design, manufacturing, and corporate presence in Japan to underscore our commitment to the Japanese market and to drive our global strategy in Asia. You can't do that long distance."[6] Domestic retrenchment thus can involve not only the loss of foreign profits but also an inability to price competitively in the home market because it no longer can take advantage of economies of scale.

MULTIPLE SOURCING ■ Once a firm has decided to produce abroad, it must determine where to do so. Although cost minimization will often dictate concentrating production in one or two plants, fear of strikes and political risks usually lead firms to follow a policy of multiple sourcing. For example, a series of strikes against British Ford in the late 1960s and early 1970s caused Ford to give lower priority to rationalization of supplies. It went for safety instead, by a policy of double sourcing. Ford since has modified this policy, but many other firms still opt for several smaller plants in different countries instead of one large plant that could take advantage of scale economies but that would be vulnerable to disruptions.

---

[6]"Whirlpool," *Fortune* (Special Advertising Section), March 18, 1993, p. S–21.

The costs of multiple sourcing are obvious; the benefits, less apparent. One benefit is the potential leverage that can be exerted against unions and governments by threatening to shift production elsewhere. For example, to reach settlement in the previously mentioned strikes against British Ford, Henry Ford II used the threat of withholding investments from England and placing them in Germany. Another, more obvious, benefit is the additional safety achieved by having several plants capable of supplying the same product.

Multiple facilities also give the firm the option of switching production from one location to another to take advantage of transient unit-cost differences arising from, say, real exchange rate changes or new labor contracts. This option is enhanced, albeit at a price, by building excess capacity into the plants.

KNOWLEDGE SEEKING ■ Some firms enter foreign markets for the purpose of gaining information and experience that is expected to prove useful elsewhere. For instance, Beecham, an English firm (now SmithKline Beecham), deliberately set out to learn from its U.S. operations how to be more competitive, first in the area of consumer products and, later, in pharmaceuticals. This knowledge proved highly valuable in competing with American and other firms in its European markets. Similarly, in late 1992, the South Korean conglomerate Hyundai moved its PC division to the United States in order to keep up with the rapidly evolving personal computer market, whose direction was set by the U.S. market.

The flow of ideas is not all one-way, however. As Americans have demanded better-built, better-handling, and more fuel-efficient small cars, Ford of Europe has become an important source of design and engineering ideas and management talent for its U.S. parent, including the hugely successful Taurus.

In industries characterized by rapid product innovation and technical breakthroughs by foreign competitors, it is imperative to track overseas developments constantly. Japanese firms excel here, systematically and effectively collecting information on foreign innovation and disseminating it within their own research and development, marketing, and production groups. The analysis of new foreign products as soon as they reach the market is an especially long-lived Japanese technique. One of the jobs of Japanese researchers is to tear down a new foreign product and analyze how it works as a base on which to develop a product of their own that will outperform the original. In a bit of a switch, Data General's Japanese operation is giving the company a close look at Japanese technology, enabling it to quickly pick up and transfer back to the United States new information on Japanese innovations in the areas of computer design and manufacturing.

More firms are building labs in Japan and hiring its scientists and engineers to absorb Japan's latest technologies. For example, Texas Instruments works out production of new chips in Japan first because, an official says, "production technology is more advanced and Japanese workers think more about quality control."[7] A firm that remains at home can be "blind-sided" by current or future competitors with new products, manufacturing processes, or marketing procedures.

Tough competition in a foreign market is a valuable experience in itself. For many industries, a competitive home marketplace has proved to be as much of a competitive advantage as cheap raw materials or technical talent. Fierce domestic competition is one reason the U.S. telecommunications industry has not lost its lead in technology, R&D, design, software, quality, and cost. Japanese and European firms are at a disadvantage in this business because they do not have enough competition in their home markets. U.S. companies have been able to engineer a great leap forward because they saw firsthand what the competition could do. Thus, for telecommunications firms such as Germany's Siemens, Japan's NEC, and France's Alcatel, a position in the U.S. market has become mandatory.

Similarly, it is slowly dawning on consumer electronics firms that to compete effectively elsewhere, they must first compete in the toughest market of all: Japan. What they learn in the process—from meeting the extraordinarily demanding standards of Japanese consumers and battling a dozen relentless Japanese rivals—is invaluable and will possibly make the difference between survival and extinction.

In contrast, protectionism has worked against Japanese pharmaceutical companies. Unlike U.S. pharmaceutical companies, which operate within a fiercely competitive home market that fosters an entrepreneurial spirit and scientific innovation, sheltered Japanese pharmaceutical companies have never had to adapt to international standards and competition, which has left them at a competitive disadvantage.

Although it may be stating the obvious to note that operating in a competitive marketplace is an important source of competitive advantage, this viewpoint appears to be a minority one today. Many companies prepared for Europe 1992 by seeking mergers, alliances, and collaboration with competitors. Some went further and petitioned their governments for protection from foreign rivals and assistance in R&D. However, to the extent that companies succeed in sheltering themselves from competition, they endanger the basis of true competitive advantage: dynamic improvement, which derives from continuous effort to enhance existing skills and learn new ones.

This point is illustrated by the sorry experience of the European film industry. In order to preserve an indigenous industry, European governments have provided subsidies

---

[7]*Wall Street Journal*, August 1, 1986, p. 6.

for local filmmakers and imposed restrictions on the showing of U.S. movies. Since 1980, however, cinema audiences for European-made films have collapsed—falling from 475 million in 1980 to 120 million in 1994. Meanwhile, the audience for American films has barely changed. In 1968, U.S. films took 35% of European box-office revenues; now they take 80%, and in some countries 90%. One reason is the very subsidies and regulations

---

ILLUSTRATION

BRIDGESTONE BUYS FIRESTONE

As noted earlier, in March 1988, Bridgestone, the largest Japanese tiremaker, bought Firestone and its worldwide tire operations. Like other Japanese companies that preceded it to the United States, Bridgestone was motivated by a desire to circumvent potential trade barriers and soften the impact of the strong yen. The move also greatly expanded Bridgestone's customer base, allowing it to sell its own tires directly to U.S. automakers, and strengthened its product line. Bridgestone excelled in truck and heavy-duty-vehicle tires, whereas Firestone's strength was in passenger-car tires. But beyond these facts, a key consideration was Bridgestone's wish to reinforce ties with Japanese auto companies that had set up production facilities in the United States. By 1992, these companies, either directly or in joint ventures with U.S. firms, had the capacity to produce about 2 million vehicles annually in the United States.

Firestone also contributed plants in Spain, France, Italy, Portugal, Argentina, Brazil, and Venezuela. Thus, Bridgestone's purchase of Firestone has firmly established the company not only in North America, but in Europe and South America as well. Formerly, it had been primarily an Asian firm, but it had come to acknowledge the need to service Japanese automakers globally by operating closer to their customers' production facilities. The increasing globalization of the automobile market has prompted vehicle producers and tiremakers alike to set up production facilities in each of the three main markets: North America, Western Europe, and Japan.

Two main factors have been responsible for this trend toward globalization: First, transport costs are high for tires, and, as a result, exporting ceased to be a viable long-term strategy for supplying distant markets. Second, shifting manufacturing overseas was the only way for the tire companies to meet the logistic challenges posed by the adoption of "just-in-time" manufacturing and inventory systems by automakers.

A series of combinations in the tire industry—including Sumitomo Rubber's purchase of Dunlop Tire's European and U.S. operations, Pirelli's acquisition of Armstrong Tire and Rubber, and Continental AG's acquisition of General Tire and Rubber and its subsequent joint venture with two Japanese tiremakers—practically forced Bridgestone to have a major presence in the important American market if it were to remain a key player in the United States and worldwide. Absent such a move, its Japanese competitors might have taken Bridgestone's share of the business of Japanese firms producing in the United States and Europe. This result would have affected its competitive stance in Japan as well.

A similar desire to increase its presence in the vital North American market was behind Michelin's 1989 acquisition of Uniroyal Goodrich. For Michelin, the addition of Uniroyal Goodrich provided entry into private-label and associate-label tire markets from which it had been absent, as well as added sales to U.S. automakers.

intended to support Europe's filmmakers; they have spawned a fragmented industry, in which producers make films to show to each other, rather than to a mass audience. In contrast, without subsidies and regulations to protect them, U.S. filmmakers have been forced to make films with global appeal rather than trying for art-house successes. Their achievement is reflected in the fact that by 1998 foreign box office accounted for more than 50% of Hollywood's revenues.

KEEPING DOMESTIC CUSTOMERS ■ Suppliers of goods or services to multinationals often will follow their customers abroad in order to guarantee them a continuing product flow. Otherwise, the threat of a potential disruption to an overseas supply line—for example, a dock strike or the imposition of trade barriers—can lead the customer to select a local supplier, which may be a domestic competitor with international operations. Hence, comes the dilemma: Follow your customers abroad or face the loss of not only their foreign but also their domestic business. A similar threat to domestic market share has led many banks, advertising agencies, and accounting, law, and consulting firms to set up foreign practices in the wake of their multinational clients' overseas expansion.

By now it should be apparent that a foreign investment may be motivated by considerations other than profit maximization and that its benefits may accrue to an affiliate far removed from the scene. Moreover, these benefits may take the form of a reduction in risk or an increase in cash flow, either directly or indirectly. Direct cash flows would include those based on a gain in revenues or a cost savings. Indirect flows include those resulting from a competitor's setback or the firm's increased leverage to extract concessions from various governments or unions (for example, by having the flexibility to shift production to another location). In computing these indirect effects, a firm must consider, of course, what would have been the company's worldwide cash flows in the absence of the investment.

20.3 DESIGNING A GLOBAL EXPANSION STRATEGY

Although a strong competitive advantage today in, say, technology or marketing skills may give a company some breathing space, these competitive advantages will eventually erode, leaving the firm susceptible to increased competition both at home and abroad. The emphasis must be on systematically pursuing policies and investments congruent with worldwide survival and growth. This approach involves five interrelated elements.

1. Awareness of Profitable Investments

It requires an awareness of those investments that are likely to be most profitable. As we have previously seen, these investments are ones that capitalize on and enhance the differential advantage possessed by the firm; that is, an investment strategy should focus explicitly on building competitive advantage. This strategy could be geared to building volume where economies of scale are all important or to broadening the product scope where economies of scope are critical to success. Such a strategy is likely to encompass a sequence of tactical projects; several may yield low returns when considered in isolation, but together they may either create valuable future investment opportunities or allow the firm to continue earning excess returns on existing investments. Proper evaluation of a sequence of tactical projects designed to achieve competitive advantage requires that the projects be analyzed jointly, rather than incrementally.

For example, if the key to competitive advantage is high volume, the initial entry into a market should be assessed on the basis of its ability to create future opportunities to build market share and the associated benefits thereof. Alternatively, market entry overseas may be judged according to its ability to deter a foreign competitor from launching a market-share battle by posing a credible retaliatory threat to the competitor's profit base. By reducing the likelihood of a competitive intrusion, foreign market entry may lead to higher future profits in the home market.

In designing and valuing a strategic investment program, a firm must be careful to consider the ways in which the investments interact. For example, where scale economies exist, investment in large-scale manufacturing facilities may be justified only if the firm has made supporting investments in foreign distribution and brand awareness. Investments in a global distribution system and a global brand franchise, in turn, are often economical only if the firm has a range of products (and facilities to supply them) that can exploit the same distribution system and brand name.

Developing a broad product line usually requires and facilitates (by enhancing economies of scope) investment in critical technologies that cut across products and businesses. Investments in R&D also yield a steady stream of new products that raises the return on the investment in distribution. At the same time, a global distribution capability may be critical in exploiting new technology.

The return to an investment in R&D is largely determined by the size of the market in which the firm can exploit its innovation and the durability of its technological advantage. As the technology-imitation lag shortens, a company's ability to fully exploit a technological advantage may depend on its being able to quickly push products embodying that technology through distribution networks in each of the world's critical national markets.

Individually or in pairs, investments in large-scale production facilities, worldwide distribution, a global brand franchise, and new technology are likely to be negative net present value projects. Together, however, they may yield a highly positive NPV by forming a mutually supportive framework for achieving global competitive advantage.

2. Selecting a Mode of Entry

This global approach to investment planning necessitates a systematic evaluation of individual entry strategies in foreign markets, a comparison of the alternatives, and selection of the optimal mode of entry. For example, in the absence of strong brand names or distribution capabilities but with a labor-cost advantage, Japanese television manufacturers entered the U.S. market by selling low-cost, private-label, black-and-white TVs.

A recent entry mode is the acquisition of a state-owned enterprise. In pursuit of greater economic efficiency or to raise cash, governments around the world are *privatizing* (selling to the private sector) many of their companies. From 1985 through 1992, governments sold off a total of $328 billion in state-owned firms. In 1992 alone, state-owned firms worth $69 billion in nearly 50 countries were sold. Many of the firms being privatized come from the same industries: airlines, utilities (telecommunications, gas, electric, water), oil, financial services (banking, insurance), and manufacturing (petro-chemicals, steel, autos). These privatizations present new opportunities for market entry in areas traditionally closed to multinationals.

3. Auditing the Effectiveness of Entry Modes

A key element is a continual audit of the effectiveness of current entry modes, bearing in mind that a market's sales potential is at least partially a function of the entry strategy. As knowledge about a foreign market increases or sales potential grows, the optimal market-penetration strategy will likely change. By the late 1960s, for example, the Japanese television manufacturers had built a large volume base by selling private-label TVs. Using this volume base, they invested in new process and product technologies, from which came the advantages of scale and quality. Recognizing the transient nature of a competitive advantage built on labor and scale advantages, Japanese companies, such as Matsushita and Sony, strengthened their competitive position in the U.S. market by investing throughout the 1970s to build strong brand franchises and distribution capabilities. The new-product positioning was facilitated by large-scale investments in R&D. By the 1980s, the Japanese competitive advantage in TVs and other consumer electronics had switched from being cost-based to one based on quality, features, strong brand names, and distribution systems.[8]

4. Using Appropriate Evaluation Criteria

A systematic investment analysis requires the use of appropriate evaluation criteria. Nevertheless, despite (or perhaps because of) the complex interactions between invest-ments or corporate policies and the difficulties in evaluating proposals, most firms still use simple rules of thumb in selecting projects to undertake. Analytical techniques are used only as a rough screening device or as a final checkoff before project approval. Although simple rules of thumb are obviously easier and cheaper to implement, there is a danger of obsolescence and consequent misuse as the fundamental assumptions underly-ing their applicability change. On the other hand, the use of the theoretically sound and recommended present value analysis is anything but straightforward. The strategic ratio-nale underlying many investment proposals can be translated into traditional capital-budgeting criteria, but it is necessary to look beyond the returns associated with the project itself to determine its true impact on corporate cash flows and riskiness. For example, an investment made to save a market threatened by competition or trade barriers must be judged on the basis of the sales that would otherwise have been lost. Also, export creation and direct investment often go hand in hand. In the case of ICI, the British chemical company, its exports to Europe were enhanced by its strong market position there in other product lines, a position due mainly to ICI's local manufacturing facilities.

---

[8] For an excellent discussion of Japanese strategy in the U.S. TV market and elsewhere, see Hamel and Prahalad, "Do You Really Have a Global Strategy?"

We saw earlier that some foreign investments are designed to improve the company's competitive posture elsewhere. For example, Air Liquide, the world's largest industrial-gas maker, opened a facility in Japan because Japanese factories make high demands of their gas suppliers and keeping pace with them ensures that the French company will stay competitive elsewhere. In the words of the Japanese unit's president, "We want to develop ourselves to be strong wherever our competitors are."[9] Similarly, a spokesperson said that Air Liquide expanded its U.S. presence because the United States is "the perfect marketing observatory."[10] U.S. electronics companies and papermakers have found new uses for the company's gases, and Air Liquide has brought back the ideas to European customers.

Applying this concept of evaluating an investment on the basis of its global impact will force companies to answer tough questions: How much is it worth to protect our reputation for prompt and reliable delivery? What effect will establishing an operation here have on our present and potential competitors or on our ability to supply competitive products, and what will be the profit impact of this action? One possible approach is to determine the incremental costs associated with, say, a defensive action such as building multiple plants (as compared with several larger ones) and then use that number as a benchmark against which to judge how large the present value of the associated benefits (e.g., greater bargaining leverage vis-à-vis host governments) must be to justify the investment.

5. Estimating the Longevity of a Competitive Advantage

The firm must estimate the longevity of its particular form of competitive advantage. If this advantage is easily replicated, both local and foreign competitors will not take long to apply the same concept, process, or organizational structure to their operations. The resulting competition will erode profits to a point where the MNC can no longer justify its existence in the market. For this reason, the firm's competitive advantage should be constantly monitored and maintained to ensure the existence of an effective barrier to entry into the market. Should these entry barriers break down, the firm must be able to react quickly and either reconstruct them or build new ones. But no barrier to entry can be maintained indefinitely; to remain multinational, firms must continually invest in developing new competitive advantages that are transferable overseas and that are not easily replicated by the competition.

20.4 ILLUSTRATION: THE JAPANESE STRATEGY FOR GLOBAL EXPANSION

In 1945, Japan was a bombed-out wreck of a nation, humiliated and forced into unconditional surrender. All through the 1950s, Japanese exports were hampered by a low-quality image. Yet less than 20 years later, Japanese companies such as Sony, Hitachi, Seiko,

---

[9]*Wall Street Journal*, November 12, 1987, p. 32.
[10]*Wall Street Journal*, February 23, 1988, p. 20.

Canon, and Toyota had established worldwide reputations equal to those of Zenith, Kodak, Ford, Philips, and General Electric.

Here are some lessons about international strategy to be learned from the Japanese, who arguably are the most successful global expansionists in history. To begin, it should be pointed out that the Japanese have invested a great deal of money and effort in quality, ever since quality gurus W. Edwards Deming and J. M. Juran crossed the Pacific in the 1950s to teach them how to manage for quality—an approach often called *total quality control* (TQC). Few U.S. companies paid much heed to Deming or Juran and TQC, but the Japanese avidly embraced their ideas. In fact, the Deming Prize is now Japan's most prestigious award for industrial quality control.

Beyond quality, the Japanese have followed a simple strategy—repeated time and again—for penetrating world markets. Whether in cars, TVs, motorcycles, or photo-copiers, the Japanese have started at the low end of the market. In each case, this market segment had been largely ignored by U.S. firms focusing on higher-margin products. At the same time, the Japanese firms invested money in process technologies and simpler product design to cut costs and expand market share. Japan's lower-cost products sold in high volume, giving the manufacturers production economies of scale that other contenders could not match. So-called **lean production**—ultra-efficient manufacturing techniques—gave Japanese industry a formidable competitive edge over U.S. rivals. One advantage was the ability to develop an array of parts and options (although as indicated in Chapter 11, variety got out of hand).

Part of the Japanese manufacturing success stems from developing new organiza-tional forms that allow them to turn out better-quality products faster and at lower cost. Companies such as Honda and Canon have taken a "concurrent engineering" approach to product development. At these companies, design engineers do not just hand off a product to manufacturing engineers with little regard for how easily it might be made or marketed; they develop it together, in teams that also include salespeople, financial analysts, customers, and suppliers to ensure that a high-quality, salable, and profitable product can be readily made at the lowest cost possible.

U.S. companies retreated to the high-end segment of each market, believing that the Japanese could not challenge them there. The incumbents' willingness to surrender the low end of the market was not entirely irrational. In each case, the incumbents had successful, established products that would be cannibalized by a vigorous response. General Motors, for example, believed that if it came up with high-quality smaller cars, sales of these cars would come at the expense of its bigger, higher-margin cars. So, it chose to hold back in responding to competitive attacks by the Japanese.

But the Japanese were not content to remain in the low-end segment of the market. Over time, they invested in new product development, built strong brand franchises and global distribution networks, and moved up-market. They amortized the costs of these investments by rapid expansion across contiguous product segments and by acceleration of the product life cycle. Here, *contiguous* refers to products that share a common tech-nology, brand name, or distribution system. In this way, the Japanese managed to take full advantage of the economies of scope inherent in core technologies, brand franchises, and distribution networks. Simultaneously, global expansion allowed them to build up production volume and garner available scale economies.

ILLUSTRATION

CANON DOESN'T COPY XEROX

The tribulations of Xerox illustrate the dynamic nature of Japanese competitive advantage.[11] Xerox dominates the U.S. market for large copiers. Its competitive strengths—a large direct sales force that constitutes a unique distribution channel, a national service network, a wide range of machines using custom-made components, and a large installed base of leased machines—have defeated attempts by IBM and Kodak to replicate its success by creating matching sales and service networks. Canon's strategy, by contrast, was simply to sidestep these barriers to entry by: (1) creating low-end copiers that it sold through office-product dealers, thereby avoiding the need to set up a national sales force; (2) designing reliability and serviceability into its machines, so users or nonspecialist dealers could service them; (3) using commodity components and standardizing its machines to lower costs and prices and boost sales volume; and (4) selling rather than leasing its copiers. By 1986, Canon and other Japanese firms had more than 90% of copier sales worldwide. And having ceded the low end of the market to the Japanese, Xerox soon found those same competitors flooding into its stronghold sector in the middle and upper ends of the market.

Canon's strategy points out an important distinction between *barriers to entry* and *barriers to imitation*.[12] Competitors, such as IBM and 3M, that tried to imitate Xerox's strategy had to pay a matching entry fee. Through competitive innovation, Canon avoided these costs and, in fact, stymied Xerox's response. Xerox realized that the more quickly it responded—by downsizing its copiers, improving reliability, and developing new distribution channels—the more quickly it would erode the value of its leased machines and cannibalize its existing high-end product line and service revenues. Hence, what were barriers to entry for imitators became barriers to retaliation for Xerox.

[11] This example appears in Gary Hamel and C. K. Prahalad, "Strategic Intent," *Harvard Business Review*, May–June 1989, pp. 63–76.

[12] This distinction is emphasized by Hamel and Prahalad, "Strategic Intent."

Historically, Japanese companies have focused on exports to service their foreign markets with the objective of their limited foreign direct investments being to secure natural resources and low-cost off-shore labor, largely in Asia. But in the 1980s, Japanese corporations changed their investment strategy and began targeting the United States and Europe. The main reason: to sidestep rising American and European protectionism. Between 1980 and 1989, America's share of Japan's cumulative foreign direct investment nearly doubled to 42% from 22%. Europe's share more than doubled to 19% from 9%.

Now, however, Japanese companies are switching their attention back to Asia, where—until recently—rapid economic growth offered better returns than mature markets in America, Europe, and even Japan. By 1993, Asia's share of new Japanese investment flows had jumped to 19% from 12% in 1989, and America's share slipped to 40% from 48% over that same period. Partly, the shift to Asia reflects the high costs of operating in Japan. Increasingly, Japanese companies are using their domestic plants to produce only the newest and the most expensive products. There, Japanese engineers fine-tune the products and assembly procedures. As competitive pressures increase and margins start shrinking the companies begin moving production offshore to take advantage of lower labor and other costs. Domestic manufacturing space and engineering expertise are then freed up for the next generation of products.

The strategic shifts in Japanese investment flows also has involved more than location. Their U.S. affiliates are steadily shifting away from local assembly in so-called screwdriver plants—designed to avoid import restrictions—to more-integrated local production, often with substantial local value added. Part of this shift reflects the much tougher competition being provided Japanese companies by their U.S. rivals.

In fact, despite the challenge posed by Japanese and other foreign competitors, many U.S. companies that rely on scientific and technical research are growing in strength. U.S. companies dominate such high-technology industries as aerospace, computers and computer peripherals, communications, chemicals, pharmaceuticals, biotechnology, and software. And U.S. companies excel in some former Japanese strongholds: Hewlett-Packard is the world leader in computer printers; AT&T dominates the cordless phone market; Motorola leads the world in cellular phones and pagers; and Motorola and Intel dominate the huge global market for microprocessors.

The American companies that have managed to succeed against their foreign competitors are those that focused on their competitive strengths and were willing to stick with markets even when profitability declined. Many of these competitive strengths are readily apparent: America's overwhelming lead in software and microprocessors, the two most profitable areas of the computer industry; the benefits of telecommunications deregulation, which spurred growth in wireless telephony; the U.S. skill in meshing software and hardware; and the strong links between American universities and American industry, which have enabled U.S. companies to dominate the pharmaceuticals and biotechnology industries.

The Japanese have had particular difficulty in "knowledge/value" industries, such as fashion, entertainment, and business and consumer services, that require a combination of technology, entrepreneurship, and creative talents such as design or the ability to tell a story. By treating knowledge/value industries as if they were strictly industrial, the Japanese have suffered a string of reversals in Hollywood (Sony's acquisition of Columbia Pictures and Matsushita's acquisition of MCA/Universal are widely regarded as financial disasters), computer software, communications, and financial services. In contrast to making cars or VCRs, knowledge/value industries are culturally intensive. Understanding the importance of taste is often more important in these fields than pure efficiency. Only a handful of Japanese companies—Nintendo and Sega come to mind—have demonstrated the ability to come up with software products that have universal appeal.

Successful companies have also learned that, as product cycles shorten, the key to profitability is not innovation by itself but innovation that is quickly embodied in new products. New products typically command a price premium based on their de facto monopoly status. At the same time, as product cycles shorten, speed to market is becoming an economic necessity, not a luxury. Indeed, without those early new-product price premiums, the company may not be able to recoup its development costs.

Here, too, American companies have learned from the Japanese. More American companies are now emulating and refining the "concurrent engineering" approach to product development. By forming teams of designers, manufacturing engineers, procurement experts, marketers, and finance people to work together on new products, companies have found that they often can dramatically compress the time it takes to develop and launch new products while slashing development and manufacturing costs as well. Using such a team, Chrysler Corporation cut the time it took to develop the Neon to 33 months from its usual four to six years while improving product quality.

ILLUSTRATION

AMERICA REGAINS THE LEAD IN SEMICONDUCTORS

In 1992, U.S. chipmakers reclaimed the global top spot from the Japanese, who grabbed it in 1986. The United States, which first commercialized the memory chip and the microprocessor in the 1960s, controlled the business well into the 1970s. Then the Japanese came on strong, targeting the highest-volume chip—the dynamic random access memory, or DRAM—and slashing prices to gain market share. By the mid-1980s, six Japanese companies dominated the DRAM market.

U.S. chipmakers got American politicians to implement a trade pact in 1986 that forced the Japanese to open their market to American chips and raise prices on their own memory chips (which increased the profitability of U.S. DRAM manufacturers). Higher prices also encouraged Korean companies to enter the field, raising capacity and eventually forcing down DRAM prices and the profitability of Japanese chipmakers.

However, what really hurt the Japanese was the fact that DRAMs became a commodity product, with commodity profit margins. In contrast, U.S. chipmakers began focusing their efforts on developing, and getting to market quickly, higher-value-added, higher-margin innovative products, such as microprocessors and custom-made circuits, which exploited American strengths in software and design. The result—an American revival in an industry that many thought the United States had lost forever.

Japanese companies still dominate those industries, such as car and television manufacturing, where incremental improvement in production technology is the critical value driver. But Japanese companies have generally proved less adept at dealing with markets where the way ahead is ambiguous. For example, for years, the focus of big Japanese computer makers such as Hitachi and Fujitsu was to beat IBM. But just when they seemed ready to catch up with IBM, IBM itself was bypassed—not by a Japanese company but by smaller and more nimble U.S. competitors, who have redefined the computer market. These companies—Compaq, Dell, Apple, and Sun—have humbled both IBM and its Japanese competitors. It remains to be seen how many Japanese companies will be as successful as Sony and Nintendo at charting new paths.

❧ 20.5 SUMMARY AND CONCLUSIONS

For many firms, becoming multinational was the end result of an apparently haphazard process of overseas expansion. However, as international operations provide a more important source of profit and as competitive pressures increase, these firms are trying to develop global strategies that will enable them to maintain their competitive edge both at home and abroad.

The key to the development of a successful strategy is to understand and then capitalize on those factors that have led to success in the past. In this chapter, we saw that the rise of the multinational firm can be attributed to a variety of market imperfections that prevent the completely free flow of goods and capital internationally. These imperfections include government regulations and controls, such as tariffs and capital controls, that impose barriers to free trade and private portfolio investment. More significant as a spawner of multinationals are market failures in the areas of firm-specific skills and information. There are various transaction, contracting, and coordinating costs involved in trying to sell a firm's managerial skills and knowledge apart from the goods it produces. To overcome these costs, many firms have created an internal market, one in which these firm-specific advantages can be embodied in the services and products they sell.

Searching for and utilizing those sources of differential advantage that have led to prior success is clearly a difficult process. This chapter sketched some of the key factors involved in conducting an appropriate global investment analysis. Essentially, such an analysis requires the establishment of corporate objectives and policies that are congruent with each other and with the firm's resources and that lead to the continual development of new sources of differential advantage as the older ones reach obsolescence.

Clearly, such a comprehensive investment approach requires large amounts of time, effort, and money; yet, competitive pressures and increasing turbulence in the international environment are forcing firms in this direction. Fortunately, the supply of managers qualified to deal with such complex multinational issues is rising to meet the demand for their services.

➤ Questions

1. Why do firms from each category below become multinational? Identify the competitive advantages that a firm in each category must have to be a successful multinational.
   a. Raw-materials seekers
   b. Market seekers
   c. Cost minimizers

2. What factors help determine whether a firm will export its output, license foreign companies to manufacture its products, or set up its own production or service facilities abroad? Identify the competitive advantages that lead companies to prefer one mode of international expansion over another.

3. Time Warner is trying to decide whether to license foreign companies to produce its films and records or to set up foreign sales affiliates to sell its products directly. What factors might determine whether it expands abroad via licensing or by investing in its own sales force and distribution network?

4. What are the important advantages of going multinational? Consider the nature of global competition.

5. Given the added political and economic risks that appear to exist overseas, are multinational firms more or less risky than purely domestic firms in the same industry? Consider whether a firm that decides not to operate abroad is insulated from the effects of economic events that occur outside the home country.

6. How is the nature of IBM's competitive advantages related to its becoming a multinational firm?

7. Goodyear Tire and Rubber Company, the world's number one tire producer before Michelin's acquisition of Uniroyal Goodrich, is competing in a global tire industry. To maintain its leadership, Goodyear has invested more than $1 billion to build the most automated tiremaking facilities in the world and is aggressively expanding its chain of wholly owned tire stores to maintain its position as the largest retailer of tires in the United States. It also has invested heavily in R&D to produce tires that are recognized as being at the cutting edge of world-class performance. Based on product innovation and high advertising expenditures, Goodyear dominates the high-performance segment of the tire market. It has captured nearly 90% of the market for high-performance tires sold as original equipment on American cars and is well represented on sporty imports. Geography has given Goodyear and other American tire manufacturers a giant assist in the U.S. market. Heavy and bulky, tires are expensive to ship overseas.
   a. What barriers to entry has Goodyear created or taken advantage of?
   b. Goodyear has production facilities throughout the world. What competitive advantages might global production provide Goodyear?

c. How do tire-manufacturing facilities in Japan fit in with Goodyear's strategy to create shareholder value?

d. How will Bridgestone's acquisition of Firestone affect Goodyear? How might Goodyear respond to this move by Bridgestone?

8. Black & Decker, the maker of small, hand-held power tools, finds that when it builds a plant in a foreign country, sales of both its locally manufactured products and its exports to that country grow. What could account for this boost in sales? Consider the likely reactions of customers, distributors, and retailers to the fact that Black & Decker is producing there.

9. OPEC nations have obviously preferred portfolio investments abroad to direct foreign investment. How does the theory of market imperfections explain this preference?

10. What was the Japanese strategy for penetrating the TV market? What similarities are there between it and the Japanese strategy for entering the U.S. car market? The photocopier market?

11. What are the benefits of having a global distribution capability?

12. How sustainable is a competitive advantage based on technology? On low-cost labor? On economies of scale? Explain.

13. What could account for the fact that most European and Japanese automakers have design studios in the United States, and especially California?

14. The value of a particular foreign subsidiary to its parent company may bear little relationship to the subsidiary's profit-and-loss statement. In addition to the manipulations described in Chapter 14, the strategic purpose or nature of a foreign unit may dictate that some of the value of the unit will show up in the form of higher profits in other affiliates.
   a. Describe three ways, aside from profit manipulations, in which the incremental cash flows associated with a foreign unit can diverge from its actual cash flows.
   b. Describe two strategic rationales for establishing and maintaining a foreign subsidiary that will lead to higher profits elsewhere in the corporation but will not be reflected in the subsidiary's profit-and-loss statement.

15. Politicians, business executives, and the media lament the sale of corporate America to foreign buyers. Recent foreign acquisitions include Firestone Tire, Pillsbury, and CBS Records. A persistent theme sounded by executives and the business press is that the depreciated dollar offers a significant financial advantage to foreign bidders for American companies. According to this argument, if the dollar has depreciated relative to, say, the yen, a Japanese company can buy a U.S. company at a discount. Evaluate this argument.

16. The chapter mentions that Hyundai shifted its PC business to the United States.
   a. What advantages might Hyundai realize from servicing the U.S. market locally as opposed to operating out of South Korea?
   b. What competitive advantages is Hyundai likely to bring to the U.S. market?

17. In 1989, the British company Beecham Group merged with the U.S. company SmithKline Beckman. What economic advantages might the two drug companies be expecting from their marriage? More generally, what economic forces underlie the ongoing process of consolidation and globalization in the world pharmaceutical industry? Consider the merger's impact in the areas of R&D, marketing, and production.

18. Hershey Foods, maker of the Hershey bar, and Gerber Products, of baby-food fame, have dominant positions in the United States and strong brand names but little presence overseas.
   a. What appeals might expanding their foreign operations hold for them?
   b. Would foreign operations increase their risks by exposing them to political and economic risks seldom, if ever, encountered in the U.S. market?
   c. What options do these companies have for going overseas?

19. Shortly after British Telecommunications announced plans to offer international communications lines to multinational companies in the United States, AT&T sought approval from the British government to sell global services to corporations with offices in Britain. Similarly, following the announcement of plans of the Stentor long-distance consortium in Canada to join forces with MCI Communications in the United States, AT&T agreed to buy a stake in Canada's MCI-like start-up Unitel Communications.
   a. What is AT&T's likely motivation for these moves?
   b. Suppose an analysis indicated that the net present values of AT&T's moves into Britain and Canada were negative. Would this result indicate that AT&T made the wrong decision? Explain.

➤ Problems

1. Suppose the worldwide profit breakdown for General Motors is 85% in the United States, 3% in Japan, and 12% in the rest of the world. Its principal Japanese competitors earn 40% of their profits in Japan, 25% in the United States, and 35% in the rest of the world. Suppose further that through diligent attention to productivity and substitution of enormous quantities of capital for labor (for example, Project Saturn), GM manages to get its automobile production costs down to the level of the Japanese.
   a. Who is likely to have the global competitive advantage? Consider, for example, the ability of GM to respond to a Japanese attempt to gain U.S. market share through a sharp price cut.
   b. How might GM respond to the Japanese challenge?
   c. Which competitive response would you recommend to GM's CEO?

2. More and more Japanese companies are moving in on what once was an exclusive U.S. preserve: making and selling the complex equipment that makes semiconductors. World sales are between $3 billion and $5 billion annually. The U.S. equipment makers already have seen their share of the Japanese market fall to 30% recently from a dominant 70% in the late 1970s. Because sales in Japan are expanding as rapidly as 50% a year, Japanese concerns have barely begun attacking the U.S. market, but U.S. experts consider it only a matter of time.
   a. What are the possible competitive responses of U.S. firms?
   b. Which one(s) would you recommend to the head of a U.S. firm? Why?

3. Airbus Industrie, the European consortium of aircraft manufacturers, buys jet engines from U.S. companies. According to a recent story in the *Wall Street Journal*, "As a result of the weaker dollar, the cost of a major component (jet engines) is declining for Boeing's biggest competitor." The implication is that the lower price of engines for Airbus gives it a competitive advantage over Boeing. Will Airbus now be more competitive relative to Boeing? Explain.

4. Nordson Co. of Amherst, Ohio, a maker of painting and glue equipment, exports nearly half its output. Customers value its reliability as a supplier. Because of an especially sharp run-up in the value of the dollar against the French franc, Nordson is reconsidering its decision to continue supplying the French market. What factors are relevant in reaching a decision?

5. Tandem Computer, a U.S. maker of fault-tolerant computers, is thinking of shifting virtually all the labor-intensive portion of its production to Mexico. What risks is Tandem likely to face if it goes ahead with this move?

6. Germany's $28 billion electronics giant, Siemens AG, sells medical and telecommunications equipment, power plants, automotive products, and computers. Siemens has been operating in the United States since 1952, but its U.S. revenues account for only about 10% of worldwide revenues. It intends to expand further in the U.S. market.
   a. According to the head of its U.S. operation, "The United States is a real testing ground. If you make it here, you establish your credentials for the rest of the world." What does this statement mean? How would you measure the benefits flowing from this rationale for investing in the United States?
   b. What other advantages might Siemens realize from a larger American presence?

7. Kao Corporation is a highly innovative and efficient Japanese company that has managed to take on and beat Proctor & Gamble in Japan. Two of Kao's revolutionary innovations include disposable diapers with greatly enhanced absorption capabilities and concentrated laundry detergent. However, Kao has had difficulty in establishing the kind of market-sensitive foreign subsidiaries that P&G has built.
   a. What competitive advantages might P&G derive from its global network of market-sensitive subsidiaries?
   b. What competitive disadvantages does Kao face if it is unable to replicate P&G's global network of subsidiaries?

◆APPENDIX 20A Corporate Stategy and Joint Ventures

The global strategies discussed in this chapter are reflected in the ownership policies of multinational firms. In particular, a company's ownership strategy—whether to have wholly owned foreign affiliates or to take on partners—appears to be related systematically to the benefits and costs of having joint venture partners. Exhibit 20A.1 lists some of the costs and benefits of entering into joint ventures.

JOINT VENTURE BENEFITS

International *joint ventures*, or strategic alliances, are cropping up everywhere. By merging their sometimes divergent skills and resources, companies can quickly establish themselves in new markets and gain access to technology that might not otherwise be available. Thus, it's not unusual to see a company such as Burroughs (now Unisys) using various partnerships to access Hitachi's technology, to package Fujitsu's high-speed facsimile machines, and to manufacture Nippon Electric's optical readers. General Motors has teamed up with Toyota and Isuzu of Japan and with Daewoo of South Korea to manufacture and sell cars and has joined with Fujitsu Fanuc of Japan to manufacture and sell robots. Similarly, AT&T has entered joint ventures with Olivetti (Italy) and N.V. Philips (Netherlands). Whatever the industry—be it chemicals, autos, pharmaceuticals, telecommunications, electronics, or even aerospace—companies find themselves in tangled webs of international consortia.

The basic advantage of joint ventures is that they typically enable a company to leverage its key capabilities to generate incremental revenue or cost savings. Joint ventures, however, typically face unusually complex problems. Maintaining them requires a daunting amount of work. With representatives of both companies sitting on the board of directors, forging a consensus can be difficult, especially when the firms involved have different expectations for the venture. Nevertheless, the advantages in

EXHIBIT 20A.1 Joint Venture Considerations

ADVANTAGES	DISADVANTAGES
Obtain:	**Disagreements over:**
Local capital	Marketing programs
Local management	Dividend policy
Assured source of raw materials	Reinvestment of earnings
Trained labor	Exports to third countries
Marketing capabilities	Sources of materials and components
Established distribution network	Transfer pricing
Technology	Management selection and remuneration
	Expansion
Aid in obtaining:	
Government approvals	**Share profits based on monopoly rents from:**
Local currency loans	Technology, marketing, and managerial capabilities
Tax incentives	
Assurances of imports	**Give up technology**
Reduce nationalistic sentiments	

terms of access to markets, low manufacturing costs, technology, and the economies of scale in product development and production have proved irresistible to many firms.

Market Access

A company with a product that it thinks might be useful to overseas markets may find formidable barriers to local entry. Such obstacles include unfamiliar language, culture, and business practice. In the case of Japan, there is also the difficulty in breaking into that nation's Byzantine distribution network.

In 1983, Armco (U.S.) and Mitsubishi (Japan) formed a joint venture to sell (and eventually manufacture) Armco's lightweight plastic composites in Japan. For Armco, the venture was a way into an otherwise impenetrable Japanese market. For Mitsubishi, the venture was a way to get Armco's materials technology. Similarly, Fuji Photo Film Co. (Japan) and Hunt Chemical Co. (U.S.) teamed up to make and sell—in Japan—photoresists, which are sensitive coatings used in the semiconductor and microelectronics industries.

Japanese firms also have found joint ventures to be of value in penetrating foreign markets. For example, several Japanese steelmakers have traded technology for access to the U.S. market. The Japanese companies and their U.S. partners include Sumitomo Metal Industries/LTV, Nippon Steel/Inland Steel, NKK/National Intergroup, NKK/Bethlehem, and Nisshin Steel/Wheeling Pittsburgh Steel.

Similarly, Japanese drug firms have found that their lack of a significant local marketing presence is their greatest hindrance to expanding in the United States. Marketing drugs in the United States requires considerable political skill in maneuvering through the U.S. regulatory process, as well as great rapport with U.S. researchers and doctors. The latter requirement means that pharmaceutical firms must develop extensive sales forces to maintain close contact with their customers. There are economies of scale here; the cost of developing such a sales force is the same whether a firm sells one product or one hundred. Thus, only firms with extensive product lines can afford a large sales force, raising a major entry barrier to Japanese drug firms trying to go it alone in the United States.

One way the Japanese drug firms have found to get around this entry barrier is to form joint ventures with U.S. drug firms, with the Japanese supplying the patents and the U.S. firms supplying the distribution network. This same strategy was followed by Novo Industri, a Danish biotechnology firm, when it linked up with Squibb Corp. of New York to sell its insulin in the United States.

In many cases, joint ventures are not just advisable for getting into foreign markets—they are mandatory. Countries such as India and Mexico require joint ventures in order to promote technology transfer from foreign to domestic firms. Union Carbide's ill-fated venture in Bhopal, India, was a result of such a requirement. Similarly, Hewlett-Packard, Apple Computer, and other U.S. computer companies formed Mexican joint ventures because they were not allowed to ship products into Mexico without such partnerships.

In recent years, many Japanese firms have set up joint ventures in the United States and Western Europe as an insurance policy against possible U.S. and European trade barriers and as a way to ease political tensions. For example, Honda Motor Co. teamed up with BL in Great Britain to design and produce cars for local sale. Similarly, Toyota formed its joint venture with GM to produce cars in California as a way to forestall tougher quotas on Japanese cars.

Technology

Improved access to technology is a powerful incentive for joint ventures. Traditionally, U.S. firms have traded technology for access to Japanese and other foreign markets. Examples cited earlier include Armco and Philip Hunt. In recent years, the trend has reversed somewhat. Kawasaki, initially weak in the technology of automation equipment, has invested to catch up with its partner, Unimation of Cincinnati, Ohio. And Elxsi, a California manufacturer of general-purpose computers, is taking advantage of foreign technology via a joint venture. Elxsi joined with Tata, a large Indian conglomerate, and the Singapore government in a start-up manufacturing and marketing firm in Singapore. Tata contributed its expertise in software engineering to the partnership.

Joint ventures are a means to make use of each other's technical strengths. Joint ventures between U.S. and Japanese firms are especially useful in linking U.S. product innovation with low-cost Japanese design and manufacturing technology. For instance, Xerox used its long-standing joint venture with Fuji Photo Film Co.—Fuji Xerox—to develop a low-cost copier equal to the competition from Ricoh and Canon.

GM set up United Motor Mfg. (Fremont, California) to build subcompacts, a 50-50 joint production venture with Toyota Motor Corp. For GM, one of the attractions of this joint venture was Japanese expertise in low-cost manufacturing. GM also hopes to learn Japanese inventory and quality-control methods. On the basis of the lessons it has already learned, especially the importance of worker participation to quality and productivity, GM has revamped its multibillion-dollar factory automation program.

In the biotechnology industry, the Japanese have entered into numerous joint ventures with U.S. R&D companies—for example, Biogen, Damon, and Genentech. The American firms supply the basic science, and the Japanese firms—particularly in the liquor, food, and chemical businesses—bring to the joint ventures their skills in brewing, fermentation, or chemical processing. These techniques are needed to make genetically engineered substances in volume.

The emerging area of multimedia is spawning numerous new joint ventures and alliances. For example, Toshiba's push into multimedia might not have been feasible without the limited partnership it formed with Time Warner. Neither company felt that alone, it could have effectively exploited the opportunities provided by multimedia. The joint venture combines Toshiba's hardware expertise with Time Warner's know-how in entertainment, film, and television. The joint venture has begun technical collaboration with a cable system in Japan and is considering future investments in both cable systems and local telephone companies. Similarly, Tele-Communications Inc. (TCI), the U.S. cable TV giant, entered into a joint venture with Sumitomo. This venture represents an attempt on the part of TCI to be part of the Japanese market—one of the world's largest markets for such services—and also one of the most difficult to break into. For their part, the Japanese partners aim to gain enough expertise to eventually create a Japanese Information Superhighway when the regulatory gridlock in Japan finally clears. It was this desire to ensure themselves a major role in Japan's Information Superhighway that led Nippon Telegraph & Telephone, Fujitsu, and Toshiba to ally themselves in 1994 with General Magic, a nascent California company that makes multimedia software.

Economies of Scale

Companies are teaming up on product development because costs have become enormous—more than a billion dollars for the design of a central-office switching system or a new mainframe computer. An example is the joint venture between AT&T and Philips, the Dutch electronics giant, to develop, manufacture, and market sophisticated telecommunications equipment outside the United States. Or companies are turning to marketing arrangements to ensure sufficient sales worldwide to justify these investments. More recently, they have been teaming up in production to gain scale advantages. In part, that's why Alfa-Romeo and Nissan Motors are jointly producing car engines in southern Italy. Similarly, Renault and International Harvester (now Navistar) agreed on joint production of parts for farm tractors as a way to cut costs and boost sales. Joint ventures can reduce development costs as well. For example, joint development of major automobile components such as car platforms, transmissions, engines, and suspensions could cut development costs for future models by as much as 40%.

CORPORATE STRATEGY AND THE OPTIMUM OWNERSHIP PATTERN

Despite the potential benefits of joint ventures, the strategies of some firms mitigate against such partnerships. Those firms that require tight control to coordinate their pricing, marketing, quality control, and production policies worldwide typically shun joint ventures. The resources that a local partner can provide (capital, for example, or marketing skills) are in abundant supply in this type of multinational and, hence, the dilution of control makes little sense. We now examine the implications for the optimum ownership pattern of four broad strategies that MNCs follow.[1]

Product Differentiation

Where marketing is used to create barriers to entry, then control over the various elements of marketing strategy is considered vital. For firms such as PepsiCo, Heinz, or Coca-Cola, bringing in local partners would likely lead to conflicts over the large advertising expenditures, channels of distribution, and pricing policies deemed optimum by headquarters. Many of these firms, however, have found that they are able to participate in joint ventures at the manufacturing stage, provided that the parent company can control quality standards and has separate wholly owned sales affiliates that market the output.

Production Rationalization

The strategy of production *rationalization* entails the concentration of production in large plants—in order to take advantage of economies of scale—and the specialization of plants in different countries in manufacturing different parts and engaging in different stages of production. This strategy requires central planning and coordination; production decisions cannot be left to individual affiliates. Having a joint venture partner is bound to cause conflicts over transfer pricing, the allocation of products and markets to each plant, and the

---

[1] See Raymond Vernon and Louis T. Wells, Jr., *Manager in the International Economy*, 3rd ed. (Englewood Cliffs, N.J.: Prentice Hall, 1976), Chapter 2.

maintenance of quality control. The Canadian farm machinery manufacturer Massey-Ferguson, for example, found that its joint ventures competed with its other affiliates for export markets. Ford eventually bought out its minority shareholders in Ford of England because of their resistance to production shifts and other necessary elements of cost minimization.

Control of Raw Materials

Firms in extractive industries typically attempt to maintain control of raw materials in order to keep them out of the hands of potential entrants into the oligopoly and to assure themselves of supplies. The high fixed and low marginal costs of production encourage the formation of joint ventures among competitors. This joint ownership creates a common cost structure in the industry and a common exposure to risk, reducing the likelihood that any one competitor will cut prices during periods of slack demand. Cutting prices would simply lead to a reallocation of the market rather than an increase in total sales and revenues because demand for most raw materials is relatively price inelastic. On the other hand, the desire to stabilize demand for the output of company-owned wells, mines, and refineries encourages wholly owned downstream facilities, such as gasoline stations and metal-working operations.

Research and Development

Some firms, such as IBM, have historically invested heavily in research and new product development within a fairly narrow product line. These companies need to maintain tight control over their marketing organizations and logistics networks in order to extract the maximum profit from their organizational capabilities; therefore, they resist joint ventures. Other firms, such as Union Carbide, follow a strategy of using high R&D expenditures to generate a diversified and innovative line of new products. Each new product line requires a different marketing strategy and, therefore, a large investment in acquiring this know-how, so these firms are more willing to trade their technology for royalty payments and equity in a joint venture with local partners. Their shortage of the management personnel and marketing skills necessary to carry their product lines to new countries makes them value the marketing capabilities and other resource contributions of their local partners more than they value control.

Over time, of course, business circumstances are likely to change and so will the costs and benefits of joint ventures. In the case of IBM, for example, its product line has widened and it faces increased competition across a broad range of its businesses. In partial response to these changed circumstances, IBM now has more than 100 joint ventures with Japanese partners. Some of these joint ventures are designed to provide IBM with expanded access to the Japanese market, while others give it access to Japanese process technologies and manufacturing capabilities or to new products that would be too costly to develop and produce on its own.

JOINT VENTURE ANALYSIS

In deciding whether to go through with a joint venture, a firm must systematically analyze the likelihood of the joint venture's success. Not all partnerships work out as planned. Some of the casualties include ventures by General Motors and Daewoo, Ampex and Toshiba, Sterling Drug and Niigata Engineering, Pentax and Honeywell, Canon and Bell & Howell, Hitachi and Singer, and Avis and Mitsubishi. However, there are also winners: Dow Chemical and Asahi, CBS and Sony, and Xerox and Fuji.

As long as a company's business strategy does not preclude a joint venture, such a partnership is most appropriate when there are strategic gaps in critical capabilities that are too expensive (or would take too long) to develop internally. Joint ventures also work best when it is desirable to access a subset of the partner's capabilities—be it market access, critical technologies, organizational skills, or production capacity—rather than take on all the baggage that comes along with an acquisition.

Even if a joint venture works on paper, however, careful advance planning is necessary for it to work in practice. Both companies must subscribe to a "prenuptial agreement" that spells out the ground rules of the partnership and their future plans and expectations. In effect, the proposed venture should be viewed as a marriage between two partners, with each partner having needs to be fulfilled by the venture and contributions to make to the venture. Before the parties enter into such an arrangement, they should determine whether there is a proper match. An initial step can involve listing the business objectives and contributions of each partner, as shown in Exhibit 20A.2.

The objectives can include gaining a distribution network, managerial expertise, or government contracts; contributions might include capital and technology. However, the match-up may not work for various reasons. For example, many Japanese-U.S. joint ventures in Japan have failed because of conflicting objectives. The U.S. firms entered into these joint ventures in hopes of using their partners' distribution networks to increase their market shares in Japan. Their Japanese partners, on the other hand, expected to gain access to U.S. technology in order to export to third-country markets, a policy at odds with U.S. corporate objectives.

Perhaps the most important lesson we can learn from the history of joint ventures is that alliances and cooperative research activity are no panacea for corporate failure to develop internally the critical skills and assets they need to compete—the *core competencies* that spawn new generations of products and enable companies to adapt quickly to changing opportunities.[2] For example, 3M has applied its skills in adhesives and coatings

EXHIBIT 20A.2 Joint Venture Analysis		
US	**NEW VENTURE**	**THEM**
We want (business objectives): — — —	⟶ ⟵	They have to offer (resources): — — —
We offer (resources): — — —	⟶ ⟵	They want (business objectives): — — —

Source: Adapted from David B. Zenoff, presentation at the University of Hawaii Advanced Management Program, Honolulu, August 1978.

---

[2] The concept and implications of core competence appear in Gary Hamel and C. K. Prahalad, "The Core Competence of the Corporation," *Harvard Business Review*, May–June 1990, pp. 79–91.

across a broad range of products and markets—bandages and dental restoratives in health care, Post-it notes and Scotch tape in office supplies, reflective highway signs, diskettes and optical disks for personal computers, and videocassettes and audiocassettes in consumer electronics. What seems to be a highly diversified portfolio of businesses turns out to rest on a few shared competencies.

Unlike physical assets, which diminish with use, core skills are enhanced as they are applied; they wither with disuse. Thus, to sustain leadership in core competencies, firms should manufacture in-house the core products—such as laser "engines" in printers or engines in cars—that embody those core competencies and provide high value added to the end products. A key advantage of in-house manufacturing is that by working with the production process on a daily basis, the firm has a better sense of the wider potential of the technology, of possible applications that it would not otherwise consider. The history of the videocassette recorder shows how production know-how can yield important technical advances.

Unfortunately, the growth options associated with investments in core products and competencies tend to be undervalued using the standard discounted cash flow analysis (Chapter 21 discusses growth options further). To compensate for this bias, firms must use an expanded net present value rule that considers the costs of *not* making such investments. Otherwise, as many American firms have discovered, they will wake up one day to find that their "partners"—who have invested in core skills and products—have metamorphosed into rivals, who now control the product and process technologies necessary to compete. To avoid this fate, companies entering collaborative arrangements must rethink their strategic goals.

Many companies enter alliances to share investment risk or to reduce the costs and risks of entering new businesses or markets on their own. Although laudable, these goals are too limiting. Rather, the primary objective should be to emerge from the alliance more competitive than upon entry.[3] This means learning new skills and capabilities from the partner. However, too many strategic alliances—especially those between Western companies and their Asian rivals—are little more than sophisticated outsourcing arrangements.

Unfortunately, as noted above, those who rent competencies from other firms instead of developing their own tend to lose control over the key value-creating activities. The immediate gain in cost savings from outsourcing can be fatal in the long term.

Collaboration can also quash innovative design and produce indistinguishable "me-too" products. For example, to deal with Japanese car makers after 1992, European producers are engaged in a variety of collaborative efforts. The danger is that this collaboration will blunt competition and remove much of the pressure for innovation.

Forced Marriages

Joint ventures that are forced on firms can be successful in foreign markets separated by trade barriers from other markets. Because the affiliate operates on a stand-alone basis, coordination with the parent company's activities elsewhere is of much less importance.

---

[3] For a discussion of how to properly structure collaborative arrangements, see Gary Hamel, Yves L. Doz, and C. K. Prahalad, "Collaborate with Your Competitors—and Win," *Harvard Business Review*, January–February 1989, pp. 133–139.

In several highly publicized cases, large U.S. multinationals have chosen to pull out of foreign countries rather than comply with government regulations that require joint ventures. Both IBM and Citibank, for example, withdrew from Nigeria, and IBM also pulled out of India. Looked at on an individual-country basis, these companies might have been better off complying with the joint venture requirements. Apparently, however, IBM and Citibank felt that to give in just once would lead other nations to demand similar equity-sharing arrangements.

An alternative to foreign direct investment (other than licensing) is to sell managerial expertise in the form of a management contract. This unbundling of services, with its attendant reduction in political and economic risks, has become more attractive to some firms. The best example would be the current sale of various types of management skills to OPEC nations.

Nevertheless, despite the risk-reducing advantages provided by management contracts and the pressure placed on them by many host governments, particularly in the Third World, to divest themselves of their foreign operations, most MNCs appear to be quite reluctant to unbundle and sell their services directly. Clearly, firms believe they can take better advantage of market imperfections and earn higher returns through direct investment.

The low prices offered by host governments may reflect the fact that these services are worth less in an unbundled form. As we have already seen, much of the value of management expertise lies in its interactions with the various organizational skills available to the firm. Unbundling these services destroys that synergy, thereby reducing their value.

QUESTIONS

1. Multinational corporations use many different entry strategies abroad, including 100% ownership of all subsidiaries, majority-owned joint ventures, minority participation in joint ventures, and the licensing of foreign companies to produce the firm's products. Discuss the characteristics of a firm and its products that might lead it to prefer 100% ownership of all of its overseas subsidiaries.

2. Motorola, the number two U.S. chipmaker, signed an agreement in 1986 with Toshiba of Japan to jointly manufacture memories and microprocessors in a new Japanese plant. The pact gives Toshiba access to Motorola's popular 32-bit microprocessor technology; Motorola gets a share in a highly efficient manufacturing facility, a source of chips, and access to the Japanese market.

   a. How can Motorola maximize the advantages of this joint venture?

   b. What are some of the risks Motorola faces?

   c. What can it do to control these risks?

3. The Airbus consortium is a joint venture among companies in four European nations—France, Germany, Spain, and Britain. Its objectives are to help preserve and enhance key technologies associated with aerospace, while creating high valued-added jobs in the allied countries. Given their objectives, what conflicts can you foresee among the partners?

◆APPENDIX 20B Strategic Implications of Europe 1992

On January 1, 1993, the Single European Market—popularly known as *Europe 1992*—was officially launched by the 12 member nations of the European Community (EC). A key element of Europe 1992 is the set of about 300 directives designed to create a single European market for goods and services.[1] Although all 300 of these directives were to have been implemented by the end of 1992, and more than 95% have been adopted to date by the EC's Council of Ministers, as of April 1994 more than 100 still had not been adopted in all 12 member nations. Nonetheless, the accomplishments so far have gone a long way toward achieving the objective of Europe 1992: to tear down barriers to trade and commerce within Europe so that European nations can achieve economic prosperity. Supporters of a single market, pointing to the example of the United States, argued that economic integration would increase efficiency and stimulate economic growth by promoting competition within the EC, thereby forcing higher productivity, and by enabling companies to attain greater economies of scale.

In a further expansion, on January 1, 1994, the members of the European Community joined with the five members of the European Free Trade Association (Austria, Sweden, Finland, Norway, and Iceland) to create the world's largest free-trade area, stretching from the Mediterranean to the Arctic, with a combined 1994 GDP of $6.6 trillion. The economic unification of Europe contains a mix of good news and bad news for corporations—including those outside as well as inside the European Community. On the one hand, companies can cash in on the purchasing power of 372 million potential customers—a market larger than the United States and Canada combined. On the other hand, they will have to comply with new laws and regulations, some of which have protectionist overtones.

LESSONS FROM U.S. DEREGULATION

Despite the enormous uncertainty surrounding Europe 1992, corporate management can look to the recent American experience with deregulation for guidance in entering uncharted territory. That experience shows clearly the competitive changes that take place when new entrants are allowed into once-restricted markets. As such, it provides valuable lessons for managers seeking to position their companies for the opening of the European market.[2]

Since 1975, the United States has deregulated various aspects of the securities industry, banking, airlines, trucking, railroads, and telecommunications. In all the industries, we can see the same set of competitive dynamics at work:

1. *The industry becomes more competitive and profitability deteriorates rapidly as strong firms expand into formerly protected markets, while many new, low-cost suppliers enter the market.* Falling profits spur staff reductions and other cost-cutting

---

[1] This appendix is based on Alan C. Shapiro, "Economic Import of Europe 1992," *Journal of Applied Corporate Finance*, Winter 1991, pp. 25–36.

[2] These lessons are elaborated in Joel A. Bleeke, "Strategic Choices for Newly Opened Markets," *Harvard Business Review*, September–October 1990, pp. 158–165.

measures. This is already happening in Europe as deregulation is exposing costly bureaucracy and duplication. In addition, the weak get weaker, and many of them fail, but the strong do not get more profitable—not right away, at least. For example, trade barriers have allowed European banks to remain highly inefficient. Deregulation will shrink bank margins dramatically. Similarly, many European industrial firms have been too insulated from market pressures, leading to serious structural problems: They have been slower than many rivals to introduce and exploit new information technologies; they tolerate expensive work rules that reduce productivity; and they bear high social-welfare costs that impose an enormous burden on doing business in Europe. As competitive pressures strip away their insulation, European firms will have to reinvent themselves to survive.

2. *The most profitable market segments come under severe price pressure as competitors flock to them.* Conversely, the least profitable segments before deregulation, which were typically cross-subsidized to hold down prices, become more attractive as the cross-subsidies are ended and many firms exit these markets. For example, the biggest changes in the banking industry will come in retail banking, which provides as much as two-thirds of European bank profits and has served to cross-subsidize expansion into deregulated wholesale and investment banking.

3. *Merger and acquisition activity accelerates.* Initially, weaker firms combine to gain the size needed to compete with the giants of the industry. However, the anticipated scale economies often do not materialize—and a wave of divestitures of at least some of the unwanted pieces obtained in these mergers typically follows. Later on, some of the strongest firms in the industry merge with each other. They also make selected acquisitions—and divestitures—to fill gaps in their product portfolio or customer segments and to focus better on their core business.

4. *Only a handful of firms survive as broad-based competitors.* Those that succeed are companies that achieve, among other things, a precise understanding of their cost structures and pricing. With such understanding, they are able to eliminate hidden cross-subsidies and create new price-service trade-offs for their customers. The rest are forced to narrow their product range to those in which they have a competitive advantage and to spin off noncore activities to survive. The result is much greater segmentation within the industry.

In short, the early years of deregulation are characterized by shakeouts, restructuring, and the consolidation of position among survivors. There is an important difference, however, between Europe 1992 and U.S. deregulation: Many of the new entrants are already mighty international competitors such as American Airlines and Toyota. Thus, the shakeout and consolidation phases are likely to be even bloodier than they were in the United States during the 1980s. Also, because low industry profitability often forecloses the option of going to the capital markets, the biggest mistake many companies can make during this period is to spend too much money on acquisitions, entry into new markets, or major capital investments—thus leaving themselves with too little cash to weather the profit drought.

U.S. multinationals stand to benefit from their experience in highly competitive markets. As a result of such experience, they have acquired a greater readiness to organize their European activities along the most economically efficient lines and to redeploy assets aggressively across national boundaries. This willingness stems in part from the painful decisions many American managers had to make during the wave of corporate restructuring in the 1980s, along with the knowledge that these decisions were the more painful for having been so long delayed. Westinghouse (now CBS), for example, cut back on mature products such as electrical equipment that it decided it could not add more value to, and focused its resources instead on such growth areas as refrigerated trucking, defense electronics, and environmental controls. In so doing, Westinghouse reduced its European workforce by about 50% since 1980.

BUSINESS STRATEGY FOR EUROPE 1992

Despite considerable uncertainty about what agreements will survive the continuing negotiations over Europe 1992, and now the European Monetary Union, companies inside and outside the European Community are acting on the assumption that these initiatives will be implemented and that they will mean stronger competition. In response, companies in industries as diverse as electrical engineering, packaged food, and insurance are entering into cross-border alliances, merging with competitors, and otherwise restructuring their operations. Most such strategies, however, represent nothing more substantial than managerial "leaps of faith"—investments of corporate time and capital whose principal aim at this point seems to be to provide companies with the flexibility to respond to whatever surprises Europe 1992 may yield.

Many of the moves are obvious ones. Those companies, such as IBM and Ford, that have acted for three decades as if Europe were one market will realize large savings from market integration. They will no longer have to make alterations in their products to meet local standards, and transportation will be quicker and cheaper.

ILLUSTRATION

THE CASE OF N.V. PHILIPS

European firms also will benefit from new opportunities for production efficiencies. Consider N.V. Philips, the Dutch electronics firm that has long operated without much regard to national borders. Giant assembly plants take in components from Philips factories across Europe and dispatch finished products to distribution centers by way of a vast trucking network. A TV factory in Belgium, for example, gets tubes from Germany, transistors from France, and plastics from Italy. In theory Philips' system of centralized manufacturing in the past should have been a model of efficiency; in practice, frontiers made it cumbersome and expensive. Trucks spent 30% of their travel time idling in lines at customs posts. To avoid shutting down assembly lines when shipments were late, factories kept extra stock on hand. Since Europe 1992, by eliminating delays at customs posts, Philips has been able to cut inventories, close warehouses, reduce clerical staff, and save several hundred million dollars a year. Also, as local standards vanish, Philips intends to shrink its vast range of washing machines, fluorescent light bulbs, and above all TV sets (Europe currently has two standards for TV reception).

In addition to opportunities for greater efficiencies in production and distribution, established multinationals also may be able to cut costs by centralizing and coordinating administrative and marketing functions. In dealing with a fragmented Europe, many multinationals have evolved into collections of unrelated national subsidiaries, each serving its own local market. However, to serve pan-European customers, these companies must develop organizations that can coordinate production, marketing, and logistics across subsidiary boundaries to present a common face to their customers. For example, IBM organizes its production by continent, but its sales by country. This structure may not suffice after European integration because of customers' increased scope for arbitrage. Perhaps, as in the United States, customers will buy where prices and sales taxes are lowest and then ship their computers. In general, companies redesigning their organizational structures to cope with post-1992 Europe must reckon with their customers' responses to the expanded choices they will have in an integrated European market.

Those U.S. firms that currently are operating in a few protected local markets, such as medical supplies, may find that when regulatory barriers fall, they will face new competitors from other European countries. Companies such as Philips, Siemens, and Thomson are committed to transforming themselves from "national champions" to global competitors. In responding to such competition, the choices available to U.S. firms include expansion in Europe through acquisition of the moderate-sized European companies that appear to be for sale; the formation of strategic alliances, such as joint ventures or cooperation agreements, for joint R&D or cross-marketing of products; or sales to competitors and withdrawal from Europe. The last may be the best option for those companies that do not realize much in the way of economies of scale. For those businesses that exhibit economies of scale, however, the greater the growth potential and the weaker the current ties between supplier and customer, the greater the opportunity to expand outward into a multinational position.

Those U.S. companies that currently are exporting to Europe may wish to consider producing there. The argument for manufacturing at home is that U.S. labor costs are generally lower and the cheap dollar makes American-made goods competitive. Yet, simply staying home may be risky if a large share of the company's sales are in Europe, and 1992 brings greater protectionism and competition from stronger European firms. Establishing production facilities abroad—through a start-up, acquisition, or joint venture—can improve relations with European customers, as well as with U.S. customers now producing abroad, by demonstrating a deeper commitment to Europe. Producing abroad also provides a hedge against exchange risk.

Finally, U.S. companies that have focused exclusively on the domestic market should reconsider their options. The thrust of Europe 1992 is to allow European companies to build a market base that gives them the scale to compete globally. Ready or not, U.S. companies in many industries are going to face aggressive competition in the U.S. market from European multinationals such as Siemens and Philips. At the same time, U.S. companies that have stayed out of Europe because the EC market was too fragmented or the local producers too well protected may find that Europe 1992 has created new opportunities for them. For example, deregulation has given American companies in industries such as telecommunications and trucking years of valuable experience with innovative product and service concepts—experience that their European counterparts lack. An integrated EC market also will provide opportunities for giant U.S.

retailers such as Circuit City, Toys "R" Us, or Wal-Mart Stores to create new international distribution networks. For decades, Europeans have endured high prices, short shopping hours, and limited selection, while high-cost producers, inefficient retailers, and shop workers' unions enjoyed all kinds of legislative protection, presenting potentially lucrative opportunities for companies such as Price/Costco, the biggest U.S. warehouse club operator, which entered Britain in late 1993.

THE MYTH OF SCALE ECONOMIES

Underlying much of the planning for a single market is the conventional wisdom—shared by many corporate executives and European governments—that, with Europe 1992, bigger will be better. In post-1992 Europe, national markets will become more like American states, and the companies that prosper will be those that learn to compete according to the time-honored American formula for success: Exploit economies of scale and build up regional brands. This view has touched off a wave of cross-border mergers, as companies seek to gain the size necessary to compete in a borderless Europe.

But if national differences across Europe run as deep as some observers maintain, Europe 1992 will pose a much more complex challenge to management than learning to think as big as Americans. For example, in both the United States and Japan the white-goods business (large appliances such as refrigerators and washing machines) is dominated by single brands that have economies of scale in both marketing and production. Europe, by contrast, has many national brands. Although the industry appears ripe for restructuring and consolidation, national brands continue to predominate. So far, widely varying national preferences, combined with the complexity of creating so many different machines, have overwhelmed economies of scale.

These difficulties are exemplified by the experiences of Philips and Sweden's Electrolux, the first firms to try to create European-wide white-goods firms. The idea was that as European lifestyles converged, Europe would become more like the United States, where a few powerful companies compete across an entire continent. But after a decade of mergers and acquisitions, the American model still does not apply to Europe. In appliances, for example, four giant firms control almost 80% of the U.S. market. In Europe, 100 appliance makers continue to battle it out. As a result of this fragmentation, although Philips and Electrolux have operations in most countries, neither firm dominates any national market, except for their small home markets and Italy (where Electrolux bought the leading local supplier, Zanussi). In recent years, the two companies have been among the least profitable European producers of washing machines. Much more profitable have been those local firms that have tenaciously defended their dominance of a single national market. Both Philips and Electrolux have found it difficult to realize economies of scale through international expansion because of differences in languages, retail systems, and consumer tastes.

Moreover, the newest technologies—flexible manufacturing, faster computers, and better telecommunications—have reduced the optimal size of many businesses, and will probably continue to do so. Computer-controlled flexible manufacturing, producing batches tailored to changing customer needs, now can be just as profitable as mass production. Indeed, smaller runs may be necessary to keep up with fast-changing and increasingly specialized markets. For example, Electrolux has developed a flexible manufacturing system that can retool quickly enough to produce its entire range of 1,000 different types of refrigerators within a week.

The view that economies of scale are less important today will not surprise those who have seen GM, with its legendary economies of scale, lose the lead in profitability not only to much smaller Japanese companies but to smaller U.S. companies such as Ford and Chrysler as well. Indeed, the restructuring of corporate America during the 1980s indicates that many companies are too large—that there are in fact significant diseconomies of scale after a certain point. Such diseconomies stem from several sources: the growth of large bureaucracies that slow decision making, the increased likelihood of *cross-subsidization* in larger companies, and the greater administrative costs (including weakened management incentives) associated with managing unrelated businesses.

For example, Philips has cross-subsidized its ventures in computers and semiconductors with profits from its protected consumer electronics and lighting divisions. Such unprofitable businesses are prime candidates for restructuring or divestiture. Philips also spawned a bloated bureaucracy, weak marketing, and operations that are highly inefficient compared with its Japanese and American rivals. Its sales per employee in 1989 were $100,000, roughly half Matsushita's and Sony's and 25% below General Electric's.

The appropriate response is to weed out unprofitable products, production facilities, and activities. This means breaking down the business into the activities involved— purchasing, manufacturing, sales, distribution, R&D—and then figuring out how much each costs and how much value each adds. It also means examining every aspect of the business—product lines, customers, organizational structure—to identify those that create value and those that destroy it. Philips began such a process in 1990. It closed factories and transferred production from high-cost Europe to cheaper Asia and chopped about one-fifth of its total workforce between 1990 and 1993. At the same time, Philips still has to decide which aspects of its consumer electronics, lighting, telecommunications equipment, and semiconductor businesses it wishes to build on and which it should drop.

ILLUSTRATION

ABB DECENTRALIZES

ABB, the result of merging two sleepy European engineering companies (Sweden's Asea and Switzerland's Brown Boveri), has become a competitive powerhouse by taking the concept of leanness to extremes. Its chairman, Percy Barnevik, turned ABB into 5,000 profit centers, each an autonomous business unit responsible for its own profitability with its own balance sheet and income statement. He then shrank headquarters drastically (the Zurich corporate staff was cut from 4,000 to 200). Seven layers of management were cut to four, and product cycle times were shortened by up to 50%.

Despite growing skepticism about the benefits of size, and the generally dismal experience of large mergers in the United States, Western European mergers exceeded $650 billion in 1998, a new record. Many of these mergers will likely fail because of the inability to bridge corporate culture differences and to make tough decisions on cost-cutting and strategy.

In short, companies should be highly selective in their pursuit of scale economies. Some computer makers, for example, have found that they can reap large economies only in R&D and component purchasing; so linking with other firms' R&D or purchasing divisions might be the best strategy. Automakers, by contrast, benefit from size in the centralized development of new engines, which explains why companies such as Renault, Peugeot, and Volvo now develop their engines jointly.

Similarly, because many of the differences between the machines sold across Europe can be achieved by combining the same parts in different ways, companies have found they can achieve economies of scale in component production, although not in producing and selling the end product. Electrolux, for example, wants to create a world-scale business in white-goods components, such as pumps and engines, as well as in the appliances themselves. It also is centralizing component purchasing and pooling research.

In sum, the surest route to pan-European efficiency is for corporate managements to subject every part of their business to a test of "critical mass." Because each activity—R&D, purchasing, manufacturing, assembly, marketing, and distribution—has its own optimal size, a merger that requires bigness across the board can be very inefficient. Megamerged firms—joined together lock, stock, and barrel—may find that they are too big in some areas, and not big enough in others. Rather, firms should focus on growing only those areas where scale economies predominate. Finally, corporate executives who associate size with competitive advantage may well have it backward: Companies become large because they are competitive; they do not become competitive because they are large.

Where economies of scale do exist, it may be preferable to realize them from worldwide sales rather than from expansion in a protected home market. Foreign competition tends to toughen companies and increase their competitiveness, improving their chances of becoming world-class competitors.

EUROPEAN MONETARY UNION AND THE SINGLE EUROPEAN MARKET

The competitive effects of a common market will be magnified by the establishment of the European Monetary Union. Greater financial transparency under EMU will reveal inefficiencies once disguised by exchange rate differentials. New competition, lured by the prospect of a huge, unified, deregulated market, is already severely pressuring Europe's banks and insurance companies, formerly pampered and protected. Hence, there has been a spate of giant European financial mergers. The result will be a radical restructuring of European financial markets, as investors and executives anticipate a more-level playing field in everything from interest rates to corporate debt ratios to price-earnings multiples in national stock markets. With a single currency, creditworthiness rather than exchange risk will become the predominant consideration of investors when they analyze debt issues. As prices across countries become easier to compare, companies will be forced by comparison shoppers to restructure inefficient operations that lead to uncompetitive prices. In addition, there will be even greater pressure for tax, regulatory, and labor policies to become more homogeneous across the different countries.

BIBLIOGRAPHY

CAVES, RICHARD E. "International Corporations: The Industrial Economics of Foreign Investment." *Economica*, February 1971, pp. 1–27.

HAMEL, GARY, and C. K. PRAHALAD. "Do You Really Have a Global Strategy?" *Harvard Business Review*, July–August 1985, pp. 139–148.

————. "Strategic Intent," *Harvard Business Review*, May–June 1989, pp. 63–76.

RUGMAN, ALAN M. "Motives for Foreign Investment: The Market Imperfections and Risk Diversification Hypothesis." *Journal of World Trade Law*, September–October 1975, pp. 567–573.

SHAPIRO, ALAN C. "Capital Budgeting and Corporate Strategy." *Midland Corporate Finance Journal*, Spring 1985, pp. 22–36.

————. "Economic Import of Europe 1992." *Journal of Applied Corporate Finance*, Winter 1991, pp. 25–36.

CAPITAL BUDGETING FOR THE MULTINATIONAL CORPORATION

Nobody can really guarantee the future. The best we can do is size up the chances, calculate the risks involved, estimate our ability to deal with them, and then make our plans with confidence.

Henry Ford II

*M*ultinational corporations evaluating foreign investments find their analyses complicated by a variety of problems that are rarely, if ever, encountered by domestic firms. This chapter examines several such problems, including differences between project and parent-company cash flows, foreign tax regulations, expropriation, blocked funds, exchange rate changes and inflation, project-specific financing, and differences between the basic business risks of foreign and domestic projects. The purpose of this chapter is to develop a framework that allows measuring, and reducing to a common denominator, the consequences of these complex factors on the desirability of the foreign investment opportunities under review. In this way, projects can be compared and evaluated on a uniform basis. The major principle behind methods proposed to cope with these complications is to maximize the use of available information while reducing arbitrary cash flow and cost of capital adjustments.

21.1 BASICS OF CAPITAL BUDGETING

Once a firm has compiled a list of prospective investments, it then must select from among them that combination of projects that maximizes the firm's value to its shareholders. This selection requires a set of rules and decision criteria that enables managers to determine, given an investment opportunity, whether to accept or reject it. The criterion of net present value is generally accepted as being the most appropriate one to use because its consistent application will lead the company to select the same investments the shareholders would make themselves, if they had the opportunity.

Net Present Value

The **net present value (NPV)** is defined as the present value of future cash flows discounted at the project's cost of capital minus the initial net cash outlay for the project. Projects with a positive NPV should be accepted; negative NPV projects should be rejected. If two projects are mutually exclusive, the one with the higher NPV should be accepted. As discussed in Chapter 18, the cost of capital is the expected rate of return on projects of similar risk. In this chapter, we take its value as given.

In mathematical terms, the formula for net present value is

$$NPV = -I_0 + \sum_{t=1}^{n} \frac{X_t}{(1 + k)^t} \tag{21.1}$$

WHERE

I_0 = the initial cash investment
X_t = the net cash flow in period t
k = the project's cost of capital
n = the investment horizon

To illustrate the NPV method, consider a plant expansion project with the following stream of cash flows and their present values:

YEAR	CASH FLOW	×	PRESENT VALUE FACTOR (10%)	=	PRESENT VALUE	CUMULATIVE PRESENT VALUE
0	−$4,000,000		1.00000		−$4,000,000	−$4,000,000
1	1,200,000		0.9091		1,091,000	−2,909,000
2	2,700,000		0.8264		2,231,000	− 678,000
3	2,700,000		0.7513		2,029,000	1,351,000

Assuming a 10% cost of capital, the project is acceptable.

The most desirable property of the NPV criterion is that it evaluates investments in the same way as the company's shareholders; the NPV method properly focuses on cash rather than on accounting profits and emphasizes the opportunity cost of the money invested. Thus, it is consistent with shareholder wealth maximization.

Another desirable property of the NPV criterion is that it obeys the **value additivity principle**. That is, the NPV of a set of independent projects is simply the sum of the

NPVs of the individual projects. This property means that managers can consider each project on its own. It also means that when a firm undertakes several investments, its value increases by an amount equal to the sum of the NPVs of the accepted projects. Thus, if the firm invests in the previously described plant expansion, its value should increase by $1,351,000, the NPV of the project.

Incremental Cash Flows

The most important and also the most difficult part of an investment analysis is to calculate the cash flows associated with the project: the cost of funding the project; the cash inflows during the life of the project; and the terminal, or ending value, of the project. Shareholders are interested in how many additional dollars they will receive in the future for the dollars they lay out today. Hence, what matters is not the project's total cash flow per period, but the **incremental cash flows** generated by the project.

The distinction between total and incremental cash flows is a crucial one. Incremental cash flow can differ from total cash flow for a variety of reasons. We now examine some of the reasons.

CANNIBALIZATION ◼ When Honda introduced its Acura line of cars, some customers switched their purchases from the Honda Accord to the new models. This example illustrates the phenomenon known as **cannibalization**, a new product taking sales away from the firm's existing products. Cannibalization also occurs when a firm builds a plant overseas and winds up substituting foreign production for parent company exports. To the extent that sales of a new product or plant just replace other corporate sales, the new project's estimated profits must be reduced by the earnings on the lost sales.

The previous examples notwithstanding, it is often difficult to assess the true magnitude of cannibalization because of the need to determine what would have happened to sales in the absence of the new product or plant. Consider Motorola's construction of a plant in Japan to supply chips to the Japanese market previously supplied via exports. In the past, Motorola got Japanese business whether it manufactured in Japan or not. But now Japan is a chip-making dynamo whose buyers no longer have to depend on U.S. suppliers. If Motorola had not invested in Japan, it might have lost export sales anyway. Instead of losing these sales to local production, however, it would have lost them to one of its rivals. The *incremental* effect of cannibalization—the relevant measure for capital-budgeting purposes—equals the lost profit on lost sales *that would not otherwise have been lost* had the new project not been undertaken. Those sales that would have been lost anyway should not be counted a casualty of cannibalization.

SALES CREATION ◼ Black & Decker, the U.S. power tool company, significantly expanded its exports to Europe after investing in European production facilities that gave it a strong local market position in several product lines. Similarly, GM's auto plants in Britain use parts made by its U.S. plants, parts that would not otherwise be sold if GM's British plants disappeared.

In both cases, an investment either created or was expected to create additional sales for existing products. Thus, *sales creation* is the opposite of cannibalization. In calculations of the project's cash flows, the additional sales and associated incremental cash flows should be attributed to the project.

OPPORTUNITY COST ■ Suppose IBM decides to build a new office building in São Paulo on some land it bought 10 years ago. IBM must include the cost of the land in calculating the value of undertaking the project. Also, this cost must be based on the current market value of the land, not the price it paid 10 years ago.

This example demonstrates a more general rule. Project costs must include the true economic cost of any resource required for the project, regardless of whether the firm already owns the resource or has to go out and acquire it. This true cost is the *opportunity cost*, the maximum amount of cash the asset could generate for the firm should it be sold or put to some other productive use. It would be foolish for a firm that acquired oil at $3 a barrel and converted it into petrochemicals to sell those petrochemicals based on $3 a barrel oil if the price of oil has risen to $30 per barrel. So, too, it would be foolish to value an asset used in a project at other than its opportunity cost, regardless of how much cash changes hands.

TRANSFER PRICING ■ By raising the price at which a proposed Ford plant in Dearborn, Michigan, will sell engines to its English subsidiary, Ford can increase the apparent profitability of the new plant, but at the expense of its English affiliate. Similarly, if Matsushita lowers the price at which its Panasonic division buys microprocessors from its microelectronics division, the latter's new semiconductor plant will show a decline in profitability.

It is evident from these examples that the transfer prices at which goods and services are traded internally can significantly distort the profitability of a proposed investment. Where possible, the prices used to evaluate project inputs or outputs should be market prices. If no market exists for the product, then the firm must evaluate the project based on the cost savings or additional profits to the corporation of going ahead with the project. For example, when Atari decided to switch most of its production to Asia, its decision was based solely on the cost savings it expected to realize. This approach was the correct one to use because the stated revenues generated by the project were meaningless, an artifact of the transfer prices used in selling its output back to Atari in the United States.

FEES AND ROYALTIES ■ Often companies will charge projects for various items such as legal counsel, power, lighting, heat, rent, research and development, headquarters staff, management costs, and the like. These charges appear in the form of fees and royalties. They are costs to the project but are a benefit from the standpoint of the parent firm. From an economic standpoint, the project should be charged only for the additional expenditures that are attributable to the project; those overhead expenses that are unaffected by the project should not be included in estimates of project cash flows.

GETTING THE BASE CASE RIGHT ■ In general, a project's incremental cash flows can be found only by subtracting worldwide corporate cash flows without the investment—the *base case*—from postinvestment corporate cash flows. To come up with a realistic base case, and thus a reasonable estimate of incremental cash flows, managers must ask the key question, "What will happen if we *don't* make this investment?" Failure to heed this question led General Motors during the 1970s to slight investment in small cars despite the Japanese challenge; small cars looked less profitable than GM's then-current mix of cars. As a result, Toyota, Nissan, and other Japanese automakers were able to expand and eventually threaten GM's base business. Similarly, many American companies—such as Kodak and Zenith—that thought overseas expansion too risky or unattractive today find their domestic competitive positions eroding. They did not adequately consider the consequences of *not* building a strong global position.

The critical error made by these and other companies is to ignore competitor behavior and assume that the base case is the status quo. But in a competitive world economy, the least likely future scenario is the status quo. A company that opts not to come out with a new product because it is afraid that the product will cannibalize its existing product line is most likely leaving a profitable niche for some other company to exploit. Sales will be lost anyway, but now they will be lost to a competitor. Similarly, a company that chooses not to invest in a new process technology because it calculates that the higher quality is not worth the added cost may discover that it is losing sales to competitors who have made the investment. In a competitive market, the rule is simple: *If you must be the victim of a cannibal, make sure the cannibal is a member of your family.*

ILLUSTRATION

INVESTING IN MEMORY CHIPS

Since 1984, the intense competition from Japanese firms has caused most U.S. semiconductor manufacturers to lose money in the memory chip business. As we saw in Chapter 20, the only profitable part of the chip business for them is in making microprocessors and other specialized chips. Why did U.S. companies continue investing in facilities to produce memory chips (the DRAMs) despite their losses in this business?

Historically, U.S. companies cared so much about memory chips because of their importance in fine-tuning the manufacturing process.

Memory chips are manufactured in huge quantities and are fairly simple to test for defects, which makes them ideal vehicles for refining new production processes. Having worked out the bugs by making memories, chip companies apply an improved process to hundreds of more complex products. Until recently, without manufacturing some sort of memory chip, it was very difficult to keep production technology competitive. Thus, making profitable investments elsewhere in the chip business was contingent on producing memory chips. As manufacturing technology has changed, diminishing the importance of memory chips as process technology drivers, U.S. chipmakers such as Intel have stopped producing DRAMs.

ACCOUNTING FOR INTANGIBLE BENEFITS ■ Related to the choice of an incorrect base case is the problem of incorporating intangible benefits in the capital-budgeting process. Intangibles such as better quality, faster time to market, quicker and less error-prone order processing, and higher customer satisfaction can have a very tangible impact on corporate cash flows, even if they cannot be measured precisely. Similarly, many investments provide intangible benefits in the form of valuable learning experiences and a broader knowledge base. For example, investing in foreign markets can sharpen competitive skills: It exposes companies to tough foreign competition; it enables them to size up new products being developed overseas and figure out how to compete with them before these products show up in the home market; and it can aid in tracking emerging technologies to transfer back home. Adopting practices, products, and technologies discovered overseas can improve a company's competitive position worldwide.

ILLUSTRATION

INTANGIBLE BENEFITS FROM INVESTING IN JAPAN

The prospect of investing in Japan scares many foreign companies. Real estate is prohibitively expensive. Customers are extraordinarily demanding. The government bureaucracy can seem impenetrable at times, and Japanese competitors fiercely protect their home market.

However, an investment in Japanese operations provides a variety of intangible benefits. More companies are realizing that to compete effectively elsewhere, they must first compete in the toughest market of all: Japan. What they learn in the process—from meeting the stringent standards of Japanese customers and battling a dozen relentless Japanese rivals—is invaluable and will possibly make the difference between survival and extinction. At the same time, operating in Japan helps a company such as IBM keep up the pressure on some of its most potent global competitors in their home market. A position in the Japanese market also gives a company an early look at new products and technologies originating in Japan, enabling it to quickly pick up and transfer back to the United States information on Japanese advances in manufacturing technology and product development. And monitoring changes in the Japanese market helps boost sales there as well.

Although the principle of incremental analysis is a simple one to state, its rigorous application is a complicated undertaking. However, this rule at least points in the right direction those executives responsible for estimating cash flows. Moreover, when estimation shortcuts or simplifications are made, it provides those responsible with some idea of what they are doing and how far they are straying from a thorough analysis.

Alternative Capital-Budgeting Frameworks

As we have just seen, the standard capital-budgeting analysis involves first calculating the expected after-tax values of all cash flows associated with a prospective investment and then discounting those cash flows back to the present, using an appropriate discount rate. Typically, the discount rate used is the **weighted average cost of capital** (WACC), where the weights are based on the proportion of the firm's capital structure accounted for by each source of capital.

AN ADJUSTED PRESENT VALUE APPROACH ∎The weighted average cost of capital is simple in concept and easy to apply. A single rate is appropriate, however, only if the financial structures and commercial risks are similar for all investments undertaken. Projects with different risks are likely to possess differing debt capacities for each project, therefore necessitating a separate financial structure. Moreover, the financial package for a foreign investment may include project-specific loans at concessionary rates or higher-cost foreign funds owing to home country exchange controls, leading to different component costs of capital for foreign investments.

The weighted average cost of capital figure can be modified, of course, to reflect these deviations from the firm's typical investment. But for some companies, such as those in extractive industries, there is no norm. Project risks and financial structure vary by country, raw material, production stage, and position in the life cycle of the project. An alternative approach is to discount cash flows at a rate that reflects only the business risks of the project and abstracts from the effects of financing. This rate, introduced as the **all-equity rate** in Chapter 18, would apply directly if the project were financed entirely by equity. The all-equity rate k^* can be used in capital budgeting by viewing the value of a project as being equal to the sum of the following components: (1) the present value of project cash flows after taxes but before financing costs, discounted at k^*; (2) the present value of the tax savings on debt financing, which is also known as the **interest tax shield**; and (3) the present value of any savings (penalties) on interest costs associated with project-specific financing.[1] This latter differential would generally be due to government regulations and/or subsidies that caused interest rates on restricted funds to diverge from domestic interest payable on unsubsidized, arm's-length borrowing. The **adjusted present value** (*APV*) with this approach is

$$APV = \begin{matrix}\text{Present value} \\ \text{of investment} \\ \text{outlay}\end{matrix} + \begin{matrix}\text{Present value} \\ \text{of operating} \\ \text{cash flows}\end{matrix} + \begin{matrix}\text{Present value} \\ \text{of interest} \\ \text{tax shield}\end{matrix} + \begin{matrix}\text{Present value} \\ \text{of interest} \\ \text{subsidies}\end{matrix}$$

$$APV = -I_0 + \sum_{t=1}^{n} \frac{X_t}{(1+k^*)^t} + \sum_{t=1}^{n} \frac{T_t}{(1+i_d)^t} + \sum_{t=1}^{n} \frac{S_t}{(1+i_d)^t} \tag{21.2}$$

WHERE

T_t = tax savings in year t due to the specific financing package
S_t = before-tax dollar (home currency) value of interest subsidies (penalties) in year t due to project-specific financing
i_d = before-tax cost of dollar (home currency) debt

The last two terms in Equation 21.2 are discounted at the before-tax cost of dollar debt to reflect the relatively certain value of the cash flows due to tax shields and interest savings (penalties). The interest tax shield in period t, T_t, equals $\tau i_d D_t$, where τ is the corporate tax rate and D_t is the incremental debt supported by the project in period t.

It should be emphasized that the all-equity cost of capital equals the required rate of return on a specific project—that is, the riskless rate of interest plus an appropriate risk premium based on the project's particular risk. Thus, k^* varies by project as project risks vary.

According to the capital asset pricing model (CAPM), the market prices only *systematic* risk relative to the market rather than total corporate risk. In other words, only interactions of project returns with overall market returns are relevant in determining project riskiness; interactions of project returns with total corporate returns can be ignored. Thus, each project has its own required return and can be evaluated without

[1] This material is based on Donald R. Lessard, "Evaluating Foreign Projects: An Adjusted Present Value Approach," in *International Financial Management*, 2nd ed., Donald R. Lessard, ed. (Boston: Warren, Gorham & Lamont), 1985.

regard to the firm's other investments. If a project-specific approach is not used, the primary advantage of the CAPM is lost—the concept of value additivity, which allows projects to be considered independently.

21.2 ISSUES IN FOREIGN INVESTMENT ANALYSIS

The analysis of a foreign project raises two additional issues other than those dealing with the interaction between the investment and financing decisions:

1. Should cash flows be measured from the viewpoint of the project or of the parent?
2. Should the additional economic and political risks that are uniquely foreign be reflected in cash-flow or discount rate adjustments?

Parent Versus Project Cash Flows

A substantial difference can exist between the cash flow of a project and the amount that is remitted to the parent firm because of tax regulations and exchange controls. In addition, project expenses such as management fees and royalties are returns to the parent company. Furthermore, the incremental revenue contributed to the parent MNC by a project can differ from total project revenues if, for example, the project involves substituting local production for parent company exports or if transfer price adjustments shift profits elsewhere in the system.

Given the differences that are likely to exist between parent and project cash flows, the question arises as to the relevant cash flows to use in project evaluation. Economic theory has the answer to this question. According to economic theory, the value of a project is determined by the net present value of future cash flows back to the investor. Thus, the parent MNC should value only those cash flows that are, or can be, repatriated net of any transfer costs (such as taxes), because only accessible funds can be used for the payment of dividends and interest, for amortization of the firm's debt, and for reinvestment.

A THREE-STAGE APPROACH ■ A three-stage analysis is recommended for simplifying project evaluation. In the first stage, project cash flows are computed from the subsidiary's standpoint, exactly as if the subsidiary were a separate national corporation. The perspective then shifts to the parent company. This second stage of analysis requires specific forecasts concerning the amounts, timing, and form of transfers to headquarters, as well as information about what taxes and other expenses will be incurred in the transfer process. Finally, the firm must take into account the indirect benefits and costs that this investment confers on the rest of the system, such as an increase or decrease in export sales by another affiliate.

ESTIMATING INCREMENTAL PROJECT CASH FLOWS ■ Essentially, the company must estimate a project's true profitability. *True profitability* is an amorphous concept, but basically it involves determining the marginal revenue and marginal costs associated with the project. In general, as mentioned earlier, incremental cash flows to the parent can be found only by subtracting worldwide parent-company cash flows (without the investment) from postinvestment parent-company cash flows. This estimating entails the following:

1. Adjust for the effects of transfer pricing and fees and royalties.
   - Use market costs/prices for goods, services and capital transferred internally.
   - Add back fees and royalties to project cash flows, because they are benefits to the parent.
   - Remove the fixed portions of such costs as corporate overhead.
2. Adjust for global costs/benefits that are not reflected in the project's financial statements. These costs/benefits include
   - Cannibalization of sales of other units
   - Creation of incremental sales by other units
   - Additional taxes owed when repatriating profits
   - Foreign tax credits usable elsewhere
   - Diversification of production facilities
   - Market diversification
   - Provision of a key link in a global service network
   - Knowledge of competitors, technology, markets, products

The second set of adjustments involves incorporating the project's strategic purpose and its impact on other units. These strategic considerations embody the factors that were discussed in Chapter 20. For example, AT&T is investing heavily in the ability to provide multinational customers with seamless global telecommunications services.

Although the principle of valuing and adjusting incremental cash flows is itself simple, it can be complicated to apply. Its application is illustrated in the case of taxes.

TAX FACTORS ■ Because only after-tax cash flows are relevant, it is necessary to determine when and what taxes must be paid on foreign-source profits. The following example illustrates the calculation of the incremental tax owed on foreign-source earning. Suppose an affiliate will remit after-tax earnings of $150,000 to its U.S. parent in the form of a dividend. Assume that the foreign tax rate is 25%, the withholding tax on dividends is 4%, and excess foreign tax credits are unavailable. The marginal rate of additional taxation is found by adding the withholding tax that must be paid locally to the U.S tax owed on the dividend. Withholding tax equals $6,000 (150,000 × 0.04), and U.S. tax owed equals $14,000. This latter tax is calculated as follows. With a before-tax local income of $200,000 (200,000 × 0.75 = 150,000), the U.S. tax owed would equal $200,000 × 0.35, or $70,000. The firm then receives foreign tax credits equal to $56,000—the $50,000 in local tax paid and the $6,000 dividend withholding tax—leaving a net of $14,000 owed the IRS. This calculation yields a marginal tax rate of 13.33% on remitted profits, as follows:

$$\frac{6,000 + 14,000}{150,000} = 0.1333$$

If excess foreign tax credits are available to offset the U.S. tax owed, then the marginal tax rate on remittances is just the dividend withholding tax rate of 4%.

Political and Economic Risk Analysis

All else being equal, firms prefer to invest in countries with stable currencies, healthy economies, and minimal political risks, such as expropriation. But all else is usually not equal, so firms must assess the consequences of various political and economic risks for the viability of potential investments.

The three main methods for incorporating the additional political and economic risks, such as the risks of currency fluctuation and expropriation, into foreign investment analysis are (1) shortening the minimum payback period, (2) raising the required rate of return of the investment, and (3) adjusting cash flows to reflect the specific impact of a given risk.

ADJUSTING THE DISCOUNT RATE OR PAYBACK PERIOD ■ The additional risks confronted abroad are usually described in general terms instead of being related to their impact on specific investments. This rather vague view of risk probably explains the prevalence among multinationals of two unsystematic approaches to account for the added political and economic risks of overseas operations. One is to use a higher discount rate for foreign operations; another, to require a shorter payback period. For instance, if exchange restrictions are anticipated, a normal required return of 15% might be raised to 20%, or a five-year payback period might be shortened to three years.

Neither of the aforementioned approaches, however, lends itself to a careful evaluation of the actual impact of a particular risk on investment returns. Thorough risk analysis requires an assessment of the magnitude and timing of risks and their implications for the projected cash flows. For example, an expropriation five years hence is likely to be much less threatening than one expected next year, even though the probability of its occurring later may be higher. Thus, using a uniformly higher discount rate simply distorts the meaning of the present value of a project by penalizing future cash flows relatively more heavily than current ones, without obviating the necessity for a careful risk evaluation. Furthermore, the choice of a risk premium is an arbitrary one, whether it is 2% or 10%. Instead, adjusting cash flows makes it possible to fully incorporate all available information about the impact of a specific risk on the future returns from an investment.

ADJUSTING EXPECTED VALUES ■ The recommended approach is to adjust the cash flows of a project to reflect the specific impact of a given risk, primarily because there is normally more and better information on the specific impact of a given risk on a project's cash flows than on its required return. The cash-flow adjustments presented in this chapter employ only expected values; that is, the analysis reflects only the first moment of the probability distribution of the impact of a given risk. Although this procedure does not assume that shareholders are risk-neutral, it does assume either that risks such as expropriation, currency controls, inflation, and exchange rate changes are unsystematic or that foreign investments tend to lower a firm's systematic risk. In the latter case, adjusting only the expected values of future cash flows will yield a lower bound on the value of the investment to the firm.

Although the suggestion that cash flows from politically risky areas should be discounted at a rate that ignores those risks is contrary to current practice, the difference is more apparent than real: Most firms evaluating foreign investments discount most likely (modal) rather than expected (mean) cash flows at a risk-adjusted rate. If an expropriation or currency blockage is anticipated, then the mean value of the probability distribution of future cash flows will be significantly below its mode. From a theoretical standpoint, of course, cash flows should always be adjusted to reflect the change in expected values caused by a particular risk; however, only if the risk is systematic should these cash flows be further discounted.

Exchange Rate Changes and Inflation

The present value of future cash flows from a foreign project can be calculated using a two-stage procedure: (1) Convert nominal foreign-currency cash flows into nominal home-currency terms, and (2) discount those nominal cash flows at the nominal domestic required rate of return. In order to properly assess the effect of exchange rate changes on expected cash flows from a foreign project, one must first remove the effect of offsetting inflation and exchange rate changes. It is worthwhile to analyze each effect separately because different cash flows may be differentially affected by inflation. For example, the **depreciation tax shield** will not rise with inflation, whereas revenues and variable costs are likely to rise in line with inflation. Or local price controls may not permit internal price adjustments. In practice, correcting for these effects means first adjusting the foreign currency cash flows for inflation and then converting the projected cash flows back into dollars using the forecast exchange rate.

ILLUSTRATION

FACTORING IN CURRENCY DEPRECIATION AND INFLATION

Suppose that with no inflation the cash flow in year 2 of a new project in France is expected to be FF 1 million, and the exchange rate is expected to remain at its current value of FF 1 = $0.20. Converted into dollars, the FF 1 million cash flow yields a projected cash flow of $200,000. Now suppose that French inflation is expected to be 6% annually, but project cash flows are expected to rise only 4% annually because the depreciation tax shield will remain constant. At the same time, because of purchasing power parity, the franc is expected to devalue at the rate of 6% annually — giving rise to a forecast exchange rate in year 2 of $0.20 \times (1 - 0.06)^2 =$ $0.1767. Then the forecast cash flow in year 2 becomes FF $1,000,000 \times 1.04^2 =$ FF 1,081,600, with a forecast dollar value of $191,119 (0.1767 \times 1,081,600).

21.3 FOREIGN PROJECT APPRAISAL: THE CASE OF INTERNATIONAL DIESEL CORPORATION

This section illustrates how to deal with some of the complexities involved in foreign project analysis by considering the case of a U.S. firm with an investment opportunity in England. International Diesel Corporation (IDC-U.S.), a U.S.-based multinational firm, is trying to decide whether to establish a diesel manufacturing plant in the United Kingdom (IDC-U.K.). IDC-U.S. expects to significantly boost its European sales of small diesel engines (40–160 hp) from the 20,000 it is currently exporting there. At the moment, IDC-U.S. is unable to increase exports because its domestic plants are producing to capacity. The 20,000 diesel engines it is currently shipping to Europe are the residual output that it is not selling domestically.

IDC-U.S. has made a strategic decision to significantly increase its presence and sales overseas. A logical first target of this international expansion is the European Community (EC). Market growth seems assured by recent large increases in fuel costs and the advent of Europe 1992 and European Monetary Union. IDC-U.S. executives believe that manufacturing in England will give the firm a key advantage with customers in England and throughout the rest of the EC.

England is the most likely production location because IDC-U.S. can acquire a 1.4-million-square-foot plant in Manchester from BL, which used it to assemble gasoline engines before its recent closing. As an inducement to locate in this vacant plant and, thereby, ease unemployment among auto workers in Manchester, the National Enterprise Board (NEB) will provide a five-year loan of £5 million ($10 million) at 3% interest, with interest paid annually at the end of each year and the principal to be repaid in a lump sum at the end of the fifth year. Total acquisition, equipment, and retooling costs for this plant are estimated to equal $50 million.

Full-scale production can begin six months from the date of acquisition because IDC-U.S. is reasonably certain it can hire BL's plant manager and about 100 other former employees. In addition, conversion of the plant from producing gasoline engines to producing diesel engines should be relatively simple.

The parent will charge IDC-U.K. licensing and overhead allocation fees equal to 7% of sales in pounds sterling. In addition, IDC-U.S. will sell its English affiliate valves, piston rings, and other components that account for approximately 30% of the total amount of materials used in the manufacturing process. IDC-U.K. will be billed in dollars at the current market price for this material. The remainder will be purchased locally. IDC-U.S. estimates that its all-equity nominal required rate of return for the project will equal 12%, based on an anticipated 3% U.S. rate of inflation and the business risks associated with this venture. The debt capacity of such a project is judged to be about 20%—that is, a debt-to-equity ratio for this project of about 1 : 4 is considered reasonable.

To simplify its investment analysis, IDC-U.S. uses a five-year capital budgeting horizon and then calculates a terminal value for the remaining life of the project. If the project has a positive net present value for the first five years, there is no need to engage in costly and uncertain estimates of future cash flows. If the initial net present value is negative, then IDC-U.S. can calculate a breakeven terminal value at which the net present value will just be positive. This breakeven value is then used as a benchmark against which to measure projected cash flows beyond the first five years.

We now apply the three-stage investment analysis outlined in the preceding section: (1) Estimate project cash flows; (2) forecast the amounts and timing of cash flows to the parent; and (3) add to, or subtract from, these parent cash flows the indirect benefits or costs that this project provides the remainder of the multinational firm.

Estimation of Project Cash Flows

A principal cash outflow associated with the project is the initial investment outlay, consisting of the plant purchase, equipment expenditures, and working-capital requirements. Other cash outflows include operating expenses, later additions to working capital as sales expand, and taxes paid on its net income.

IDC-U.K. has cash inflows from its sales in England and other EC countries. It also has cash inflows from three other sources:

- The tax shield provided by depreciation and interest charges
- Interest subsidies
- The terminal value of its investment, net of any capital gains taxes owed upon liquidation

Recapture of working capital is not assumed until eventual liquidation because this working capital is necessary to maintain an ongoing operation after the fifth year.

INITIAL INVESTMENT OUTLAY ■ Total plant acquisition, conversion, and equipment costs for IDC-U.K. were previously estimated at $50 million. The plant and equipment will be depreciated on a straight-line basis over a five-year period, with a zero salvage value.

Of the $50 million in net plant and equipment costs, $10 million will be financed by NEB's loan of £5 million at 3%. The remaining $40 million will be supplied by the parent in the form of equity capital.

Working-capital requirements—comprising cash, accounts receivable, and inventory—are estimated at 30% of sales, but this amount will be partially offset by accounts payable to local firms, which are expected to average 10% of sales. Therefore, net investment in working capital will equal approximately 20% of sales. The transfer price on the material sold to IDC-U.K. by its parent includes a 25% contribution to IDC-U.S.'s profit and overhead. That is, the variable cost of production equals 75% of the transfer price. Lloyds Bank is providing an initial working-capital loan of £1.5 million ($3 million). All future working-capital needs will be financed out of internal cash flow. Exhibit 21.1 summarizes the initial investment.

FINANCING IDC-U.K. ■ Based on the information just provided, IDC-U.K.'s initial balance sheet, in both pounds and dollars, is presented in Exhibit 21.2. The debt ratio (debt to total assets) for IDC-U.K. is 33:53, or 62%. Note that this debt ratio

EXHIBIT 21.1 Initial Investment Outlay in IDC-U.K. (£1 = $2)

	£ (MILLIONS)	$ (MILLIONS)
Plant purchase and retooling expense	17.5	35
Equipment		
Supplied by parent (used)	2.5	5
Purchased in the United Kingdom	5	10
Working capital		
Bank financing	1.5	3
Total initial investment	£26.5	$53

EXHIBIT 21.2 Initial Balance Sheet of IDC-U.K. (£1 = $2)

	£ (MILLIONS)	$ (MILLIONS)
Assets		
Current assets	1.5	3
Plant and equipment	25	50
Total assets	26.5	53
Liabilities		
Loan payable (to Lloyds)	1.5	3
Total current liabilities	1.5	3
Loan payable (to NEB)	5	10
Loan payable (to IDC-U.S.)	10	20
Total liabilities	16.5	33
Equity	10	20
Total liabilities plus equity	£26.5	$53

could vary from 25%, if the parent's total investment was in the form of equity, all the way up to 100%, if the parent provided all of its $40 million investment for plant and equipment as debt. In other words, as discussed in Chapter 18, an affiliate's capital structure is not independent; rather, it depends on its parent's investment policies.

As discussed in the previous section (see Equation 21.2), the tax shield benefits of interest write-offs are represented separately. Assume that IDC-U.K. contributes $10.6 million to its parent's debt capacity (0.2 × $53 million), the dollar market rate of interest for IDC-U.K. is 8%, and the U.K. tax rate is 40%. This calculation translates into a cash flow in the first and subsequent years equal to $10,600,000 × 0.08 × 0.40, or $339,000. Discounted at 7%, this cash flow provides a benefit equal to $1.4 million over the next five years.

INTEREST SUBSIDIES ■ Based on a 5% anticipated rate of inflation in England and on an expected annual 2% depreciation of the pound relative to the dollar, the market rate on the pound loan to IDC-U.K. would equal about 10%. Thus, the 3% interest rate on the loan by the National Enterprise Board represents a 7% subsidy to IDC-U.K. The cash value of this subsidy equals £350,000 (£5,000,000 × 0.07), or approximately $700,000 annually for the next five years, with a present value of $2.6 million.[2]

SALES AND REVENUE FORECASTS ■ At a profit-maximizing price of £250 per unit in the first year ($490 at the projected year 1 exchange rate), demand for diesel engines in England and the other Common Market countries is expected to increase by 10% annually, from 60,000 units in the first year to 88,000 units in the fifth year. It is assumed here that purchasing power parity holds with no lag and that real prices remain constant in both absolute and relative terms. Hence, the sequences of nominal pound prices and

---

[2] The exact present value of this subsidy is given by the difference between the present value of debt service on the 3% loan discounted at 10% and the face value of the loan.

exchange rates, reflecting anticipated annual rates of inflation equaling 5% and 3% for the pound and dollar, respectively, are

	YEAR					
	0	1	2	3	4	5
Price (pounds)	—	250	278	308	342	380
Exchange rate (dollars)	2.00	1.96	1.92	1.89	1.85	1.82

It is also assumed here that purchasing power parity holds with respect to the euro and other currencies of the various EC countries to which IDC-U.K. exports. These exports account for about 60% of total IDC-U.K. sales. Disequilibrium conditions in the currency markets or relative price changes can be dealt with using an approach similar to that taken in the exposure measurement example (Spectrum Manufacturing) in Chapter 10.

In the first year, although demand is at 60,000 units, IDC-U.K. can produce and supply the market with only 30,000 units (owing to the six-month start-up period). Another 20,000 units are exported by IDC-U.S. to its English affiliate at a unit transfer price of £250, leading to no profit for IDC-U.K. Because these units would have been exported anyway, IDC-U.K. is not credited from a capital budgeting standpoint with any profits on these sales. IDC-U.S. ceases its exports of finished products to England and the EC after the first year. From year 2 on, IDC-U.S. is counting on an expanding U.S. market to absorb the 20,000 units. Based on these assumptions, IDC-U.K.'s projected sales revenues are shown in Exhibit 21.3, line C.

In nominal terms, IDC-U.K.'s pound sales revenues are projected to rise at a rate of 15.5% annually, based on a combination of the 10% annual increase in unit demand and the 5% annual increase in unit price ($1.10 \times 1.05 = 1.155$). Dollar revenues will increase at about 13% annually, due to the anticipated 2% annual pound devaluation.

PRODUCTION COST ESTIMATES ■Based on the assumptions that relative prices will remain constant and that purchasing power parity will hold continually, variable costs of production, stated in real terms, are expected to remain constant, whether denominated in pounds or in dollars. Hence, the pound prices of both labor and material sourced in England and components imported from the United States are assumed to increase by 5% annually. Unit variable costs in the first year are expected to equal £140, including £30 ($60) in components purchased from IDC-U.S.

In addition, the license fees and overhead allocations, which are set at 7% of sales, will rise at an annual rate of 15.5% because pound revenues are rising at that rate. With a full year of operation, initial overhead expenses would be expected to equal £1,100,000. Actual overhead expenses incurred, however, are only £600,000 because the plant does not begin operation until midyear. These expenses are partially fixed, so their rate of increase should be about 8% annually.

The plant and equipment, valued at £25 million, can be written off over five years, yielding an annual depreciation charge against income of £5 million. The cash flow associated with this tax shield remains constant in nominal pound terms but declines in nominal dollar value by 2% annually. With a 3% rate of U.S. inflation, its real value is, therefore, reduced by 5% annually, the same as its loss in real pound terms.

EXHIBIT 21.3 Present Value of IDC-U.K.: Project Viewpoint

	0	1	2	3	4	5	5+
				YEAR			
A. Sales (units)		30,000	66,000	73,000	80,000	88,000	
B. Price per unit (£)		250	263	276	289	304	
C. Sales revenue (£ millions)		7.5	17.3	20.1	23.2	26.7	
D. Variable cost per unit (£)		140	147	154	162	170	
E. Total variable cost (£ millions)		4.2	9.7	11.3	13.0	15.0	
F. Licensing fees and royalties (0.07 × line C, in £ millions)		0.5	1.2	1.4	1.6	1.9	
G. Overhead expenses (£ millions)		0.6*	1.2	1.3	1.4	1.5	
H. Depreciation (£ millions)		5.0	5.0	5.0	5.0	5.0	
I. Total expenses (E + F + G + H, in £ millions)		10.3	17.1	19.0	21.0	23.3	
J. Profit before tax (C − I, in £ millions)		−2.8	0.2	1.2	2.2	3.4	
K. U.K. corporate income taxes @ 40% = 0.40 × J**		0.0	0.0	0.0	0.3	1.4	
L. Net profit after tax (J − K, in £ millions)		−2.8	0.2	1.2	1.9	2.0	
M. Terminal value for IDC-U.K. [2.7 × (L + H), for year 5, in £ millions]							19.0
N. Initial investment, including working capital (£ millions)	−26.5						
O. Working capital investment at 20% of revenue (0.2 × C, in £ millions)		1.5	3.5	4.0	4.6	5.3	
P. Required addition to working capital (line O for year t − line O for year t − 1; t = 2, . . . , 5, in £ millions)		0.0	2.0	0.6	0.6	0.7	
Q. IDC–U.K. net cash flow (L + H + M + N − P, in £ millions)	−26.5	2.2	3.3	5.6	6.3	6.3	19.0
R. £ exchange rate ($)	$2.00	$1.96	$1.92	$1.89	$1.85	$1.82	$1.82
S. IDC-U.K. cash flow (Q × R, in $ millions)	−53.0	4.3	6.3	10.6	11.6	11.5	34.5
T. Present value factor at 12%	1.0	0.8929	0.7972	0.7118	0.6355	0.5674	0.5674
U. Present value (S × T, in $ millions)	−53.0	3.8	5.0	7.5	7.4	6.5	19.6
V. Cumulative present value ($ millions)	−$53.0	−$49.2	−$44.2	−$36.7	−$29.3	−$22.8	−$3.2

*Represents overhead for less than one full year.

**Loss carryforward from year 1 of £2.8 eliminates tax for years 2 and 3 and reduces tax for year 4.

Annual production costs for IDC-U.K. are estimated in Exhibit 21.3, lines D-I. It should be realized, of course, that some of these expenses are, like depreciation, a non-cash charge or, like licensing fees, a benefit to the overall corporation.

Total production costs rise less rapidly each year than the 15.5% annual increase in nominal revenue. This situation is due both to the fixed depreciation charge and to the semifixed nature of overhead expenses. Thus, the profit margin should increase over time.

PROJECTED NET INCOME ■ Net income for years 1 through 5 is estimated on line L of Exhibit 21.3. The effective tax rate on corporate income faced by IDC-U.K. in England is estimated to be 40%. The £2.8 million loss in the first year is applied against income in years 2, 3, and 4, reducing corporate taxes owed in those years.

ADDITIONS TO WORKING CAPITAL ■ One of the major outlays for any new project is the investment in working capital. IDC-U.K. begins with an initial investment in working capital of £1.5 million ($3 million). Working-capital requirements are projected at a constant 20% of sales. Thus, the necessary investment in working capital will increase by 15.5% annually, the rate of increase in pound sales revenue. These calculations are shown on lines O and P of Exhibit 21.3.

TERMINAL VALUE ■ Calculating a terminal value is a complex undertaking, given the various possible ways to treat this issue. Three different approaches are pointed out. One approach is to assume that the investment will be liquidated after the end of the planning horizon and to use this value. However, this approach just takes the question one step further: What would a prospective buyer be willing to pay for this project? The second approach is to estimate the market value of the project, assuming that it is the present value of remaining cash flows. Again, though, the value of the project to an outside buyer may differ from its value to the parent firm, owing to parent profits on sales to its affiliate, for instance. The third approach is to calculate a breakeven terminal value at which the project is just acceptable to the parent and then use that as a benchmark against which to judge the likelihood of the present value of future cash flows exceeding that value.

Most firms try to be quite conservative in estimating terminal values. IDC-U.K. calculates a terminal value based on the assumption that the market value of the project will be 2.7 times the net cash flow in year 5 (net income plus depreciation), or £19 million.

ESTIMATED PROJECT PRESENT VALUE ■ We are now ready to estimate the net present value of IDC-U.K. from the viewpoint of the project. As shown in Exhibit 21.3, line V, the NPV of project cash flows equals −$3.2 million. Adding to this amount the $2.6 million value of interest subsidies and the $1.4 million present value of the tax shield on interest payments yields an overall positive project net present value of $0.8 million. The estimated value of the interest tax shield would be correspondingly greater if this analysis were to incorporate benefits derived over the full 10-year assumed life of the project, rather than including benefits from the first five years only. Over 10 years, the present value of the tax shield would equal $2.4 million, bringing the overall project net present value to $1.8 million. The latter approach is the conceptually correct one.

Despite the favorable net present value for IDC-U.K., it is unlikely a firm would undertake an investment that had a positive value only because of interest subsidies or the interest tax shield provided by the debt capacity of the project. However, this is exactly what most firms do if they accept a marginal project, using a weighted cost of capital. Based on the debt capacity of the project and its subsidized financing, IDC-U.K. would have a weighted cost of capital of approximately 10%. At this discount rate, IDC-U.K. would be marginally profitable.

It would be misleading, however, to conclude the analysis at this point without recognizing and accounting for differences between project and parent cash flows and their impact on the worth of investing in IDC-U.K. Ultimately, shareholders in IDC-U.S. will benefit from this investment only to the extent that it generates cash flows that are, or can be, transferred out of England. The value of this investment is now calculated from the viewpoint of IDC-U.S.

Estimation of Parent Cash Flows

From the parent's perspective, additional cash outflows are recorded for any taxes paid to England or the United States on remitted funds. IDC-U.S. has additional cash inflows as well. It receives licensing and overhead allocation fees each year for which it incurs no additional expenses. If it did, the expenses would have to be charged against the fees. IDC-U.S. also profits from exports to its English affiliate.

LOAN PAYMENTS ■ IDC-U.K. first will make all necessary loan repayments before paying dividends. Specifically, IDC-U.K. will repay the £1.5 million working-capital loan from Lloyds at the end of year 2 and NEB's loan of £5 million at the end of the fifth year. Their dollar repayment costs are estimated at $2.9 million and $9.3 million, respectively, based on the forecasted exchange rates. These latter two loan repayments are counted as parent cash inflows because they reduce the parent's outstanding consolidated debt burden and increase the value of its equity by an equivalent amount. Assuming that the parent would repay these loans regardless, having IDC-U.K. borrow and repay funds is equivalent to IDC-U.S. borrowing the money, investing it in IDC-U.K., and then using IDC-U.K.'s higher cash flows (because it no longer has British loans to service) to repay IDC-U.S.'s debts.

REMITTANCES TO IDC-U.S. ■ IDC-U.K. is projected to pay dividends equal to 100% of its remaining net cash flows after making all necessary loan repayments. It also pays licensing and overhead allocation fees equal, in total, to 7% of gross sales. On both of these forms of transfer, the English government will collect a 10% withholding tax. These remittances are shown in Exhibit 21.4. IDC-U.S., however, will not owe any further tax to

EXHIBIT 21.4 Dividends and Fees and Royalties Received by IDC-U.S. (U.S. $ Millions)

	YEAR					
	1	*2*	*3*	*4*	*5*	*5+*
A. Net cash flow to IDC-U.K. (from Exhibit 21.3, line S)	4.3	6.3	10.6	11.6	11.5	34.5
B. Loan repayments by IDC-U.K.		2.9			9.3	
C. Dividend paid to IDC-U.S. (A − B)	4.3	3.3	10.6	11.6	2.2	34.5
D. Fees and royalties (Exhibit 21.3, line F × line G)	1.0	2.3	2.7	3.0	3.4	15.5*
E. Withholding tax paid to England @ 10% = .10 × (C + D)	0.5	0.6	1.3	1.5	0.6	5.0
F. Net income received by IDC-U.S. (C + D − E, in $ millions)	$4.8	$5.1	$11.9	$13.1	$5.1	$45.0
G. Exchange rate	$1.96	$1.92	$1.89	$1.85	$1.82	$1.82

*Estimated present value of future fees and royalties. These were not incorporated in the terminal value figure of $25 million.

EXHIBIT 21.5 Net Cash Flows from Exports to IDC-U.K.

	YEAR					
	1	*2*	*3*	*4*	*5*	*5+*
A. Sales (units)	30,000	66,000	73,000	80,000	88,000	88,000
B. Components purchased from IDC–U.S.						
1. Unit price ($)	60.0	61.8	63.7	65.6	67.5	67.5
2. Total export revenue	1.8	4.1	4.6	5.2	5.9	5.9
(A × B1 in $ millions)						
C. After-tax cash flow	$0.3	$0.7	$0.8	$0.9	$1.0	$1.0
(0.165 × B2 in $ millions)						

the IRS because the company is assumed to have excess foreign tax credits. Otherwise, IDC-U.S. would have to pay U.S. corporate income taxes on the dividends and fees it receives, less any credits for foreign income and withholding taxes already paid. In this case, IDC-U.K. losses in the first year, combined with the higher British corporate tax rate, will assure that IDC-U.S. would owe minimal taxes to the IRS even if it did not have any excess foreign tax credits.

EARNINGS ON EXPORTS TO IDC-U.K. ▪ With a 25% margin on its exports, and assuming it has sufficient spare-parts manufacturing capacity, IDC-U.S. has incremental earn-

EXHIBIT 21.6 Present Value of IDC-U.K.: Parent Viewpoint (U.S. $ Millions)

				YEAR			
	0	*1*	*2*	*3*	*4*	*5*	*5+*
A. Cash inflows							
1. Loan repayments by IDC-U.K.			2.9			9.3	
(from Exhibit 21.4, line B)							
2. Dividends paid to IDC-U.S.		4.3	3.3	10.6	11.6	2.2	34.5
(from Exhibit 21.4, line C)							
3. Fees and royalties paid to		1.0	2.3	2.7	3.0	3.4	15.5
IDC-U.S. (from Exhibit 21.4, line D)							
4. Net cash flows from exports		0.3	0.7	0.8	0.9	1.0	4.1*
(from Exhibit 21.5, line C)							
5. Total cash inflows		5.6	9.3	14.0	15.4	15.9	54.1
B. Cash outflows							
1. Plant and equipment	50						
2. Working capital	3						
3. Withholding tax paid to U.K.		0.5	0.6	1.3	1.5	0.6	5.0
(from Exhibit 21.4, line E)							
4. Total cash outflows	53	0.5	0.6	1.3	1.5	0.6	5.0
C. Net cash flow (A5 − B4)	−53	5.1	8.7	12.7	14.0	15.3	49.1
D. Present-value factor at 12%	1.0	0.8929	0.7972	0.7118	0.6355	0.5674	0.5674
E. Present value (C × D)	−53	4.5	6.9	9.0	8.9	8.7	27.9
F. Cumulative present value	−$53	−$48.5	−$41.5	−$32.5	−$23.6	−$14.9	$12.9
($ millions)							

*Estimated present value of future earnings on export sales to IDC-U.K.

ings on sales to IDC-U.K. equaling 25% of the value of these shipments. After U.S. corporate tax of 35%, IDC-U.S. generates cash flows valued at 16.5% (25% × 65%) of its exports to IDC-U.K. These cash flows are presented in Exhibit 21.5.

ESTIMATED PRESENT VALUE OF PROJECT TO IDC-U.S. ■ In Exhibit 21.6, all the various cash flows are added up, net of tax and interest subsidies on debt; and their present value is calculated at $12.9 million. Adding the $5 million in debt-related subsidies ($2.4 million for the interest tax shield and $2.6 million for the NEB loan subsidy) brings this value up to $17.9 million. It is apparent that, despite the additional taxes that must be paid to England and the United States, IDC-U.K. is more valuable to its parent than it would be to another owner on a stand-alone basis. This situation is primarily due to the various licensing and overhead allocation fees received and the incremental earnings on exports to IDC-U.K.

LOST SALES ■ There is a circumstance, however, that can reverse this conclusion. This discussion has assumed that IDC-U.S. is now producing at capacity and that the 20,000 diesels currently being exported to the EC can be sold in the United States, starting in year 2. Should this assumption not be the case (that is, should 20,000 units of IDC-U.K. sales just replace 20,000 units of IDC-U.S. sales), then the project would have to be charged with the incremental cash flow that IDC-U.S. would have earned on these lost exports. We now see how to incorporate this effect in a capital-budgeting analysis.

Suppose the incremental after-tax cash flow per unit to IDC-U.S. on its exports to the EC equals $180 at present and that this contribution is expected to maintain its value in current dollar terms over time. Then, in nominal dollar terms, this margin grows by 3% annually. If we assume lost sales of 20,000 units per year, beginning in year 2 and extending through year 10, and a discount rate of 12%, the present value associated with these lost sales equals $19.5 million. The calculations are presented in Exhibit 21.7. Subtracting the present value of lost sales from the previously calculated present value of $17.9 million yields a net present value of IDC-U.K. to its parent equal to −$1.6 million (−$6.6 million ignoring the interest tax shield and subsidy).

EXHIBIT 21.7 Value of Lost Export Sales

	YEAR								
	2	*3*	*4*	*5*	*6*	*7*	*8*	*9*	*10*
A. Lost unit sales	20,000	20,000	20,000	20,000	20,000	20,000	20,000	20,000	20,000
B. Cash flow per unit*	185.4	191.0	196.7	202.6	208.7	214.9	221.4	228.0	234.9
C. Total cash flow from exports (A × B)	3.7	3.8	3.9	4.1	4.2	4.3	4.4	4.6	4.7
D. Present value factor at 12%	0.7972	0.7118	0.6355	0.5674	0.5066	0.4523	0.4039	0.3606	0.3220
E. Present value (C × D)	3.0	2.7	2.5	2.5	2.1	1.9	1.8	1.6	1.5
F. Cumulative present value	$3.0	$5.7	$8.2	$8.2	$12.6	$14.5	$16.3	$18.0	$19.5

*The figures in this row grow by 3% each year. So, 185.4 = 180(1.03), and so on.

This example points up the importance of looking at incremental cash flows generated by a foreign project, rather than total cash flows. An investment that would be marginally profitable on its own, and quite profitable when integrated with parent activities, becomes unprofitable when taking into account earnings on lost sales.

21.4 POLITICAL RISK ANALYSIS

It is apparent from the figures in Exhibit 21.5 that IDC-U.S.'s English investment is quite sensitive to the potential political risks of currency controls and expropriation. The net present value of the project does not turn positive until well after its fifth year of operation (assuming there are no lost sales). Should expropriation occur or exchange controls be imposed at some point during the first five years, it is unlikely that the project will ever be viable from the parent's standpoint. Only if compensation is sufficiently great in the event of expropriation, or if unremitted funds can earn a return reflecting their opportunity cost to IDC-U.S. with eventual repatriation in the event of exchange controls, can this project still be viable in the face of these risks.

The general approach recommended previously for incorporating political risk in an investment analysis usually involves adjusting the cash flows of the project (rather than its required rate of return) to reflect the impact of a particular political event on the present value of the project to the parent. This section shows how these cash flow adjustments can be made for the cases of expropriation and exchange controls.

Expropriation

The extreme form of political risk is **expropriation**. Expropriation is an obvious case where project and parent company cash flows diverge. The approach suggested here examines directly the impact of expropriation on the present value of the project to the parent. In this section, we examine the technique of adjusting expected cash flows to show how expropriation and currency controls affect the value of specific projects.

UNITED FRUIT COMPANY CALCULATES THE CONSEQUENCES OF EXPROPRIATION Suppose that United Fruit Company (UFC) is worried that its banana plantation in Honduras will be expropriated during the next 12 months.[3] The Honduran government has promised, however, that compensation of $100 million will be paid at the year's end if the plantation is expropriated. UFC believes that this promise would be kept. If expropriation does not occur this year, it will not occur any time in the foreseeable future. The plantation is expected to be worth $300 million at the end of the year. A wealthy Honduran has just offered UFC $128 million for the plantation. If UFC's risk-adjusted discount rate is 22%, what is the probability of expropriation at which UFC is just indifferent between selling now or holding onto its plantation?

Exhibit 21.8 displays UFC's two choices and their consequences. If UFC sells out now, it will receive $128 million today. Alternatively, if it chooses to hold on to the plantation, its property will be worth $300 million if expropriation does not occur and worth only $100 million in the event the Honduran government expropriates its plantation and compensates UFC. If the probability of expropriation is p, then the expected end-of-year value of the plantation to UFC (in millions of dollars) is $100p + 300(1 - p) = 300 - 200p$. The present value of the

amount, using UFC's discount rate of 22%, is $(300 - 200p)/1.22$. Setting this equal to the $128 million offer by the wealthy Honduran yields a value of $p = 72\%$. In other words, if the probability of expropriation is at least 72%, UFC should sell out now for $128 million. If the probability of expropriation is less than 72%, it would be more worthwhile for UFC to hold on to its plantation.

EXHIBIT 21.8 United Fruit Company's Choices (U.S. $ Millions)

	EXPROPRIATION	NO EXPROPRIATION	EXPECTED PRESENT VALUE
Sell out now	128	128	128
Wait	100	300	$[100p + 300(1 - p)]/1.22$

[3] Illustration suggested by Richard Roll.

Blocked Funds

The same method of adjusting expected cash flows can be used to analyze the effects of various exchange controls. In any discussion of **blocked funds**, it must be pointed out that if all funds are expected to be blocked in perpetuity, then the value of the project is zero.

ILLUSTRATION BRASCAN CALCULATES THE CONSEQUENCES OF CURRENCY CONTROLS On January 1, 1981, the Indonesian electrical authority expropriated a power-generating station owned by Brascan, Inc., a Canadian operator of foreign electric facilities.[4] In compensation, a perpetuity of C$50 million will be paid annually at the end of each year. Brascan believes, however, that the Indonesian Central Bank may block currency repatriations during the calendar year 1983, allowing only 75% of each year's payment to be

repatriated (and no repatriation of reinvestments from the other 25%). Assuming a cost of capital of 20% and a probability of currency blockage of 40%, what is the current value (on January 1, 1981) of Indonesia's compensation?

Exhibit 21.9 displays the two possibilities and their consequences for the cash flows Brascan expects to receive. If currency controls are not imposed, Brascan will receive C$50 million annually, with the first payment due December 31, 1981. The present value of this stream of cash equals C$250 million (50/0.2). Alternatively, if controls are imposed, Brascan will receive C$50 million at the end of the first two years and C$37.5 million (50 × 0.75) on each December 31 thereafter. The present value of these cash flows is C$206.6 million [50/1.2 + 50/(1.2)^2 + (37.5/0.2)/(1.2)^2].[5] Weighting these present values by the probability that each will come to pass yields an expected present value (in millions of Canadian dollars) of 0.6 × 250 + 0.4 × 206.6 = C$232.6 million.

EXHIBIT 21.9 Cash Flows to Brascan (C$ Millions)

	CASH FLOW AT YEAR END		
	1981	1982	1983 and On
Currency controls	50.0	50.0	37.5
No currency controls	50.0	50.0	50.0

---

[4] Illustration suggested by Richard Roll.

[5] As of the end of year 2 the $37.5 million annuity beginning in year 3 has a present value equal to 37.5/0.2. The present value of this annuity as of today equals [37.5/0.2]/(1.2)^2.

21.5 GROWTH OPTIONS AND PROJECT EVALUATION

The discounted cash flow (DCF) analysis presented so far treats a project's expected cash flows as given at the outset. This approach presupposes a static approach to investment decision making: It assumes that all operating decisions are set in advance. In reality, though, the opportunity to make decisions contingent on information to become available in the future is an essential feature of many investment decisions.

Consider the decision of whether to reopen a gold mine. The cost of doing so is expected to be $1 million. There are an estimated 40,000 ounces of gold remaining in the mine. If the mine is reopened, the gold can be removed in one year at a variable cost of $390 per ounce. Assuming an expected gold price in one year of $400/ounce, the expected profit per ounce mined is $10. Clearly, the expected cash inflow (ignoring taxes) of $400,000 next year ($10 × 40,000) is far below that necessary to recoup the $1 million investment in reopening the mine, much less to pay the 15% yield required on such a risky investment. However, intuition — which suggests a highly negative project NPV of −$652,174 (−$1,000,000 + 400,000/1.15) — is wrong in this case. The reason is that the cash flow projections underlying the classical DCF analysis ignore the option *not* to produce gold if it is unprofitable to do so.

Here is a simple example that demonstrates the fallacy of always using expected cash flows to judge an investment's merits. Suppose there are only two possible gold prices next year: $300/ounce and $500/ounce, each with probability 0.5. The expected gold price is

EXHIBIT 21.10 The Gold Mine-Operating Decision

Today	Gold price	Next year	Cash flow

Close mine

$300/ounce
($p = 0.5$)

$0 × 40,000 = $0

Invest
$1 million

Expected
cash flow

$4,400,000 × 0.5 + 0 × 0.5 = $2,200,000

$500/ounce
($p = 0.5$)

$110 × 40,000 = $4,400,000

Mine gold

$400/ounce, but this expected price is irrelevant to the optimal mining decision rule: Mine gold if, and only if, the price of gold at year's end is $500/ounce. Exhibit 21.10 shows the cash flow consequences of that decision rule. Closure costs are assumed to be zero.

Incorporating the mine owner's option *not* to mine gold when the price falls below the cost of extraction reveals a positive net present value of $913,043 for the decision to reopen the gold mine:

$$\frac{\text{NPV of gold mine}}{\text{investment}} = -\$1,000,000 + \frac{\$2,200,000}{1.15}$$

$$= \$913,043$$

As the example of the gold mine demonstrates, the ability to alter decisions in response to new information may contribute significantly to the value of a project. Such investments bear the characteristics of options on securities and should be valued accordingly. As we saw in the case of foreign exchange, a call option gives the holder the right, but not the obligation, to buy a security at a fixed, predetermined price (called the exercise price) on or before some fixed future date. By way of analogy, the opportunities a firm may have to invest capital to increase the profitability of its existing product lines and benefit from expanding into new products or markets may be thought of as **growth options**.[6] Similarly, a firm's ability to capitalize on its managerial talent, experience in a particular product line, its brand name, technology, or its other resources may provide valuable but uncertain future prospects.

Growth options are of great importance to multinational firms. Consider the value of IDC-U.K.'s production and market positions at the end of its planning horizon. IDC-U.S. may increase or decrease the diesel plant's output depending on current market conditions, expectations of future demand, and relative cost changes, such as those due to

---

[6]A good discussion of growth options is contained in W. Carl Kester, "Today's Options for Tomorrow's Growth," *Harvard Business Review*, March–April 1984, pp. 153–160.

currency movements. The plant can be expanded; it can be shut down, then reopened when production and market conditions are more favorable; or it can be abandoned permanently. Each decision is an option from the viewpoint of IDC-U.S. The value of these options, in turn, affects the value of the investment in IDC-U.K.

Moreover, by producing locally, IDC-U.S. will have an enhanced market position in the EC that may enable the company to expand its product offerings at a later date. The ability to exploit this market position depends on the results of IDC-U.S.'s R&D efforts and the shifting pattern of demand for its products. In all these cases, the optimal operating policy depends on outcomes that are not known at the project's inception.

Similarly, the investments that many Western firms are now considering in Eastern Europe also can be thought of as growth options. Some view investments there as a way to gain entry into a potentially large market. Others see Eastern Europe as an underdeveloped area with educated and skilled workers but low wages and view such investments as a low-cost backdoor to Western European markets. In either case, companies who invest there are buying an option that will pay off in the event that Eastern European markets boom or that Eastern European workers turn out to be much more productive with the right technology and incentives than they were under communism. Other investments are undertaken, in part, to gain knowledge that can later be capitalized on elsewhere. For example, in announcing its plans to build in Alabama, Mercedes officials said the factory would serve as a laboratory for learning to build cars more efficiently. Similarly, several Regional Bell Operating Companies (RBOCs) such as SBC Communications and USWest Communications have established operations in Great Britain to learn how—and whether—to provide services such as cable TV, combined cable TV and telephone service, and personal communications services (PCS) that were prohibited to RBOCs in the U.S. market. By failing to take into account the benefits of operating flexibility, learning, and potentially valuable add-on projects, the traditional DCF will tend to understate project values.

The problem of undervaluing investment projects using the standard DCF analysis is particularly acute for strategic investments. Many strategically important investments, such as investments in R&D, factory automation, a brand name, or a distribution network, provide growth opportunities because they are often only the first link in a chain of subsequent investment decisions.

Valuing investments that embody discretionary follow-up projects requires an expanded net present value rule that considers the attendant options. More specifically, the value of an option to undertake a follow-up project equals the expected project NPV using the conventional discounted cash-flow analysis plus the value of the discretion associated with undertaking the project. This relation is shown in Exhibit 21.11. Based on the discussion of currency options in Chapter 6, the latter element of value (the discretion to invest or not invest in a project) depends on the following.

1. *The length of time the project can be deferred*: The ability to defer a project gives the firm more time to examine the course of future events and to avoid costly errors if unfavorable developments occur. A longer time interval also raises the odds that a positive turn of events will dramatically boost the project's profitability and turn even a negative NPV project into a positive one.

EXHIBIT 21.11 Valuing a Growth Option to Undertake a Follow-Up Project

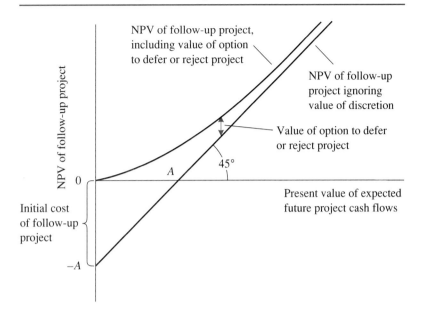

2. *The risk of the project*: Surprisingly, the riskier the investment, the more valuable is an option on it. The reason is the asymmetry between gains and losses. A large gain is possible if the project's NPV becomes highly positive, whereas losses are limited by the option not to exercise when the project NPV is negative. The riskier the project, the greater the odds of a large gain without a corresponding increase in the size of the potential loss. Thus, growth options are likely to be especially valuable for MNCs because of the large potential variation in costs and the competitive environment.

3. *The level of interest rates*: Although a high discount rate lowers the present value of a project's future cash flows, it also reduces the present value of the cash outlay needed to exercise an option. The net effect is that high interest rates generally raise the value of projects that contain growth options.

4. *The proprietary nature of the option*: An exclusively owned option is clearly more valuable than one that is shared with others. The latter might include the chance to enter a new market or to invest in a new production process. Shared options are less valuable because competitors can replicate the investments and drive down returns. For the multinational firm, though, most growth options arise out of its intangible assets. These assets, which take the form of skills, knowledge, brand names, and the like, are difficult to replicate and so are likely to be more valuable.

ILLUSTRATION

FORD GIVES UP ON SMALL CAR DEVELOPMENT

In late 1986, Ford gave up on small-car development in the United States and handed over the job to Japan's Mazda. Although seemingly cost-effective in the short run (Ford should save about $500 million in development costs for one car model alone), such a move—which removed a critical mass from Ford's own engineering efforts—could prove dangerous in the longer term. Overcoming engineering obstacles unique to subcompact cars—for example, the challenges of miniaturization—enhances engineers' skills and allows them to apply innovations to all classes of vehicles. By eroding its technological base, Ford may have yielded the option of generating ideas that can be applied elsewhere in its business. Moreover, the cost of reentering the business of in-house design can be substantial. The abandonment option is not one to be exercised lightly.

Some American consumer-electronics companies, for example, are learning the penalties of ceding major technologies and the experiences that come from working with these technologies on a day-to-day basis. Westinghouse Electric, (now CBS) after quitting the development and manufacture of color television tubes in 1976, recently decided to get back into the color-video business. However, because it lost touch with the product, Westinghouse has been able to reenter only by way of a joint venture with Japan's Toshiba.

Similarly, RCA and other U.S. manufacturers several years ago conceded to the Japanese development of videocassette recorders and laser video disk players. Each technology has since spawned entirely new, popular product lines—from video cameras to compact disk players—in which U.S. companies are left with nothing to do beyond marketing Japanese-made goods.

To take another example, RCA and Westinghouse first discovered the principles of liquid crystal displays (LCDs) in the 1960s. But while the Americans did not follow up with investment and development, Japanese companies did. Sharp, Seiko, and Casio used LCDs in calculators and digital watches. That gave them knowledge of the technology so that later, when laptop computers developed needs for sophisticated graphics and color pictures, Japanese manufacturers could deliver increasingly capable screens.

Even those companies that merely turn to outside partners for technical help could nevertheless find their skills atrophying over the years as their partners handle more of the complex designing and manufacturing. Such companies range from Boeing, which has enlisted three Japanese firms to help engineer a new plane, to Honeywell, which is getting big computers from NEC. The corresponding reduction in in-house technological skills decreases the value of the option these firms have to develop and apply new technologies in novel product areas.

❧ 21.6 SUMMARY AND CONCLUSIONS

Capital budgeting for the multinational corporation presents many elements that rarely, if ever, exist in domestic capital budgeting. The primary thrust of this chapter has been to adjust project cash flows instead of the discount rate to reflect the key political and economic risks that MNCs face abroad. Tax factors also are incorporated via cash-flow adjustments. Cash-flow adjustments are preferred on the pragmatic grounds that there is available more and better information on the effect of such risks on future cash flows than on the required discount rate. Furthermore, adjusting the required rate of return of a project to reflect incremental risk does not usually allow for adequate consideration of the time pattern and magnitude of the risk being evaluated. Using a uniformly higher discount rate to reflect additional risk involves penalizing future cash flows relatively more heavily than present ones.

This chapter showed how these cash-flow adjustments can be carried out by presenting a lengthy numerical example. It also discussed the significant differences that can exist between project and parent cash flows and showed how these differences can be accounted for when estimating the value to the parent firm of a foreign investment. The chapter also pointed out that failure to take into account the options available to managers to adjust the scope of a project will lead to a downward bias in estimating project cash flows. These options include the possibility of expanding or contracting the project or abandoning it, the chance to employ radical new process technologies by utilizing skills developed from implementing the project, and the possibility of entering the new lines of business to which a project may lead.

➤ Questions

1. A foreign project that is profitable when valued on its own will always be profitable from the parent firm's standpoint. True or false. Explain.

2. What are the principal cash outflows associated with the IDC-U.K. project?

3. What are the principal cash inflows associated with the IDC-U.K. project?

4. In what ways do parent and project cash flows differ on the IDC-U.K. project? Why?

5. Suppose the real value of the pound declines. How would this decline likely affect the economics of the IDC-U.K. project?

6. Describe the alternative ways to treat the interest subsidy provided by the British government.

7. Suppose England raised its corporate tax rate by 1 percentage point. How would this increase affect the economics of the IDC-U.K. project?

8. Why are IDC-U.S. earnings on exports to IDC-U.K. credited to the project?

9. Why are loan repayments by IDC-U.K. to Lloyds and NEB treated as a cash inflow to the parent company?

10. Under what circumstances should IDC-U.S. earnings on lost export sales to the United Kingdom and the rest of the EC countries be treated as a cost of the project?

11. Under what circumstances should these lost export earnings be ignored when evaluating the project?

12. How sensitive is the value of the project to the threat of currency controls and expropriation? How can the financing be structured to make the project less sensitive to these political risks?

13. What options does investment in the new British diesel plant provide to IDC-U.S.? How can these options be accounted for in the traditional capital-budgeting analysis?

14. Should the cost of capital for the IDC-U.K. project be higher, lower, or the same as the cost of capital for a similar project to manufacture and sell diesel engines in the United States? Explain.

15. Early results on the Lexus, Toyota's upscale car, showed it was taking the most business from customers changing from BMW (15%), Mercedes (14%), Toyota (14%), General Motors' Cadillac (12%), or Ford's Lincoln (6%). With what in the auto business is considered a high percentage of sales coming from its own customers (14%), how badly is Toyota hurting itself with the Lexus?

16. Comment on the following statement that appeared in *The Economist* (August 20, 1988, p. 60): "Those oil producers that have snapped up overseas refineries—Kuwait, Venezuela, Libya and, most recently, Saudi Arabia—can feed the flabbiest of them with dollar-a-barrel crude and make a profit. . . . The majority of OPEC's existing overseas refineries would be scrapped without its own cheap oil to feed them. Both Western European refineries fed by Libyan oil (in West Germany and Italy) and Kuwait's two overseas refineries (in Holland and Denmark) would almost certainly be idle without it."

17. Some economists have stated that too many companies are not calculating the cost of *not* investing in new technology, world-class manufacturing facilities, or market position overseas. What are some of these costs? How are these costs related to the notion of growth options discussed in the chapter?

18. In December 1989, General Electric spent $150 million to buy a controlling interest in Tungsram, the Hungarian state-owned light bulb maker. Even in its best year, Tungsram earned less than a 4% return on equity (based on the price GE paid).

a. What might account for GE's decision to spend so much money to acquire such a dilapidated, inefficient manufacturer?

b. A Hungarian lighting worker earns about $170 a month in Hungary, compared with about $1,700 a month in the United States. Do these figures indicate that Tungsram will be a low-cost producer? Explain.

➤ Problems

1. Suppose a firm projects a $5 million perpetuity from an investment of $20 million in Spain. If the required return on this investment is 20%, how large does the probability of expropriation in year 4 have to be before the investment has a negative NPV? Assume that all cash inflows occur at the end of each year and that the expropriation, if it occurs, will occur just before the year 4 cash inflow or not at all. There is no compensation in the event of expropriation.

2. Suppose a firm has just made an investment in France that will generate $2 million annually in depreciation, converted at today's spot rate. Projected annual rates of inflation in France and in the United States are 7% and 4%, respectively. If the real exchange rate is expected to remain constant and the French tax rate is 50%, what is the expected real value (in terms of today's dollars) of the depreciation charge in year 5, assuming that the tax write-off is taken at the end of the year?

3. Jim Toreson, chairman and CEO of Xebec Corporation, a Sunnyvale, California, manufacturer of disk-drive controllers, is trying to decide whether to switch to offshore production. Given Xebec's well-developed engineering and marketing capabilities, Toreson could use offshore manufacturing to ramp up production, taking full advantage of both low-wage labor and a grab bag of tax holidays, low-interest loans, and other government largess. Most of his competitors seem to be doing it. The faster he follows suit, the better off Xebec would be according to the conventional discounted cash-flow analysis, which shows that switching production offshore is clearly a positive NPV investment. However, Toreson is concerned that such a move would entail the loss of certain intangible strategic benefits associated with domestic production.

a. What might be some strategic benefits of domestic manufacturing for Xebec? Consider the fact that its customers are all U.S. firms and that manufacturing technology—particularly automation skills—is key to survival in this business.

b. What analytic framework can be used to factor these intangible strategic benefits of domestic manufacturing (which are intangible costs of offshore production) into the factory location decision?

c. How would the possibility of radical shifts in manufacturing technology affect the production location decision?

d. Xebec is considering producing more-sophisticated drives that require substantial customization. How does this possibility affect its production decision?

e. Suppose the Taiwan government is willing to provide a loan of $10 million at 5% to Xebec to build a factory there. The loan would be paid off in equal annual installments over a five-year period. If the market interest rate for such an investment is 14%, what is the before-tax value of the interest subsidy?

f. Projected before-tax income from the Taiwan plant is $1 million annually, beginning at the end of the first year. Taiwan's corporate tax rate is 25%, and there is a 20% dividend withholding tax. However, Taiwan will exempt the plant's income from corporate tax (but not withholding tax) for the first five years. If Xebec plans to remit all income as dividends back to the United States, how much is the tax holiday worth?

g. An alternative sourcing option is to shut down all domestic production and contract to have Xebec's products built for it by a foreign supplier in a country such as Japan. What are some of the potential advantages and disadvantages of foreign contracting vis-à-vis manufacturing in a wholly owned foreign subsidiary?

➤ Bibliography

LESSARD, DONALD R. "Evaluating Foreign Projects: An Adjusted Present Value Approach." In *International Financial Management*, 2nd. ed., edited by Donald R. Lessard. New York: John Wiley & Sons, 1985.

SHAPIRO, ALAN C. "Capital Budgeting for the Multi-national Corporation." *Financial Management*, Spring 1978, pp. 7–16.

_____. "International Capital Budgeting." *Midland Journal of Corporate Finance*, Spring 1983, pp. 26–45.

THE MEASUREMENT AND MANAGEMENT OF POLITICAL RISK

People say they want clarification of the rules of the game, but I think it isn't very clear what isn't clear to them.

Adolfo Hegewisch Fernandez,
Mexico's Subsecretary for Foreign Investment

Potential investors don't want flexibility, they want fixed rules of the game.

John Gavin, U.S. Ambassador to Mexico

*I*n recent years, there has been a significant increase in developing and developed countries alike in the types and magnitudes of political risks that multinational companies have faced. Although **expropriation**[1] is the most obvious and extreme form of political risk, there are other significant political risks including currency or trade controls, changes in tax or labor laws, regulatory restrictions, and requirements for additional local production. The common denominator of such risks is not hard to identify: government intervention into the workings of the economy that affects, for good or ill, the value of the firm. Although the consequences usually are adverse, changes in the political environment can provide opportunities as well. The imposition of quotas on autos from Japan, for example, was undoubtedly beneficial to U.S. automobile manufacturers.

The purpose of this chapter is to provide a framework that can facilitate a formal assessment of political risk and its implications for corporate decision making. Both international banks and nonbank multinationals analyze political risk, but from different

---

[1] The terms *expropriation* and *nationalization* are used interchangeably in this book and refer specifically to the taking of foreign property, with or without compensation.

perspectives. This chapter takes the perspective of nonbank MNCs, who analyze political risk in order to determine the investment climate in various countries. Political risk assessments may be used in investment analyses to screen out countries that are excessively risky or to monitor countries in which the firm is currently doing business to determine whether new policies are called for.

The basic approach to managing political risk presented here involves two key steps: (1) identifying political risk and its likely consequences, and (2) developing policies in advance to cope with the possibility of political risk, which basically involves strengthening a company's bargaining position in any confrontation with government policies. The experiences in Chile of Kennecott and of Anaconda, detailed in the last section of this chapter, illustrate this approach.

22.1 MEASURING POLITICAL RISK

Despite the near-universal recognition among multinational corporations, political scientists, and economists of the existence of **political risk**, there is no unanimity yet about what constitutes that risk and how to measure it. The two basic approaches to viewing political risk are from a country-specific perspective and a firm-specific perspective. The former perspective depends on *country risk analysis*, whereas the latter depends on a more micro approach.

A number of commercial and academic political-risk forecasting models are available today. These models normally supply country risk indices that attempt to quantify the level of political risk in each nation. Most of these indices rely on some measure(s) of the stability of the local political regime.

Political Stability

Measures of political stability may include the frequency of changes of government, the level of violence in the country (for example, violent deaths per 100,000 population), number of armed insurrections, conflicts with other states, and so on. The basic function of these stability indicators is to determine how long the current regime will be in power and whether that regime also will be willing and able to enforce its foreign investment guarantees. Most companies believe that greater political stability means a safer investment environment.

A basic problem in many Third World countries is that the local actors have all the external trappings of genuine nation-states—United Nations-endorsed borders, armies, foreign ministries, flags, currencies, and national airlines—but they are nothing of the kind. They lack social cohesion, political legitimacy, and the institutional infrastructures that are necessary for economic growth.

Economic Factors

Other frequently used indicators of political risk include economic factors, such as inflation, balance-of-payments deficits or surpluses, and the growth rate of per capita GDP. The intention behind these measures is to determine whether the economy is in good

THREATS TO THE NATION-STATE

From Canada to the former Czechoslovakia, from India to Ireland, and from South Africa to the former Soviet Union, political movements centered around ethnicity, national identity, and religion are reemerging to contest some of the most fundamental premises of the modern nation-state. In the process, they are reintroducing ancient sources of conflict so deeply submerged by the Cold War that they seemed almost to have vanished from history's equation.

The implications of this resurgence of national, ethnic, and religious passions are profound.

- A host of modern nation-states—from Canada to Lebanon to Iraq—are beginning to crumble, while others—such as Yugoslavia, Czechoslovakia, and Somalia—have already disintegrated, because the concept of the "melting pot," the idea that diverse and even historically hostile peoples could readily be assimilated under larger political umbrellas in the name of modernization and progress, has failed them. Even in the strongest nations, including the United States, the task of such assimilation has proved difficult and the prognosis is for even greater tension in the years ahead.

- After 70 years on the road to nowhere, the Soviet Union finally arrived in 1991. Now, turmoil in the states making up the former Soviet Union and parts of China threaten to blow apart the last remnants of an imperial age that began more than 500 years ago. The turbulent dismantling of nineteenth-century European empires after World War II may be matched by new waves of disintegration within the former Soviet and Chinese Communist empires, with incalculable consequences for the rest of the world. Stretching from the Gulf of Finland to the mountains of Tibet and beyond, the sheer scale of the potential instability would tax the world's capacity to respond. Ethnic unrest could spill into neighboring countries, old border disputes could reignite, and, if the central governments tried to impose order with force, civil wars could erupt within two of the world's largest nuclear powers.

- Around the world, fundamentalist religious movements have entered the political arena in a direct challenge to one of the basic principles of the modern age: that governments and other civic institutions should be predominantly secular and religion confined to the private lives of individuals and groups. Since the end of the Middle Ages, when religion dominated not just government but every aspect of society, the pervasive trend in the past 500 years has been to separate church and state. Now, in many parts of the world, powerful movements—reacting against the secular quality of modern public culture and the tendency of traditional values to be swept aside in periods of rapid change—are insisting on a return to God-centered government. One consequence of this trend is to make dealings between states and groups more volatile. As the United States learned with the Arab-Israeli conflict, the Iranian revolution, and the Persian Gulf war, disputes are far harder to manage when governments root their positions in religious principle.

Paradoxically, at the same time that many states and societies are fragmenting over religion, ethnicity, and national culture, their people nourish hopes of achieving economic progress by allying themselves to one or another of the new trade blocs—Europe, North America, Pacific Rim—now taking shape. Yet in many cases such dreams will not materialize. Civil strife and dogmatic politics hold little allure for foreign investors; bankers lend money to people whose first priority is money, not religion or ethnic identity. The challenge for business is to create profitable opportunities in a world that is simultaneously globalizing and localizing.

shape or requires a quick fix, such as expropriation to increase government revenues or currency inconvertibility to improve the balance of payments. In general, the better a country's economic outlook, the less likely it is to face political and social turmoil that will inevitably harm foreign companies.

Subjective Factors

More-subjective measures of political risk are based on a general perception of the country's attitude toward private enterprise: whether private enterprise is considered a necessary evil to be eliminated as soon as possible or is actively welcomed. The attitude toward multinationals is particularly relevant and may differ from the feeling regarding local private ownership. Consider, for example, the Soviet Union and other Eastern European countries that have actively sought products, technology, and even joint ventures with Western firms while refusing to tolerate (until recently) domestic free enterprise. In general, most countries probably view foreign direct investment in terms of a cost/benefit trade-off and are not either for or against it in principle.

An index that tries to incorporate all these economic, social, and political factors into an overall measure of the business climate, including the political environment, is the Profit Opportunity Recommendation (POR) rating, shown in Exhibit 22.1. The scores of

EXHIBIT 22.1 Profit Opportunity Recommendation Rankings, 1997

Low Risk (70–100)	POR Composite Score	High Risk (40–54)	
Switzerland	82	Chile	53
Singapore	79	Saudi Arabia	53
Japan	76	South Africa	52
Taiwan (ROC)	76	Czech Republic	51
Netherlands	74	Italy	50
Germany	71	Thailand	49
Norway	71	Turkey	46
Austria	70	Egypt	45
		Hungary	45
Moderate Risk (55–69)		Indonesia	45
United States	69	Poland	45
Belgium	66	Argentina	43
France	63	Colombia	43
Sweden	63	India	43
Ireland	62	Philippines	43
Denmark	61	Greece	43
United Kingdom	61	Iran	42
Finland	59	Kazakhstan	42
Portugal	59	Vietnam	42
Spain	59	Brazil	41
Malaysia	58	Mexico	40
Australia	57	Peru	40
China (PRC)	57	Russia	40
South Korea	57	Venezuela	40
Canada	55		
		Prohibitive Risk (0–39)	
		Pakistan	39
		Ukraine	38
		Ecuador	37

Source: BERI S.A., Washington, D.C., 1997.

countries listed on the POR scale are based on an aggregation of the subjective assessments of a panel of experts.

POLITICAL RISK AND UNCERTAIN PROPERTY RIGHTS ■ Models such as POR are useful insofar as they provide an indication of the general level of political risk in a country. From an economic standpoint, political risk refers to uncertainty over **property rights**. If the government can expropriate either legal title to property or the stream of income it generates, then political risk exists. Political risk also exists if property owners may be constrained in the way they use their property. This definition of political risk encompasses government actions ranging from outright expropriation to a change in the tax law that alters the government's share of corporate income to laws that change the rights of private companies to compete against state-owned companies. Each action affects corporate cash flows and hence the value of the firm.

ILLUSTRATION

KOMINEFT RHYMES WITH THEFT

In early 1995, Komineft, a Russian oil company, instituted a 3-for-2 stock split, but didn't tell shareholders.

It also said that only those investors on the registry in May 1984 would get the new shares, thereby diluting by a third the stakes of those who had bought shares afterward, which included most foreign buyers. Komineft, although conceding there was a problem, insisted it had done nothing wrong. It was probably right. The few Russian laws there are to protect shareholders' interests are often contradictory, allowing Russian companies to ignore shareholder rights and get away with it. Although Komineft later reversed its position on the stock split, episodes such as this one have taken most of the luster off of investing in Russian companies.

Other investors in Russia have had similar experiences with vague and shifting property rights. For example, in February 1995, Texaco's $45 billion deal to drill in the Russian Arctic hit a last-minute snag when a regional production association suddenly demanded a 50% share of the profits. In addition, oil companies have suffered from arbitrary changes in tax and export laws despite the fact that when Western oil companies registered their joint ventures in 1991, the Soviet government assured them they would be able to export 100% of production tax-free. However, as the government struggled over cash shortages, the oil companies' "rights" disappeared. As a result, every foreign oil company operating in the former Soviet Union is scaling back its operations there or pulling out altogether.

The key questions that companies should ask in assessing the degree of political risk they face in a country, particularly one undergoing a political and economic transition, are as follows:

- Has economic reform become institutionalized, thereby minimizing the chance of abrupt policy changes that would adversely affect an investment's value?
- Are the regulatory and legal systems predictable and fair? Constant rule changes involving foreign ownership, taxes, currency controls, trade, or contract law raise investment risk.
- Is the government reasonably competent, maintaining the value of its currency and preserving political stability?

ILLUSTRATION POLITICAL RISK IN VENEZUELA

According to the *Wall Street Journal* (December 31, 1980, p. 10), when Venezuela's oil income quadrupled in 1973, a high government official declared, "Now we have so much money that we won't need any new foreign investment." President Carlos Perez calculated that the country was rich enough to buy machinery from abroad and set up factories without any foreign participation. He overlooked the fact that Venezuela did not have enough skilled technicians to run and maintain sophisticated equipment. Despite an orgy of buying foreign-made machinery, economic growth stalled as companies were left with equipment they could not operate.

By 1980, President Luis Herrera, who succeeded Perez, recognized the mistake and invited foreigners back. But once shunned, foreigners did not rush back, particularly because the Herrera administration sent out such mixed signals that investors could not be certain how sincere the welcome was or how long it would last.

Consider the uncertainty faced by a foreign investor trying to decipher government policy from the following statements. On the one hand, the Superintendent of Foreign Investment insisted that the government had had "a change of heart" and declared that the country needed "new capital and new technology in manufacturing, agro-industry, and construction of low-cost housing." On the other hand, the head of the powerful Venezuelan Investment Fund, which invests much of the country's oil income, was downright hostile to foreign investors. He said, "Foreign investment is generally unfavorable to Venezuela. Foreign investors think Venezuela is one big grab bag, where they can come and pick out whatever goodies they want." Still another official, the president of the Foreign Trade Institute, said, "It is a grave error to think that foreign investment can contribute to the transfer of technology, capital formation, the development of managerial capacity, and equilibrium in the balance of payments."

The collapse of oil prices brought the return of Carlos Perez as a free-market reformer, who instituted needed economic changes in Venezuela. However, the Venezuelan government failed to pursue political reforms in tandem with economic reforms, and the economic reforms themselves worsened existing economic inequality and undermined President Perez's hold on power. After Perez was forced out of office on corruption charges in 1994, his successor, President Rafael Caldera, spent more than 10% of the nation's GDP to bail out weak banks (financed by running the central bank's printing presses overtime), stopped all privatizations, appointed the country's best-known Marxist economist to the board of the central bank, imposed price and currency controls, and spoke out strongly against the private sector. The result was high inflation and a massive outflow of capital, prompting an 87% devaluation of the Venezuelan bolivar.

CAPITAL FLIGHT ■ One good indicator of the degree of political risk is the seriousness of capital flight. **Capital flight** refers to the export of savings by a nation's citizens because of fears about the safety of their capital. By its nature, capital flight is difficult to measure accurately because it is not directly observed in most cases. Nevertheless, one can usually infer the capital outflows, using balance-of-payments figures—particularly the entry labeled "errors and omissions." These estimates indicate that capital flight represents an enormous outflow of funds from developing countries.

EXHIBIT 22.2 Latin American Flight Capital:
Estimated Assets Held Abroad at Year End, 1988

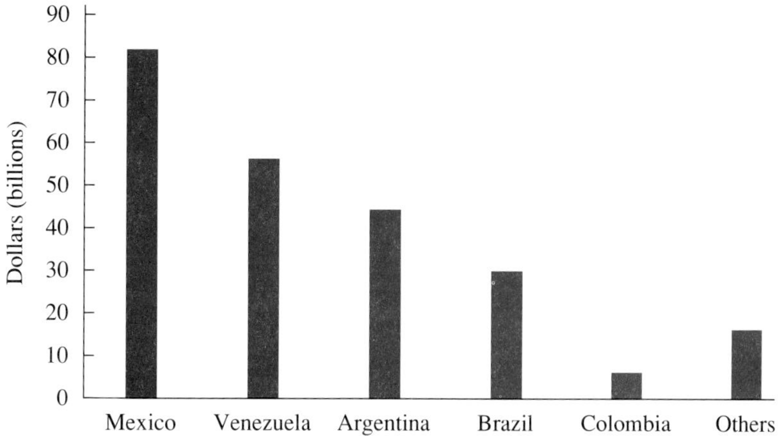

Between 1974 and 1982, Argentina borrowed $32.6 billion. Estimates of the level of capital flight from Argentina over the same period range from $15 billion to more than $27 billion.[2] These estimates would mean that capital flight amounted to between half and four-fifths of the entire inflow of foreign capital to Argentina. For Venezuela, the inflow was $27 billion over the same period, with capital flight estimated at between $12 billion and $22 billion. The inflow for Mexico was $79 billion between 1979 and 1984; the outflow has been estimated at between $26 billion and $54 billion for the same period. Other debtor countries, such as Nigeria and the Philippines, have also had large capital outflows. As conditions in many of these countries worsened, capital flight continued. As of the end of 1988, Morgan Guaranty estimated that Latin Americans held $243 billion in assets abroad, far exceeding the amount of loans to the region held by U.S. banks. And this excludes assets taken abroad before 1977. Exhibit 22.2 shows the breakdown of the $243 billion figure by country.

Capital flight occurs for several reasons, most of which have to do with inappropriate economic policies. These reasons include government regulations, controls, and taxes that lower the return on domestic investments. In countries where inflation is high and domestic inflation hedging is difficult or impossible, investors may hedge by shifting their savings to foreign currencies deemed less likely to depreciate. They may also make the shift when domestic interest rates are artificially held down by their governments, or when they expect a devaluation of an overvalued currency.

Perhaps the most powerful motive for capital flight is political risk. In unstable political regimes (and in some stable ones), wealth is not secure from government seizure, especially when changes in regime occur. Savings may be shifted overseas to protect them. For example, the citizens of Hong Kong, which was turned over to communist

---

[2] These figures come from Steven Plaut, "Capital Flight and LDC Debt," *FRBSF Weekly Letter*, Federal Reserve Bank of San Francisco, January 8, 1988.

EXHIBIT 22.3 Private Capital Returns to Latin America

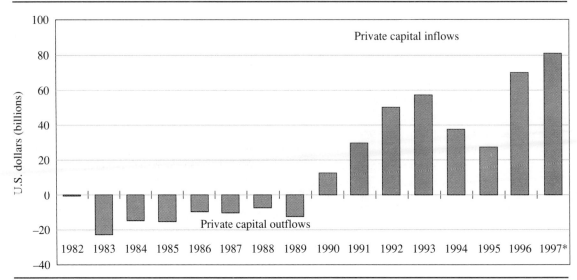

*Estimated as of August 1997.

Source: Bank for International Settlements and Inter-American Development Bank.

China on July 1, 1997, responded to the anticipated change in regime by sending large sums of money abroad in advance.

Common sense dictates that if a nation's own citizens do not trust the government, then investment there is unsafe. After all, residents presumably have a better feel for conditions and government intentions than do outsiders. Thus, when analyzing investment or lending opportunities, multinational firms and international banks must bear in mind the apparent unwillingness of the nation's citizens to invest and lend in their own country.

What is needed to halt capital flight are tough-minded economic policies—the kind of policies that make investors want to put their money to work instead of taking it out. As we shall see in the next section, these policies include cutting budget deficits and taxes, removing barriers to investment by foreigners, selling off state-owned enterprises, allowing for freer trade, and avoiding currency overvaluations that virtually invite people to ship their money elsewhere before the official exchange rate falls.

Such policies are now being employed in Latin America, with predictable results. As seen in Exhibit 22.3, beginning in 1990, capital flight has been reversed, with private capital flooding back into Latin America. These capital inflows helped fuel the extraordinary performance of the Latin American stock markets discussed in Chapter 19.

The Micro Approach

Despite the increased sophistication of political risk models such as POR, however, their usefulness remains problematic. For one thing, political instability by itself does not necessarily contribute to political risk. Changes of government in Latin America, for example, are quite frequent; yet most multinationals continue to go about their business relatively undisturbed.

The most important weakness of these indices, however, lies in their assumption that all firms in a country face the same degree of political risk. This assumption is manifestly untrue, as is indicated by the empirical evidence on the post-World War II experiences of American and British MNCs. The data clearly show that, except in those countries that went communist, companies differ in their susceptibilities to political risk depending on their industry, size, composition of ownership, level of technology, and degree of vertical integration with other affiliates.[3] For example, the extreme form of political risk, **expropriation** (or creeping expropriation), is more likely to occur in the extractive, utility, and financial service sectors of an economy than in the manufacturing sector. Moreover, some firms may benefit from the same event that harms other firms. For instance, a company that relies on imports will be hurt by trade restrictions, whereas an import-competing firm may well be helped.

Because political risk has a unique meaning for, and impact on, each firm, it is doubtful that any index of generalized political risk will be of much value to a company selected at random. The specific operating and financial characteristics of a company will largely determine its susceptibility to political risk and, hence, the effects of that risk on the value of its foreign investment.

In terms of the large majority of countries, expropriation appears to be used as a fairly selective instrument of policy, with the actions taken both limited in scope and, from the government's perspective, rational. Rarely do governments, even revolutionary ones, expropriate foreign investments indiscriminately. In general, the greater the perceived benefits to the host economy and the more expensive its replacement by a purely local operation, the smaller the degree of risk to the MNC. This selectivity suggests that firms can take actions to control their exposure to political risk.

22.2 COUNTRY RISK ANALYSIS

We now examine in more detail some of the economic and social factors that contribute to the general level of risk in the country as a whole—termed **country risk**. The primary focus here is on how well the country is doing economically. As noted earlier, the better a nation's economic performance, the lower the likelihood that its government will take actions that adversely affect the value of companies operating there.

Fiscal Irresponsibility

To begin, fiscal irresponsibility is one sign of a country likely to be politically risky because it will probably have an insatiable appetite for money. Thus, one country risk indicator is the government deficit as a percentage of gross domestic product. The higher

---

[3] See, for example, studies by J. Frederick Truitt, "Expropriation of Foreign Investment: Summary of the Post-World War II Experience of American and British Investors in Less Developed Countries," *Journal of International Business Studies*, Fall 1970, pp. 21–34; Robert G. Hawkins, Norman Mintz, and Michael Provissiero, "Government Takeovers of U.S. Foreign Affiliates," *Journal of International Business Studies*, Spring 1976, pp. 3–15; and David Bradley, "Managing Against Expropriation," *Harvard Business Review*, July–August, 1977, pp. 75–83.

this figure, the more the government is promising to its citizens relative to the resources it is extracting in payment. This gap lowers the possibility that the government can meet its promises without resorting to expropriations of property.

Controlled Exchange Rate System

The economic problems presented by a fiscally irresponsible government are compounded by having a *controlled exchange rate system*, where currency controls are used to fix the exchange rate. A controlled rate system goes hand in hand with an overvalued local currency, which is the equivalent of taxing exports and subsidizing imports. The risk of tighter currency controls and the ever-present threat of a devaluation encourage capital flight. Similarly, multinational firms will try to repatriate their local affiliates' profits rather than reinvest them. A controlled rate system also leaves the economy with little flexibility to respond to changing relative prices and wealth positions, exacerbating any unfavorable trend in the nation's terms of trade.

Wasteful Government Spending

Another indicator of potential political risk is the amount of unproductive spending in the economy. To the extent that capital from abroad is used to subsidize consumption or is wasted on showcase projects, the government will have less wealth to draw on to repay the nation's foreign debts and is more likely to resort to exchange controls, higher taxes, and the like. In addition, funds diverted to the purchase of assets abroad (capital flight) will not add to the economy's dollar-generating capacity unless investors feel safe in repatriating their overseas earnings.

Resource Base

The resource base of a country consists of its natural, human, and financial resources. Other things equal, a nation with substantial natural resources, such as oil or copper, is a better economic risk than is one without those resources. However, typically, all is not equal. Hence, nations such as South Korea or Taiwan turn out to be better risks than resource-rich Argentina or Brazil. The reason has to do with the quality of human resources and the degree to which these resources are allowed to be put to their most efficient use.

A nation with highly skilled and productive workers, a large pool of scientists and engineers, and ample management talent will have many of the essential ingredients needed to pursue a course of steady growth and development. Two further factors are necessary: (1) a stable political system that encourages hard work and risk taking by allowing entrepreneurs to reap the rewards (and bear the losses) from their activities, and (2) a free-market system that ensures that the prices people respond to correctly signal the relative desirability of engaging in different activities. In this way, the nation's human and natural resources will be put to their most efficient uses. The evidence by now is overwhelming that free markets bring wealth and that endless state meddling brings waste. The reason is simple: Unlike a government-controlled economy, free markets do not tolerate and perpetuate mistakes.

Country Risk and Adjustment to External Shocks

Recent history shows that the impact of external shocks is likely to vary from nation to nation; some countries deal successfully with these shocks, and others succumb to them. The evidence suggests that domestic policies play a critical role in determining how effectively a particular nation will deal with external shocks. Asian nations, for example, successfully coped with falling commodity prices, rising real interest rates, and rising exchange rates because their policies promoted timely internal and external adjustment, as is manifest in relatively low inflation rates and small current-account deficits.

The opposite happened in Latin America, where most countries accepted the then-prevalent ideology that growth is best promoted by an **import-substitution development strategy** characterized by extensive state ownership, controls, and policies to encourage import substitution. Many of these countries took over failing private businesses, nationalized the banks, protected domestic companies against imports, ran up large foreign debts, and heavily regulated the private sector. Whereas the "East Asian Tigers"—Hong Kong, South Korea, Taiwan, and Singapore—tested their ability to imitate and innovate in the international marketplace, Latin American producers were content with the exploitation of the internal market, charging prices that were typically several times the international price for their goods. The lack of foreign competition has contributed to long-term inefficiency among Latin American manufacturers.

In addition, by raising the cost of imported materials and products used by the export sector, the Latin American import-substitution development strategies worsened their international competitive position, leaving the share of exports in GDP far below that of other LDCs. Moreover, state expenditures on massive capital projects diverted resources from the private sector and exports. Much of the investment went to inefficient state enterprises, leading to wasted resources and large debts.

The decline in commodity prices and the simultaneous rise in real interest rates should have led to reduced domestic consumption. However, fearing that spending cuts would threaten social stability, Latin American governments delayed cutting back on projects and social expenditures. The difference between consumption and production was made up by borrowing overseas, thereby enabling their societies to temporarily enjoy artificially high standards of living.

Latin American governments also tried to stimulate their economies by increasing state spending, fueled by high rates of monetary expansion. This response exacerbated their difficulties because the resulting high rates of inflation combined with their fixed exchange rates to boost real exchange rates substantially and resulted in higher imports and lower exports. Moreover, the overvalued exchange rates, interest rate controls, and political uncertainties triggered massive capital flight from the region—estimated at up to $100 billion during the two-year period of 1981 and 1982. The result was larger balance-of-payments deficits that necessitated more foreign borrowing and higher debt-service requirements. Moreover, in an attempt to control inflation, the Latin American governments imposed price controls and interest rate controls. These controls led to further capital flight and price rigidity. Distorted prices gave the wrong signals to the residents, sending consumption soaring and production plummeting.

The centralization of economic power in much of the Third World turned the state into a huge patronage machine and spawned a swollen and corrupt bureaucracy to administer the all-encompassing rules and regulations. Avaricious elites, accountable to no one, used the labyrinthine controls and regulations to enrich themselves and further the interests of their own ethnic groups or professional class at the expense of national economic health and well-being.

Many of these countries are finally rejecting earlier policies and trying to stimulate the private sector and individual initiative. The experience in countries that are reforming their economies—including the former Soviet Union, China, and India, as well as much of Latin America—illustrates that it is far easier to regulate and extend the state's reach than to deregulate and retrench. When the state becomes heavily involved in the economy, many special interests from the state bureaucracy, business, labor, and consumer groups come to rely on state benefits. Of course, they actively oppose reforms that curtail their subsidies.

The process of reform is greatly complicated by egalitarian ideologies that deprecate private success while justifying public privilege and by the pervasiveness of the state, which distorts the reward pattern and makes it easier to get rich by politics than by industry, by connections than by performance. These ideologies tend to take investment and the provision of productive capacity for granted (they like capital but dislike capitalists), while being generally more concerned with redistributing the benefits and mitigating the costs of economic progress than with protecting its foundations.

The message is clear. In evaluating a nation's riskiness, it is not sufficient to identify factors—such as real interest shocks or world recession—that would systematically affect the economies of all foreign countries to one extent or another. It is necessary also to determine the susceptibility of the various nations to these shocks. This determination requires a focus on the financial policies and development strategies pursued by the different nations.

Some of the common characteristics of country risk are:

- A large government deficit relative to GDP
- A high rate of money expansion, especially if it is combined with a relatively fixed exchange rate
- Substantial government expenditures yielding low rates of return
- Price controls, interest rate ceilings, trade restrictions, and other government-imposed barriers to the smooth adjustment of the economy to changing relative prices
- High tax rates that destroy incentives to work, save, and invest
- Vast state-owned firms run for the benefit of their managers and workers
- A citizenry that demands, and a political system that accepts, government responsibility for maintaining and expanding the nation's standard of living through public-sector spending and regulations (The less stable the political system, the more important this factor will likely be.)
- The absence of basic institutions of government—a well-functioning legal system, reliable regulation of financial markets and institutions, and an honest civil service

Alternatively, indicators of a nation's long-run economic health include the following:

- *A structure of incentives that rewards risk taking in productive ventures*: People have clearly demonstrated that they respond rationally to the incentives they face, given the information and resources available to them. This statement is true whether we are talking about shopkeepers in Nairobi or bankers in New York. A necessary precondition for productive investment to take place is secure legal rights to own and sell at least some forms of property. If property is not secure, people have an incentive to consume their resources immediately, or transfer them overseas, lest they be taken away. Low taxes are also important because they encourage productive efforts and promote savings and investment.

■ *A legal structure that stimulates the development of free markets*: Wealth creation is made easier by stable rules governing society and fair and predictable application of laws administered by a judicial system free of corruption. Such a legal structure, which replaces official whim with the rule of law, combined with a system of property rights and properly enforceable contracts, facilitates the development of free markets. The resulting market price signals are most likely to contain the data and provide the incentives that are essential to making efficient use of the nation's resources. Free markets, however, do more than increase economic efficiency. By quickly rewarding success and penalizing failure, they also encourage successful innovation and economic growth. Conversely, the lack of a rule of law and a well-defined commercial code, as in Russia, hampers the development of a market economy by making it difficult to enforce contracts and forcing businesses to pay protection money to thugs.

■ *Minimal regulations and economic distortions*: Complex regulations are costly to implement and waste management time and other resources. Moreover, reduced government intervention in the economy lowers the incidence of corruption. After all, why bribe civil servants if their ability to grant economic favors is minimal? Instead, the way to succeed in an unregulated economy is to provide superior goods and services to the market.

■ *Clear incentives to save and invest*: In general, when there are such incentives, when the economic rules of the game are straightforward and stable, when property rights are secure, taxes on investment returns are low, and when there is political stability, a nation's chances of developing are maximized.

■ *An open economy*: Free trade not only increases competition and permits the realization of comparative advantage, it also constrains government policies and makes them conform more closely to those conducive to increases in living standards and rapid economic growth.

The sorry economic state of Eastern Europe dramatically illustrates the consequences of pursuing policies that are the exact opposite of those recommended. Thus, the ability of the Eastern European countries to share in the prosperity of the Western world depends critically on reversing the policies they followed under communist rule.

Similarly, investors assessing the prospects for Western nations would also do well to recognize the benefits of markets and incentives. Government economic intervention in the form of subsidies and regulations is appealing, but governments make poor venture capitalists and for a simple reason: Industries not yet born do not have lobbyists, whereas old and established ones have lots of them. The net result is that the bulk of subsidies go to those industries resisting change. For example, Germany spends more on subsidies for powerful smokestack industries such as shipbuilding and coal mining than to support basic research. It also continues to shelter the state-owned telecommunications monopoly, Deutsche Telekom, and its suppliers, such as Siemens, from most foreign competition; government sheltering in turn dampens innovation across a range of industries. When government does invest in high-tech industries, it usually botches the job. Examples are legion, including the tens of billions of wasted tax dollars spent to develop European software, semiconductor, computer, and aerospace industries.

THE 1948 WEST GERMAN ERHARD REFORMS

At the end of World War II, the German economy lay in ruins. Industrial output in 1948 was one-third its 1936 level because of a massive disruption in production and trade patterns. Aside from the devastation caused by the war, economic disruption was aggravated by wartime money creation, price controls, and uncertainty about economic policy. Each day vast, hungry crowds traveled to the countryside to barter food from farmers; an extensive black market developed; and cigarettes replaced currency in many transactions.

In June 1948, Ludwig Erhard, West Germany's economic czar, announced an extensive reform package. This package created a new currency, the Deutsche mark (DM), and dismantled most price controls and rationing ordinances. It also implemented a restrictive monetary policy, lowered tax rates, and provided incentives for investment.

Erhard's reforms almost immediately established sound and stable macroeconomic conditions and led to the German "economic miracle." Consumer prices initially rose by 20%, but inflation then subsided to an average annual rate of about 1% between 1949 and 1959. Goods that had been hoarded or sold only in the black market flooded the market. Industrial production increased 40% in the second half of 1948 and then tripled over the next 10 years. Real GDP and productivity also grew rapidly. Although unemployment rose from 3% in the first half of 1948 to more than 10% in the first half of 1950, it then disappeared over the next eight years. In 1958, the DM became convertible.

Economic reform could not have produced such dramatic results if West Germany had not had key structural elements already in place. It had the legal framework necessary for a market economy, many intact businesses, and skilled workers and managers. Marshall Plan money helped, but absent the reforms, any aid would have been wasted.

As the West German example illustrates, realism demands that nations—especially those in the Third World and Eastern Europe—come to terms with their need to rely more on self-help. The most successful economies, such as Hong Kong, South Korea, and Taiwan, demonstrate the importance of aligning domestic incentives with world market conditions. As a result of their market-oriented policies, Asian nations have had remarkable economic success over the past three decades as reflected in their strong economic growth and rising standards of living. Their recent problems just reinforce the importance of such policies: When Asian nations substituted the visible hand of state intervention for the invisible hand of the market in allocating capital, they helped create the financial crisis that rocked Asia in 1997. The evidence indicates that the more distorted the prior economic policies, the more severe the crisis, with Indonesia, South Korea, Thailand, and Malaysia being hit much harder than Hong Kong, Taiwan, and Singapore.

Like it or not, nations must make their way in an increasingly competitive world economy that puts a premium on self-help and has little time for the inefficiency and pretension of *statism*—that is, the substitution of state-owned or state-guided enterprises for the private sector—as the road to economic success. Statism destroys initiative and leads to economic stagnation. Free enterprise is the road to prosperity.

This recognition—that they cannot realize the benefits of capitalism without the institutions of capitalism—is dawning in even the most socialist countries of Europe,

Asia, and Latin America. For example, in 1989, Vietnamese families were given the right to work their own land and sell their output at market prices. Within a year, rice production rose so dramatically that Vietnam went from the edge of famine to the world's third-largest rice exporter.

Market-oriented reform of Eastern European and LDC economic policies lies at the heart of any credible undertaking to secure these nations' economic and financial rehabilitation. The first and most critical step is to cut government spending. In practical terms, cutting government spending means reducing the bloated public sectors that permeate most Latin American and communist countries.

Reform of the public sector has probably gone the furthest in Latin America, where, shocked by the severe miseries of the 1980s (known in Latin America as the *Lost Decade*), many of these countries abandoned the statism, populism, and protectionism that have crippled their economies since colonial times. Chile and Colombia have embarked on fairly comprehensive reform programs, emphasizing free markets and sound money, and, despite some backsliding and significant problems with corruption, Mexico has made surprisingly good headway (see Exhibit 22.4 for a summary of the changes in Mexico's economic policies). We saw in Chapter 2 that Argentina also has begun radical reform of its economy, highlighted by *privatizing*—returning to the private sector from the public sector—major activities and galvanizing the private sector by deregulation and the elimination of protectionism. And Poland has instituted a radical set of market reforms as well. The next big challenge for all these countries is to revamp the entire civil service, including police, regulatory agencies, judiciaries, and all the other institutions necessary for the smooth functioning of a market economy.

Yet in the best of circumstances, structural reform meets formidable political obstacles: labor unions facing job and benefit losses, bureaucrats fearful of diminished power and influence (not to mention their jobs), and local industrialists concerned about increased competition and reduced profitability. All are well aware that the benefits of restructuring are diffuse and materialize only gradually, while they must bear the costs immediately.

Despite these obstacles, reducing state subsidies on consumer goods and to inefficient industries, removing trade barriers and price controls, and freeing interest rates and the exchange rate to move to market levels is probably the most straightforward and workable solution to economic stagnation. These actions, if implemented, can increase

---

EXHIBIT 22.4 Mexico's Economic Policies: Then and Now

OLD MEXICAN MODEL	NEW MEXICAN MODEL
Large budget deficits	Fiscal restraint
Rapid expansion of money supply	Monetary discipline
Nationalization	Privatizations
Restrict foreign direct investment	Attract foreign direct investment
High tax rates	Tax reform
Import substitution	Trade liberalization
Controlled currency	End currency controls
Price and interest rate controls	Prices and interest rates set by market
Government dominates economy	Reduced size and scope of government

ILLUSTRATION

STRATEGIES FOR
EASTERN **E**UROPEAN
ECONOMIC **S**UCCESS

Eastern Europe has the basic ingredients for successful development: an educated workforce, low wages, and proximity to large markets. However, the key to creating a dynamic market economy is to mobilize the energies and savings of the populace on the broadest possible scale; without this support any reform package is doomed to failure. Unfortunately, under communism, people had no incentive to take risks, and, thus, took no initiative. State-owned firms were focused on the bureaucrats who gave them orders rather than the customers who bought from them, toward output rather than profit, toward the social welfare of their employees rather than efficiency. Market signals became even more muffled because the state banking system continued to finance loss-makers, no matter how hopeless, so that jobs were not jeopardized. The result was no innovation, an insatiable demand for investment capital, no concern with profitability, no consumer orientation, and low-quality merchandise. Here is a thumbnail sketch, based on these considerations, of the factors that will influence Eastern Europe's chances of economic success.

■ *Prerequisites*: Privatization of bloated state enterprises to force efficiency and customer responsiveness;[4] market prices to signal relative scarcity and opportunity cost; private property, to provide incentives; and a complete revamping of the legal, financial, and administrative institutions that govern economic activity to permit enforceable contracts and property rights. The creation of these capitalist institutions is critical, for without them changes made today could easily be reversed tomorrow. These legal institutions would also go a long way to reducing the corruption that is rampant in these societies.

■ *Economic reforms*: Decontrolling prices; eliminating subsidies and restrictions on international trade; creating strong, convertible currencies, so that people can receive something for their efforts; permitting bankruptcy, so that assets and people can be redeployed; doing away with regulations of small businesses; introducing a free capital market; and demonopolizing state enterprises through privatization to introduce competition and boost productivity. Privatization and deregulation should also improve the economic infrastructure, a pressing need because Eastern Europe currently is plagued by dilapidated state-owned transport, power, and telecommunications systems.

These are not mere details. Completing such tasks will embroil the region in all the wrangles about wealth distribution and the size and role of the state that Western countries have spent gener-

---

[4]For a discussion of privatization programs around the world and their performance, see William L. Megginson, Robert C. Nash, and Matthias van Randenborgh, "The Financial and Operating Performance of Newly Privatized Firms: An International Empirical Analysis," University of Georgia working paper, April 1993.

output by making the economy more efficient, can reduce consumption, and, thereby, can increase the quantity of goods available for export. They will also discourage capital flight and stimulate domestic savings and investment.

The simple truth is that a nation's success is not a function of the way its government harnesses resources, manages workers, or distributes wealth. Rather, economic success depends on the ability and willingness of a nation's people to work hard and take risks in the hopes of a better life. From this perspective, the state's best strategy is to provide basic stability—and little else—thereby permitting the humble to rise and the great to fall.

ations trying to resolve. And there is another complicating factor. For persons born since 1945, the habits that constitute the tradition of private property, markets, and creativity have been blotted out. Two generations in Eastern Europe have never experienced private property, free contracts, markets, or inventive enterprise; the skills are gone. Whether and how fast these skills and habits can be resurrected remain open questions. A worrying trend is the recent electoral success throughout Eastern Europe of ex-communists who played on the populace's envy over the economic success of a few fellow citizens. Nonetheless, given the prerequisites and reforms outlined above, here are four strategies aimed at getting Eastern Europe off to a fast start.

1. *Deregulating agriculture* (especially farmers' access to markets, ownership of land, and decontrol of prices) offers the best hope for quickly easing food shortages and for containing food price increases as subsidies are eliminated. Communal ownership of land stifles initiative. Raising efficiency in agriculture is critical to successful development. Vietnam provides a good example of the potential gains from freeing up farmers.

2. *Privatizing small businesses*, on the broadest possible scale, is desperately needed in order to create new job opportunities for workers displaced by the inevitable restructuring of heavy industry. Small businesses have the greatest potential for harnessing individual initiative and creating new jobs quickly and are an indispensable part of the infrastructure of any dynamic market economy. Any owner is better than the state, so getting the job done is more important than how it is done.

3. *Manufacturing low-technology goods* plays a key role in increasing skill levels and in disseminating technology throughout the economy. The region's current comparative advantage lies in low-technology manufactured goods. East Europeans cannot be expected immediately to produce cars, computers, and consumer electronics of sufficient quality to compete with the West; moving up the ladder to more sophisticated products will take considerable time and lots of Western investment and expertise.

4. *Direct investment in local production* is a preferred strategy because none of the countries in the region can afford a huge influx of imports. Western companies should target relatively cheap, everyday products of less-than-premium quality. Living standards in Eastern Europe and the former Soviet Union are too low to warrant mass purchases of anything but the most basic Western goods.

Perhaps a fifth strategy is speed. The lesson of reform so far is that fortune favors the brave. Those countries that opted for shock therapy—Poland, the Czech Republic, and the Baltic states of Estonia, Latvia, and Lithuania—have had the smallest overall output declines and the fastest subsequent growth. In contrast, the gradualist countries—Hungary, Bulgaria, and Romania—have had delayed economic recovery and increased social costs. In the end, these countries spared their citizens none of the hardships that gradualism was supposed to avoid.

22.3 MANAGING POLITICAL RISK

After the firm has analyzed the political environment of a country and assessed its implications for corporate operations, it then must decide whether to invest there and, if so, how to structure its investment to minimize political risk. The key point remains that political risk is not independent of the firm's activities; the configuration of the firm's investments will, in large measure, determine its susceptibility to changing government policies.

Preinvestment Planning

Given the recognition of political risk, an MNC can follow at least four separate, though not necessarily mutually exclusive, policies: (1) avoidance, (2) insurance, (3) negotiating the environment, and (4) structuring the investment.

AVOIDANCE ■ The easiest way to manage political risk is to avoid it, and many firms do so by screening out investments in politically uncertain countries. However, inasmuch as all governments make decisions that influence the profitability of business, all investments, including those made in the United States, face some degree of political risk. For example, U.S. steel companies have had to cope with stricter environmental regulations requiring the expenditure of billions of dollars for new pollution control devices, and U.S. oil companies have been beleaguered by so-called windfall profit taxes, price controls, and mandatory allocations. Thus, risk avoidance is impossible.

The real issue is the degree of political risk a company is willing to tolerate and the return required to bear it. A policy of staying away from countries considered to be politically unstable ignores the potentially high returns available and the extent to which a firm can control these risks. After all, companies are in business to take risks, provided these risks are recognized, intelligently managed, and provide compensation.

INSURANCE ■ An alternative to risk avoidance is insurance. Firms that insure assets in politically risky areas can concentrate on managing their businesses and forget about political risk—or so it appears. Most developed countries sell *political risk insurance* to cover the foreign assets of domestic companies. The coverage provided by the U.S. government through the **Overseas Private Investment Corporation** (OPIC) is typical. Although its future has been in doubt several times—some U.S. citizens believe it is an instrument of U.S. imperialism, and others feel that the government has no right to subsidize a service that would otherwise be provided by private enterprise—OPIC has managed to survive because of the general belief that this program, by encouraging U.S. direct investment in less-developed countries, helps these countries to develop.

The OPIC program provides U.S. investors with insurance against loss due to the specific political risks of expropriation, currency inconvertibility, and political violence—that is, war, revolution, or insurrection. By the end of 1995, OPIC was insuring $11.8 billion in U.S. investments overseas. To qualify, the investment must be a new one or a substantial expansion of an existing facility and must be approved by the host government. Coverage is restricted to 90% of equity participation. For very large investments or for projects deemed especially risky, OPIC coverage may be limited to less than 90%. The only exception is institutional loans to unrelated third parties, which may be insured for the full amount of principal and interest.

Similar OPIC political risk protection is provided for leases. OPIC's insurance provides lessors with coverage against loss due to various political risks, including the inability to convert into dollars local currency received as lease payments.

OPIC also provides business income coverage (BIC), which protects a U.S. investor's income flow if political violence causes damage that interrupts operation of the foreign enterprise. For example, an overseas facility could be bombed and partially or totally destroyed. It may take weeks or months to rebuild the plant, but during the rebuilding process the company still must meet its interest and other contractual payments and pay skilled workers in order to retain their services pending the reopening of the business. BIC allows a business to meet its continuing expenses and to make a normal profit during the period its operations are suspended. This is similar to the business interruption insurance available from private insurers for interruptions caused by nonpolitical events.

Another special program run by OPIC provides coverage for U.S. exporters of goods and services. OPIC insures against arbitrary drawings by a government buyer of bid, performance, advance payment and other guaranties (usually issued in the form of standby letters of credit). An arbitrary drawing is one not justified by the terms of the contract. OPIC also will insure a contractor's assets against loss due to political violence or confiscation. Protection in the event of contractual disputes is offered as well.

Premiums are computed for each type of coverage on the basis of a contractually stipulated maximum insured amount and a current insured amount that may, within the limits of the contract, be elected by the investor on a yearly basis. The current insured amount represents the insurance actually in force during any contract year.

The difference between the current insured amount and maximum insured amount for each coverage is called the *standby amount*. The major portion of the premium is based on the current insured amount, with a reduced premium rate being applicable to the standby amount. For expropriation and war coverage, the insured must maintain current coverage at a level equal to the amount of investment at risk.

The cost of the coverage varies by industry and risk insured. These costs, which are listed in Exhibit 22.5, are not based solely on objective criteria; they also reflect subsidies geared to achieving certain political aims, such as fostering development of additional energy supplies.

The only private insurer of consequence against expropriation risks is Lloyd's of London. There are several possible reasons why a large-scale private market for expropriation insurance has failed to develop. One major barrier to entry is the magnitude of the potential expropriation losses. A $1 billion claim may bankrupt a private insurer, unless that loss is only a small percentage of its total operations. However, the large loss factor can be dealt with by reinsuring most of the risks with other insurance firms and private investors. This reinsuring is done routinely in the insurance field.

Other possible obstacles to private insurers contemplating the sale of expropriation insurance are the problems of adverse selection and adverse incentives. **Adverse selection** refers to the possibility that only high-risk multinationals will seek insurance. This problem can be dealt with in several ways. The ways include adjusting premiums in accord with the perceived risks, screening out certain high-risk applicants, and providing premium reductions for firms engaged in activities that are likely to reduce expropriation risks.

The problem of **adverse incentives**—which is another term for **moral hazard**—is that by reducing the riskiness of certain activities, insurance may prompt firms to engage in activities with a higher probability of expropriation. Firms may undertake investments that were previously too risky and neglect certain policies responsive to the host country's needs. If the local affiliate is in financial difficulty, the parent can also take actions that would increase the possibility of expropriation or else collude with the host government to expropriate the affiliate—in much the same way that the owner of a failing business might commit arson to collect on the fire insurance. In effect, purchasing political risk

EXHIBIT 22.5 OPIC Insurance Fees: Annual Base Rates Per $100 of Coverage

Coverage	MANUFACTURING/SERVICES PROJECTS		NATURAL RESOURCE PROJECTS OTHER THAN OIL AND GAS	
	Current	(Standby)	Current	(Standby)
Inconvertibility	$0.30	$(0.25)	$0.30	$(0.25)
Expropriation	0.60	(0.25)	0.90	(0.25)
Political violence[1]				
Business income	0.45	(0.25)	0.45	(0.25)
Assets	0.60	(0.25)	0.60	(0.25)
Interference with operations	—	—	—	—
Disputes	—	—	—	—
Bid Bonds	—	—	—	—
Performance, advance payment, and other guaranties	—	—	—	—

[1] Discounted rates may be available for combined business income and assets political violence coverage.
[2] Covered amount is the amount of disbursed principal plus accrued interest, less principal paid to date.

Source: Program Handbook, Reprinted with permission from the Overseas Private Investment Corp. (OPIC).

insurance is equivalent to purchasing a put option on the project. The MNC will seek to exercise this put option—which effectively involves selling its foreign project to the insurance company for the amount of coverage—whenever the market value of the project falls below the insurance claim.

The problem of adverse incentives can be coped with to a certain extent by *coinsurance*—forcing the purchaser to self-insure part of the losses—and by refusing to pay off on a claim if it can be shown that the insured firm caused the expropriation. Although these are not ideal solutions, the problems of adverse selection and adverse incentives should not pose major deterrents to the establishment of a viable market for expropriation insurance.

The most important barrier preventing private competition with Lloyd's is likely to be the existence of OPIC and other government-operated expropriation schemes. By offering subsidized insurance, these government plans have made private expropriation insurance unprofitable. As in other instances of market distortion, it is in the MNC's best interest to buy insurance when it is priced at a below-market rate. The rate is uniform across countries (for example, 0.6% for manufacturing operations), so it clearly pays to insure in risky nations (mostly LDCs) and not insure in low-risk nations. Thus, government plans are faced with the aforementioned problem of adverse selection.

Even with subsidized rates, there are two fundamental problems with relying on insurance as a protection from political risk. First, there is an asymmetry involved. If an investment proves unprofitable, it is unlikely to be expropriated. Because business risk is not covered, any losses must be borne by the firm itself. On the other hand, if the investment proves successful and is then expropriated, the firm is compensated only for the value of its assets. This result is related to the second problem: Although the economic value of an investment is the present value of its future cash flows, only the capital investment in assets is covered by insurance. Thus, although insurance can provide partial protection from political risk, it falls far short of being a comprehensive solution.

EXHIBIT 22.5 (Continued)

OIL AND GAS PROJECTS			INSTITUTIONAL LOANS AND LEASES		CONTRACTORS AND EXPLORERS GUARANTY COVERAGE	
Exploration	Development/ Production	(Standby)	Covered Amount[2]	(Undisbursed Principal)	Current	(Standby)
$0.30	$0.30	$(0.25)	$0.45	$(0.20)	$0.30	$(0.25)
0.40	1.50	(0.25)	0.40–0.90	(0.20)	0.60	(0.25)
			0.40–0.70	(0.20)		
—	—		—	—	—	—
0.75	0.75	(0.25)	—	—	0.60	(0.25)
0.40	0.40	(0.25)	—	—	—	—
—	—	—	—	—	0.70	(0.25)
—	—	—	—	—	0.50	(0.25)
—	—	—	—	—	0.60	(0.25)

NEGOTIATING THE ENVIRONMENT ■ In addition to insurance, therefore, some firms try to reach an understanding with the host government before undertaking the investment, defining rights and responsibilities of both parties. Also known as a **concession agreement**, such an understanding specifies precisely the rules under which the firm can operate locally.

In the past, these concession agreements were quite popular among firms investing in less-developed countries, especially in colonies of the home country. They often were negotiated with weak governments. In time, many of these countries became independent or their governments were overthrown. Invariably, the new rulers repudiated these old concession agreements, arguing that they were a form of exploitation.

Concession agreements still are being negotiated today, but they seem to carry little weight among Third World countries. Their high rate of obsolescence has led many firms to pursue a more active policy of political risk management.

STRUCTURING THE INVESTMENT ■ Once a firm has decided to invest in a country, it then can try to minimize its exposure to political risk by increasing the host government's cost of interfering with company operations. This action involves adjusting the operating policies (in the areas of production, logistics, exporting, and technology transfer) and the financial policies to closely link the value of the foreign project to the multinational firm's continued control. In effect, the MNC is trying to raise the cost to the host government of exercising its ever-present option to expropriate or otherwise reduce the local affiliate's value to its parent.[5]

---

[5] Arvind Mahajan, "Pricing Expropriation Risk," *Financial Management*, Winter 1990, pp. 77–86, points out that when a multinational firm invests in a country, it is effectively writing a call option to the government on its property. The aim of political risk management is to reduce the value to the government of exercising this option.

One key element of such a strategy is keeping the local affiliate dependent on sister companies for markets or supplies or both. Chrysler, for example, managed to hold on to its Peruvian assembly plant even though other foreign property was being nationalized. Peru ruled out expropriation because of Chrysler's stranglehold on the supply of essential components. Only 50% of the auto and truck parts were manufactured in Peru. The remainder—including engines, transmissions, sheet metal, and most accessories—were supplied from Chrysler plants in Argentina, Brazil, and Detroit. In a similar instance of vertical integration, Ford's Brazilian engine plant generates substantial exports, but only to other units of Ford. Not surprisingly, the data reveal no expropriations of factories that sell more than 10% of their output to the parent company.[6]

Similarly, by concentrating R&D facilities and proprietary technology, or at least key components thereof, in the home country, a firm can raise the cost of nationalization. This strategy will be effective only if other multinationals with licensing agreements are not permitted to service the nationalized affiliate. Another element of this strategy is establishing a single, global trademark that cannot be legally duplicated by a government. In this way, an expropriated consumer-products company would sustain significant losses by being forced to operate without its recognized brand name.

Control of transportation—including shipping, pipelines, and railroads—has also been used at one time or another by the United Fruit Company and other multinationals to gain leverage over governments. Likewise, sourcing production in multiple plants reduces the government's ability to hurt the worldwide firm by seizing a single plant and, thereby, it changes the balance of power between government and firm.

Another defensive ploy is to develop external financial stakeholders in the venture's success. This defense involves raising capital for a venture from the host and other governments, international financial institutions, and customers (with payment to be provided out of production) rather than employing funds supplied or guaranteed by the parent company. In addition to spreading risks, this strategy will elicit an international response to any expropriation move or other adverse action by a host government. A last approach, particularly for extractive projects, is to obtain unconditional host government guarantees for the amount of the investment that will enable creditors to initiate legal action in foreign courts against any commercial transactions between the host country and third parties if a subsequent government repudiates the nation's obligations. Such guarantees provide investors with potential sanctions against a foreign government, without having to rely on the uncertain support of their home governments.

Operating Policies

After the multinational has invested in a project, its ability to further influence its susceptibility to political risk is greatly diminished but not ended. It still has at least five different policies that it can pursue with varying chances of success: (1) planned divestment, (2) short-term profit maximization, (3) changing the benefit/cost ratio of expropriation, (4) developing local stakeholders, and (5) adaptation.

---

[6] Bradley, "Managing Against Expropriation," pp. 75–83.

PLANNED DIVESTMENT ■ Several influential authors, notably Raul Prebisch and Albert Hirschman, have suggested that multinational firms phase out their ownership of foreign investments over a fixed time period by selling all or a majority of their equity interest to local investors.[7] Such *planned divestment*, however, may be difficult to conclude to the satisfaction of all parties involved. If the buyout price was set in advance and the investment proved unprofitable, the government probably would not honor the purchase commitment. Moreover, with the constant threat of expropriation present during the bargaining, it is unlikely that a fair price could be negotiated.

SHORT-TERM PROFIT MAXIMIZATION ■ Confronted with the need to divest itself wholly or partially of an equity position, the multinational corporation may respond by attempting to withdraw the maximum amount of cash from the local operation. By deferring maintenance expenditures, cutting investment to the minimum necessary to sustain the desired level of production, curtailing marketing expenditures, producing lower-quality merchandise, setting higher prices, and eliminating training programs, the firm can maximize cash generation for the short term, regardless of the effects of such actions on longer-run profitability and viability. This policy, which almost guarantees that the company will not be in business locally for long, is a response of desperation. Of course, the behavior is likely to accelerate expropriation if such was the government's intention—perhaps even if it were not the government's intention originally.

Hence, the firm must select its time horizon for augmenting cash outflow and consider how this behavior will affect government relations and actions. Surprisingly, most politicians do not seem to appreciate how strongly their rhetoric affects corporate decisions. In effect, government rhetoric about the evils of multinationals becomes a self-fulfilling prophecy because these threats induce more and more myopic corporate behavior.

The secondary implications of the short-term profit maximization strategy must be evaluated as well. The unfriendly government can be replaced by one more receptive to foreign investment (as occurred in Chile), or the multinational firm may want to supply the local market from affiliates in other countries. In either case, an aggressive tactic of withdrawing as much as possible from the threatened affiliate will probably be considered a hostile act and will vitiate all future dealings between the MNC and the country. Moreover, it is unlikely that a firm can get away with this behavior for long. Other governments will be put on notice and begin taking closer and more skeptical looks at the company's actions in their countries.

One alternative to this indirect form of divestment is to do nothing and hope that even though the local regime can take over an affiliate (with minor cost), it will choose not to do so. This wish is not necessarily in vain because it rests on the premise that the country needs foreign direct investment and will be unlikely to receive it if existing operations are expropriated without fair and full compensation. However, this strategy is essentially passive, resting on a belief that other multinationals will hurt the country (by withholding potential investments) if the country nationalizes local affiliates. Whether

---

[7] Raul Prebisch, "The Role of Foreign Private Investment in the Developing of Latin America," Sixth Annual Meeting of the IA-ECOSOC, June 1969; and Albert O. Hirschman, "How to Divest in Latin America, and Why," *Essays in International Finance*, No. 76, Princeton University, November 1969.

ILLUSTRATION

BEIJING JEEP

After the United States restored diplomatic relations with China in 1979, Western businesses rushed in to take advantage of the world's largest undeveloped market. Among them was American Motors Corporation (AMC). In 1983, AMC and Beijing Automotive Works formed a joint venture called the Beijing Jeep Company to build and sell jeeps in China.[8] The aim of Beijing Jeep was to first modernize the old Chinese jeep, the BJ212, and then replace it with a "new, second-generation vehicle" for sale in China and overseas. Because it was one of the earliest attempts to combine Chinese and foreign forces in heavy manufacturing, Beijing Jeep became the "flagship" project other U.S. firms watched to assess the business environment in China. Hopes were high.

AMC viewed this as a golden opportunity: Build jeeps with cheap labor and sell them in China and the rest of the Far East. The Chinese government wanted to learn modern automotive technology and earn foreign exchange. Most important, the People's Liberation Army wanted a convertible-top, four-door jeep, so that Chinese soldiers could jump in and out and open fire from inside the car.

That the army had none of these military vehicles when they entered Tienanmen Square in 1989 has to do with the fact that this jeep could not be made from any of AMC's existing jeeps. However, in signing the initial contracts, the two sides glossed over this critical point. They also ignored the realities of China's economy. For managers and workers, productivity was much lower than anybody at AMC had ever imagined. Equipment maintenance was minimal. Aside from windshield solvents, spare-tire covers, and a few other minor parts, no parts could be manufactured in China. The joint venture, therefore, had little choice but to turn the new Beijing jeep into the Cherokee Jeep, using parts kits imported from the United States. The Chinese were angry and humiliated not to be able to manufacture any major jeep components locally.

They got even angrier when Beijing Jeep tried to force its Chinese buyers to pay half of the Cherokee's $19,000 sticker price in U.S. dollars. With foreign exchange in short supply, the Chinese government ordered its state agencies, the only potential customers, not to buy any more Cherokees and refused to pay the $2 million that various agencies owed on 200 Cherokees already purchased.

The joint venture would have collapsed right then had Beijing Jeep not been such an important symbol of the government's modernization program. After deciding it could not let the venture fail, China's leadership arranged a bailout. The Chinese abandoned their hopes of making a new military jeep, and AMC gained the right to convert renminbi (the Chinese currency) into dollars at the official (and vastly overvalued) exchange rate. With this right, AMC realized it could make more money by replacing the Cherokee with the old, and much cheaper to build, BJ212s. The BJ212s were sold in China for renminbi, and these profits were converted into dollars.

It was the ultimate irony: An American company that originally expected to make huge profits by introducing modern technology to China and by selling its superior products to the Chinese found itself surviving, indeed thriving, by selling the Chinese established Chinese products. AMC succeeded because its venture attracted enough attention to turn the future of Beijing Jeep into a test of China's open-door policy.

---

[8] This example is adapted from Jim Mann, *Beijing Jeep: The Short, Unhappy Romance of American Business in China* (New York: Simon and Shuster, 1989).

this passive approach will succeed is a function of how dependent the country is on foreign investment to realize its own development plans and the degree to which economic growth will be sacrificed for philosophical or political reasons.

A more active strategy is based on the premise that expropriation is basically a rational process—that governments generally seize property when the economic benefits outweigh the costs. This premise suggests two maneuvers characteristic of active political-risk management: Increase the benefits to the government if it does not nationalize a firm's affiliate, and increase the costs if it does.

CHANGING THE BENEFIT/COST RATIO ■ If the government's objectives in an expropriation are rational—that is, based on the belief that economic benefits will more than compensate for the costs—the multinational firm can initiate a number of programs to reduce the perceived advantages of local ownership and, thereby, diminish the incentive to expel foreigners. These steps include establishing local R&D facilities, developing export markets for the affiliate's output, training local workers and managers, expanding production facilities, and manufacturing a wider range of products locally as substitutes for imports. It should be recognized that many of the foregoing actions lower the cost of expropriation and, consequently, reduce the penalty for the government. A delicate balance must be observed.

Realistically, however, it appears that those countries most liable to expropriation view the benefits—real, imagined, or both—of local ownership as more important than the cost of replacing the foreign investor. Although the value of a subsidiary to the local economy can be important, its worth may not be sufficient to protect it from political risk. Thus, one aspect of a protective strategy must be to engage in actions that raise the cost of expropriation by increasing the negative sanctions it would involve. These actions include control over export markets, transportation, technology, trademarks and brand names, and components manufactured in other nations. Some of these tactics may not be available once the investment has been made, but others still may be implemented. However, an exclusive focus on providing negative sanctions may well be self-defeating by exacerbating the feelings of dependence and loss of control that often lead to expropriation in the first place. Where expropriation appears inevitable, with negative sanctions only buying more time, it may be more productive to prepare for negotiations to establish a future contractual-based relationship.

DEVELOPING LOCAL STAKEHOLDERS ■ A more positive strategy is to cultivate local individuals and groups who have a stake in the affiliate's continued existence as a unit of the parent MNC. Potential stakeholders include consumers, suppliers, the subsidiary's local employees, local bankers, and joint venture partners.

Consumers worried about a change in product quality or suppliers concerned about a disruption in their production schedules (or even a switch to other suppliers) brought about by a government takeover may have an incentive to protest. Similarly, well-treated local employees may lobby against expropriation.[9] Local borrowing could help give local

---

[9] French workers at U.S.-owned plants, satisfied with their employers' treatment of them, generally stayed on the job during the May 1968 student-worker riots in France, even though most French firms were struck.

bankers a stake in the health of the MNC's operations if any government action threatened the affiliate's cash flows and, thereby, jeopardized loan repayments.

Having local private investors as partners seems to provide protection. One study found that joint ventures with local partners have historically suffered only a 0.2% rate of nationalization, presumably because this arrangement establishes a powerful local voice with a vested interest in opposing government seizure.[10]

The shield provided by local investors may be of limited value to the MNC, however. The partners will be deemed to be tainted by association with the multinational. A government probably would not be deterred from expropriation or enacting discriminatory laws because of the existence of local shareholders. Moreover, the action can be directed solely against the foreign investor, and the local partners can be the genesis of a move to expropriate to enable them to acquire the whole of a business at a low or no cost.

ADAPTATION ■ Today, some firms are trying a more radical approach to political risk management. Their policy entails adapting to the inevitability of potential expropriation and trying to earn profits on the firm's resources by entering into licensing and management agreements. For example, oil companies whose properties were nationalized by the Venezuelan government received management contracts to continue their exploration, refining, and marketing operations. These firms have recognized that it is not necessary to own or control an asset such as an oil well to earn profits. This form of arrangement may be more common in the future as countries develop greater management abilities and decide to purchase from foreign firms only those skills that remain in short supply at home. Firms that are unable to surrender control of their foreign operations because these operations are integrated into a worldwide production-planning system or some other form of global strategy are also the least likely to be troubled by the threat of property seizure, as was pointed out in the aforementioned Chrysler example.

22.4 ILLUSTRATION: KENNECOTT AND ANACONDA IN CHILE

Most raw material seekers active in the Third World have found themselves under considerable pressure either to divest or to enter into minority joint ventures in order to avoid outright expropriation by host governments submerged by the tidal wave of economic nationalism. The tale of the involvement of two U.S. copper MNCs in Chile illustrates how political risk can be managed ex ante through a policy of multilateral entrapment.[11]

Both Kennecott and Anaconda had long held and operated substantial copper mines in Chile, but they had radically different outlooks on Chile's future. Kennecott relied on the giant mine of El Teniente for 30% of its world output, but it invested mini-

[10] Bradley, "Managing Against Expropriation," pp. 75–83.

[11] Adapted from Theodore H. Moran, *The Politics of Dependence: Copper in Chile* (Princeton, N.J.: Princeton University Press, 1974) by Laurent L. Jacque and Alan C. Shapiro.

mally above depreciation to keep production slightly increasing. Between 1945 and 1965 it did not attempt to develop any new mining sites. However, in 1964, under pressure from President Frei to expand and modernize its operation at El Teniente, Kennecott initiated an ambitious capital expenditures plan aimed at increasing copper production from 180,000 metric tons to 280,000 metric tons per year. The expansion plan was to be financed by selling a 51% interest in the mine for $80 million to the Chilean government in exchange for a ten-year management contract. In addition, further financing was sourced from the U.S. Eximbank ($110 million to be paid back over a period of 10 to 15 years) and the Chilean Copper Corporation ($24 million). In exchange for agreeing to a minority position in the newly created joint venture, Kennecott obtained a special reassessment of the book value of the El Teniente property (from $69 million to $286 million) and a dramatic reduction in taxes from 80% to 44% on its share of the profits.

Kennecott did not commit one cent to the new mine; it also developed a multinational web of stakeholders in the project. Kennecott began by insuring its equity sale to the Chilean government (reinvested in the mine) with the U.S. Agency for International Development (AID) and ensured that the Eximbank loan was unconditionally guaranteed by the Chilean state. In addition, any disputes between Kennecott and Chile would be submitted to the law of the state of New York. Kennecott also raised $45 million for the new joint venture by writing long-term contracts for the future output (literally mortgaging copper still in the ground) with European and Asian customers. Finally, collection rights on these contracts were sold to a consortium of European banks ($30 million) and Mitsui & Co., the Japanese trading company ($15 million).

Anaconda, by contrast, had been bullish on Chile all along. Having invested heavily in its own name from 1945 to 1965 in new mines and the modernization of old ones, it refused voluntary divestiture and was eventually forced to sell 51% of its Chilean holdings to the state in 1969. Although it had partial coverage of its holdings with AID before its forced divestiture, Anaconda had allowed the policy to lapse after 1969.

The real test of these different strategies came in 1971, when the defiant new Marxist government of President Allende assumed power. Both firms shortly fell prey to Chilean vengeance as Allende fulfilled his pledge to expropriate without compensation the foreign interests in Chilean copper. Kennecott received compensation from OPIC (the successor organization to AID) of $80 million plus interest—an amount that surpassed the book value of its pre-1964 holdings and that was eventually reimbursed to the U.S. government by Chile as a condition for rolling over the Chilean debt. Kennecott, on its own, was using the unconditional guarantee initially extracted from the Frei government for the original sale amount to obtain a writ of attachment in the U.S. federal courts against all Chilean property within the courts' jurisdiction, including the jets of Lanchile when they landed in New York. Faced with these actions, the Allende government assumed all debt obligations that the joint venture had contracted with Eximbank and the consortium of European banks and Mitsui. In effect, Kennecott had been freed from any further international obligations, financial and otherwise.

Anaconda was expropriated without compensation from either the Chilean government—because it had no leverage, being the sole investor—or from OPIC—because it had failed to insure against political risk. The only recourse left to Anaconda's board of directors was to fire its entire management—which the board did.

❧ 22.5 SUMMARY AND CONCLUSIONS

Country risk analysis is the assessment of factors that influence the likelihood that a country will have a healthy investment climate. A favorable political risk environment depends on the existence of a stable political and economic system in which entrepreneurship is encouraged and free markets predominate. Under such a system, resources are most likely to be allocated to their highest-valued uses, and people will have the greatest incentive to take risks in productive ventures.

Regardless of whether a nation has a high or low country risk rating, however, individual foreign investments typically are subject to differing degrees of political risk, based on their time pattern of benefits. The major benefits to a host country from a foreign investment usually appear at the beginning. Over time, the added benefits become smaller and the costs more apparent. Unless the firm is continually renewing these benefits—by introducing more products, say, or by expanding output and developing export markets—it is likely to be subject to increasing political risks. The common attitude of government is to ignore the past and ask what a firm will do for it in the future. In a situation where the firm's future contributions are unlikely to evoke a favorable government reaction, the firm had best concentrate on protecting its foreign investments by raising the costs of nationalization.

➤ Questions

1. When investing in a copper mine in Peru, how can a U.S. mining firm such as Kennecott reduce its political risk?

2. Generally speaking, what is the most appropriate means of managing expropriation risk?

3. Can avoiding politically risky countries eliminate a company's political risk? Explain.

4. What factors affect the degree of political risk faced by a firm operating in a foreign country?

5. How might a government budget deficit lead to inflation?

6. What political realities underlie a government budget deficit?

7. What are some indicators of country risk? Of country health?

8. What obstacles do Third World countries such as Argentina, Brazil, and Ghana face in becoming developed nations with strong economies?

9. What can we learn about economic development and political risk from the contrasting experiences of East and West Germany, North and South Korea, and China and Taiwan, Hong Kong, and Singapore?

10. What role do property rights and the price system play in national development and economic efficiency?

11. What indicators would you look for in assessing the political riskiness of an investment in Eastern Europe?

12. Exhibit 22.4 describes some economic changes that have been instituted by Mexico in recent years.
    a. What are the likely consequences of those changes?
    b. Who are the winners and the losers from these economic changes?

13. What is the link between a controlled exchange rate system and political risk?

14. Milton Friedman has suggested that public-sector firms in Latin American countries should simply be given away, possibly to their employees. How do you think workers would feel about being given (for free) ownership of the public sector firms that employ them? Explain.

➤ Problems

1. Comment on the following statement discussing Mexico's recent privatization. "Mexican state companies are owned in the name of the people but are run and now privatized to benefit Mexico's ruling class."

2. Between 1981 and 1987, direct foreign investment in the Third World plunged by more than 50%. The World Bank is concerned about this decline and wants to correct it by improving the investment climate in Third World countries. Its solution:

Create a Multilateral Investment Guarantee Agency (MIGA) that will guarantee foreign investments against expropriation at rates to be subsidized by Western governments.

a. Assess the likely consequences of MIGA on both the volume of Western capital flows to Third World nations and the efficiency of international capital allocation.

b. How will MIGA affect the probability of expropriation and respect for property rights in Third World countries?

c. Is MIGA likely to improve the investment climate in Third World nations?

d. According to a senior World Bank official (*Wall Street Journal*, December 22, 1987, p. 20), "There is vastly more demand for political risk coverage than the sum total available." Is this a valid economic argument for setting up MIGA? Explain.

e. Assess the following argument made on behalf of MIGA by a State Department memo: "We should avoid penalizing a good project [by not providing subsidized insurance] for bad government policies over which they have limited influence . . . Restrictions on eligible countries [receiving insurance subsidies because of their doubtful investment policies] will decrease MIGA's volume of business and spread of risk, making it harder to be self-sustaining." (Quoted in the *Wall Street Journal*, December 22, 1987, p. 20.)

3. By the year 2000, China aims to boost its electricity-generating capacity by more than half. To do that, it is planning on foreigners' investing at least $20 billion of the roughly $100 billion tab. However, Beijing has informed investors that, contrary to their expectations, they will not be permitted to hold majority stakes in large power-plant or equipment-manufacturing ventures. In addition, Beijing has insisted on limiting the rate of return that foreign investors can earn on power projects. Moreover, this rate of return will be in local currency without official guarantees that the local currency can be converted into dollars and it will not be permitted to rise with the rate of inflation. Beijing says that if foreign investors fail to invest in these projects, it will raise the necessary capital by issuing bonds overseas. However, these bonds will not carry the "full faith and credit of the Chinese government."

a. What problems do you foresee for foreign investors in China's power industry?

b. What options do potential foreign investors have to cope with these problems?

c. How credible is the Chinese government's fallback position of issuing bonds overseas to raise capital in lieu of foreign direct investment?

4. You have been asked to head up a special presidential commission on the Russian economy. Your first assignment is to assess the economic consequences of the following six policies and suggest alternative policies that may have more favorable consequences.

a. Under the current Russian system, any profits realized by a state enterprise are turned over to the state to be used as the state sees fit. At the same time, shortfalls of money do not constrain enterprises from consuming resources. Instead, the state bank automatically advances needy enterprises credit, at a zero interest rate, to buy the inputs they need to fulfill the state plan and to make any necessary investments.

b. The Russian fiscal deficit has risen to an estimated 13.1% of GDP. This deficit has been financed almost exclusively by printing rubles. At the same time, prices are controlled for most goods and services.

c. Russian enterprises are allocated foreign exchange to buy goods and services necessary to accomplish the state plan. Any foreign exchange earned must be turned over to the state bank.

d. In an effort to introduce a more market-oriented system, the government has allowed some Russian enterprises to set their own prices on goods and services. However, other features of the system have not been changed: Each enterprise is still held accountable for meeting a certain profit target; only one state enterprise can produce each type of good or service; and individuals are not permitted to compete against state enterprises.

e. Given the disastrous state of Russian agriculture, the Russian government has permitted some private plots on which anything grown can be sold at unregulated prices in open-air markets. Because of their success, the government has recently expanded this program, giving Russian farmers access to much more acreage. At the same time, a number of Western nations are organizing massive food shipments to Russia to cope with the current food shortages.

f. The United States and other Western nations are considering instituting a Marshall Plan for Eastern Europe that would involve massive loans to Russia and other Eastern European nations in order to prop up the reform governments.

5. The president of Mexico has asked you to advise him on the likely economic consequences of the

following five policies designed to improve Mexico's economic environment. Describe the consequences of each policy, and evaluate the extent to which these proposed policies will achieve their intended objective.

a. Expand the money supply to drive down interest rates and stimulate economic activity.

b. Increase the minimum wage to raise the incomes of poor workers.

c. Impose import restrictions on most products to preserve the domestic market for local manufacturers and, thereby, increase national income.

d. Raise corporate and personal tax rates from 50% to 70% to boost tax revenues and reduce the Mexican government deficit.

e. Fix the nominal exchange rate at its current level in order to hold down the cost to Mexican consumers of imported necessities (assume that inflation is currently 100% annually in Mexico).

6.* The president of Brazil has just appointed you to work with the country's cabinet ministers to launch a radical restructuring of the Brazilian economy. Inflation is running at more than 1,000% annually, and the federal government is running a deficit in excess of 10% of GDP (the U.S. deficit is about 3% of GDP). To finance the deficit, the government has incurred huge debts, both internally and externally. In your initial discussions with the cabinet ministers, you realize that there is considerable disagreement about a number of specific program proposals. Your job is to assess the issues and the relative merits of the proposed policies.

a. The Governor of the Banco do Brasil, Brazil's central bank, wants to cease its purchases of government bonds issued by the Ministry of Finance to fund the ongoing federal budget deficit. The Banco do Brasil has acquired 50% to 60% of all government bonds issued in the past several years with money expressly created for that purpose. In other words, it has been monetizing the deficit. Other cabinet ministers are afraid that this policy will lead to higher interest rates and wonder how the deficit can be financed otherwise.

b. The Minister of Infrastructure has proposed that his ministry will begin privatizing the hundreds of state-owned enterprises under his administration. These enterprises include virtually all of Brazil's steel industry, mining industry, electric utilities, the telephone company, national oil

company, chemical companies, and a wide range of manufacturing companies. Opponents claim that this move will lead to massive unemployment and the bankruptcy of vital national industries.

c. The Minister of Political Economy has proposed that Brazil enter into free trade agreements with its Latin American neighbors. This policy would involve eliminating all tariffs, duties, and fees on imports. A number of other government leaders oppose this move, because the Brazilian market is larger and generally more protected than those of its neighbors. They feel that opening the border would expose Brazil to rapid growth in imports that exceed any incremental export activity.

d. The Minister of Finance has proposed creating a new tax on consumption and lowering income tax rates. His concern is that Brazil's personal savings rate has been close to zero over the past several years. He believes that increased savings will help to dampen inflation, lower interest rates on the federal debt, and promote exports. Critics of this proposal argue that the vast majority of Brazil's population are living very near the poverty line and that a consumption tax would be highly regressive (hit the poor relatively harder than the rich). It also would tend to dampen domestic demand, the principal engine of economic growth in Brazil.

e. The Minister of Labor has proposed raising the minimum wage to raise the income of poor workers and, thereby, offset the effects of restructuring on them. Other cabinet members are concerned about the effects this policy will have on employment and competitiveness.

f. The Central Bank has proposed that it replace the current fixed exchange rate system with a freely floating exchange rate system. Critics of this proposal argue that floating the cruzeiro will devalue the currency and raise the cost of living (by boosting the price of imported necessities) for Brazilians.

g. In order to reduce the money supply and, thereby, suppress inflation, the Minister of Finance has proposed freezing all bank accounts. Depositors will be able to withdraw only the cruzeiro equivalent of about $1,000. Other cabinet ministers are concerned about possible adverse consequences of such a freeze.

---

*Developed by William H. Davidson and Alan C. Shapiro.

➤ Bibliography

BRADLEY, DAVID. "Managing Against Expropriation." *Harvard Business Review*, July–August 1977, pp. 75–83.

KOBRIN, STEPHEN J. "Political Risk: A Review and Reconsideration." *Journal of International Business Studies*, Spring–Summer 1979, pp. 67–80.

KRUEGER, ANNE O. "Asian Trade and Growth Lessons." *American Economic Review*, May 1990, pp. 108–112.

LANDES, DAVID S. "Why Are We So Rich and They So Poor?" *American Economic Review*, May 1990, pp. 1–13.

SHAPIRO, ALAN C. "The Management of Political Risk." *Columbia Journal of World Business*, Fall 1981, pp. 45–56.

STOBAUGH, ROBERT B. "How to Analyze Foreign Investment Climates." *Harvard Business Review*, September–October 1969, pp. 100–108.

VAN AGTAMAEL, A.W. "How Business Has Dealt with Political Risk." *Financial Executive*, January 1976, pp. 26–30.

Case Studies

CASE V.1

THE INTERNATIONAL MACHINE CORPORATION

The International Machine Corporation (IMC) is a large, well-established manufacturer of a wide variety of food processing and packaging equipment. Total revenue for last year was $12 billion, of which 45% was generated outside of the United States. IMC has subsidiaries in 23 countries, with licensing arrangements in 8 others.

The management of IMC is currently contemplating the establishment of a subsidiary in Mexico. IMC has been exporting products to Mexico for several years, and its international division believes there is sufficient demand for the product and that a Mexican investment might be appropriate at this time. More important, management believes that the Mexican market is expanding, that the economy is growing, and that producing such products locally appears to be consistent with the national aspirations of the Mexican government.

Mexican inflation is projected to be 20% annually, and the U.S. inflation rate is expected to be 10% annually. The current exchange rate is $1 = Ps 7.2 and is expected to remain fixed in real terms over the life of the investment. The following list contains details of the contemplated investment.

A. Initial investment
  1. It is estimated that it would take one year to purchase and install plant and equipment.
  2. Imported machinery and equipment will cost $9 million. No import duties will be levied by the Mexican government. With a small allowance for banking fees, the bill will come to Ps 65.5 million.
  3. The plant would be set up on government-owned land that will be sold to the project for Ps 6.5 million.
  4. IMC plans to maintain effective control of the subsidiary with ownership of 60% of equity. The remaining 40% is to be distributed widely among Mexican financial institutions and private investors. Accordingly, IMC needs to invest U.S. $6 million in the project.

B. Working capital
  1. The company plans to maintain 5% of annual sales as a minimum cash balance.
  2. Accounts receivable are estimated to be 73 days of annual sales.
  3. Inventory is estimated to be 20% of annual sales.
  4. Accounts payable are estimated to be 10% of annual sales.
  5. Other payables are estimated to be 5% of annual sales.
  6. Licensing and overhead allocation fees are paid annually at the end of the year.

C. Sales volume
  1. Sales volume for the first year is estimated to be 200 units.
  2. Selling price in the first year will be Ps 458,000 per unit.
  3. Unit sales growth of 10% is expected during the project life.
  4. An annual price increase of 20% is expected.

D. Cost of goods sold
  1. The U.S. parent company is expected to provide parts and components adding up to Ps 59,000 per unit in the first year of operation. These costs (in U.S. dollars) are expected to rise on an average of 10% annually, in line with the projected U.S. inflation rate.

Source: This is an edited version of "The International Machine Corporation: An Analysis of Investment in Mexico," by Vinod B. Bavishi, University of Connecticut, and Haney A. Shawkey, State University of New York at Albany. Permission to use this case was provided by Professors Bavishi and Shawkey.

2. Local material and labor costs are expected to be Ps 137,000 per unit, with an annual rate of increase of 20%.
3. Manufacturing overhead (without depreciation) is expected to be Ps 9.2 million the first year of operation. An average rate of increase of 15% is expected.
4. Depreciation of manufacturing equipment is to be computed on a straight-line basis, with a projected life of 10 years and zero salvage value to be assumed.

E. Selling and administrative costs
1. The variable portion of selling and administrative costs are expected to equal 10% of annual sales revenue.
2. Semifixed selling costs are expected to equal 5% of the first year's sales. These costs will then rise at 15% annually.

F. Licensing and overhead allocation fees
1. The parent company will levy Ps 23,000 per unit as licensing and overhead allocation fees, payable at year end in U.S. dollars.
2. This fee will increase 20% per year to compensate for Mexican inflation.

G. Interest expense
1. Local borrowings can be obtained for working capital purposes at 15%. Borrowing will occur at the end of the year with the full year's interest budgeted in the following year.
2. Any excess funds can be invested in Mexican marketable securities with an annual rate of return of 15%. Investment will be made at the end of the year, with the full year's interest to be received in the following year.

H. Income taxes
1. Corporate income taxes in Mexico are 42% of taxable income.
2. Withholding taxes on licensing and overhead allocation fees are 20%.
3. The parent company's effective U.S. tax rate is 35%, which is the rate used in analyzing investment projects. It can be assumed that the parent company can take appropriate credits for taxes paid to, or withheld by, the Mexican government.

I. Dividend payments
1. No dividends will be paid for the first three years.
2. Dividends equal to 70% of earnings will be paid to the shareholders, beginning in the fourth year.

J. Terminal payment
It is assumed that, at the end of the tenth year of operation, IMC's share of net worth in the Mexican subsidiary will be remitted in the form of a terminal payment.

K. Parent company's capital structure
1. Domestic debt equals U.S. $1 billion with an average before-tax cost of 12%. The cost of new long-term debt is estimated at 14% before tax.
2. An amount equivalent to $600 million of parent debt is denominated in various foreign currencies, and after adjusting for previous exchange gains/losses the cost (or effective cost) of this debt has averaged 16%.
3. Shareholder equity (capital, surplus, and retained earnings) equals U.S. $1.5 billion. The company plans to pay U.S. $3.20 in dividends per share during the coming year. Over the last 10 years, earnings and dividends have grown at a compounded rate of 7%. The market price of common stock was $40, and number of shares outstanding were 60 million as of last December 31.

L. Exports lost
At present IMC is exporting about 25 units per year to Mexico. If IMC decides to establish the Mexican subsidiary, it is expected that the after-tax effects on income due to the lost exports sales would be $648,000, $742,000, and $930,000 in the first three years of operation, respectively. IMC assumes it cannot count on these export sales for more than three years because the Mexican government is determined to see that such machinery is manufactured locally in the near future.

1. Should IMC make this investment?
2. What is IMC's required rate of return for this project?
3. What factors and assumptions are critical to your project analysis?

CASE V.2

EURO DISNEYLAND

It is 1987, and in a muddy sugar-beet field 20 miles east of Paris, Walt Disney Co. is creating Euro Disneyland. By the time of its scheduled opening in 1992, Euro Disneyland (EDL) is expected to cost FF 15 billion ($2.5 billion based on an exchange rate of FF 6 = $1). Most signs point to the park's success. Two million Europeans already visit its American parks every year. And Paris demographics look great: In two hours, 17 million people could drive to the park and 310 million people could fly to the park.

Disney's risk appears to be modest. It has invested $350 million in planning the park but has put up just $145 million for 49% of EDL's equity. Public investors will pay $1 billion for the other 51% in a stock offering in October 1989. The public company, through its traded shares, will give Europeans a chance to participate in the success of the project once the gates open in 1992. (When the stock begins trading on the Paris Bourse in October 1989, Disney's stake will be valued at $1 billion—an $855 million gain in value.)

Even if profits are weak, Disney will rake in fat management fees. And it could clean up just on the land: It has rights to buy 4,800 acres from the government at just $7,500 an acre—compared with $750,000 an acre for similar land in the area. Disney can resell chunks to other developers for any price it can get. The acreage should jump in value once high-speed rail lines from Paris and London (via the Chunnel, the tunnel under the English Channel) are in place.

But there is risk nonetheless. The most critical variable is attendance. Many experts think surprises may await Disney. They doubt that attendance will meet Disney's expectations. Others think that the crowds will come but will spend less than Disney projects. Disney faces other unknowns as well. MCA/Universal, which will be bought by Matsushita in 1990, is thinking of building a park in London, which would cannibalize Euro Disneyland's attendance. There's also the grim winter weather, which prompts European parks to close until spring. Then there's the challenge of training 12,000 Europeans, half of them French, to be Disney "cast members." Bowing to French individualism, Disney will relax its personal grooming code a bit. Disney may also decide to change its ban on booze if customers call loudly enough for wine and beer. Disney claims that its experience with Tokyo Disneyland, its last major development project, shows that it can deal with non-American culture and bad weather. Tokyo Disneyland was completed on time and within 1% of budget. It has also been a huge commercial success.

FINANCING

In March 1987, Disney and the French government sign a "Master Agreement" for Euro Disneyland. In accordance with that agreement, Disney forms a holding company to control development of the entire site. It pays $145 million for 49% of the holding company's shares and sells 51% of EDL to European investors for a little over $1 billion; of the latter shares, around half are sold to the French.

The holding company set up as an SCA (*Societé Commanditeé par Actions*), a unique French corporate form that is very similar to an American limited partnership. Disney is the *gerant*, or general partner. The SCA structure allows Disney to control management, even with a minority shareholding. Thus, even though the holding company owns EDL, Disney will manage it and collect an estimated $35 million a year in royalties on sales of admission tickets, food, and souvenirs.

The master agreement is basically an inducement for Disney to bring Euro Disneyland to Paris rather than to Spain's Mediterranean coast. The inducements include the following:

- Loans of up to FF 4.8 billion are available from a French government agency. The loans carry a fixed interest rate of 7.85%, in contrast to a normal commercial rate of 9.25%.
- EDL can use accelerated depreciation to write off the construction costs of its extravaganza over a 10-year period.
- The French government will invest $350 million in park-related infrastructure. This includes sewer and telephone trunk lines, subway and road links to Paris, and a new line of its 156-mile-per-hour *train a grande vitesse* (TGV) that will link EDL to the Chunnel (and thence to London), Brussels, and Lyons.
- Disney is allowed to buy 4,800 acres of land at 1971 prices. The average cost is FF 11.1 per square meter compared with FF 1,000 per square meter for development land in the Paris suburbs.
- The French government agrees to cut the value-added tax (VAT) on ticket sales to just 7% instead of the usual 18.5%.

Disney will structure the $2.5 billion Euro Disneyland project so that the operating losses created during construction can be used, along with the accelerated depreciation benefits, as tax shelters. These tax benefits will be sold to a group of French companies for $200 million. Equity contributions include the $1.15 billion stock sale plus Disney's $350 million investment in project planning. The remaining $800 million of project financing will come from loans subsidized by the French government. The repayment schedule on these loans is as follows (in FF millions):

1992–1996	1997	1998	1999	2000	2001
0	FF 960	FF 960	FF 960	FF 960	FF 960

Euro Disneyland will also require about $115 million in working capital initially, and this amount is expected to grow at the rate of sales revenue. The working capital will be financed by French franc bank loans carrying an interest rate expected to average about 9.5% annually.

FINANCIAL PROJECTIONS

Disney's projections assume a minimum 1992 attendance rate at Euro Disneyland of 11 million visitor-days and maximum of 16 million. These numbers compare with 1988 attendance at the domestic U.S. parks of 25.1 million for Orlando's Disney World and 13 million for Anaheim's Disneyland. In projections of future years' attendance, a conservative 3% annual growth rate is reasonable. The range of EDL attendance figures based on these assumptions is shown in Exhibit V2.1.

The next step is to forecast individual expenditures at the park on a daily basis, for admission as well as for food and merchandise. Disney's projections assume that each visitor will spend FF 78.6 on merchandise, FF 59.5 on food and beverages, and FF 5.5 on parking and other items (e.g., stroller rentals). Admission fees are estimated at FF 144.4 apiece.

These estimates are based on 1989 francs. French inflation is expected to increase these figures by 5% each year. As of late 1989, the French franc:dollar exchange rate was about FF 6 = $1, but this rate could obviously vary considerably. For example, U.S. inflation is expected to average about 4% annually, about 1% below the expected French inflation rate.

Euro Disneyland will also collect "participation fees" from various corporate sponsors, such as Kodak and Renault. These fees are payment for the privilege of sponsoring specific attractions in return for promotional considerations (e.g., Kraft's "The Land") and are expected to approximate $35 million in 1992.

EDL's pretax operating margin is expected to be about 35%. That money will not all flow into the hands of EDL shareholders. EDL must pay interest on its debt and French tax. The effective tax rate is estimated at 55%. In addition, EDL must pay Disney royalties, a base management fee, and an incentive-based bonus fee. Under the master agreement, Disney will collect 10% of the revenues generated by ticket sales and 5% of all expenditures on food, beverage, and merchandise. Disney also can collect significantly higher fees from EDL if the park exceeds certain operating cash flow (OCF) targets. Specifically, Disney will collect 30% of the park's OCF in the range of FF 1.4–2.1 billion; 40% of all OCF between FF 2.1 and 2.8 billion; and 50% of all OCF above FF 2.8 billion. Disney will receive no incentive fee if the park's OCF is less than FF 1.4 billion. In this case, the operating cash flow will equal operating income as Disney has sold its depreciation tax benefits.

In November 1989, Walt Disney Co. announced plans to add a movie studio theme park as the second gated attraction to Euro Disneyland. The movie studio theme park is expected to open in 1996. In addition to strengthening Disney's film production capabilities in Europe (where the movement to enforce geographic quotas continues to gain momentum), the second attraction could add as much as FF 3.3 billion to 1996 operating revenues.

EXHIBIT V2.1 Projected Number of Visitor Days at Euro Disneyland (Millions)

1992	1993	1994	1995	1996
11	11.3	11.7	12.0	12.4
12	12.4	12.7	13.1	13.5
13	13.4	13.8	14.2	14.6
14	14.4	14.9	15.3	15.8
15	15.5	15.9	16.4	16.9
16	16.5	17.0	17.5	18.0

Source: Liz Buyer, "EuroDisney: As Close to Risk Free as a Deal Can Get," *The Journal of European Business*, March/April 1992, p. 27. Reprinted with permission of *Journal of European Business* and Faulkner and Gray.

QUESTIONS

1. These questions are related to the $800 million (FF 4.8 billion at FF 6 = $1) in French government-subsidized loans.
   a. What is the value to Disney of the French government's loan subsidies?
   b. What exchange risk is this project subject to from the standpoint of Disney? How can financing be used to mitigate this exchange risk?
   c. Suppose it turns out that having $800 million in franc financing actually adds to Disney's economic exposure. How should this affect Disney's willingness to accept the full amount of financing offered by the French government?

2. Based on purchasing power parity, project the dollar:franc exchange rate from 1989 through 1996.

3. What is the range of projected dollar net income for Euro Disneyland for the years 1992–1996?

4. Suppose the terminal value at the end of 1996 is estimated at seven times net income for 1996. Using a 15% cost of capital, what is the range of net present values of Euro Disneyland as a stand-alone project?

5. Using the same 15% cost of capital, what is the range of net present values of Walt Disney's investment in Euro Disneyland?

6. Should Walt Disney go ahead with this project? What other factors might you consider in estimating the value of Euro Disneyland to Walt Disney?

◆ GLOSSARY OF KEY WORDS AND TERMS IN INTERNATIONAL FINANCE

Acceptance A time draft that is accepted by the drawee. Accepting a draft means writing *accepted* across its face, followed by an authorized person's signature and the date. The party accepting a draft incurs the obligation to pay it at maturity.

Accounting Exposure The change in the value of a firm's foreign-currency-denominated accounts due to a change in exchange rates.

Act of State Doctrine This doctrine says that a nation is sovereign within its own borders and its domestic actions may not be questioned in the courts of another nation.

Adjusted Present Value The net present value of a project using the all-equity rate as a discount rate. The effects of financing are incorporated in separate terms.

Administrative Pricing Rule IRS rules used to allocate income on export sales to a foreign sales corporation.

Advance Pricing Agreement (APA) Procedure that allows the multinational firm, the IRS, and the foreign tax authority to work out, in advance, a method to calculate transfer prices.

Adverse Incentives *See* moral hazard.

Adverse Selection The possibility that only the highest-risk customers will seek insurance.

African Development Bank (AFDB) The AFDB makes or guarantees loans and provides technical assistance to member states for various development projects.

Agency Costs Costs that stem from conflicts between managers and stockholders and between stockholders and bondholders.

All-Equity Beta The beta associated with the unleveraged cash flows of a project or company.

All-Equity Rate The discount rate that reflects only the business risks of a project and abstracts from the effects of financing.

All-In Cost The effective interest rate on a loan, calculated as the discount rate that equates the present value of the future interest and principal payments to the net proceeds received by the borrower; it is the internal rate of return on the loan.

American Depository Receipt (ADR) A certificate of ownership issued by a U.S. bank as a convenience to investors in lieu of the underlying foreign corporate shares it holds in custody.

American Option An option that can be exercised at any time up to the expiration date.

American Shares Securities certificates issued in the United States by a transfer agent acting on behalf of the foreign issuer. The certificates represent claims to foreign equities.

American Terms Method of quoting currencies; it is expressed as the number of U.S. dollars per unit of foreign currency.

Appreciation *See* revaluation.

Arab Fund for Economic and Social Development (AFESD) A multilateral Arab fund that actively searches for projects in Arab League countries and then assumes responsibility for project implementation.

Arbitrage The purchase of securities or commodities on one market for immediate resale on another in order to profit from a price discrepancy.

Arm's-Length Price Price at which a willing buyer and a willing unrelated seller would freely agree to transact (i.e., a market price).

Asiacurrency (or Asiadollar) Market Offshore financial market located in Singapore that channels investment dollars to a number of rapidly growing Southeast Asian countries and provides deposit facilities for those investors with excess funds.

Asian Development Bank (ADB) The ADB guarantees or makes direct loans to member states and private ventures in Asian/Pacific nations and helps to develop local capital markets by underwriting securities issued by private enterprises.

Ask The price at which one can sell a currency. Also known as the offer price.

Atlantic Development Group for Latin America (ADELA) An international private investment company dedicated to the socioeconomic development of Latin America. Its objective is to strengthen private enterprise by providing capital and entrepreneurial and technical services.

At-the-Money An option whose exercise price is the same as the spot exchange rate.

Back-to-Back Loan An intercompany loan, also known as a *fronting loan* or *link financing*, that is channeled through a bank.

Baker Plan A plan by U.S. Treasury Secretary James Baker under which 15 principal middle-income debtor countries (the "Baker countries") would undertake growth-oriented structural reforms, to be supported by increased financing from the World Bank and continued lending from commercial banks.

Balance of Payments Net value of all economic transactions—including trade in goods and services, transfer payments, loans, and investments—between residents of the same country and those of all other countries.

Balance of Trade Net flow of goods (exports minus imports) between countries.

Balance-Sheet Exposure *See* accounting exposure.

Bank Capital The equity capital and other reserves available to protect bank depositors against credit losses.

Bank Draft A draft addressed to a bank; *See also* draft.

Bank Loan Swap *See* debt swap

Bank for International Settlements (BIS) Organization headquartered in Basle that acts as the central bank for the industrial countries' central banks. The BIS helps central banks manage and invest their foreign exchange reserves and also holds deposits of central banks so that reserves are readily available.

Banker's Acceptance Draft accepted by a bank; *See also* draft.

Barrier Option *See* knockout option.

Basis Point One hundred basis points equal one percent of interest.

Basis Swap Swap in which two parties exchange floating interest payments based on different reference rates.

Bearer Securities Securities that are unregistered.

Bear Spread A currency spread designed to bet on a currency's decline. It involves buying a put at one strike price and selling another put at a lower strike price.

Beggar-Thy-Neighbor Devaluation A devaluation that is designed to cheapen a nation's currency and thereby increase its exports at others' expense and reduce imports. Such devaluations often led to trade wars.

Beta A measure of the systematic risk faced by an asset or project. Beta is calculated as the covariance between returns on the asset and returns on the market portfolio divided by the variance of returns on the market portfolio.

Bid The price at which one can buy a currency.

Bid-Ask Spread The difference between the buying and selling rates.

Bill of Exchange *See* bank draft.

Bill of Lading A contract between a carrier and an exporter in which the former agrees to carry the latter's goods from port of shipment to port of destination. It is also the exporter's receipt for the goods.

Black Market An illegal market that often arises when price controls or official rationing lead to shortages of goods, services, or assets.

Black-Scholes Option Pricing Model The most widely used model for pricing options. Named after its creators, Fischer Black and Myron Scholes.

Blocked Currency A currency that is not freely convertible to other currencies due to exchange controls.

Brady Bonds New government securities issued under the Brady Plan whose interest payments were backed with money from the International Monetary Fund.

Brady Plan Plan developed by U.S. Treasury Secretary Nicholas Brady in 1989 that emphasized LDC debt relief through forgiveness instead of new lending. It gave banks the choice of either making new loans or writing off portions of their existing loans in exchange for Brady Bonds.

Branch A foreign operation incorporated in the home country.

Bretton Woods System International monetary system established after World War II under which each government pledged to maintain a fixed, or pegged, exchange rate for its currency vis-à-vis the dollar or gold. As one ounce of gold was set equal to $35, fixing a currency's gold price was equivalent to setting its exchange rate relative to the dollar. The U.S. government pledged to maintain convertibility of the dollar into gold for foreign official institutions.

Bull Spread A currency spread designed to bet on a currency's appreciation. It involves buying a call at one strike price and selling another call at a higher strike price.

Call Provision Clause giving the borrower the option of retiring the bonds before maturity should interest rates decline sufficiently in the future.

Capital Account Net result of public and private international investment and lending activities.

Capital Asset Pricing Model (CAPM) A model for pricing risk. The CAPM assumes that investors must be compensated for the time value of money plus systematic risk, as measured by an asset's beta.

Capital Flight The transfer of capital abroad in response to fears of political risk.

Capital Market Imperfections Distortions in the pricing of risk, usually attributable to government regulations and asymmetries in the tax treatment of different types of investment income.

Capital Market Integration The situation that exists when real interest rates are determined by the global supply and global demand for funds.

Capital Market Segmentation The situation that exists when real interest rates are determined by local credit conditions.

Capital Productivity The ratio of output (goods and services) to the input of physical capital (plant and equipment).

Cash Flow Exposure Used synonymously with economic exposure, it measures the extent to which an exchange rate change will change the value of a company through its impact on the present value of the company's future cash flows.

Cash Pooling *See* pooling.

Central Bank The nation's official monetary authority.

Chicago Mercantile Exchange (CME) The largest market in the world for trading standardized futures and options contracts on a wide variety of commodities, including currencies and bonds.

CHIPS *See* Clearing House Interbank Payments System.

Clean Draft A draft unaccompanied by any other papers; it is normally used only for nontrade remittances.

Clean Float *See* free float.

Cleanup Clause A clause inserted in a bank loan requiring the company to be completely out of debt to the bank for a period of at least 30 days during the year.

Clearing House Interbank Payments System (CHIPS) A computerized network for transfer of international dollar payments, linking about 140 depository institutions that have offices or affiliates in New York City.

Commercial Invoice A document that contains an authoritative description of the merchandise shipped, including full details on quality, grades, price per unit, and total value, along with other information on terms of the shipment.

Commercial Paper (CP) A short-term unsecured promissory note that is generally sold by large corporations on a discount basis to institutional investors and to other corporations.

Compensating Balance The fraction (usually 10% to 20%) of an outstanding loan balance that a bank requires borrowers to hold on deposit in a noninterest-bearing account.

Concession Agreement An understanding between a company and the host government that specifies the rules under which the company can operate locally.

Consignment Under this selling method, goods are only shipped, but not sold, to the importer. The exporter (consignor) retains title to the goods until the importer (consignee) has sold them to a third party. This arrangement is normally made only with a related company because of the large risks involved.

Consular Invoice An invoice, which varies in its details and information requirements from nation to nation, that is presented to the local consul in exchange for a visa.

Controlled Foreign Corporation (CFC) A foreign corporation whose voting stock is more than 50% owned by U.S. stockholders, each of whom owns at least 10% of the voting power.

Convertible Bonds Fixed-rate bonds that are convertible into a given number of shares prior to maturity.

Corporate Governance The means whereby companies are controlled.

Correspondent Bank A bank located in any other city, state, or country that provides a service for another bank.

Cost of Equity Capital The minimum rate of return necessary to induce investors to buy or hold a firm's stock. It equals a basic yield covering the time value of money plus a premium for risk.

Countertrade A sophisticated form of barter in which the exporting firm is required to take the countervalue of its sale in local goods or services instead of in cash.

Country Risk General level of political and economic uncertainty in a country affecting the value of loans or investments in that country. From a bank's standpoint, it refers to the possibility that borrowers in a country will be unable to service or repay their debts to foreign lenders in a timely manner.

Coupon Swap Swap in which one party pays a *fixed rate* calculated at the time of trade as a spread to a particular Treasury bond, and the other side pays a *floating rate* that resets periodically throughout the life of the deal against a designated index.

Covered Interest Arbitrage   Movement of short-term funds between two currencies to take advantage of interest differentials with exchange risk eliminated by means of forward contracts.

Covered Interest Differential   The difference between the domestic interest rate and the hedged foreign interest rate.

Cross-Default Clause   Clause in a loan agreement which says that a default by a borrower to one lender is a default to all lenders.

Cross-Hedge   Hedging exposure in one currency by the use of futures or other contracts on a second currency that is correlated with the first currency.

Cross-Rate   The exchange rate between two currencies, neither of which is the U.S. dollar, calculated by using the dollar rates for both currencies.

Currency Arbitrage   Taking advantage of divergences in exchange rates in different money markets by buying a currency in one market and selling it in another.

Currency Call Option   A financial contract that gives the buyer the right, but not the obligation, to buy a specified number of units of foreign currency from the option seller at a fixed dollar price, up to the option's expiration date.

Currency Collar   A contract that provides protection against currency moves outside an agreed-upon range. It can be created by simultaneously buying an out-of-the-money put option and selling an out-of-the-money call option of the same size. In effect, the purchase of the put option is financed by the sale of the call option.

Currency Controls   *See* exchange controls.

Currency Futures Contract   Contract for future delivery of a specific quantity of a given currency, with the exchange rate fixed at the time the contract is entered. Futures contracts are similar to forward contracts except that they are traded on organized futures exchanges and the gains and losses on the contracts are settled each day.

Currency of Denomination   Currency in which a transaction is stated.

Currency Option   A financial contract that gives the buyer the right, but not the obligation, to buy (call) or sell (put) a specified number of foreign currency units to the option seller at a fixed dollar price, up to the option's expiration date.

Currency of Determination   Currency whose value determines a given price.

Currency Put Option   A financial contract that gives the buyer the right, but not the obligation, to sell a specified number of foreign currency units to the option seller at a fixed dollar price, up to the option's expiration date.

Currency Risk Sharing   An agreement by the parties to a transaction to share the currency risk associated with the transaction. The arrangement involves a customized hedge contract imbedded in the underlying transaction.

Currency Spread   An options position created by buying an option at one strike price and selling a similar option at a different strike price. It allows speculators to bet on the direction of a currency at a lower cost than buying a put or a call option alone but at the cost of limiting the position's upside potential.

Currency Swap   A simultaneous borrowing and lending operation whereby two parties exchange specific amounts of two currencies at the outset at the spot rate. They also exchange interest rate payments in the two currencies. The parties undertake to reverse the exchange after a fixed term at a fixed exchange rate.

Current Account   Net flow of goods, services, and unilateral transactions (gifts) between countries.

Current Assets   Short-term assets, including cash, marketable securities, accounts receivable, and inventory.

Current Exchange Rate   Exchange in effect today.

Current Liabilities   Short-term liabilities, such as accounts payable and loans expected to be repaid within one year.

Current/Noncurrent Method   Under this currency translation method, all of a foreign subsidiary's current assets and liabilities are translated into home currency at the current exchange rate while noncurrent assets and liabilities are translated at the historical exchange rate (that is, at the rate in effect at the time the asset was acquired or the liability incurred).

Currency Controls   See exchange controls.

Current Rate Method   Under this currency translation method, all foreign currency balance-sheet and income items are translated at the current exchange rate.

Cylinder   The payoff profile of a currency collar created through a combined put purchase and call sale.

Debt Relief   Reducing the principal and/or interest payments on LDC loans.

Debt Swap A set of transactions (also called a debt-equity swap) in which a firm buys a country's dollar bank debt at a discount and swaps this debt with the central bank for local currency that it can use to acquire local equity.

Depreciation *See* devaluation.

Depreciation Tax Shield The value of the tax write-off on depreciation of plant and equipment.

Derivatives Contracts that derive their value from some underlying asset (such as a stock, bond, or currency), reference rate (such as a 90-day Treasury bill rate), or index (such as the S&P 500 stock index). Popular derivatives include swaps, forwards, futures, and options.

Devaluation A decrease in the spot value of a currency.

Direct Quotation A quote that gives the home currency price of a foreign currency.

"Dirty" Float *See* managed float.

Discounting A means of borrowing against a trade or other draft. The exporter or other borrower places the draft with a bank or other financial institution and, in turn, receives the face value of the draft less interest and commissions.

Discount Basis Means of quoting the interest rate on a loan under which the bank deducts the interest in advance.

Doctrine of Sovereign Immunity Doctrine that says a nation may not be tried in the courts of another country without its consent.

Documentary Draft A draft accompanied by documents that are to be delivered to the drawee on payment or acceptance of the draft. Typically, these documents include the bill of lading in negotiable form, the commercial invoice, the consular invoice where required, and an insurance certificate.

Domestic International Sales Corporation (DISC) A domestic U.S. corporation that receives a tax incentive for export activities.

Double Dip Lease A cross-border lease in which the disparate rules of the lessor's and lessee's countries let both parties be treated as the owner of the leased equipment for tax purposes.

Down-and-In Option An option that comes into existence if and only if the currency weakens enough to cross a preset barrier.

Down-and-Out Call A knockout option that has a positive payoff to the option holder if the underlying currency strengthens but is canceled if it weakens sufficiently to hit the outstrike.

Down-and-Out Put A knockout option that has a positive payoff if the currency weakens but is canceled if it weakens beyond the outstrike.

Draft An unconditional order in writing—signed by a person, usually the exporter, and addressed to the importer—ordering the importer or the importer's agent to pay, on demand (sight draft) or at a fixed future date (time draft), the amount specified on its face. Accepting a draft means writing *accepted* across its face, followed by an authorized person's signature and the date. The party accepting a draft incurs the obligation to pay it at maturity.

Dragon Bond Debt denominated in a foreign currency, usually dollars, but launched, priced, and traded in Asia.

Drawdown The period over which the borrower may take down the loan.

Dual Currency Bond Bond that has the issue's proceeds and interest payments stated in foreign currency and the principal repayment stated in dollars.

Dual Syndicate Equity Offering An international equity placement where the offering is split into two tranches—domestic and foreign—and each tranche is handled by a separate lead manager.

EAFE Index The Morgan Stanley Capital International Europe, Australia, Far East Index, which reflects the performance of all major stock markets outside of North America.

Economic Exposure The extent to which the value of the firm will change due to an exchange rate change.

Economies of Scale Situation in which increasing production leads to a less-than-proportionate increase in cost.

Economies of Scope Scope economies exist whenever the same investment can support multiple profitable activities less expensively in combination than separately.

Edge Act Corporation A subsidiary, located in the United States, of a U.S. bank that is permitted to carry on international banking and investment activities.

Effective Interest Rate The true cost of a loan; it equals the annual interest paid divided by the funds received.

Efficient Frontier The set of portfolios that has the smallest possible standard deviation for its level of expected return and has the maximum expected return for a given level of risk.

Efficient Market One in which new information is readily incorporated in the prices of traded securities.

Electronic Trading System A system used in the foreign exchange market that offers automated matching. Traders can enter buy and sell orders directly into their terminals on an anonymous basis, and these prices will be visible to all market participants. Another trader, anywhere in the world, can execute a trade by simply hitting two buttons.

Emerging Markets A term that encompasses the stock markets in all of South and Central America; all of the Far East with the exception of Japan, Australia, and New Zealand; all of Africa; and parts of Southern Europe, as well as Eastern Europe and countries of the former Soviet Union.

Equity-Related Bonds Bonds that combine features of the underlying bond and common stock. The two principal types of equity-related bonds are convertible bonds and bonds with equity warrants.

Equity Warrants Securities that give their holder the right to buy a specified number of shares of common stock at a specified price during a designated time period.

Euro New currency created for members of the European Monetary Union and issued by the European Central Bank. Coins and bills denominated in the euro will be unavailable until 2002.

Eurobank A bank that accepts and makes loans in Eurocurrencies.

Eurobond A bond sold outside the country in whose currency it is denominated.

Eurobond Market Market in which Eurobonds are issued and traded.

Euro-Commercial Paper (Euro-CP) Euronotes that are not underwritten.

Eurocurrency A currency deposited in a bank outside the country of its origin.

Eurocurrency Market The set of banks that accept deposits and make loans in Eurocurrencies.

Eurodollar A U.S. dollar on deposit outside the United States.

Eurodollar Future A cash-settled futures contract on a three-month, $1,000,000 Eurodollar deposit that pays LIBOR.

Euroequity Issue A syndicated equity offering placed throughout Europe and handled by one lead manager.

Euro-Medium Term Note (Euro-MTN) A nonunderwritten Euronote issued directly to the market. Euro-MTNs are offered continuously rather than all at once like a bond issue. Most Euro-MTN maturities are under five years.

Euronote A short-term note issued outside the country of the currency it is denominated in.

Europe, Australia, Far East (EAFE) Index An index put out by Morgan Stanley Capital International which tracks the performance of the 20 major stock markets outside of North America.

European Bank for Reconstruction and Development (EBRD) A development bank supposed to finance the privatization of Eastern Europe.

European Central Bank The central bank for the European Monetary Union. It has the sole power to issue a single European currency called the **euro**.

European Currency Unit (ECU) A composite currency, consisting of fixed amounts of 12 European currencies.

European Investment Bank (EIB) A development bank that offers funds for certain public and private projects in European and other nations associated with the Common Market.

European Monetary System (EMS) Monetary system formed by the major European countries under which the members agree to maintain their exchange rates within a specific margin around agreed-upon, fixed central exchange rates. These central exchange rates are denominated in currency units per ECU.

European Monetary Union Eleven members of the European Community who have joined together to establish a single central bank (the European Central Bank) that issues a common European currency (usually referred to as the euro).

European Option An option that can be exercised only at maturity.

European Terms Method of quoting currencies; it is expressed as the number of foreign currency units per U.S. dollar.

European Union Organization of fifteen European nations whose purpose is to promote economic harmonization and tear down barriers to trade and commerce within Europe.

Exchange Controls Restrictions placed on the transfer of a currency from one nation to another.

Exchange-Rate Mechanism (ERM) Arrangement at the heart of the European Monetary System which allows each member of the EMS to determine a mutually agreed-on central exchange rate for its currency; each rate is denominated in currency units per ECU.

Exchange Risk The variability of a firm's (or asset's) value that is due to uncertain exchange rate changes.

Exercise Price The price at which an option is exercised. Also known as **strike price**.

Export-Import Bank (Eximbank) U.S. government agency dedicated to facilitating U.S. exports, primarily through subsidized export financing.

Exposure Netting Offsetting exposures in one currency with exposures in the same or another currency, where exchange rates are expected to move in such a way that losses (gains) on the first exposed position should be offset by gains (losses) on the second currency exposure.

Expropriation The taking of foreign property, with or without compensation, by a government.

Factor Specialized buyer, at a discount, of company receivables.

FASB No. 8 *See* Statement of Financial Accounting Standards No. 8

FASB No. 52 *See* Statement of Financial Accounting Standards No. 52.

FedWire The Federal Reserve' network for transferring fed funds.

Fiat Money Nonconvertible paper money.

Financial Accounting Standards Board Organization in the United States that sets the rules that govern the presentation of financial statements and resolves other accounting issues.

Financial Channels The various ways in which funds, allocated profits, or both are transferred within the multinational corporation.

Financial Deregulation The dismantling of various regulations that restrict the nature of financial contracts that consenting parties, such as borrowers and lenders, could enter into.

Financial Economics A discipline that emphasizes the use of economic analysis to understand the basic workings of financial markets, particularly the measurement and pricing of risk and the intertemporal allocation of funds.

Financial Innovation The process of segmenting, transferring, and diversifying risk to lower the cost of capital as well as creating new securities that avoid various tax and regulatory costs.

Fisher Effect States that the nominal interest differen-

tial between two countries should equal the inflation differential between those countries.

Fixed Exchange Rate An exchange rate whose value is fixed by the governments involved.

Fixed-Rate Bonds Bonds that have a fixed coupon, set maturity date, and full repayment of the principal amount at maturity.

Floating Currency A currency whose value is set by market forces.

Floating Exchange Rate An exchange rate whose value is determined in the foreign exchange market.

Floating-Rate Bond A bond whose interest rate is reset every three to six months, or so, at a fixed margin above a mutually agreed-upon interest rate "index" such as LIBOR for Eurodollar deposits, the corresponding Treasury bill rate, or the prime rate.

Foreign Bank Market That portion of domestic bank loans supplied to foreigners for use abroad.

Foreign Bond Market That portion of the domestic bond market that represents issues floated by foreign companies or governments.

Foreign Credit Insurance Association The FCIA is a cooperative effort of Eximbank and a group of approximately 50 of the leading marine, casualty, and property insurance companies that administers the U.S. government's export-credit insurance program. FCIA insurance offers protection from political and commercial risks to U.S. exporters: The private insurers cover commercial risks, and the Eximbank covers political risks.

Foreign Direct Investment The acquisition abroad of physical assets such as plant and equipment, with operating control residing in the parent corporation.

Foreign Equity Market That portion of the domestic equity market that represents issues floated by foreign companies.

Foreign Exchange Brokers Specialists in matching net supplier and demander banks in the foreign exchange market.

Foreign Exchange Market Intervention Official purchases and sales of foreign exchange that nations undertake through their central banks to influence their currencies.

Foreign Exchange Risk *See* exchange risk.

Foreign Market Beta A measure of foreign market risk that is derived from the capital asset pricing model and is calculated relative to the U.S. market in the same way that individual asset betas are calculated.

Foreign Sales Corporation (FSC) A special type of corporation created by the Tax Reform Act of 1986 that is designed to provide a tax incentive for exporting U.S.-produced goods.

Foreign Tax Credit Home country credit against domestic income tax for foreign taxes already paid on foreign-source earnings.

Forfaiting The discounting—at a fixed rate without recourse—of medium-term export receivables denominated in fully convertible currencies (e.g., U.S. dollar, Swiss franc, Deutsche mark).

Forward Contract Agreement between a bank and a customer (which could be another bank) that calls for delivery, at a fixed future date, of a specified amount of one currency against dollar payment; the exchange rate is fixed at the time the contract is entered into.

Forward Differential The forward discount or premium on a currency expressed as an annualized percentage of the spot rate.

Forward Discount A situation that pertains when the forward rate expressed in dollars is below the spot rate.

Forward Forward A contract that fixes an interest rate today on a future loan or deposit.

Forward Market Hedge The use of forward contracts to fix the home currency value of future foreign currency cash flows. Specifically, a company that is long a foreign currency will sell the foreign currency forward, whereas a company that is short a foreign currency will buy the currency forward.

Forward Premium A situation that pertains when the forward rate expressed in dollars is above the spot rate.

Forward Rate The rate quoted today for delivery at a fixed future date of a specified amount of one currency against dollar payment.

Forward Rate Agreement (FRA) A cash-settled, over-the-counter forward contract that allows a company to fix an interest rate to be applied to a specified future interest period on a notional principal amount.

Free Float An exchange rate system characterized by the absence of government intervention. Also known as a **clean float**.

Functional Currency As defined in FASB No. 52, an affiliate's functional currency is the currency of the primary economic environment in which the affiliate generates and expends cash.

Fundamental Analysis An approach to forecasting asset prices that relies on painstaking examination of the macroeconomic variables and policies that are likely to influence the asset's prospects.

Funds Adjustment A hedging technique designed to reduce a firm's local currency accounting exposure by altering either the amounts or the currencies (or both) of the planned cash flows of the parent or its subsidiaries.

Futures Contracts Standardized contracts that trade on organized futures markets for specific delivery dates only.

Futures Option An option contract calling for delivery of a standardized IMM futures contract in the currency rather than the currency itself.

G-5 Nations The United States, France, Japan, Great Britain, and Germany.

G-7 Nations The G-5 nations plus Italy and Canada.

Global Fund A mutual fund that can invest anywhere in the world, including the United States.

Gold Standard A system of setting currency values whereby the participating countries commit to fix the prices of their domestic currencies in terms of a specified amount of gold.

Good Funds Funds that are available for use.

Government Budget Deficit A closely watched figure that equals government spending minus taxes.

Growth Options The opportunities a company may have to invest capital so as to increase the profitability of its existing product lines and benefit from expanding into new products or markets.

Hard Currency A currency expected to maintain its value or appreciate.

Hedge To enter into a forward contract in order to protect the home currency value of foreign-currency-denominated assets or liabilities.

Herstatt Risk Named after a German bank that went bankrupt, this is the risk that a bank will deliver currency on one side of a foreign exchange deal only to find that its counterparty has not sent any money in return. Also known as **settlement risk**.

Historical Exchange Rate In accounting terminology, it refers to the rate in effect at the time a foreign currency asset was acquired or a liability incurred.

Home Bias The tendency of investors to hold domestic assets in their investment portfolios.

Hyperinflationary Country Defined in FASB No. 52 as one that has cumulative inflation of approximately 100% or more over a three-year period.

IFC Emerging Markets Index An index of developing country stock markets published by the International Finance Corporation.

Implied Volatility The volatility that, when substituted in the Black-Scholes option pricing formula, yields the market price of the option.

Import-Substitution Development Strategy A development strategy followed by many Latin American countries and other LDCs that emphasized import substitution—accomplished through protectionism—as the route to economic growth.

Indexed Bond Bond that pays interest tied to the inflation rate. The intent is to fix the real interest rate on the bond.

Indirect Quotation A quote that gives the foreign currency price of the home currency.

Inflation Change in the general level of prices.

Inflation Risk Refers to the divergence between actual and expected inflation.

Inter-American Development Bank (IADB) A source of long-term capital in Latin America.

Intercompany Loan Loan made by one unit of a corporation to another unit of the same corporation.

Interbank Market The wholesale foreign exchange market in which major banks trade currencies with each other.

Intercompany Transaction Transaction, such as a loan, carried out between two units of the same corporation.

Interest Rate/Currency Swap Swap that combines the features of both a currency swap and an interest rate swap. It converts a liability in one currency with a stipulated type of interest payment into one denominated in another currency with a different type of interest payment.

Interest Rate Parity A condition whereby the interest differential between two currencies is (approximately) equal to the forward differential between two currencies.

Interest Rate Swap An agreement between two parties to exchange interest payments for a specific maturity on an agreed upon principal amount. The most common interest rate swap involves exchanging fixed interest payments for floating interest payments.

Interest Tax Shield The value of the tax write-off on interest payments (analogous to the depreciation tax shield).

International Bank for Reconstruction and Development (IBRD) Also known as the World Bank, the IBRD is owned by its member nations and makes loans at nearly conventional terms to countries for projects of high economic priority.

International Banking Facility (IBF) A bookkeeping entity of a U.S. financial institution that is permitted to conduct international banking business (such as receiving foreign deposits and making foreign loans) largely exempt from domestic regulatory constraints.

International Diversification The attempt to reduce risk by investing in more than one nation. By diversifying across nations whose economic cycles are not perfectly in phase, investors can typically reduce the variability of their returns.

International Finance Subsidiary A subsidiary incorporated in the United States (usually in Delaware) whose sole purpose was to issue debentures overseas and invest the proceeds in foreign operations, with the interest paid to foreign bondholders not subject to U.S. withholding tax. The elimination of the corporate withholding tax has ended the need for this type of subsidiary.

International Fisher Effect States that the interest differential between two countries should be an unbiased predictor of the future change in the spot rate.

International Fund A mutual fund that can invest only outside the United States.

International Monetary Fund (IMF) International organization created at Bretton Woods, N.H., in 1944 to promote exchange rate stability, including the provision of temporary assistance to member nations trying to defend their currencies against transitory phenomena.

International Monetary Market (IMM) The IMM is a market created by the Chicago Mercantile Exchange for the purpose of trading currency futures.

International Monetary System The set of policies, institutions, practices, regulations, and mechanisms that determine the rate at which one currency is exchanged for another.

In-the-Money An option that would be profitable to exercise at the current price.

Intrinsic Value The value of an option that is attributable to its being in-the-money. An out-of-the money option has no intrinsic value.

Inverse Floaters Floating-rate notes with coupons that move opposite to the reference rate.

Investment Banker A financing specialist who assists organizations in designing and marketing security issues.

Irrevocable Letter of Credit An L/C that cannot be revoked without the specific permission of all parties concerned, including the exporter.

J-Curve Theory that says a country's trade deficit will initially worsen after its currency depreciates because higher prices on foreign imports will more than offset the reduced volume of imports in the short run.

Keiretsu The large industrial groupings—often with a major bank at the centers—that form the backbone of corporate Japan.

Knockout Option An option that is similar to a standard option except that it is canceled—that is, knocked out—if the exchange rate crosses, even briefly, a pre-defined level called the outstrike. If the exchange rate breaches this barrier, the holder cannot exercise this option, even if it ends up in-the-money. Also known as **barrier options**.

Law of One Price The theory that exchange-adjusted prices on identical tradeable goods and financial assets must be within transaction costs of equality worldwide.

L/C *See* letter of credit.

LDC Debt Swap *See* debt swap.

LDC Debit-Equity Swap *See* debt swap.

Leading and Lagging A means of shifting liquidity by accelerating (leading) and delaying (lagging) international payments by modifying credit terms, normally on trade between affiliates.

Lean Production Ultra-efficient manufacturing techniques pioneered by Japanese industry that gave it a formidable competitive edge over U.S. rivals.

Lender of Last Resort Official institution that lends funds to countries or banks that get into financial trouble. It is designed to avert the threat of a financial panic.

Letter of Credit A letter addressed to the seller, written and signed by a bank acting on behalf of the buyer, in which the bank promises to honor drafts drawn on itself if the seller conforms to the specific conditions contained in the letter.

Line of Credit An informal agreement that permits a company to borrow up to a stated maximum amount from a bank. The firm can draw down its line of credit when it requires funds and pay back the loan balance when it has excess cash.

Link Financing *See* back-to-back loan.

Liquidity The ability to readily exchange an asset for goods or other assets at a known price, thereby facilitating economic transactions. It is usually measured by the difference between the rates at which dealers can buy and sell that asset.

Listed Options Option contracts with prespecified terms (amount, strike price, expiration date, fixed maturity) that are traded on an organized exchange.

Loan Syndication Group of banks sharing a loan.

Lock Box A postal box in a company's name to which customers remit their required payments.

London Interbank Bid Rate (LIBID) The rate paid by one bank to another for a deposit.

London Interbank Offer Rate (LIBOR) The deposit rate on interbank transactions in the Eurocurrency market.

Long Position In the foreign exchange market, it means that one has more assets in a particular currency than liabilities.

Look-Thru A method for calculating U.S. taxes owed on income from controlled foreign corporations that was introduced by the Tax Reform Act of 1986.

Louvre Accord Named for the Paris landmark where it was negotiated, this accord called for the G-7 nations to support the falling dollar by pegging exchange rates within a narrow, undisclosed range, while they also moved to bring their economic policies into line.

Maastricht Criteria Tough standards on inflation, currency stability, and deficit spending established in the Maastricht Treaty that European nations must meet in order to join EMU.

Maastricht Treaty Agreement under which the EC nations would establish a European Monetary Union with a single central bank having the sole power to issue a single European currency called the euro.

Market Efficiency A market in which the prices of traded securities readily incorporate new information.

Managed Float Also known as a "dirty" float, this is a system of floating exchange rates with central bank intervention to reduce currency fluctuations.

Market Risk Premium The difference between the required return on a particular stock market and the risk-free interest rate.

Marking to Market A daily settlement feature in which profits and losses of futures contracts are paid over every day at the end of trading. More generally, this term refers to pricing assets at their market value rather than their book value.

Monetary/Nonmonetary Method Under this translation method, monetary items (for example, cash, accounts payable and receivable, and long-term debt) are translated at the current rate while nonmonetary items (for example, inventory, fixed assets, and long-term investments) are translated at historical rates.

Monetary Union A group of states that join together to have a single central bank that issues a common currency.

Money Market Hedge The use of simultaneous borrowing and lending transactions in two different currencies to lock in the home currency value of a foreign currency transaction.

Moral Hazard The tendency to incur risks that one is protected against.

Morgan Stanley Capital International World Index An internationally diversified index of developed country stock markets that combines the EAFE index with the U.S. market index.

Multicurrency Clause This clause gives a Eurocurrency borrower the right to switch from one currency to another when the loan is rolled over.

Multinational Cash Mobilization A system designed to optimize the use of funds by tracking current and near-term cash positions and redeploying those funds in an efficient manner.

Multinational Financial System The aggregate of the internal transfer mechanisms available to the MNC to shift profits and money among its various affiliates. These transfer mechanisms include transfer price adjustments, leading and lagging interaffiliate payments, dividend payments, fees and royalties, intercompany loans, and intercompany equity investments.

Nationalization The taking of property, with or without compensation, by a government.

Net Liquidity Balance The change in private domestic borrowing or lending that is required to keep payments in balance without adjusting official reserves. Nonliquid, private, short-term capital flows and errors and omissions are included in the balance; liquid assets and liabilities are excluded.

Netting *See* exposure netting or payments netting.

No-Arbitrage Condition The relationship between exchange rates such that profitable arbitrage opportunities don't exist. If this condition is violated on an ongoing basis, we would wind up with a money machine.

Nominal Exchange Rate The actual exchange rate; it is expressed in current units of currency.

Nominal Interest Rate The price quoted on lending and borrowing transactions. It is expressed as the rate of exchange between current and future units of currency unadjusted for inflation.

Nonrecourse Method of borrowing against an asset, such as a receivable, under which the lender assumes all the credit and political risks except for those involving disputes between the transacting parties. The lender has no recourse to the borrower.

Note Issuance Facility (NIF) A facility provided by a syndicate of banks that allows borrowers to issue short-term notes, which are then placed by the syndicate providing the NIF. Borrowers usually have the right to sell their notes to the bank syndicate at a price that yields a prearranged spread over LIBOR.

Notional Principal A reference amount against which the interest on a swap is calculated.

Official Reserves Holdings of gold and foreign currencies by official monetary institutions.

Official Reserve Transactions Balance The adjustment required in official reserves to achieve balance-of-payments equilibrium.

Offshore Finance Subsidiary A wholly owned affiliate incorporated overseas, usually in a tax-haven country, whose function is to issue securities abroad for use in either the parent's domestic or foreign business.

Open Account Selling This selling method involves shipping goods first and billing the importer later.

Open Interest The number of futures contracts outstanding at any one time.

Open-Market Operation Purchase or sale of government securities by the monetary authorities to increase or decrease the domestic money supply.

Operating Exposure Degree to which an exchange rate change, in combination with price changes, will alter a company's future operating cash flows.

Optimum Currency Area Largest area in which it makes sense to have only one currency. It is defined as that area for which the cost of having an additional currency—higher costs of doing business and greater currency risk—just balances the benefits of another currency—reduced vulnerability to economic shocks associated with the option to change the area's exchange rate.

Option A financial instrument that gives the holder the right—but not the obligation—to sell (put) or buy (call) another financial instrument at a set price and expiration date.

Out-of-the-Money An option that would not be profitable to exercise at the current price.

Outright Rate Actual forward rate expressed in dollars per currency unit, or vice versa.

Outsourcing The practice of purchasing a significant percentage of intermediate components from outside suppliers.

Outstrike Pre-defined exchange rate at which a knock-out option is canceled if the spot rate crosses this price even temporarily.

Overdraft A line of credit against which drafts (checks) can be drawn (written) up to a specified maximum amount.

Overseas Private Investment Corporation (OPIC) Agency of the U.S. government that provides political risk insurance coverage to U.S. multinationals. Its purpose is to encourage U.S. direct investment in less-developed countries.

Over-the-Counter Market (OTC) A market in which the terms and conditions on contracts are negotiated between the buyer and seller, in contrast to an organized exchange where contractual terms are fixed by the exchange.

Over-the-Counter Options Option contracts whose specifications are generally negotiated as to the amount, exercise price, underlying instrument, and expiration. They are traded by commercial and investment banks.

Parallel Loan Simultaneous borrowing and lending operation usually involving four related parties in two different countries.

Payments Netting Reducing fund transfers between affiliates to only a netted amount. Netting can be done on a bilateral basis (between pairs of affiliates) or on a multilateral basis (taking all affiliates together).

Pegged Currency A currency whose value is set by the government.

Peso Problem Reference to the possibility that during the time period studied investors anticipated significant events that didn't materialize, thereby invalidating statistical inferences based on data drawn from that period.

Plaza Agreement A coordinated program agreed to in September 1985 that was designed to force down the dollar against other major currencies and thereby improve American competitiveness.

Political Risk Uncertain government action that affects the value of a firm.

Pooling Transfer of excess affiliate cash into a central account (pool), usually located in a low-tax nation, where all corporate funds are managed by corporate staff.

Possession Corporation A U.S. corporation operating in a U.S. possession. Such companies are entitled to certain tax breaks.

Price Elasticity of Demand Percentage change in the quantity demanded of a particular good or service for a given percentage change in price.

Price-Specie-Flow Mechanism Adjustment mechanism under the classical gold standard whereby disturbances in the price level in one country would be wholly or partly offset by a countervailing flow of specie (gold coins) that would act to equalize prices across countries and automatically bring international payments back in balance.

Private Export Funding Corporation (PEFCO) Company that mobilizes private capital for financing the export of big-ticket items by U.S. firms by purchasing at fixed interest rates the medium- to long-term debt obligations of importers of U.S. products.

Private Placements Securities that are sold directly to only a limited number of sophisticated investors, usually life insurance companies and pension funds.

Privatization The act of returning state-owned or state-run companies back to the private sector, usually by selling them off.

Product Cycle The time it takes to bring new and improved products to market. Japanese companies have excelled in compressing product cycles.

Property Rights Rights of individuals and companies to own and utilize property as they see fit and to receive the stream of income that their property generates.

Protectionism Protecting domestic industry from import competition by means of tariffs, quotas, and other trade barriers.

Purchasing Power Parity The notion that the ratio between domestic and foreign price levels should equal the equilibrium exchange rate between domestic and foreign currencies.

Put-Call Option Interest Rate Parity A parity condition that relates put and call currency options prices to the interest differential between two currencies and, by extension, to their forward differential.

Quota Government regulation specifying the quantity of particular products that can be imported to a country.

Range Forward *See* currency collar.

Rational Expectations The idea that people rationally anticipate the future and respond to what they see ahead.

Real Exchange Rate The spot rate adjusted for relative price level changes since a base period.

Real Interest Rate The nominal interest rate adjusted for expected inflation over the life of the loan; it is the exchange rate between current and future goods.

Real or Inflation-Adjusted Exchange Rate Measured as the nominal exchange rate adjusted for changes in relative price levels.

Regional Fund A mutual fund that invests in a specific geographic areas overseas, such as Asia or Europe.

Regulatory Arbitrage The process whereby the users of capital markets issue and trade securities in financial centers with the lowest regulatory standards and, hence, the lowest costs.

Reinvoicing Center A subsidiary that takes title to all goods sold by one corporate unit to another affiliate or to a third-party customer. The center pays the seller and in turn is paid by the buyer.

Reporting Currency The currency in which the parent firm prepares its own financial statements; that is, U.S. dollars for a U.S. company.

Revaluation An increase in the spot value of a currency.

Revolving Credit Agreement A line of credit under which a bank (or syndicate of banks) is *legally committed* to extend credit up to the stated maximum.

Revolving Underwriting Facility (RUF) A note issuance facility that includes underwriting services. The RUF gives borrowers long-term continuous access to short-term money underwritten by banks at a fixed margin.

Right of Offset Clause that gives each party to a swap or parallel loan arrangement the right to offset any nonpayment of principal or interest with a comparable nonpayment.

Risk Arbitrage The process that leads to equality of risk-adjusted returns on different securities, unless market imperfections that hinder this adjustment process exist.

Roll Over Date Date on which the interest rate on a floating-rate loan is reset based on current market conditions.

Rule 144A Rule adopted by the Securities and Exchange Commission (SEC) in 1990 that allows qualified institutional investors to trade in unregistered private placements, making them a closer substitute for public issues.

Same-Day Value Funds that are credited and available for use the same day they are transferred.

Samurai Bonds Yen bonds sold in Japan by a non-Japanese borrower.

Section 482 United States Department of Treasury regulations governing transfer prices.

Securitization The matching up of borrowers and lenders wholly or partly by way of the financial markets. This process usually refers to the replacement of nonmarketable loans provided by financial intermediaries with negotiable securities issued in the public capital markets.

Seigniorage The profit to the central bank from money creation; it equals the difference between the cost of issuing the money and the value of the goods and services that money can buy.

Secondary Market The market where investors trade securities already bought.

Settlement Risk *See* Herstatt risk.

Shogun Bonds Foreign currency bonds issued within Japan by Japanese corporations.

Shogun Lease Yen-based international lease.

Short Position In the foreign exchange market, it means that one has more liabilities in a particular currency than assets.

Sight Draft A draft that must be paid on presentation or else dishonored.

Single-Country Fund A mutual fund that invests in individual countries outside the United States, such as Germany or Thailand.

Smithsonian Agreement After the currency turmoil of August 1971, the United States agreed in December 1971 to devalue the dollar to 1/38 of an ounce of gold, and other countries agreed to revalue their currencies by negotiated amounts vis-à-vis the dollar.

Society for Worldwide Interbank Financial Telecommunications (SWIFT) A dedicated computer network to support funds transfer messages internationally between more than 900 member banks worldwide.

Soft Currency A currency expected to depreciate.

Sovereign Risk The risk that the country of origin of the currency a bank is buying or selling will impose foreign exchange regulations that will reduce or negate the value of the contract; also refers to the risk of government default on a loan made to it or guaranteed by it.

Special Drawing Rights (SDR) A new form of international reserve assets, created by the IMF in 1967, whose value is based on a portfolio of widely used currencies.

Spot Rate The price at which foreign exchange can be bought or sold with payment set for the same day.

Statement of Financial Accounting Standards No. 8 This is the currency translation standard previously in use by U.S. firms.

Statement of Financial Accounting Standards No. 52 This is the currency translation standard currently in use by U.S. firms. It basically mandates the use of the current rate method.

Statistical Discrepancy A number on the balance-of-payments account that reflects errors and omissions in collecting data on international transactions.

Step-Down Note A debt instrument with a high coupon in earlier payment periods and a lower coupon in later payment periods.

Step-Up Note A callable debt issue that features one or more increases in a fixed rate or a step-up in a spread over LIBOR during the life of the note.

Sterilized Intervention Foreign exchange market intervention in which the monetary authorities have insulated their domestic money supplies from the foreign exchange transactions with offsetting sales or purchases of domestic assets.

Strike Price *See* exercise price.

Structured Notes Interest-bearing securities whose interest payments are determined by reference to a formula set in advance and adjusted on specified reset dates.

Subpart F Special category of foreign-source "unearned" income that is currently taxed by the IRS whether or not it is remitted back to the United States.

Subsidiary A foreign-based affiliate that is a separately incorporated entity under the host country's law.

Swap A foreign exchange transaction that combines a spot and a forward contract. More generally, it refers to a financial transaction in which two counterparties agree to exchange streams of payments over time, such as in a currency swap or an interest rate swap.

Swap Rate The difference between spot and forward rates expressed in points (e.g., $0.0001 per pound sterling or DM 0.0001 per dollar).

SWIFT *See* Society for Worldwide Interbank Financial Telecommunications.

Systematic (Nondiversifiable) Risk Marketwide influences that affect all assets to some extent, such as the state of the economy.

Systematic Risk That element of an asset's risk that cannot be eliminated no matter how diversified an investor's portfolio.

Target-Zone Arrangement A monetary system under which countries pledge to maintain their exchange rates within a specific margin around agreed-upon, fixed central exchange rates.

Tariff A tax imposed on imported products. It can be used to raise revenue, to discourage purchase of foreign products, or some combination of the two.

Tax Arbitrage The shifting of gains or losses from one tax jurisdiction to another to profit from differences in tax rates.

Tax Haven A nation with a moderate level of taxation and/or liberal tax incentives for undertaking specific activities such as exporting.

Tax Reform Act of 1986 A 1986 law involving a major overhaul of the U.S. tax system.

Technical Analysis An approach that focuses exclusively on past price and volume movements—while totally ignoring economic and political factors—to forecast future asset prices.

Temporal Method Under this currency translation method, the choice of exchange rate depends on the underlying method of valuation. Assets and liabilities valued at historical cost (market) are translated at the historical rate (current rate).

Term Loan A straight loan, often unsecured, that is made for a fixed period of time, usually 90 days.

Terms of Trade The weighted average of a nation's export prices relative to its import prices.

Time Draft A draft that is payable at some specified future date and as such becomes a useful financing device.

Time Value The excess of an option's value over its intrinsic value.

Total Dollar Return The dollar return on a nondollar investment, which includes the sum of any dividend/interest income, capital gains (losses), and currency gains (losses) on the investment.

Trade Acceptance A draft accepted by a commercial enterprise; See draft.

Trade Draft A draft addressed to a commercial enterprise; See draft.

Transaction Exposure The extent to which a given exchange rate change will change the value of foreign-currency-denominated transactions already entered into.

Transfer Price The price at which one unit of a firm sells goods or services to an affiliated unit.

Translation Exposure *See* accounting exposure.

Treasury Workstation A software package implemented on a computer that treasurers use to run their cash management systems.

Triangular Currency Arbitrage A sequence of foreign exchange transactions, involving three different currencies, that one can use to profit from discrepancies in the different exchange rates.

Unbiased Nature of the Forward Rate (UFR) States that the forward rate should reflect the expected future spot rate on the date of settlement of the forward contract.

Underwriting The act by investment bankers of purchasing securities from issuers for resale to the public.

United Currency Options Market (UCOM) Market set up by the Philadelphia Stock Exchange in which to trade currencies.

Universal Banking Bank practice, especially in Germany, whereby commercial banks perform not only investment banking activities but also take major equity positions in companies.

Unsterilized Intervention Foreign exchange market intervention in which the monetary authorities have not insulated their domestic money supplies from the foreign exchange transactions.

Unsystematic (Diversifiable) Risk Risks that are specific to a given firm, such as a strike.

Up-and-In Option An option that comes into existence if and only if the currency strengthens enough to cross a preset barrier.

Up-and-Out Option An option that is canceled if the underlying currency strengthens beyond the outstrike.

Value-Added Tax Method of indirect taxation whereby a tax is levied at each stage of production on the value added at that specific stage.

Value Additivity Principle The principle that the net present value of a set of independent projects is simply the sum of the NPVs of the individual projects.

Value-at-Risk (VAR) A calculation which allows a financial institution to estimate the maximum amount it might expect to lose in a given time period with a certain probability.

Value Date The date on which the monies must be paid to the parties involved in a foreign exchange transaction. For spot transactions, it is set as the second working day after the date on which the transaction is concluded.

Value-Dating Refers to when value (credit) is given for funds transferred between banks.

Virtual Currency Option A foreign currency option listed on the Philadelphia Stock Exchange which is settled in U.S. dollars rather than in the underlying currency.

Weighted Average Cost of Capital (WACC) The required return on the funds supplied by investors. It is a weighted average of the costs of the individual component debt and equity funds.

Working Capital The combination of current assets and current liabilities.

World Bank *See* International Bank for Reconstruction and Development.

Yankee Bonds Dollar-denominated foreign bonds sold in the United States.

Yankee Stock Issues Stock sold by foreign companies to U.S. investors.

◆ INDEX

799

SYMBOLS AND ACRONYMS

a_h	Expected real return on home currency loan
a_f	Expected real return on a foreign currency loan
ADR	American depository receipt
APV	Adjusted present value
B/L	Bill of lading
β	Beta coefficient, a measure of an asset's riskiness
β^*	All-equity beta
β_e	Levered β
C_t	Local currency cash flows in period t
C	Cost
$C(E)$	Price of a foreign currency call option
d	Amount of currency devalution
D	Forward discount
D_f	Amount of foreign currency debt
e_t	Nominal exchange rate at time t
e'_t	Real exchange rate at time t
E	(a) Exercise price on a call option or (b) Amount of equity
E_f	Foreign subsidiary retained earnings
f_t	t-period forward exchange rate
g	(a) Expected dividend growth rate or
	(b) Expected rate of foreign currency appreciation against the dollar
HC	Home currency
i_f	(a) Expected rate of foreign inflation per period or
	(b) Before-tax cost of foreign debt
i_h	Expected rate of home country inflation per period
i_d	Before-tax cost of domestic debt
I_0	Initial investment
IRPT	Interest rate parity theory
k	Cost of capital
k_0	Weighted cost of capital